To Professor Alec Walters

With warmest regards

Jerry Kemp.

COMPREHENSIVE CHEMICAL KINETICS

COMPREHENSIVE

Section 1. THE PRACTICE AND THEORY OF KINETICS

Volume 1 The Practice of Kinetics

Volume 2 The Theory of Kinetics

Volume 3 The Formation and Decay of Excited Species

Section 2. HOMOGENEOUS DECOMPOSITION AND ISOMERISATION REACTIONS

Volume 4 Decomposition of Inorganic and Organometallic Compounds

Volume 5 Decomposition and Isomerisation of Organic Compounds

Section 3. INORGANIC REACTIONS

Volume 6 Reactions of Non-metallic Inorganic Compounds

Volume 7 Reactions of Metallic Salts and Complexes, and Organometallic Compounds

Section 4. ORGANIC REACTIONS (6 volumes)

Volume 9 Addition and Elimination Reactions of Aliphatic Compounds

Volume 10 Ester Formation and Hydrolysis and Related Reactions

Volume 13 Reactions of Aromatic Compounds

Section 5. POLYMERISATION REACTIONS (2 volumes)

Section 6. OXIDATION AND COMBUSTION REACTIONS (2 volumes)

Section 7. SELECTED ELEMENTARY REACTIONS (2 volumes)

Additional Sections

HETEROGENEOUS REACTIONS

SOLID STATE REACTIONS

KINETICS AND TECHNOLOGICAL PROCESSES

CHEMICAL KINETICS

EDITED BY

C. H. BAMFORD

M.A., Ph.D., Sc.D. (Cantab.), F.R.I.C., F.R.S.

Campbell-Brown Professor of Industrial Chemistry,

University of Liverpool

AND

C. F. H. TIPPER

Ph.D. (Bristol), D.Sc. (Edinburgh)

Senior Lecturer in Physical Chemistry,

University of Liverpool

VOLUME 7

REACTIONS OF METALLIC SALTS AND COMPLEXES, AND ORGANOMETALLIC COMPOUNDS

ELSEVIER PUBLISHING COMPANY

AMSTERDAM - LONDON - NEW YORK

1972

ELSEVIER PUBLISHING COMPANY
335 JAN VAN GALENSTRAAT
P.O. BOX 211, AMSTERDAM, THE NETHERLANDS

AMERICAN ELSEVIER PUBLISHING COMPANY, INC.
52 VANDERBILT AVENUE
NEW YORK, NEW YORK 10017

LIBRARY OF CONGRESS CARD NUMBER 76-151731
ISBN 0-444-40861

WITH 35 ILLUSTRATIONS AND 228 TABLES

PRINTED IN THE NETHERLANDS

COMPREHENSIVE CHEMICAL KINETICS

Contributors to Volume 7

D. BENSON — Department of Chemistry,
Widnes Technical College,
Widnes, Lancs., England

L. J. CSÁNYI — Institute of Inorganic and Analytical Chemistry,
József Attila University,
Szeged, Hungary

T. J. KEMP — School of Molecular Sciences,
University of Warwick,
Coventry, England

C. H. LANGFORD — Department of Chemistry,
Carleton University,
Ottawa, Canada

M. PARRIS — Department of Chemistry,
Carleton University,
Ottawa, Canada

P. J. PROLL — Department of Chemistry,
Widnes Technical College,
Widnes, Lancs., England

Preface

The rates of chemical processes and their variation with conditions have been studied for many years, usually for the purpose of determining reaction mechanisms. Thus, the subject of chemical kinetics is a very extensive and important part of chemistry as a whole, and has acquired an enormous literature. Despite the number of books and reviews, in many cases it is by no means easy to find the required information on specific reactions or types of reaction or on more general topics in the field. It is the purpose of this series to provide a background reference work, which will enable such information to be obtained either directly, or from the original papers or reviews quoted.

The aim is to cover, in a reasonably critical way, the practice and theory of kinetics and the kinetics of inorganic and organic reactions in gaseous and condensed phases and at interfaces (excluding biochemical and electrochemical kinetics, however, unless very relevant) in more or less detail. The series will be divided into sections covering a relatively wide field; a section will consist of one or more volumes, each containing a number of articles written by experts in the various topics. Mechanisms will be thoroughly discussed and relevant non-kinetic data will be mentioned in this context. The methods of approach to the various topics will, of necessity, vary somewhat depending on the subject and the author(s) concerned.

It is obviously impossible to classify chemical reactions in a completely logical manner, and the editors have in general based their classification on types of chemical element, compound or reaction rather than on mechanisms, since views on the latter are subject to change. Some duplication is inevitable, but it is felt that this can be a help rather than a hindrance.

Section 3 deals with reactions in which at least one of the reactants is an inorganic compound. Many of the processes considered also involve organic compounds, but autocatalytic oxidations and flames, polymerisation and reactions of metals themselves and of certain unstable ionic species, *e.g.* the solvated electron, are discussed in later sections. Where appropriate, the effects of low and high energy radiation are considered, as are gas and condensed phase systems but not fully heterogeneous processes or solid reactions. Rate parameters of individual elementary steps, as well as of overall reactions, are given if available.

In volume 7 reactions of metallic salts, complexes and organometallic compounds are covered. Isomerisation and group transfer reactions of "inert" metal complexes and certain organometallics (not involving a change in oxidation state) are considered first, followed by oxidation–reduction processes (a) between different valency states of the same metallic element (b) between salts of different metals and (c) involving "covalent" inorganic or organic compounds. Finally, induced reactions are discussed separately.

The Editors desire to record their sincere appreciation of the continuing advice and support from the members of the Advisory Board.

Liverpool
October, 1971

C. H. BAMFORD
C. F. H. TIPPER

Contents

Chapter 4 (T. J. Kemp)

Oxidation–reduction reactions between covalent compounds and metal ions 274

Chapter 1

Reactions of Inert Complexes and Metal Organic Compounds

C. H. LANGFORD AND M. PARRIS

1. Introduction

This chapter is concerned with the simplest reactions of inert transition metal complexes. Fig. 1 shows a typical compound. This is Co(III) coordinated to six NH_3 molecules to form a triply positive cation $[Co(NH_3)_6]^{3+}$. It is indicated in Fig. 1 to be in aqueous solution where water molecules occupy positions in what

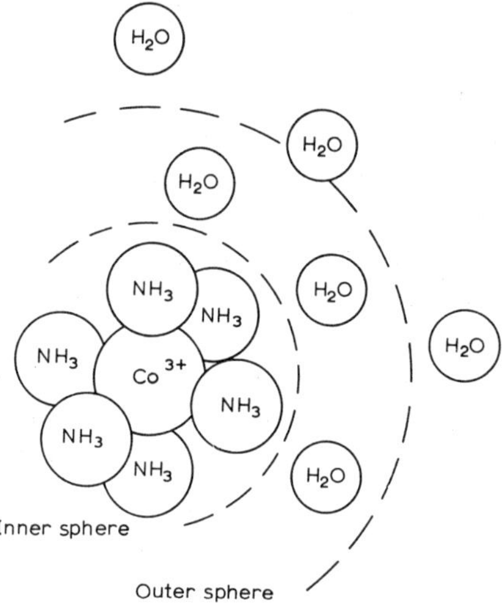

Fig. 1. The $Co(NH_3)_6{}^{3+}$ ion in aqueous solution. The "inner sphere" contains six ammonia ligands strongly bonded to Co(III). The "outer sphere" contains several water molecules.

may be called the *outer* or *second* coordination sphere. The reactions in question are typified by the replacement of one of the NH_3 molecules by a water molecule from the outer sphere. "Inert" will be considered to mean any complex whose reactions occur slowly enough for conventional experimental kinetic techniques to be applied; in general this means half times longer than 10 sec. The inert complexes have sometimes been called "robust" but this term seems to be more suggestive of thermodynamic stability than kinetic non-lability[1].

This chapter is restricted to treatment of the inert complexes because these provide an extraordinary range of experimental results to examine. The richness of results arises because a wide variety of closely related structures are synthetically available when ligand substitution does not occur too rapidly. It is also of theoretical importance to consider reactions of inert complexes because the range of closely related structures allows careful examination of small structure changes and correspondingly detailed information about the mechanisms of reactions. The basic ideas about the mechanistic pathways of ligand substitution that arise from study of inert complexes also serve as an excellent starting point for analysis of all ligand substitutions including fast ones. In fact, two cases to be given detailed treatment, Co(III) and Pt(II), have served admirably as paradigms for the general study of ligand substitution. The systems to be considered here include the octahedral complexes of Co(III), Rh(III), Cr(III), Ru(III) and Ir(III) which contain principally amine ligands and square planar complexes of Pt(II), Pd(II) and Au(III) which contain similar ligands. For lack of a more precise characterization of the situation we shall describe these as complexes of the "harder" ligands in the sense of Pearson's distinction between hard and soft acids and bases[2]. Consideration will also be given to some organometallic systems, principally the metal carbonyls. These, by contrast with the other inert complexes, essentially involve ligands which are soft bases in the Pearson classification. A cautious reading of the evidence suggests separating consideration of these two types of complex since their reactions occur under quite different circumstances (*e.g.* type of solvent), but there is actually no strong evidence that the major generalizations about mechanism do not extend over both classes of system.

There are essentially two distinct types of experiment which have been utilised by students of the mechanisms of ligand substitution. One type involves the detailed analysis of the rate law governing the reaction (including stereochemistry) and this type can yield (under favorable circumstances) insight into the number of elementary steps of the reaction and the stoichiometric composition of the transition state[3]. However, the experimental rate law for a single reaction cannot reveal the energetic role played by the various groups in the complex. The value of the rate coefficient itself reveals the energy difference between two points along the reaction coordinate, the initial state and the transition state[4]. The role that any particular substituent (or component of the environment) plays in determining that energy difference may only be assessed from a variation of the group (or factor of the environment). Thus, the energetics of activation must be inferred from a second type of experiment in which rates of a series of reactions presumed to have related mechanisms are compared. It is this question of the energetics of activation which demands our consideration first.

In a ligand substitution reaction, two groups must always receive attention. There is a bond to the leaving group to be broken and a bond to the entering group to be formed. The relative importance of these two processes provides a basic dichotomy for the classification of substitutions. If a reaction rate is sensitive to the nature of the entering group, it is clear that the energy of the bond being formed is important to the activation process; and its influence must be in the nature of an entering group assistance to activation. If otherwise, then the entering group could not influence the rate, and the minimum requirement for a substitution reaction is that the bond to the leaving group be broken.

The question of the sensitivity of the rate of substitution to the nature of the entering ligand provides a basis for the mechanistic dichotomy. A reaction which is clearly insensitive to the nature of the entering group must reach its transition state principally by the internal accumulation of the energy to break the bond to the leaving group within the ground state complex because any significant assistance (in the sense of formation of a new bond) should be *selective* with respect to the ligand assisting. Such a reaction, insensitive to the nature of the entering ligand, will be said to have a *dissociative mode* of activation. Reactions which are sensitive to the nature of the entering group will be characterized as having an *associative* mode of activation because the assistance of the entering ligand does play an important part in the determination of the free energy of activation, although not necessarily to the exclusion of the leaving ligand. As the discussions of particular cases will suggest, the dichotomy between *dissociative* and *associative* activation does encompass the secondary effects on the energy of activation by other groups in the complex or by the factors of the environment of the complex (*e.g.* solvent effects).

A discussion of ligand exchange reactions of organometallic compounds associated with oxidation–reduction processes leading to free-radical formation will be found in Volume 14 (Free-radical polymerization).

2. Stoichiometric mechanisms

So far nothing has been said about rate laws. This has been intentional, for there is no simple relationship between the modes of activation and the concentration dependencies that determine the rate law. The form of the rate law for a reaction, that is, the dependence of rate on concentrations over the widest possible variations, tests hypotheses concerning the stoichiometric composition of the transition state and the number of elementary steps of the reaction. These aspects may be called the stoichiometric mechanism[3]. There is enough variation in activation mode among reactions of similar stoichiometric mechanism to recommend classifying reactions in two separate ways, first according to stoichiometric mechanism, then according to activation mode. Moreover, these two distinguishing classifications follow the natural evolution in experiment. Stoichiometric mechanism is inferred from analysis of rate laws; mode of activation is inferred from consideration of relative rates.

Note that the three transition states, ***a***, ***b*** and ***c***, in Fig. 3 all contain both the

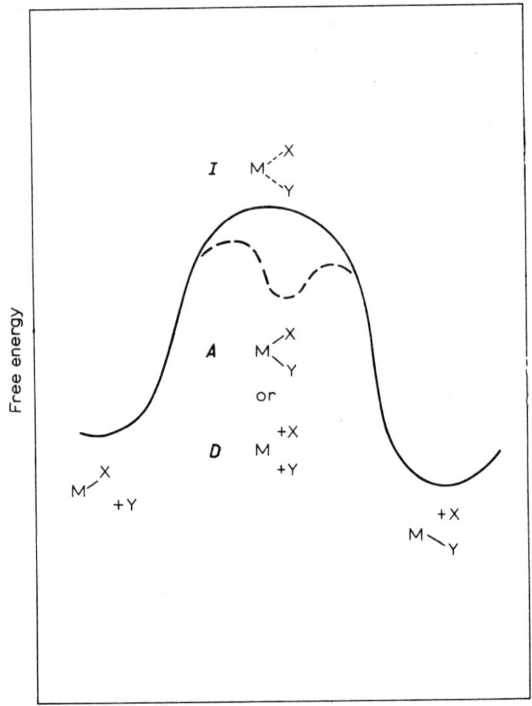

Fig. 2. Free energy *vs.* reaction coordinate for the concerted (*I*) and two-step (*A, D*) reactions.

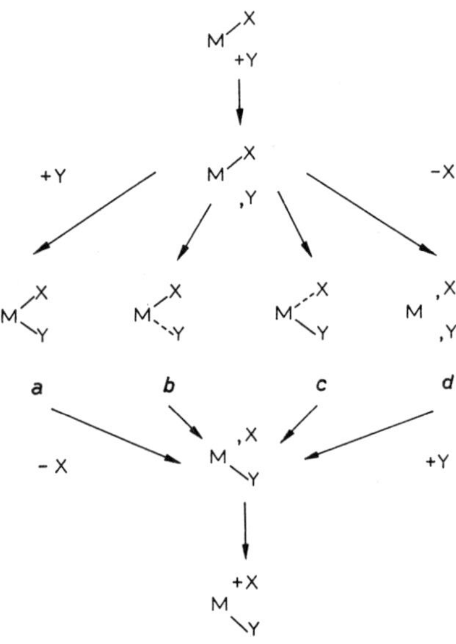

Fig. 3. Possible transition states in the reaction MX+Y → MY+X.

entering and leaving ligands, Y and X in addition to the remaining metal complex moiety. All three of these must be reached *via* an initial encounter with the entering groups. Since we consider here only "slow" reactions of non-labile complexes we might write a simple mechanism in all three cases as

$$\mathrm{MX+Y} \overset{K}{\rightleftharpoons} \mathrm{MX, Y} \qquad \text{(fast equilibrium)}$$

$$\mathrm{MX, Y} \overset{k}{\rightarrow} \mathrm{MY+X} \qquad \text{(slow)} \tag{1}$$

Here MX, Y designates an *outer sphere* or *second sphere* complex. There is every reason to suppose that formation and dissociation of MX, Y occurs at rates approaching the diffusional-control limit so that the slow conversion to MY is a negligible perturbation on the equilibrium of the first step. There is a similarity here the Langmuir, the Michaelis–Menten and the Lindemann–Hinshelwood schemes.

Two common limiting forms of the rate law for mechanism (1) are encountered experimentally. In the event that the equilibrium constant, K, for outer sphere complexation is small in relation to the concentration of MX and Y, the rate law becomes

$$-\frac{\mathrm{d[MX]}}{\mathrm{d}t} = k_{\mathrm{obs}}[\mathrm{MX}][\mathrm{Y}] \tag{2}$$

where k_{obs}, the second-order rate coefficient, is kK. If K is large, the outer sphere equilibration may become saturated and the rate law reduce to

$$-\frac{\mathrm{d[MX]}}{\mathrm{d}t} = k'_{\mathrm{obs}}[\mathrm{MX}] \tag{3}$$

where k'_{obs} now equals k, the rate coefficient for the slow step. It has not always been realised that these two forms may arise from the same mechanism. Choosing, as is usual, Y to be in excess, a general expression under mechanism (1) may be written

$$-\frac{\mathrm{d[MX]}}{\mathrm{d}t} = \frac{kK[\mathrm{Y}]}{1+K[\mathrm{Y}]}\cdot[\mathrm{MX}] \tag{4}$$

Recall that three transition states might be considered as falling within the pattern (1). Transition state ***a*** of Fig. 3 involves strong binding of both X and Y. In this case, it is quite possible that an *intermediate* of increased coordination number is formed during the reaction. Since the initial attack of Y determines the stereochemical course of any reaction obeying the rate law (4) there is no simple way to identify an intermediate of increased coordination number save its actual accumulation to the extent that it can be detected. But, should such an intermediate be detected, the pattern of reaction clearly differs in a qualitative way from others following the scheme (1). A substitution involving an intermediate of increased coordination number is a reaction with at least three elementary steps: (*i*) encounter, (*ii*) addition of entering group, (*iii*) loss of leaving group. Recognising that such a path must involve an ***a*** transition state it may be usefully labelled an ***A***, associative, stoichiometric pathway.

The other two transition states that are consistent with mechanism (1) are not expected to involve an intermediate. They may accomplish an *interchange* between the outer and inner coordination spheres in a single elementary step. Such processes may be felicitously labelled *interchange*, ***I***, reactions. Note that interchange reactions may have either ***a*** or ***d*** transition states. That is, there may be $\boldsymbol{I_a}$ or $\boldsymbol{I_d}$ reactions. It is interesting to note precisely how these cases are distinguished. If a wide enough concentration range can be studied so that the rate coefficient, k, for the slow step can be found, it is a straightforward matter. However, interpretation of information from the range covered by equation (2) must normally be attempted. It must then be determined whether variations in the observed second-order rate coefficients as Y is varied reflect only variations in the outer sphere association constant K, or if there are significant variations in k. If variations in k_{obs} are attributable to K only, k is constant and the reaction employs the ***d*** activation mode. Definite information concerning values of K is unfortunately scarce.

Simple examples of ***A***, $\boldsymbol{I_a}$ and $\boldsymbol{I_d}$ reactions are seen to have the same rate law. There remains another important mechanism that differs. This is the pathway through an intermediate of reduced coordination number that is possible if a transition state like ***d*** of Fig. 3 occurs. This path may be called *dissociative*, ***D***. The mechanism may be represented as

$$\begin{aligned} &\mathrm{MX} \underset{k_x}{\overset{k_{-x}}{\rightleftharpoons}} \mathrm{M+X} \quad \text{(slow)} \\ &\mathrm{M+Y} \xrightarrow{k_y} \mathrm{MY} \quad \text{(fast)} \end{aligned} \tag{5}$$

Mechanism (5) gives rise to the rate law

$$-\frac{\mathrm{d[MX]}}{\mathrm{d}t} = k_{-x}[\mathrm{MX}] \tag{6}$$

as long as the rate of capture of the intermediate M by Y is large compared to its rate of recapture by X. If recapture by X becomes important compared to reaction with Y the expression becomes

$$-\frac{\mathrm{d[MX]}}{\mathrm{d}t} = \frac{k_{-x}[\mathrm{Y}]}{(k_x/k_y)+[\mathrm{Y}]} \cdot [\mathrm{MX}] \tag{7}$$

which is not easily distinguished experimentally from expression (4) for mechanism (1). Table 1 summarizes the mechanistic categories described and Fig. 2 indicates their relationships on an energy–reaction progress diagram.

TABLE 1

CLASSIFICATION OF LIGAND SUBSTITUTION MECHANISMS

Mode of activation	*Stoichiometric mechanism*		
	Intermediate of increased coordination number	*One-step process*	*Intermediate of reduced coordination number*
Associative activation	***A***	$\boldsymbol{I_a}$	
Dissociative activation		$\boldsymbol{I_d}$	***D***

The categorization just described was proposed recently[3]. In most of the literature, substitution reactions have been characterized according to the scheme introduced by Hughes and Ingold for organic reactions. This scheme has been critically refined by Basolo and Pearson[1]. Following Basolo and Pearson, the following rough equivalence may be listed

$$S_N2\ (\text{lim}) = \boldsymbol{A}$$

$$S_N2 = \boldsymbol{I_a}$$

$$S_N1 = \boldsymbol{I_d}$$

$$S_N1\ (\text{lim}) = \boldsymbol{D}$$

S_N denotes "substitution, nucleophilic" and 1 or 2 the molecularity of the process. The designation S_N1 seems an objectionable usage since it describes as "unimolecular" a process which requires the entering group as a stoichiometric component of the transition state and suggests a too sharp distinction from the S_N2 or $\boldsymbol{I_a}$ process. This could be deemed reasonable if molecularity were defined as "the number of ligands changing covalence" but such a definition is probably no longer an operational one, and it seems unfortunate to diminish the clear operational significance which "molecularity" has for gas-phase reactions.

3. Reactions of Co(III) complexes

The most extensively studied family of non-labile complexes is the cobalt(III) ammine series. These are octahedral systems and all those to be considered are low spin d^6 systems. The subtle variations that can be achieved synthetically make

References pp. 52–55

them especially attractive for the systematic kineticist and the reactions of Co(III) systems have served as the model for development of most of the key concepts about substitution reactions of octahedral systems.

A good starting place for discussion is the rate law for substitution reactions in acidic aqueous solutions. One typical reaction is

$$Co(NH_3)_5Cl^{2+} + H_2O \rightarrow Co(NH_3)_5OH_2{}^{3+} + Cl^- \quad (8)$$

The rate law for this solvolytic process is

$$-\mathrm{d}\frac{[Co(NH_3)_5Cl^{2+}]}{\mathrm{d}t} = k[Co(NH_3)_5Cl^{2+}] \quad (9)$$

which is of course expected in this case from almost any mechanistic picture. When the solvent is a potential ligand, the perpetual encounter between solvent molecules and complex severely limits the cases in which the rate equation is directly informative. However, one limit of mechanism is available. The general reaction (where Y is an arbitrary anionic ligand)

$$Co(NH_3)_5Cl^{2+} + Y^- \rightarrow Co(NH_3)_5Y^{2+} + Cl^- \quad (10)$$

and related ones, in which Cl^- is replaced by another leaving group, or the NH_3 ligands are replaced by others inert to substitution, all follow a rate law similar to equation (9)[5-8]. Significantly, the concentration of Y^- does not appear in the rate law in any case except when $Y^- = OH^-$. Overlooking the exception of OH^- for the moment, Y^- is not a stoichiometric participant of transition state(s) leading to its entry and there must be an *intermediate* in the reactions. There are two choices of pathway, *viz.*

$$\begin{aligned} &\text{Co}-\text{X} \xrightarrow{\text{slow}} \text{Co} + \text{X} \\ &\text{Co} + \text{Y} \xrightarrow{\text{fast}} \text{Co}-\text{Y} \end{aligned} \quad (11)$$

$$\begin{aligned} &\text{Co}-\text{X} + \text{H}_2\text{O} \xrightarrow{\text{slow}} \text{Co}-\text{OH}_2 + \text{X} \\ &\text{Co}-\text{OH}_2 + \text{Y} \xrightarrow{\text{fast}} \text{Co}-\text{Y} + \text{H}_2\text{O} \end{aligned} \quad (12)$$

The first possibility, (11), clearly is concerned with a *dissociative* mode of activation (***d***). The second, (12), might be *associative* led by water attack. But, this interpretation of (12) involves commitment to the proposition that no nucleophile (possibly excepting OH^-) has been discovered which is better than water, *i.e.* the associative attack must always involve water in the first instance. The proposition is unattractive and there is an appealing alternative. Pathway (12) might also represent a case of *dissociative* activation without a stable intermediate. Since water will be the predominant component of the second coordination sphere a dissociative interchange reaction (I_d) would lead to the aquo complex as the immediate product. The main point at the moment is that either of the attractive interpretations of the absence of [Y] in rate laws leads to a model involving the *dissociative* mode of activation. It is interesting that pathway (12) has been clearly demonstrated to be common in Co(III) chemistry[6, 9]. In the Co(III) case, as in all others in which reaction with solvent predominates, it is fruitful to adopt the tentative view that activation is *dissociative* and seek support for this from other criteria.

It is pertinent, then, to seek a dependence of substitution rates on (*i*) leaving group, (*ii*) solvent, (*iii*) steric crowding, (*iv*) charge, (*v*) nature of non-labile substituents including stereochemistry, consistent with this picture of the activation mode. If these tests generally support ***d*** modes it will be desirable to examine rate laws closely to attempt a distinction between ***D*** and $\boldsymbol{I_d}$ stoichiometric pathways.

In a dissociative process the reaction rate is expected to decrease as the strength of the metal to leaving ligand bond increases. This trend is generally observed in Co(III) ammine complexes. As can be seen in Table 2, a partial leaving group order is

$$NO_3^-, I^- > Br^- > Cl^- > F^- > NO_2^- > N_3^-$$

This is to be compared to Yatsimirskii's bond energy order[10] estimated for the gas phase

$$NO_2^- > NCS^- > I^-, Cl^- > Br^- > NO_3^-$$

Perhaps the closest definition of the role of the leaving group emerges from correlation of rates with equilibria in reactions of the family $Co(NH_3)_5X^{2+}$. In a dissociative mode the leaving group in the transition state strongly resembles the leaving group in the product state. If X is an anionic ligand, in the transition state it should resemble the free anion. The activation free energy should respond to changes in leaving group in much the same way as the free energy difference for the overall reaction responds. A linear free energy relationship (see Vol. 2, Chapter 4) is suggested between the activation energy $\Delta G^{\ddagger}$ and the free energy of reaction ΔG^0 of the form $\Delta(\Delta G^{\ddagger}) = \beta\Delta(\Delta G^0)$, where $\Delta(\Delta G)$ denotes change in the free energy quantity with change of X. In the ideal *dissociative* case, β would be unity. This has been realised for the $Co(NH_3)_5X^{2+}$ family[11] as shown in Fig. 4.

A persuasive reason for preferring the dissociative over the associative interpretation of equation (12) has emerged from recent work on reactions of Co(III)

TABLE 2

RATES OF ACID HYDROLYSIS OF SOME Co(III) COMPLEXES AT 250 °C

The labile ligand is italicized.

*Complex**	$k(sec^{-1})$	$\Delta H^{\ddagger}(kcal)$	$\Delta S^{\ddagger}(eu)$	*Ref.*
$Co(NH_3)_5\ \mathit{OP(OCH_3)_3}$	2.5×10^{-4}	—	—	13
$Co(NH_3)_5\ \mathit{NO_3}^{2+}$	2.7×10^{-5}	25.5	+6	14
$Co(NH_3)_5\ \mathit{I}^{2+}$	8.3×10^{-6}	—	—	14
$Co(NH_3)_5\ \mathit{Br}^{2+}$	6.3×10^{-6}	23.5	−4	14
$Co(NH_3)_5\ \mathit{OH_2}^{3+}$	5.8×10^{-6}	27	+6	15
$Co(NH_3)_5\ \mathit{Cl}^{2+}$	1.7×10^{-6}	23	−9	14
$Co(NH_3)_5\ \mathit{SO_4}^{+}$	1.2×10^{-6}	19	−24	14
$Co(NH_3)_5\ \mathit{OPO_3H_2}^{2+}$	2.6×10^{-7}	—	—	13
$Co(NH_3)_5\ \mathit{NO_2}^{2+}$	1.15×10^{-8}	—	—	16
$Co(NH_3)_5\ \mathit{NCS}^{2+}$	5.0×10^{-10}	31	0	17
$Co(NH_3)_5\ \mathit{OH}^{2+}$	very slow	—	—	14
$Co(NH_3)_6^{3+}$	$\sim 10^{-10}$	—	—	18
trans $Coen_2OH\ \mathit{Cl}^{+}$	1.6×10^{-3}	26.2	+20	18, 19
cis $Coen_2OH\ \mathit{Cl}^{+}$	1.2×10^{-2}	23.1	+10	18, 19
trans $Coen_2Br\mathit{Cl}^{+}$	4.5×10^{-5}	24.9	+2	18, 19
cis $Coen_2Br\mathit{Cl}^{+}$	1.4×10^{-4}	23.5	+14	18, 19
trans $Coen_2\mathit{Cl_2}^{+}$	3.5×10^{-5}	26.2	+14	18, 19
cis $Coen_2\mathit{Cl_2}^{+}$	2.4×10^{-4}	21.5	−5	18, 19
trans $Coen_2N_3\mathit{Cl}^{+}$	2.2×10^{-4}	22.5	0	18, 19
cis $Coen_2N_3\mathit{Cl}^{+}$	2.0×10^{-4}	21.3	−4	18, 19
trans $Coen_2NCS\mathit{Cl}^{+}$	5×10^{-8}	30.2	+9	18, 19
cis $Coen_2NCS\mathit{Cl}^{+}$	1.1×10^{-5}	20.1	−14	18, 19
trans $Coen_2NH_3\mathit{Cl}^{2+}$	3.4×10^{-7}	23.2	−11	18, 19
cis $Coen_2NH_3\mathit{Cl}^{2+}$	5×10^{-7}	24.5	−6	18, 19
trans $Coen_2OH_2\mathit{Cl}^{2+}$	2.5×10^{-6}	—	—	20
cis $Coen_2OH_2\mathit{Cl}^{2+}$	1.6×10^{-6}	—	—	20
trans $Coen_2CN\mathit{Cl}^{+}$	8.2×10^{-5}	22.5	−2	21, 19
trans $Coen_2NO_2\mathit{Cl}^{+}$	9.8×10^{-4}	20.9	−2	18, 19
cis $Coen_2NO_2\mathit{Cl}^{+}$	1.1×10^{-4}	21.8	−3	18, 19
trans $Coen_2NO_2\mathit{Br}^{+}$	4.0×10^{-3}	—	—	18, 19
trans $Coen_2\mathit{Br_2}^{+}$	1.2×10^{-6}	—	—	18
trans $Co(NH_3)_4\mathit{Cl_2}^{+}$	1.8×10^{-3}	—	—	23
cis $Co(NH_3)_4\mathit{Cl_2}^{+}$	fast	—	—	23

* en represents ethylene diamine.

complexes in solvents other than water. Work of Tobe, Watts, Langford and their respective collaborators[12] has demonstrated that these solvents, dimethyl formamide, dimethyl sulphoxide, dimethyl acetamide and methanol also function as "preferred" nucleophiles. This reinforces the suggestion that it is the high solvent concentration and not solvent nucleophilicity that is important. Furthermore, it is found that in these solvents some direct replacement by an anion may be observed but that such replacement is always associated with ion pair formation and that reaction rates show very little sensitivity to the *nature* of the entering ion[12].

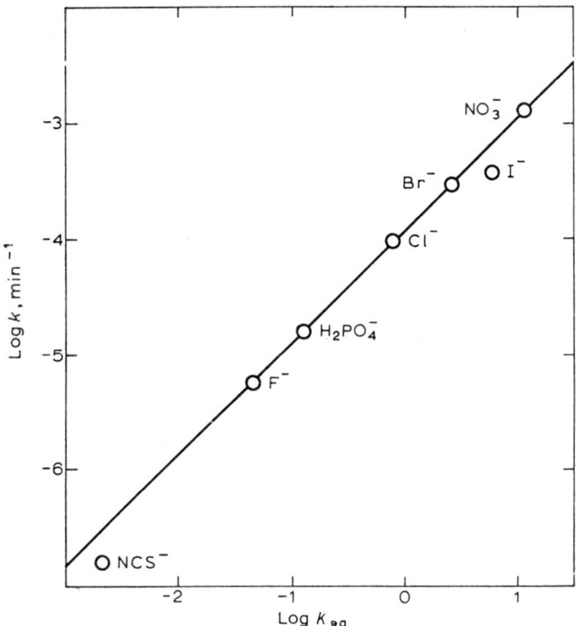

Fig. 4. Linear free-energy relationship for the reaction, $Co(NH_3)_5OH_2^{3+} + X^- \rightleftarrows Co(NH_3)_5X^{2+} + H_2O$. Log k (rate coefficient) *vs.* log K (equilibrium constant).

The interpretation of the next factor, steric crowding, is quite straightforward if its effects can be isolated. Steric crowding should inhibit an *associative* reaction but accelerate a *dissociative* one. It is frequently difficult to isolate the steric effects for a reaction in solution since the structure variations that result in crowding of the reaction site may also modify the surrounding solvent structure in an uncontrollable way or be associated with important electronic effects. There is at least one clear cut experiment concerned with the steric effect on a Co(III) complex that supports the dissociative mode. This is the comparison of the acid hydrolysis rates for *d, l* and *meso* $Cobn_2Cl_2^+$ (bn = 2,3-diaminobutane). It can be seen that the methyl groups on the chelate rings must be opposed in the *meso* form and staggered in the *d, l* form. The *meso* form hydrolyzes about thirty times faster[24]. The other available data are consistent with this suggestion of acceleration by steric crowding and its implication of dissociation.

The effect of overall charge on the complex is perhaps even more difficult to isolate than the steric effects. Each change of charge type is accompanied by an important change in the electronic arrangement about the metal which the discussion (below, p. 12) of electronic effects of non-labile ligands shows to be quite important. However, an assessment of the rates for the 2^+ and 3^+ charged species cited in Table 2 suggests that substitution rates at Co(III) decrease as the overall charge on the complex is increased. This conforms to expectation for the dissociative model when the leaving group is anionic. For the associative model, one might expect opposite or at least very small charge effects (see the discussion of Pt(II), p. 20).

In this catalogue of structure variation experiments to test the hypothesis of a dissociative activation mode, the last is the role of non-labile ligands. This question has been examined using the hydrolysis reactions of the family of complexes *cis* and *trans* $Coen_2ACl^+$ where Cl^- is the leaving group and A is a variable non-labile substituent (see rates in Table 2). The first approach taken to the analysis of the data was to classify the ligands A as electron donors or electron acceptors on the basis of organic chemical precedent and then to plot the observed 25° rate coefficients as a function of decreasing electron donor–increasing electron acceptor properties[25]. The two branches of the curve (Fig. 5) were given a two-

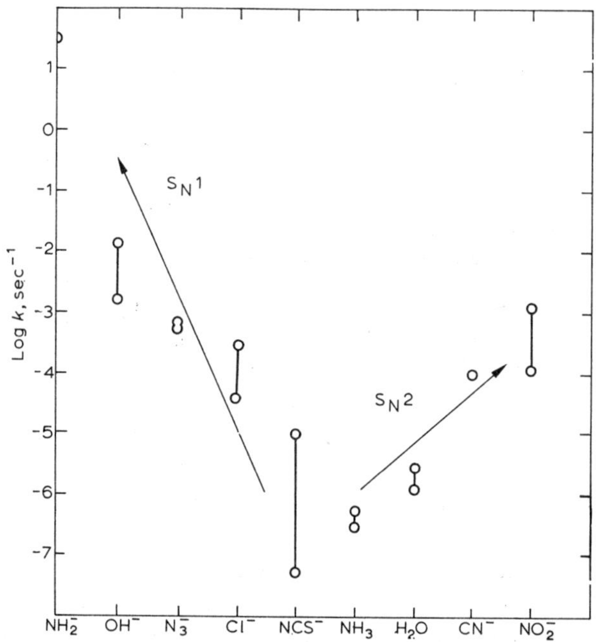

Fig. 5. Rates of hydrolysis of a series of $Coen_2ACl^+$ complexes. The abscissa represents the electron-donating or -accepting power of A.

mechanism interpretation. Good electron donors were supposed to replenish the depleted electron density at Co(III) in a dissociative transition state. Electron acceptors (NO_2^-, CN^-) were supposed to drain away the excess electron density at Co(III) in an associative transition state. A difficulty for this attractive hypothesis is that there is no correspondingly simple pattern in the values of $\Delta H^\ddagger$. The extended effort to obtain some evidence for nucleophilic discrimination in the reactions supposed to involve an associative transition state have been reviewed[12,26]. No direct support has emerged for the postulation of an associative transition state. Fortunately, there is an alternative account of the situation. The dissociative transition state is an incipient five coordinate complex. In the extreme case there are two possible geometries, trigonal bipyramidal and square pyramidal and different ligands may stabilize geometrically different transition states. Note (Fig. 6) that the square pyramidal form cannot lead to *cis–trans* isomerization

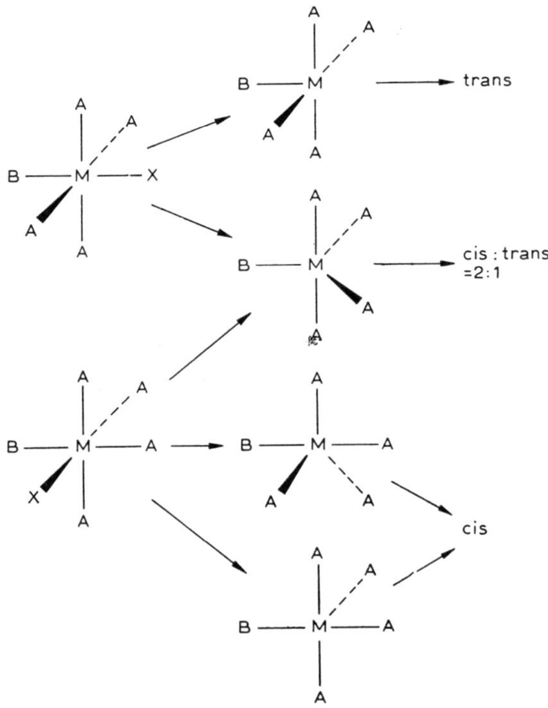

Fig. 6. Stereochemical changes accompanying the dissociative reaction $MA_4BX \rightarrow MA_4B \rightarrow MA_4BY$.

whereas the trigonal bipyramidal form can. A fairly satisfactory correlation between stereochemical rearrangement emerges following the suggestion[19] that the formation of a strongly trigonal form is accompanied by a positive $\Delta S^{\ddagger}$. The two-geometry uniformly dissociative model seems to give the most consistent account of the effects of non-labile ligands.

The hypothesis of dissociative activation in Co(III) reactions stands the available tests well. It is therefore profitable to attempt to distinguish the ***D*** from the $\boldsymbol{I_d}$ pathways. Fig. 7 summarizes the two pathways consistent with ***d*** activation, and the general methods for establishing the stoichiometric mechanism $\boldsymbol{I_d}$ are illustrated by the example of $Co(NH_3)_5OH_2^{3+}$.

First, a case against the ***D*** path may be constructed. A knowledge of the rate

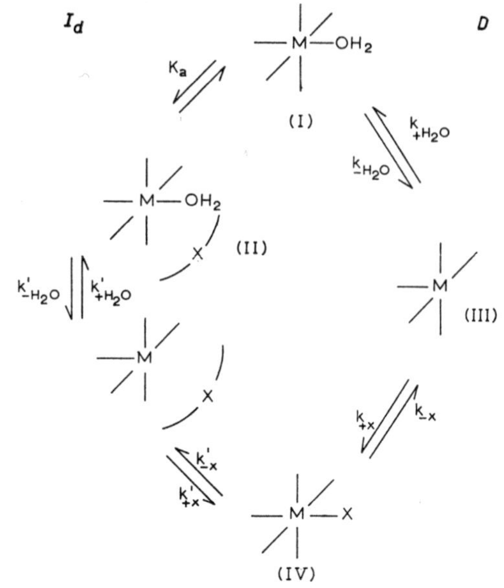

Fig. 7. The I_d *vs.* the D mechanism.

of water exchange of $Co(NH_3)_5OH_2^{3+}$ and the rate of hydrolysis of $Co(NH_3)_5X^{2+}$ under concentration conditions where the reaction goes to completion gives k_{-H_2O} and k_{-X}. These may be combined with the overall equilibrium constant for the reaction to give the ratio k_{+H_2O}/k_{+X}, the competition ratio for the intermediate $Co(NH_3)_5^{3+}$. A number of these competition ratios were calculated by Haim and Taube[27] including the case $X = SCN^-$. The assumed mechanism may be tested by "generating the intermediate" from a different source and checking the competition ratio by evaluati ng the immediate product distribution. This was done by Pearson and Moore[6] who "generated the intermediate" by hydrolysis of the labile nitrato and bromo pentaamines of Co(III) in the presence of a large concentration of SCN^- ion. In conflict with the explicit predictions from the D mechanism[27], they found no evidence for capture of the "intermediate" by thiocyanate. Unless very small concentrations of Br^- or NO_3^- in solution affect the reactivity of the intermediate, it is necessary to conclude that it is not there.

Now we can proceed to assemble the positive evidence for the I_d path (I → II → IV, Fig. 7). Once the outer sphere complex, (II), is formed, all replacements of water should occur at the same rate, k'_{-H_2O}. If the "ion pairing" constant K_a is known, or a limiting rate of anion entry corresponding to saturation of the association is observable, the rates of conversion of (II) into (IV) may be compared for various X. All should be equal to k'_{-H_2O} if the activation mode is d, but they will not equal the rate of water exchange which was identified with k_{-H_2O} on the D path. The reason is that species (II) has a number of solvent molecules in its outer coordination sphere as well as the ligand X. Even if the rate of dissociation of water is unaffected by the presence of X in the neighborhood, the most probable result of water loss will be water exchange and not X entry. Thus on the $\boldsymbol{I_d}$ path we expect all ion pairs to show closely similar rates of conversion to aniono complexes, but we expect these rates to lie below the solvent exchange rate by the appropriate statistical factor for the population of the outer sphere. Recently [9,13] the rates of formation of $Co(NH_3)_5X$ from $Co(NH_3)_5OH_2 \ldots X$ have been reported relative to the water exchange rate for $X = SO_4^{=}, Cl^-, SCN^-$, and $H_2PO_4^-$. The values are 0.24, 0.21, 0.16 and 0.13, respectively. The values span a range of a factor of two which must be admitted to be a little larger than the experimental uncertainty and also easily within the differences among the anions in their probability of occupancy of the crucial outer sphere site adjacent to the leaving water molecule. All are nearly a factor of five below the water exchange rate. These results conform neatly to the $\boldsymbol{I_d}$ predictions.

Examining the relationship between the probability that a ligand occupies an outer sphere site, the rate of ligand incorporation, and the solvent exchange rate, appears to be the most general method for identifying the $\boldsymbol{I_d}$, pathway. It has been applied to several other systems recently. Rates of anion entry into *cis*-$[Coen_2NO_2(DMSO)]^{2+}$ in dimethyl sulphoxide (DMSO) have been compared[28] to the DMSO exchange rate obtained from the deuterium tracer NMR experiments of Lantzke and Watts[29]. The pattern is very akin to that for $Co(NH_3)_5OH_2^{3+}$ in water. Similarly the rate of sulphate anation in the pair *cis*-$[Coen_2(OH_2)_2]^{3+} \ldots SO_4^{2-}$ has been found to be 0.25 times the water exchange rate of the free ion[29a].

Several authors have suggested that the $\boldsymbol{I_d}$ pathway may prove to be the most common mechanism in substitution reactions of octahedral complexes generally. However, the $\boldsymbol{D}$ path can be clearly demonstrated in some cases including at least two examples from Co(III) chemistry. The path (I → III → IV, Fig. 7) through the five coordinate intermediate would lead, in the case of rate studies in the presence of excess anionic ligand, to observed first-order rate constants governed by equation (13)

$$k_{obs} = \frac{k_{-H_2O}[X^-] + k_{+H_2O}k_{+X}/k_{-X}}{k_{+H_2O}/k_{+X} + [X^-]} \tag{13}$$

This form of $[X^-]$ dependence was observed by Haim *et al.*[30,31] in studies of the anation of $Co(CN)_5OH_2^{2-}$. Equation (13) predicts a limiting rate at high $[X^-]$ equal to the solvent exchange rate, and allows substantial variation in reactivity of X^- groups. These features are realised in the $Co(CN)_5OH_2^{2-}$ system. The reactivity order toward the intermediate $Co(CN)_5^{2-}$ is: $OH^- > I_3^- > NH_3 > SCN^- >$ thiourea $> NH_3 > Br^- > S_2O_3^{2-} > NCO^- > H_2O$, spanning about four orders of magnitude. A parallel case[32] has evolved for the intermediate

$Co(NH_3)_4SO_3^+$. The order of reactivity toward this five coordinate species is: $OH^- > NO_2^- > CN^- > NH_3 > H_2O$ and spans six orders of magnitude.

These two five coordinated Co(III) species are, to date, the only ones clearly established from detailed knowledge of the rate law for substitution (a situation very clearly subject to change). Strongly suggestive evidence for others has been accumulated from another approach. Loeliger and Taube[33] and Sargeson *et al.*[34] have examined reactions of a complex where there is a distribution of products (*i.e.* ^{16}O and ^{18}O aquo complexes or stereoisomerically different complexes). They argue that a constant product ratio strictly independent of the nature of the leaving group implies product formation *after* the leaving group is removed from the scene of reaction and the existence of an intermediate. The key word here is "strictly". Leaving group effects may often be quite subtle. In $\boldsymbol{I_d}$ processes, for example, they would appear only to the extent that they modified outer sphere populations. Some persuasive indication of $\boldsymbol{D}$ reaction has been presented for some of the so-called induced aquations of Co(III) amine complexes. Induced aquations include Hg^{2+}-catalyzed halide loss and rapid azide loss catalyzed by nitrous acid.

Before leaving Co(III) chemistry we must consider the base hydrolysis reaction

$$Co(NH_3)_5X + OH^- \rightarrow Co(NH_3)_5OH + X^- \tag{14}$$

The usual rate law is typified by

$$-d\frac{[Co(NH_3)_5X]}{dt} = k[Co(NH_3)_5X][OH^-] \tag{15}$$

It is important to appreciate that the values of k in equation (15) are often quite large when compared to the rates summarized in Table 2. Some of these values appear in Table 3.

The simplest interpretation of equation (15) would assume a nucleophilic attack on Co(III) by OH^-. This, however, would put OH^- in an extraordinary category of nucleophilicity. Garrick[38] was the first to note that an alternative explanation for the role of OH^- was available. In the alternative, the conjugate base of the initial complex ammine is presumed to be formed in small amount and to function as the actual reactive species

$$\begin{aligned} Co(NH_3)_5Cl^{2+} + OH^- &\rightleftharpoons Co(NH_3)_4(NH_2)Cl^+ + H_2O \quad \text{(fast equilibrium)} \\ Co(NH_3)_4(NH_2)Cl^+ &\rightarrow Co(NH_3)_4(NH_2)^{2+} + Cl^- \quad \text{(slow)} \\ Co(NH_3)_4(NH_2)^{2+} + H_2O &\rightarrow Co(NH_3)_5OH^{2+} \quad \text{(fast)} \end{aligned} \tag{16}$$

The mechanism allows for a slow step analogous to the $\boldsymbol{d}$ process observed in

TABLE 3

RATES OF BASE HYDROLYSIS OF SOME Co(III) COMPLEXES

The labile ligand is italicized.

*Complex**	k_{OH}*(l. mole⁻¹.sec⁻¹)*	*T(°C)*	E_a*(kcal. mole⁻¹)*	*ΔS(eu)*	*Ref.*
$Co(NH_3)_5I^{2+}$	23	25	29	+42	35
$Co(NH_3)_5Br^{2+}$	7.5	25	28	+40	35
$Co(NH_3)_5Cl^{2+}$	0.85	25	29	+36	35
$Co(NH_3)_5N_3^{2+}$	3×10^{-4}	25	33	+35	35
$Co(NH_3)_5NO_2^{2+}$	4.2×10^{-6}	25	38	+30	35
trans $Coen_2Cl_2^+$	85	0	23.2	—	36
trans $Coen_2OHCl^+$	0.017	0	22.8	—	36
trans $Coen_2NO_2Cl^+$	0.080	0	24.4	—	36
cis $Coen_2Cl_2^+$	15.1	0	24.6	—	36
cis $Coen_2OHCl^+$	0.37	0	22.4	—	36
cis $Coen_2NO_2Cl^+$	0.03	0	23.1	—	36
trans Co(*d, l*-bn)$_2Cl_2^+$	2100	25	—	—	37
trans Co(*Meso*-bn)$_2Cl_2^+$	9800	25	—	—	37

* bn represents 2,3-diaminobutane.

acid hydrolysis, and the extensive evidence of a parallel between acid and base hydrolytic reactivity has been reviewed[39]. It has also been established by measurement of H–D exchange on the ammine ligands[40,41] that the conjugate base can be formed sufficiently rapidly. The most telling experiments, though, are those that establish that product formation occurs in a step independent of the initial hydroxide involvement. Green and Taube[42] have shown that the $^{16}O/^{18}O$ isotope fractionation factors in base hydrolysis are more easily explained assuming incorporation of O from H_2O than from OH^-, and Sargeson *et al.*[43] have shown that other anions (Y^-) can effectively compete with OH^- in the base catalyzed pathway to give products $Co(NH_3)_5Y^{2+}$ in addition to $Co(NH_3)_5OH^{2+}$. Thus, there seems little doubt that direct OH^- attack on Co(III) is excluded.

The reason for high reactivity on the base-catalyzed pathway remains something of a puzzle. The simplest, but not entirely convincing interpretation, suggests that the NH_2^- ligand functions as an electron donor similar to OH^-. Archer[44] has suggested, on spectroscopic and stereochemical grounds, that the conjugate base species may be labile because it is a *high spin* d^6 complex. This view is rendered more attractive by Watt and Knifton's recent report[45] of a paramagnetic solid Co(III) conjugate base species isolated from liquid NH_3. Gillard[46] has made the interesting suggestion that OH^- may not function as a base but as an electron donor, to produce a transient OH radical and a labile Co(II) species.

4. Cr(III), Rh(III), Ru(III), Ir(III) and Pt(IV) complexes

The related octahedral non-labile complexes which have received some attention will be grouped together simply because there is much less information available than there is with respect to Co(III) complexes. Reactions of Pt(IV) complexes are very slow and attempts to isolate simple thermal substitutions from Pt(II)-catalyzed redox pathways and photochemical reactions have not yet been very successful[47–49]. Information on Ir(III) complexes is very limited to date and probably not adequate for mechanistic analysis[50, 51]. A reasonable account of Cr(III), Rh(III) and Ru(III) behaviour in acid solutions may be constructed adhering to the postulate of ***d*** activation, although experimental results are less comprehensive than in the Co(III) case and the assignment of activation mode is less secure. Reactions of all three also shed important light on the base hydrolysis pathway. Some important rate coefficients for these systems are collected in Table 4.

Leaving-group orders for acid hydrolyses of both the $Cr(NH_3)_5X^{2+}$ and $Cr(OH_2)_5X^{2+}$ systems are parallel to those for Co(III), suggesting the ***d*** activation mode. This point is supported by a correlation between $\Delta S^{\ddagger}$ for hydrolysis of $Cr(OH_2)_5X^{2+}$ and S^0_{hydr}, the entropy of hydration of the ions X^-, discovered by Swaddle and King[55]. This suggests that X^- functions as a solvated anion in the transition state. The evidence from charge dependence in Cr(III), indicated in Table 4, is also consistent with a ***d*** process. Detailed evidence on steric crowding of Cr(III) is lacking but it has been noted[66] that $Co(CH_3NH_2)_5Cl^{2+}$ hydrolyzes faster than $Co(NH_3)_5Cl^{2+}$ whereas $Cr(CH_3NH_2)_5Cl^{2+}$ hydrolyzes more slowly than $Cr(NH_3)_5Cl^{2+}$. This seems discordant.

One of the earliest suggestions of what we characterize as the $\boldsymbol{I_d}$ mechanism emerged from a study by Jones *et al.*[67] of reactions of $Cr(NH_3)_5Br^{2+}$ ion *paired* with various organic anions. They found that the initial entering group was water even though the final thermodynamic product was the complex of the organic anion. Fairly strong evidence for the stoichiometric $\boldsymbol{I_d}$ pathway was presented by Duffy and Earley[68] who established that reactions of $Cr(NH_3)_5OH_2^{3+}$ with SCN^-, Cl^- and $H_2{}^{18}O$ follow a pattern very similar to the reactions of $Co(NH_3)_5OH_2^{3+}$ (see p. 11). It has been shown[69] that hydrolysis of $Cr(OH_2)_5I^{2+}$ in the presence of Cl^- produces 10–20 % $Cr(OH_2)_5Cl^{2+}$. This could be explained by the formation of the ***D*** intermediate $Cr(OH_2)_5{}^{3+}$ which reacts selectively with Cl^-. Unfortunately, a much more likely explanation can be given because of the demonstration[70] that I^- has a strong *trans*-labilizing effect in $Cr(OH_2)_5I^{2+}$. The observed product probably arises from rapid hydrolysis of $Cr(OH_2)_4ClI^+$.

Perhaps the most telling piece of information about Rh(III) reactions in aqueous solutions is a study[71] of the reactions

$$trans\ Rhen_2Cl_2{}^+ + 2\,X^- \rightarrow trans\ Rhen_2X_2{}^+ + 2\,Cl^- \tag{17}$$

as a function of the entering group, X. The observations are summarized in Table 5. The small differences very probably reflect secondary factors and there is no evidence for nucleophilic attack over this quite varied set of nucleophiles. The

TABLE 4

RATES OF HYDROLYSIS REACTIONS OF SOME OCTAHEDRAL COMPLEXES AT 25 °C

The labile ligand is italicized

A. Acid hydrolysis

Complex	$k(sec^{-1})$	$E_a(kcal.mole^{-1})$	*Ref.*
$Cr(NH_3)_5I^{2+}$	1.0×10^{-3}	21	52
$Cr(NH_3)_5Br^{2+}$	6.8×10^{-5}	24	53
$Cr(NH_3)_5Cl^{2+}$	7.3×10^{-6}	24	53
$Cr(NH_3)_6^{3+}$	1.0×10^{-7}	26	54
$Cr(OH_2)_5F^{2+}$	6.2×10^{-10}	29	55
$Cr(OH_2)_5Cl^{2+}$	2.8×10^{-7}	25	55
$Cr(OH_2)_5Br^{2+}$	3.1×10^{-6}	24	56
$Cr(OH_2)_5I^{2+}$	8.4×10^{-5}	23	55
$Cr(OH_2)_5N_3^{2+}$	4.6×10^{-8}	33	55
trans $Cr(OH_2)_4Cl_2^+$	8.3×10^{-5}	27	57
cis $Cr(OH_2)_4Cl_2^+$	4.3×10^{-5}	—	57
trans $Cren_2Cl_2^+$	2.2×10^{-5}	23	58
cis $Cren_2Cl_2^+$	3.3×10^{-4}	21	58
trans $Cren_2OH_2Cl^{2+}$	$<10^{-6}$	—	58
cis $Cren_2OH_2Cl^{2+}$	$\sim3\times10^{-5}$	—	58
$Cr(NCS)_6^{3-}$	5.3×10^{-5}	—	59
$Ru(NH_3)_5Cl^{2+}$	7.0×10^{-7}	23	60
$Ru\ Cl_6^{3-}$	~1	—	61
trans $Ru(OH_2)_3Cl_3$	2.1×10^{-6}	—	61
$Ru(OH_2)_5Cl^{2+}$	$\sim10^{-8}$	—	61
$Rh(NH_3)_5Br^{2+}$	10^{-8}	26	62
$Ir(NH_3)_5Br^{2+}$	2×10^{-10}	27	51
$Rh(OH_2)Cl_5^{2-}$	3×10^{-4}	—	63
$Ir(OH_2)Cl_5^{2-}$	$\sim10^{-6}$	—	64
$IrCl_6^{3-}$	9.4×10^{-6}	30	64

B. Base hydrolysis

Complex	$k_{OH}(l.mole^{-1}sec^{-1})$	$E_a(kcal.mole^{-1})$	*Ref.*
$Cr(NH_3)_5I^{2+}$	3.6	27	52
$Cr(NH_3)_5Br^{2+}$	6.9×10^{-2}	26	52
$Cr(NH_3)_5Cl^{2+}$	1.7×10^{-3}	27	52
$Ru(NH_3)_5Cl^{2+}$	4.9	—	60
$Rh(NH_3)_5I^{2+}$	7.3×10^{-5}	33	65
$Rh(NH_3)_5Br^{2+}$	3.4×10^{-4}	31	65
$Rh(NH_3)_5Cl^{2+}$	4.1×10^{-4}	29	65

pattern is very similar to the indiscriminate reactivity of Co(III) complexes and very probably means that the initial product of substitution is *normally* the aquo compound. This would again recommend provisional assignment of the dissociative mode of activation. However, the evidence available from other effects on Rh(III) reactivity does not support the hypothesis as clearly as is the case in Co(III) chemistry. Steric crowding does result in faster reactions, but the effect

References pp. 52–55

is smaller[71]. Of course, this may be as expected for the larger Rh(III) ion. Limited data[71] on the effect of changing charge on the ion is contrary to expectation, for the more highly-charged species react the faster. It is not here at all clear that only overall charge effects are being measured.

TABLE 5

RATES OF: *trans* Rhen $_2Cl_2^+ + 2X_- \rightarrow$ *trans* Rhen $_2X_2^+ + 2Cl^-$ AT 80 °C

X	$k \times 10^5 (sec^{-1})$	X	$k \times 10^5 (sec^{-1})$
OH^- (0.1 M)	5.1	I^- (0.05 M)	5.1
NO_2^- (0.1 M)	4.2	$^{36}Cl^-$ (0.01 M)	4.0
NO_2^- (0.05 M)	4.2	Thiourea (0.1 M)	4.9
I^- (0.1 M)	5.2	NH_3 (5 M)	4.0

A very striking feature of Rh(III) chemistry is illustrated in Table 5. This is the reduced importance of the hydroxide-dependent base hydrolysis. Comparison of data in Table 4 with those in Table 3 reveals that base hydrolysis is also less important in Cr(III) than in Co(III) but that it is an important feature of Ru(III) chemistry. Some clue to this puzzling situation is provided by the observation[60] that $Ru(NH_3)_5Cl^{2+}$ undergoes H–D exchange more than 100 times faster than $Co(NH_3)_5Cl^{2+}$, $Rh(NH_3)_5Cl^{2+}$ or $Cr(NH_3)_5Cl^{2+}$. There could be, it seems, more of the conjugate base species present in the Ru(III) system. In fact it has been estimated that the reactivity for the conjugate base species follows the order: Co(III) $\gg$ Rh(III) $\sim$ Cr(III) $>$ Ru(III)[72]. There does not as yet appear to be a simple explanation for this order of reactivity. One more interesting fact deserves note. Base hydrolysis is important in reactions of $Rh(NH_3)_5Cl^{2+}$ (see Table 4) but is undetected in reactions of *trans* $Rhen_2pyCl^{2+}$. An ammine group *trans* to the leaving group appears to be required.

5. Complexes of Pt(II), Pd(II), Au(III) and Rh(I)

To this point the complexes considered have shared the coordination number six and approximate octahedral geometry. It has been argued that they also share the dissociative reaction mode. There are examples of reactions both with and without intermediates of reduced (that is, 5) coordination, but the insensitivity to entering ligands is a consistent feature. It will be interesting, shortly, to see if the dissociative pattern persists in more or less "organometallic" octahedral systems but first we shall give some attention to the non-labile square planar systems.

As we turn to complexes with only four groups bound in a planar arrangement and two potential coordination positions open it should not be too surprising to encounter a change of mechanism. That such a change does occur is revealed from the rate law which was discovered by Rich and Taube[73] for the radio-chloride exchange reaction

$$AuCl_4^- + {}^{36}Cl^- \rightarrow Au^{36}Cl_4^- + Cl^- \qquad (18)$$

The rate law for exchange is

$$\text{rate} = (k_2 + k_2[{}^{36}Cl^-])[AuCl_4^-] \qquad (19)$$

and incorporates a first-order term independent of the entering ligand plus a second-order term *dependent upon the entering ligand.* This rate law has been found to be quite general for substitution reactions of square planar systems. The most detailed experiments have been carried out on Pt(II) complexes.

The new feature is the k_2 term in the rate law. Fig. 8 shows how this depends on the nature of the entering ligand for the particular reaction

$$Pt(dien)Br^+ + Y \rightarrow Pt(dien)Y + Br^- \qquad (20)$$

(dien = 1,4,7-triazaheptane,diethylenetriamine; a tridentate ligand). The figure shows apparent first-order rate coefficients in the presence of excess Y as functions of Y, values of k_2 ranging from "0" and 8.8×10^{-4} l.mole^{-1}.sec^{-1} for OH^- and Cl^- to 4.3×10^{-1} and 8.3×10^{-1} l.mole^{-1}.sec^{-1} for SCN^- and thiourea. The pathway characterized by k_2 is clearly *selective* with respect to entering groups and must be *associative*. This result raises the interesting question of what com-

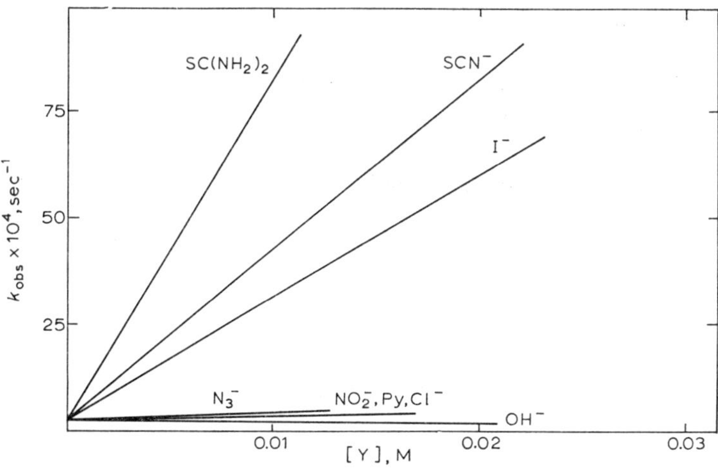

Fig. 8. Relative reactivities of different entering groups Y with $PtdienBr^+$ in aqueous solution at 25 °C. Data from ref. 74.

References pp. 52–55

prises a good nucleophile for Pt(II). Apparently OH^-, which is both a strong base and a good nucleophile for attack at carbon is quite unreactive toward Pt(II). Of course, reactivity of nucleophilic reagents is quite sensitively dependent upon the nature of the substrate. A good deal of effort has been directed toward the exploration of the correlation of nucleophilic reactivity with easily measurable characteristics of the nucleophile. A two-parameter approach based on the basicity (toward H^+) and polarizability of the nucleophile meets with considerable success. Nucleophilicity toward Pt(II) is clearly most dependent upon polarizability, but it should be emphasized that the exact connection between the "bulk" polarizability of a species and its nucleophilic reactivity is far from clear.

Nucleophilic reactivity toward Pt(II) complexes may be conveniently systematized *via* linear free energy relationships[75] established between reactions of *trans* $Ptpy_2Cl_2$ (py = pyridine) with various nucleophiles and reactions of other Pt(II) complexes with the same nucleophiles. First, each nucleophile is characterized by a nucleophilicity parameter, n^0_{Pt} derived from its reactivity toward the common substrate, *trans* $Ptpy_2Cl_2$. Reactivity toward other Pt(II) substrates is then quite satisfactorily represented by an equation of the form (21), wherein k_Y is the value of k_2 in the reaction with nucleophile Y

$$\log k_Y = sn^0_{Pt} + \log k_s \tag{21}$$

The constant *s*, characteristic of the substrate complex, reflects its sensitivity to variation in nucleophilicity as assessed by the $Ptpy_2Cl_2$ reaction. It is called the nucleophilic discrimination factor (NDF). The intercept log k_s turns out to be related to the value of the k_1 term in the rate law for the solvent in question. Some representative ligands involved in attack on Pt(II) complexes may be listed in order of decreasing n^0_{Pt} as follows[75]

$$(C_6H_5)_3P > S{=}C(NH_2)_2 > SeCN^- > CN^- > SCN^- > I^- > Br^- > Py$$
$$> NH_3 > Cl^- > CH_3O^-$$

Some values of *s* (or NDF) fall in the order[76]

$$\textit{trans}\ Pt[(C_2H_5)_3P]_2Cl_2 > Ptpy_2Cl_2 > Pten_2Cl_2 > Pt\ dien\ Br^+$$
$$> Pt\ dien\ Cl^+ > Pt\ dien\ OH_2^{2+}$$

The k_2 term in the rate law for square planar substitution is very clearly connected with an associative mechanism. The k_1 term may be also. Consider the pathway (S = solvent)

$$\begin{aligned} L_3Pt\text{–}X + S &\rightarrow L_3Pt\text{–}S + X \quad \text{(slow)} \\ L_3Pt\text{–}S + Y &\rightarrow L_3Pt\text{–}Y + S \quad \text{(fast)} \end{aligned} \tag{22}$$

This would accomplish substitution of Y for X by a potentially associative solvent attack followed by a fast replacement of solvent. This interpretation of the k_1 path is strongly supported by two lines of evidence.

First, the order of k_1 values in different solvents is quite reasonably interpreted as a nucleophilicity order[77, 78] (*e.g.* $(CH_3)_2SO > H_2O \sim CH_3NO_2 > C_2H_5OH$) and second, the k_1 rate is (as is the k_2) greatly reduced by steric crowding[77, 79]. Of course, a nucleophilic attack by solvent is a very likely process, *a priori*. In any solvent, the solvent itself will be the *poorest* nucleophile that can be studied since poorer ones will not effectively compete. Thus the k_s term of equation (21) corresponds to the k_1 value. The aquo intermediate of scheme (22) has been trapped by using reactions in the presence of OH^-, a poor nucleophile but good base[86].

In the reactions of Pt(II) complexes three ligands, the entering, the leaving and the *trans* group, are found to have a large influence upon the rate. As we have seen, the entering group is quite important and that fact establishes associative activation. It is probably not surprising that the leaving group is also important. In fact, leaving group variations span the same wide range of rates as entering groups. A typical replacement rate series is[81]

$$NO_3^- > OH_2 > Cl^- > Br^- > I^- > N_3^- > SCN^- > NO_2^- > CN^-$$

H_2O departs about 10^5 times as fast as CN^-. A good attacking group is a poor leaving group with very few exceptions. OH^- is notable, for it is a very poor nucleophile for Pt(II) but is only very slowly replaced. In the main, lability also corresponds to thermodynamic instability, for example in the series

rates: $Pt(dien)I^+ < Pt(dien)Br^+ < Pt(dien)Cl^+$

formation constants: $Pt(dien)I^+ > Pt(dien)Br^+ > Pt(dien)Cl^+$

The third ligand which plays a major role in determining rates of substitution at Pt(II) is the ligand which is *trans* to the leaving group. Its effect is as spectacular as the effect of the directly involved entering and leaving ligands, and, since first being recognized and explored by Werner[82] and Chernyaev[83], it has fascinated students of Pt chemistry. The *trans* effect has been crucial to the control of synthesis in the square planar series. A fairly constant order of labilizing effect of *trans* ligands can be given. If we could study the rate of release of a single leaving group being replaced by a single entering group with the *trans* group varied from CN^- to OH^-, the rate would decrease by a factor of about 10^6. A *trans* effect order[84] is given in Table 6. Inspection of the table reveals that a good *trans* labilizing ligand corresponds to a good nucleophile and a poor leaving group in the great majority of cases.

The complementary roles of *trans*, entering, and leaving ligands when contrasted with the much smaller role of variable *cis* ligands suggest an explicit model for

References pp. 52–55

TABLE 6

Trans EFFECT ORDERING OF LIGANDS

Relative rates

very large	>	*large*	>	*moderate*	>	*small*
CO CN^- >C=C<	>	NO_2^- $SC(NH_2)_2$ I^- SCN^-	>	Br^- Cl^-	>	Py NH_3 OH^-
		$PMe_3 > PEt_3$, $H^- > CH_3^- > C_6H_5^-$				

the transition state. Associative attack at Pt(II) probably is well represented by the scheme

(23)

In scheme (23) the transition state is concerned, as approximating a trigonal bipyramidal five-coordinate species, with the leaving, entering and *trans* ligands sharing the trigonal plane. It is significant that the model also accounts for the fact that substitutions at Pt(II) occur with *retention* of geometrical isomerism[85] (*cis* → *cis* and *trans* → *trans*).

Finally, it is noteworthy that the associative activation model suggested for reactions of Pt(II) species is strongly supported by interpretation of the activation energetics. The characteristic finding[86] is a low value of the activation enthalpy and a highly negative activation entropy. These combine to suggest that formation of the transition state is accompanied by a net increase in bonding.

If all Pt(II) complexes are activated associatively, it becomes plausible to suggest that the five-coordinate species in scheme (23) is not merely a transition state but also an intermediate. Unfortunately, this question of stoichiometric mechanism cannot be resolved by asking for evidence in the rate law for competitive reactions with the intermediate. The only way to look for the intermediate directly is to search for its accumulation. Such accumulation could be evidenced by departure from the second-order rate law at high ligand concentration if the step forming the intermediate becomes rapid compared to its subsequent breakdown. Such kinetic behaviour has been observed once in a rather unusual Pt(II) complex containing sulphur donor atom ligands[87]. Indirect evidence suggesting the intermediate comes from the preparation of several stable five-coordinate Pt(II) complexes[88], such as $Pt(SnCl_3)_5^{3-}$, $PtH(SnCl_3)_4^{3-}$ and $Pt[(C_2H_5)_3P]_2H(SnCl_3)_2^-$. The generality of the occurrence of the five-coordinate intermediate must be regarded as unknown.

This section so far has concentrated on the evidence for ***a*** and possibly ***A*** reactivity at Pt(II). Little has been said of the other square planar systems. This is because much more data are available on Pt(II) and where data for other complexes exist, they exhibit the pattern made familiar by Pt(II). The rate coefficients for reactions of $AuCl_4^-$ are larger[73] than those for $PtCl_4^{2-}$ suggesting increased bonding by the entering group in the transition state. It is also true that square planar Pd(II) and Ni(II) complexes react faster than the corresponding Pt(II) systems. Probably this reflects easier achievement of five-coordination. Axial perturbations in solution are more pronounced for these square planar systems. Some rates of reaction of complexes with pyridine are compared in Table 7. The Rh(I) and Ir(I) square planar complexes $Rh(CO)_2$(*p*-anisidine)Cl and $Ir(CO)_2$ (*p*-toluidine)Cl undergo second-order exchange with ^{14}CO at −80 °C in ethanol, with k_2 approx. 2 l.mole^{-1}.sec^{-1} in both cases.

TABLE 7

RATES OF REACTIONS OF ANALOGUES OF Pt(II) COMPLEXES WITH PYRIDINE

Complex	*[Py] (M)*	*Solvent*	*Temp* (°C)	k_{obs}(*sec*$^{-1}$)	*Ref.*
Ptdien SCN^+	0.00010	H_2O	25	6.2×10^{-8}	81
Pddien SCN^+	0.00123	H_2O	25	4.2×10^{-2}	81
Ptdien NO_2^+	0.00592	H_2O	25	5.0×10^{-8}	81
Pddien NO_2^+	0.00124	H_2O	25	3.3×10^{-2}	81
trans-$Pt[(C_2H_5)_3P]_2$(*o*-tolyl)Cl	0.0062	C_2H_5OH	−25	1.7×10^{-5}	90
trans-$Pd[(C_2H_5)_3P]_2$(*o*-tolyl)Cl	0.0062	C_2H_5OH	−40	5.8×10^{-3}	90
trans-$Ni[(C_2H_5)_3P]_2$(*o*-tolyl)Cl	0.0062	C_2H_5OH	−65	1.6×10^{-2}	90

6. Complexes with B class ligands: the binary carbonyls

The characteristic of these complexes in general is that ligands very often possess vacant (π-bonding) orbitals. It applies particularly to those unsaturated ligands, such as CN, CO and hydrocarbons with relatively low energy (and vacant) antibonding orbitals, less particularly to the ligand atoms with relatively low energy (and vacant) *d* orbitals (those which do not belong to the first short period). We refer to the *A*, *B* dichotomy described by Chatt *et al.*[91] and the more general division into "hard" and "soft" acids or bases developed by Pearson[2]. Oxidation numbers are low, in the most typical cases being zero and very often negative. In a purely formal sense, this follows from the above, in that the ligand can accommodate some fraction of the formal negative charge on the metal. More exactly, π^* molecular orbitals in the complex will always be lowered in energy by the presence of such ligands[92]. Coordination numbers are generally low and covalent contributions to the total metal-ligand bond appear to be high.

These four characteristics find expression in a symbiotic principle[93], that is to say, that ligand-metal association tends to form well-matched systems in the sense that B class ligands combine preferentially with B class metals and even promote association with other B class ligands. The main virtue of a B class ligand appears to lie in its minimizing interelectron repulsion, so promoting all of the above characteristics. There is an independent criterion of this effect, namely the position of the ligand in the nephelanxetic series[94]. We will here be concerned primarily with organometallic rather than coordination compounds, generally, that is, uncharged complexes with weakly basic ligands in the Brønsted sense. Thus neither complexes nor ligands will require a particularly polar solvent environment which might interfere with kinetic interpretation. Against this advantage must be set, firstly, the ambiguity which may arise in the oxidation number of the metal and, secondly, the possibility that ligand reaction or rearrangement may occur without normal substitution on the metal, and this may dominate the observations. The stereochemistry of organometallic compounds is more varied[95] than that of coordination compounds, and it is evident that more unorthodox substrates are open to investigation. However, the kinetic information available to date concerns compounds of conventional, especially octahedral and tetrahedral, stereochemistry. The greatest body of information concerns the carbonyl complexes – in a sense these may perhaps be characteristic of the whole group of B type compounds, as the ammines have been conveniently taken as characteristic of the A type compounds.

Some initial assessment of the metal–carbonyl bond may first be made. The simpler neutral binary carbonyls fall conveniently into three eighteen-electron classes, *viz.* the octahedral chromium group hexacarbonyls, the trigonal bipyramidal iron group pentacarbonyls and tetrahedral nickel tetracarbonyl. In addition, dimers or polymers with metal–metal bonds are formed by the elements of groups VII and VIII while carbonylate anions and some cations are formed which are structurally akin to the three basic monomeric types. Table 8 indicates some of these formal relationships between the different carbonyls. Replacement of CO by other ligands, anionic such as the halides or neutral such as the group V ligands, hydrocarbons and the like, or a change in the charge on the complex or a change in its geometry will result in a change in the nature of the M–C bond. This has been adequately reviewed[96]: briefly, the total M–C bond owes its peculiar

TABLE 8

SOME BINARY CARBONYL COMPOUNDS

			$Fe(CO)_4^{2-}$	$Co(CO)_4^-$	$Ni(CO)_4$
				$Co_2(CO)_8$	
		$Mn(CO)_5^-$	$Fe(CO)_5$		
		$Mn_2(CO)_{10}$			
$V(CO)_6^-$	$Cr(CO)_6$	$Mn(CO)_6^+$			
$V(CO)_6$					

stability to two synergic contributions[97], firstly, a σ-bond using a predominantly *sp* orbital of carbon and, secondly, a π-bond using a π^*, predominantly anti-bonding C–O, orbital. In very similar systems the force constant of the C–O bond might be expected to be inversely related to that of the M–C bond simply because a greater contribution of metal to ligand π bonding (an increase in M–C bond order) necessarily results in a smaller net π contribution to the C–O bond. *A priori*, there is no good reason why force constant should relate to bond strength in other than very similar systems, but a like evaluation may be applied cautiously to the substituted carbonyls and, indeed, is a very useful method of monitoring directional or other effects of carbonyl substitution. For the moment, the three series of isoelectronic carbonyls in Table 9 demonstrate fairly adequately that increase in the metal negative charge or decrease in its oxidation number increases the M–C bond order. Again, there is no reason why a low bond order should necessarily relate to a high rate of substitution unless the activation step is one of essentially complete dissociation.

TABLE 9

ν_{CO}(cm^{-1}) FOR SOME CARBONYL COMPOUNDS

			$Fe(CO)_4^{2-}$	$Co(CO)_4^-$	$Ni(CO)_4$
			1786	1886	2057
		$Mn(CO)_5^-$	$Fe(CO)_5$		
		1895	2034		
		1863	2014		
$V(CO)_6^-$	$Cr(CO)_6$	$Mn(CO)_6^+$			
1859	1985	2090			
	$Mo(CO)_6$				
	1990				
	$W(CO)_6$				
	1980				

* Data taken from refs. 98, 99, 100.

The ligand replacement reactions in the B type complexes superficially resemble those of the A type, with one important difference, the frequent appearance of a second-order term. In many cases thermal exchange or substitution in the binary carbonyls is a first-order reaction, whether it takes place in the gas phase or in solution. Only octahedral $V(CO)_6$, tetrahedral $Ni(CO)_4$ and closely allied $Co_2(CO)_8$ exchange relatively rapidly[103]. $V(CO)_6$, the only example of a paramagnetic binary carbonyl, must fall into a special category, for, if metal to carbon π donation does contribute some part to the overall bond in the carbonyls, then $V(CO)_6$ is the only one which in this respect falls short of a full complement of six $d\pi$ electrons. The very rapid exchange of all CO's in $Co_2(CO)_8$ may take place *via* a partial one-ended dissociation of a bridging CO together with a rapid randomization of the remaining non-bridged ligands[101,102]. This is compatible with an interpretation of the results for $Fe_3(CO)_{12}$ and the structurally similar $Co_4(CO)_{12}$ (ref. 105) wherein all CO's exchange at the same rate[104]. For $Mn_2(CO)_{10}$ and $Fe(CO)_5$, containing no such bridging CO's the rates of exchange are exceedingly slow.

The rapid first-order exchange rate[108] for $Ni(CO)_4$ is to be contrasted with the much slower rates[130] for the carbonylate anions $Co(CO)_4^-$ and $Fe(CO)_4^{2-}$, and this is quite in accordance with a higher M–C bond order in these two latter compounds, provided that the primary reaction step is a dissociative one. The group VI hexacarbonyls exchange CO several orders of magnitude more slowly than $Ni(CO)_4$ by a first-order process in the gas phase[106,107]. There is very good evidence that the photochemical substitutions of these hexacarbonyls proceed *via* a dissociative step[109]. The thermal substitution of $Mo(CO)_6$ with aromatic and olefinic ligands[110] follows first-order kinetics, but at much higher concentrations of the more nucleophilic amines and phosphines[111,112] a second-order term becomes apparent. The slow exchange[103] of CO in $Fe(CO)_5$ is very much accelerated in aqueous acid[118] and this is considered to be due to formation of $[Fe(CO)_5H]^+$. Either the labilizing influence of the H ligand is responsible for increasing the rate of substitution or else the positive charge on the complex, reducing the M–C bond order, has this effect (for an alternative interpretation, see p. 38). In any case, *substitution* of the CO in $Fe(CO)_5$ is also an exceedingly slow reaction[119]. Although a second-order dependence in $Ni(CO)_4$ substitution has not been detected, yet the first-order rate coefficient (triphenyl phosphine as entering ligand) does show a strong dependence upon the nature of the solvent[113]. Generally, saturated solvents lead to slower reactions, higher activation enthalpies and higher entropies. Table 10 gives some typical values for rate parameters of exchange and substitution, first-order for the tetrahedral carbonyls, first and second-order for the octahedral carbonyls.

It has been pointed out that the types of solvents which are used here, are not generally such as would enter into strong association with the substrate. The molecularity of the substitution reaction may then stand more chance of being an operational concept. Amongst the binary carbonyls, the only systems which have been extensively studied have been nickel tetracarbonyl and the hexacarbonyls of group VI. For the former, the observation of a first-order rate is at least consistent with a rate-determining dissociation of one carbonyl ligand followed by reaction of the intermediate with whichever nucleophile should be available.

$$Ni(CO)_4 \xrightarrow{\text{slow}} Ni(CO)_3 + CO$$

$$Ni(CO)_3 + L \xrightarrow{\text{fast}} Ni(CO)_3L \tag{24}$$

The reaction has been considered[108] as an ideal, or limiting S_N1 dissociation[116],

TABLE 10

RATES OF THERMAL SUBSTITUTION IN SOME BINARY CARBONYLS

$M(CO)_n + L \rightarrow M(CO)_{n-1}L + CO$

Part 1. *Rate* $= k_1[M(CO)_n]$

Compound	*L*	*Solvent*	*Temp.* (°C)	k_1 *(sec⁻¹)*	$\Delta H^‡$ *(kcal. mole⁻¹)*	$\Delta S^‡$ *(eu)*	*Ref.*
$Ni(CO)_4$	$C^{18}O$	Hexane	0.5	2.45×10^{-4}	24.3	14	108
$Ni(CO)_4$	$(C_6H_5)_3P$	Hexane	0.5	2.45×10^{-4}	24.2	13	108
$Ni(CO)_4$	$(C_6H_5)_3P$	Cyclohexane	25	6.87×10^{-3}	26.6	20.9	113
$Ni(CO)_4$	$(C_6H_5)_3P$	Toluene	25	1.94×10^{-2}	20.4	2.0	113
$Co(CO)_4^-$	^{14}CO	H_2O	60	very slow	—	—	130
$Fe(CO)_4^{2-}$	^{14}CO	H_2O	60	very slow	—	—	130
$Cr(CO)_6$	^{14}CO	Gas	117	2×10^{-5}	38.7(E_a)	—	106
$Mo(CO)_6$	^{14}CO	Gas	116	7.5×10^{-5}	30.8(E_a)	—	107
$W(CO)_6$	^{14}CO	Gas	142	2.6×10^{-6}	39.8(E_a)	—	107
$Mo(CO)_6$	Benzene	*n*-Decane-cyclohexane	112	2×10^{-5}	—	—	110
$Mo(CO)_6$	Mesitylene	*n*-Decane-cyclohexane	112	2.75×10^{-5}	—	—	110
$Mo(CO)_6$	Toluene	*n*-Decane-cyclohexane	112	1.74×10^{-5}	—	—	110
$Mo(CO)_6$	Hexamethyl benzene	*n*-Decane-cyclohexane	112	1.34×10^{-4}	—	—	110
$Mo(CO)_6$	Cyclo-heptatriene	*n*-Decane-cyclohexane	106	5.6×10^{-5}	30.0(E_a)	—	110
$Mo(CO)_6$	Norbor-nadiene	*n*-Decane-cyclohexane	112	3×10^{-4}	—	—	110
$Cr(CO)_6$	$(n\text{-But})_3P$	Decalin	130.7	1.38×10^{-4}	40.2	22.6	112
$Mo(CO)_6$	$(EtO)_3P$	Decalin	112.0	2.13×10^{-4}	31.7	6.7	112
$W(CO)_6$	$(C_6H_5)_3P$	Decalin	165.7	1.15×10^{-4}	39.9	13.8	112

Part 2. *Rate* $= k_2[M(CO)_n][L] + k_1[M(CO)_n]$

Compound	*L*	*Solvent*	*Temp.* (°C)	k_1 *(sec⁻¹)*	$\Delta H^‡$ *(kcal. mole⁻¹)*	$\Delta S^‡$ *(eu)*	*Ref.*
$Cr(CO)_6$	$(n\text{-But})_3P$	Decalin	130.7	8.54×10^{-5}	25.5	−14.3	112
$Mo(CO)_6$	$(n\text{-But})_3P$	Decalin	112.0	2.05×10^{-4}	21.7	−14.9	112
$W(CO)_6$	$(n\text{-But})_3P$	Decalin	165.7	7.1×10^{-5}	29.2	−6.9	112
$Cr(CO)_6$	$(C_2H_5O)_3P$	Decalin	130.7	4.5×10^{-5}	—	—	112
$Mo(CO)_6$	$(C_2H_5O)_3P$	Decalin	112.0	6.96×10^{-4}	—	—	112
$W(CO)_6$	$(C_2H_5O)_3P$	Decalin	165.7	1.7×10^{-4}	—	—	112
$Cr(CO)_6$	$(C_6H_5)_3P$	Decalin	130.7	4.31×10^{-5}	—	—	112
$Mo(CO)_6$	$(C_6H_5)_3P$	Decalin	112.0	1.77×10^{-4}	—	—	112
$W(CO)_6$	$(C_6H_5)_3P$	Decalin	165.7	8.88×10^{-5}	—	—	112

leading to either exchange or substitution, in competition but summing to the same total rate. There are for example strong correlations between rates and activation parameters for the CO exchange and the substitution. However, it has been shown[113] that there is also a strong solvent dependence of the first-order rate coefficient for triphenylphosphine substitution into nickel tetracarbonyl, from which considerable contribution of solvent to the transition state may be inferred. Those solvents (electron-releasing, aromatic) which promote the reaction are those which generally form the most stable complexes of, for example, the group VI metals and which, in $Ni(CO)_4$, decrease the CO bond order[114]. There are no strong arguments for either dissociative or associative activation modes. On the one hand, there is no evidence for the required intermediate $Ni(CO)_3$ and on the other it is difficult to envisage a direct attack upon the Ni in $Ni(CO)_4$. There is, however, an alternative possibility presented below (see p. 31).

In the case of the hexacarbonyls, the rate-expression contains not only the same type of first-order term but in addition one second-order overall. For good entering groups (but not CO, for example) the rate expression contains a term strictly first-order in both the complex and the entering nucleophile. The first-order rates of CO exchange are practically identical with the rates of substitution in hydrocarbon solvents, but there is nevertheless some acceleration in ether (THF, dioxan) solutions. This solvent-dependence is not so well-marked[110] as in the case of nickel tetracarbonyl. The second-order rate of substitution very strongly depends upon the basicity[115] of the entering nucleophile

$$(n\,\mathrm{But})_3P > (C_2H_5O)_3P > C_2H_5C(CH_2O)_3P > (C_6H_5)_3P > (C_6H_5O)_3P > (C_6H_5)_3As$$

These observations are, indeed, consistent with an associative activation, generation of a seven-coordinated intermediate (easier in the case of Mo and W than for Cr because their larger sizes produce less steric hindrance) by attack taking place directly upon the metal atom, that is, with an *A* or a limiting S_N2 mechanism[116], accompanied by a reaction sequence involving dissociative activation similar to scheme (24) above, *viz.*

$$M(CO)_6 + L \xrightarrow{\text{slow}} M(CO)_6L$$

$$M(CO)_6L \xrightarrow{\text{fast}} M(CO)_5L + CO \qquad (25)$$

The experimental results are equally consistent with an initial attack taking place upon one of the coordinated ligands. A faster nucleophilic attack at the metal by stronger bases is unexpected in view of the fact that these are electronically saturated compounds. For this reason base attack upon the coordinated CO should not be excluded from consideration as a possible mode of activation, and it is worth noting here that oxygen exchange but not CO exchange occurs[117] in aqueous solutions of $Re(CO)_6{}^+$. There will be available empty orbitals, largely located on the ligands and electron donation into these will promote a loss of the ligands. In effect, it is likely that the C of the CO ligand bears a net positive charge and as a result is subject to nucleophilic attack

$$\text{M–CO} + \text{X} \rightarrow \text{M–}\overset{\displaystyle \text{X}}{\overset{|}{\text{CO}}} \rightarrow \text{M–X} + \text{CO} \tag{26}$$

In the general case, an incoming nucleophile would be expected to be favoured by (*i*) a high basicity consistent with (*ii*) a high polarizability, and the metal complex to favour its approach if (*iii*) it contains electron-acceptive, or B class ligands. An interpretation of the available data may be essayed on these lines. The infrared data upon $Ni(CO)_4$ are consistent with a weakening of the C–O bond[114], and it would be of interest to examine the solvent effect upon the Ni–C bond.

7. Complexes with B class ligands: the substituted carbonyls

There has been considerable success in accounting for the grosser changes in the carbonyl bond character upon substitution by evaluating only the relative π-bonding capacity of CO and the substituting ligands. Analyses of IR spectral changes[120] allowed assessment of the contribution of these substituting ligands to the limited π-electron content in certain of the carbonyls. The methods have usually ignored the inductive (or σ-bonding) dependence of the spectral changes and to a large measure this appears justified when limited to the tetrahedral complexes, where π or σ interactions between all ligands must be equivalent, to a first approximation. It is not, however, justified in the case of the octahedral complexes, for in this geometry there will be a distinct difference between the mutual π and σ interactions of *cis* and *trans* ligands. An evaluation of the π and σ contributions to total bond-order in a series of octahedral carbonyl complexes[121] has shown that these are indeed cooperative rather than complementary in proportion, and that good π-accepting ligands, such as PF_3, NO, CO (the B class ligands), are *ipso facto* better σ-donors. Conversely, the poor π-accepting ligands, amines, ethers (the A class ligands), are not exceptionally good σ-donors either in these types of compounds. The intermediate ligands may be placed in a rational sequence of π-acceptance ($PX_3 > P(OR)_3 > PR_3 > NR_3$). Since the ligand-metal bond which is eventually to be broken upon reaction will be extended in the transition state even of an ***A*** reaction it might perhaps be expected that σ-bonding, being less distance-dependent than π-bonding, will contribute relatively more to a transition state than to a ground state. Accordingly, arguments based upon bond-orders which have been derived from average force-constants can only be used with reticence, particularly with regard to any supposed dissociative reactions.

It may be concluded fairly generally, though, that CO, virtually at the head of the π-accepting ligand series[122], should usually gain, thermodynamically, upon substitution of CO by a ligand lower in the series. In $Ni(CO)_4$, say, the tetrahedral arrangement ensures that a maximum of two $d\,\pi$ orbitals may be mixed in with the π^* ligand orbitals. Up to two CO groups are readily replaced; thereafter further substitution is difficult. In the case of the octahedral carbonyls, up to three CO's may be so replaced, for there are now three $d\,\pi$ orbitals of suitable symmetry. It has been pointed out, though, that π-acceptance by the CO ligand is overall possibly less than has been supposed[123] and in any case the argument is less strong for $M(CO)_4$ than for $M(CO)_6$ since d-π^* overlap is likely to be less effective for the tetrahedral than for the octahedral compounds[124].

Kinetic experiments have largely been limited to series of compounds based upon the simple tetrahedral and octahedral structures of nickel tetracarbonyl and the group VI hexacarbonyls respectively. Replacement of CO by neutral ligands produces substrates of use in the assessment of directional and non-directional influences without introducing the complications of altered oxidation number, charge type, etc. Replacement of CO by NO produces the same net effect as a reduction to carbonylate anion but without altering the charge, and the resulting nitrosyl carbonyl may be looked upon as a carbonylate with, attached, the very electron-acceptive NO^+ ligand. The replacement of CO by halide has the opposite effect, tantamount to oxidation of the metal and generation of a carbonyl halide isoelectronic with its left-neighbouring carbonyl in the Periodic Table. Olefins behave as neutral, exceptionally polarizable (B type) ligands, and aromatic compounds (similar ligands) form those exceptionally stable products in which three CO ligands have been replaced by a single arene. Table 11 shows some of the relations between groups of carbonyl derivatives.

Exchange and some substitutions in $Ni(CO)_4$ have been shown[108] to proceed by first-order processes with virtually the same activation parameters, consistent with a rate-determining dissociation. In the series $Co(CO)_3NO$ through Mn(CO)

TABLE 11

SOME RELATED TYPES OF SUBSTITUTED CARBONYL COMPLEXES *

$Cr(CO)_6$	$Mn(CO)_6^+$	$Mn(CO)_4NO$	$Fe(CO)_5$	$Mn(CO)(NO)_3$	$Fe(CO)_2(NO)_2$
				and $Co(CO)_3NO$	$Ni(CO)_4$
$Cr(CO)_5X^-$	$Mn(CO)_5X$			$Co(CO)_2(NO)L$	$Ni(CO)_3L$
$Cr(CO)_5L$	$Mn(CO)_4LX$				$Ni(CO)_2L_2$
$Cr(CO)L_2$					
$Cr(CO)_4(L–L)$					
$ArCr(CO)_3$	$CpMn(CO)_3$				

* L = neutral monodentate ligand; L–L = neutral bidentate ligand; X = halide ligand; Ar = aromatic ligand; Cp = cyclopentadiene.

$(NO)_3$ the ^{14}CO exchange is again first order although there is some possibility of a confusing gas-phase reaction for the former[125]. A first-order reaction between $Co(CO)_3NO$ and $(C_6H_5)_3As$ apparently occurs, but the remaining tetrahedral nitrosyl carbonyls (the *pseudo*-nickel carbonyls) react according to a second-order rate law. Tables 12 and 13 contain some rate data for these nitrosyl compounds and show that the rate coefficient generally increases with the basicity of the entering ligand. Those rates of CO exchange which have been measured are slower than the second-order substitutions except for the doubtful case of $Co(CO)_3NO$, and it has appeared reasonable to conclude that the substitutions are associative. There are, indeed, many examples known of pentacoordinated complexes containing B type ligands[155]. In the nitrosyl carbonyls, the formal oxidation number falls: Co(−I), Fe(−II), Mn(−III), and it has been argued that this would cause an increase in the M–C bond order and so reduce the possibility for dissociative reaction. The decrease in rates in the series $(n\text{-But})_3P > (C_6H_5)_3P > (C_6H_5)_3As$ is consistent with either smaller basicity or, equivalently here, polarizability.

TABLE 12

SECOND-ORDER RATE COEFFICIENTS ($l.\ mole^{-1}sec^{-1}$) FOR THE REACTIONS

$$M(CO)_{4-x}(NO)_x + L \rightarrow M(CO)_{3-x}(NO)_xL + CO$$

Compound	*L = (n-But)$_3$P*	*$(C_6H_5)_3P$*	*$(C_6H_5)_3As$*	*Ref.*
$Co(CO)_3(NO)^a$	9×10^{-2}	10^{-3}	2.3×10^{-6}	125
$Fe(CO)_2(NO)_2{}^a$	2.6×10^{-1}	10^{-3}	slow	126
$Mn(CO)(NO)_3{}^b$	>200	~54	~5.6	127

^{a} at 25 °C; ^{b} at 22 °C.

TABLE 13

SOME DIFFERENCES IN KINETIC BEHAVIOUR BETWEEN $Fe(CO)_2(NO)_2$ AND $Co(CO)_3(NO)$* AT 25 °C. $M(CO)_{4-x}(NO)_x + L \rightarrow M(CO)_{3-x}(NO)_xL + CO$

L	*M*	*Solvent*	*$k_1(sec^{-1})$*	*$k_2(l.mole^{-1}.sec^{-1})$*
$(C_6H_5)_3P$	Fe	toluene	0	10^{-3}
$(C_6H_5)_3P$	Co	toluene	0	10^{-3}
$(C_6H_5)_3P$	Fe	THF	10^{-3}	1.6×10^{-3}
$(C_6H_5)_3P$	Co	THF	0	1.3×10^{-3}
Py	Fe	toluene	0	5.6×10^{-3}
Py	Co	toluene	0	4×10^{-5}
$(C_6H_5)_3As$ (0.1 *M*)	Fe	toluene	k_{obs} =	$1.5\times10^{-6}\ sec^{-1}$
$(C_6H_5)_3As$ (0.1 *M*)	Co	toluene	k_{obs} =	$10^{-3}\ sec^{-1}$

* Data from ref. 170.

In the case of the nickel compounds, further substitution which is first-order takes place[120]

$$Ni(CO)_3L + L' \rightarrow Ni(CO)_2LL' + CO \quad (27)$$

or

$$Ni(CO)_2L_2 + L' \rightarrow Ni(CO)_2LL' + L \quad (28)$$

While the rate is independent of both the concentration and the nature of L′, it varies with L as: $L = Cl_3P > (C_6H_5)_3P > (n\text{-But})_3P > (C_2H_5O)_3P > > (C_6H_5O)_3P$. The decrease in rate parallels an increase in basicity (σ-donation) and/or a decrease in nephelauxetic effect (π-acceptance) with the exception, that is, of the phosphite, which appears an anomalously stable compound on both counts.

Further second-order substitutions can take place also in the cobalt series of compounds

$$Co(NO)(CO)_2L + L \rightarrow Co(NO)(CO)L_2 + CO$$
$$L = (CH_3O)_3P > (n\text{-But})_3P \gg (C_6H_5)_3P \quad (29)$$

or

$$Co(NO)(CO)_2L + L' \rightarrow Co(NO)(CO)_2L' + L$$
$$L' = (n\text{-But})_3P > (C_6H_5)_3P > (C_6H_5O)_3P$$
$$L = CH_3C(CH_2O)_3P > (CH_3O)_3P > (C_6H_5)_3P \quad (30)$$

Generally, the more basic and more polarizable nucleophiles enter into the complexes more readily, while, as in the case of the nickel compounds, initial presence of these same ligands aids the reaction. Steric hindrance in entering or already present ligands markedly reduces the rates. The best entering groups appear to be small, polarizable bases in which σ-donating power appears to be of more immediate importance than polarizability, while the most readily labilized complexes are those of the same types of ligands in which low polarizability becomes more important. These two criteria are no more than general indications however. It has been previously mentioned that, in a tetrahedral complex, the presence of the NO ligand could quite readily reduce any dissociative contribution relative to associative, and this could be explained quite adequately in terms of an increase in the M–C bond order; this is supported by IR data with regard to the C–O bond order. However, it would not be expected to *increase* the (nucleophile's) associative contribution by, in effect, reducing the positive charge upon the metal. An alternative explanation may be that the presence of polarizable non-labile ligands is essential in generating the presumably trigonal bipyramidal associative intermediate. The NO, or any other ligand of good π-accepting plus σ-donating ability, would then be in the position of a trans-labilizing group in square-planar substitution. It should be noted here that second-order kinetics are observed[128] for the rapid exchange of tertiary phosphines (L) with $Ni(II)L_2X_2$ and $Co(II)L_2X_2$ ($X = Cl > Br > I$).

For the octahedral compounds replacement of carbonyl by other ligands has effects upon substitution kinetics which are somewhat similar to those for the tetrahedral compounds but more apparent. Whereas for tetrahedral complexes increased lability only fairly generally follows upon replacement of CO by the better π acceptors, in the case of the octahedral complexes a fairly distinct division may be made between the strongly labilizing substituents (which are A ligands) and the non-labilizing (which are B ligands). Table 14 shows some of the types of ligands which fall into these two groups.

Considerable investigation of the octahedral carbonyl complexes has been carried out. To a certain degree this is because definitive evidence for associative substitution in the case of type A complexes has been conspicuously lacking whereas for the type B compounds there seem to be several well-substantiated examples. A general summary of the main types of octahedral substitutions which have been kinetically examined is given in Table 15.

TABLE 14

SOME OCTAHEDRAL COMPOUNDS WHICH UNDERGO SLOW* SUBSTITUTION

$$M(CO)_5X + L \rightarrow M(CO)_4LX + CO$$

M	*X*	*L*	*Ref.*
Mn	CO	^{14}CO	130
Mn	$SO_2CH_2C_6H_5$	$(C_6H_5)_3P$, C_5H_5N	131
Mn	$AuP(C_6H_5)_3$	^{14}CO	132
Cr	CNC_6H_5	^{14}CO	107
Mo	R_3P	R_3P	110
W	CO	^{14}CO	107

SOME OCTAHEDRAL COMPOUNDS WHICH UNDERGO FAST* SUBSTITUTION

$$M(CO)_5X + L \rightarrow M(CO)_4LX + CO \text{ or } M(CO)_4X_2 + 2\,L, L\text{–}L \rightarrow M(CO)_4L_2, M(CO)_4(L\text{–}L) + 2\,X$$

M	*X*	*L, L–L*	*Ref.*
Mn	Br	$(C_6H_5)_3P$	133
Mn	H	CO	134
Cr	Cl, Br, I	$(C_6H_5)_3P$	135
Mo	*cis* $(pyridine)_2$	dipyridyl	136
W	dipyridyl	$(RO)_3P$	137

* Relative to the parent reaction, X = L = CO.

TABLE 15

SUBSTITUTION REACTIONS OF SOME OCTAHEDRAL CARBONYL COMPOUNDS

Reactants	*Product*	*Solvent*	*Order*	*Metal*	*Inert ligand*	*Entering ligand*	*Leaving ligand*	*Ref.*
$M(CO)_6 + L$	$M(CO)_5L$	Decalin	1	$\Delta H^{\dagger}$: Mo < Cr, W	—	—	—	142
			2	$\Delta H^{\dagger}$: W < Cr < Mo	—	$R_3P > (RO)_3P > A_3P$		142
$M(CO)_5X^- + L$	*cis*$M(CO)_4LX^-$	Diglyme	1	Mo > Cr > W	Cl > Br > I	—	—	135
	$M(CO)_5L$	Diglyme	2	W, Mo > Cr	—	$(p-ClC_6H_4)_3P$ $> (p-FC_6H_4)_3P$ $> (C_6H_5)_3P$	Br > Cl, I	135
cis$M(CO)_4LX^-$ $+L$	$M(CO)_4L_2$ (Cr: *tr*, Mo: *cis*			Cr > Mo > W = 0	—	$(C_6H_5)_3P$ $> (p-FC_6H_4)_3P$ $> (p-ClC_6H_4)_3P$	I > Br > Cl	135
$M(CO)_5X + L$	*cis*$M(CO)_4LX$	*c*-Hexane CCl_4 $> C_6H_6$ $> CHCl_3$ > Acetone $> C_6H_5NO_2$	1	Mn > Re	Cl > Br > I	—	—	146, 147
cis$M(CO)_4LX$	*cis*$M(CO)_3$ LL^1X				Pyr, $C_6H_5NH_2$ $> (C_6H_5)_3P$ $> (C_6H_5)_3As$ $> R_3P > (C_6H_5O)_3P$ Cl > Br > I	—	—	147, 133
cis, tr$M(CO)_4L_2$ +(N–N) or (P–P)	$M(CO)_4$(N–N) or $M(CO)_4$(P–P)	Toluene Benzene	1	$\Delta H^{\dagger}$: Mo < W	—	—	*cis*: PCl_3 > Pyr $> (C_6H_5)_3As$ $> (C_6H_5)_3P$ $> Cl_2(C_6H_5)P$ $> (C_6H_5)_3Sb$ *tr*: $(C_6H_5)_3P$ $> (C_6H_5O)_3P$	138, 139

TABLE 15 (continued)

Reactants	Product	Solvent	Order	Metal	Inert ligand	Entering ligand	Leaving ligand	Ref.
$M(CO)_4(N-N)$ +L	*cis*$M(CO)_3$ (N–N)L	$(CH_2Cl)_2$ > C_6H_5Cl	1	—	Increases with pK_a(N–N) = o-phen, dipyr	—	—	140, 142
	cis$M(CO_3$ (N–N)L (M = Mo, W only)				—		—	
	cis$M(CO)_4L_2$ (M = Mo, W only N–N = dipyr, 1,2-diamino-2Me-propane only)	$(CH_2Cl)_2$ = C_6H_5Cl	2	ΔH†: Mo > W	Decreases with pK_a(N–N) = *o*-phen, dipyr	R_3P > $(RO)_3P$ > $(C_6H_5)_3P$ > $(C_6H_5O)_3P$	Diamine > dipyr > *o*-phen = 0	141, 143
$M_2(CO)_8$+L	$M(CO)_3XL_2$ (M = Mn only)	CCl_4 = C_6H_5Cl $(CH_2Cl)_2$	1	—	Cl > Br > I	—	Cl > Br > I	149
	cis$M(CO)_4LX$		2	Re > Mn		Increases with pK_a(L) –Picoline > Pyr > $(C_6H_5)_3P$ > p-Cl-Pyr		148, 149
$Mo(CO)_4$ (cyclo-1,5-C_8H_{12}+2L or (N–N) or (P–P)	$Mo(CO)_4L_2$ or $Mo(CO)_4(N-N)$ or $Mo(CO)_4(P-P)$	C_6H_6 $CHCl_3$ $(CH_2Cl)_2$	1	—	—	*o*-phen > $((C_6H_5)_2PCH_2)_2$ > $(R_2PCH_2)_2$ > $(C_6H_5)_2P$ > dipy	—	138, 144
			2	—	—	R_3P > Picoline > Pyr		138, 144
$M(CO)_4$ (2,5-dithiahexane)+L	*cis, tr*$M(CO)_4L_2$	$(CH_2Cl)_2$	2	Cr > Mo; ΔH†: Mo < Cr; ΔS†: Cr(+) > Mo(—)		$(RO)_3P$ > $(C_6H_5O)_3P$		145

References pp. 52–55

The two-term rate law for substitution reactions of the group VI hexacarbonyls[140] has been previously mentioned (see p. 29) and it will be useful to summarize the evidence for associative activation in this case.

i. There is reasonably good agreement between the rate of ^{14}CO exchange in the gas phase and the *first*-order rate of substitution in decalin, suggesting that this term represents a dissociative reaction.

ii. $\Delta H^{\ddagger}_2$ is significantly higher for Cr than for either Mo or W, and this is in agreement with the generally observed order of stability for those group VI compounds which are seven-coordinated.

iii. $\Delta S^{\ddagger}$ is positive for the first-order rate, negative for the second-order and these values are expected for, respectively, dissociative and associative activations.

Substitution in the pentacarbonyl manganese and rhenium compounds ($XM(CO)_5$, where X is formally anionic) is invariably more rapid than in the $M(CO)_6^+$ cations. The reactions are complicated by the further replacement steps but, in the first instance, produce *cis* $Mn(CO)_4XL$ at a rate independent of the nature and concentration of the entering ligand (*i.e.* first-order) and largely determined by the nature of X[142]. Usually, the rate of substitution is very much faster than the *exchange* in $M(CO)_6^+$ and when X is a halogen the rates are demonstrably[134, 137] the same. Ligands which particularly labilize the CO's are those of type A, of relatively high polarizing power but low polarizability, that is the halides and NO_3^-, while no significant acceleration is noted for the type B ligands. There is one especially interesting exception, for $HMn(CO)_5$ and $HRe(CO)_5$ both undergo extremely rapid ^{14}CO exchange. The same effect is noticed in $HCo(CO)_4$ and $HFe(CO)_5^+$ and it has been suggested[154] that the extreme labilizing ability of H, which is unexpected since H is an extreme type B ligand[2], lies in its being able to take part in an insertion (migration) reaction[153]

$$HMn(CO)_5 \rightleftarrows (HCO)Mn(CO)_4$$
$$(HCO)Mn(CO)_4 + CO \rightleftarrows (HCO)Mn(CO)_5 \tag{31}$$

For the group VI pentacarbonyl halide anions a first-order substitution of carbon monoxide by phosphine is also observed[135] to produce *cis* $M(CO)_4LX^-$, the rate of the reaction decreasing (Cl > Br > I) with the non-labile ligand in the same way as for the Mn, Re carbonyl halides. In this case there is a concurrent second-order substitution of the halide to produce $M(CO)_5L$. It should be noted that the order Cl > Br > I is unexpected if a simple ***D*** process is the mechanism of the first-order reaction, for, on the basis of π-bonding capacity (I > Br > Cl), the M–C bond strength should be the greater for Cl than for Br, I. Again, π-bonding arguments rarely give satisfactory predictions concerning reaction rates (see p. 26). Thus

$$M(CO)_5X^- + A_3P \rightarrow M(CO)_4(A_3P)X^- + CO \tag{32}$$

First-order: M = Mo > Cr > W, X = Cl > Br > I

$$M(CO)_5X^- + A_3P \rightarrow M(CO)_5(A_3P) + X^- \quad (33)$$

Second-order: M = W, Mo > Cr. X = Br > Cl, I

$$A = p\text{-}ClC_6H_5 > p\text{-}FC_6H_5 > C_6H_5$$

It is surprising that the second-order rate coefficients decrease as the basicity of the entering ligand increases; it appears that ligand nucleophilicity is compounded of basicity and micropolarizability in no simple manner.

The products of this first stage of the reaction, *cis* Mo or $Cr(CO)_4LX^-$ and *cis* Mn or $Re(CO)_4LX$ undergo further substitution. The Mn, Re compounds behave according to first-order kinetics while the Cr, Mo compounds show second-order kinetics to form *trans* $Cr(CO)_4L_2$ or *cis* $Mo(CO)_4L_2$, rate coefficients increasing this time as expected, with the basicity of the entering ligand L.

Complexity in the manganese and rhenium pentarbonyl halides' substitution arises from the fact that these decompose in inert solvents to form the halogen-bridged dimers $[M(CO)_4X]_2$. Both monomers and dimers react with phosphines, arsines, pyridine, aniline etc. to give the disubstituted compounds $M(CO)_3XL_2$. Thus three final products or any mixture of them may be obtained, depending upon the precise reaction conditions

$$M(CO)_5X \xrightarrow{-CO} [M(CO)_4X]_2 \quad (34)$$

$$(M(CO)_4X)_2 \xrightarrow[-CO]{+L} cis\ M(CO)_4LX \quad (35)$$

$$M(CO)_5X \xrightarrow[-CO]{+L} cis\ M(CO)_4LX \quad (36)$$

$$M(CO)_4LX \xrightarrow[-CO]{+L} cis\ M(CO)_3L_2X \quad (37)$$

One factor which promotes all reactions is the electronegativity of X(Cl > Br > I), although an extensive study of reaction (34) has not been carried out. Release of CO, reaction (37), is greatly accelerated when L is a basic or a N-donor ligand, $(n\text{-But})_3P$ rather than $(C_6H_5)_3P$ and aniline rather than CNC_6H_5, $(n\text{-ButO})_3P$. For a given donor atom, the rate increases with the bulk of the remainder of the ligand. The reactions

$$Mn(CO)_5X \rightarrow cis\ Mn(CO)_4LX$$
$$(\text{or } Mn(CO)_3LL'X) \quad (38)$$

are first-order[1,33] However, the manganese dimers $[Mn(CO)_4X]_2$ undergo cleavage

$$[Mn(CO)_4X]_2 + 4\ L \rightarrow 2\ Mn(CO)_4XL_2 \tag{39}$$

where L = pyridine or substituted picolines, according to a two-term rate law, and the corresponding rhenium dimers $[Re(CO)_4X]_2$ with the same ligands (*plus* $(C_6H_5)_3P$) follow second-order kinetics to yield $Re(CO)_4XL$.

The simplest overall interpretation of these data is in terms of a rate-determining dissociation. Entropies of activation are positive and the solvent-dependence for a better leaving group (Cl) is less marked than for a worse one (Br) in the case of reaction (38)[147]. For the dimeric carbonyls, $[M(CO)_4X]_2$, bridge-breaking, essentially the same dissociation, could result in a rapid pre-equilibrium. If this were followed by a second dissociative step, then the kinetics could be first-order (as for Mn), while a rate-determining entry of L could produce second-order kinetics (as for Re).

$$M(CO)_4X_2M(CO)_4 \xrightarrow{\text{fast}} M(CO)_4XM(CO)_4X \tag{40}$$

$$M(CO)_4XM(CO)_4X \xrightarrow{\text{slow}} 2\ M(CO)_4X \xrightarrow[\text{fast}]{+2L} 2\ M(CO)_4LX \tag{41}$$

$$M(CO)_4XM(CO)_4X \xrightarrow{+L} M(CO)_4LXM(CO)_4X \xrightarrow[\text{fast}]{+L} 2\ M(CO)_4LX \tag{42}$$

There is no real indication of the part which the solvent plays in these reactions. The intermediate generated in the fast pre-equilibrium (40) may, for example, be more or less solvated (as also the species $M(CO)_4X$). The effects of varying the solvent are small, however.

In the case of replacement of CO from the group VI carbonyl compounds there is additional evidence to the effect that the type A ligands labilize CO whereas the type B do not, but rather promote a second-order reaction. For the group VII octahedral compounds there is no strong evidence in favour of an associative activation step, except when interpretation is obscured by subsequent or concurrent reaction (but see ref. 146). There is, however, good reason to believe that such an associative reaction does occur in certain of the group VI compounds.

The rate of replacement of CO by phosphines and phosphites

$$M(CO)_5L + L' \rightarrow cis\ M(CO)_4LL' + CO \tag{43}$$

does not generally differ greatly from that in the unsubstituted carbonyl when L = R_3P, $(RO)_3P$ or CNC_6H_5[110]. When L = halide or pyridine, that is a type A ligand, the substitution is very much faster[135] but it is still first-order, independent of concentration and nature of the entering nucleophile. There is, indeed, a second-order *halide* substitution accompanying *CO* substitution but the seemingly important point is that isocyanide, phosphine or phosphite do not have a labilizing influence upon the CO whereas pyridine and the halides do so. This is more convincingly brought out by the reactions which $M(CO)_4B$ compounds undergo, where B is a bidentate ligand, *viz.* a substituted *ortho*-phenanthroline, substituted dipyridyl, 1,2-diamino-2-methyl-propane, 2,5-dithiahexane or 1,5-*cyclo*octadiene. Along this series we see a fall in group basicity and an increase in polarizability. At one extreme, the *o*-phenanthrolines are "harder" ligands than CO, while at the other, the di-olefin is of comparable "softness". This series might perhaps be extended to the sets of π-(aromatic ligand) $Mo(CO)_3$[150] and 1,3,5-*cyclo*-heptatriene $M(CO)_3$ compounds in which the trend continues to completely second-order kinetics and replacement, not of CO, but of the hydrocarbon.

When the bidentate B is dipyridyl or a substituted *o*-phenanthroline and M is Cr, only a first-order substitution is observed

$$Cr(CO)_4(N\text{–}N)+L \rightarrow \textit{cis}\ Cr(CO)_3L(N\text{–}N)+CO \tag{44}$$

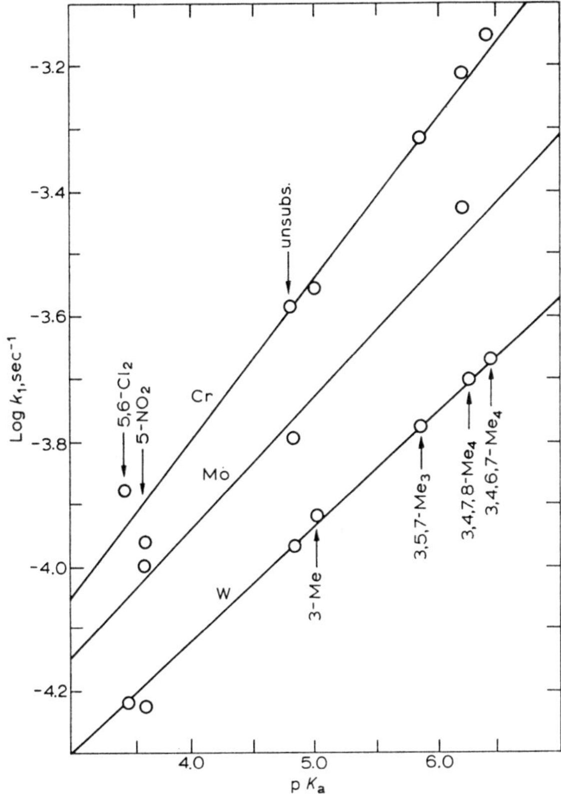

Fig. 9. Linear free-energy relationship for the reaction $M(CO)_4$ *o*-phen+$P(OCH_2)_3CCH_3$ → $M(CO)_3$ *o*-phen$P(OCH_2)_3CCH_3$+CO(M = Cr, Mo,W). Log k_1 *vs.* pK_a *o*-phen.

References pp. 52–55

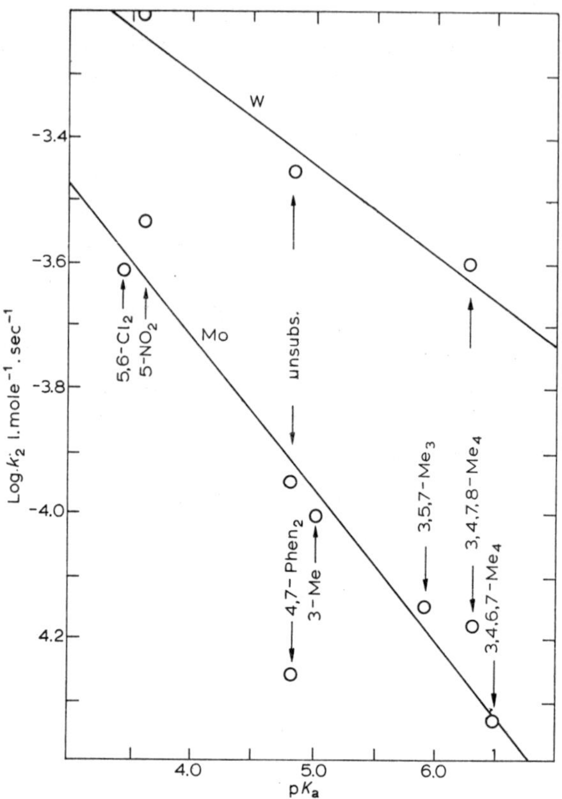

Fig. 10. Linear free-energy relationship for the reaction
$M(CO)_4$ *o*-phen+$P(OCH_2)_3CCH_3$ → $M(CO)_3$ *o*-phen$P(OCH_2)_3CCH_3$+CO(M = Mo, W).
Log k_2 *vs.* pK_a *o*-phen.

The same first-order replacements are seen when M is Mo or W, somewhat slower than in the case of Cr, but still much faster than for the hexacarbonyls. The rate increases with the pK_a of the inert ligand (N–N) and Fig. 9 shows the linear free-energy relation between log k_1 and pK_a. The relative orders would not have been expected on the basis of any π-bonding effects since increasing back-donation to CO would increase the M–C bond order. This increase in M–C bond order is supported by a decrease in ν_{CO} with increasing *o*-phenanthroline basicity. The same consideration applies for the pentacarbonyl halide anions where the first-order rates decrease (Cl > Br > I), unexpectedly as the halide polarizability increases.

When M is Mo or W, the substitution reaction follows a two-term rate expression and in addition there is some replacement of the dipyridyl or 1,2-diamino-2-methyl propane but not of *o*-phenanthroline. The proportion of bidentate replacement to CO replacement is practically independent of the concentration of the entering group[141]. There is now a *decrease* in the (second-order) rate coefficient with the pK_a of the inert *o*-phenanthroline (Fig. 10) and an increase with what may be judged to be the nucleophilicity of the entering phosphine or phosphite.

It is apparent that the existence of a second-order term in the rate expression does not of itself offer any proof of associative or dissociative activation, for there are two possible alternative mechanisms compatible. These are:

(*i*) An *association* of L to give a seven-coordinated intermediate, followed by a rate-determining loss of either CO or of (N–N), *viz.*

$$M(CO)_4(N{-}N) + L \rightleftharpoons M(CO)_4L(N{-}N) \xrightarrow{-CO} M(CO)_3L(N{-}N) \quad (45)$$

$$M(CO)_4L(N{-}N) \xrightarrow[-(N{-}N)]{+L} M(CO)_4L_2 \quad (46)$$

(*ii*) A *dissociation* of one end of the bidentate ligand, followed by a rate-determining entry of L, *viz.*

$$M(CO)_4\left(\begin{matrix}N\\N\end{matrix}\right) \rightleftharpoons M(CO)_4(N{-}N) \xrightarrow[-CO]{+L} M(CO)_3(N{-}N)L \quad (47)$$

$$M(CO)_4(N{-}N) \xrightarrow[-(N{-}N)]{+L} M(CO)_4L_2 \quad (48)$$

Alternative (*ii*) corresponds to the $[Re(CO)_4X]_2$ case, equations (41) and (42) above. However, it was here favoured largely because no second-order term was observed for the $Re(CO)_5X$[142] and $Re(CO)_4LX$[152] substitution. In the case of $Mo(CO)_5Py$, expected to be closely similar to $Mo(CO)_4dipy$, a second-order dependence has been observed.

8. Theoretical considerations

In the final section of this chapter, we shall attempt to give a brief rationalization of the regularities and peculiarities of the reactions of non-labile complexes which have been discussed in the previous sections. The theoretical framework in which the discussion will be conducted is that of molecular orbital theory (MOT). The MOT is to be preferred to alternative approaches for it allows consideration of all of the semi-quantitative results of crystal field theory without sacrifice of interest in the bonding system in the complex. In this enterprise we note the apt remark[156]: "Kinetics is like medicine or linguistics, it is interesting, it is useful, but it is too early to expect to understand much of it". The electronic theory of reactivity remains in a fairly primitive state. However, theoretical considerations may not safely be ignored. They have proved a valuable stimulus to incisive experiment.

Two remarkably successful generalizations from the preceding pages deserve attention first of all. Octahedral complexes show a pronounced tendency to react

References pp. 52–55

dissociatively. In contrast, square planar complexes show a very pronounced tendency to associative reaction. If nothing else were operative, these tendencies might be anticipated on steric grounds alone, since the planar complexes have two open faces for attack, these "access" routes being closed in the octahedral systems. It may be that this steric difference is of prime importance but there is at least one significant and relevant difference in electronic structure.

Figs. 11 and 12 show typical MO diagrams for square planar and octahedral complexes. Inspection reveals that the metal p_z orbital (z is the axial direction) in a square planar complex is involved in the π bonding system and available for σ bonding in the transition state. This is a feature shared by nucleophilic substitution at square planar complexes with the spectacularly associative nucleophilic aromatic substitutions. The octahedral complexes discussed in this chapter

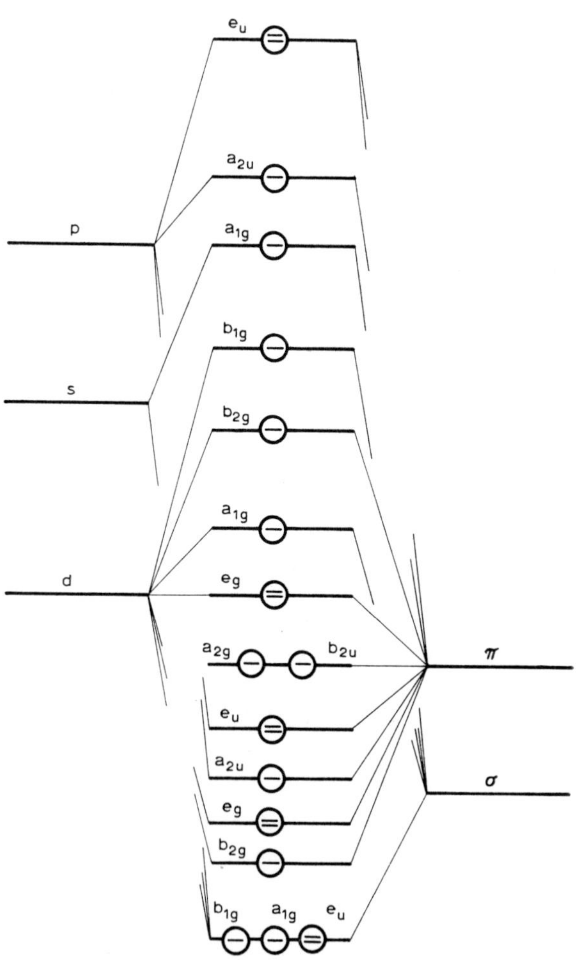

Fig. 11. MO diagram for a square planar complex.

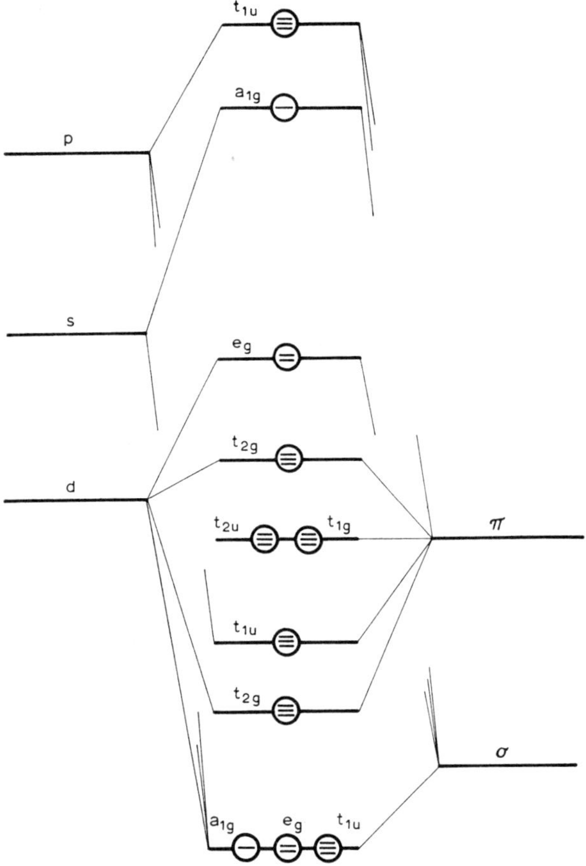

Fig. 12. MO diagram for an octahedral complex.

stand in sharp contrast. The metal p_z orbital is part of the σ bonding system in the ground state and, as long as there are three or more d electrons the $d\pi$ orbitals are filled and not readily available for σ bonding in the transition state. Orgel[157] noted several years ago that an associative mechanism would be quite plausible if there were fewer than three d electrons. So far, there has been no definitive evidence in support of this suggestion, although there are some indications in the results of fast reaction kineticists on the behaviour of V(III)[158, 159]. When the ligands can act as strong π acceptors then the situation may become modified. π donors, that is, halide ions, amines, oxy ligands can only destabilise the d orbitals of π symmetry. π accepting ligands on the whole may considerably stabilise these, to a degree which depends markedly upon the matching (or lack of it) of metal ($d\pi$) and ligand (π^*) orbitals. If there is good matching, particularly in the cases of CO and NO, to a less degree perhaps PF_3, the filled orbitals will concentrate into the internuclear regions leaving in the region of the ligand atom a net positive charge. The orbital system of the octahedral B-type complexes should be qualitatively similar to that of the square-planar d^8 complexes.

Moving from consideration of general trends to specific phenomena, perhaps the most fruitful problem for the theoretician of ligand substitution is the remarkable *trans* labilizing effect encountered in Pt(II) chemistry. The molecular orbital theory of this effect developed by Gray[160] serves admirably as a starting point for a more general molecular orbital theory of the effects of the non-labile ligands in a complex. If we appeal to the observation that the leaving group, the entering group and the group *trans* to the leaving group play similar roles in square planar substitutions, we may with reason postulate a transition state approaching a trigonal bipyramid structure. The theoretical problem becomes comparison of bonding in the square planar ground state and the trigonal bipyramid activated state.

First the σ bonding may be analyzed. Of the four metal valence orbitals involved in strong σ bonding in a square planar complex, only the p orbitals have *trans* directional properties. The *trans* group and the leaving group must share the same p orbital and a *trans* ligand with a strong σ interaction with the p orbital must weaken the bonding to the leaving group. Moreover, when the complex is converted from the square planar form to the activated trigonal bipyramid an additional p orbital (the p_z) is involved in σ bonding to the entering, leaving and *trans* ligands (see Fig. 13). It follows that the *trans* ligand which had approximately a half-share in a metal p orbital in the ground state may have approximately a two-thirds share in a metal p orbital in this transition state. The energy difference between ground state and transition state should be relatively small for good ligand (σ) to metal (p) donors located *trans* to the leaving group. Calculations[160] have shown that ligands such as H^-, PR_3 and CH_3^- have unusually large overlap integrals with the Pt $6p$ orbital. It is probably safe to conclude that the large *trans* effect of these ligands is a consequence of the σ bonding effect described. Other *trans* labilizing ligands must also depend for their effectiveness to a large extent on this effect.

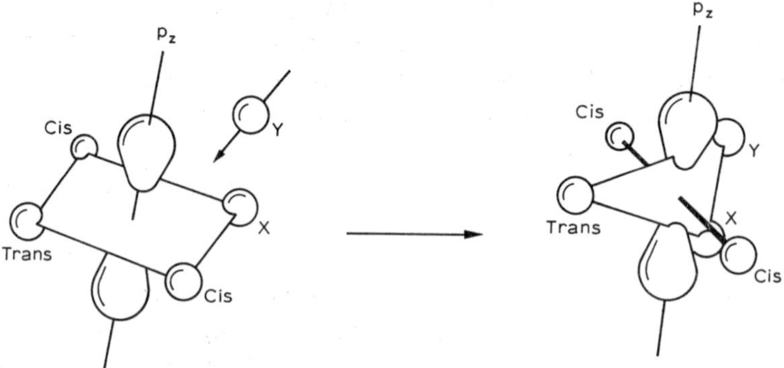

Fig. 13. Change in the metal p_z orbital structure in square- planar substitution.

Some ligands may supplement this σ effect with one operating through the π bonding system. Three d orbitals have proper symmetry for π interaction, xy, zx and yz. Assuming coordinates chosen as before, it is clear that the xz and yz interact only with a pair of *trans* disposed ligands in the square planar ground state. For example, the xz orbital may be shared between the leaving and *trans* ligands. In the trigonal bipyramid, four d orbitals are of π symmetry (xz, yz, x^2–y^2 and xy) and *all* are shared in π interaction with ligands in the trigonal plane, *viz.* the *trans*, entering and leaving ligands. Now, in d^8 complexes (all of the square planar cases we have discussed) all of the π antibonding orbitals derived from the metal d orbitals are filled. In this case, it is a great advantage for the *trans* ligand to have empty, reasonably low-lying (π) orbitals which can contribute to delocalization of the charge in the antibonding π orbitals as a new ligand (the entering ligand) is attached at the metal. Thus the effect of a good π acceptor *trans* ligand is to lower the activation energy for the reaction. Table 16 gives

TABLE 16

RELATIVE σ- AND π-*trans* EFFFECTS OF LIGANDS

(See ref. 3, p. 27)

Estimated σ-effect order
$H^- > PR_3 > -SCN^- > I^-, CH_3^-, CO, CN^- > Br^- > Cl^- > NH_3 > OH^-$
Estimated π-effect order
$H_2C = CH_2, CO > CN^- > -NO_2^- > -SCN^- > I^- > Br^- > Cl^- > NH_3 > OH^-$

theoretical estimates of the σ and π *trans* effect orders as derived from overlap integral calculations[160]. It can be seen from comparison of these orders with the experimental order of *trans* effect that both σ and π effects are important. Earlier emphasis had been placed on π effects but it is important to notice not only that a very large part of the experimental *trans* effect order can be discussed using the σ order alone, but that some ligands (*e.g.* H^-) absolutely require consideration of the σ effect.

The MOT discussion of the *trans* effect immediately illuminates the part played by entering and leaving groups. The ligand which functions effectively in the *trans* position should do so also in the other two positions in the trigonal plane; the entering and the leaving positions. There should be a parallel between the *trans* effect series and both the order of *effectiveness* (nucleophilicity) of entering groups and the order of *inertness* of leaving groups, and these correlations are, of course, well known.

Now, consider leaving group orders in general. The order of inertness at Pt(II) (which corresponds to the theoretically comprehensible *trans* effect series) is largely opposite to the order of ligands in the well-known spectrochemical series[161].

Those ligands which are "strong field" in the sense of producing a large splitting between the $d\pi$ antibonding (xz, yz, xy) and $d\sigma$ antibonding (x^2–y^2, z^2) orbitals of an octahedral complex are frequently ones readily displaced from square planar complexes of Pt(II). This circumstance is the opposite of that observed in the octahedral complexes of Co(III) and Cr(III). In these latter two cases the ligands *high* in the spectrochemical series are the ones which are least readily replaced.

The spectrochemical series reflects the splitting of molecular orbitals derived from the metal d orbitals and a large ligand field splitting is expected in the event that strong σ bonding occurs between metal d and ligand σ orbitals. However, strong σ bonding, ligand to metal p orbital, does not make a direct contribution to the ligand field splitting. In the π bonding framework of the complex the most important effects are directly reflected in the ligand field splitting. Interaction of filled metal $d\pi$ orbitals with "lone pair" electrons on the ligands in orbitals of π symmetry makes the $d\pi$ orbitals antibonding. As this interaction *increases* in importance, the ligand field splitting *decreases*. Conversely, for ligands with low-lying empty π orbitals (*e.g.* the π^* levels of CO and CN^-) the interaction with the $d\pi$ orbitals of the metal lowers these levels and *increases* the ligand field splitting. We can expect, then, that this easily accessible spectroscopic parameter might provide a general guide to the role of a ligand in d orbital bonding of both σ and π types but little direct information about the important matter of σ bonding involving s and p orbitals. More detailed analyses of optical and magnetic resonance spectra promise parameters offering more direct guidance concerning the role of the s and p metal orbitals. Some recent work is suggestive[124, 162–164].

Returning to the leaving ligand orders, that order observed for the third transition series metal, Pt(II), could be explained on the hypothesis that changes in σ bonding to metal p orbitals are more important than changes in σ bonding to d orbitals as the leaving group is changed (this hypothesis is implicit in the *trans* effect theory). Correlation of lability at Cr(III) and Co(III) centres with the spectrochemical series suggests importance of d orbital σ bonding for the lighter first transition series metals. Only one later period metal octahedral system has been studied extensively enough for a definite statement concerning leaving group orders. In the Rh(III) complexes the lability order $Cl^- > Br^- > I^-$ has been established[65]. This is opposite to the Co(III) order and opposite to that expected from the ligand field splitting as discussed above. It does correspond, however, to the lability order for the three ligands at Pt(II) centres and suggests that this second transition series metal reflects changes in σ binding through metal p orbitals in its lability order.

As the *trans* effect theory indicates, there should be some relationship between lability of a ligand and its role as a "labilizing" group in another position in a complex. In an octahedral complex reacting *via* a dissociative mode of activation, the transition state has five strongly bound ligands. This state will be stabilized if there should be present within the complex a group which binds strongly with σ orbitals formerly utilized by the leaving group. That is, in octahedral as in square planar complexes, a non-labile group should function to labilize other groups. In a very rough sense the data given above on the systems $Coen_2ACl^+$ where A is a variable group influencing the rate of loss of Cl^- could be construed to agree with this proposition. The labilizing order was

$$NH_2^- \gg OH^- > NO_2^- \sim CN^- \sim Cl^- > NCS^-$$

There is perhaps a significant π bonding effect, yet to be considered, but there is at least one more immediately obvious point revealed in the results. The leaving group orders suggested a dominant role for d orbital σ binding in Co(III). $d\sigma$ orbitals have four-fold symmetry and there should be no great difference between the influences of a ligand *cis* or *trans* to the leaving group. There should be no general *trans* effect in Co(III) or in Cr(III) chemistry. However, there should appear a *trans* effect in octahedral Rh(III) chemistry if p orbital discrimination amongst ligands is now of great importance. Recent work[165, 166] has established a distinct *trans* effect in Rh(III) amine complexes. With several constant leaving groups, the *trans* effect order is: $I^- > Br^- > Cl^-$, that is, as for Pt(II). It has also been observed[71, 167] that the presence of an amine group *trans* to the leaving ligand in a Rh(III) complex results in hydroxide catalysed (base) hydrolysis but that an amine in the *cis* position alone is insufficient. Apparently NH_2^- labilizes effectively only when *trans* to the leaving group. It is understandable, then, that extended search for clear cut *trans* effects in octahedral complexes is finally bearing fruit only as studies turn to the second and third transition series metals.

Having some theory of the σ effect, we now consider π effects in the dissociative mode substitution reactions of octahedral complexes. It has already been pointed out that ligands which have filled orbitals of π symmetry interact with metal $d\pi$ orbitals making them function as antibonding levels. This will destabilize ground states and make such ligands (*e.g.* Cl^-, OH_2, OH^-) rather better leaving groups than might be anticipated from consideration of a σ effect alone. Although this effect should correlate with spectrochemical position it is difficult to distinguish the role of σ from π bonding in determining lability. When these ligands with filled π orbitals function (as non-labile ligands) in modifying the reactivity of other ligands they play a role opposite to that "expected" from their position in the lability series. This is because the same π antibonding interaction which weakens the bond favours the formation of a dissociative transition state. Although this may be simply phrased in the suggestion that "lone pair" electrons (of π symmetry) can be supplied to the increasingly electron deficient metal, a more precise formulation may be the correlation noted[168] between the labilizing effect of such ligands and their position low in the nephelauxetic series. Relatively, stabilization of the transition state follows from reduction of the interelectron repulsions brought about by donor ligands.

Basolo and Pearson[169] were among the first to point out that this particular type of ligand will function as π donor in a stereospecific fashion in the dissociative transition state. Fig. 14 shows how an electron-filled *p* orbital on the ligand can overlap directly with a *p* orbital being depleted by the departure of a *cis* ligand without any rearrangement. On the other hand the same ligands *trans* to one another cannot form orbital overlap of σ orbitals unless there is rearrangement toward trigonal bipyramid geometry as shown in Fig. 15. These considerations are, in all probability, the explanation for the greater reactivity of *cis* $Coen_2ACl^+$ as compared to *trans* $Coen_2ACl^+$ when A is Cl^-, Br^- or OH^-. The suggestion is supported by the observation that it is the complex of this type that undergoes isomerization in the course of acid hydrolysis. All *cis* isomers of the $Coen_2ACl^+$

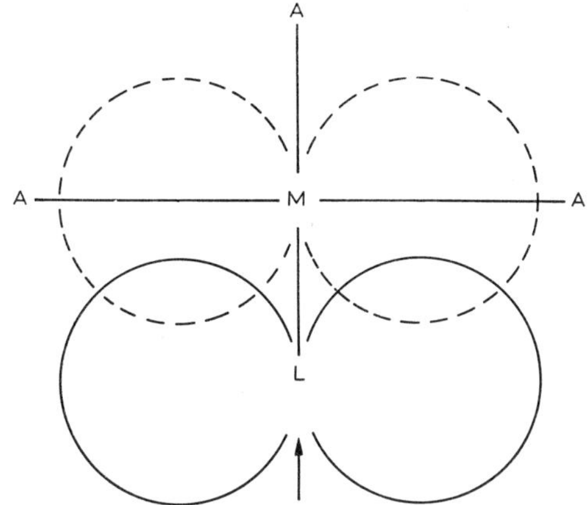

Fig. 14. The overlap of (ligand) *p* orbital with a vacant *cis* (metal) *p* orbital without rearrangement.

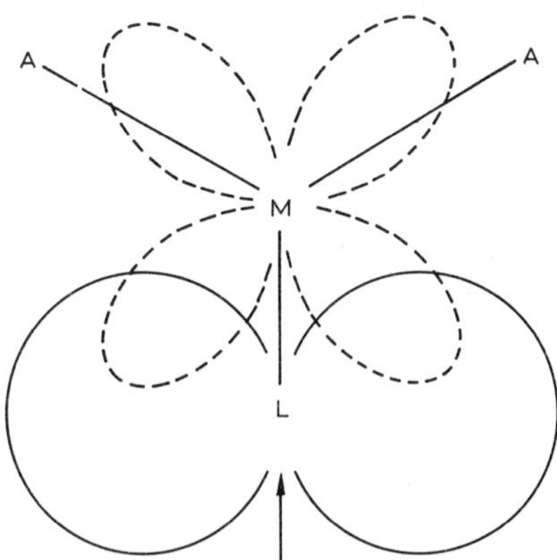

Fig. 15. The overlap of (ligand) *p* orbital with a vacant orbital upon rearrangement of the *trans* compound to a trigonal bipyramid.

system yield *cis* $Coen_2AOH_2^{2+}$ and of the *trans* $Coen_2ACl^+$ system only those complexes with A = OH^-, Cl^-, Br^- and NCS^- lead to any *cis* product.

Finally, it is important to turn to those ligands which are π acceptors, those with low-lying empty orbitals that bond with metal $d\pi$ orbitals. This interaction increases their ligand field splitting, advancing their position in the spectrochemical series, and at the same time decreases lability by adding additional bonding above the σ bonding. It is again true that there is inverse correlation between position in the spectrochemical series and lability. It is also again true that the π interactions do not play the same role when the ligand functions to labilize another group. To a first approximation, π bonding does not stabilize nor seriously destabilize a *dissociative* transition state, and one expects such ligands (*e.g.* NO_2^-, CN^-, CO) to produce labilizing effects principally within the σ bonding framework. These could well be large (see the σ *trans* effect series for Pt(II) in Table 16) but they will be *smaller* than anticipated from the ligands' spectroschemical position or lack of lability. The present suggestion is supported by the fact that NO_2^- and CN^- as ligands A in $Coen_2ACl^+$ are more effective in labilizing Cl^- from the *trans* position. It is not improbable that this small *trans* effect arises through σ bonding with metal p orbitals. A theoretical assessment of σ and π contributions in octahedral systems (such as the Co(III) complexes) remains to be made, yet the lability and labilizing orders seem consistent with a theory emphasizing σ bonding modifiied by some important π effects.

Some assessments of σ and π overlap populations for the simpler carbonyl systems have been made[124, 164]. For these, in which the metal is in an abnormally low oxidation state it is probable that the main variable is within the π system while the σ system remains relatively constant for a given geometry. Details differ, but it appears that metal d to ligand π^* overlap is more effective for the octahedral than for the tetrahedral systems (for these latter, the metal p orbital is more effective). The most interesting single result is that, in the octahedral systems the carbon site shows promise of some electrophilic character. The π system is effectively removed from the metal such that at least one cogent argument against nucleophilic attack directly at the metal is lessened, and it has been pointed out[164] that nucleophilic attack directly at the carbon followed by an internal ligand migration is not at all unlikely. It is worthy of note that this two-site concept has recently been convincingly used[170] with regard to some of the anomalies in the $Fe(CO)_2(NO)_2$ substitution kinetics. There is a spectacular solvent effect for, in THF or methanol, the relative orders of rates for nucleophiles are very different from those in toluene or dichloromethane (see Table 13). Particularly, $(C_6H_5)_3As$ and CO, very poor reagents in toluene, are much more effective in THF. In addition, nitrogen bases, quite ineffective for $Co(CO)_3NO$, are more effective for $Fe(CO)_2(NO)_2$ than are the phosphines, which are good ligands for $Co(CO)_3NO$. One interpretation of these findings is in terms of initiating solvent or other A-type nucleophilic attack at carbon followed by substitution of CO. This will give rise to first-order kinetics if the solvent is a good nucleophile and second-order kinetics if the entering ligand should be

$$\mathrm{M{-}C{-}O + S \rightarrow M{-}\underset{\substack{| \\ S}}{C}{-}O} \qquad \text{(slow)} \tag{49}$$

$$\mathrm{M{-}\underset{\substack{| \\ S}}{C}{-}O + B \rightarrow M{-}B + CO + S} \qquad \text{(fast)} \tag{50}$$

While it is too early to venture a generalization extending to all of the B-type compounds, specifically why "harder" ligands appear to exert a labilizing influence in the octahedral compounds, these are some indications that it is not in order to consider ligands in these B complexes merely as extensions of those in the A compounds. Reactivity of the coordinated ligand, its rearrangement and migration may prove to be an important feature.

REFERENCES

1 F. Basolo and R. G. Pearson, *Mechanisms of Inorganic Reactions*, Wiley, New York, 1967.
2 R. G. Pearson, *J. Am. Chem. Soc.*, 85 (1963) 3533.
3 C. H. Langford and H. B. Gray, *Ligand Substitution Processes*, Benjamin, New York, 1966.
4 S. Glasstone, K. S. Laidler and H. Eyring, *The Theory of Rate Processes*, McGraw-Hill, New York, 1941.
5 F. Basolo, B. D. Stone and R. G. Pearson, *J. Am. Chem. Soc.*, 76 (1954) 3079.
6 R. G. Pearson and J. W. Moore, *Inorg. Chem.*, 3 (1964) 1334.
7 A. W. Adamson and R. G. Wilkins, *J. Am. Chem. Soc.*, 76 (1954) 3379.
8 G. W. Ettle and C. H. Johnson, *J. Chem. Soc.*, (1940) 1490.
9 C. H. Langford and W. R. Muir, *J. Am. Chem. Soc.*, 87 (1967) 3141.
10 K. B. Yatsimirskii and L. Pankora, *Zh. Obshch. Khim.*, 19 (1949) 611;
K. B. Yatsimirskii, *Zh. Obshch. Khim.*, 20 (1950) 1408.
11 C. H. Langford, *Inorg. Chem.*, 4 (1965) 265.
12 M. L. Tobe, *Rec. Chem. Progr.*, 27 (1966) 79; *Advan. Chem.*, 49 (1965) 1 and references therein.
13 H. Taube and W. Schmidt, *J. Am. Chem. Soc.*, 80 (1958) 2642.
14 See the documented table in ref. 1, p. 164.
15 H. R. Hunt and H. Taube, *J. Am. Chem. Soc.*, 80 (1958) 2642.
16 G. C. Halor, *J. Chem. Soc. A*, (1966) 1.
17 D. L. Gay and G. C. Lalor, *J. Chem. Soc. A*, (1966) 1179.
18 M. L. Tobe, *Sci. Progr.*, 48 (1960) 484. See also P. J. Staples and M. L. Tobe, *J. Chem. Soc.*, (1960) 4803;
M. E. Baldwin, S. C. Chan and M. L. Tobe, *J. Chem. Soc.*, (1961) 4637;
S. C. Chan and M. L. Tobe, *J. Chem. Soc.* ,(1963) 5200.
19 M. L. Tobe, *Inorg. Chem.*, 7 (1968) 1260.
20 S. C. Chan, *J. Chem. Soc.*, (1963) 5137.
21 S. C. Chan and M. L. Tobe, *J. Chem. Soc.*, (1963) 514.
22 C. H. Langford and M. L. Tobe, *J. Chem. Soc.*, (1963) 506.
23 R. G. Pearson, C. R. Boston and F. Basolo, *J. Phys. Chem.*, 59 (1955) 304.
24 R. G. Pearson, C. R. Boston and F. Basolo, *J. Am. Chem. Soc.*, 75 (1953) 3089.
25 C. K. Ingold, R. S. Nyholm and M. L. Tobe, *Nature*, 187 (1960) 477.
26 Ref. 1, pp. 79–84.

27 A. Haim and H. Taube, *Inorg. Chem.*, 2 (1963) 1199.
28 W. R. Muir and C. H. Langford, *Inorg. Chem.*, 7 (1968).
29 I. R. Lantzke and D. W. Watts, *J. Am. Chem. Soc.*, 89 (1967) 815.
29aR. Murry and C. Barraclough, *J. Chem. Soc.*, (1965) 7047.
30 A. Haim and W. K. Wilmarth, *Inorg. Chem.*, 1 (1962) 573.
31 A. Haim, R. J. Grassi and W. K. Wilmarth, *Advan. Chem.*, 49 (1965) 31.
32 J. Halpern, R. A. Palmer and L. M. Blakely, *J. Am. Chem. Soc.*, 88 (1966) 2877.
33 D. A. Loeliger and H. Taube, *Inorg. Chem.*, 5 (1966) 1376.
34 D. A. Buckingham, I. I. Olson and A. M. Sargeson, *Inorg. Chem.*, 6 (1967) 1807.
35 G. C. Lalor and J. Lang, *J. Chem. Soc.*, (1963) 5620; S. C. Chan, K. Y. Hui, J. Miller and W. S. Tang, *J. Chem. Soc.*, (1967) 3207.
36 S. C. Chan and M. L. Tobe, *J. Chem. Soc.*, (1962) 4531; (1963) 514.
37 R. G. Pearson, R. E. Meeker and F. Basolo, *J. Am. Chem. Soc.*, 78 (1956) 709.
38 E. J. Garrick, *Nature*, 139 (1937) 507.
39 Ref. 1, p. 177 ff.
40 F. Basolo, J. W. Palmer and R. G. Pearson, *J. Am. Chem. Soc.*, 82 (1960) 1037.
41 J. W. Palmer and F. Basolo, *J. Phys. Chem.*, 64 (1960) 788.
42 M. Green and H. Taube, *Inorg. Chem.*, 2 (1963) 948.
43 D. A. Buckingham, I. I. Olson and A. M. Sargeson, *J. Am. Chem. Soc.*, 88 (1966) 5443.
44 R. D. Archer, *Advan. Chem.*, 62 (1967) 542.
45 G. W. Watt and J. F. Knifton, *Inorg. Chem.*, 7 (1968) 1159.
46 R. D. Gillard, *J. Chem. Soc. A*, (1967) 917.
47 N. M. Nikolaeva, S. V. Ptitsyn and E. D. Pastakova, *Russ. J. Inorg. Chem.*, 10 (1965) 574.
48 R. C. Johnson and E. R. Berger, *Inorg. Chem.*, 4 (1965) 1262.
49 B. Corain and A. J. Poe, *J. Chem. Soc. A*, (1967) 1633.
50 A. A. El-Awady, E. J. Bounsall and C. S. Garner, *Inorg. Chem.*, 6 (1967) 79.
51 A. B. Lamb and L. T. Fairhall, *J. Am. Chem. Soc.*, 45 (1923) 378.
52 M. A. Levine, T. P. Jones, W. E. Harris and W. J. Wallace, *J. Am. Chem. Soc.*, 83 (1961) 2453.
53 A. Ogard and H. Taube, *J. Am. Chem. Soc.*, 80 (1958) 1084.
54 J. Bjerrum and G. C. Lamm, *Acta Chem. Scand.*, 9 (1955) 216.
55 T. F. Swaddle and E. L. King, *Inorg. Chem.*, 3 (1964); 4 (1965) 532.
56 F. A. Guthrie and E. L. King, *Inorg. Chem.*, 3 (1964) 916.
57 H. B. Johnson and W. C. Reynolds, *Inorg. Chem.*, 2 (1963) 468.
58 D. J. MacDonald and C. S. Garner, *J. Am. Chem. Soc.*, 83 (1961) 4152; *Inorg. Chem.*, 1 (1962) 20.
59 S. Behrendt and C. H. Langford, to be published.
60 J. A. Broomhead, F. Basolo and R. G. Pearson, *Inorg. Chem.*, 3 (1964) 826.
61 R. E. Connick, *Advances in the Chemistry of Coordination Compounds, Ed.*, S. Kirscher, MacMillan, New York, 1961, p. 15.
62 H. B. Lamb, *J. Am. Chem. Soc.*, 61 (1939) 699.
63 J. V. Rund, F. Basolo and R. G. Pearson, *Inorg. Chem.*, 3 (1964) 658.
64 J. A. Poulsen and C. S. Garner, *J. Am. Chem. Soc.*, 84 (1962) 2032.
65 A. W. Bushnell, G. C. Lalor and E. A. Moelwyn-Hughes, *J. Chem. Soc. A*, (1966) 719.
66 M. Parris, *J. Chem. Soc. A*, (1967) 583.
67 T. P. Jones, W. E. Harris and W. J. Wallace, *Can. J. Chem.*, 39 (1959) 2371.
68 N. V. Duffy and J. E. Earley, *J. Am. Chem. Soc.*, 89 (1967) 816.
69 M. Ardon, *Inorg. Chem.*, 4 (1965) 372.
70 P. Moore, F. Basolo and R. G. Pearson, *Inorg. Chem.*, 5 (1966) 223.
71 S. A. Johnson, F. Basolo and R. G. Pearson, *J. Am. Chem. Soc.*, 85 (1963) 1781.
72 Ref. 3, Chap. 3.
73 R. L. Rich and H. Taube, *J. Phys. Chem.*, 58 (1954) 1.
74 H. B. Gray, *J. Am. Chem. Soc.*, 84 (1962) 1548.
75 U. Belluco, L. Cattalini, F. Basolo, R. G. Pearson and A. Turco, *J. Am. Chem. Soc.*, 87 (1965) 241.

76 U. Belluco, *Coord. Chem. Rev.*, 1 (1966) 111.
77 R. G. Pearson, H. B. Gray and F. Basolo, *J. Am. Chem. Soc.*, 82 (1960) 787.
78 R. S. Drago, U. A. Mode and J. G. Kay, *Inorg. Chem.*, 5 (1966) 2050.
79 R. Wenquo, *Master's thesis, Northwestern University, Evanston, Ill.*, 1965.
80 H. B. Gray and R. J. Olcott, *Inorg. Chem.*, 1 (1962) 481.
81 F. Basolo, H. B. Gray and R. G. Pearson, *J. Am. Chem. Soc.*, 82 (1960) 4200.
82 A. Werner, *Z. Anorg. Allgem. Chem.*, 3 (1893) 267.
83 I. I. Chernyaev, *Ann. Inst. Platine* U.S.S.R., 4 (1926) 261; *Izv. Sektora Platiny i Drug. Blagorodn. Metal. Inst. Obshch. i Neorgan. Khim. Akad. Nauk SSSR*, 5 (1927) 118.
84 Ref. 2, p. 24.
85 Ref. 2, p. 22.
86 U. Belluco, R. E. Hove, F. Basolo, R. G. Pearson and A. Turco, *Inorg. Chem.*, 5 (1966) 591.
87 J. P. Fackler, W. C. Seidel and J. A. Fetchin, *J. Am. Chem. Soc.*, 90 (1968) 2707.
88 R. D. Cramer, A. V. Lindsey, C. T. Prewitt and U. G. Stollberg, *J. Am. Chem. Soc.*, 87 (1965) 685.
89 A. Wojcicki and H. B. Gray, *Proc. Chem. Soc.*, (1960) 358.
90 F. Basolo, J. Chatt, H. B. Gray, R. G. Pearson and B. L. Shaw, *J. Chem. Soc.*, (1961) 2207.
91 S. Ahrland, J. Chatt and N. R. Davies, *Quart. Rev.*, 12 (1958) 265.
92 H. B. Gray and N. A. Beach, *J. Am. Chem. Soc.*, 85 (1963) 2922.
93 C. K. Jorgensen, *Inorg. Chem.*, 8 (1964) 1201.
94 C. K. Jorgensen, *Absorption Spectra and Chemical Bonding in Complexes*, Addison-Wesley, Reading, Mass., 1962.
95 M. R. Churchill and R. Masob, *Advan. Organometal. Chem.*, 5 (1967) 93.
96 J. Chatt, P. L. Pauson and L. M. Venanzi, *Organometallic Chemistry, Ed. H. Zeiss, Reinhold*, New York, 1960, Chap. 10.
97 L. E. Orgel, *An Introduction to Transition Metal Ion Chemistry*: *Ligand Field Theory*, Wiley, New York, 1960.
98 F. A. Cotton and C. S. Kraihanzel, *J. Am. Chem. Soc.*, 84 (1962) 4432.
99 W. F. Edgell, W. E. Wilson and R. Summitt, *Spectrochim. Acta*, 19 (1963) 863.
100 W. Beck and R. E. Nitzshmann, *Z. Naturforsch.*, 17b (1962) 577.
101 F. Basolo and A. Wojcicki, *J. Am. Chem. Soc.*, 83 (1961) 520.
102 K. Noack, *Helv. Chim. Acta*, 47 (1964) 1064.
103 D. F. Keeley and R. E. Johnson, *Inorg. Nucl. Chem.*, 11 (1959) 33.
104 G. Getini, O. Gambino, E. Sappa and G. A. Vaglio, *Ric. Sci.*, 37 (1967) 430.
105 C. H. Wei and L. F. Dahl, *J. Am. Chem. Soc.*, 88 (1966) 1821.
106 G. Pajaro, F. Calderazzo and R. Ercoli, *Gazz. Chim. Ital.*, 90 (1960) 1486.
107 G. Cetini and O. Gambino, *Atti Accad. Sci. Torino*, 97 (1963) 1197.
108 J. P. Day, F. Basolo, R. G. Pearson, L. F. Kangas and P. M. Henry, *J. Am. Chem. Soc.*, 90 (1968) 1925.
109 I. W. Stolz, G. R. Dobson and R. K. Sheline, *J. Am. Chem. Soc.*, 85 (1963) 1013.
110 H. Werner and R. Prinz, *J. Organometal. Chem.*, 5 (1966) 79.
111 R. J. Angelici and J. R. Graham, *J. Am. Chem. Soc.*, 88 (1966) 3658.
112 J. R. Graham and R. J. Angelici, *Inorg. Chem.*, 6 (1967) 2082.
113 R. J. Angelici and B. E. Leach, *J. Organometal. Chem.*, 11 (1968) 203.
114 G. Bor, *Spectrochim. Acta*, 18 (1962) 817.
115 C. A. Streuli, *Anal. Chem.*, 32 (1960) 985.
116 Ref. 1, p. 128.
117 E. L. Muertterties, *Inorg. Chem.*, 4 (1965) 1841.
118 F. Basolo, A. T. Brault and A. J. Poe, *J. Chem. Soc.*, (1964) 676.
119 E. E. Siefert and R. J. Angelici, J. *Organometal. Chem.*, 8 (1967) 374.
120 L. R. Meriwether and M. L. Fiene, *J. Am. Chem. Soc.*, 81 (1959) 4200.
121 W. A. G. Graham, *Inorg. Chem.*, 7 (1968) 315.
122 W. D. Horrocks and R. C. Taylor, *Inorg. Chem.*, 2 (1963) 723.
123 G. R. Dobson, *Inorg. Chem.*, 4 (1965) 1673.

124 A. F. Schreinev and T. L. Brown, *J. Am. Chem. Soc.*, 90 (1968) 3366.
125 E. M. Thorsteinson and F. Basolo, *J. Am. Chem. Soc.*, 88 (1966) 3929.
126 Ref. 1, p. 572.
127 W. Wawersik and F. Basolo, *J. Am. Chem. Soc.*, 89 (1967) 4626.
128 L. H. Pignolet and W. D. Horrocks, *J. Am. Chem. Soc.*, 90 (1968) 922.
129 E. M. Thorsteinson and F. Basolo, *Inorg. Chem.*, 5 (1966) 1691.
130 W. Hieber and K. Wollmann, *Chem. Ber.*, 95 (1962) 1552.
131 F. A. Hartmann and A. Wojcicki, *J. Am. Chem. Soc.*, 88 (1966) 844.
132 S. Breitschaft and F. Basolo, *J. Am. Chem. Soc.*, 88 (1966) 2702.
133 R. J. Angelici and F. Basolo, *Inorg. Chem.*, 2 (1963) 728.
134 P. S. Braterman, R. W. Hamill and H. D. Keasz, *J. Am. Chem. Doc.*, 89 (1967) 2851.
135 A. D. Allen and P. F. Barrett, *Can. J. Chem.*, 46 (1968) 1655.
136 M. Graziano, F. Zingales and U. Belluco, *Inorg. Chem.*, 6 (1967) 1582.
137 R. J. Angelici, F. Basolo and A. J. Poe, *J. Am. Chem. Soc.*, 85 (1963) 2215.
138 F. Zingales, F. Canziani and F. Basolo, *J. Organometal. Chem.*, 7 (1967) 461.
139 M. Graziani, F. Zingales and U. Belluco, *Inorg. Chem.*, 6 (1967) 1582.
140 R. J. Angelici and J. R. Graham, *J. Am. Chem. Soc.*, 87 (1965) 5586.
141 J. R. Graham and R. J. Angelici, *J. Am. Chem. Soc.*, 87 (1965) 5590.
142 R. J. Angelici and J. R. Graham, *Inorg. Chem.*, 6 (1967) 988.
143 J. R. Graham and R. J. Angelici, *Inorg. Chem.*, 6 (1967) 992.
144 F. Zingales, M. Graziani and U. Belluco, *J. Am. Chem. Soc.*, 89 (1967) 256.
145 G. C. Faber and G. R. Dobson, *Inorg. Chem.*, 7 (1968) 584.
146 F. Zingales, M. Graziani, F. Faraone and U. Belluco, *Inorg. Chim. Acta*, 1 (1967) 172.
147 R. J. Angelici and F. Basolo, *J. Am. Chem. Soc.*, 84 (1962) 2495.
148 F. Zingales, U. Sartorelli, F. Canziani and M. Raneglia, *Inorg. Chem.*, 6 (1967) 154.
149 F. Zingales and U. Sartorelli, *Inorg. Chem.*, 6 (1967) 1243.
150 A. Pidcock, J. D. Smith and B. W. Taylor, *J. Chem. Soc. A*, (1967) 872.
151 A. Pidcock and B. W. Taylor, *J. Chem. Soc. A*, (1967) 877.
152 F. Zingales and U. Sartorelli, *Inorg. Chem.*, 6 (1967) 1246.
153 R. F. Heck, *Advan. Organometal. Chem.*, 4 (1966) 243.
154 Ref. 1, p. 546.
155 E. L. Muertterties and R. A. Schunn, *Quart. Rev. (London)*, 20 (1966) 245.
156 C. K. Jorgensen, *Structure and Bonding*, 3 (1967) 106.
157 L. E. Orgel, *J. Chem. Soc.*, (1952) 4756.
158 B. R. Baker, N. Sutin and T. J. Welch, *Inorg. Chem.*, 6 (1967) 1948.
159 W. Kruse and D. Thusius, *Inorg. Chem.*, 7 (1968) 364.
160 Ref. 3, p. 24 ff.
161 Ref. 94, p. 109.
162 H. B. Gray and J. Alexander, *J. Am. Chem. Soc.*, 90 (1968)
163 L. M. Venanzi, *Chemistry in Britain*, (1968) 162.
164 K. G. Caulton and R. F. Fenske, *Inorg. Chem.*, 7 (1968) 1273.
165 H. L. Bolt, E. J. Bounsall and H. J. Poe, *J. Chem. Soc. A*, (1966) 1275.
166 H. L. Bolt and H. J. Poe, *J. Chem. Soc., A*, (1967) 205.
167 U. Klabunde, quoted in ref. 1, p. 187.
168 Ref. 3, p. 72.
169 See, for example, ref. 1, p. 173 ff. for a contemporary account.
170 D. E. Morris and F. Basolo, *J. Am. Chem. Soc.*, 90 (1968) 2536.

Chapter 2

Reactions in Solution Between Various Metal Ions of the Same Element in Different Oxidation States

P. J. PROLL

1. Introduction

This chapter attempts to survey the studies which have been made on the various electron transfer reactions, occurring between metal ions (of the same element) in homogeneous solution. These reactions include the types known as exchange reactions

$$Co^{3+} + Co^{2+} \rightleftharpoons Co^{2+} + Co^{3+}$$

$$NpO_2^{2+} + NpO_2^{+} \rightleftharpoons NpO_2^{+} + NpO_2^{2+}$$

$$Mn(CN)_6^{3-} + Mn(CN)_6^{4-} \rightleftharpoons Mn(CN)_6^{4-} + Mn(CN)_6^{3-}$$

mutual oxidation–reduction reactions

$$Np(V) + Np(III) \rightleftharpoons 2\ Np(IV)$$

$$Cr(VI) + 3\ Cr(II) = 4\ Cr(III)$$

$$Mn(VII) + 4\ Mn(II) = 5\ Mn(III)$$

catalysed substitutions and isomerisations

$$Pt(NH_3)_5Cl^{3+} + Cl^- \xlongequal{Pt(II)} Pt(NH_3)_4Cl_2^{2+} + NH_3$$

$$Cr(H_2O)_5Cl^{2+} + H_2O \xlongequal{Cr(II)} Cr(H_2O)_6^{3+} + Cl^-$$

$$CrNC^{2+} \xlongequal{Cr(II)} CrCN^{2+}$$

and electron transfer reactions of the type

$$Fe(III)X + Fe(II)Y \rightleftharpoons Fe(II)X + Fe(III)Y$$

$$IrCl_6^{2-} + IrBr_6^{3-} \rightleftharpoons IrCl_6^{3-} + IrBr_6^{2-}$$

$$CoCl^{2+} + Co^{2+} \rightleftharpoons Co^{3+} + CoCl^{+}$$

The latter are often included with the exchange reaction type. Both inner- and outer-sphere reaction mechanisms can occur in these electron transfer reactions.

The isotopic method is the most widely used experimental technique for the study of exchange reactions. A suitable radioactive tracer is used to label one or other of the oxidation states of the metal ion; the extent of the exchange reaction is then followed by the analysis of the amount of radioactivity present in one or both of the oxidation states after mixing reactants, quenching the reaction and separating the reactants at various times. Depending on the exchange rate, either a rapid mixing and quenching technique or the more usual slower technique is employed. The rates of the exchange reactions are then obtained from plots of $\ln(1-F)$ *versus* time (where F is the fraction exchanged at time t) either from the slopes of the above plots or from the evaluation of the time of half exchange ($t_{\frac{1}{2}}$) from these plots and the application of the McKay[1] equation,

$$\text{Rate} = \frac{[\text{X}][\text{Y}]}{[\text{X}]+[\text{Y}]} \cdot \frac{0.693}{t_{\frac{1}{2}}} = \frac{-\ln(1-F)}{t} \cdot \frac{[\text{X}][\text{Y}]}{[\text{X}]+[\text{Y}]}$$

where X and Y are the two exchanging species. The rates of exchange obtained for a wide variety of conditions are then used to obtain the rate law, rate coefficient and mechanism in the usual manner. In some systems non-linear McKay plots have been found and a modified treatment of exchange data has to be employed. Separation of the two oxidation states of the metal ion can be achieved using the physical methods of ion migration, ion exchange, diffusion, solvent extraction, and various chemical methods – usually involving precipitation of one of the exchanging species. The possibility of separation induced exchange, which can lead to erroneous results, is usually investigated and corrections applied, if necessary, to the exchange data. More than one separation method should, if possible, be used.

Exchange reactions can be sometimes investigated by the techniques of polarimetry, nuclear magnetic resonance and electron spin resonance. The optical activity method requires polarimetric measurements on the rate of racemization in mixtures of *d*-X (or *l*-X) and *l*-Y (or *d*-Y).

The reactions

$$d\text{-X} + l\text{-Y} \rightleftharpoons d\text{-Y} + l\text{-X}$$

$$d\text{-X} \rightleftharpoons l\text{-X}$$

$$l\text{-Y} \rightleftharpoons d\text{-Y}$$

occur. The various kinetic possibilities resulting have been observed and discussed in papers by Eichler and Wahl[2] and Im and Busch[3,4]. The nuclear magnetic resonance method requires measurements on a suitable spectral line (either proton or metal resonance). From either the position or the broadening, brought about by the presence of a paramagnetic exchanging species, of the resonance line of the diamagnetic exchanging species an estimate of the exchange rate coefficient can be obtained. An electron spin resonance method has also been used but has met with little success to date in metal ion–metal ion exchange systems. The theory underlying these resonance methods has been presented in the literature. Relevant papers include those by McConnell and Weaver[5], McConnell and Berger[6], Bruce *et al.* [7], Ward and Weissman[8], Dietrich and Wahl[9], and Larsen and Wahl[10].

In the study of reactions of the types other than exchange mentioned previously, the usual technique involves the spectrophotometric examination of reaction mixtures. The absorbance changes that occur, at a suitable wavelength where only one species (either reactant or product) absorbs, as the reaction proceeds are measured (manually or recorded). Treatment of the data *via* the Beer–Lambert law enables rate coefficients and laws to be found in the usual manner. Stopped flow and temperature jump techniques have been used for very rapid reactions.

In this chapter, the results that have been obtained to date by the various techniques are reviewed. Sections 2 to 7 deal with metal elements in the same order as their periodic group classification, with the exception of the transition, rare earth and actinide elements which are dealt with in sections 8, 9 and 10, respectively.

Previously many excellent reviews on these and related topics have appeared in the literature. Among these are articles by Sutin[11], Stranks and Wilkins[12], Halpern[13], Vlcek[14], Amphlett[15] and Zwolinski *et al.* [16].

2. Copper, silver and gold

2.1 THE EXCHANGE REACTION BETWEEN Cu(II) AND Cu(I)

McConnell and Weaver[1], using the NMR line width method, have obtained a value for the observed rate coefficient, for the reaction in 12 M hydrochloric acid media, of 5×10^7 l.mole^{-1}.sec^{-1}. The width of the ^{63}Cu NMR line from copper(I) (~ 1 M) was observed in the presence of copper(II) (10^{-4} to 10^{-2} M). In this media the exchanging species are probably $CuCl_4^{2-}$ and $CuCl_3^{2-}$. Optical interaction effects have been observed in mixtures of Cu(I) and Cu(II) in chloride media[2].

2.2 THE EXCHANGE REACTION BETWEEN Ag(II) AND Ag(I)

Gordon and Wahl[1] have used the radioisotope ^{110}Ag as a tracer for a study of the exchange of silver between Ag(II) and Ag(I) in acidic media. The precipitation of $Ag(phen)_2(ClO_4)_2$, brought about by the addition of *o*-phenanthroline, formed the basis of the separation method. The experimental data were obtained using a flow apparatus with reaction times of < 10 sec. Owing to the instability of silver-(II), media 6 M in H^+ and temperatures in the range −14.8 to 11.4 °C were employed. From measurements of the half-times ($t_{\frac{1}{2}}$) and the generalised expression

$$k_{obs} = \frac{\ln 2}{t_{\frac{1}{2}}[Ag(I)+Ag(II)]}[Ag(I)]^{1-a}[Ag(II)]^{1-b}$$

where a and b are orders of the reaction with respect to Ag(I) and Ag(II), a rate law

$$\text{rate} = k_{obs}[Ag(II)]^2$$

was obtained. The value of k_{obs} (0.2 °C) for media 5.87 M in $HClO_4$ was 1020 $l.mole^{-1}.sec^{-1}$. An Arrhenius plot led to an activation energy of 12.5 $kcal.mole^{-1}$ with an entropy of activation of −1 $cal.deg^{-1}.mole^{-1}$.

Variation of the hydrogen-ion concentration, over the range 5.87 to 3.94 M at a constant ionic strength of 5.87 M (ClO_4^-), led to the conclusion that the full rate law was

$$\text{rate} = k'[Ag(II)]^2/[H^+]^4$$

The mechanism for this exchange was postulated as

$$Ag^{2+} + xH_2O \rightleftharpoons Ag(OH)_x^{(2-x)+} + xH^+ \quad \text{rapid}$$

$$2\,Ag(OH)_x^{(2-x)+} \rightleftharpoons Ag(III) + Ag(I)$$

with the value of x possibly equal to 2.

Previously, Bruno and Santoro[2] had found complete exchange occurred between the ions $Ag(dipy)_2^{2+}$ or $Ag(phen)_2^{2+}$ and Ag^+, in nitric acid solution, within the separation time.

2.3 THE EXCHANGE REACTION BETWEEN Au(III) AND Au(I)

Under conditions where the dismutation reaction is slow the exchange between Au(III) and Au(I) has been shown to proceed at a measurable rate; at 0 °C in 0.09 M HCl, an exchange half-time of about 2 min was observed. The isotopic method (^{198}Au) and a separation method based on the precipitation of dipyridine-chloroaurate(III) was used to obtain data. An acceleration in the exchange rate was observed as the HCl concentration was increased[1].

2.4 THE EXCHANGE REACTION BETWEEN Au(III) AND Au(II)

No direct studies have been made on this exchange system, but Rich and Taube[1] have been able to propose a limit of $> 1.7 \times 10^6$ l.mole^{-1}.sec^{-1} at 0 °C for the rate coefficient k'_2 of the reaction

$$\mathrm{Au^{II}_{Cl}} + \mathrm{AuCl_4^-} \xrightarrow{k'_2}$$

as a result of their studies on the exchange of Cl^- with $AuCl_4^-$ in the presence of Fe(II).

2.5 THE DISPROPORTIONATION OF Au(II)

A limit has been proposed by Rich and Taube[1] for the rate coefficient k'_3 of the process

$$2\,\mathrm{Au(II)} \xrightarrow{k'_3} \mathrm{Au(I)} + \mathrm{Au(III)}$$

of $> 10^8$ l.mole^{-1}.sec^{-1} at 0 °C.

3. Mercury

3.1 THE EXCHANGE REACTION BETWEEN Hg(II) AND Hg(I); THE DISPROPORTIONATION REACTION OF Hg(II)

The earliest reported studies on the exchange, by King[1], and Haissinsky and Cottin[2], utilised the isotopic method (^{205,203}Hg) and showed the rapid nature of the process. Complete exchange was observed within 2 min at room temperature in acidic nitrate media[2]. Similar results were obtained with acidic perchlorate media[1]. The separation methods used the insolubility of Hg(I) as (*a*) chloride[1,2], (*b*) chromate[1] and (*c*) sulphate[1].

Wolfgang and Dodson[3] in a further study suggested a method of separation using the picrolonic acid precipitation of Hg(I). In perchlorate media even with reactant concentrations as low as 10^{-5} *M*, complete exchange was always observed by these workers, except in the presence of added CN^- ($\sim$ [Hg(II)]) when the kinetics could be examined. With reactant concentrations in the ranges Hg(I) 6×10^{-3} to 7×10^{-2} *M*, and Hg(II) $\sim 1.6 \times 10^{-3}$ to 1.6×10^{-2} *M* the rate law found was

$$\text{rate} = k_{\text{obs}}[\mathrm{Hg(II)}][\mathrm{Hg(I)}]$$

The observed rate coefficient (k_{obs}) was found to be dependent on the ratio $[CN^-]/[Hg(II)]$ as was the amount of zero time exchange. At 0 °C, for $[CN^-] = [Hg(II)]$, a value of 5.7×10^{-3} l.mole^{-1}.sec^{-1} was calculated, and the overall activation energy obtained was 14 kcal.mole^{-1} over the range 0 to 30 °C. The rate of exchange was found to be independent of ionic strength, hydrogen ion concentration and surface area and was not affected by light, oxygen or colloidal mercury.

Wolfgang and Dodson suggest that the exchange process in the absence of cyanide is controlled by a rapid dismutation reaction

$$Hg_{2(aq)}^{2+} \rightarrow Hg_{(aq)}^{2+} + Hg_{(aq)}$$

and have calculated a theoretical rate coefficient of $\sim 3.2 \times 10^5$ sec^{-1} at 25 °C. In the presence of cyanide, however, a new rate-determining step, involving a neutral Hg(II)–CN complex, possibly

$$Hg_2^{2+} + HgCN(ClO_4) \rightarrow$$

is present. King[4] and Adamson[5] have pointed out that the equilibria

$$Hg^{2+} + CN^- \rightleftharpoons HgCN^+$$

and

$$Hg^{2+} + HgCN^+ \rightleftharpoons HgCN^+ + Hg^{2+}$$

must not be rapid for the above mechanisms to be valid, and the observed zero-time exchange should be related to the concentration of free Hg^{2+}.

Wolfgang and Dodson[6] have since made a study of the exchange reaction between Hg^{2+} and $HgCN^+$ using ^{203}Hg and IO_3^- precipitation of the species Hg^{2+}. They have proposed a mechanism involving the steps

$$Hg^{2+} + HgCN^+ \rightarrow HgCN^+ + Hg^{2+}$$

$$Hg(CN)_2 + Hg^{2+} \underset{k_2}{\overset{k_1}{\rightleftharpoons}} HgCN^+ + HgCN^+ \qquad (K)$$

to account for the experimental observations on the slow exchange. A value for the equilibrium constant K of 20 at 25 °C has been obtained. With the aid of this value of K, it was found possible to calculate the zero-time exchange (Z_0) mentioned previously. Good agreement was obtained between the calculated and observed values of Z_0. The rate coefficients k_1 and k_2 have values (at 0 °C, $\mu = 0.12\ M$) of 1.7×10^{-2} and 8.3×10^{-4} l.mole^{-1}.sec^{-1}, respectively, both activation energies being 17 kcal.mole^{-1}. For the exchange $HgCN^+$–Hg(I), a rate coefficient (0 °C) of 7.0×10^{-3} l.mole^{-1}.sec^{-1} and activation entropy of -18 cal.deg^{-1}.mole^{-1} have been calculated.

Peschanski[7], during the course of a further study on this exchange in the presence of cyanide, has obtained results which are in fair agreement with those above.

3.2 THE EXCHANGE REACTION BETWEEN Hg(II) AND Hg(I) IN NON-AQUEOUS MEDIA

Peschanski[1], using the isotopic method (^{205}Hg), has found complete exchange (0 °C) in methanol and various other non-aqueous media. The separation methods used were, (*a*) paper and column chromatography, (*b*) paper electrophoresis, and (*c*) precipitation of Hg(I) with chloride. In the presence of cyanide ions, however, less than complete exchange could be observed. Zero-time exchange was again found to vary in the same manner as for aqueous media. Similar effects were observed in the presence of chloride ions.

4. Thallium

4.1 THE EXCHANGE REACTION BETWEEN Tl(III) AND Tl(I)

Although attempts have been made to study this reaction using ThC as an indicator[1,2], the stability of ThC has restricted these investigations. In 1948 the production of ^{204,206}Tl enabled more detailed studies to be made by Harbottle and Dodson[3] and by Prestwood and Wahl[4]. In preliminary reports, these authors presented data for the exchange in perchloric[3,4], hydrochloric[3] and nitric[4] acid media, obtained using separation methods involving precipitation of (*a*) Tl(I) as chromate[3] or bromide[4] and (*b*) Tl(III) as hydroxide[4]. A rate law

$$\text{rate} = k_{\text{obs}}[\text{Tl(III)}][\text{Tl(I)}]$$

was observed for each of the above three media. In 0.4 M $HClO_4$ k_{obs} (49.5 °C) has a value of 5.6×10^{-4} l.mole^{-1}.sec^{-1} and an associated activation energy[3] of 12 kcal.mole^{-1}.

Prestwood and Wahl[5], in a report of a more detailed study conducted at an ionic strength of 3.68 M (ClO_4^-) and temperatures in the range 10 to 50 °C, have presented results consistent with a more complicated rate expression

$$\text{rate} = (k_1 + k_2'[\text{H}^+]^{-1})[\text{Tl(III)}][\text{Tl(I)}]$$

(plots of k_{obs} *versus* $[H^+]^{-1}$ were linear with a positive intercept). This was interpreted in terms of two exchange pathways (rate coefficients k_1 and k_2), *viz.*

$$Tl^{3+} + H_2O \rightleftharpoons TlOH^{2+} + H^+ \qquad K_2$$

$$Tl^{3+} + Tl^+ \xrightarrow{k_1}$$

$$TlOH^{2+} + Tl^+ \xrightarrow{k_2}$$

with $k_2 > k_1$ and K_2 having a small value. The values of k_1 and $k'_2 (= k_2 K_2)$ at 25 °C are 4.3×10^{-5} l.mole^{-1}.sec^{-1} and 2×10^{-6}sec^{-1}, respectively, with corresponding activation energies of 17.6 and 10.3 kcal.mole^{-1}. The thallium concentration employed in this work was 5×10^{-3} to 5×10^{-2} M.

Harbottle and Dodson[6], have presented data obtained during a study of the exchange in perchlorate media, under constant high ionic strength conditions ($\mu = 6.0$ M). From results obtained with Tl(III) 1.4×10^{-4} to 1.7×10^{-2} M, Tl(I) 1×10^{-4} to 2×10^{-2} M and H^+ 3×10^{-1} to 6.0 M, it was concluded that only the step defined by k_2 was operative and the extent of hydrolysis was large (plots of k_{obs}^{-1} *versus* $[H^+]$ were linear with a positive intercept). On this basis, a value for k'_2 (25 °C) of 2.6×10^{-5} l.mole^{-1}.sec^{-1} and an associated activation energy and entropy of 14.7 kcal.mole^{-1} and -32 cal.deg^{-1}.mole^{-1} were calculated.

Dodson[7], in an attempt to resolve the difference in interpretation, carried out further experiments at various acidities and constant ionic strengths at 50 °C. He was able to show that the data given by Prestwood and Wahl[5] and by Harbottle and Dodson[6] were in good agreement with his own, and the additional data he obtained at $\mu = 2.0$, 4.0 and 6.0 M could be interpreted in terms of exchange *via* $TlOH^{2+}$ and Tl^+. The rate was found to decrease with increasing ionic strength and a revised activation energy for the process involving k_2 of 16 kcal.mole^{-1} was given.

In 1955 Rossotti[8], utilising the data available at that time on the hydrolysis of Tl(III)[9] and Tl(I)[10] was able to show that the interpretation given by Prestwood and Wahl[5] of their data ($\mu = 3.68$ M) was feasible. Estimates were made of k_1 and k_2 (25 °C, $\mu = 3.68$ M) of 4.28×10^{-5} and 3.4×10^{-4} l.mole^{-1}.sec^{-1}, respectively.

In 1961, Roig and Dodson[11] carried out a further study of the exchange in perchlorate media under identical conditions (25 °C, $\mu = 3.0$ M) to those in the Tl(III) hydrolysis study[9]. The isotope ^{202}Tl was used, with a separation procedure based on extracting Tl(III) from reaction mixtures with either methyl isobutyl ketone or diethyl ether. The exchange was examined in the absence of light, and a correction procedure to eliminate the catalytic effects of traces of chloride ions was used since Tl(III) concentrations of $\sim 10^{-4}$ M were necessary at the very low acidities employed. Using the known values of the first and second hydrolysis constants of Tl(III) (K_2 and K_3)

References pp. 142–152

$$TlOH^{2+} + H_2O \rightleftharpoons Tl(OH)_2^+ + H^+ \qquad K_3$$

the values of the rate coefficients k_1, k_2, and k_3 for the steps

$$Tl^{3+} + Tl^+ \xrightarrow{k_1}$$

$$TlOH^{2+} + Tl^+ \xrightarrow{k_2}$$

$$Tl(OH)_2^+ + Tl^+ \xrightarrow{k_3}$$

were calculated to be 7.0×10^{-5}, 2.5×10^{-5}, and $< 2.8 \times 10^{-5}$ l.mole^{-1}.sec^{-1}, respectively. Activation parameters obtained for the first step (k_1) were $E_a = 17.4$ kcal.mole^{-1} and $\Delta S^{\ddagger} = -21$ cal.deg^{-1}.mole^{-1}.

Gilks and Waind[12] have examined the exchange in heavy water and have found the exchange rate to be about two-thirds of that in ordinary water. A similar dependence, at constant ionic strength, on D^+ as on H^+ was observed and the rate of exchange was found to decrease as the ionic strength was increased. The isotope ^{204}Tl and chloroplatinate precipitation of Tl(I) was employed in this study. Waind[13] has discussed the role of water molecules in these exchange reactions.

The accelerating effect of nitrate ions (over the range 10^{-1} to 1.0 M, at $\mu = 3.68$ M) has also been studied[5]. An additional pathway involving the species $TlNO_3{}^{2+}$, *viz.*

$$Tl^{3+} + NO_3^- \rightleftharpoons TlNO_3^{2+} \qquad K_4$$

$$TlNO_3^{2+} + Tl^+ \xrightarrow{k_4}$$

was thought to occur at $[NO_3^-] < 0.4$ M. At higher $[NO_3^-]$, pathways involving $Tl(NO_3)_2^+$ were also thought likely. The full rate law found by Prestwood and Wahl[5] was

$$\text{rate} = (k_1 + k_2'[H^+]^{-1} + k_4'[NO_3])[Tl(III)][Tl(I)]$$

The value of $k_4'(= k_4 K_4)$ at 25 °C is 4.45×10^{-4} l^2.mole^{-2}.sec^{-1}, and the associated activation energy is 16 kcal.mole^{-1}.

The effect of added chloride on the exchange at constant ionic strength and acidity has been investigated by Harbottle and Dodson[6]. For Tl(III) $\sim 6.8 \times 10^{-3}$ to 6.8×10^{-4} M and Tl(I) $\sim 2 \times 10^{-2}$ M, $Cl^- < 3.5 \times 10^{-2}$ M produced a decrease in the rate whereas $Cl^- > 3.5 \times 10^{-2}$ M produced an increase. A minimum in the rate was found to occur at a ratio $[Cl^-]/[Tl(III)] \sim 1.5$. The exchange pathways believed to occur are

$$TlCl^{2+} + Tl^{+} \xrightarrow{k_5}$$

$$TlCl_2^{+} + Tl^{+} \xrightarrow{k_6}$$

$$TlCl_4^{-} + TlCl_3^{2-} \xrightarrow{k_7}$$

$$TlCl_4^{-} + TlCl_4^{3-} \xrightarrow{k_8}$$

and it was concluded that k_5 and k_6 were less than k_1 or k_2 and that k_7 and k_8 were larger than k_1 or k_2. For the exchange in 0.4 *M* HCl ($\mu = 6.0$ *M*) at temperatures 31.8 to 41.8 °C, an overall activation energy 29.6 kcal.mole^{-1} was found.

Gilks *et al.*[14] have since confirmed the above observations and have found a similar effect to operate in deuterated solvent. At high $[Cl^-]$ the rate in D_2O is equal to that in H_2O.

Challenger and Masters[15] have made some observations on the exchange in 0.4 *M* H_2SO_4 solution using concentrations of Tl(III) 8.5×10^{-4} to 1.7×10^{-3} *M* and of Tl(I) 1.7×10^{-3} to 8.5×10^{-3} *M*, with the tracer ^{204}Tl and a chloroplatinate precipitation of Tl(I). The rate law

$$\text{rate} = k_{obs}[Tl(III)][Tl(I)]$$

was obeyed and led to a calculated value of k_{obs} (25 °C) 1.98×10^{-2} l.mole^{-1}.sec^{-1} with an associated activation energy (0 to 44.3 °C) and entropy of 13.8 kcal.mole^{-1} and -22 cal.deg^{-1}.mole^{-1}, respectively.

Brubaker and Mickel[16] later reported results, obtained in a more detailed study for sulphate media of ionic strength of 3.68 *M* with the species Tl(III), Tl(I), H^+ and $SO_4{}^{2-}$ in the concentration ranges 1.8×10^{-3} to 1.08×10^{-2} *M*, 1.0×10^{-3} to 2.0×10^{-3} *M*, 4.5×10^{-1} to 2.90 *M* and 7×10^{-3} to 7×10^{-1} *M*, respectively. The chromate precipitation method was used. The experimental data at 24.9 °C were found to be consistent with a detailed rate equation

$$\text{rate} = \frac{0.0345 + k_9'[H^+]^2[SO_4^{2-}] + k_{10}'[H^+]^2[SO_4^{2-}]^3[Tl(III)][Tl(I)]}{[H^+]^2 + K_2[H^+] + K_2K_3 + K_9[H^+]^2[SO_4^{2-}](1 + K_{10}[SO_4^{2-}])}$$

where K_2 has been defined previously and K_3, K_9, and K_{10} are given by

$$K_3 = \frac{[TlO^+][H^+]}{[TlOH^{2+}]}\,;\; K_9 = \frac{TlSO_4^+}{[Tl^{3+}][SO_4^{2-}]} \text{ and } K_{10} = \frac{[TlSO_4^-]}{[Tl^+][SO_4^{2-}]}$$

k_9' and k_{10}' are products of exchange rate coefficients and equilibrium constants and have numerical values (25 °C) of 0.012 and 0.2, respectively. Measurements on the exchange at other temperatures were made and the exchange rate was found to be dependent on the ionic strength.

In a further study, Brubaker *et al.*[17] have reported on the effects of the addition of chloride ion to the sulphate exchange system at virtually constant ionic strength (3.68 M), sulphate and hydrogen-ion concentrations. For the concentration ratio $[Cl^-]/[Tl(III)]$ of 9.2×10^{-2} to 9.5 at 24.9 °C results analogous to the effect observed in perchlorate media[6] were obtained. The minimum in the rate corresponded to a ratio of ~ 2.5. Results were also presented for the conditions, constant $[Cl^-]$ and variable $[SO_4^{2-}]$ and $[H^+]$ ($\mu = 3.68\ M$). Brubaker *et al.*[17] have suggested that the exchange paths most likely to occur in sulphate media are

$$TlSO_4^+ + Tl^+ \xrightarrow{k_9}$$

$$Tl(SO_4)_2^- + TlSO_4^- \xrightarrow{k_{10}}$$

for which activated complexes involving sulphate bridges are likely. Wiles[18], has made a further study of this system in the presence of small amounts of added sulphate and has observed a first order dependence on this anion. Results at higher concentrations of added sulphate agree with those mentioned previously. A sulphate bridge activated state was again proposed.

The effect of the addition of cyanide to the exchange reaction in perchlorate media has been reported by Penna-Franca and Dodson[19]. An effect similar to the addition of chloride was observed with a minimum occurring at a ratio $[CN^-]/[Tl(III)]$ of 3.5 ($\mu = 0.5\ M$, 30 °C). It was concluded that the pathways involving Tl^+ and $TlCN^{2+}$ or $Tl(CN)_2^+$ are slow whereas the pathways involving Tl^+ or cyanide complexes of Tl(I) [$TlCN$, $Tl(CN)_2^-$, $Tl(CN)_3^{2-}$] and $Tl(CN)_3$ or $Tl(CN)_4^-$ are more rapid.

Dodson *et al.*[20] have also studied in detail the effect of added bromide on the exchange reaction using precipitation and extraction separations and ^{204}Tl and ^{202}Tl as indicators. The variation in the rate of exchange was found to be governed by the law

$$\text{rate} = k_1[Tl^{3+}][Tl^+]+k_2[TlOH^{2+}][Tl^+]+k_{11}[TlBr_2^+] \\ +k_{12}[TlBr_3]+k_{13}[TlBr_4^-][Tl^+]+k_{14}[TlBr_4^-][TlBr_2^-]$$

for added bromide $< 0.2\ M$. The rate coefficients have values ($\mu = 0.5\ M$, 30 °C): k_{11} and k_{12}, 2.2×10^{-6} and 1.25×10^{-6} sec^{-1}; k_{13} and k_{14}, 1.28×10^{-3} and 6.6×10^{-7} l.mole^{-1}.sec^{-1}, respectively. Plots of the observed rate of exchange *versus* $[Br^-]$ exhibit a maximum followed by a minimum and a further increase at high bromide ion concentrations. For the step

$$TlBr_4^- + TlBr_2^- \xrightarrow{k_{14}}$$

the possibility of a symmetrical activated complex was suggested and the steps having rate coefficients k_{11} and k_{12} were proposed as

$$TlBr_2^+ \xrightarrow{k_{11}} Tl^+ + Br_2$$

$$TlBr_3 \xrightarrow{k_{12}} Tl^+ + Br_2 + Br^-$$

Brubaker and Andrade[21] have made a study of the effects of acetic and succinic acids on the exchange; a reduction in the exchange rate was observed†.

Some conclusions on the exchange reaction have been made as a result of studies on the redox reactions between Tl(I) and Ce(IV)[22] and between V(IV) and Tl(III)[23]. Gryder and Dorfman[22] have pointed out that Tl(II) cannot be invoked for both the exchange between Tl(III) and Tl(I) and the reaction of Tl(I) with Ce(IV) which is slower than the exchange. Sykes[24] has shown, on the basis of a mechanism proposed for the reaction between Tl(III) and V(IV), that Tl(II) cannot be involved in the exchange reaction as this would lead to Tl(I) accelerating the above redox reaction. It is interesting to note that as long ago as 1949, McConnell and Davidson[25] reported the lack of interaction absorbance phenomenon between Tl(III) and Tl(I). This might be expected if Tl(II) was absent. Radiation-induced exchange reactions which are more rapid than the thermal exchange reactions are thought to involve Tl(II)[15,26].

Challenger and Masters[15] have made a detailed study of the X-ray-induced exchange reaction in 0.4 M H_2SO_4 solution, over the ranges Tl(III) 2×10^{-3} to 7×10^{-3} M and Tl(I) 1.4×10^{-6} to 7×10^{-6} M, with X-ray intensities (I) 1.7×10^{19} to 4×10^{20} eV.l^{-1}.min^{-1}. The rate law observed was

$$\text{rate} = k_{obs}[\text{Tl(III)}]^{0.5}I^{0.67}$$

The exchange mechanism proposed involves the steps

$$\text{Tl(I)} + \text{OH} \rightarrow \text{Tl(II)} + \text{OH}^-$$

$$\text{Tl(III)} + \text{H} \rightarrow \text{Tl(II)} + \text{H}^+$$

$$\text{Tl(II)} + \text{Tl(I)} \rightarrow \text{exchange}$$

$$\text{Tl(II)} + \text{Tl(III)} \rightarrow \text{exchange}$$

$$2\,\text{Tl(II)} \rightarrow \text{Tl}^+ + \text{Tl}^{3+}$$

Radiation of wavelength 253.7 mμ has also been used to induce exchange. Stranks and Yandell[26], on the basis of their observations, have proposed a similar

† The exchange reaction in the presence of $CH_2ClCOOH$, $CHCl_2COOH$ and CCl_3COOH has recently been examined by McGregor and Wiles[27].

mechanism

$$Tl^{3+} + h\nu \rightleftharpoons Tl^{2+} + OH + H^+$$

$$TlOH^{2+} + h\nu \rightleftharpoons Tl^{2+} + OH$$

$$Tl^{+} + OH \rightarrow Tl^{2+} + OH^-$$

$$Tl^{3+} + Tl^{2+} \xrightarrow{k_{15}} \text{exchange}$$

$$Tl^{+} + Tl^{2+} \xrightarrow{k_{16}} \text{exchange}$$

$$2\,Tl^{2+} \rightarrow Tl^{+} + Tl^{3+}$$

The value of the ratio k_{15}/k_{16} (at 25 °C) is 1.2 (see also Stranks and Yandell[28]).

5. Tin and lead

5.1 THE EXCHANGE REACTION BETWEEN Sn(IV) AND Sn(II) IN AQUEOUS MEDIA

In aqueous solution this exchange has been studied, in the absence of oxygen, in chloride[1,2] and sulphate[3] media. The isotope ^{113}Sn was used as the indicator and the separation of Sn(IV) and Sn(II) was achieved by the formation of the insoluble salts caesium hexachlorostannate(IV)[1-3] and stannous oxalate[3].

Davidson *et al.*[1] have investigated the exchange reaction in chloride media, where the rate law obeyed is

$$\text{rate} = k_{obs}[\text{Sn(IV)}][\text{Sn(II)}]$$

A value for k_{obs} of 9.2×10^{-3} l.mole^{-1}.sec^{-1} was calculated from data obtained at 25.2 °C for 10 M HCl solution with Sn(IV) and Sn(II) in the ranges 6×10^{-3} to 3.2×10^{-2} M and 9×10^{-3} to 5.7×10^{-2} M, respectively. The value of k_{obs} was found to be dependent on the concentration of HCl. An overall activation energy of 10.8 kcal.mole^{-1} was obtained. These authors, following the observation of Whitney and Davidson[4], have studied in some detail the interaction observed in mixtures of Sn(IV) and Sn(II) and have concluded that an equilibrium

$$SnCl_6^{2-} + SnCl_4^{2-} \rightleftharpoons Sn_2Cl_{10}^{4-}$$

occurs. Exchange could occur *via* this equilibrium.

Craig and Davidson[2] have investigated the photochemical exchange, using light of wavelengths ~ 365 mμ (a wavelength where the interaction complex absorbs strongly) from a high intensity source. Uranyl oxalate, which has similar absorption characteristics to the complex ($Sn_2Cl_{10}^{4-}$), was used for the actinometer. Quantum yields for the process were found to be ~ 0.2. It was concluded that exchange occurs *via* an activated complex which is unsymmetrical.

The exchange in sulphate media, where a similar rate law is obeyed and the exchange rate is slower than in chloride media, has been studied by Gordon and Brubaker[3]. At $\mu = 4.98\ M$, k_{obs} (25 °C, $[H^+] = 3.99\ M$ and $[SO_4^{2-}] = 0.99\ M$) has a value 1.7×10^{-5} l.mole^{-1}.sec^{-1}. The effect of variation in the hydrogen ion (3.30 to 5.0 M) and sulphate ion (0.70 to 1.40 M) concentrations at constant ionic strength ($LiClO_4$ and Li_2SO_4) was found to conform to a relationship

$$k_{obs} = k'[H^+]^{-2} + k''[SO_4^{2-}][H^+]^{-1}$$

suggesting the involvement of two exchange pathways. The possible steps proposed were

$$SnO(SO_4)_2^{2-} + SnOH^+ \rightarrow$$

$$Sn(OH)(SO_4)_2^- + Sn(OH)SO_4^- \rightarrow$$

$$Sn(OH)_2SO_4 + Sn(OH)SO_4^- \rightarrow$$

The main species of Sn(IV) and Sn(II) in sulphate media were thought to be $SnOH^+$ and $Sn(SO_4)_2$. The overall activation energy (over the range 25–50 °C, in 3.0 M H_2SO_4 with $\mu = 4.98\ M$) was calculated as 19 kcal.mole^{-1}.

The addition of chloride ion was found to increase the rate of reaction and, if present in the range 0–0.5 M, to lead to abnormal orders with respect to Sn(IV) and Sn(II). Above 0.5 M Cl^- a rate law

$$\text{rate} = k_{obs}[\text{Sn(IV)}][\text{Sn(II)}][Cl^-]$$

was obeyed, with k_{obs} (25 °C, 1 M $[Cl^-]$) having a value 1.23×10^{-3} l^2.mole^{-2}.sec^{-1} with an associated overall activation energy of 22.6 kcal.mole^{-1}. The exchange path

$$SnCl_6^{2-} + SnCl_4^{2-} \rightarrow$$

was concluded to occur[3]. At high temperatures, McKay plots were found to show deviations from linearity for exchange in presence or absence of chloride, suggesting more than one exchange path is operative.

Interaction absorbance data have also been reported by Gordon and Brubaker[3] for mixtures of Sn(IV) and Sn(II) in sulphate media. Exchange *via* interaction dimers, Sn(IV)–Sn(II), seems likely.

5.2 THE EXCHANGE REACTION BETWEEN Sn(IV) AND Sn(II) IN NON-AQUEOUS MEDIA

The exchange of ^{121}Sn between $SnCl_4$ and $SnCl_2$ in anhydrous methanol and ethanol in the absence of oxygen has been investigated by Meyer *et al.*[1,2]. From the linearity of plots of $t_{\frac{1}{2}}$ *versus* the inverse of the total tin concentration, which was varied over the range $\sim 4\times10^{-2}$ to $\sim 3\times10^{-1}$ M, a rate law

$$\text{rate} = k_{obs}[\text{Sn(IV)}][\text{Sn(II)}]$$

was suggested. The kinetic parameters found were: (*a*) methanol solution[1]; k_{obs} (33 °C) 7.12×10^{-5} l.mole^{-1}.sec^{-1}, activation energy (over range 33–47 °C) 21 kcal.mole^{-1} and activation entropy -11.5 cal.deg^{-1}.mole^{-1}; (*b*) ethanol solution[2]; k_{obs} (25 °C) 6.03×10^{-5} l.mole^{-1}.sec^{-1}, activation energy (over the range 25–42 °C) 23.7 kcal.mole^{-1} and activation entropy -0.4 cal.deg^{-1}.mole^{-1}. The separation of Sn(II) and Sn(IV) was based on the insolubility of the oxalate of Sn(II) in alcoholic media.

The mechanism proposed for ethanol media, which is similar to that for aqueous media, was

$$SnCl_4+SnCl_2 \rightleftharpoons Sn_2Cl_6 \qquad \text{rapid}$$

$$\text{or } SnCl_3^+ + SnCl_3^- \rightleftharpoons Sn_2Cl_6 \qquad \text{rapid}$$

$$Sn_2Cl_6 \rightarrow SnCl_2+SnCl_4$$

Presumably in methanol solution a similar mechanism is operative.

5.3 THE EXCHANGE REACTION BETWEEN Pb(IV) AND Pb(II) IN AQUEOUS MEDIA

Investigations by Zintl and Ranch[1], suggested that, in aqueous alkali, the oxyanions of lead (plumbate and plumbite) do not exchange at room temperature. This has been confirmed by Fava[2], who detected no exchange in 7 M KOH over a period of ten days at room temperature, but found measurable exchange at temperatures in the range 57 to 100 °C with reactant concentrations $\sim 2\times10^{-2}$ M. The barium plumbate separation method was used with the tracer Ra D.

The corresponding overall activation energy obtained was 33 kcal.mole^{-1}. At 100 °C, the measured half-time for exchange was 13 min.

5.4 THE EXCHANGE REACTION BETWEEN Pb(IV) AND Pb(II) IN NON-AQUEOUS MEDIA

Exchange between Pb(IV) and Pb(II) as acetates in acetic acid solution has been observed[1] using the isotopic method (Th B). However, Evans *et al.* [2], using Ra D as the indicator, with reactant concentrations in the range 10^{-2} to 10^{-1} *M*, concluded that there was no exchange at 80 °C in a period of several hours. The reaction was carried out in a dry-box system and anhydrous acetic acid was used as the solvent. Three separation procedures were used: (*a*) conversion of the lead species to the plumbate and plumbite anions, followed by precipitation of barium plumbate (a method used previously for the aqueous exchange system), (*b*) precipitation of lead(IV) as the dioxide, and (*c*) rapid cooling and consequent crystallisation of the lead(IV). Method (*b*) was reported to bring about induced exchange.

6. Arsenic and antimony

6.1 THE EXCHANGE REACTION BETWEEN As(V) AND As(III)

Wilson and Dickenson[1] observed no exchange, over a period of three hours at 100 °C, between arsenate and arsenite ions in media ranging from aqueous acid to aqueous alkali. Martin *et al.*[2] have found similar results for the exchange between arsenate and thioarsenite ions in aqueous media. However, in liquid ammonia exchange occurred between ammonium arsenate and arsenic trisulphide[2]. The isotopic method was used[1,2].

6.2 THE EXCHANGE REACTION BETWEEN Sb(V) AND Sb(III) IN AQUEOUS MEDIA

Absorbance measurements on mixtures of Sb(V) and Sb(III) in hydrochloric acid media, have led Whitney and Davidson[1,2] to propose a dimeric equilibrium

$$SbCl_6^- + SbCl_4^- \rightleftharpoons Sb_2Cl_{10}^{2-}$$

Observations on Sb(V) solutions, in the same media, have shown that the absorbance is dependent upon (*a*) time, and (*b*) hydrochloric acid concentration[1-3]. This has been shown to be due to reactions

$$SbCl_6^- + x\,H_2O \rightleftharpoons SbCl_{6-x}(OH)_x + x\,H^+ + x\,Cl^-$$

References pp. 142–152

Neumann[3] has measured values of the equilibrium constants ($x = 1$ and 2) in media 6–11 M HCl. Neumann and Ramette[4] have obtained data on the rate of hydrolysis of Sb(V) using spectrophotometry.

Bonner[5], using the isotopic method (^{124}Sb) and either precipitation of the oxinate of Sb(III) or an isopropyl ether extraction of Sb(V), obtained the first results on this exchange reaction. In media ~ 6 M HCl a complex rate law, *viz.*

$$\text{rate} = k'[\text{Sb(III)}]^{0.6}[\text{Sb(V)}]^{1.1}[\text{H}^+]^4[\text{Cl}^-]^9$$

was found to be consistent with the experimental data. Concentrations were varied over the ranges: Sb(III) = Sb(V), 8×10^{-4} to 4×10^{-2} M; Cl^-, 5.4 to 6.1 M; H^+, 4.7 to 6.1 M. The value of k' (25 °C) was calculated as 2.4×10^{-14} $1^{13.7}$. mole$^{-13.7}$.sec^{-1} with the overall activation energy ~ 27 kcal.mole^{-1}.

Neumann and Brown[6], have suggested that the only species of Sb(V) that can exchange with Sb(III) is the ion $SbCl_6^-$. Numerous reasons for this assumption have been discussed. These authors have shown that the observed half-times of exchange in media 6–12 M with respect to HCl[5-7] can be satisfactorily predicted on the basis of a mechanism

$$\text{B} \rightleftharpoons \text{SbCl}_6^- \quad (1)$$

$$\text{SbCl}_6^- + \text{SbCl}_3 \rightleftharpoons \text{SbCl}_3 + \text{SbCl}_6^- \quad (2)$$

where B represents species of Sb(V) which are present in equilibrium with $SbCl_6{}^-$, in which either reaction (1) or reaction (2) can be rate determining depending on the conditions. In concentrated HCl the order of the reaction with respect to [Sb(III)] was demonstrated to be unity. Various transition states, involving chloride bridges, have been suggested[6].

Cheek *et al.* [8], have since published data for HCl media showing (*a*) a maximum exchange rate (~ 9 M HCl), (*b*) non-linear exchange-time plots (~ 7 M HCl), (*c*) an order in Sb(III) of 1.0 (11.8 M HCl) and 0.9 (9.5 M HCl), (*d*) an activation energy which depends on the acidity, and (*e*) details of rate dependence upon chloride and hydrogen ion concentrations. The observed rate coefficient (25 °C) and activation energy are 5.7×10^{-3} l.mole^{-1}.sec^{-1} and 17.2 kcal.mole^{-1}, respectively, for media 11.8 M in HCl. These authors conclude that exchange occurs between Sb(III) and two Sb(V) species which are present in slow equilibrium with each other.

Kambara *et al.*[9] have reported data for the exchange in media 0.8 to 4.0 M with respect to HCl, the highest rate being observed at 2.0 M HCl. Added lithium chloride and the previous treatment of reactants was also found to affect the rate of exchange.

Bonner and Goishi[10] have reviewed the complex exchange kinetics and the hydrolysis reactions of Sb(V), and have made a further study of the exchange in the region ~ 7 *M* HCl. The exchange data were treated according to a reaction scheme in which Sb(III) could exchange with two forms of Sb(V) which are themselves in a slowly established hydrolysis equilibrium. For a medium 7.04 *M* in HCl, the values of k_1 and k_2 the rate coefficients for exchange between Sb(III) and the two forms of Sb(V), are 2.32×10^{-2} l.mole^{-1}.sec^{-1} and $\leqq 6.7 \times 10^{-5}$ l.mole^{-1}.sec^{-1}, respectively, at 24.95 °C. Energy of activation for the step with coefficient k_1 is 17.2 kcal.mole^{-1}, with an entropy of activation of -10.3 cal.deg^{-1}.mole^{-1}. Some evidence was also obtained for a photochemical effect which only shows itself at low [Sb(III)]. Bonner and Goishi have also pointed out that the interaction dimer[2] and the Sb(III)-catalysed hydrolysis of Sb(V)[4] must have unsymmetrical activated complexes, which are different from each other and from the exchange activated complex.

Turco and Faroane[11] and Turco[12] have investigated the effect of bromide ions on the Sb(V)–Sb(III) exchange reaction in 3.15 *M* HCl media and have found a complex rate law[11]

$$\text{rate} = k_{\text{obs}}[\text{Sb(V)}]^{1.1}[\text{Sb(III)}]^{0.15}[\text{H}^+]^{4.2}[\text{Br}^-]^{3.5}$$

which was attributed to the exchange occurring *via* two pathways, one independent of Sb(III) (the main pathway) and one involving both Sb(V) and Sb(III). A simplified rate law

$$\text{rate} = k_3'[\text{Sb(V)}] + k_4'[\text{Sb(V)}][\text{Sb(III)}]$$

was also found to fit the experimental data. At low [Sb(III)] and [Br$^-$] (7.5×10^{-4} to 1×10^{-2} *M* and 1.0 *M*, respectively), only the first term in the above rate law was observed. The value of k_3' (22 °C) is 1.3×10^{-4} sec^{-1}. The step postulated to account for this exchange was

$$\text{Sb(V)Br}_n\text{(X)} \rightleftharpoons \text{Sb(III)Br}_{(n-2)}\text{(X)} + \text{Br}_2$$

where X represents an hydroxide or chloride ion.

Brubaker and Sincius[13, 14] have reported the exchange reaction as not occurring in sulphate media ([H$^+$], 3 to 12 *M*) unless chloride ions are present. McKay plots deviating from linearity were observed for added chloride ion 0.4 to 6.0 *M*, and complicated dependencies on the ions H$^+$, SO_4^{2-} and Sb(V) were also found. For mixtures of Sb(III) and Sb(V), the principle of additivity of absorbance was obeyed in sulphate media.

Turco[15] has also detected little exchange between Sb(III) and Sb(V) in alkaline media (1.8 *M* KOH) at room temperature.

6.3 THE EXCHANGE REACTION BETWEEN Sb(V) AND Sb(III) IN NON-AQUEOUS MEDIA

Barker and Kahn[1] have made a detailed study of the exchange in carbon tetrachloride media using the isotope ^{124}Sb to label the Sb(III) species. The reaction was carried out in sealed ampoules covered with Al foil in the presence of an atmosphere of He or Ar gas. The separation method used involved complexing the Sb(V) with fluoride (brought about by addition of ethanol, HCl and HF) followed by precipitation of the Sb(III) with H_2S and finally addition of boric acid and HCl, removal of the CCl_4, and treatment with H_2S to remove the Sb(V). Zero-time exchange was ~ 5 %.

Concentration ranges employed in this study were Sb(III), 1.5×10^{-2} to 1.2×10^{-1} *M*, and Sb(V), 5.5×10^{-3} to 6.7×10^{-2} *M*, the chloride salts being used. The experimental data led to a rate law

$$\text{rate} = k_1[SbCl_5]+k_2'[SbCl_5]^2[SbCl_3]$$

with values of the coefficients k_1 and k_2' (at 50.1 °C) 1.6×10^{-7} sec^{-1} and 1.8×10^{-4} l^2.mole^{-2}.sec^{-1}. Activation energies for the processes represented by k_1 and k_2' were calculated as 19 and 15 kcal.mole^{-1}, respectively, from data obtained over the range 50.1 to 81 °C. The rate of exchange was found to be affected by light, but not by increase in surface area.

A mechanism involving complex chloride species, *viz.*

$$SbCl_5 \overset{k_1}{\rightleftharpoons} SbCl_3+Cl_2$$

$$2\ SbCl_5 \rightleftharpoons Sb_2Cl_{10} \qquad \text{(rapid)}$$

$$Sb_2Cl_{10}+SbCl_3 \rightleftharpoons SbSb_2Cl_{13} \qquad \text{(rapid)}$$

$$SbSb_2Cl_{13} \overset{k_2}{\rightarrow} SbCl_3+SbSbCl_{10}$$

was proposed to account for these observations. A suggestion was made concerning the structure of the complex ($SbSb_2Cl_{13}$).

Price and Brubaker[2,3] have examined the catalytic effect of hydrogen chloride on the rate of exchange in the same media. Over a range of HCl concentrations, 6×10^{-3} to 6×10^{-2} *M*, the data obtained was consistent with a rate law

$$\text{rate} = k_1[SbCl_5]+k_2'[SbCl_5]^2[SbCl_3]+k_3[HSbCl_6][SbCl_3]$$

k_3 (50 °C) has a value 2.07×10^{-5} l.mole^{-1}.sec^{-1} and is the rate coefficient for the exchange pathway

$$HSbCl_6+SbCl_3 \xrightarrow{k_3}$$

It is of interest to note that the ^{32}P exchange between PCl_5 and PCl_3 also proceeds, in CCl_4 media, *via* a dissociation step[4]

$$PCl_5 \overset{k_4}{\rightleftharpoons} PCl_3 + Cl_2$$

with a rate law

$$\text{rate} = k_4[PCl_5]$$

While the phosphorus exchange is ~ 130 times faster than the antimony exchange for the same basic process the energies of activation (16 and 19 kcal.mole^{-1}) are similar. No light sensitivity was observed in the ^{32}P exchange.

6.4 THE Sb(III)-CATALYSED HYDROLYSIS OF Sb(V)

Neumann and Ramette[1] have found the hydrolysis of Sb(V) to be catalysed by Sb(III). An activated complex, which must be unsymmetrical, has been proposed[2].

7. Tellurium

7.1 THE EXCHANGE REACTION BETWEEN Te(VI) AND Te(IV)

The exchange between the acids H_6TeO_6 and H_2TeO_3 has been investigated using the radioisotopes ^{127}Te, ^{129}Te[1] and ^{132}Te[2]. No exchange was detected over periods of many hours at 95 °C in acidic chloride or perchlorate solution up to 6 *M* with respect to H^+. Separation was achieved using the precipitation of the tellurite anion which occurs at pH 4.20.

8. Transition metals

8.1 VANADIUM AND TANTALUM

8.1.1 The exchange reaction between V(III) and V(II)

King and Garner[1], using the isotopic method (^{48}V), made an attempt to study this exchange reaction in both aqueous sulphate and perchlorate media. The separation methods tried were (*a*) ion exchange, and (*b*) precipitation of the vanadium (III) with ammonia, after first complexing the vanadium (II) in aqueous ethanol as $V(dipy)_3^{2+}$. Complete exchange was observed within minutes at 2 °C, in the absence of oxygen, with reactant concentrations of ~ 10^{-1} *M*. Evidence was obtained that the exchange of V^{3+} with $V(dipy)_3^{2+}$ is slow.

References pp. 142–152

Krishnamurty and Wahl[2] tried a number of separation methods and eventually used a modified dipyridyl–ammonia separation to obtain kinetic data for this exchange. The zero-time exchange lay between 25 and 50 %, depending on the conditions; the activity of the $V(dipy)_3^{2+}$ ion was measured. The rate law obtained for perchlorate media was

$$\text{rate} = k_{obs}[V(III)][V(II)]$$

At 25 °C, at an ionic strength of 2.0 M and $[H^+]$ of 1 M, k_{obs} was found to have a value of 1.4×10^{-2} l.mole^{-1}.sec^{-1}. The overall activation energy and entropy were calculated as 13.2 kcal.mole^{-1} and -25 cal.deg^{-1}.mole^{-1}, respectively.

The rate of exchange was found to be dependent on the hydrogen-ion concentration (up to 5×10^{-1} M) in a manner which led to these authors suggesting that the exchange pathways

$$V^{3+} + V^{2+} \xrightarrow{k_1}$$

$$VOH^{2+} + V^{2+} \xrightarrow{k_2}$$

exist. Analysis of the exchange data obtained at an ionic strength 2.0 M led to values of k_1 and k_2K_2 of 1.02×10^{-2} l.mole^{-1}.sec^{-1} and 3.5×10^{-3} sec^{-1}, respectively, where K_2 refers to

$$V^{3+} + H_2O \rightleftharpoons VOH^2 + H^+ \qquad K_2$$

The estimated value of k_2 is 1.75 l.mole^{-1}.sec^{-1}.

Addition of chloride in the range 5×10^{-3} to 3×10^{-2} M, at constant acidity and ionic strength, was found to increase the rate of exchange. This was interpreted in terms of an exchange pathway involving the VCl^{2+} ion, *viz.*

$$V^{3+} + Cl^- \rightleftharpoons VCl^{2+} \qquad K_3$$

$$VCl^{2+} + V^{2+} \xrightarrow{k_3}$$

The term k_3K_3 has a value of 1.4 l^2.mole^{-2}.sec^{-1} in 2 M perchlorate media of constant acidity. Other pathways involving chloride are also possible.

8.1.2 The exchange reaction between V(IV) and V(III)

Furman and Garner[1] have measured the rate of this exchange reaction in aqueous perchlorate media using the isotope ^{48}V as indicator. Numerous methods of separation were tried, kinetic data being obtained using an ion exchange technique. The rate law obeyed, over the concentrations range V(IV), 1×10^{-2} to 5×10^{-2} *M*, and V(III), 8×10^{-3} to 5×10^{-2} *M*, and in the absence of oxygen, was

$$\text{rate} = k_{obs}[V(IV)][V(III)]$$

The observed rate coefficient k_{obs} was found to be dependent on the hydrogen-ion concentration (plots of k_{obs} *versus* $[H^+]^{-1}$ were linear passing through the origin), suggesting that only an acid dependent pathway for exchange occurs. The pathway suggested was

$$VO^{2+} + VOH^{2+} \xrightarrow{k_1}$$

Using the available data for the equilibrium

$$V^{3+} + H_2O \rightleftharpoons VOH^{2+} + H^+ \qquad K_1$$

k_1 was estimated as 14 l.mole^{-1}.sec^{-1} at 25 °C and ionic strength of 2.5 *M*. The calculated activation energy and entropy obtained for this step were 10.7 kcal. mole^{-1} and -24 cal.deg^{-1}.mole^{-1}.

The presence of thiocyanate ions appears to catalyse this exchange but no detailed measurements have been made.

8.1.3 The exchange reaction between V(V) and V(IV)

Tewes *et al.*[1], using both precipitation and extraction separation techniques, observed essentially complete exchange within the separation time (1 min) during an isotopic study (^{48}V) of this exchange. The media ranged from 0.3 *M* perchloric acid to 7 *M* hydrochloric acid; reactant concentrations were $\sim 10^{-2}$ *M*.

More recently Giuliano and McConnell[2], using an NMR technique based on line width changes of the ^{51}V line, have made some measurements of the exchange rate. The rate law was found to be

$$\text{rate} = k_{obs}[V(V)]^2[V(IV)]$$

in chloride–perchlorate media. At concentrations 3 *M* Cl^- and 6.5 *M* H^+, the value of k_{obs} is 1.5×10^6 l^2.mole^{-2}.sec^{-1}. The rate of exchange was found to be dependent on the hydrogen and chloride ion concentrations.

References pp. 142–152

The mechanism proposed was

$$2\,V(V) \rightleftharpoons (V(V))_2$$

$$(V(V))_2 + V(IV) \xrightarrow{k_1} (V(V))_2 + V(IV)$$

with the acid dependence being possibly due to the reaction

$$2\,VO_2^+ + 2\,H^+ \rightleftharpoons V_2O_3^{4+} + H_2O$$

Concentrations of V(IV) and V(V) used in this study were $\sim 1.5 \times 10^{-2}$ and 2×10^{-1} *M*, respectively.

8.1.4 Reactions between vanadium ions

A detailed study of the reaction

$$VO^{2+} + V^{2+} + 2\,H^+ = 2\,V^{3+} + H_2O$$

in perchlorate media, in the absence of air, has been made by Newton and Baker[1] using a spectrophotometric method, a wavelength of 760 mμ (absorption by V(IV)) being used to obtain rate data. The rate law, obeyed over a range of conditions, was

$$-d[V(IV)]/dt = -d[V(II)]/dt = k_{obs}[V(IV)][V(II)]$$

with no retardation being produced by addition of V(III). A small but definite increase in the rate coefficient (k_{obs}) was observed as the hydrogen-ion concentration was increased at constant ionic strength. This effect led to a more detailed rate expression

$$\text{rate} = (k' + k''[H^+])[VO^{2+}][V^{2+}]$$

and the conclusion was that the main step was

$$VO^{2+} + V^{2+} \rightarrow$$

with the possibility of a step

$$VO^{2+} + V^{2+} + H^+ \rightarrow$$

or a medium effect operating. For the process represented by k' the activation enthalpy and entropy were 12.3 kcal.mole^{-1} and -16.5 cal.deg^{-1}.mole^{-1}, and for that represented by k'', 9.8 kcal.mole^{-1} and -31.6 cal.deg^{-1}.mole^{-1}.

During this study, an intermediate absorbing at 425 mμ was detected and shown in a further study[2] to be a dimer (VOV^{4+}), with nearly two-thirds of the V(IV)–V(II) reaction proceeding *via* this species in an inner-sphere step, the remainder reacting *via* an outer-sphere pathway. The mechanism proposed for the reaction was

$$VO^{2+}+V^{2+}+2\,H^{+} \rightarrow 2\,V^{3+}+H_2O \quad (k_1)$$

$$VO^{2+}+V^{2+} \rightarrow VOV^{4+} \quad (k_2)$$

$$VOV^{4+}+H^{+} \rightarrow V^{3+}+VOH^{2+} \quad (k_3)$$

$$VOH^{2+}+H^{+} \rightleftharpoons V^{3+}+H_2O \quad \text{(rapid)}$$

From spectrophotometric measurements (at 425 mμ) on the appearance and disappearance of dimer, values for the rate coefficients k_2 and k_3 were found to be (at 0 °C and ionic strength 1.0 M) 6.7×10^{-2} and 3.3×10^{-1} l.mole^{-1}.sec^{-1}, respectively. Under the same conditions k_1 was calculated to be 3.9×10^{-2} l.mole^{-1}.sec^{-1}.

Both sulphate and chloride ions were found to accelerate the reaction under constant ionic strength conditions and, although medium effects may operate, pathways involving chloride and sulphate ions are possible. For the sulphate ion addition Newton and Baker[1] conclude that the step

$$VO^{2+}+V^{2+}+SO_4^{2-} \rightarrow$$

is always present. Olver and Ross[3], however, during a study of the catalytic polarographic reduction of V(III), were able to obtain data suggesting the dependence of the rate of this reaction on $[HSO_4^-]$ rather than on $[SO_4^{2-}]$ as above, and suggested a pathway involving this ion. The rate law proposed was

$$\text{rate} = k'[VO^{2+}][V^{2+}]+k_4[VOHHSO_4^{2+}][V^{2+}]$$

Daugherty and Newton[4] have examined the reaction between VO_2^+ and V^{3+} and have found their results to be consistent with a rate law

$$-\mathrm{d}[VO_2^+]/\mathrm{d}t = (k_4'[H^+]^{-2}+k_5'[H]^{-1}+k_6+k_7'[H^+])[V(III)][V(V)]$$

the major part of the reaction proceeding *via* a pathway associated with the coefficient k_5.

References pp. 142–152

The reaction

$$V^{2+}+VO_2^+ +2H^+ = V^{3+}+VO^{2+}+H_2O$$

has been investigated by Espenson and Krug[5], using the stopped-flow technique. Results indicating a rate law

$$-d[VO_2^+]/dt = (k_8+k_9'[H^+])[V^{2+}][VO_2^+]$$

were obtained. The values of the rate coefficients k_8(l.mole^{-1}.sec^{-1}) and k_9'(l^2.mole^{-2}.sec^{-1}), at 25 °C and $\mu = 1.01\ M$, with the associated activation enthalpies (kcal.mole^{-1}) and entropies (cal.deg^{-1}.mole^{-1}) in parentheses are 2.58×10^3 (1.9, -36.8) and 2.16×10^3 (1.8 and -37), respectively.

8.1.5 Reactions between tantalum cluster ions

Espenson and McCarley[1] have estimated the rate coefficient for the reaction

$$Ta_6Cl_{12}^{4+}+Ta_6Cl_{12}^{2+} = 2\ Ta_6Cl_{12}^{3+}$$

from spectrophotometric observations on mixtures of the two reactants to be greater than 10^5 l.mole^{-1}.sec^{-1}.

For the reaction

$$Ta_6Br_{12}^{2+}+Ta_6Br_{12}^{4+} = 2Ta_6Br_{12}^{3+}$$

estimates[2,3] for the rate coefficient of $\geqslant 5\times10^7$ and 6.8×10^7 l.mole^{-1}.sec^{-1} have been recently reported.

For the reaction

$$Ta_6Cl_{12}^{2+}+Ta_6Br_{12}^{3+} = Ta_6Cl_{12}^{3+}+Ta_6Br_{12}^{2+}$$

the value of the rate coefficient proposed[2] was $\geqslant 10^8$ l.mole^{-1}.sec^{-1}.

8.2 CHROMIUM, MOLYBDENUM AND TUNGSTEN

8.2.1 The exchange reaction between Cr(III) and Cr(II)

The earliest estimate of the rate coefficient for the exchange in perchlorate media was obtained by Plane and Taube[1] from data obtained on the catalytic effect of

Cr^{2+}_{aq} on the $Cr(H_2O)_6^{3+}$–H_2O exchange reaction. A value of 4.7×10^{-4} l.mole^{-1}.sec^{-1} was proposed.

Anderson and Bonner[2] made the first detailed kinetic study on the exchange using the isotopic method (^{51}Cr) and a separation method based on the conversion of Cr(II) into Cr(III) oxalate and an ion-exchange treatment. To prevent oxidation of Cr(II) during exchange a hydrogen atmosphere was maintained over the reaction mixture. The rate law found to be obeyed for the concentration ratio range Cr(III)/Cr(II) of between 3.3×10^{-2} and 2.0 in perchlorate media was

$$\text{rate} = k_{obs}[\text{Cr(III)}][\text{Cr(II)}]$$

From data obtained at $\mu \sim 1.0$ *M* with $[H^+]$ varying over the range 0.213 to 1.0 *M*, k_{obs} was found to be given by

$$k_{obs} = k_1 + k_2'[H^+]^{-1}$$

which was considered by Anderson and Bonner to arise from the exchange pathways

$$Cr^{3+} + Cr^{2+} \xrightarrow{k_1}$$

$$CrOH^{2+} + Cr^{2+} \xrightarrow{k_2}$$

The species $CrOH^{2+}$ being formed by the hydrolysis

$$Cr^{3+} + H_2O \rightleftharpoons CrOH^{2+} + H^+ \quad K_2$$

with k_2' given by the product K_2k_2. On this basis k_1 and k_2 were evaluated as 2×10^{-5} and 7.0 l.mole^{-1}.sec^{-1}, respectively (at 24.5 °C $\mu \sim 1.0$ *M*). The value obtained for the overall activation energy was 22 kcal.mole^{-1}. For the second step (k_2) a hydrogen atom transfer mechanism was suggested.

The effect of chloride ion on the exchange was found by these workers to be very small, whereas Plane and Taube[1] had estimated a rate coefficient about five times larger in the presence of 10^{-2} *M* chloride ion than in perchlorate solution. Van der Straaten and Aten[3] have studied the exchange in media 1 *M* with respect to HCl and have estimated a rate coefficient $\geqq 3.0 \times 10^{-2}$ l.mole^{-1}.sec^{-1}. The isotopic method (^{51}Cr) and a separation procedure based on the precipitation of Cr(II) as the acetate complex was used.

The exchange reactions

$$CrX^{2+} + Cr^{2+} \rightleftharpoons Cr^{2+} + CrX^{2+}$$

where X is, F^-, Cl^-, Br^-, N_3^-, NCS^-, CN^-, and $H_2PO_2^-$ have been investigated[4–8].

The radioisotope ^{51}Cr was used to tag the Cr^{2+} species and the separation of Cr^{2+} from CrX^{2+} was achieved using an ion-exchange method, after oxidation of the Cr^{2+} to Cr^{3+} with Fe(III) or oxygen. The reactions were carried out in the absence of oxygen in perchlorate media. For the systems involving chloride, fluoride[4], and azide[6], King *et al.* have found a rate law

$$\text{rate} = k_3[CrX^{2+}][Cr^{2+}]$$

to be obeyed over a range of conditions including variation in the $[H^+]$. Values of the rate coefficient k_3 (l.mole^{-1}.sec^{-1}), conditions in parentheses, are, for $X = F^-, Cl^-$, and N_3^- 2.5×10^{-3}, 9.1 (0 °C, $\mu = 1.0\ M$) and 1.3 (0 °C, $\mu = 0.5\ M$), respectively. Assuming a similar rate law when $X = Br^-$ and NCS^- the calculated rate coefficients were > 60 (0 °C, $\mu = 1.0\ M$) and 1.8×10^{-4} (27 °C, $\mu = 1.0\ M$) l.mole^{-1}.sec^{-1}, respectively. For $X = Cl^-$ a value of 8.3 l.mole^{-1}sec^{-1} (0 °C, $\mu \sim 1.0\ M$) has been obtained by Taube and King[5]. Activation parameters have been obtained for the systems CrF^{2+}–Cr^{2+} and CrN_3^{2+}–Cr^{2+}; these are $\Delta H^{\ddagger} = 13.7$ and 9.6 kcal.mole^{-1} and $\Delta S^{\ddagger} = -20$ and -22.8 cal.deg^{-1}.mole^{-1}, respectively[4,6].

The transition states proposed for these exchange systems involve anion bridges[4,5]. For the system $CrNCS^{2+}$–Cr^{2+}, Ball and King[4] suggest that either a nitrogen bridged transition state or a two stage process

$$CrNCS^{2+} + Cr^{2+} \rightarrow Cr^{2+} + CrSCN^{2+}$$

$$CrSCN^{2+} + Cr^{2+} \rightarrow Cr^{2+} + CrNCS^{2+}$$

occurs.

Espenson *et al.*[7,8] have found for $X = H_2PO_2^-$ and CN^- a hydrogen ion dependent rate law

$$\text{rate} = k_{obs}[CrX^{2+}][Cr^{2+}]$$

where the observed rate coefficients are given by $k_{obs} = k_4'[H^+]^{-1}$ for $X = H_2PO_2^-$, (over the range of concentrations, $CrH_2PO_2^{2+}$ 3.7×10^{-3} to 4.0×10^{-2} M, Cr^{2+} 1.8×10^{-3} to 2.5×10^{-2} M, and H^+ 2.6×10^{-2} to 1.0 M) and by $k_{obs} = k_5 + k_6'[H^+]^{-1}$ for $X = CN^-$ (over the concentration ranges, $CrCN^{2+}$ 4.5×10^{-3} to 2.6×10^{-2} M, Cr^{2+} 1.0×10^{-2} to 1.9×10^{-2} M, and H^+ 1.2×10^{-2} to 3.5×10^{-1} M).

For the hypophosphito complex, k_4' has a value 6.11×10^{-4} sec^{-1} (25 °C, $\mu = 1.0\ M$) with the associated activation enthalpy and entropy values of 19.7 kcal.mole^{-1} and -7.2 cal.deg^{-1}.mole^{-1}.

For the exchange involving the monocyanochromium(III), values of k_5 and k_6' are 7.7×10^{-2} l.mole^{-1}.sec^{-1} and 4.2×10^{-3} sec^{-1} at 25 °C and $\mu = 1.0\ M$, respectively. The activation parameters corresponding to these steps are $\Delta H_5^{\ddagger}$ 9.3 kcal.mole^{-1}, $\Delta H_{6'}^{\ddagger}$ 17.2 kcal.mole^{-1}, $\Delta S_5^{\ddagger}$ -32 cal.deg^{-1}.mole^{-1} and $\Delta S_{6'}^{\ddagger}$ -8 cal.deg^{-1}.mole^{-1}. For the steps associated with k_4' and k_6' transition states having bridges involving the ions X^- and OH^- have been proposed. For the pathway associated with k_5 a mechanism

$$CrCN^{2+} + Cr^{2+} \rightarrow Cr^{2+} + CrNC^{2+}$$

$$CrNC^{2+} \rightarrow CrCN^{2+}$$

$$CrNC^{2+} + Cr^{2+} \rightarrow CrCN^{2+} + Cr^{2+}$$

has been suggested.

Snellgrove and King[9] have examined the exchange reaction

$$cis\text{-}Cr(N_3)_2^+ + Cr^{2+} \xrightarrow{k_7}$$

and have suggested two azide groups participate in the transition state bridging. The isotopic method (^{51}Cr) and ion-exchange separation of reactants (after treatment with H_2O_2) was used. The observed second-order rate coefficient has a value 60 l.mole^{-1}.sec^{-1} (0 °C, $\mu = 0.5\ M$). The effect of increasing the hydrogen ion concentration was very small.†

Stranks[10] has reported the activation energy and entropy for the exchange system $Cr(urea)_6^{3+}$–$Cr(urea)_6^{2+}$, to be 13 kcal.mole^{-1} and -40 cal.deg^{-1}.mole^{-1}.

8.2.2 The exchange reaction between Cr(VI) and Cr(III)

The early studies on the exchange system, carried out by Muxart *et al.*[1], Menker and Garner[2], and Burgus and Kennedy[3], showed the exchange to be slow in sulphate[1,2], perchlorate[2], nitrate[3], and hydroxide media[2,3]. The isotopic method (^{51}Cr) was used with separation of the Cr(III) and Cr(VI) being achieved by the precipitation of chromic oxide[2] and lead chromate[3]. Some evidence was obtained for the retardation of the exchange by hydrogen ions.

Altman and King[4] made the first detailed study of this system, using Cr(III) solutions containing only the monomeric species and Cr(VI) solutions which had been allowed to age. The isotopic method and lead chromate precipitation separation were used to obtain kinetic data at a temperature of 94.8 °C. Over the range of concentrations, Cr(VI) 2.3×10^{-4} to 8.4×10^{-2} M, Cr(III) 1.8×10^{-3} to

† Data for the exchange of ^{51}Cr between Cr^{2+} and the Cr(III) species $Cr(OAc)_2^+$ and *cis*-$Cr(Ox)_2^-$ have also been reported[11,12].

8.5×10^{-2} *M*, H^+ 4×10^{-2} to 8.3×10^{-1} *M*, a rate law

$$\text{rate} = (k' + k''[H^+]^{-2})[Cr^{3+}]^{4/3}[H_2CrO_4]^{2/3}$$

was obtained, with values of k' and k'' ($\mu = 0.91$ *M* and 94.8 °C) of 1.4×10^{-5} l.mole^{-1}.sec^{-1} and 6.6×10^{-7} mole.l^{-1}.sec^{-1}, respectively. The exchange was thought to occur *via* a slow step, either

$$Cr(III) + Cr(V) \rightleftharpoons Cr(V) + Cr(III)$$

or

$$Cr(III) + Cr(V) \rightleftharpoons Cr(IV) + Cr(IV)$$

with the other steps

$$Cr(V) + Cr(VI) \rightleftharpoons Cr(VI) + Cr(V)$$

$$Cr(III) + Cr(VI) \rightleftharpoons Cr(IV) + Cr(V)$$

also occurring.

8.2.3 *The reaction between Cr(VI) and Cr(II)*

Hegedus and Haim[1] have examined the reaction of Cr(VI) with Cr(II)

$$3\,Cr^{2+} + Cr(VI) = 2\,Cr^{3+} + Cr_2(OH)_2^{4+}$$

using the isotopic method (^{51}Cr) in an attempt to verify the reaction mechanism proposed by Ardon and Plane[2], *viz.*

$$Cr(VI) + Cr^{2+} \rightarrow Cr(V) + Cr^{3+}$$

$$Cr(V) + Cr^{2+} \rightarrow Cr(IV) + Cr^{3+}$$

$$Cr(IV) + Cr^{2+} \rightarrow Cr_2(OH)_2^{4+}$$

Separation of the products was achieved using an ion-exchange method, the reaction taking place in sealed vessels with the Cr(VI) labelled (^{51}Cr). Under various conditions only 40 % of the ^{51}Cr appears in the product $Cr_2(OH)_2^{4+}$. This led these workers to propose a modified mechanism in which the step involving Cr(V) proceeds in two ways

$$Cr(V) + Cr(II) \rightarrow Cr(IV) + Cr^{3+}$$

$$Cr(V) + Cr(II) \rightarrow Cr^{3+} + Cr(IV)$$

the former step being about four times faster than the latter. No kinetic measurements have been made.

8.2.4 *Cr(II)-catalysed substitution and isomerisation reactions of Cr(III)*

A number of aquation reactions of the type

$$Cr(III)X = Cr^{3+} + X^-$$

have been studied and found to be catalysed by Cr(II), usually in the form Cr^{2+}_{aq}. The reactions are usually carried out in a sealed container, in perchlorate media and in the presence of an inert gas which prevents atmospheric oxidation of the Cr^{2+}; the rate data are usually obtained from spectrophotometric measurements. Studies have been made on the complex ions Cr(III)X for X = Cl^- (Taube and King[1], Adin and Sykes[2], and Pennington and Haim[3]), Br^- (Adin *et al.*[4], and Pennington and Haim[3]), I^- (Adin *et al.*[4], and Pennington and Haim[3]),† F^- (Adin *et al.*[4]), N_3^- (Doyle *et al.*[5]), and NH_3 (Espenson and Carlyle[6]). The wavelengths used in the absorbance studies were 608 mμ (Cl^-), 622 mμ (Br^-), 306 or 475 mμ (I^-), 408 mμ (F^-), 433 mμ (N_3^-), and 504 to 522 mμ (NH_3).†† Since these aquation reactions can also occur *via* pathways which are not catalysed by Cr(II), terms from both catalysed and non-catalysed reactions appear in the rate law, which is usually of the form

$$d[Cr(III)X]/dt = (k_1 + k_2'[H^+]^{-1} + k'[Cr^{2+}])[Cr(III)X]$$

The rate coefficients k_1 and k_2' relate to the non-catalysed process and k' is the apparent rate coefficient for the catalysed pathways. For the systems where X = Cl^-, Br^- and I^-, k' is given by k_3' $[H^+]^{-1}$, and a reaction pathway

$$CrX(OH)^+ + Cr^{2+} \xrightarrow{k_3}$$

was postulated, with the species $Cr(X)(OH)^+$ produced *via* the equilibrium

$$CrX^{2+} + H_2O \rightleftharpoons Cr(X)(OH)^+ + H^+ \quad K_3$$

and $k_3' = k_3K_3$. Values of the observed rate coefficient k_3' (sec^{-1}) are 3.2×10^{-4} (Cl^-)[5], 1.74×10^{-3} (Br^-)[4], 2.13×10^{-2} (I^-)[4] (at 25 °C, μ = 2.0 *M*) and $\sim 4 \times 10^{-4}$ (Cl^-), $\sim 2 \times 10^{-3}$ (Br^-), and 2.2×10^{-2} (I^-), (μ = 1.0 *M*, 25 °C)[3]. Activation parameters ($\Delta H^{\ddagger}$ and $\Delta S^{\ddagger}$) of 20.2 kcal.mole^{-1}, −7.4 cal.deg^{-1}.mole^{-1}

† See also Pennington and Haim[23].
†† The reaction for X = CN^- has also been investigated[24].

(Cl^-)[6], 18.2 kcal.mole^{-1}, -10 cal.deg^{-1}.mole^{-1} (Br^-)[4], and 17 kcal.mole^{-1}, -9.4 cal.deg^{-1}.mole^{-1} (I^-)[4] were also obtained. For the reaction

$$CrCl^{2+} + Cr^{2+} \underset{k_{-3a}}{\overset{k_{3a}}{\rightleftharpoons}} Cr^{2+} + Cl^- + Cr^{3+} \quad K_{3_a}$$

the related value of K_{3a} has been determined and an estimate of k_{3a} (at 20 °C, $\mu = 2.0\ M$) of 20 l.mole^{-1}.sec^{-1} has been made[2]. The observed rate coefficient k_{-3a} has been evaluated as 3.4×10^{-3} l^2. mole^{-2}. sec^{-1} ($[H^+] = 1.0 \times 10^{-1}\ M$ and $Cl^- = 1.0\ M$ at 40 °C)[2].

Sykes *et al.*[2,4] have proposed a transition state involving a hydroxide bridge for the processes defined by k_3.

For the systems $X = F^-$, N_3^-, and NH_3, two catalysed pathways were indicated by the rate law, which took the same form as previously, with

$$k' = (k_3'[H^+]^{-1} + k_4)$$

Values of k_3' and k_4 (at 25 °C and $\mu = 2.0\ M$) obtained were, 8.6×10^{-6} sec^{-1} and 4.4×10^{-5} l.mole^{-1}.sec^{-1} (F^- at 25 °C)[4], 3.4×10^{-5} sec^{-1} and 5.2×10^{-4} l.mole^{-1}.sec^{-1} (N_3^- at 40 °C)[5] and 5.9×10^{-5} sec^{-1} and 2.4×10^{-5} l.mole^{-1}.sec^{-1} (NH_3 at 25 °C)[6], respectively. Activation parameters calculated by Espenson and Carlyle[6] for the $CrNH_3^{3+}$–Cr^{2+} system are ($\Delta H^\ddagger$ and $\Delta S^\ddagger$) 21.6 kcal.mole^{-1}, -5.6 cal.deg^{-1}.mole^{-1} (k_3') and 13.9 kcal.mole^{-1}, -33 cal.deg^{-1}.mole^{-1} (k_4).† Transition state complexes proposed for the steps associated with k_3 are analogous to those given previously for the chloride, bromide and iodide complex ions. For the pathway defined by k_4, transition states $[CrOHCrXH^{4+}]^\ddagger$ and $[CrH_2OCrX^{4+}]^\ddagger$ ($X = F^-$ or N_3^-)[4,5] and[6] $[CrH_2OCrNH_3{}^{5+}]^\ddagger$ have been suggested.

For the reaction between CrN_3^{2+} and Cr^{2+} it was found necessary to allow for the reaction

$$2\,Cr^{2+} + 2\,H^+ + HN_3 = 2\,Cr^{3+} + NH_3 + N_2$$

before evaluation of the rate data from the absorbance measurements[5].

Aquation reactions of some disubstituted aquo ions of Cr(III) have also been found to be catalysed by Cr^{2+}, *viz.*

$$CrX_2^+ + Cr^{2+} = CrX^{2+} + Cr^{2+} + X^-$$

Kinetic studies on the reactions (in the absence of oxygen, in perchlorate media) have been made[7-10] for the ions CrX_2^+ where $X = Cl^-$, F^-, N_3^- and CN^-; the rate data was obtained using spectrophotometric measurements for the

† Deutsch and Taube[25] and Nordmeyer and Taube[26] have recently obtained data for the similar Cr(III)X systems, X = acetate[25] and nicotinamide[26].

chloride (*cis*-, 245 to 260 mμ), (*trans*-, 448 or 635 mμ)[7], azide (*cis*-, 275 mμ)[9] and cyanide (*cis*-, 460 mμ)[10] complex ions and the isotopic method (^{51}Cr) with ion-exchange separation for the fluoride complex ion[8]. For these reactions a transition state, involving the ion X forming a single bridge[7-9], has been suggested by the form of the rate law

$$-d[CrX_2^+]/dt = k''[CrX_2^+][Cr^{2+}]$$

applicable to the catalysed pathway. The non-catalysed aquation reaction was found to be very much slower than the above reactions.

For the ion $CrCl_2^+$, Espenson and Slocum[7] have obtained values of k'' (l.mole^{-1}.sec^{-1}), at 25 °C, $\mu = 1.0$ M, of 401 (*trans*-) and 288 (*cis*-) with corresponding activation parameters ($\Delta H^{\ddagger}$, kcal.mole^{-1} and $\Delta S^{\ddagger}$, cal.deg^{-1}.mole^{-1}) of 4.9 and -30 (*trans*-) and 5.7 and -28.2 (*cis*-), respectively. The value of k'' was found to be independent of $[H^+]$ in the case of the *trans*-isomer only. Previously, Taube and Meyers[11] had estimated a value of 166 l.mole^{-1}.sec^{-1} at 2 °C (media 1 M in $HClO_4$) and Johnson and Reynolds[12] had obtained a value of ~ 100 l.mole^{-1}.sec^{-1} for *cis*-$CrCl_2^+$). For the reverse reaction

$$CrCl^{2+} + Cr^{2+} + Cl^- = CrCl_2^+ + Cr^{2+}$$

Taube and King[1] have obtained a rate coefficient of (50.2 l^2.mole^{-2}.sec^{-1} at 0 °C (media 1 M in $HClO_4$).

Haim[9] has found a value of k'' of 7.6 l.mole^{-1}sec^{-1} at 25 °C (media 0.2 M in $HClO_4$) for the aquation of *cis*-$Cr(N_3)_2^+$. Corresponding activation parameters of 8.1 kcal.mole^{-1} and -27 cal.deg^{-1}.mole^{-1} were also obtained. For the aquation of *cis*-$Cr(CN)_2^+$, Birk and Espenson[10] have reported a value of 4.19 l.mole^{-1}.sec^{-1} at 25 °C ($\mu = 1.0$ M).

Chia and King[8] have evaluated k'' as 1×10^{-2} l.mole^{-1}.sec^{-1} at 25 °C (*cis*-CrF_2^+) and 1.5×10^{-3} l.mole^{-1}.sec^{-1} at 25 °C (*trans*-CrF_2^+). The activation enthalpy and entropy for the aquation of *cis*-CrF_2^+ are 13 kcal.mole^{-1} and -24 cal.deg^{-1}.mole^{-1}, respectively.

The species $Cr(CN)_3$ also undergoes a catalysed aquation

$$Cr(CN)_3 + Cr^{2+} = Cr^{2+} + CrCN^{2+} + 2\,CN^-$$

with a second order rate coefficient of 4.56 l.mole^{-1}.sec^{-1} at 25 °C ($\mu = 1.0$ M)[10].†

Cr^{2+} also catalyses some substitution reactions of the species $Cr(NH_3)_5X^{2+}$, *viz.*

$$Cr(NH_3)_5X^{2+} + 5\,H^+ = CrX^{2+} + 5\,NH_4^+$$

† For the similar system involving the species $Cr(Ox)_3^{3-}$ see Hutchital[27].

where X is the fluoride, chloride, bromide or iodide group. Ogard and Taube[13], using a spectrophotometric method (detecting the appearance of the product CrX^{2+}) have obtained rate data for these reactions, which were carried out in perchlorate media in the absence of oxygen. The rate law found was

$$-\mathrm{d}[Cr(NH_3)_5X^{2+}]/\mathrm{d}t = k_0[Cr(NH_3)_5X^{2+}]+k_5[Cr(NH_3)_5X^{2+}][Cr^{2+}]$$

where k_0 relates to the non-catalysed reactions

$$Cr(NH_3)_5X^{2+} = Cr(NH_3)_5^{3+}+X^-$$

and k_5 (the rate coefficient for the catalysed steps) has values (l.mole^{-1}.sec^{-1} at 25 °C, $\mu = [H^+] = 1.0\ M$) of 2.7×10^{-4} (F^-), 5.1×10^{-2} (Cl^-), 0.32 (Br^-), and ~ 5.5 (I^-). For the species $Cr(NH_3)_5Cl^{2+}$ k_5 was found not to be affected by variation in $[H^+]$ but was reduced by change in the solvent from H_2O to D_2O. Activation parameters of 13.4, 11.1, 8.5 kcal.mole^{-1} and −30, −23, −33 cal. deg^{-1}.mole^{-1} were calculated for the species $Cr(NH_3)_5F^{2+}$, $Cr(NH_3)_5Cl^{2+}$ and $Cr(NH_3)_5Br^{2+}$, respectively. The reaction $(NH_3)_5CrOHCr(NH_3)_4Cl^{4+}+5H^+ = (NH_3)_5Cr^{2+}+5NH_4^+ +CrCl^{2+}$ has been found by Hoppenjans *et al.*[28] to be catalysed by Cr^{2+}.

Cannon[14] has made a study of the Cr(II) catalysed reaction

$$Cr(NH_3)_5Cl^{2+}+n\,OAc^-+5\,H^+ = Cr(OAc)_n^{(3-n)}+5\,NH_4^+ +Cl^-$$

which occurs in acetate buffer. Absorbance measurements at 593 mμ were used to obtain rate data. The rate law observed was

$$\mathrm{d}[Cr(OAc)_3]/\mathrm{d}t = k'''[Cr(NH_3)_5Cl^{2+}][Cr(II)]^{\frac{1}{2}}$$

from which it was concluded that the mechanism was

$$Cr_2(OAc)_6^{2-} \rightleftharpoons 2\,Cr(OAc)_3^- \qquad K_6$$

$$Cr(OAc)_3^- +Cr(NH_3)_5Cl^{2+} \xrightarrow{k_6} Cr(OAc)_3Cl^- +Cr(II)+5\,NH_4^+$$

$$Cr(OAc)_3Cl^- \rightarrow Cr(OAc)_3+Cl^-$$

$$Cr(OAc)_3 \rightarrow \text{polymer}$$

$$Cr(II)+3\,OAc^- \rightleftharpoons Cr(OAc)_3^-$$

Using the value of K_6, obtained from spectrophotometric measurements, k_6 was calculated as 1.2 l.mole^{-1}.sec^{-1} (25 °C, $\mu = 1.0\ M$) with an associated activation energy of 12 kcal.mole^{-1}. For the reaction

$$Cr_2(OAc)_6^{2-} + Cr(NH_3)_5Cl^{2+} \xrightarrow{k_7}$$

an estimate of $k_7 \leqq 0.13$ l.mole^{-1}.sec^{-1} at 45 °C ($\mu = 1.0\ M$) was made.

De Chant and Hunt[15,16] have recently reported some kinetic data on the reactions

$$cis\text{- or } trans\text{-}Cr(NH_3)_4Cl^{2+} + 4\ H^+ = CrCl^{2+} + 4\ NH_4^+$$

which are also catalysed by Cr^{2+}. In perchlorate media in the absence of oxygen a rate law

$$\text{rate} = k_8[Cr(NH_3)_4Cl^{2+}][Cr^{2+}]$$

was found. In the presence of chloride ions the rate law became

$$\text{rate} = (k_8 + k_9'[Cl^-])[Cr(NH_3)_4Cl^{2+}][Cr^{2+}]$$

Values of the kinetic parameters k_8, $\Delta H^\ddagger$ and $\Delta S^\ddagger$ are 1.16 l.mole^{-1}.sec^{-1} at 25 °C ($\mu = 1.0\ M$), 9.7 kcal.mole^{-1} and 26 cal.deg^{-1}.mole^{-1} for the *trans*-isomer and 1.14×10^{-1} l.mole^{-1}.sec^{-1}, 10.6 kcal.mole^{-1} and -27.6 cal.deg^{-1}.mole^{-1} for the *cis*-species. Both pathways associated with k_8 and k_9' are thought to occur *via* chloride bridge transition states.

Pennington and Haim[17] have investigated the similar Cr^{2+}-catalysed aquations of the *cis*- and *trans*-$Cr(en)_2Cl^{2+}$, and the *cis*- and *trans*-$Cr(en)_2Cl_2^+$ species of Cr(III), in perchlorate media, *viz.*

$$cis\text{- or } trans\text{-}Cr(en)_2Cl_2^+ + Cr^{2+} + 2\ H^+ = Cr^{2+} + 2\ enH^+ + Cl^- + CrCl^{2+}$$

$$cis\text{- or } trans\text{-}Cr(en)_2Cl^{2+} + Cr^{2+} + 2\ H^+ = Cr^{2+} + 2\ enH^+ + CrCl^{2+}$$

In the absence of oxygen the rate law obtained from spectrophotometric measurements at wavelengths 382 mμ (*trans*-isomers) and 510 or 512 mμ (*cis*-isomers) was found to be of the form

$$-d[X]/dt = k_{10}[X][Cr^{2+}] + k_{11}'[X]$$

where X represents the Cr(III) species mentioned and k_{10} and k_{11}' are the rate coefficients for the catalysed and non-catalysed pathways, respectively. (Only for the *cis*-isomers was the term involving k_{11}' large enough, under experimental conditions, to appear.) Values of the kinetic parameters k_{10} (l.mole^{-1}.sec^{-1} at 25 °C, $\mu = 1.0\ M$), $\Delta H^\ddagger$ (kcal.mole^{-1}) and $\Delta S^\ddagger$(cal.deg^{-1}.mole^{-1}) are 1.16×10^{-2},

10.2 and -33 for *cis*-$Cr(en)_2Cl^{2+}$, 1.95×10^{-2}, 9.0 and -36 for *cis*-$Cr(en)_2Cl_2^+$, 4.34×10^{-1}, 8.0 and -33 for *trans*-$Cr(en)_2Cl^{2+}$ and 5.09×10^{-1}, 8.7 and -31 for *trans*-$Cr(en)_2Cl_2^+$, respectively. A chloride bridge transition state was again suggested.

The Cr(II)-catalysed substitution reaction between Cr(III) and *N*-methyliminodiacetic acid H_2L

$$Cr(III) + L^{2-} = CrL^+$$

has been studied by Cannon and Earley[18]. A rate law

$$d[CrL^+]/dt = k''''[Cr^{3+}][Cr^{2+}][HL^-][H^+]^{-2}$$

was found to account for the experimental data. At 25 °C and $\mu \sim 1.0$ *M*, k'''' has a value 2.2×10^{-4} sec^{-1}. The mechanism proposed by these workers was

$$H_2L \rightleftharpoons HL^- + H^+$$

$$HL^- \rightleftharpoons L^{2-} + H^+$$

$$Cr^{2+} + L^{2-} \rightleftharpoons CrL$$

$$Cr^{3+} + H_2O \rightleftharpoons CrOH^{2+} + H^+$$

$$CrL + CrOH^{2+} \rightarrow CrL^+ + Cr^{2+} + OH^-$$

which was also concluded to occur in the presence of *N*-phenyliminodiacetic acid. A spectrophotometric technique (546 mμ) was used to obtain rate data.†

Cr(II) in the form of the EDTA complex has been found to undergo a very fast reaction with the Cr(III) species CrX^{2+}(X = F, Cl, or Br). The products are Cr(III) EDTA complex ions[19]. However, the kinetic parameters for the reaction between the Cr(III) complex with EDTA and the Cr(II) complex with *trans*-1,2-diaminocyclohexanetetraacetate have recently been reported by Wilkins and Yelin[29], as 3.0×10^3 l.mole^{-1}.sec^{-1}, at 25 °C, 5.0 kcal.mole^{-1} and -26 cal.deg^{-1}.mole^{-1}.

Cr(II) catalysed isomerisations of the ions $CrNC^{2+}$ and $CrSCN^{2+}$

$$CrNC^{2+} + Cr^{2+} = Cr^{2+} + CrCN^{2+}$$

$$CrSCN^{2+} + Cr^{2+} = Cr^{2+} + CrNCS^{2+}$$

have been studied using spectrophotometric techniques at wavelengths of 520 mμ

† Investigations of the reactions $CrI^{2+} + X \underset{Cr^{2+}}{=} CrX^{2+} + I^-$ where $X = F^-, Cl^-$ or Br^-, have also been reported[23].

and 260 to 350 mμ, respectively[20-22]. For the ion $CrSCN^{2+}$ two catalysed pathways have been proposed (k_{12}, k_{13})

$$CrSCN^{2+} + Cr^{2+} \xrightarrow{k_{12}}$$

$$CrSCN^{2+} + H_2O \rightleftharpoons Cr(SCN)(OH)^{+} + H^{+} \quad K_{13}$$

$$Cr(SCN)(OH)^{+} + Cr^{2+} \xrightarrow{k_{13}}$$

the observed rate coefficient being given by

$$k_{obs} = k_{12} + k_{13} K_{13}[H^{+}]^{-1}$$

Values of k_{12} and $k_{13}K_{13}$ at 25 °C (μ = 1.0 M) were calculated by Haim and Sutin[21] as 40 l.mole^{-1}.sec^{-1} and 2 sec^{-1}, respectively.

For the species $CrNC^{2+}$, Birk and Espenson[22] have found a rate law

$$-\mathrm{d}\ln[CrNC^{2+}]/\mathrm{d}t = k_{obs} = k'_{14} + k'_{15}[Cr^{2+}]$$

where the terms k'_{14} and k'_{15} $[Cr^{2+}]$ relate to the non-catalysed and catalysed processes. The value of k'_{15} is 5.7×10^{-1} l.mole^{-1}.sec^{-1} at 15 °C (μ = 1.0 M, $[H^{+}] = 4 \times 10^{-1}$ M). Hutchital[27] has recently obtained kinetic data for the reaction *trans*-$Cr(Ox)_2^-$ $\overset{Cr^{2+}}{=\!=}$ *cis*-$Cr(Ox)_2^-$.

8.2.5 The exchange reaction between Mo(V) and Mo(IV)

The exchange of Mo between the anions $Mo(CN)_8^{3-}$ and $Mo(CN)_8^{4-}$ has been investigated by the isotopic method (^{99}Mo) and the separation methods (*a*) precipitation of $Mo(CN)_8^{4-}$ with either ethanol or cadmium ions, and (*b*) precipitation of $Mo(CN)_8^{3-}$ with tetraphenylarsonium chloride. Complete exchange was observed by Wolfgang[1] even with reactant concentrations $\sim 5 \times 10^{-5}$ M. An estimate of the rate coefficient at 2 °C of $>10^3$ l.mole^{-1}.sec^{-1} has been suggested.

More recently, a value of 3×10^4 l.mole^{-1}.sec^{-1} has been calculated for the exchange rate coefficient at 10 °C and zero ionic strength by Campion *et al.*[2] using the Marcus theory and rate coefficients for the reactions

$$Mo(CN)_8^{3-} + Os(dipy)_3^{2+} \underset{k_r}{\overset{k_f}{\rightleftharpoons}} Mo(CN)_8^{4-} + Os(dipy)_3^{3+}$$

k_f and k_r for the above reactions were measured by the temperature jump method.

References pp. 142–152

8.2.6 The exchange reaction between W(V) and W(IV)

Using the radio-isotope ^{185}W as the indicator, and a direct injection technique the exchange between the anions octacyanotungstate(V) and octacyanotungstate (IV) has been investigated by Goodenow and Garner[1]. Tetraphenylarsonium chloride was used to precipitate the $W(CN)_8^{3-}$ from reaction mixtures. In the absence of light, in acidic (HCl) or alkaline (KOH) media with reactant concentrations in the range 10^{-3} to 10^{-4} *M*, complete exchange was observed at temperatures ~ 1 °C. On this basis a rate coefficient of $> 4 \times 10^4$ l.mole^{-1}.sec^{-1} has been proposed.

From a study of the paramagnetic line broadening of aqueous (10^{-2} *M*) solutions of $W(CN)_8^{3-}$ by $W(CN)_8^{4-}$ a limit for $k \leqslant 4 \times 10^8$ l.mole^{-1}.sec^{-1} has been set by Weissman and Garner[2].

8.3 MANGANESE

8.3.1 The exchange reaction between Mn(II) and Mn(I)

Using nuclear magnetic resonance line-broadening measurements (^{55}Mn), Matteson and Bailey[1] have studied the exchange between the isonitrile complex ions of Mn(II) and Mn(I), $(RNC)_6Mn^{2+}$ and $(RNC)_6Mn^+$ (where R is either the ethyl or tertiary butyl group) in the solvents acetonitrile, ethanol and dimethylsulphoxide. In acetonitrile, the evaluated rate coefficients (l.mole^{-1}.sec^{-1}) for the exchanges at 7 °C, are 6.4×10^5 (R = ethyl) and 4.0×10^4 (R = *t*-butyl). Activation parameters, $\Delta H^{\ddagger}$ and $\Delta S^{\ddagger}$, of 1.7 kcal.mole^{-1} and -25 cal.deg^{-1}.mole^{-1} (R = ethyl) and 4.6 kcal.mole^{-1} and -21 cal.deg^{-1}.mole^{-1} (R = *t*-butyl) have also been reported. The exchange in the solvent dimethylsulphoxide was found to be more rapid than that in ethanol or acetonitrile (see also Matteson and Bailey[2]).

8.3.2 The exchange reaction between Mn(III) and Mn(II)

Polissar[1] has observed 100 % exchange between Mn(III), as the oxalato ion $Mn(Ox)_2^-$, and manganous sulphate. Adamson[2], using manganic chloride as the source of Mn(III), has observed the exchange to be incomplete in a time ~ 15 sec with reactant concentrations ~ 10^{-3} *M*. Both workers[1,2] used a separation technique based on the precipitation of Mn(IV), present *via* the equilibrium

$$2\,Mn(III) \rightleftharpoons Mn(II) + Mn(IV)$$

as the oxide, with perchloric acid solutions as reaction media and the isotope ^{54}Mn as the indicator.

Diebler and Sutin[3], using perchlorate salts of Mn(III) and Mn(II) and the isotopic method, have also observed complete exchange, in media 6 M in $HClO_4$, within a time of one min. Two additional separation techniques were employed: extraction of the Mn(III) into benzene as the complex dibutylphosphate and precipitation of the Mn(II) as the salt $MnNH_4PO_4$. However, using Marcus theory, Diebler and Sutin[3], have been able to calculate a value for the exchange rate coefficient at 25 °C (3 M in $HClO_4$) of 3×10^{-4} l.mole^{-1}.sec^{-1}, from data obtained during a study of the reactions of Co(III) with Mn(II) and of Mn(III) with numerous Fe(II) complexes.

Adamson[4] has reported the exchange between the manganese anions $Mn(CN)_6^{3-}$ and $Mn(CN)_6^{4-}$ to be measurable.

The kinetic parameters for the reaction between the Mn(III) complex with EDTA and the Mn(II) complex with *trans*-1,2-diaminocyclohexanetetraacetate have recently been reported by Wilkins and Yelin[5] as 1.2 l.mole^{-1}.sec^{-1}, 7.1 kcal.mole^{-1} and -34 cal.deg^{-1}.mole^{-1}.

8.3.3 The exchange reaction between Mn(VII) and Mn(VI)

This exchange in alkaline media has been studied, using the isotopic method (^{54}Mn and ^{56}Mn), by many workers. Libby[1], using the separation afforded by the insolubility of barium manganate, has found 100 % exchange. Hornig *et al.*[2], attempted to study the exchange using pyridine extraction of the permanganate but observed 100 % exchange, as did Adamson[3], who in addition to the above separation methods, used a method based on the precipitation of manganese dioxide. Bonner and Potratz[4], using reactant concentrations $\sim 10^{-4}$ to 10^{-5} M and extraction of Mn(VII) as the triphenylsulphonium salt with either ethylenedichloride or chloroform, were able to suggest a limit of > 1500 l.mole^{-1}.sec^{-1} for the rate coefficient at 0 °C for the exchange in 2 M NaOH.

Sheppard and Wahl[5] were the first workers to make rate measurements on this exchange reaction. They used triphenylsulphonium bromide and tetraphenylarsonium chloride as the separating agents; both reagents remove the permanganate anion. The radio-isotope ^{54}Mn was used to label the manganate ion. In a further report, Sheppard and Wahl[6] give details of their special reaction vessel and the quenching solution. The rate law obeyed is

$$\text{rate} = k_{obs}[MnO_4^-][MnO_4^{2-}]$$

where k_{obs} has a value of 710 l.mole^{-1}.sec^{-1} at 0.1 °C for 0.16 M NaOH. The energy and entropy of activation were calculated as 10.5 kcal.mole^{-1} and -9 cal.deg^{-1}.mole^{-1}, respectively.

The addition of various cations (Cs^+, K^+, Na^+ and Li^+) was found to alter the value of k_{obs}, the highest value being observed in the presence of Cs^+; various anions were also added but only slight effects were noted. The suggestion of a cation bridge transition state was made.

A detailed investigation of the catalysis by the caesium cation has been made by Gjertsen and Wahl[7], who were able to show from the linear plots of k_{obs} *versus* $[Cs^+]$ that the rate law could be rewritten as

$$\text{rate} = (k' + k''[Cs^+])[MnO_4^-][MnO_4^{2-}]$$

At an ionic strength of 0.16 M the values of k' and k'' (0 °C) were 7.1×10^2 l.mole^{-1}.sec^{-1} and 1.2×10^4 l^2.mole^{-2}.sec^{-1}, respectively. Sodium hydroxide was used to maintain the ionic strength constant.

Meyers and Sheppard[8] have since made a study of this exchange using the nuclear magnetic line-broadening method (^{55}Mn). From the results obtained the rate law

$$\text{rate} = k_{obs}[MnO_4^-][MnO_4^{2-}]$$

was confirmed and, as shown in previous work[6], the value of k_{obs} was found to depend on both the concentration and nature of the cation. In media 0.57 M with respect to KOH, k_{obs} (20 °C) has a value 5.6×10^3 l.mole^{-1}.sec^{-1} ,with an associated activation energy of 8.3 kcal.mole^{-1}.

Britt and Yen[9], using the pulsed nuclear resonance technique, have obtained exchange data comparable to those obtained by the isotopic method. An observed rate coefficient at 0 °C and $\mu = 1.06$ M of 1.23×10^3 l.mole^{-1}.sec^{-1} has been calculated for the exchange in the presence of 1.0 M Na^+.

8.3.4 The exchange reaction between Mn(VII) and Mn(III)

Polissar[1] was unable to detect exchange between Mn(VII), as MnO_4^-, and Mn(III), as the complex ion $Mn(Ox)_2^-$, using the isotopic method.

8.3.5 The exchange reaction between Mn(VII) and Mn(II)

Using the isotope ^{56}Mn as a means of labelling the oxidation states of manganese, Polissar[1] has made a preliminary study of the exchange reaction in perchlorate media between Mn(VII) and Mn(II). He concluded that no measurable exchange occurred between these two oxidation states ($\sim 10^{-2}$ M) in a time of 15 min. The reactants were separated by the addition of sodium hydroxide which precipitated manganese dioxide. Adamson[2], however, using reactant concentrations

$\sim 10^{-3}$ *M*, has observed a slow exchange in media 3 M with respect to $HClO_4$. The experimental results suggested a complicated rate law of approximately the form

$$\text{rate} = k_{\text{obs}}[MnO_4^-]^{1/3}[Mn^{2+}]^{4/3}[H^+]^{4/3}$$

with k_{obs} having a value $\sim 7.2 \times 10^{-4}$ $l^2.\text{mole}^{-2}.\text{sec}^{-1}$ at 25 °C. Adamson suggested that the mechanism was

$$MnO_4^- + 3\,Mn^{2+} + 6\,H^+ \rightleftharpoons MnO^{2+} + 3\,Mn^{3+} + 3\,H_2O \qquad K_1$$

$$Mn^{3+} + H_2O \rightleftharpoons MnO^+ + 2H^+ \qquad K_2$$

$$Mn^{4+} + H_2O \rightleftharpoons MnO^{2+} + 2\,H^+ \qquad K_3$$

$$MnO^+ + MnO^{2+} \xrightarrow{k_1} \text{exchange}$$

$$MnO^{2+} + H_2O \rightarrow MnO_2(s) + 2\,H^+$$

$$MnO^{2+} + Mn^{2+} + 2\,H^+ \rightleftharpoons 2\,Mn^{3+} + H_2O$$

$$Mn^{4+} + Mn^{2+} \rightleftharpoons 2\,Mn^{3+}$$

which leads to a theoretical rate expression

$$\text{rate} = (k_1^2 K_1\, K_2^2/3)^{\frac{1}{2}}[MnO_4^-]^{\frac{1}{2}}[Mn^{2+}]^{\frac{3}{2}}[H^+]$$

which was considered to be in satisfactory agreement with the observations.

Happe and Martin[3], using the radio-isotope ^{54}Mn and a separation method based on the precipitation of the ion MnO_4^- with tetraphenylarsonium nitrate, have observed a slow exchange in nitrate media. The experimental results, obtained over the ranges $[HNO_3]$ 1 to 2 *M*, $[MnO_4^-]$ 2.7×10^{-4} to 5.7×10^{-4} *M* and $[Mn^{2+}]$ 8.9×10^{-4} to 1.35×10^{-3} *M*, show some agreement with the rate law found by Adamson[2].

8.3.6 The reaction of Mn(VII) and Mn(II)

The reaction of the permanganate and manganous ions can take place under conditions where either $MnO_2(s)$ or Mn(III), which decomposes slowly, are the main products, *viz.*

$$3\,Mn^{2+} + 2\,MnO_4^- + 2\,H_2O = 5\,MnO_2(s) + 4\,H^+$$

$$4\,Mn^{2+} + MnO_4^- + 8\,H^+ = 5\,Mn^{3+} + 4\,H_2O$$

The reaction to produce the hydrated manganese dioxide has been studied by

References pp. 142–152

many workers[1-3]. The reaction in the presence of oxalate[4] or periodate[5] ions has also been investigated. Oxalato and iodato complexes of Mn(III) are formed as intermediates in the reactions.

Rosseinsky and Nicol[6] have investigated in detail the reaction in highly acidic perchlorate media which leads to the production of Mn^{3+}. Using [Mn(II)] in the range 2×10^{-2} to 1×10^{-1} M with $[MnO_4^-] \sim 10^{-4}$ M the rate law, found from absorbance measurements at a wavelength of 525 mμ, is

$$-d[MnO_4^-]/dt = k_{obs}[MnO_4^-][Mn(II)]^2$$

The variation in the value of the observed rate coefficient k_{obs} with acidity was found to be of the form

$$k_{obs} = k_1' + k_2'[H^+]$$

At a constant ionic strength of 3.31 M, values of the rate coefficients k_1' and k_2' at 24.4 °C are 2.04 l^2. $mole^{-2}.sec^{-1}$ and 3.05 $l^3.mole^{-3}.sec^{-1}$, respectively, with the corresponding activation parameters of -0.6 and $+0.2$ $kcal.mole^{-1}$ and -59 and -56 $cal.deg^{-1}.mole^{-1}$.

8.4 IRON, RUTHENIUM AND OSMIUM

8.4.1 The exchange reaction between Fe(III) and Fe(II) in aqueous media

The earliest attempts to measure the rate of exchange between ferrous and ferric ions in aqueous media utilised the diffusion separation technique. Little agreement was obtained by the different workers[1-4]. Diffusion separation factors, found to be ~ 0.5[1], ~ 1.4[2], ~ 3.5[3] and ~ 1.2[4], illustrate the difficulty of the technique. The isotopes used to label the iron were either ^{55}Fe or ^{59}Fe, and exchange was found to be complete in hours[2,4] or many days[1,3] in perchlorate media.

Silverman and Dodson[5] made the first detailed isotopic study of this exchange system using the separation afforded by the addition of 2,2^1-dipyridyl at pH 5, followed by the precipitation of the ferric iron with either ammonia or 8-hydroxyquinoline. Dodson[6], using this separation method, had previously obtained an overall rate coefficient of 16 $l.mole^{-1}.sec^{-1}$ at 23 °C for 0.4 M perchloric acid media. The exchange in perchlorate and perchlorate–chloride media was found to conform to a rate law, first order with respect to both total ferrous and ferric ion concentrations, with an observed rate constant (k_{obs}) dependent on the hydrogen-ion concentration, *viz.*

$$k_{obs} = k_1 + k_2'[H^+]^{-1}$$

This dependence was interpreted in terms of an $[H^+]$-independent pathway

$$Fe^{3+} + Fe^{2+} \xrightarrow{k_1}$$

and an $[H^+]$-dependent pathway

$$FeOH^{2+} + Fe^{2+} \xrightarrow{k_2}$$

the $FeOH^{2+}$ being produced by the rapid hydrolysis reaction

$$Fe^{3+} + H_2O \rightleftharpoons FeOH^{2+} + H^+ \qquad K_{2W}$$

From intercepts and slopes of plots of k_{obs} *versus* $[H^+]^{-1}$, values of k_1 and $k_2K_{2W}(k_2')$ were evaluated as 3.3 l.mole^{-1}.sec^{-1} and 3.88 sec^{-1} at a temperature of 21.6 °C. Using known values of K_{2W} (1.43×10^3 M at 21.6 °C), k_2 was calculated as 2700 l.mole^{-1}.sec^{-1}. The activation energies obtained for the steps defined by k_1 and k_2 were 9.9 and 7.4 kcal.mole^{-1}, respectively, with corresponding entropies of activation −25 and −18 cal.deg^{-1}.mole^{-1}. Reasonable agreement with these results has since been reported by many workers during the course of their studies.

For the step defined by k_2 a hydrogen-atom transfer mechanism has been suggested by Dodson[7]. Eimer *et al.*[8], using the same technique, observed no effect on the rate as oxygen present was varied from 4×10^{-7} to 2×10^{-2} M, thus eliminating any oxygen dependent mechanism. Hudis and Dodson[9] have observed a reduction in both rate coefficients k_1 and k_2 when heavy water was used as the solvent. (At 7.1 °C in D_2O, $k_1 = 0.7$ l.mole^{-1}.sec^{-1} and $k_2 = 765$ l.mole^{-1}.sec^{-1}). It was concluded that a hydrogen-atom transfer occurred in both pathways. Fukushima and Reynolds[10] have reinvestigated the deuterium isotope effect. Using values of the equilibrium constant K_{2D} for

$$Fe^{3+} + D_2O \rightleftharpoons FeOD^{2+} + D^+ \qquad K_{2D}$$

obtained by spectrophotometric measurements under the same conditions as their isotope exchange measurements, these workers obtained values of the rate coefficients k_{1D} and k_{2D} (25 °C)

$$Fe^{3+} + Fe^{2+} \xrightarrow{k_{1D}}$$

$$FeOD^{2+} + Fe^{2+} \xrightarrow{k_{2D}}$$

for the exchange in $DClO_4$ media at $\mu = 0.5$ M of 1.0 l.mole^{-1}.sec^{-1} and 3.04×10^3 l.mole^{-1}.sec^{-1}, respectively. For the second step (k_{2D}) the activation enthalpy obtained was 11.5 kcal.mole^{-1} with the entropy −4 cal.deg^{-1}.mole^{-1}.

References pp. 142–152

Reynolds and Lumry[11] have discussed the role of water in this exchange and have suggested, for both steps, a mechanism involving water bridges.

Horne[12] has studied the kinetics of exchange in aqueous perchlorate media at temperatures down to −78 °C by the isotopic method (^{59}Fe) and dipyridyl separation. The same rate law in these ice media as in aqueous solution was observed, although the acid dependence was small. Horne concluded that the same exchange mechanism occurs in solid and liquid solvent. Evidence for a "Grotthus-type" mechanism has been summarised[13].

8.4.2 The effect of inorganic ions on the exchange reaction between Fe(III) and Fe(II)

The effect of the addition of inorganic ions has been investigated using the isotopic method (^{55}Fe or ^{59}Fe) and the 2,2′-dipyridyl separation. The rate law in the presence of inorganic anions is given by

$$\text{rate} = k_{obs}[\text{Fe(III)}][\text{Fe(II)}]$$

where k_{obs} is the observed second-order rate coefficient, which is related to the rate coefficients for the individual reaction pathways and the formation constants of the species involved in these pathways. The latter constants must be known for the evaluation of the rate coefficients and constant ionic strength conditions must be maintained. The reaction pathways

$$Fe^{3+} + Fe^{2+} \xrightarrow{k_1}$$

$$FeOH^{2+} + Fe^{2+} \xrightarrow{k_2}$$

operative in perchlorate media, have to be taken into account.

The effect of chloride ions was investigated first by Silverman and Dodson[1]. These authors observed an increase in k_{obs} as the concentration of chloride ion was increased from 0 to 0.55 *M* in perchlorate media of constant acidity. The rate expression found to fit the experimental data was

$$\text{rate} = k_1[Fe^{3+}][Fe^{2+}] + k_2[FeOH^{2+}][Fe^{2+}] + k_3[FeCl^{2+}][Fe^{2+}] + k_4[FeCl_2^+][Fe^{2+}]$$

where k_3 and k_4 are the rate coefficients for the reactions

$$FeCl^{2+} + Fe^{2+} \xrightarrow{k_3}$$

$$FeCl_2^+ + Fe^{2+} \xrightarrow{k_4}$$

From the estimated values of K_3 and K_4

$$Fe^{3+} + Cl^- \rightleftharpoons FeCl^{2+} \qquad K_3 = 3.01 \text{ l.mole}^{-1}$$

$$FeCl^{2+} + Cl^- \rightleftharpoons FeCl_2^+ \qquad K_4 = 0.94 \text{ l.mole}^{-1}$$

for the exchange conditions, values of k_3 and k_4 were calculated as 29 and 51 l.mole^{-1}.sec^{-1}, respectively, at 20 °C ($\mu = 0.55$ M). Entropies and energies of activation associated with the steps involving chloride (k_3 and k_4) are -24 and -20 cal.deg^{-1}.mole^{-1} and 8.8 and 9.7 kcal.mole^{-1}, respectively.

Sutin *et al.*[2] have made a detailed study of the chloride-catalysed paths in deuterated water media, the exchange occurring more slowly in this solvent than in water. Accurate values of K_3, in both water and heavy water, were obtained by spectrophotometry and enabled more precise rate parameters to be calculated. For aqueous media, values are 22.8 l.mole^{-1}.sec^{-1} for k_3 ($\mu = 0.50$ M and 20 °C), 11.5 kcal.mole^{-1} (activation energy), and -15 cal.deg^{-1}.mole (activation entropy); for deuterated solvent the corresponding values are 9.1 l.mole^{-1}.sec^{-1}, 13.2 kcal.mole^{-1}, and -10.2 cal.deg^{-1}.mole^{-1}.

Sutin *et al.*[3] have recently re-examined the pathways of exchange involving chloride and have evaluated the rate coefficients as $k_3 = 57.6$ and $k_4 = 159$ l. mole^{-1}.sec^{-1} at 25 °C and $\mu = 3.0$ M. These workers have pointed out that step with coefficient k_3 can occur in two ways

$$FeCl^{2+} + Fe^{2+} \overset{k'_3}{\rightleftharpoons} FeCl^+ + Fe^{3+}$$

for which a value $k'_3 = 12.1$ l.mole^{-1}.sec^{-1} ($\mu = 3.0$ M and 25 °C) has been obtained[3], and

$$FeCl^{2+} + Fe^{2+} \overset{k''_3}{\rightleftharpoons} FeCl^{2+} + Fe^{2+}$$

with k_3 being the sum of 2 k'_3 and k''_3. Calculations have led to the evaluation of k''_3 as 33.4 l.mole^{-1}.sec^{-1} ($\mu = 3.0$ M and 25 °C). Horne[4] has shown that chloride paths are unimportant at low temperatures in ice media.

Hudis and Wahl[5] have examined the effect of fluoride ion on the exchange rate and have found it necessary to include three fluoride exchange paths

$$FeF^{2+} + Fe^{2+} \xrightarrow{k_5}$$

$$FeF_2^+ + Fe^{2+} \xrightarrow{k_6}$$

$$FeF_3 + Fe^{2+} \xrightarrow{k_7}$$

Values obtained for the rate coefficients k_5, k_6, and k_7 (l.mole^{-1}.sec^{-1} at 0 °C and $\mu = 0.5$ M) were 9.7, 2.5 and $\sim$ 0.5, with activation parameters for the first two steps (k_5 and k_6) of $\Delta S^{\ddagger}$ -21 and -22 cal.deg^{-1}.mole^{-1} and E_a 9.1 and 9.5 kcal.mole^{-1}, respectively. A hydrogen-atom transfer mechanism was proposed for these steps.

Menashi *et al.*[6] have examined the exchange when only the FeF^{2+} of the Fe(III) fluoro-species was present and when the rate of formation of FeF^{2+} and the rate of the ^{59}Fe exchange are of the same order of magnitude. The results obtained led these workers to conclude that exchange occurs *via* an activated complex $[Fe_2F^{4+}]^{\ddagger}$.

Thiocyanate ion also provides additional pathways for exchange, *viz.*

$$FeSCN^{2+} + Fe^{2+} \xrightarrow{k_8}$$

$$Fe(SCN)_2^{+} + Fe^{2+} \xrightarrow{k_9}$$

The rate law found by Laurence[7], for the exchange in the presence of thiocyanate ion in the range 3.2×10^{-5} to 8.6×10^{-2} M was

$$\text{rate} = k_1[Fe^{3+}][Fe^{2+}] + k_2[FeOH^{2+}][Fe^{2+}] + k_8[FeSCN^{2+}][Fe^{2+}] + k_9[Fe(SCN)_2^{+}][Fe^{2+}]$$

At 25 °C and ionic strength 0.5 M, values of the rate coefficients k_8 and k_9 obtained were 41.5 and 7.6 l.mole^{-1}.sec^{-1}, respectively, with the associated activation energies for these pathways of 7.9 and 8.6 kcal.mole^{-1}. Conocchioli and Sutin[8] have recently shown that, since $k_8 = k_{12} + 2k_{11} + 2k_{10}$, where the rate coefficients k_{10}, k_{11} and k_{12} are for the reactions

$$FeNCS^{2+} + Fe^{2+} \overset{k_{10}}{\rightleftharpoons} FeNCS^{+} + Fe^{3+}$$

$$FeNCS^{2+} + Fe^{2+} \overset{k_{11}}{\rightleftharpoons} Fe^{2+} + FeSCN^{2+}$$

$$FeNCS^{2+} + Fe^{2+} \overset{k_{12}}{\rightleftharpoons} Fe^{2+} + FeNCS^{2+}$$

the value of $(k_{12} + 2k_{11})$ is $\sim$ 30 l.mole^{-1}.sec^{-1}. The value of k_{10} used was 10.5 l.mole^{-1}.sec^{-1} and k_8 was estimated to be 51.6 l.mole^{-1}.sec^{-1} under the same conditions (μ = 3.0 M and 25 °C).

Horne and Axelrod[9] have suggested that this exchange may not be first order with respect to each iron oxidation state at high thiocyanate concentrations. A lower value for the rate coefficient k_8 and a slightly higher activation energy than that quoted previously[7] have been reported[8, 9].

An azide catalysed path

$$FeN_3^{2+} + Fe^{2+} \xrightarrow{k_{13}}$$

has also been found by Bunn *et al.*[10,11], who have evaluated, from the kinetic data and the measured formation constant of FeN_3^{2+}, the rate coefficient k_{13} as 4.75×10^3 l.mole^{-1}.sec^{-1} at 10 °C and $\mu = 0.55$ M. The rate coefficient in D_2O was found to be lower than in H_2O. Arrhenius plots were linear in the range 0–13 °C, but not above the higher temperature, and led to calculated activation energies of about 14 kcal.mole^{-1} for both solvents. Positive activation entropies, of 7.0 and 8.5 cal.deg^{-1}.mole^{-1} for H_2O and D_2O media, were calculated for this pathway. A reaction mechanism of the inner-sphere type was postulated, *viz.*

$$Fe^{3+}N_3^- + Fe^{2+} \underset{k_{-14}}{\overset{k_{14}}{\rightleftharpoons}} Fe^{3+}N_3^-Fe^{2+}$$

$$Fe^{3+}N_3^-Fe^{2+} \underset{k_{-15}}{\overset{k_{15}}{\rightleftharpoons}} Fe^{2+}N_3^-Fe^{3+}$$

$$Fe^{2+}N_3^-Fe^{3+} \underset{k_{-16}}{\overset{k_{16}}{\rightleftharpoons}} Fe^{2+} + {}^-N_3Fe^{3+}$$

with the relative magnitudes of k_{-14} and k_{15} accounting for the curvature of the Arrhenius plot.

The presence of phosphoric acid also provides an additional pathway

$$Fe(HPO_4)^+ + Fe^{2+} \xrightarrow{k_{17}}$$

Sheppard and Brown[12] have evaluated the rate coefficient k_{17} as 4180 l.mole^{-1}.sec^{-1} with 0.53 M $HClO_4$ at 20 °C. The overall activation energy and entropy corresponding to the term $K_{17}k_{17}$, where K_{17} is the equilibrium constant of the reaction

$$Fe^{3+} + HPO_4^{2-} \rightleftharpoons Fe(HPO_4)^+ \qquad K_{17}$$

were calculated as 15 kcal.mole^{-1} and 6 cal.deg^{-1}.mole^{-1}.

The presence of bromide ions also leads to alternative pathways for exchange, *viz.*

$$FeBr^{2+} + Fe^{2+} \xrightarrow{k_{18}}$$

$$FeBr_2^+ + Fe^{2+} \xrightarrow{k_{19}}$$

Horne[13] has obtained a value for k_{18} at 0 °C ($\mu = 0.55$ M) of 4.9 l.mole^{-1}.sec^{-1} with a corresponding activation energy of 8.0 kcal.mole^{-1}. The activation energy found for the second step (k_{19}) was 9.6 kcal.mole^{-1}.

Four recent kinetic studies have been made of the exchange in the presence of the sulphate ion[12,14–16], in perchlorate media, some measure of agreement being observed between the results of the various workers.

Reynolds and Fukushima[14] have made an extensive study of this system and have interpreted their results, obtained with varying sulphate and hydrogen-ion concentrations, in terms of the exchange paths

$$Fe^{3+} + Fe^{2+} \xrightarrow{k_1}$$

$$FeOH^{2+} + Fe^{2+} \xrightarrow{k_2}$$

$$FeSO_4^+ + Fe^{2+} \xrightarrow{k_{20}}$$

$$Fe(SO_4)_2^- + Fe^{2+} \xrightarrow{k_{21}}$$

$$FeOH^{2+} + FeSO_4 \xrightarrow{k_{22}}$$

At an ionic strength of 0.75 M the calculated values of the rate coefficients k_{20}, k_{21}, and k_{22} (l.mole^{-1}.sec^{-1} at 25 °C) are 5.86×10^2, 2.07×10^4 and 1.4×10^6, respectively. Although the rate of the step associated with k_{20} was found to vary with the ionic strength of the medium, k_{21} and k_{22} would not appear to depend on this property. In this study the concentration ranges used were Fe(II) 6.2×10^{-7} to 3.16×10^{-5} M, Fe(III) 2×10^{-7} to 1.1×10^{-6} M, SO_4^{2-} up to 8×10^{-3} M, and H^+ 3.8×10^{-2} to 1.25×10^{-1} M.

Sheppard and Brown[12] have estimated a value of k_{20} of 9.80×10^2 l.mole^{-1}.sec^{-1} (at 28 °C and $\mu = 1.0$ M) from data obtained with total iron $\sim 10^{-4}$ M and sulphate ion up to 8×10^{-2} M. Activation energy and entropy for the combined term $K_{20} k_{20}$, where K_{20} refers to

$$Fe^{3+} + SO_4^{2-} \rightleftharpoons FeSO_4^+ \qquad K_{20}$$

were calculated as 13.5 kcal.mole^{-1} and -2 cal.deg^{-1}.mole^{-1}.

Willix[15] has estimated from his kinetic data a value of k_{20} of 360 l.mole^{-1}.sec^{-1} (at 25.7 °C, $\mu = 1.0$ M) and has evaluated the activation energy and entropy as 8.3 kcal.mole^{-1} and -18.7 cal.deg^{-1}.mole^{-1} for this pathway, using data on the temperature dependence of the equilibrium K_{20}.

Bächmann and Lieser[16] have estimated rate coefficients (at 25 °C, $\mu = 1.0$ M) of 2.9×10^2 and 1.8×10^4 l.mole^{-1}.sec^{-1} for the steps defined by k_{20} and k_{21}. The corresponding activation energies obtained for these steps were 13.8 and 15 kcal.mole^{-1}. The entropy for the first step (k_{20}) was evaluated as -1.2 cal. deg^{-1}.mole^{-1}.

Suggestions that the sulphate catalysed paths may involve a mechanism with a sulphate-bridged activated complex, as opposed to a hydrogen-atom transfer mechanism, have been made[14].

8.4.3 The effect of organic ligands on the exchange reaction between Fe(III) and Fe(II)

The effect of some organic acids on the exchange in perchlorate media has been investigated[1-3]. Fumaric[1,3], benzoic and *o*-phthalic acids[1] have been shown to cause little or no alteration in the rate, whereas acetic, succinic, carbolic[1], oxalic[1,2] and tartaric acids[3] have an accelerating influence. The isotopic method (^{59}Fe) and the 2,2′-dipyridyl separation have been used. The rate law observed was[1-3]

$$\text{rate} = k_{obs}[\text{Fe(III)}][\text{Fe(II)}]$$

At constant ionic strength and acidity (~ 0.55 *M*), with oxalic acid (H_2Ox) up to 9×10^{-2} *M*, and at temperatures in the range 0.02 °C to 20.57 °C, the rate data were found by Horne[1] to be consistent with an expression

$$\text{rate} = R_0 + k_3[\text{FeOx}^+][\text{Fe}^{2+}] + k_4[\text{Fe(Ox)}_2^-][\text{Fe}^{2+}]$$

where R_0 represents the exchange rate at zero oxalate ion concentration. Values of k_3 and k_4 were calculated as 2140 and 4250 l.mole^{-1}.sec^{-1}, respectively, at 20 °C. For the step associated with k_3 the entropy and energy of activation were estimated as -14 cal.deg^{-1}.mole^{-1} and 9.2 kcal.mole^{-1}. Sheppard and Brown[2] have reported values of k_3 (20 °C) of 3300 l.mole^{-1}.sec^{-1}, and of the entropy of activation of 28 cal.deg^{-1}.mole^{-1} and the energy of activation of 21 kcal.mole^{-1} for the overall process associated with $k_3 K_3$ where K_3 refers to the equilibrium

$$\text{Fe}^{3+} + \text{Ox}^{2-} \rightleftharpoons \text{FeOx}^+$$

McAuley and Brubaker[3] have found that for added tartaric acid (H_2Tar) up to 5×10^{-3} *M*, at temperatures in the range 0 to 10 °C with constant acidity (0.11 *M*) and ionic strength 0.55 *M*, rate data were consistent with an expression

$$\text{rate} = R_0 + k_5[\text{FeHTar}^{2+}][\text{Fe}^{2+}] + k_6[\text{Fe(HTar)}_2^+][\text{Fe}^{2+}]$$

The term defined by k_5 is relatively unimportant, since plots of k_{obs} *versus* $[H_2Tar]^2$ were linear. The pathways for exchange in the presence of tartaric[3] and oxalic acids[1] are

$$\text{FeOx}^+ + \text{Fe}^{2+} \xrightarrow{k_3}$$

$$\text{Fe(Ox)}_2^- + \text{Fe}^{2+} \xrightarrow{k_4}$$

$$\text{FeHTar}^{2+} + \text{Fe}^{2+} \xrightarrow{k_5}$$

$$\text{Fe(HTar)}_2^+ + \text{Fe}^{2+} \xrightarrow{k_6}$$

References pp. 142–152

in addition to those normally operative in aqueous media. The values of k_5 and k_6 could not be calculated owing to lack of data on the formation constants of $Fe(HTar)_2^+$ and $FeTar^{2+}$.

Chakrabarty *et al.*[4] have measured the exchange in the presence of 8-quinolinol (HQ), using a similar method. The rate law at high acidities (pH < 2) and low concentrations of Fe^{3+} was found to be

$$\text{rate} = R_0 + k_7[FeQ^{2+}][Fe^{2+}]$$

where k_7 is the rate coefficient for the pathway

$$FeQ^{2+} + Fe^{2+} \xrightarrow{k_7}$$

At 25 °C and $\mu = 0.5$ *M* the value of k_7 is 450 l.mole^{-1}.sec^{-1}.

100 % exchange of iron, between the ethylenediaminetetraacetate complexes of

iron(III) and iron(II) (FeY^{2-} and FeY^-), has been observed within the time required (15 sec) for either, precipitation, or ion exchange separation of the species of iron[5]. Reynolds *et al.*[6] have since investigated the exchange of Fe^{2+} with FeY^-, which they have found to be measurable and to follow a rate law

$$\text{rate} = k_{obs}[H^+]^{1.7}[Fe^{2+}]^{0.7}[FeY^-]$$

These workers have concluded that, in the exchange of FeY^{2-} and FeY^-, a dissociation step cannot take part in the mechanism of exchange.

Attempts to measure the exchange between 1,10-phenanthroline and substituted 1,10-phenanthroline complexes of iron (II) and iron (III) have led to a lower limit being placed on the rate coefficients. Eimer and Medalia[7], using the isotopic method (^{55}Fe) and either perchlorate precipitation or extraction using a chloroform solution of camphorsulphonic acid of the iron(II) complex, observed 100 % exchange in 15 sec with the 5,6-dimethylphenanthroline complexes in 1 *M* sulphate media at 0 °C. Eichler and Wahl[8], using similar separation methods, but with a direct injection technique, observed complete exchange, between the phenanthroline complexes at 0 °C, in 3.0 *M* sulphuric acid, in less than 0.07 sec. This was calculated to correspond to a rate coefficient in excess of 10^5 l.mole^{-1}.sec^{-1}. Using the optical activity method, these two workers placed a limit > 160 l.mole^{-1}.sec^{-1} on the same rate coefficient.

In heavy water 3 *M* with respect to D_2SO_4, a lower limit of $k > 10^5$ l.mole^{-1}.sec^{-1} at 26 °C has been proposed by Dietrich and Wahl[9] and Larsen and Wahl[10], as a result of an NMR investigation of the position of the proton absorption bands of mixtures of the iron (III) complexes and iron (II) complexes. This limit applies to the exchange between iron (III) and iron (II) complexes of 1,10phenanthroline[9], 4,7-dimethyl-, 3,4,7,8-tetramethyl-, 3,5,6,8-tetramethyl- and 5,6-dimethyl-substituted phenanthroline complexes[10].

For the phenanthroline and 4,7- and 5,6-dimethylphenanthroline complexes a further estimate of $k > 3 \times 10^7$ l.mole^{-1}.sec^{-1} has been proposed, on the basis of NMR line broadening measurements, by Larsen and Wahl[11].

The isotopic method has been used in conjunction with a flow apparatus by Stranks[12], to measure the exchange between the cyclopentadienyl complexes of iron (III) and iron (II) in methanol. Separation was based on the insolubility of $Fe(C_5H_5)^+$ in petroleum ether at -80 °C. Using Fe(II) and Fe(III) $\sim 10^{-4}$ M and short reaction times ($\sim$ msec), a rate coefficient 8.7×10^5 l.mole^{-1}.sec^{-1} at -75 °C was obtained. The rate of exchange in the presence of chloride ions and inert electrolytes was found to be more rapid. Calculations using Marcus Theory showed reasonable agreement with the experimental observations. In deuterated acetone, line broadening measurements have led to an estimate[9] of this rate coefficient of $> 10^5$ l.mole^{-1}.sec^{-1} at 26 °C.

8.4.4 The exchange reaction between Fe(III) and Fe(II) in non-aqueous and mixed solvents

An isotopic investigation (^{59}Fe) of the exchange of iron (III) and iron (II) as perchlorates in absolute ethanol, with chromatographic separation on an alumina column, led to complete exchange being observed within the separation time[1]. However, Horne[2], using the 2,2′-dipyridyl separation method and the isotope ^{55}Fe has observed a decrease in the exchange rate as water is replaced by acetone, methanol, ethanol and propanol. The rate tending to zero in the absence of water. In aqueous ethanol the rate law was found to be

$$\text{rate} = k_{\text{obs}}[\text{Fe(III)}][\text{Fe(II)}]$$

with activation parameters at low water concentrations of about 10 kcal.mole^{-1} ($\Delta H^{\ddagger}$) and -20 cal.deg^{-1}.mole^{-1} ($\Delta S^{\ddagger}$). Increase in the ionic strength led to an increase in the rate, whereas increase in $HClO_4$ concentration brought about a reduction. The rate of exchange has also been reported to be much slower in isopropanol[3] and nitromethane[4] than in aqeous media. Various reasons for the slow rate in alcohol media have been put forward[2,3].

Menashi *et al.*[5] have made a detailed isotopic study (^{59}Fe) of the exchange in dimethylsulphoxide, in the absence of oxygen, using the 2,2′-dipyridyl separation. With Fe(II) and Fe(III) in the ranges 1×10^{-5} to 2.8×10^{-4} M and 1×10^{-5} to 1×10^{-4} M, respectively, the rate data was found to conform to a first order dependence on both oxidation states of iron. The value of the observed rate coefficient (20 °C), in the presence of 2.5×10^{-2} M H_2O and 2×10^{-5} M $HClO_4$, at $\mu = 0.2$ M ($NaClO_4$), was 18 l.mole^{-1}.sec^{-1}.

References pp. 142–152

The effect of variation in the perchloric acid concentration was found to be complex. At concentrations lower than 7×10^{-6} M, there was a rapid decrease in the observed rate coefficient as the perchloric acid was increased; above 7×10^{-6} M no change occurred in the rate as the $[HClO_4]$ was altered. Perchlorate ion concentration had no effect on the rate of exchange. The effect of the addition of water was found to be dependent on the perchloric acid concentration: either an increase ($HClO_4 = 2\times10^{-5}$ M), or a decrease followed by an increase ($HClO_4 = 2\times10^{-2}$ M) being observed. Ionic strength effects have also been studied. The overall activation enthalpy and entropy values obtained[5] were 9.6 kcal.mole^{-1} and -20 cal.deg^{-1}.mole^{-1}, respectively.

It is thought that exchange can occur through the species $Fe(DMSO)_6^{3+}$ and $Fe(DMSO)_6^{2+}$ possibly *via* an inner sphere mechanism; the exchange occurs in the absence of water.

Wada and Reynolds[6] have investigated the effect of chloride ion, over the range 1×10^{-4} to 6×10^{-4} M, on this exchange system. A rate law at constant $HClO_4$

$$\text{rate} = \left(\frac{k_0+k'[Cl^-]}{1+K_1[Cl^-]}\right)[Fe(II)][Fe(III)]$$

was found, which was interpreted in terms of the reaction mechanism

$$Fe^{2+}+Fe^{3+} \xrightarrow{k_0} Fe^{3+}+Fe^{2+}$$

$$FeCl^{+}+Fe^{3+} \xrightarrow{k_1} Fe^{3+}+FeCl^{+}$$

$$Fe^{2+}+Cl^{-} \overset{K_1}{\rightleftharpoons} FeCl^{+} \qquad (K_1 = 1.4\times10^{-3}\ \text{l.mole}^{-1})$$

On this basis, a value of the rate coefficient k_1 was estimated as 3.2×10^{2} l.mole^{-1}.sec^{-1} (at 20 °C and $\mu = 0.1$ M). The energy and entropy of activation corresponding to the term k_1K_1 were calculated as 8.7 kcal.mole^{-1} and -0.5 cal.deg^{-1}.mole^{-1}, respectively.

8.4.5 The exchange reaction between hexacyanoferrate (III) and hexacyanoferrate (II)

The earliest attempts to measure the rate of this exchange reaction met with little success, 100 % exchange being found to occur during the time necessary for mixing and separating the two species of iron[1-4]. The radioactive indicator used was usually a mixture of isotopes (^{59}Fe and ^{55}Fe), and the media ranged from molar aqueous hydrochloric acid to 0.05 M sodium hydroxide. The separation methods tried included (*a*) precipitation of the ion $Fe(CN)_6^{4-}$ as the insoluble salts $KCeFe(CN)_6$[1], $Pb_2Fe(CN)_6$[2] and $KFe_2(CN)_6$[3] and (*b*) physical methods based on diffusion, ion exchange and electrophoresis[4].

Wahl and Deck[5] were able to obtain an estimate of an assumed second-order rate coefficient ($\sim 10^3$ l.mole^{-1}.sec^{-1} at 4 °C) using a separation procedure based on the extraction of $Fe(CN)_6^{3-}$ by a chloroform solution of Ph_4AsCl, in the presence of the ions $Co(CN)_6^{3-}$ and $Ru(CN)_6^{4-}$, to reduce the exchange between the iron species in the two liquid phases. A similar estimate was obtained using a precipitation method in the presence of the carrier $Ru(CN)_6^{4-}$. A direct injection technique was used as short reaction times were necessary. Wahl[6] has reviewed the large induced exchanges occurring in the chemical separation methods. The extraction procedure when the carriers $Co(CN)_6^{3-}$ and $Ru(CN)_6^{4-}$ are present provides the most satisfactory method of separation[6].

From the variation in the line width of the single ^{14}N NMR line, observed with aqueous mixtures of $K_4Fe(CN)_6$ and $K_3Fe(CN)_6$ of about 0.5 *M* with respect to each species of iron, a rate coefficient (9.2×10^4 l.mole^{-1}.sec^{-1} at 32 °C) has been calculated[7,8]. An associated entropy and energy of activation of -24.3 cal.deg^{-1}.mole^{-1} and 4.2 kcal.mole^{-1}, respectively, has also been found[8]. At temperatures above 40 °C the activation energy may be larger than that mentioned above.† Results confirming the positive catalytic effects of various inorganic cations (H^+, Li^+, Na^+, K^+, Rb^+, Cs^+, NH_4^+, Mg^{2+}, Ca^{2+} and Sr^{2+}), reported earlier[9], were also obtained and the possibility of a cation bridge mechanism was discussed.

Wahl *et al.*[9,10] have completed the first detailed isotopic study of this exchange reaction and have shown paths catalysed by cations (H^+, K^+, Ca^{2+}, Ba^{2+}, Ph_4As^+[9] and various tetraalkylammonium ions[10]) occur. The first order with respect to both $Fe(CN)_6^{4-}$ and $Fe(CN)_6^{3-}$ was confirmed[10], *viz.*

$$\text{rate} = k_{obs}[Fe(CN)_6^{3-}][Fe(CN)_6^{4-}]$$

The dependence of the exchange rate on the [cation] is complicated as a result of complexing, *viz.*

$$M^+ + Fe(CN)_6^{4-} \rightleftharpoons MFe(CN)_6^{3-}$$
$$M^+ + Fe(CN)_6^{3-} \rightleftharpoons MFe(CN)_6^{2-}$$

and the exchange paths involving all species of iron which occur, *viz.*

$$Fe(CN)_6^{4-} + Fe(CN)_6^{3-} \xrightarrow{k_0}$$
$$MFe(CN)_6^{3-} + Fe(CN)_6^{3-} \xrightarrow{k_{1a}}$$
$$Fe(CN)_6^{4-} + MFe(CN)_6^{2-} \xrightarrow{k_{1b}}$$
$$MFe(CN)_6^{3-} + MFe(CN)_6^{2-} \xrightarrow{k_2}$$

† See, however, Loewenstein and Ron[12].

For aqueous potassium hydroxide (10^{-2} M) the value of the observed rate coefficient k_{obs} is 226 l.mole^{-1}.sec^{-1} at 0.1 °C. Over the temperature range 0.1 to 20.6 °C an activation energy of 6.0 kcal.mole^{-1} was calculated. At zero cation concentration k_0 (0.1 °C), obtained by extrapolation, has a value of 6.0 l.mole^{-1}. sec^{-1} with an associated activation energy and entropy of 9.1 kcal.mole^{-1} and -24 cal.deg^{-1}.mole^{-1}, respectively.

Wahl *et al.*[10] have suggested that the rate coefficient obtained by the NMR method[7,8] at cation concentrations $\sim 5 \times 10^{-1}$ M is that for the step k_2 in the above mechanism. There is reasonable agreement between the data obtained by these two different procedures. A comparison has been made of the experimental results with those obtained from the Marcus theory[10].

For the exchange reactions between $Fe(CN)_5N_3^{3-}$ and $Fe(CN)_5N_3^{4-}$, $Fe(CN)_5NH_3^{2-}$ and $Fe(CN)_5NH_3^{3-}$, $Fe(CN)_5P(C_6H_5)_3^{2-}$ and $Fe(CN)_5P(C_6H_5)_3^{3-}$, and $Fe(dipy)(CN)_4^{-}$ and $Fe(dipy)(CN)_4^{2-}$, Stasiw and Wilkins[11] have recently evaluated rate coefficients of 3×10^3, 1×10^5, 1×10^5 and 4×10^7 l.mole^{-1}.sec^{-1}, respectively, at 25 °C.

8.4.6 *Reactions of Fe(III) with Fe(II)*

A stopped flow technique coupled with spectrophotometric analysis of the iron (II) complex formed has been used to investigate[1,2] the reactions of some organic complexes of iron(III) with the ion Fe^{2+}. The iron(III) was complexed with 1,10-phenanthroline[1], various substituted 1,10-phenanthrolines (5-methyl-, 5-nitro-, 5-chloro-, 5-phenyl-, 5,6-dimethyl-, 4,7-diphenyl-, and 3,4,7,8-tetramethyl-) and 2,2′-dipyridine, 4,4′-dimethyl-2,2′-dipyridine, and 2,2′,2″-tripyridine[2]. The wavelengths used for the analysis lay in the region 500–552 mμ.

For the reaction $Fe(phen)_3^{3+}$–Fe^{2+} in 0.5 M $HClO_4$ a rate coefficient of 3.7×10^4 l.mole^{-1}.sec^{-1} at 25 °C was calculated by Sutin and Gordon[1]. The entropy and energy of activation values obtained were -37 cal.deg^{-1}.mole^{-1} and 0.8 kcal.mole^{-1}. Little acid dependence was observed. The rate coefficients (l.mole. sec^{-1} at 25 °C in 0.5 M $HClO_4$) for the other systems[2] are : $Fe(5\text{-Me-phen})_3^{3+}$–$Fe^{2+}$, 2×10^4; $Fe(5\text{-NO}_2\text{-phen})_3^{3+}$–$Fe^{2+}$, 1.1×10^6; $Fe(5\text{-Cl-phen})_3^{3+}$–$Fe^{2+}$, 2.1×10^5; and $Fe(5,6\text{-Me}_2\text{-phen})_3^{3+}$–$Fe^{2+}$, 7.8×10^3. The rate coefficients for these reactions are between 7 and 9 times larger in 0.5 M H_2SO_4. For the systems $Fe(5\text{-ph-phen})_3^{3+}$–$Fe^{2+}$, $Fe(4,7\text{-ph}_2\text{-phen})_3^{3+}$–$Fe^{2+}$ and $Fe(3,4,7,8\text{-Me}_4\text{-phen})_3^{3+}$–$Fe^{2+}$, the rate coefficients with 0.5 M H_2SO_4 at 25 °C are 3.2×10^5, 3.3×10^4, and 1.9×10^3 l.mole^{-1}.sec^{-1}, respectively. Ford-Smith and Sutin[2] suggest that electron transfer occurs when the ferrous ion is among the phenanthroline groupings of the iron(III) complex.

For the iron(III) complexes (*a*) $Fe(dipy)_3^{3+}$ (*b*) $Fe(4,4'\text{-Me}_2\text{-dipy})_3^{3+}$ and (*c*) $Fe(tripy)_2^{3+}$ the rate coefficients (l.mole^{-1}.sec^{-1}) with perchloric and sulphuric acids (0.5 M) are[2] (a) 2.7×10^4 and 2.2×10^5 (b) 6.0×10^2 and 5.9×10^3 and (c) 8.5×10^4 and 7.4×10^5. Estimates of second order rate coefficients in excess of 10^8 l.mole^{-1}.sec^{-1}, have been made by Gordon *et al.*[3] for the reactions between $Fe(phen)_3^{3+}$ and $Fe(CN)_6^{4-}$, $Fe(4,4'\text{-}Me_2\text{-}phen)_3^{3+}$ and $Fe(3,4,7,8\text{-}Me_4\text{-}phen)_3^{2+}$, and $Fe(phen)_3^{3+}$ and $Fe(4,4'\text{-}Me_2\text{-}dipy)_3^{2+}$ at 25 °C in 0.5 M sulphuric or perchloric acids.

The same technique has been used to measure the rate coefficients for the reactions of ferricyanide ions with ferrohemoglobin[4] and ferrocytochrome *c* (ref. 5). At 25 °C, $\mu = 0.1$ M and pH 6, the values are 7×10^4 and 1.6×10^7 l.mole^{-1}.sec^{-1}, respectively.

Stasiw and Wilkins[6] have made an investigation of the reactions between $Fe(CN)_6^{4-}$ and $Fe(CN)_5NH_3^{2-}$, $Fe(CN)_5H_2O^{2-}$, $Fe(CN)_5P(C_6H_5)_3^{2-}$ and $Fe(dipy)(CN)_4^-$. The values of the rate coefficients (l.mole^{-1}.sec^{-1}) with the activation enthalpies (kcal.mole^{-1}) and entropies (cal.deg^{-1}.mole^{-1}) and ionic strength in parentheses, reported by these workers are: 7×10^3 (3.3, -32; $\mu = 0.05$ M), 1×10^3 (3.2, -33; $\mu = 0.02$ M), 8×10^4 (3.3, -25; $\mu = 0.05$ M), and 8×10^6, respectively. For the reactions between $Fe(CN)_6^{3-}$ and $Fe(CN)_5N_3^{4-}$, and $Fe(dipy)(CN)_4^{2-}$ the corresponding values are 8×10^4 (1.8, -30; $\mu = 0.05$ M) and 2×10^4, respectively.

The kinetic parameters for the reaction between the Fe(III) complex with EDTA and the Fe(II) complex with *trans*-1,2-diaminocyclohexanetetraacetate have recently been reported, by Wilkins and Yelin[7], as 3×10^4 l.mole^{-1}.sec^{-1}, 4.0 kcal.mole^{-1} and -25 cal.deg^{-1}.mole^{-1}.

8.4.7 The Fe(II)-catalysed aquation of Fe(III)

Sutin *et al.*[1-3] have found iron(II) to catalyse the aquation of the iron(III) ions, $FeCl^{2+}$ (refs. 1, 2) and $FeNCS^{2+}$ (ref. 3), in perchlorate media. A flow technique, with spectrophotometric detection at 336 mμ (disappearance of $FeCl^{2+}$) and 460 mμ (disappearance of $FeNCS^{2+}$), was used to obtain rate data. The rate law

$$\frac{-\mathrm{d}\,[FeX^{2+}]}{\mathrm{d}t} = k_{obs}([FeX^{2+}]-[FeX^{2+}]_{eq})$$

where X is either the chloride or thiocyanate ions, was used to treat the data obtained. The observed rate coefficient k_{obs} is related to the reactions

$$FeX^{2+} \rightleftharpoons Fe^{3+} + X^-$$

$$FeX^{+} \rightleftharpoons Fe^{2+} + X^-$$

$$FeX^{2+} + Fe^{2+} \underset{k_{-1}}{\overset{k_1}{\rightleftharpoons}} FeX^{+} + Fe^{3+}$$

References pp. 142–152

and in the case of the chlorocomplex, following the study of the effect of hydrogen ion[2], the additional reactions,

$$FeCl^{2+} + H_2O \rightleftharpoons ClFeOH^+ + H^+ \qquad (K_2)$$

$$ClFeOH^+ \rightleftharpoons FeOH^+ + Cl^-$$

$$ClFeOH^+ + Fe^{2+} \underset{k_{-2}}{\overset{k_2}{\rightleftharpoons}} FeCl^+ + FeOH^{2+}$$

At a constant ionic strength of 3.0 M, $Mg(ClO_4)_2$, and a temperature of 25 °C, k_1 (NCS^-) has a value of 10.5 l.mole^{-1}.sec^{-1} (at constant $[H^+] = 1.80$ M) and k_1 (Cl^-) of 6.2 l.mole^{-1}.sec^{-1}. The value of the composite term k_2K_2 (Cl^-) is 14.8 sec^{-1} under the same conditions. The step associated with k_2 is thought likely to involve a transition state with a hydroxide bridge.

8.4.8 The reaction between Fe(IV) and Fe(II)

A reaction between iron (IV) and iron (II)

$$Fe(IV) + Fe^{2+} \rightarrow [Fe(III)]_2$$

has been postulated to occur in the oxidation of iron(II) by various two-equivalent oxidants[1].

8.4.9 The exchange reaction between Ru(VII) and Ru(VI)

The exchange between the ruthenium anions RuO_4^- and RuO_4^{2-} in aqueous hydroxide media has been found rapid. A limit for the rate coefficient at 0 °C of $> 1.7 \times 10^3$ l.mole^{-1}.sec^{-1} has been proposed by Luoma and Brubaker[1]. The isotopic method (^{106}Ru), and separation procedures based on the precipitation of the RuO_4^{2-} or RuO_4^- species with barium or tetraphenylarsonium ions, respectively, were used. Attempts to use an ESR technique failed.

8.4.10 Ru(II)-catalysed substitution reactions of Ru(III)

Ru(II) in the form of $Ru(NH_3)_5^{2+}$ catalyses the substitution reactions[1]

$$Ru(NH_3)_5OH_2^{3+} + Cl^- \xrightarrow{k'_1}$$

$$Ru(NH_3)_5Cl^{2+} + H_2O \xrightarrow{k'_2}$$

At an ionic strength 0.1 M the values of the catalytic rate coefficients k'_1 and k'_2 (25 °C) were estimated from spectrophotometric measurements to be 4×10^3 $l^2.mole^{-2}.sec^{-1}$ and 2×10^2 $l.mole^{-1}.sec^{-1}$, respectively.

8.4.11 The exchange reaction between Os(III) and Os(II)

The exchange between *l*-$Os(dipy)_3^{3+}$ and *d*-$Os(dipy)_3^{2+}$ has been studied by the optical activity method by Dwyer and Garfas[1], and Eichler and Wahl[2]. The former authors reporting complete exchange in 95 sec at 5 °C for reactants $\sim 5 \times 10^{-4}$ M, while the latter authors report complete exchange in a time < 15 sec at 4 °C under comparable conditions.

Eichler and Wahl[2] have attempted an isotopic study (^{191}Os and ^{185}Os) of the exchange reaction between $Os(dipy)_3^{3+}$ and $Os(dipy)_3^{2+}$ using a direct injection technique so that reaction times $\sim 7 \times 10^{-2}$ sec were possible. With total osmium $\sim 10^{-5}$ M in aqueous sulphate media at 0 °C complete exchange was observed. The separation methods used were, (*a*) perchlorate precipitation (in presence of iron(II) carrier) and (*b*) extraction with *p*-toluenesulphonic acid in nitromethane, of the osmium(II) complex. A lower limit of 1×10^5 $l.mole^{-1}.sec^{-1}$ was placed on the rate coefficient (0 °C, 3.0 M H_2SO_4). Dietrich and Wahl[3] using the line broadening effect produced by $Os(dipy)_3^{3+}$ on the NMR spectrum of $Os(dipy)_3^{2+}$ have been able to propose a value of $> 5 \times 10^4$ $l.mole^{-1}.sec^{-1}$ at 6 °C in D_2O (0.14 M $[Cl^-]$ and 5×10^{-3} M $[D^+]$).

Campion *et al.*[4] have calculated a value for the exchange coefficient (10 °C, zero ionic strength) of 1×10^7 $l.mole^{-1}.sec^{-1}$ from the observed rate coefficients for the reactions

$$Os(dipy)_3^{2+} + Mo(CN)_8^{3-} \rightleftharpoons Os(dipy)_3^{3+} + Mo(CN)_8^{4-}$$

obtained by a temperature jump method, and the application of the Marcus theory.

8.5 COBALT, RUTHENIUM AND IRIDIUM

8.5.1 The exchange reaction between Co(III) and Co(II) in aqueous media; the effect of inorganic anions

The earliest investigation of the exchange reaction between the aquated ions of Co(III) and Co(II) was carried out by Hoshowsky *et al.*[1], using the isotopic method (^{60}Co). When sulphate salts ($\sim 10^{-2}$ M) were employed, complete exchange was observed between the two oxidation states of cobalt, in a time of less than two min. Two separation methods were employed: (*a*) adsorption on an alumina column, and (*b*) precipitation of the Co(III) as the cobaltinitrite.

References pp. 142–152

The first rate measurements on this system were made by Bonner and Hunt[2] who, in a preliminary study, were able to show the exchange in perchlorate media was overall second-order. The observed rate coefficient (0 °C, 1 *M* in $HClO_4$) was 0.77 $l.mole^{-1}.sec^{-1}$. The separation method involved complexing with EDTA at pH 8 before extracting the Co(II), after addition of thiocyanate ions, with methyl isobutyl ketone. In a further article[3] these two authors report in detail their studies on this system. Overall activation energy and entropy values of 13.2 $kcal.mole^{-1}$ and -13 $cal.deg^{-1}.mole^{-1}$ are given. From results obtained at varying acidity and constant ionic strength, Bonner and Hunt[3] concluded that the rate law could be of the form

$$\text{rate} = (k_1 + k_2'[H^+]^{-1})[Co(III)][Co(II)]$$

where k_1 and k_2' have values of 0.83 $l.mole^{-1}.sec^{-1}$ and 0.14 sec^{-1}, respectively, at 3.2 °C and $\mu = 1.0$ *M*. The rate of exchange was found to be influenced by the addition of nitrate or sulphate ions and ionic strength but not by the presence of chloride ions, oxygen, light or platinum surfaces. In D_2O the rate was found to be much slower than in H_2O.

Shankar and De Souza[4, 5], in a similar study using the isotopic method (^{60}Co), employed mainly ethylenediamine and ammonia to stop the reaction and a separation based on the extraction of Co(II) (as the thiocyanate) with amyl alcohol–ether mixtures. These two workers were able to prove that the exchange was accurately first order in both Co(II) and Co(III) over a wide range of conditions. Results in general agreement with those mentioned previously[3] were obtained for the effect of ionic strength, heavy water and hydrogen ion concentration on the exchange reaction. The mechanism suggested involved the two exchange pathways

$$Co^{3+} + Co^{2+} \xrightarrow{k_1}$$

$$CoOH^{2+} + Co^{2+} \xrightarrow{k_2}$$

with the $CoOH^{2+}$ species being produced *via* the equilibrium

$$Co^{3+} + H_2O \rightleftharpoons CoOH^{2+} + H^+ \qquad K_2$$

From a detailed study of the exchange, at various temperatures (in the range 0 to 20 °C) and acidities at a constant ionic strength of $\mu = 1.0$ *M*, the kinetic parameters were calculated[5]. k_1 and k_2' ($k_2' = k_2K_2$) have values of 0.48 $l.mole^{-1}.sec^{-1}$ and 0.22 sec^{-1}, respectively, at 0 °C. For the exchange pathway associated with k_1, values of the activation enthalpy and entropy of 12.6 $kcal.mole^{-1}$ and -14 $cal.deg^{-1}.mole^{-1}$, respectively, were reported. For the second pathway

(k_2) the authors estimated values of 11.6 kcal.mole^{-1} and -7 cal.deg^{-1}.mole^{-1}, using thermodynamic parameters found by Sutcliffe and Weber[6] for the equilibrium. Comparisons have been made of the activation parameters obtained for this exchange system with those obtained for similar systems[5].

Habib and Hunt[7] have continued the study of this reaction, obtaining further data with special reference to the effects of ionic strength, sulphate and hydrogen-ion concentrations. From data obtained on the dependence of the rate on the $[H^+]$ at various temperatures, values of the kinetic parameters differing slightly from those above[5] have been obtained. Values of $\Delta H_1^{\ddagger}$ and $\Delta H_2^{\ddagger}$ and $\Delta S_1^{\ddagger}$ and $\Delta S_2^{\ddagger}$ (at $\mu = 1.0$ M) obtained were 11.8, 5.3 kcal.mole^{-1} and -17 and -31 cal.deg^{-1}.mole^{-1}, respectively. The value of k_2 was estimated as $\sim 6.7 \times 10^3$ l.mole^{-1}.sec^{-1} at 18 °C, $\mu = 1.0$ M.

The effect of the addition of sulphate and fluoride ions were found by these workers[7] to increase the rate of exchange; addition of acetate and trifluoroacetate ions produced relatively minor changes. For the addition of sulphate ions, a rate law

$$\text{rate} = (k_1 + k_2'[H^+]^{-1} + k_3'[SO_4^{2-}])[Co(III)][Co(II)]$$

was found to fit the experimental observations, suggesting a pathway

$$CoSO_4^+ + Co^{2+} \xrightarrow{k_3}$$

with k_3' being given by k_3K_3, where K_3 is the constant for the equilibrium

$$Co^{3+} + SO_4^{2-} \rightleftharpoons CoSO_4^+$$

At $\mu = 0.5$ M the value of k_3' is 5.2×10^2 l^2.mole^{-2}.sec^{-1} (0.15 °C); the corresponding activation enthalpy and entropy values are 20.5 kcal.mole^{-1} and 28.5 cal.deg^{-1}.mole^{-1}.

Shankar and De Souza[8] have also recently investigated the effect of the additions of various anions to this system in both water and heavy water solvent. Fluoride was found to have very little influence on the exchange rate while acetate, nitrate and sulphate ions produced an increase. For the addition of sulphate ions an estimate of the rate coefficient k_3 of ~ 20 l.mole^{-1}.sec^{-1} (at 14 °C and $\mu = 2.0$ M) was made. For the addition of nitrate and acetate, values of the coefficients k_4' and k_5' (where $k_4' = k_4K_4$ and $k_5' = k_5K_5$), *viz.*

$$CoNO_3^{2+} + Co^{2+} \xrightarrow{k_4}$$

$$CoOAc^{2+} + Co^{2+} \xrightarrow{k_5}$$

$$Co^{3+} + NO_3^- \rightleftharpoons CoNO_3^{2+} \qquad K_4$$

$$Co^{3+} + OAc^- \rightleftharpoons CoOAc^{2+} \qquad K_5$$

at 0 °C are 2.83×10^4 ($\mu = 1.0$ M) and 0.762 $1.^2$mole^{-2}.sec^{-1} ($\mu = 2.0$ M), respectively.

Shankar and De Souza[8] also found that the rate of exchange was increased by the addition of silver ions owing to the occurrence of equilibria such as

$$Co^{3+} + Ag^+ \rightleftharpoons Co^{2+} + Ag^{2+}$$

Conocchioli *et al.*[9], from spectrophotometric observations on the equilibrium

$$Co^{3+} + Cl^- \rightleftharpoons CoCl^{2+}$$

have been able to evaluate the rate coefficients for the analogous reactions

$$CoCl^{2+} + Co^{2+} \underset{k_7}{\overset{k_6}{\rightleftharpoons}} Co^{3+} + CoCl^+$$

They found values of 1.0 l.mole^{-1}.sec^{-1} for k_6 and 26 1^2.mole^{-2}.sec^{-1} for k_7K_7 (at 25 °C, $\mu = 3.0$ M) where K_7 is the constant for the equilibrium

$$Co^{2+} + Cl \rightleftharpoons CoCl^+$$

The exchange of ^{60}Co between the complex anions, 12-tungstocobaltate(III) ($CoO_4W_{12}O_{36}{}^{5-}$) and 12-tungstocobaltate(II) ($CoO_4W_{12}O_{36}{}^{5-}$), in aqueous chloride media has been studied by Rasmussen and Brubaker[10]. Tetrabutylammonium iodide in acetate media was used to give a separation by precipitation of the Co(III). The rate showed a first order dependence on Co(II) and Co(III) concentrations but was independent of the H^+ concentration. At $\mu = 1.02$ M (LiCl) the rate coefficient has a value of 1.01 l.mole^{-1}.sec^{-1} at 0 °C. Variation in the ionic strength and dielectric constant was found to affect the kinetic parameters. At zero ionic strength, the rate coefficient has a value of 4.5×10^{-3} l.mole^{-1}.sec^{-1} (0 °C). The activation energy of the reaction was reported as 18 kcal.mole^{-1}.

The rate coefficient for the exchange reaction

$$Co(CN)_6^{3-} + Co(CN)_5^{3-} \longrightarrow$$

has been evaluated as $< 1 \times 10^{-5}$ l.mole^{-1}.sec^{-1} at 25 °C by Birk and Halpern[11] using data obtained by Adamson[12].

8.5.2 Exchange reactions involving complexes of Co(III) and Co(II) with ammonia and organic ligands

The early work[1-4] on exchange reactions involving the complex hexammino species $Co(NH_3)_6^{3+}$ and $Co(NH_3)_6^{2+}$ led to the conclusion that the ^{60}Co exchanges between $Co(NH_3)_6^{3+}$ and $Co(NH_3)_6^{2+}$, and $Co(NH_3)_6^{3+}$ and Co^{2+} were extremely slow. The reactions were studied in the absence of oxygen, which catalyses the exchange, and in media varying between acid and aqueous ammonia. The separation method used was the extraction of Co(II), as the thiocyanate complex, with a mixed organic solvent.

Subsequently, Stranks[5] has investigated a series of cobaltammines with regard to the electron exchange between cobalt(III) and cobalt(II), using the isotopic method ^{60}Co. The systems studied included the Co(III) complexes and ion-pairs $Co(NH_3)_6^{3+}$, $Co(NH_3)_6^{3+} \cdot OH^-$, $Co(NH_3)_6^{3+} \cdot Cl^-$, $Co(NH_3)_5OH^{2+}$, *trans*-Co $(NH_3)_4(OH)_2^+$ and *cis*-$Co(NH_3)_4(OH)_2^+$ with Co(II) complexes of the type $Co(NH_3)_n^{2+}$ where n has values 3, 4, 5 and 6.† All exchanges refer to perchlorate media in the absence of oxygen. Energies of activation $\sim$ 13.5 kcal.mole^{-1} and activation entropies between -28 and -37 cal.deg^{-1}.mole^{-1} have been found for these exchanges. The rate coefficients (l.mole^{-1}.sec^{-1} at 64.5 °C, $\mu = 1.0$ M) evaluated from necessary kinetic and ion-pair association data are for the exchange of $Co(NH_3)_n^{2+}$ with $Co(NH_3)_6^{3+} \cdot OH^-$ 5.5×10^{-3}, with $Co(NH_3)_6^{3+} \cdot Cl^-$ 7.3×10^{-4}, with $Co(NH_3)_5OH^{2+}$ 9.0×10^{-4}, with *trans*-$Co(NH_3)_4(OH)_2^+$ 4.2×10^{-3} and with *cis*-$Co(NH_3)_4(OH)_2^+$ 2.5×10^{-2}. Under the same conditions the rate coefficient for the exchange $Co(NH_3)_6^{3+}$–$Co(NH_3)_n^{2+}$ is $< 1.7 \times 10^{-10}$ l.mole^{-1}.sec^{-1}. The mechanism proposed by Stranks was

$$Co(NH_3)_6^{3+} + H_2O \rightleftharpoons Co(NH_3)_6^{3+} \cdot OH^- + H^+$$

$$Co(NH_3)_6^{3+} \cdot OH^- + Co(NH_3)_n^{2+} \longrightarrow Co(NH_3)_6^{2+} + Co(NH_3)_n^{3+} + OH^-$$

involving a hydrogen or hydroxo bridge activated complex. A similar transition state was thought likely to occur for the $Co(NH_3)_6^{3+} \cdot Cl^-$–$Co(NH_3)_n^{2+}$ exchange.

Appleman *et al.*[6] have investigated the exchange of ^{60}Co between the species $Co(NH_3)_5OH^{2+}$ and Co(II), $Co(NH_3)_n^{2+}$ and $Co(NH_3)_{(n-1)}OH^+$ where n has values between 0 and 6, in aqueous ammonia. All kinetic data was obtained using a separation procedure based on the precipitation of the salt $Co(NH_3)_5H_2OHgCl_5$. Light and oxygen were excluded from the reaction vessels. A rate law of the form

$$\begin{aligned} \text{rate} &= k[Co(NH_3)_5OH^{2+}][Co^{2+}][NH_3]^x \\ &= k_{obs}[Co(NH_3)_5OH^{2+}][Co(NH_3)_x^{2+}] \end{aligned}$$

was found from the data obtained over the range of $[NH_3]$ 0.092 to 0.167 M, [Co(II)] in the range 1.5×10^{-6} to 9×10^{-3} M with [Co(III)] $\sim 4 \times 10^{-3}$ M. The value of x was found to vary with the ammonia concentration. For the exchange occurring under conditions when $x = 5$ the value of k_{obs} (35.5 °C, $\mu = 2.28$ M) is 4.4×10^{-4} l.mole^{-1}.sec^{-1} with corresponding activation parameters of 15.6 kcal.

† See also Biradar *et al.*[15].

mole^{-1} ($\Delta H^{\ddagger}$) and -23.5 cal.deg^{-1}.mole^{-1} ($\Delta S^{\ddagger}$).† No exchange was detected between $Co(NH_3)_5OH^{2+}$ and Co^{2+} over a long period of time.

Lewis *et al.*[4] have studied the exchange between $Co(en)_3^{3+}$ and $Co(NH_3)_6^{2+}$ and between $Co(NH_3)_6^{3+}$ and $Co(en)_3^{2+}$ occurring in the absence of oxygen. The observed rate coefficients for the exchanges at $\mu = 0.98$ M are $< 1.2 \times 10^{-7}$ l.mole^{-1}.sec^{-1} (45 °C) and 2.0×10^{-2} l.mole^{-1}.sec^{-1} (25 °C), respectively. The exchange between $Co(en)_3^{3+}$ and $Co(en)_3^{2+}$ was also investigated, using the isotopic method, by these workers, who found catalytic effects were produced by oxygen and various added solids. The rate law showed a near first-order dependence on $Co(en)_3^{2+}$ and an order of between 0.58 and 1.0, with respect to $Co(en)_3^{3+}$. The activation energy of the process was found to depend on the ionic strength of the medium and varied from 12.75 to 15.75 kcal.mole^{-1}. At 25 °C, $\mu = 0.98$ M a rate coefficient of 5.2×10^{-4} l.mole^{-1}.sec^{-1} was calculated.

Dwyer and Sargeson[7] have used the optical activity method in their study of the exchange between the tris-(ethylenediamine) complexes of Co(III) and Co(II)

$$d\text{-}Co(en)_3^{3+} + Co(en)_3^{2+} \rightleftharpoons l\text{-}Co(en)_3^{3+} + Co(en)_3^{2+}$$

The cobalt(II) complex, which is optically unstable, was formed in reaction mixtures by the addition of ethylenediamine (excess) to Co^{2+}. The rate law obtained from the racemization data was

$$\text{rate} = k_1[\text{Co(III)}][\text{Co(II)}] + k''[\text{Co(III)}][\text{en}]$$

indicating that racemization occurs *via* two paths only one of which (associated with k_1) is related to the electron transfer process. At 25 °C and $\mu = 0.98$ M, k_1 has a value 7.7×10^{-5} l.mole^{-1}.sec^{-1}, the corresponding values of the activation energy and entropy are 14.1 kcal.mole^{-1} and -32 cal.deg^{-1}.mole^{-1}. Plots of log k_1 *versus* $\mu^{\frac{1}{2}}$ were found to be linear. No specific anion effects were observed.

Stranks[8] has also studied the exchange between tris- and bis-ethylenediamine complexes of Co(III) and Co(II). The isotopic method was used. At low ionic strengths (ClO_4^-) a rate law

$$\text{rate} = k_2[\text{Co(III)}][\text{Co(II)}]$$

was indicated by the kinetic data. The same rates of the exchange were found between $Co(en)_3^{3+}$ and $Co(en)_n^{2+}$ when n had values of 1, 2 or 3. Similar results were obtained for the exchange between $Co(en)_2^{3+}$ and $Co(en)_n^{2+}$. Hydroxide and other anions were found to catalyse the exchange reactions. The rate coefficients (l.mole^{-1}.sec^{-1}) at 50 °C and $\mu = 0.2$ M for the various exchange systems, were

† Recently, Williams and Hunt[16] have published further data for this exchange system.

evaluated by Stranks as: $Co(en)_3^{3+}$–$Co(en)_n^{2+}$, 1.4×10^{-4}; $Co(en)_3^{3+}\cdot OH^-$–$Co(en)_n^{2+}$, 9.2×10^{-4}; $Co(en)_3^{3+}Cl^-$–$Co(en)_n^{2+}$, 5×10^{-4}; $Co(en)_3^{3+}\cdot Br^-$–$Co(en)_n^{2+}$ 3.0×10^{-4}; $Co(en)_3^{3+}\cdot I^-$–$Co(en)_n^{2+}$, 2.0×10^{-4}; $Co(en)_3^{3+}\cdot SO_4^{2-}$–$Co(en)_n^{2+}$, 1.3×10^{-5} $Co(en)_2^{3+}$–$Co(en)_n^{2+}$, 2.7×10^{-4}; $Co(en)_2OH^{2+}$–$Co(en)_n^{2+}$, 9.7×10^{-4}, and $Co(en)_2(OH)_2^+$–$Co(en)_n^{2+}$, 2.0×10^{-3}. For all these processes the activation energies are ~ 13.8 kcal.mole^{-1} with activation entropies in the range -27 to -32 cal.deg^{-1}.mole^{-1}. These are similar to the values obtained for the analogous ammonia complexes[5]. No exchange has been detected between $Co(en)_3^{3+}$ and Co^{2+} in aqueous solution over a period of 24 hours[9].

The exchange of ^{60}Co between the EDTA complexes of Co(II) (CoY^{2-}) and Co(III) (CoY^-), in aqueous perchlorate and nitrate media at pH = 2, has been investigated by Adamson and Vorres[10]. Using a ion-exchange separation method, a rate law

$$\text{rate} = k_3'[CoY^-][CoY^{2-}]$$

was found to be obeyed over a range of concentrations, CoY^{2-} 1.6×10^{-2} to 1.2×10^{-1} M and CoY^{2-} 1.8×10^{-2} to 8.6×10^{-2} M. At 85 °C, pH = 2, k_3' has a value 2.1×10^{-4} l.mole^{-1}.sec^{-1}. The activation parameters are 22 kcal.mole^{-1} ($\Delta H^\ddagger$) and -17 cal.deg^{-1}.mole^{-1} ($\Delta S^\ddagger$). Increase in ionic strength and surface area and exposure to light produce small increases in the rate of exchange. The mechanism proposed was a direct electron transfer process: steps involving dissociation of CoY^- and CoY^{2-} were not considered feasible.

Im and Busch[11] have made a further study of this reaction. They used the optical activity method, investigating the rate of racemization of d-CoY^-, in the presence of CoY^{2-} which is optically unstable and catalyses the process, *viz.*

$$d\text{-}CoY^- \rightleftharpoons l\text{-}CoY^- \qquad k$$

$$d\text{-}CoY^- + CoY^{2-} \rightleftharpoons CoY^{2-} + l\text{-}CoY^- \qquad k_3$$

The equation $-\ln(\alpha_t/\alpha_0) = k_{obs}t$ was used to treat the experimental data ($k_{obs} = k_3'[CoY^{2-}]+2k$) and it was found that k_3' (the apparent rate coefficient) had a value of 1.5×10^{-4} l.mole^{-1}.sec^{-1} at 85 °C (pH = 2.0). Variation of the solvent from H_2O to D_2O did not bring about any change in the rate of racemization, whereas increase in the pH of the medium decreased the rate. This was explained in terms of the formation of the species $Co(H\text{–}Y)^-$

$$Co(H\text{–}Y)^- \rightleftharpoons CoY^{2-} + H^+ \qquad K_4$$

and the two pathways

$$d\text{-}CoY^- + Co(H\text{–}Y)^- \xrightarrow{k_4}$$

$$d\text{-}CoY^- + CoY^{2-} \xrightarrow{k_5}$$

The rate coefficients are related by the expression

$$k_3' = k_4\{H^+/(K_4+H^+)\}+k_5\{K_4/(K_4+H^+)\}$$

Values of k_4 and k_5 (l.mole^{-1}.sec^{-1}) at 100 °C and the corresponding activation parameters ($\Delta H^{\ddagger}$ and $\Delta S^{\ddagger}$ in parentheses) are 8.0×10^{-4} (24 kcal.mole^{-1}, -9 cal. deg^{-1}.mole^{-1}) and 1.4×10^{-4} (20 kcal.mole^{-1} and -21 cal.deg^{-1}.mole^{-1}).

Im and Busch[12], using the optical activity method, have also made a study of the exchange of the propylenediaminetetraacetate complexes of Co(III) and Co(II) ($CoPY^-$ and $CoPY^{2-}$). This system can be treated by the McKay equation. Optical rotation data was obtained at temperatures between 80 and 100°C and various pH's between 2.0 and 7.0. This was again found to be consistent with exchange *via* two pathways (k_6 and k_7)

$$\text{L-Co(H-}d\text{-PY)}^- \rightleftharpoons \text{L-Co(}d\text{-PY)}^{2-}+H^+$$

$$\text{D-Co(}l\text{-PY)}^- + \text{L-Co(H-}d\text{-PY)}^- \xrightarrow{k_6} \text{D-Co(H-}l\text{-PY)}^- + \text{L-Co(}d\text{-PY)}^-$$

$$\text{D-Co(}l\text{-PY)}^- + \text{L-Co(}d\text{-PY)}^{2-} \xrightarrow{k_7} \text{D-Co(}l\text{-PY)}^{2-} + \text{L-Co(}d\text{-PY)}^-$$

D and L, and *d* and *l*, refer to the rotational properties of the compound and the ligand, respectively. Values of the rate coefficients k_6 and k_7 at 100 °C are 7.0×10^{-4} and 2.0×10^{-4} l.mole^{-1}.sec^{-1}, respectively.

Baker *et al.*[13], have obtained some kinetic data on the exchange reactions between the 1,10-phenanthroline, 2,2′-dipyridine and 2,2′, 2″-tripyridine complexes of Co(III) and Co(II) in both H_2O and D_2O solution. The isotopic method (^{60}Co) and separations involving either extraction of Co(II) with organic solvents (*n*-hexanol–ether or saturated sodium acetate in *n*-hexanol) or precipitation of Co(III) as $Co(phen)_3(I_3)_3$ were used. For the 1,10-phenanthroline system, a rate law

$$\text{rate} = k_{obs}[\text{Co(III)}][\text{Co(II)}]^y$$

was observed, where y has a value of ~ 1 in the presence of perchlorate or nitrate ions and $\frac{1}{2}$ in the presence of chloride ions. The values of the observed rate coefficients at 0 °C ($\mu = 0.1$ M), which are dependent on the anion concentration, are 1.1 l.mole^{-1}.sec^{-1} (ClO_4^-), 4.0 l.mole^{-1}.sec^{-1} (NO_3^-) and 1.2×10^{-2} l$^{\frac{1}{2}}$. mole$^{-\frac{1}{2}}$.sec^{-1} (Cl^-). The mechanism postulated invokes the existence of ion-pairs of cobalt(III) and the anion concerned. The rates of exchange in D_2O were little different from those in H_2O. Data have recently been published concerning this exchange system for media containing polymeric electrolytes[15].

For the exchange between $Co(dipy)_3^{3+}$–$Co(dipy)_3^{2+}$ and $Co(tripy)_2^{3+}$–$Co(tripy)_2^{2+}$, second-order rate coefficients (0 °C) in perchlorate media of ~ 12 and 1.6 l.mole^{-1}.sec^{-1}, respectively, were obtained. Previously, Ellis *et al.*[14] had estimated values for the exchange rate coefficients (20 °C) of 18.7 and 4.46 l. mole^{-1}.sec^{-1} for the 1,10-phenanthroline and dipridine systems.

8.5.3 *The exchange reaction between Co(III) and Co(II) in non-aqueous media*

Grossman and Garner[1] have investigated the exchange reaction between the species $Co(NH_3)_6^{3+}$ and $Co(NH_3)_6^{2+}$, in the form of their nitrate salts, using the isotope ^{60}Co with anhydrous liquid ammonia as solvent. The reaction was carried out in a sealed system at temperatures of 25° and 45 °C, with a separation procedure based on ether–pentanol extraction of the thiocyanate complex of Co(II). With Co(III) and Co(II) $\sim 3\times10^{-3}$ *M*, observed rate coefficients of 6×10^{-5} (25 °C) and 7×10^{-4} (45 °C) l.mole^{-1}.sec^{-1} were obtained. The activation energy for the exchange process is 23 kcal.mole^{-1}. Grossman and Garner[1] have pointed out that the exchange mechanism could involve the dissociation of the species $Co(NH_3)_6^{3+}$, since the rate of ammonia exchange with this species is of the same order of magnitude.

West[2] has observed very slow exchange reactions in pyridine solvent between Co(II) in the form of $Co(OAc)_2$ and Co(III) in the forms tris-(1-nitroso-2-naphthol)cobalt(III) and tris-(acetylacetone)cobalt(III).

Baker *et al.*[3] have obtained some evidence for a decreasing rate of exchange, as water is replaced by acetone, with the system $Co(phen)_3^{3+}$–$Co(phen)_3^{2+}$.

8.5.4 *Co(II)-catalysed substitution reactions of Co(III)*

Numerous cyanide substitution reactions of cobalt(III) complexes, $Co^{III}(NH_3)_5X$ and $Co^{III}(CN)_5X$ have been investigated (mainly by Halpern *et al.*[1,3,4]) and found[1-4] to be catalysed by Co(II) in the form of $Co(CN)_5^{3-}$. Two reactions of the pentammine complexes can occur, *viz.*

$$Co^{III}(NH_3)_5X+5\,CN^- = Co^{III}(CN)_5X+5\,NH_3$$

and

$$Co^{III}(NH_3)_5X+6\,CN^- = Co(CN)_6^{3-}+5\,NH_3+X$$

the former by an inner-sphere mechanism with the group X acting as a bridge, *viz.*

$$Co(CN)_5^{3-}+Co^{III}(NH_3)_5X \xrightarrow{k_1} [(CN)_5Co^{II}\text{-}X\text{-}Co^{III}(NH_3)_5]^{\ddagger}$$

$$\text{rate} = k_1[Co^{III}(NH_3)_5X][Co(CN)_5^{3-}]$$

and the latter reaction *via* an outer-sphere mechanism

$$Co(CN)_5^{3-} + CN^- \rightleftharpoons Co(CN)_6^{4-} \qquad K_2$$

$$Co(CN)_6^{4-} + Co^{III}(NH_3)_5X \xrightarrow{k_2} [(CN)_5Co^{II}(CN)(X)Co^{III}(NH_3)_5]^{\ddagger}$$

$$\text{rate} = k_2 K_2[Co^{III}(NH_3)_5X][Co(CN)_5^{3-}][CN^-]$$

The nature of the group X determines the type of reaction which is the most important. For X = azide, thiocyanate, hydroxide, chloride[1], bromide and iodide[2] the inner-sphere bath operates while for X = ammonia or oxyanions (including carboxylates)[1] the main pathway is the outer-sphere reaction. For X = fluoride or nitrite the concentration of the cyanide ion present determines which is the major reaction pathway.

Halpern *et al.*[1], from data recorded using a stopped flow technique with spectrophotometric detection, have obtained values of k_1 and $k_2 K_2$ for each of the above-mentioned anions.

For the Co(III) complex $Co(NH_3)_5NO_2^{2+}$, Halpern and Nakamura[3] have obtained spectrophotometric evidence for the inner-sphere reaction occurring *via* $Co(CN)_5ONO^{3-}$ which isomerises to give the product $Co(CN)_5NO_2^{3-}$. The species $Co(NH_3)_5CN^{2+}$ also reacts in this manner to give $Co(CN)_5NC^{3-}$ and finally $Co(CN)_6^{3-}$.

Birk and Halpern[4] have studied the Co(II), $Co(CN)_5^{3-}$, catalysed substitution reactions

$$Co^{III}(CN)_5X + CN^- = Co(CN)_6^{3-} + X$$

which occur *via* an inner-sphere mechanism and an activated complex $[X(NC)_4Co^{III}\text{–}CN\text{–}Co^{II}(CN)_5]^{\ddagger}$ to give $Co(CN)_5NC^{3-}$ and finally $Co(CN)_6^{3-}$. Absorbance measurements in the region of λ_{max} of the species $Co^{III}(CN)_5X$ were used to obtain rate data. The rate law observed was

$$\text{rate} = k_3[Co^{III}(CN)_5X][Co(CN)_5^{3-}]$$

where the rate coefficients k_3 decrease in the order X = H_2O, Br^-, Cl^-, SCN^-, I^-, N_3^-, OH^- and SO_3^{2-}. Values of k_3 for each of these ions have been determined. Limits for the rate coefficients of the outer-sphere reactions, involving the ion $Co(CN)_6^{4-}$, have also been set by these workers. For the chloride and iodide complexes an additional term, k_4 $[Co^{III}(CN)_5X]$, was required in the rate law to account for a non-catalysed pathway.

Farina and Wilkins[5] have published data relating to the reactions: $Co(tripy)^{2+}$ with $Co(dipy)_3^{3+}$, $Co(dipy)_2^{3+}$, $Co(dipy)^{3+}$, $Co(phen)_3^{3+}$, $Co(phen)_2^{3+}$, $Co(phen)^{3+}$,

$Co(3,5,6,8\text{-}Me_4\text{-}phen)_3^{3+}$, Co^{3+} and $CoOH^{2+}$; $Co(py)_4Cl_2^+$ with $Co(tripy)_2^{2+}$, $Co(dipy)_3^{2+}$ and $Co(phen)_3^{2+}$; $Co(phen)^{2+}$ with $Co(dipy)^{3+}$, and $Co(dipy)_3^{2+}$ with $Co(tripy)_2^{3+}$. Values of the rate coefficients (at 0 °C, l.mole^{-1}.sec^{-1}) with the corresponding activation enthalpies (kcal.mole^{-1}) and entropies (cal.deg^{-1}.mole^{-1}) and ionic strengths in parentheses are: 6.4×10 (8.4, −19; $\mu = 0.05\ M$), 1.3×10^4 (4.2, −24; $\mu = 0.05\ M$), 6.8×10^2 (10.8, −6; $\mu = 0.05\ M$), 2.8×10^2 (6.9, −21; $\mu = 0.05\ M$), 3×10^4 (4.3, −22; $\mu = 0.05\ M$), 1.4×10^3 (10.9, −4; $\mu = 0.05\ M$), 6.8×10 (10.6, −12; $\mu = 0.02\ M$), 7.4×10^4 (3.4, −23; $\mu = 1.0\ M$), 1.7×10^4 (5.1, −20; $\mu = 0.027\ M$), 3.0×10^5 (5.2, −14; $\mu = 0.00016\ M$), 1.1×10^4 (5.2, −20; $\mu = 0.00016\ M$), 9.1×10^3 (8.4, −12; $\mu = 0.00016\ M$), 6×10^{-2} (8.7, −30; $\mu = 1.55\ M$), and 2.7×10 (8.7, −20; $\mu = 0.05\ M$), respectively.

8.5.5 The reaction of Co(IV) with Co(II)

To account for their experimental observations on the reaction of lead(IV) acetate with cobalt(II) acetate in anhydrous acetic acid solution, Sutcliffe *et al.*[1] have proposed a mechanism involving the Co(IV) steps, *viz.*

$$\mathrm{Co(IV)+Co(II)} \xrightarrow{2k_1} \mathrm{2\ Co(III)}$$

$$\mathrm{Co(IV)+Pb(II)} \overset{k_2}{\rightleftharpoons} \mathrm{Co(II)+Pb(IV)}$$

Whilst no direct measurements have been made on these Co(IV) reactions, the ratio $(k_2/2\,k_1)$ has been evaluated[2] as 0.15 at 25 °C.

8.5.6 Rh(I)-catalysed substitution reactions of Rh(III)

Rund *et al.*[1] have suggested that the pyridine substitution reaction

$$\mathrm{RhCl_5^{2-} + 4\ py} = trans\text{-}\mathrm{Rh\ py_4\ Cl_2^+ + 3\ Cl^-}$$

proceeds *via* a pathway involving Rh(I), *viz.*

$$\mathrm{Rh(I) + 4\ py} \longrightarrow \mathrm{Rh\ py_4^+}$$

$$\mathrm{Rh\ py_4^+ + RhCl_5^{2-}} \longrightarrow \mathrm{(Rh\ py_4\text{–}Cl\text{–}RhCl_4)^-}\ \text{slow}$$

$$\mathrm{(Rh\ py_4\text{–}Cl\text{–}RhCl_4)^-} \longrightarrow trans\text{-}\mathrm{Rh\ py_4Cl^{2+} + Rh(I)}$$

$$trans\text{-}\mathrm{Rh\ py_4Cl^{2+} + Cl^-} \longrightarrow trans\text{-}\mathrm{Rh\ py_4Cl_2^+}$$

Other workers[2,3] have proposed steps involving catalysis by species other than Rh(I) for some similar reactions.

References pp. 142–152

8.5.7 *The exchange reaction between Ir(IV) and Ir(III)*

The isotopic method (^{192}Ir) has been used to investigate the exchanges between the anions $IrCl_6^{2-}$ and $IrCl_6^{3-}$. The separation methods used by Sloth and Garner[1] were: (*a*) extraction of $IrCl_6^{2-}$ with either 2-butanone or acetone–chloroform, and (*b*) precipitation of the iridium (IV) as the ammonium salt. These authors observed complete exchange in less than 42 sec, with reactant concentrations $\sim 10^{-4}$ *M*, at 1 °C in the absence of light with 1 *M* HCl as solvent. This corresponds to a rate coefficient in excess of 290 l.mole^{-1}.sec^{-1}.

Recently, Hurwitz and Kustin[2] have reinvestigated this exchange reaction using the same isotopic procedure and the 2-butanone separation method, in conjunction with a stopped flow apparatus. A rate coefficient of 2.3×10^5 l.mole^{-1}.sec^{-1} was obtained for the conditions, temperature 25 °C and ionic strength 0.1 *M*. Application of the Marcus theory to results obtained for the reaction

$$IrCl_6^{2-} + Fe(5,6\text{-}Me_2\text{-phen})_3^{2+} \rightleftharpoons IrCl_6^{3-} + Fe(5,6\text{-}Me_2\text{-phen})_3^{3+}$$

by the temperature jump method[3] had led previously tc an estimate for this rate coefficient of 2.5×10^4 l.mole^{-1}.sec^{-1}.

8.5.8 *The reaction between complexes of Ir(IV) and Ir(III)*

Hurwitz and Kustin[1] have investigated the reactions

$$IrCl_6^{2-} + IrBr_6^{3-} \underset{k_2}{\overset{k_1}{\rightleftharpoons}} IrCl_6^{3-} + IrBr_6^{2-}$$

using the temperature jump method, with spectrophotometric detection (495 mμ). With potassium nitrate media ($\mu = 0.1$ *M*) values of k_1 and k_2 (10 °C) are 1.1×10^7 and 1.5×10^6 l.mole^{-1}.sec^{-1}, respectively, with energies of activation of 5.7 (k_1) and 7.5 (k_2) kcal.mole^{-1}. The rate coefficients were found to depend on the ionic strength of the medium and possibly on the size of the cation present. Hurwitz and Kustin[1] have also obtained a value for the exchange rate coefficient (Marcus theory) from this data of 1×10^5 l.mole^{-1}.sec^{-1}.

8.6 PLATINUM

8.6.1 *The exchange reaction between Pt(IV) and Pt(II); Pt(II)-catalysed substitution reactions of Pt(IV)*

Two types of mechanism have been postulated to explain the experimental observations on this exchange system.

Rich and Taube[1], using the isotopic method and a separation achieved by precipitation of the ion $PtCl_6^{2-}$ as Cs_2PtCl_6, have found Pt exchange in 10^{-2} *M* HCl media, at 25 °C, to occur in minutes in the presence of UV light and in many hours under normal conditions. On this basis and using additional information obtained from experiments on chloride exchange with the ions $PtCl_4^{2-}$ and $PtCl_6^{2-}$, a mechanism involving a Pt(III) species, *viz.*

$$PtCl_4^{2-} + PtCl_6^{2-} \rightleftharpoons 2\,PtCl_5^{2-}$$

$$PtCl_6^{2-} \xrightarrow{h\nu} PtCl_5^{2-} + Cl$$

$$PtCl_5^{2-} + PtCl_4^{2-} \longrightarrow \text{exchange}$$

$$PtCl_5^{2-} + PtCl_6^{2-} \longrightarrow \text{exchange}$$

was proposed to account for the experimental observations. A similar chain mechanism has been postulated by McCarley *et al.*[2], for the platinum exchange system, $Pt(en)Br_4$–$Pt(en)Br_2$, *viz.*

$$Pt(en)Br_4 \xrightarrow{h\nu} Pt(en)Br_3 + Br$$

$$Pt(en)Br_3 + Pt(en)Br_2 \longrightarrow \text{exchange}$$

$$Pt(en)Br_3 + Pt(en)Br_4 \longrightarrow \text{exchange}$$

$$Pt(en)Br_3 + Br \longrightarrow Pt(en)Br_4$$

in *N*, *N'*-dimethylformamide solution in the presence of light. Both of the above exchange systems show inhibitions by $IrCl_6^{2-}$.

Basolo *et al.*[3,4] have found a rate law

$$\text{rate} = k'[\text{Pt(IV)}][\text{Pt(II)}][\text{Cl}^-]$$

for the exchange of chloride ions with *trans*-$Pt(en)_2Cl_2^{2+}$ in the presence of $Pt(en)_2^{2+}$ in the absence of light; k' has a value 15 $l^2.mole^{-2}.sec^{-1}$ at 25 °C. The mechanism postulated was

$$Pt(en)_2^{2+} + Cl^- \rightleftharpoons Pt(en)_2Cl^+$$

$$Pt(en)_2Cl_2^{2+} + Pt(en)_2Cl^+ \rightleftharpoons [Cl(en)_2Pt\text{–}Cl\text{–}Pt(en)_2Cl]^{3+}$$

$$[Cl(en)_2Pt\text{–}Cl\text{–}Pt(en)_2Cl]^{3+} \rightleftharpoons Pt(en)_2Cl^+ + Pt(en)_2Cl_2^{2+}$$

which, as Basolo *et al.*[3,4] have pointed out, provides a pathway for platinum exchange. This has since been confirmed[5,6].

Cox *et al.*[5], using the isotope ^{195}Pt to label the species and tetraphenyl boron precipitation of platinum(II), $Pt(en)_2^{2+}$, as the separation method, have found the above rate law to be applicable over the ranges [Pt(II)] 8×10^{-4} to 5×10^{-3} *M*, [Pt(IV)] 5×10^{-4} to 5×10^{-3} *M* and [Cl^-] 1×10^{-3} to 1.8×10^{-2} *M*. A value of k' (25 °C) of 12.1 $l^2.mole^{-2}.sec^{-1}$ was obtained with the activation entropy and enthalpy of -20 $cal.deg^{-1}.mole^{-1}$ and 10.1 $kcal.mole^{-1}$, respectively. The activation energy found[6] for chloride exchange was 11.5 $kcal.mole^{-1}$. Using ^{14}C to label the species $Pt(en)_2^{2+}$, similar confirmation has been obtained for this exchange mechanism; k' (25 °C) has a value[6] of 16 $l^2.mole^{-2}.sec^{-1}$ for media $\sim 9 \times 10^{-3}$ *M* with respect to H^+.

Basolo *et al.*[6], have found similar platinum(II)-catalysed chloride exchange reactions with other Pt(IV) complexes, including *cis*- and *trans*-$Pt(NH_3)_4Cl_2^{2+}$, $Pt(NH_3)_5Cl^{3+}$ and *trans*-$Pt(NH_3)_3Cl_3^+$. These reactions proceed by the chloride bridge mechanism above and the apparent rate coefficients (k' $l^2.mole^{-2}.sec^{-1}$, 25 °C) for platinum exchange, which was concluded to occur *via* this pathway, are: *trans*-$Pt(NH_3)_4Cl_2^{2+}$–$Pt(NH_3)_4^{2+}$, ~ 6.3; *cis*-$Pt(NH_3)_4Cl_2^{2+}$–$Pt(NH_3)_4^{2+}$, $\sim 2.5 \times 10^{-3}$; $Pt(NH_3)_5Cl^{3+}$–$Pt(NH_3)_4^{2+}$, $\sim 6 \times 10^{-4}$; *trans*-$Pt(NH_3)_3Cl_3^+$–$Pt(NH_3)_4^{2+}$, ~ 2.0; and *trans*-$Pt(NH_3)_3Cl_3^+$–$Pt(NH_3)_3Cl^+$, ~ 3.0. No exchange was detected for the systems *trans*-$Pt(tetrameen)_2Cl_2^{2+}$ with either $Pt(en)_2^{2+}$ or $Pt(tetrameen)_2^{2+}$ where chloride bridges cannot form (en = ethylenediamine and tetrameen = tetramethyl en).

The systems (pn = propylenediamine)

$$Pt(en)_2^{2+} + trans\text{-}Pt(l\text{-}pn)_2X_2^{2+} \rightleftharpoons trans\text{-}Pt(en)_2X_2^{2+} + Pt(l\text{-}pn)_2^{2+}$$

where X is Cl, Br or OH, have been investigated by polarimetry. The platinum exchange rate increases in the order $OH^- < Cl^- < Br^-$, and the reaction proceeds *via* the bridged mechanism[6].

Johnson and Basolo[7] have found for the reactions

$$Pt(l\text{-}pn)_2^{2+} + trans\text{-}Pt(en)_2Cl_2^{2+} + 2\,X^- \rightleftharpoons trans\text{-}Pt(l\text{-}pn)_2X_2^{2+} + Pt(en)_2^{2+} + 2\,Cl^-$$

where X^- is Br^-, CNS^-, CN^-, Cl^-, CNO^- and NO_2^-, half times of minutes, whereas for $X^- = OH^-$, SO_4^{2-}, ClO_4^-, $C_2H_3O_2^-$, F^-, and NO_3^- the half times are hours. These reactions, which involve exchange and substitution, are also thought to occur *via* a bridged dimer mechanism. Polarimetry was again used to obtain the experimental results.

Ellison *et al.*[8] have reported that the substitution of one nitrite ion for one chloride ion in the Pt(IV) species, *trans*- and *cis*-$Pt(NH_3)_4Cl_2^{2+}$ and *trans*-$Pt(en)_2Cl_2^{2+}$, is catalysed by Pt(II). In the absence of light, a rate law,

$$\text{rate} = k'[Pt(IV)][Pt(II)][NO_2^-]$$

and a mechanism similar to that proposed for the chloride exchange reactions with Pt(IV) was thought to occur. The values of k' ($l^2.mole^{-2}.sec^{-1}$) at 50 °C are: *trans*-$Pt(en)^2Cl_2^{2+}$, 10.1; *cis*-$Pt(NH_3)_4Cl_2^{2+}$, 2.1×10^{-3}; and *trans*-$Pt(NH_3)_4Cl$, 3.25. The species *trans*-$Pt(tetrameen)_2Cl_2$ was found not to undergo substitution.

Other platinum(II) catalysed substitution reactions which have been reported are

$$trans\text{-}Pt(NH_3)_4Cl_2^{2+} + Br^- = trans\text{-}Pt(NH_3)_4ClBr^{2+} + Cl^- \quad (1)$$

$$trans\text{-}Pt(NH_3)_4ClBr^{2+} + Br^- = trans\text{-}Pt(NH_3)_4Br_2^{2+} + Cl^- \quad (2)$$

$$trans\text{-}Pt(NH_3)_4Br_2^{2+} + Cl^- = trans\text{-}Pt(NH_3)_4BrCl^{2+} + Br^- \quad (3)$$

$$trans\text{-}Pt(NH_3)_4BrCl^{2+} + Cl^- = trans\text{-}Pt(NH_3)_4Cl_2^{2+} + Br^- \quad (4)$$

$$trans\text{-}Pt(NH_3)_4Cl_2^{2+} + NH_3 = Pt(NH_3)_5Cl^{3+} + Cl^- \quad (5)$$

$$Pt(NH_3)_5I^{3+} + Cl^- = trans\text{-}Pt(NH_3)_4ICl^{2+} + NH_3 \quad (6)$$

$$Pt(NH_3)_5I^{3+} + Br^- = trans\text{-}Pt(NH_3)_4IBr^{2+} + NH_3 \quad (7)$$

$$Pt(NH_3)_5I^{3+} + I^- = trans\text{-}Pt(NH_3)_4I_2^{2+} + NH_3 \quad (8)$$

$$Pt(NH_3)_5Br^{3+} + Br^- = trans\text{-}Pt(NH_3)_4Br_2^{2+} + NH_3 \quad (9)$$

$$Pt(NH_3)_5Cl^{3+} + Cl^- = trans\text{-}Pt(NH_3)_4Cl_2^{2+} + NH_3 \quad (10)$$

for which rate data has been obtained using spectrophotometric methods, by Rettew and Johnson[9] for reactions (1), (2), (3) and (4), Johnson and Berger[10] for reaction (5) and Mason and Johnson[11] for reactions (6), (7), (8), (9) and (10). Values of the apparent rate coefficients ($l^2.mole^{-2}.sec^{-1}$) at 25 °C for reactions (1) to (10), with the corresponding activation enthalpies ($kcal.mole^{-1}$) and entropies ($cal.deg^{-1}.mole^{-1}$) and ionic strength in parentheses, are : 108(8, −24; $\mu = 0.2\ M$), 1.9×10^4 (3, −30; $\mu = 0.2\ M$), 6.3 (11, −20; $\mu = 0.2\ M$), 4.2×10^3 (6, −22; $\mu = 0.2\ M$), 1.21 (6, −37; $\mu = 0.375\ M$), 5.6×10^2 (11, −10; $\mu = 0.016\ M$), 1.2×10^4 (8, −15; $\mu = 0.016\ M$), 3.9×10^2 (6, −29; $\mu = 0.016\ M$), 12 (10, −19; $\mu = 3.2\ M$) and 1.2×10^{-3} (18, −13; $\mu = 3.2\ M$), respectively. In all the above reactions a halide-bridged species participates. Johnson and Berger[16] have made some rate measurements on the similar catalysed system

$$trans\text{-}Pt(NH_3)_4pyCl^{3+} + Cl^- \rightleftharpoons trans\text{-}Pt(NH_3)_4Cl^{2+} + py$$

In dimethylformamide solution, platinum exchange between $Pt(en)Br_2$ and $Pt(en)Br_4$ has been found to be light-sensitive and catalysed by bromide ions. In the absence of light, a value $< 1.33 \times 10^{-2}$ $l.mole^{-1}.sec^{-1}$ was obtained for the observed rate coefficient (25 °C) from data obtained by the isotopic method (^{195}Pt) and separation using an ether precipitation of $Pt(en)Br_2$. The bridge mechanism was thought to occur[2].

In methanol solution reactions of the type (diars = *o*-phenylene*bis*dimethylarsine)

$$trans\text{-}Pt(PEt_3)_2X_4 + Pt(diars)_2^{2+} + Y^- = Pt(diars)_2XY^{2+} + trans\text{-}Pt(PEt_3)_2X_2 + X^-$$

where X^- and Y^- are either chloride or bromide ions, also occur *via* a halide bridge mechanism. Peloso and Ettore[12], who have made a study of these reactions using a spectrophotometric method, have found a rate law

$$\text{rate} = k_1' K[Pt(diars)_2^{2+}][Pt(PEt_3)_2X_4][Y^-]/(1+K[Y^-])$$

where k_1' is an apparent rate coefficient and K the constant of the equilibrium

$$Pt(diars)_2^{2+} + Y^- \rightleftharpoons Pt(diars)_2Y^+$$

to be applicable. Values of k_1' at 30 °C ($\mu = 0.0385\ M$) and the associated activation parameters ($\Delta H^\ddagger$ and $\Delta S^\ddagger$) obtained by these workers are: 4.14×10^{-1} l.mole^{-1}.sec^{-1}, 6.4 kcal.mole^{-1} and -39 cal.deg^{-1}.mole^{-1}($X^- = Cl^-$, $Y^- = Cl^-$) 7.65 l.mole^{-1}.sec^{-1}, 5.3 kcal.mole^{-1} and -41 cal.deg^{-1}.mole^{-1} ($X^- = Cl^-$, $Y^- = Br^-$); 2.65×10^3 l.mole^{-1}.sec^{-1}, 3.4 kcal.mole^{-1} and -32 cal.deg^{-1}.mole^{-1} ($X^- = Br^-$, $Y^- = Cl^-$); and 3.9×10^3 l.mole^{-1}.sec^{-1}, 3.8 kcal.mole^{-1} and -30 cal.deg^{-1}.mole^{-1} ($X^- = Br^-$, $Y^- = Br^-$). The mechanism proposed was

$$trans\text{-}Pt(PEt_3)_2X_4 + Pt(diars)_2Y^+ \rightleftharpoons X_3(PEt_3)_2Pt\text{–}X\text{–}Pt(diars)_2Y^+$$

$$X_3(PEt_3)_2Pt\text{–}X\text{–}Pt(diars)_2Y^+ \rightarrow trans\text{-}Pt(PEt_3)_2X_2 + Pt(diars)_2XY^{2+} + X^-$$

The species *trans*-$Pt(PEt_3)_2X_2$ and $Pt(diars)_2XY^{2+}$ can undergo non-catalysed substitution by the ion Y^-. Reactions with $Y^- = I^-$ have also been studied; the final products are $Pt(PEt_3)_2I_2$ and I_3^-.

Reports have recently appeared in the literature on studies of the catalysed reactions

$$trans\text{-}Pt(dien)NH_3Cl_2^{2+} + 2Br^- = trans\text{-}Pt(dien)NH_3Br_2^{2+} + 2Cl^- \quad (11)$$

$$trans\text{-}Pt(CN)_4Br_2^{2-} + Cl^- = trans\text{-}Pt(CN)_4ClBr^{2-} + Br^- \quad (12)$$

$$trans\text{-}Pt(en)(NO_2)_2Cl_2 + 2Br^- = trans\text{-}Pt(en)(NO_2)_2Br_2 + 2Cl^- \quad (13)$$

$$trans\text{-}Pt(NO_2)_4Br_2^{2-} + Cl^- = trans\text{-}Pt(NO_2)_4ClBr^{2-} + Br^- \quad (14)$$

by Mason[13], reactions (12) and (14), and Syamal and Johnson[14], reactions (11) and (13); and the reactions

$$trans\text{-}Pt(dien)BrCl_2^+ + Pt(dien)Br^+ + 2Br^- = trans\text{-}Pt(dien)Br_3^+ + Pt(dien)Br^+ + 2Cl^- \quad (15)$$

$$trans\text{-}Pt(dien)NO_2Cl_2^+ + Pt(dien)Br^+ + 2Br^- = trans\text{-}Pt(dien)Br_3^+ + Pt(dien)NO_2^+ + 2Cl^- \quad (16)$$

$$trans\text{-}Pt(NH_3)_4Cl_2^{2+} + Pt(dien)Br^+ + 2Br^- = trans\text{-}Pt(dien)Br^+ + Pt(NH_3)_4^{2+} + 2Cl^- \quad (17)$$

$$trans\text{-}Pt(NH_3)_4Cl_2^{2+} + Pt(dien)NO_2^+ + 2Br^- = trans\text{-}Pt(dien)NO_2Br_2^+ + Pt(NH_3)_4^{2+} + 2Cl^- \quad (18)$$

$$trans\text{-}Pt(NH_3)_4Cl_2^{2+} + Pt(dien)NH_3^{2+} + 2Br^- = trans\text{-}Pt(dien)NH_3Br_2^{2+} + Pt(NH_3)_4^{2+} + 2Cl^- \quad (19)$$

$$trans\text{-}Pt(dien)NH_3Cl_2^{2+} + Pt(dien)Br^+ + 2Br^- = trans\text{-}Pt(dien)Br_3^+ + Pt(dien)NH_3^{2+} + 2Cl^- \quad (20)$$

by Bailey and Johnson[15]. For reactions (13) to (20) a rate law

$$\text{rate} = k'[\text{Pt(IV)}][\text{Pt(II)}][\text{X}]$$

where X is either the chloride or bromide ion, has been obtained[13,14]. For reactions (11) and (12) the rate laws obtained were

$$\text{rate} = (k'[\text{Br}]+k''[\text{Br}]^2)[\text{Pt(IV)}][\text{Pt(II)}]$$

and

$$\text{rate} = (k_1+k'[\text{Cl}^-])[\text{Pt(IV)}][\text{Pt(II)}]$$

respectively.

For reactions (11)–(20) the values of the rate coefficients (k') ($l^2.mole^{-2}.sec^{-1}$) with the associated activation enthalpy ($kcal.mole^{-1}$) and entropy ($cal.deg.^{-1}mole^{-1}$) and ionic strength in parentheses are: 3.0×10^2 (4.2, −33; $\mu = 0.1\ M$)[14], 7.2×10 (2.5, −42; $\mu = 1.01\ M$)[13], 1.64×10^2 (5.8, −29; $\mu = 0.1\ M$)[13], 7.6×10^{-1} (−1.6, −64; $\mu = 1.01\ M$)[14] at 25 °C, and 2.0×10^2 (6.2, −27; $\mu = 0.2\ M$), 9.3×10^2 (5.0, −28; $\mu = 0.2\ M$), 1.3×10^2 (7.6, −23; $\mu = 0.2\ M$), 1.2×10^2 (8.4, −21; $\mu = 0.2\ M$), 4.0×10^2 (5.9, −27; $\mu = 0.2\ M$), and 1.7×10^2 (8.0, −21; $\mu = 0.2\ M$)[15] at 24.2 °C, respectively. The corresponding values associated with the rate coefficients k_1 and k'' (at 25 °C) are 7.8 $l.mole^{-1}.sec^{-1}$ (−0.5,

-56; $\mu = 1.01$ M)[13] and 1.7×10^3 $l^3.mole^{-3}.sec^{-1}$ (5.9, -24; $\mu = 0.1$ M)[14], respectively.

Data has also been reported concerning some Pt(II) catalysed reactions for aqueous methanol and aqueous dioxane media[14].

9. Cerium and europium

9.1 THE EXCHANGE REACTION BETWEEN Ce(IV) AND Ce(III)

Ion migration[1] and diffusion separation[2] techniques, in conjunction with the isotopic method (^{141}Ce and ^{144}Ce) have led to complete exchange being observed in both sulphate and perchlorate media. However, in 6 M nitric acid solution with a separation involving an ether extraction of Ce(IV) an observed rate coefficient (25 °C) of 0.27 $l.mole^{-1}.sec^{-1}$ and an activation energy of 13.4 $kcal.mole^{-1}$ have been reported[2].

As a result of a more detailed study, Gryder and Dodson[3] have found, for media $\sim$ 6 M in (a) nitrate and (b) perchlorate, an order with respect to Ce(III) (1.7×10^{-3} to 1.9×10^{-2} M) of unity. The order with respect to Ce(IV) (4.25×10^{-3} to 4.25×10^{-2} M), lay between zero and one in perchlorate media, and was about 0.90 in nitrate media. For perchlorate solution the rate law obtained was

$$\text{rate} = k'[\text{Ce(III)}] + k_1'[\text{Ce(IV)}][\text{Ce(III)}]$$

For media 6.18 M [$HClO_4$] at 0 °C, k' and k_1' have values $\sim 8.3 \times 10^{-5}$ sec^{-1} and 6.7×10^{-3} $l.mole^{-1}.sec^{-1}$ with activation energies for these steps (k' and k_1') of 19.4 and 16.8 $kcal.mole^{-1}$, respectively. The mechanism proposed involved the exchange steps

$$\text{Ce(IV)} + \text{Ce(III)} \rightarrow$$

$$\text{Ce(IV)} + \text{Ce(III)}^* \rightarrow$$

and an excitation process, which depends on the environment

$$\text{Ce(III)} + \text{S} \rightleftharpoons \text{Ce(III)}^* + \text{S}$$

From the dependence of the rate on [H^+] it was concluded that hydrolysed species are involved in the excitation step. For nitrate media where the step associated with k_1' predominates under the experimental conditions, the effect of variation in the [H^+] (1–6 M) at $\mu = 6.18$ M was studied and results were interpreted on the basis of a rate law

$$\text{rate} = k_1'[\text{Ce(IV)}][\text{Ce(III)}]$$

The variation in the rate coefficient k_1', *viz.*

$$k_1' = k_1 + k_1''[H^+]^{-2}$$

was consistent with two exchange pathways, one involving fully aquated ions and one hydrolysed species. At 0 °C k_1 and k_1'' have values of 2.5×10^{-2} l.mole^{-1}.sec^{-1} and 0.3 mole.l^{-1}.sec^{-1} with activation energies of 7.7 and 24 kcal.mole^{-1}, respectively.

Using a separation method based on the extraction of Ce(IV) by butyl phosphate Parchen and Duke[4] have made a further study of this exchange reaction. In perchlorate media ($\sim$ 6 M), at 0 °C, with Ce(IV) and Ce(III) in the ranges 8×10^{-4} to 10^{-2} M and 8×10^{-4} to 8×10^{-3} M, respectively, the exchange data indicated a rate law

$$\text{rate} = k_1'[\text{Ce(III)}][\text{Ce(IV)}] + k_2'[\text{Ce(III)}][\text{Ce(IV)}]^2$$

Variation in the hydrogen-ion concentration, 5.85 to 2.04 M at $\mu = 5.9$ M ($NaClO_4$), gave data in accordance with the participation of hydrolysed species of Ce(IV), the rate at low concentrations of H^+ being much faster than at high $[H^+]$. Two pathways were suggested

$$\text{Ce(OH)}_3^+ + \text{Ce(III)} \xrightarrow{k_1}$$

$$\text{CeOCeOH}^{5+} + \text{Ce(III)} \xrightarrow{k_2}$$

It is interesting to note that a dimer of Ce(IV) has also been invoked to account for observations on this exchange system at a platinum surface[5]. The rate of exchange between the fully aquated ions of Ce(IV) and Ce(III) was concluded to be relatively slow[4].

Observations, made by Sigler and Masters[6], on the exchange in 0.4 M H_2SO_4 solution, have led to a rate law

$$\text{rate} = k_{\text{obs}}[\text{Ce(IV)}][\text{Ce(III)}]$$

which was obeyed over the ranges Ce(IV), 10^{-4} to 10^{-3} M and Ce(III), 5×10^{-4} to 1.1×10^{-2} M. k_{obs} at 0 °C has a value $\sim 4.2 \times 10^{-1}$ l.mole^{-1}.sec^{-1}. These authors have also reported their results for the exchange reaction induced by hydrogen peroxide. A mechanism

$$H_2O_2 + \text{Ce(IV)} \rightleftharpoons HO_2 + \text{Ce(III)} + H^+$$

$$HO_2 + \text{Ce(IV)} \rightarrow \text{Ce(III)} + O_2 + H^+$$

References pp. 142–152

was proposed. The isotopic method (^{144}Ce) and *n*-butylphosphate separations were used.

Challenger and Masters[7] have found an X-ray induced cerium exchange reaction, which is faster than the normal exchange reaction, occurs in nitric and sulphuric acid media.

In nitrate media (~ 6 *M*), fluoride ion has a catalytic effect on the exchange reaction between Ce(IV) and Ce(III). Hornig and Libby[8] have made a detailed study of this effect, over the range of added KF, 0 to 8.4×10^{-4} *M*, and have concluded that a pathway involving a monofluoro complex occurs, possibly involving a fluoride-bridged activated complex.

9.2 THE EXCHANGE REACTION BETWEEN Eu(III) AND Eu(II)

The isotope ^{152}Eu has been used as a tracer in the study of this exchange system[1,2]. In view of the instability of the Eu(II) species the absence of both light and oxygen are necessary. Separation of the two species, Eu(III) and Eu(II), was achieved using a hydroxide precipitation of the trivalent europium, brought about by the addition of aqueous ammonia. Observations on the exchange in perchlorate media were found to be obscured by the ClO_4^- oxidation of Eu(II). However, in the presence of chloride ions the exchange rate could be measured in perchlorate-containing media. A rate law for constant ionic strength (2.0 *M*)

$$\text{rate} = k_{obs}[\text{Eu(III)}][\text{Eu(II)}][\text{Cl}^-]$$

was found to be obeyed. The concentration ranges employed were Eu(III) 3.5×10^{-2} to 1.2×10^{-1} *M*, Eu(II) 2.6×10^{-2} to 6.8×10^{-2} *M* and Cl^- 9×10^{-2} to 1.86 *M*. The value of k_{obs} ($\mu = 2.0$ *M* and 39.4 °C) is 1.8×10^{-3} $l^2.mole^{-2}.sec^{-1}$, almost independent of hydrogen-ion concentration. The associated activation energy obtained over the range 32 to 50 °C was 20.8 $kcal.mole^{-1}$. Meier and Garner[2] have proposed the exchange pathway

$$EuCl^{2+} + Eu^{2+} \rightarrow$$

which is much more rapid than any involving the fully aquated or hydrolysed ions of Eu(III) and Eu(II), to occur.

10. Uranium, neptunium, plutonium and americium

10.1 EXCHANGE REACTIONS BETWEEN URANIUM IONS

The exchange reaction between U(VI) and U(IV) in aqueous sulphate[1], chloride[2] and perchlorate[3] media and in various mixed aqueous–organic solvents containing hydrochloric acid[4-7] has been studied. The isotopic method (with ^{233}U as the tracer) was used with separations of the two oxidation states achieved by (*a*) precipitation of the U(IV) as either the cupferron complex[4,7] or as the fluoride UF^2 [4,5,6] and (*b*) aqueous extraction of U(IV) after the addition of thenoyltrifluoroacetone in benzene[3].

Betts[1] obtained evidence for the exchange in sulphate media being catalysed by light of wavelengths ~ 340 mμ, the region of absorption of U(VI). In the presence of light a rate law, approximately

$$\text{rate} = k'[\text{U(VI)}]^{\frac{1}{2}}[\text{U(IV)}]^{\frac{1}{2}}[\text{H}^+]^{-\frac{1}{2}}$$

and activation energy of 8.5 kcal.mole^{-1} were found.

Rona[2] has obtained results on the exchange in chloride media over the ranges [U(VI)] 2.4×10^{-2} to 1.5×10^{-1} *M*, [U(IV)] 2.5×10^{-2} to 6.4×10^{-2} *M* and $[H^+]$ 2.5×10^{-2} to 1.4×10^{-1} *M*, which are consistent with a rate law

$$\text{rate} = k''[\text{U(VI)}][\text{U(IV)}]^2[\text{H}^+]^{-\frac{1}{2}}$$

The rate was found to be unaffected by light and either $[Cl^-]$ or $[ClO_4^-]$. The overall activation energy obtained was 33.4 kcal.mole^{-1}. The mechanism of exchange suggested by Rona[2] was

$$U^{4+} + H_2O \rightleftharpoons UOH^{3+} + H^+ \qquad (K)$$

$$UOH^{3+} + UO_2^{2+} + 2\,H_2O \rightleftharpoons \left(\begin{array}{c}\text{O=U–O–U–OH}\\ \quad|\qquad\;|\\ \quad\text{OH}\quad\text{OH}\end{array}\right)^{3+} + 2\,H^+$$

$$\left(\begin{array}{c}\text{O=U–O–U–OH}\\ \quad|\qquad\;|\\ \quad\text{OH}\quad\text{OH}\end{array}\right)^{3+} + UOH^{3+} \rightarrow \text{exchange}$$

Masters and Schwartz[3] have made a study of the exchange in perchlorate media and have found two exchange pathways operate. One pathway involves exchange *via* the process

$$\text{U(VI)} + \text{U(IV)} \rightleftharpoons 2\text{U(V)}$$

with an activation energy of 38 kcal.mole^{-1}, the other *via* the pathway observed for hydrochloric acid media with an activation energy of ~ 28 kcal.mole^{-1}. For the former pathway a rate law

$$\text{rate} = k_{\text{obs}}[\text{U(VI)}][\text{U(IV)}]$$

where $k_{\text{obs}} = k'''([\text{H}^+]^3 + K[\text{H}^+]^2)^{-1}$, was found to be obeyed over the range of

References pp. 142–152

conditions [U(VI)] 2×10^{-3} to 4×10^{-2} *M*, [U(IV)] 1×10^{-4} to 5×10^{-3} *M*, $[H^+]$ 6×10^{-2} to 2×10^{-1} *M*, at temperatures of 25.1 to 47.1 °C. At 25.1 °C and $\mu = 2.0$ *M*, k''' has a value of 2.13×10^{-7} $mole^2.l^{-2}.sec^{-1}$. Values of the activation parameters of 37.5 $kcal.mole^{-1}$ and -36 $cal.deg^{-1}.mole^{-1}$ were calculated. Ultraviolet radiation was found to increase the exchange rate in perchlorate media.

Amis *et al.*[4-7], have found with various mixtures of aqueous hydrochloric acid with ethanol[4,5], acetone[6], or ethylene glycol[7], exchange dependences with respect to U(VI), U(IV) and H^+ which vary with the organic solvent content. Values of 2.70, 0 and -1.26 (100 % ethanol) and 0.85, 0.66 and -0.53 (100 % ethylene glycol) have been reported for the orders with respect to U(VI), U(IV) and H^+. With aqueous ethanol the activation energy was also found to depend upon the solvent content; light was found to have no effect on the exchange rate[5]. Changes in the mechanism are thought to occur[4,6,7].

The rate coefficient for the exchange between U(VI) and U(V) has been estimated as 52 $l.mole^{-1}.sec^{-1}$, at 25 °C in 1.0 *M* $HClO_4$, by Gordon and Taube[8], from observations of the U(V)-catalysed water exchange reaction of aqueous U(VI).

10.2 REACTIONS BETWEEN URANIUM IONS

Numerous studies have been made on the forward reaction of the equilibrium,

$$2\,U(V) \rightleftharpoons U(VI) + U(IV)$$

using polarographic[1-7], potentiometric[8,9] and spectrophotometric techniques[10-12]. A majority of the investigations led to the rate law

$$-d[U(V)]/dt = k_1'[H^+][U(V)]^2$$

and a mechanism

$$UO_2^+ + H^+ \rightleftharpoons UOOH^{2+}$$

$$UO_2^+ + UOOH^{2+} \longrightarrow UO_2^{2+} + UOOH^+$$

$$UOOH^+ \longrightarrow U(IV)$$

The value of the rate coefficient k_1' ($l^2.mole^{-2}.sec^{-1}$) has been calculated as 130 ($\mu = 0.4$ *M*)[1], 1800 ($\mu = 3.8$ *M*)[2], 156 ($\mu = 0.5$ *M*)[4], 260 ($\mu = 0.5$ *M*, D_2O)[4], 143 ($\mu = 0.4$ *M*)[5], 436 ($\mu = 2.1$ *M*)[6], 192 ($\mu = 1.0$ *M*)[7], 417 ($\mu = 2.0$ *M*)[9], 150 ($\mu = 0.5$ *M*)[11] and 135 ($\mu = 1.0$ *M*)[11] at 25 °C in perchlorate media. Imai[6] has also reported that addition of the anions Cl^-, Br^-, I^- and NO_3^- accelerates this reaction. Activation parameters $\Delta H^{\ddagger}$ (kcal.mole^{-1}) and $\Delta S^{\ddagger}$ (cal.deg^{-1}.mole^{-1}) for $\mu = 2.1$ *M* of 10.2 and -17 (perchlorate), 12.7 and -2.7 (0.1 *M* added chloride) and 14.8 and $+0.4$ (1 *M* added bromide) have been calculated. For chloride media, corresponding values of k_1' (l^2.mole^{-2}.sec^{-1}) of 175 ($\mu = 0.5$ *M*)[2] and $\sim$ 100 ($\mu = 0.15$ *M*)[3] at 25 °C have been obtained. With 0.5 *M* sulphate media a value of 1200 l^2.mole^{-2}.sec^{-1} at 30.4 °C has been found[8], with the corresponding activation energy 9.5 kcal.mole^{-1}.

Newton and Baker[12], using wavelengths of 737 mμ or 648 mμ (U(IV) absorption), have obtained similar overall rate coefficients to those obtained by polarography[5,6]. However, they have found U(VI) to retard the disproportionation reaction due to the presence of an additional process

$$\mathrm{U(VI)+U(V)} \rightleftharpoons \mathrm{U_2(XI)} \qquad K_2$$

$$\mathrm{U_2(XI)+U(V)} \xrightarrow{k_2} \mathrm{2\,U(VI)+U(IV)}$$

The dimer U_2(XI), $U_2O_4^{3+}$, was detected by its absorption at 737 mμ. Analysis of spectrophotometric data using the rate law

$$-\mathrm{d[U(V)]/d}t = (k_1'[\mathrm{H^+}]+k_2 K_2[\mathrm{U(VI)}])[\mathrm{U(V)}]^2/(1+K_2[\mathrm{U(VI)}])^2$$

led to values of k_1' (l^2.mole^{-2}.sec^{-1}), k_2 (l.mole^{-1}.sec^{-1}) and K_2 (l.mole^{-1}) at 25.1 °C and $\mu = 2.0$ *M* with [H$^+$] 5×10^{-2} to 2×10^{-1} *M* of $\sim$ 500, $\sim$ 14 and $\sim$ 16.7, respectively. The values of these coefficients show some dependence on [H$^+$]; various reasons for this effect, including medium effects, have been fully discussed by these workers. Activation parameters for the step associated with k_1' of 11 kcal.mole^{-1} ($\Delta H^{\ddagger}$) and -11 cal.deg^{-1}.mole^{-1} ($\Delta S^{\ddagger}$) have been evaluated.

The reverse reaction of the disproportionation equilibrium has been investigated by Masters and Schwartz[13]. It provides a pathway for the exchange of U(VI) and U(IV) as has been previously mentioned.

10.3 EXCHANGE REACTIONS BETWEEN NEPTUNIUM IONS

The exchange reaction between Np(VI) and Np(V) has been investigated using the isotopic method (^{239}Np) and an extraction separation (Np(VI) with tributylphosphate or thenoyltrifluoroacetone in toluene)[1-3]. Cohen *et al.*[1] have found, for the exchange in perchlorate media, a rate law

$$\mathrm{rate} = k_{\mathrm{obs}}[\mathrm{Np(VI)}][\mathrm{Np(V)}]$$

to be obeyed over the ranges [Np(VI)] 1.8×10^{-5} to 9.27×10^{-5} *M*, [Np(V)]

2.6×10^{-5} to 1.04×10^{-4} M, and $[H^+]$ 0.32 to 0.99 M. At $\mu = 1.0$ M and a temperature of 0 °C, k_{obs} has a value of ~ 30 l.mole^{-1}.sec^{-1}, with an associated energy and entropy of activation of 8.3 kcal.mole^{-1} and -24 cal.deg^{-1}.mole^{-1}, respectively.

Cohen *et al.*[2] and Sullivan *et al.*[3], in further articles, report some data obtained under higher ionic strength conditions and have suggested that k_{obs} depends on $[H^+]$ in the manner[3]

$$k_{obs} = k_1 + k_2'[H^+]$$

Values obtained for the coefficients k_1 and k_2' at 4.5 °C ($\mu = 3.0$ M) were 73.5 l.mole^{-1}.sec^{-1} and 15 l^2.mole^{-2}.sec^{-1}, respectively, with corresponding activation energies of 12 and 15 kcal.mole^{-1}.

In deuterated solvent the rate of exchange was found to be lower than in aqueous media; the ratio $k_{obs}(H_2O)/k_{obs}(D_2O)$ was found to be dependent on the acid concentration. Sullivan *et al.*[3] have suggested a hydrogen atom transfer process and a water bridging process

$$NpO_2^+ + H^+ \rightleftharpoons NpO_2H^{2+} \qquad K_2$$

$$NpO_2H^{2+} + NpO_2^{2+} \xrightarrow{k_2} \text{exchange}$$

$$NpO_2^{2+} + NpO_2^+ + H_2O \xrightarrow{k_1} \text{exchange}$$

to account for the two exchange pathways.

Cohen *et al.*[2] have studied the effect of the addition of both chloride and nitrate ions (up to 3 M at $\mu = 3.0$ M) on this exchange. Whereas nitrate ion was found to have vertually no effect on the exchange rate, chloride ion had an accelerating influence. For exchange in the presence of chloride ions rate data was found to be fitted by an expression

$$\text{rate} = k_0'[NpO_2^{2+}][NpO_2^+] + k_3[NpO_2Cl^+][NpO_2^+] + k_4[NpO_2Cl_2]\,[NpO_2^+]$$

with values of the rate coefficients k_0', k_3 and k_4 at 0 °C ($\mu = 3.0$ M, $[H^+] = 3.0$ M) of 88.8, 215, and 86 l.mole^{-1}.sec^{-1}, respectively. The energies and entropies of activation were found to have values of 10.6 kcal.mole^{-1}, -12.6 cal.deg^{-1}.mole^{-1} (k_0'), 15.4 kcal.mole^{-1}, 6.7 cal.deg^{-1}.mole^{-1} (k_3) and 15.3 kcal.mole^{-1}, 5 cal.deg^{-1}.mole^{-1} (k_4). A chlorine-atom transfer mechanism has been discussed.

Cohen *et al.*[4], have also made some observations on the exchange in water–sucrose and water–ethylene glycol mixed solvents containing perchloric acid (0.106 M). Over a range of dielectric constant 68 to 88, no alteration in the exchange rate was observed.

Sullivan *et al.*[5], have studied the perchlorate media exchange reaction between Np(V) and Np(IV) using the isotopic method (^{239}Np tracer) and a separation based on the extraction of Np(IV) with thenoyltrifluoacetone in benzene. Data obtained over the ranges [Np(V)] 5×10^{-3} to 4.9×10^{-2} *M*, [Np(IV)] 3.9×10^{-3} to 3.2×10^{-2} *M* and [H$^+$] 4×10^{-2} to 4.5×10^{-1} *M* were found to fit a rate expression

$$\text{rate} = k'_5[\text{Np(V)}]^2[\text{H}^+] + k'_6[\text{Np(V)}]^{\frac{1}{2}}[\text{Np(IV)}]^{\frac{3}{2}}[\text{H}^+]^{-2}$$

The rate coefficients k'_5 and k'_6 at 25 °C ($\mu = 1.2$ *M*) have values and corresponding activation parameters of 1.10×10^{-5} l^2.mole^{-2}.sec^{-1}, 17.6 kcal. mole^{-1}, -22 cal.deg^{-1}.mole^{-1} and 6.45×10^{-8} mole.l^{-1}.sec^{-1}, 36.8 kcal.mole^{-1}, 32 cal.deg^{-1}mole^{-1}, respectively. Some surface catalysis was found; the above results relate to teflon reaction vessels. The processes suggested by these authors were, for the path associated with k'_5,

$$2\,\text{NpO}_2^+ + \text{H}^+ \rightleftharpoons \text{NpO}_2^{2+} + \text{NpO}_2\text{H}^+$$

$$\text{NpO}_2\text{H}^+ + \text{Np(OH)}_2^{2+} \longrightarrow \text{exchange}$$

and for the path associated with k'_6

$$\text{Np}^{4+} + \text{NpO}_2^+ \rightleftharpoons \text{NpO}_2^{2+} + \text{Np}^{3+}$$

$$\text{Np(OH)}_2^{2+} + \text{NpO}_2^{2+} \longrightarrow \text{exchange}$$

with the species Np(OH)_2^{2+} produced *via* the equilibria

$$\text{Np}^{4+} + \text{H}_2\text{O} \rightleftharpoons \text{NpOH}^{3+} + \text{H}^+$$

$$\text{NpOH}^{3+} + \text{H}_2\text{O} \rightleftharpoons \text{Np(OH)}_2^{2+} + \text{H}^+$$

The above processes are related to the disproportionation reaction of Np(V) which has also been investigated. However exchange rate constants and activation parameters are different from those predicted from disproportionation data[5].

10.4 REACTIONS BETWEEN NEPTUNIUM IONS

The reaction between Np(V) and Np(III)

$$\text{NpO}_2^+ + \text{Np}^{3+} \overset{k_1}{\rightleftharpoons} 2\,\text{Np}^{4+}$$

in the absence of oxygen in perchlorate media has been studied by Hindeman *et al.*[1], using a spectrophotometric method (723 mμ, Np(IV) absorption). The rate law governing the reaction

$$-\mathrm{d}[\mathrm{NpO_2^+}]/\mathrm{d}t = k_1'[\mathrm{NpO_2^+}][\mathrm{Np^{3+}}][\mathrm{H^+}]$$

was found to be obeyed over the ranges $[NpO_2^+]$ 2.5×10^{-4} to 1.0×10^{-3} M, $[Np^{3+}]$ 2.98×10^{-4} to 1.04×10^{-3} M and $[H^+]$ 1.08×10^{-1} to 1.98 M. At 25 °C and $\mu = 2.0$ M, k_1' has a value 43.2 $l^2.mole^{-2}.sec^{-1}$ (H_2O solution) and 34.3 $l^2.mole^{-2}.sec^{-1}$ (D_2O solution). The activation energies obtained in this study were 6.52 (H_2O) and 5.62 (D_2O) $kcal.mole^{-1}$. The mechanism suggested by these workers involves reaction of the species $Np(O)(OH)^{2+}$, which is produced by the equilibrium

$$\mathrm{NpO_2^+ + H^+ \rightleftharpoons Np(O)(OH)^{2+}} \qquad K_1$$

and Np^{3+}. The equilibrium reaction between Np(VI) and Np(IV)

$$\mathrm{Np(VI) + Np(IV)} \underset{k_4}{\overset{k_3}{\rightleftharpoons}} 2\ \mathrm{Np(V)}$$

in aqueous perchlorate, ethylene glycol–water perchlorate and sulphate media has also been investigated by Hindeman *et al.*[2-5]. The rate law controlling the forward reaction (k_3) was found from spectrophotometric data [724 mμ for Np(IV) or 983 mμ for Np(V)] for perchlorate media to be

$$-\mathrm{d}[\mathrm{Np(IV)}]/\mathrm{d}t \text{ or } +\mathrm{d}[\mathrm{Np(V)}]/\mathrm{d}t = k_{\mathrm{obs}}[\mathrm{NpO_2^{2+}}][\mathrm{Np^{4+}}]^a$$

where $k_{obs} = k_3'[H^+]^{-2}$, and a has a value of $\frac{1}{3}$ in ethylene glycol–perchlorate media[5] and 1 in aqueous perchlorate media[2]. The rate was also found to be dependent on the ethylene glycol concentration. In media 12.1 M ethylene glycol k_3' has a value of 1.58×10^{-1} $mole^{\frac{5}{3}}.l^{-}.^{\frac{5}{3}}sec^{-1}$ (25 °C); in aqueous perchlorate the corresponding value is 4.5×10^{-2} $mole.l^{-1}.sec^{-1}$ (at 25 °C, $\mu = 2.0$ M), with activation parameters ($\Delta H^\ddagger$ and $\Delta S^\ddagger$) of 24.6 $kcal.mole^{-1}$ and 17.8 $cal.deg^{-1}.mole^{-1}$. Using the data above and that available on the equilibrium, Hindeman *et al.*[2], obtained values of k_4' (at 25 °C, $\mu = 1.0$ M) and associated activation parameters of 9.57×10^{-9} $l^3.mole^{-3}.sec^{-1}$, 17.1 $kcal.mole^{-1}$ and -38.1 $cal.deg^{-1}.mole^{-1}$ for the reverse reaction, the rate law suggested being

$$-\mathrm{d}[\mathrm{Np(V)}]/\mathrm{d}t = k_4'[\mathrm{Np(V)}]^2[\mathrm{H^+}]^2$$

The mechanism proposed by these authors was

$$Np^{4+}+H_2O \rightleftharpoons NpO^{2+}+2\,H^+ \quad \text{or}$$

$$Np^{4+}+2\,H_2O \rightleftharpoons Np(OH)_2^{2+}+2\,H^+$$

$$NpO_2^{2+}+NpO^{2+} \longrightarrow NpO_2^{+}+NpO^{3+}$$

$$NpO^{3+}+H_2O \rightleftharpoons NpO_2^{+}+2\,H^+ \quad \text{or}$$

$$NpO_2^{2+}+Np(OH)_2^{2+} \longrightarrow 2\,Np(O)(OH)^{2+}$$

$$Np(O)(OH)^{2+} \rightleftharpoons NpO_2^{+}+H^+$$

A reinvestigation of the rate dependence on $[H^+]$ has since been made[3] and it was concluded that the relation

$$k_{obs} = k_3'[H^+]^{-2}+k_3''[H^+]^{-3}$$

was more accurate. Values of k_3' and k_3'' at 24.9 °C (μ = 2.2 M) obtained were 4.27×10^{-2} mole.l^{-1}.sec^{-1} and 5.04×10^{-3} mole^2.l^{-2}.sec^{-1}, respectively.

For sulphate media much more complex rate laws for both the forward and reverse steps of the reaction were found by Sullivan *et al.*[4], the observed rate coefficient (k_{obs}) being dependent on both $[H^+]$ and $[HSO_4^-]$. For the forward reaction

$$-d[Np(IV)]/dt = k_{obs}[Np(VI)][Np(IV)]$$
$$= k_3'[NpO_2^{2+}][Np^{4+}][H^+]^{-2}+(k_7[H^+]^{-2}+k_8[H^+]^{-3})$$
$$(k_5[NpO_2^{2+}][NpSO_4^{2+}]+k_6[NpO_2SO_4][NpSO_4^{2+}])$$

The values of the coefficients k_3', k_5, k_6, k_7 and k_8 at 25 °C (μ = 2.2 M) are 4.48×10^{-2} mole.l^{-1}.sec^{-1}, 7.12×10^{-2} l.mole^{-1}.sec^{-1}, 0.119 l.mole^{-1}.sec^{-1}, 3.6 mole^2.l^{-2} and 1.7 mole^3.l^{-3}, respectively. The value of the overall activation energy for the reaction in sulphate media is 23 kcal.mole^{-1}.

For the reverse process the rate law obtained (sulphate media) was

$$-d[Np(V)]/dt = k'[Np(V)]^2[H^+]^{0.09}[HSO_4^-]^{1.32}$$
$$= (k_9[HSO_4^-]+k_{10}[HSO_4^-]^2)(k_{11}+k_{12}[H^+])[Np(V)]^2$$

Values of k_9, k_{10}, k_{11} and k_{12} at 25 °C (μ = 2.2 M) are 5×10^{-3} l^2.mole^{-2}.sec^{-1}, 1.46×10^{-3} l^3.mole^{-3}.sec^{-1}, 7.39×10^{-1} and 1.35×10^{-1} mole.l^{-1}, respectively. Possible transition states have been discussed[4].

The reactions have also been studied in deuterated media[3], the corresponding values of the terms k_3' and k_3'' are 8.56×10^{-3} mole.l^{-1}.sec^{-1} and 1.02×10^{-3} mole^2.l^{-2}.sec^{-1} (at 24.9 °C, μ = 2.2 M perchlorate media); an overall activation

References pp. 142–152

energy for the reaction of 26.85 kcal.mole^{-1} was also calculated from the experimental data. For the forward reaction in either perchlorate or sulphate media, the reaction rate ratio (H_2O: D_2O) lies between 4.33 and 5.65 depending on the temperature. For the back reaction the same ratio has a value of 0.4, and the values of the terms k_9 and k_{10} at 25 °C (μ = 2.0 M) for deuterated sulphate media are 1.17×10^{-2} l^2.mole^{-2}.sec^{-1} and 5.8×10^{-3} l^3.mole^{-3}.sec^{-1}, respectively[3].

10.5 EXCHANGE REACTIONS BETWEEN PLUTONIUM IONS

Keenan[1,2] has made an investigation of the exchange reaction between Pu(IV) and Pu(III) in perchlorate media. The isotopic method was used with an α energy analyser to separate the tracer activity (^{238}Pu) from that normally present from the major constituent (^{239}Pu). Tributylphosphate extraction of the Pu(IV) formed the basis of the separation method. It was shown that the rate law has the approximate form

$$\text{rate} = k_{obs}[\text{Pu(IV)}][\text{Pu(III)}]$$

For 0.5 M perchloric acid k_{obs} has a value 1.8×10^2 l.mole^{-1}.sec^{-1} at 0 °C[1]. From results obtained at μ = 2.0 M, [H^+] = 0.40 to 2.0 M, Keenan concluded that the pathways operative were

$$Pu^{4+} + Pu^{3+} \xrightarrow{k_1}$$

$$PuOH^{3+} + Pu^{3+} \xrightarrow{k_2}$$

the hydrolysed species $PuOH^{3+}$, being produced *via* the equilibrium

$$Pu^{4+} + H_2O \rightleftharpoons PuOH^{3+} + H^+ \qquad K_2$$

Using known values of the constant K_2, the values of k_1 and k_2 were calculated as 1.8×10^2 and 1.3×10^4 l.mole^{-1}.sec^{-1}, respectively, at 0 °C and μ = 2.0 M. The activation energies and entropies obtained for the k_1 step were 7.7 kcal.mole^{-1} and -31 cal.deg^{-1}.mole^{-1}; for the k_2 step values of 2.8 kcal.mole^{-1} and -32 cal.deg^{-1}.mole^{-1}were found.

10.6 REACTIONS BETWEEN PLUTONIUM IONS

Rabideau *et al.*[1,2] have used a spectrophotometric method [wavelengths, 600 mμ for (Pu(III)[1] and 830 mμ for (Pu(VI)[2]] to investigate the reactions

$$PuO_2^{2+} + Pu^{3+} \underset{k_2}{\overset{k_1}{\rightleftharpoons}} PuO_2^{+} + Pu^{4+}$$

The rate law

$$-d[PuO_2^{2+}]/dt = -d[Pu^{3+}]/dt = k_1(1-K/K^*)[PuO_2^{2+}][Pu^{3+}]$$

(where K and K^* are the values of $[Pu^{3+}][PuO_2^{2+}]/[PuO_2^{+}][Pu^{4+}]$ at equilibrium and at any time t, respectively) was found to be obeyed in both aqueous and deuterated perchlorate media. The calculated values of k_1 at 25 °C, ($\mu = 1.0$ M), which were found not to depend on $[H^+]$, are 2.66 $l.mole^{-1}.sec^{-1}$ (H_2O) and 1.49 $l.mole^{-1}.sec^{-1}$ (D_2O). The values of k_2 under the same conditions after correction for hydrolysis effects were calculated as 37.4 $l.mole^{-1}.sec^{-1}$ (H_2O) and 56.5 $l.mole^{-1}.sec^{-1}$ (D_2O). Activation parameters ($\Delta H^{\ddagger}$ and $\Delta S^{\ddagger}$) were: 4.82 $kcal.mole^{-1}$ and -40.4 $cal.deg^{-1}.mole^{-1}$ (k_1), 13.6 $kcal.mole^{-1}$ and -5.7 $cal.deg^{-1}.mole^{-1}$ (k_2) for H_2O solution; 4.5 $kcal.mole^{-1}$ and -42.6 $cal.deg^{-1}.mole^{-1}$ (k_1), 11.8 $kcal.mole^{-1}$ and -11 $cal.deg^{-1}.mole^{-1}$ (k_2) for D_2O solution. Neither of the rate coefficients (k_1, k_2) was affected by the addition of chloride ions.

In acetic acid media in the presence of 0.5 M $HClO_4$, a modified reaction

$$Pu(VI) + 2\,Pu(III) = 3\,Pu(IV)$$

occurs, owing to the instability of Pu(V) in this media. Alexi *et al.*[3] have used a spectrophotometric method to make a brief investigation and have suggested that the steps

$$Pu(VI) + Pu(III) \rightleftharpoons Pu(IV) + Pu(V)$$

$$Pu(V) + Pu(III) \rightarrow 2\,Pu(IV)$$

are amongst the processes occurring. The reaction appears to depend on the concentration of the Pu present.

From spectrophotometric studies on 0.5 M HCl solutions of plutonium containing Pu(V), Connick[4] attempted to investigate the disproportionation reaction of this species

$$2\,Pu(V) \underset{k'_4}{\overset{k'_3}{\rightleftharpoons}} Pu(IV) + Pu(VI)$$

From his results, Connick concluded that the reactions also occurring were

$$Pu(V) + Pu(III) \underset{k'_6}{\overset{k'_5}{\rightleftharpoons}} 2\,Pu(IV)$$

$$Pu(V) + Pu(IV) \underset{k'_1}{\overset{k'_2}{\rightleftharpoons}} Pu(VI) + Pu(III)$$

References pp. 142–152

and he was able to obtain values of the rate coefficients k'_5 and k'_3 at 25 °C of 5.8×10^{-2} and $< 2 \times 10^{-3}$ l.mole^{-1}.sec^{-1}, respectively; in conjunction with equilibrium data, estimates of k'_6 and k'_4 for the same conditions, of 3.5×10^{-4} and $< 1.2 \times 10^{-6}$ l.mole^{-1}.sec^{-1}, respectively, were made.

Rabideau[5], using EMF measurements to obtain rate data, has since reinvestigated the disproportionation reaction of Pu(V), using solutions containing very small amounts of Pu(IV) and Pu(III). He was able to conclude from the dependence of the rate coefficient (k'_3) on the first power of $[H^+]$ that the steps and equilibrium present were

$$PuO_2^+ + H^+ \rightleftharpoons Pu(O)OH^{2+}$$

$$PuO_2^+ + Pu(O)OH^{2+} \longrightarrow PuO_2^{2+} + Pu(O)OH^+$$

$$Pu(O)OH^+ \longrightarrow Pu(IV)$$

The values of k'_3 obtained from this work are 4.0×10^{-3} l.mole^{-1}.sec^{-1} (at 25 °C, $\mu = 1.0$ M, $[H^+] = 1.0$ M, in H_2O media) and 5.25×10^{-3} l.mole^{-1}.sec^{-1} (for D_2O media under the same conditions). For aqueous media, activation parameters of $\Delta H^{\ddagger} = 19$ kcal.mole^{-1} and $\Delta S^{\ddagger} = -5.8$ cal.deg^{-1}.mole^{-1} were calculated.

Rabideau[6] has also investigated, using EMF measurements, the overall reaction

$$3\ Pu(IV) \rightleftharpoons Pu(VI) + 2\ Pu(III)$$

which is governed by the rate law

$$-d[Pu(IV)]/dt = 3\ k'_6(1 - K^*/K)[Pu(IV)]^2$$

(where K and K^* are the values of $[Pu(III)]^2[Pu(VI)]/[Pu(IV)]^3$ at equilibrium and at time t, respectively) and involves the reactions defined by k'_6, k'_5, k'_2, and k'_1. From results obtained with perchlorate media of various acidities (k'_6 dependent on $[H^+]^3$) it was concluded that the equilibria and pathways operating were

$$Pu^{4+} + H_2O \rightleftharpoons PuOH^{3+} + H^+$$

$$PuOH^{3+} + H_2O \rightleftharpoons Pu(OH)_2^{2+} + H^+$$

$$PuOH^{3+} + Pu(OH)_2^{2+} \longrightarrow Pu^{3+} + PuO_2^+ + H_3O^+$$

$$PuO_2^+ + Pu^{4+} \rightleftharpoons Pu^{3+} + PuO_2^{2+}$$

The value of k'_6 at 25 °C ($\mu = 1.0$ M) obtained was 2.7×10^{-5} l.mole^{-1}.sec^{-1}. Previously, Connick and McVey[7], using data available at that time for this system had calculated a value of k'_6 of 2.5×10^{-5} l.mole^{-1}.sec^{-1} for approximately the same conditions.

Rabideau and Cowan[8] have made a similar investigation of this reaction in hydrochloric acid media and have found identical features. At 25 °C (media 1 M in HCl, $\mu = 1.0\ M$) k'_6 has the value 1.5×10^{-4} l.mole^{-1}.sec^{-1} with associated activation parameters ($\Delta H^{\ddagger}$ and $\Delta S^{\ddagger}$) of 39 kcal.mole^{-1} and 53 cal.deg^{-1}.mole^{-1}. These results show good agreement with those calculated previously by Connick McVey[7] ($k'_6 = 9.8 \times 10^{-5}$ l.mole^{-1}.sec^{-1}, $\Delta H^{\ddagger} = 40$ kcal.mole^{-1} and $\Delta S^{\ddagger} = 60$ cal.deg^{-1}.mole^{-1}) for similar conditions.

10.7 EXCHANGE REACTIONS BETWEEN AMERICIUM IONS

Keenan *et al.*[1] have carried out a brief survey on the exchange reactions

$$AmO_2^+ + Am^{3+} \rightarrow$$

$$AmO_2^{2+} + AmO_2^+ \rightarrow$$

In perchlorate media the latter reaction was found to be rapid (100 % exchange in < 60 sec) and the former reaction slow (half times of > 200 hours at 100 °C). The isotopic method was used (^{242}Am).

10.8 REACTIONS BETWEEN AMERICIUM IONS

Am(V), AmO_2^+, undergoes disproportionation and redox reactions in aqueous acidic media, which can be followed by spectrophotometry at wavelengths of ~ 812 mμ for Am(III), ~ 715 mμ for Am(V) and ~ 992 mμ for Am(VI). The earlier work[1-5] on solutions of Am(V) in hydrochloric[1], nitric[5], sulphuric[5], and perchloric acid[2-5] media, which were hindered by α-radiation reactions (^{241}Am), led to the conclusion that the rate law

$$-d[AmO_2^+]/dt = k_{obs}[AmO_2^+]^2[H^+]^4$$

was obeyed[1,3-5].

In 1962, Coleman[6], using the relatively stable ^{243}Am, was able to study the reaction of Am(V) in various media in more detail and found a rate law

$$-d[AmO_2^+]/dt = k'_1[AmO_2^+]^2[H^+]^2 + k''_1[AmO_2^+]^2[H^+]^3$$

In perchlorate media, values of k'_1 and k''_1 at 75.7 °C ($\mu = 2.0\ M$) are 7×10^{-4} l^3.mole^{-3}.sec^{-1} and 4.6×10^{-4} l^4.mole^{-4}.sec^{-1}, respectively. The rate increased in the solvent order $HClO_4 < HNO_3 < HCl < H_2SO_4$. No rate measurements in dibutylphosphoric acid media were possible.

References pp. 142–152

The stoichiometric equation found for the overall reaction was

$$3\ \mathrm{Am(V)} = 2\ \mathrm{Am(VI)} + \mathrm{Am(III)}$$

with the probable steps

$$2\ \mathrm{Am(V)} \rightleftharpoons \mathrm{Am(VI)} + \mathrm{Am(IV)}$$

$$\mathrm{Am(IV)} + \mathrm{Am(V)} \rightarrow \mathrm{Am(III)} + \mathrm{Am(VI)}$$

The reaction between Am(VI) and Am(III) has been detected[6] but no kinetic data has been reported.

Americium(IV) has been found to undergo similar reactions, *viz.*

$$2\ \mathrm{Am(IV)} \rightleftharpoons \mathrm{Am(V)} + \mathrm{Am(III)}$$

$$\mathrm{Am(IV)} + \mathrm{Am(V)} \rightarrow \mathrm{Am(III)} + \mathrm{Am(VI)}$$

Penneman *et al.*[7], have studied these reactions in acidic sulphate, nitrate and perchlorate media, and have found that only in sulphate media is the latter step relatively important. The disproportionation reaction rate coefficient has been estimated as $> 1.03 \times 10^{-7}$ l.mole^{-1}.sec^{-1} at 0 °C (media 5×10^{-2} M in HNO_3).

REFERENCES

1. *Introduction*

1 H. A. C. McKay, *Nature*, 42 (1938) 997.
2 E. Eichler and A. C. Wahl, *J. Am. Chem. Soc.*, 80 (1958) 4145.
3 Y. A. Im and D. H. Busch, *J. Am. Chem. Soc.*, 83 (1961) 3357.
4 Y. A. Im and D. H. Busch, *J. Am. Chem. Soc.*, 83 (1961) 3362.
5 H. M. McConnell and H. E. Weaver, *J. Chem. Phys.*, 25 (1956) 307.
6 H. M. McConnell and S. B. Berger, *J. Chem. Phys.*, 27 (1957) 230.
7 C. R. Bruce, R. E. Norberg and S. I. Weissman, *J. Chem. Phys.*, 24 (1956) 473.
8 R. L. Ward and S. I. Weissman, *J. Am. Chem. Soc.*, 79 (1957) 2086.
9 M. W. Dietrich and A. C. Wahl, *J. Chem. Phys.*, 38 (1963) 1591.
10 D. W. Larsen and A. C. Wahl, *J. Chem. Phys.*, 41 (1964) 908.
11 N. Sutin, *Ann. Rev. Nucl. Sci.*, 12 (1962) 285.
12 D. R. Stranks and R. G. Wilkins, *Chem. Rev.*, 57 (1957) 743.
13 J. Halpern, *Quart. Rev.* (London), 15 (1961) 207.
14 A. A. Vlcek, *Chemie*, 9 (1957) 305.
15 C. B. Amphlett, *Quart. Rev.* (London), 8 (1954) 219.
16 B. J. Zwolinski, R. J. Marcus and H. Eyring, *Chem. Rev.*, 55 (1955) 157.

2.1 *The exchange reaction between Cu(II) and Cu(I)*

1 H. M. McConnell and H. E. Weaver, *J. Chem. Phys.*, 25 (1956) 307.
2 H. M. McConnell and N. Davidson, *J. Am. Chem. Soc.*, 72 (1950) 3168

2.2 *The exchange reaction between Ag(II) and Ag(I)*

1 B. M. Gordon and A. C. Wahl, *J. Am. Chem. Soc.*, 80 (1958) 273.
2 M. Bruno and V. Santoro, *Ric. Sci.*, 26 (1956) 3072.

2.3 *The exchange reaction between Au(III) and Au(I)*

1 A. Turco and G. Sordillo, *Gazz. Chim. Ital.*, 85 (1955) 977.

2.4 *The exchange reaction between Au(III) and Au(II)*

1 R. L. Rich and H. Taube, *J. Phys. Chem.*, 58 (1954) 6.

2.5 *The disproportionation of Au(II)*

1 R. L. Rich and H. Taube, *J. Phys. Chem.*, 58 (1954) 6.

3.1 *The exchange reaction between Hg(II) and Hg(I); the disproportionation reaction of Hg(II)*

1 E. L. King, *J. Am. Chem. Soc.*, 71 (1949) 3553.
2 M. Haissinsky and M. Cottin, *J. Chim. Phys.*, 46 (1949) 476.
3 R. L. Wolfgang and R. W. Dodson, *J. Phys. Chem.*, 56 (1952) 872.
4 E. L. King, *J. Phys. Chem.*, 56 (1952) 876.
5 A. W. Adamson, *J. Phys. Chem.*, 56 (1952) 876.
6 R. L. Wolfgang and R. W. Dodson, *J. Am. Chem. Soc.*, 76 (1954) 2004.
7 D. Peschanski, *J. Chim. Phys.*, 50 (1953) 640.

3.2 *The exchange reaction between Hg(II) and Hg(I) in non-aqueous media*

1 D. Peschanski, *J. Chim. Phys.*, 50 (1953) 634.

4..1 *The exchange reaction between Tl(III) and Tl(I)*

1 J. Zirkler, *Z. Physik.*, 97 (1934) 410; 98 (1935) 75; 99 (1936) 669; *Z. Physik. Chem.*, A187 (1940) 103.
2 V. Majer, *Z. Physik. Chem.*, A179 (1937) 51.
3 G. Harbottle and R. W. Dodson, *J. Am. Chem. Soc.*, 70 (1948) 880.
4 R. J. Prestwood and A. C. Wahl, *J. Am. Chem. Soc.*, 70 (1948) 880.
5 R. J. Prestwood and A. C. Wahl, *J. Am. Chem. Soc.*, 71 (1949) 3137.
6 G. Harbottle and R. W. Dodson, *J. Am. Chem. Soc.*, 73 (1951) 2442.
7 R. W. Dodson, *J. Am. Chem. Soc.*, 75 (1953) 1795.
8 F. J. C. Rossotti, *J. Inorg. Nucl. Chem.*, 1 (1955) 159.
9 G. Biedermann, Arkiv Kemi, 5 (1953) 441.
10 R. P. Bell and J. H. B. George, *Trans. Faraday Soc.*, 49 (1953) 619.
11 E. Roig and R. W. Dodson, *J. Phys. Chem.*, 65 (1961) 2175.
12 S. Gilks and G. M. Waind, *Discussions Faraday Soc.*, 29 (1960) 102.
13 G. M. Waind, *Discussions Faraday Soc.*, 29 (1960) 135.
14 S. Gilks, T. Rodgers and G. M. Waind, *Trans. Faraday Soc.*, 57 (1961) 1371.
15 G. E. Challenger and B. J. Masters, *J. Am. Chem. Soc.*, 78 (1956) 3012.
16 C. H. Brubaker and J. P. Mickel, *J. Inorg. Nucl. Chem.*, 4 (1957) 55.
17 C. H. Brubaker, K. O. Groves, J. P. Mickel and C. P. Knop, *J. Am. Chem. Soc.*, 79 (1957) 4641.
18 D. R. Wiles, *Can. J. Chem.*, 36 (1958) 167.
19 E. Penna-Franca and R. W. Dodson, *J. Am. Chem. Soc.*, 77 (1955) 2651.
20 L. G. Carpenter, M. H. Ford-Smith, R. P. Bell and R. W. Dodson, *Discussions Faraday Soc.*, 29 (1960) 92.

21 C. H. Brubaker and C. Andrade, *J. Am. Chem. Soc.*, 81 (1959) 5282.
22 J. W. Gryder and M. C. Dorfman, *J. Am. Chem. Soc.*, 83 (1961) 1254.
23 W. C. E. Higginson, D. R. Rosseinsky, J. B. Stead and A. G. Sykes, *Discussions Faraday Soc.*, 29 (1960) 49.
24 A. G. Sykes, *J. Chem. Soc.*, (1961) 5549.
25 H. McConnell and N. Davidson, *J. Am. Chem. Soc.*, 71 (1949) 3845.
26 D. R. Stranks and J. R. Yandell, *Exchange Reactions*, I.A.E.A., Vienna, 1965, p. 83.
27 R. G. McGregor and D. R. Wiles, *J. Chem. Soc.* A, (1970) 323.
28 D. R. Stranks and J. K. Yandell, *J. Phys. Chem.*, 73 (1969) 840.

5.1 *The exchange reaction between Sn(IV) and Sn(II) in aqueous media*

1 C. I. Browne, R. P. Craig and N. Davidson, *J. Am. Chem. Soc.*, 73 (1951) 1946.
2 R. P. Craig and N. Davidson, *J. Am. Chem. Soc.*, 73 (1951) 1951.
3 G. Gordon and C. H. Brubaker, *J. Am. Chem. Soc.*, 82 (1960) 4448.
4 J. E. Whitney and N. Davidson, *J. Am. Chem. Soc.*, 71 (1949) 3809.

5.2 *The exchange reaction between Sn(IV) and Sn(II) in non-aqueous media*

1 E. G. Meyer and M. A. Melnick, *J. Phys. Chem.*, 61 (1957) 367.
2 E. G. Meyer and M. Kahn, *J. Am. Chem. Soc.*, 73 (1951) 4950.

5.3 *The exchange reaction between Pb(IV) and Pb(II) in aqueous media*

1 E. Zintl and A. Ranch, *Ber.*, 57B (1924) 1743.
2 A. Fava, *J. Chim. Phys.*, 50 (1953) 403.

5.4 *The exchange reaction between Pb(IV) and Pb(II) in non-aqueous media*

1 G. V. Hevesy and L. Zechmeister, *Ber.*, 53B (1920) 415.
2 E. A. Evans, J. L. Huston and T. H. Norris, *J. Am. Chem. Soc.*, 74 (1952) 4986.

6.1 *The exchange reaction between As(V) and As(III)*

1 J. N. Wilson and R. G. Dickenson, *J. Am. Chem. Soc.*, 59 (1937) 1358.
2 M. Martin, P. Daudel, R. Daudel and P. Magnier, *Compt. Rend.*, 224 (1947) 195.

6.2 *The exchange reaction between Sb(V) and Sb(III) in aqueous media*

1 J. E. Whitney and N. Davidson, *J. Am. Chem. Soc.*, 69 (1947) 2076.
2 J. E. Whitney and N. Davidson, *J. Am. Chem. Soc.*, 71 (1949) 3809.
3 H. M. Neumann, *J. Am. Chem. Soc.*, 76 (1954) 2611.
4 H. M. Neumann and R. W. Ramette, *J. Am. Chem. Soc.*, 78 (1956) 1848.
5 N. A. Bonner, *J. Am. Chem. Soc.*, 71 (1949) 3909.
6 H. M. Neumann and H. Brown, *J. Am. Chem. Soc.*, 78 (1956) 1843.
7 C. H. Cheek, *Ph. D. Thesis*, Washington University, 1953.
8 C. H. Cheek, N. A. Bonner and A. C. Wahl, *J. Am. Chem. Soc.*, 83 (1961) 80.
9 T. Kambara, K. Yamaguchi and S. Yasuba, *Exchange Reactions*, I.A.E.A., Vienna, 1965, p. 101.
10 N. A. Bonner and W. Goishi, *J. Am. Chem. Soc.*, 83 (1961) 85.
11 A. Turco and G. Faroane, *Ric. Sci.*, 25 (1955) 2887.
12 A. Turco, *Gazz. Chim. Ital.*, 88 (1958) 365.
13 C. H. Brubaker and J. A. Sincius, *Abstr. Paper 139*th *Meeting Am. Chem. Soc.*, Missouri, 1961.
14 C. H. Brubaker and J. A. Sincius, *J. Phys. Chem.*, 65 (1961) 867.
15 A. Turco, *Gazz. Chim. Ital.*, 28A (1953) 231.

6.3 *The exchange reaction between Sb(V) and Sb(III) in non-aqueous media*

1 F. B. Barker and M. Kahn, *J. Am. Chem. Soc.*, 78 (1956) 1317.
2 K. R. Price and C. H. Brubaker, *Exchange Reactions*, I.A.E.A., Vienna, 1965, p. 113.
3 K. R. Price and C. H. Brubaker, *Inorg. Chem.*, 4 (1965) 1351.
4 W. E. Becker and R. E. Johnson, *J. Am. Chem. Soc.*, 79 (1957) 5157.

6.4 *The Sb(III)-catalysed hydrolysis of Sb(V)*

1 H. M. Neumann and R. W. Ramette, *J. Am. Chem. Soc.*, 78 (1956) 1848.
2 N. A. Bonner and W. Goishi, *J. Am. Chem. Soc.*, 83 (1961) 85.

7.1 *The exchange reaction between Te(VI) and Te(IV)*

1 M. Haissinsky and M. Cottin, *Anal. Chim. Acta*, 3 (1949) 226.
2 M. W. Hanson and T. C. Hoering, *J. Phys. Chem.*, 61 (1957) 699.

8.1.1 *The exchange reaction between V(III) and V(II)*

1 W. R. King and C. S. Garner, *J. Am. Chem. Soc.*, 74 (1952) 3709.
2 K. V. Krishnamurty and A. C. Wahl, *J. Am. Chem. Soc.*, 80 (1958) 5921.

8.1.2 *The exchange reaction between V(IV) and V(III)*

1 S. G. Furman and C. S. Garner, *J. Am. Chem. Soc.*, 74 (1952) 2333.

8.1.3 *The exchange reaction between V(V) and V(IV)*

1 H. A. Tewes, J. B. Ramsey and C. S. Garner, *J. Am. Chem. Soc.*, 72 (1950) 2422.
2 C. R. Giuliano and H. M. McConnell, *J. Inorg. Nucl. Chem.*, 9 (1959) 171.

8.1.4 *Reactions between vanadium ions*

1 T. W. Newton and F. B. Baker, *J. Phys. Chem.*, 68 (1964) 228.
2 T. W. Newton and F. B. Baker, *Inorg. Chem.*, 3 (1964) 569.
3 J. W. Olver and J. W. Ross, *J. Phys. Chem.*, 66 (1962) 1699.
4 N. A. Daugherty and T. W. Newton, *J. Phys. Chem.*, 68 (1964) 612.
5 J. H. Espenson and L. A. Krug, *Inorg. Chem.*, 8 (1969) 2633.

8.1.5 *Reactions between tantalum cluster ions*

1 J. H. Espenson and R. E. McCarley, *J. Am. Chem. Soc.*, 88 (1966) 1063.
2 J. H. Espenson and D. J. Boone, *Inorg. Chem.*, 7 (1968) 636.
3 N. Winograd and T. Kuwana, *J. Am. Chem. Soc.*, 92 (1970) 224.

8.2.1 *The exchange reaction between Cr(III) and Cr(II)*

1 R. A. Plane and H. Taube, *J. Phys. Chem.*, 56 (1952) 33.
2 A. Anderson and N. A. Bonner, *J. Am. Chem. Soc.*, 76 (1954) 3826.
3 H. van der Straaten and A. H. W. Aten, *Rec. Trav. Chim.*, 73 (1954) 89.
4 D. L. Ball and E. L. King, *J. Am. Chem. Soc.*, 80 (1958) 1091.
5 H. Taube and E. L. King, *J. Am. Chem. Soc.*, 76 (1954) 4053.
6 R. Snellgrove and E. L. King, *Inorg. Chem.*, 3 (1964) 288.
7 K. A. Schroder and J. H. Espenson, *J. Am. Chem. Soc.*, 89 (1967) 2548.
8 J. P. Birk and J. H. Espenson, *J. Am. Chem. Soc.*, 90 (1968) 2266.
9 R. Snellgrove and E. L. King, *J. Am. Chem. Soc.*, 84 (1962) 4609.

10 D. R. Stranks, *Discussions Faraday Soc.*, 29 (1960) 79.
11 E. Deutsch and H. Taube, *Inorg. Chem.*, 7 (1968) 1532.
12 D. H. Hutchital, *Inorg. Chem.*, 9 (1970) 486.

8.2.2 *The exchange reaction between Cr(VI) and Cr(III)*

1 R. Muxart, P. Daudel, R. Daudel and M. Haissinsky, *Nature*, 159 (1947) 538.
2 H. E. Menker and C. S. Garner, *J. Am. Chem. Soc.*, 71 (1949) 371.
3 W. H. Burgus and J. W. Kennedy, *J. Chem. Phys.*, 18 (1950) 97.
4 C. Altman and E. L. King, *J. Am. Chem. Soc.*, 83 (1961) 2825.

8.2.3 *The reaction between Cr(VI) and Cr(II)*

1 L. S. Hegedus and A. Haim, *Inorg. Chem.*, 6 (1967) 664.
2 M. Ardon and R. A. Plane, *J. Am. Chem. Soc.*, 81 (1959) 3197.

8.2.4 *Cr(II)-catalysed substitution and isomerisation reactions of Cr(III)*

1 H. Taube and E. L. King, *J. Am. Chem. Soc.*, 76 (1954) 4053.
2 A. Adin and A. G. Sykes, *J. Chem. Soc.*, A (1966) 1518.
3 D. E. Pennington and A. Haim, *J. Am. Chem. Soc.*, 88 (1966) 3450.
4 A. Adin, J. Doyle and A. G. Sykes, *J. Chem. Soc. A*, (1967) 1504.
5 J. Doyle, A. G. Sykes and A. Adin, *J. Chem. Soc. A*, (1968) 1314.
6 J. H. Espenson and D. W. Carlyle, *Inorg. Chem.*, 5 (1966) 586.
7 J. H. Espenson and S. G. Slocum, *Inorg. Chem.*, 6 (1967) 906.
8 Y. Chia and E. L. King, *Discussions Faraday Soc.*, 29 (1960) 109.
9 A. Haim, *J. Am. Chem. Soc.*, 88 (1966) 2325.
10 J. P. Birk and J. H. Espenson, *J. Am. Chem. Soc.*, 90 (1968) 2266.
11 H. Taube and H. Meyers, *J. Am. Chem. Soc.*, 76 (1954) 2103.
12 H. B. Johnson and W. L. Reynolds, *Inorg. Chem.*, 2 (1963) 468.
13 A. E. Ogard and H. Taube, *J. Am. Chem. Soc.*, 80 (1958) 1084.
14 R. D. Cannon, *J. Chem. Soc. A*, (1968) 1098.
15 M. J. de Chant and J. B. Hunt, *J. Am. Chem. Soc.*, 89 (1967) 5988.
16 M. J. de Chant and J. B. Hunt, *J. Am. Chem. Soc.*, 90 (1968) 3695.
17 D. E. Pennington and A. Haim, *Inorg. Chem.*, 5 (1966) 1887.
18 R. D. Cannon and J. E. Earley, *J. Chem. Soc. A*, (1968) 1102.
19 R. F. N. Thorneley, B. Kipling and A. G. Sykes, *J. Chem. Soc. A*, (1968) 2847.
20 A. Haim and N. Sutin, *J. Am. Chem. Soc.*, 87 (1965) 4210.
21 A. Haim and N. Sutin, *J. Am. Chem. Soc.*, 88 (1966) 434.
22 J. P. Birk and J. H. Espenson, *J. Am. Chem. Soc.*, 90 (1968) 1153.
23 D. E. Pennington and A. Haim, *Inorg. Chem.*, 6 (1967) 2138.
24 J. P. Birk and J. H. Espenson, *Inorg. Chem.*, 7 (1968) 991.
25 E. Deutsch and H. Taube, *Inorg. Chem.*, 7 (1968) 1532.
26 F. Nordmeyer and H. Taube, *J. Am. Chem. Soc.*, 90 (1968) 1162.
27 D. H. Hutchital, *Inorg. Chem.*, 9 (1970) 486.
28 D. W. Hoppenjans, J. B. Hunt and L. Penzhorn, *Inorg. Chem.*, 7 (1968) 1467.
29 R. G. Wilkins and R. E. Yelin, *Inorg. Chem.*, 7 (1968) 2667.

8.2.5 *The exchange reaction between Mo(V) and Mo(IV)*

1 R. L. Wolfgang, *J. Am. Chem. Soc.*, 74 (1952) 6144.
2 R. Campion, N. Purdie and N. Sutin, *J. Am. Chem, Soc.*, 85 (1963) 3528.

8.2.6 *The exchange reaction between W(V) and W(IV)*

1 E. L. Goodenow and C. S. Garner, *J. Am. Chem. Soc.*, 77 (1955) 5272.
2 S. I. Weissman and C. S. Garner, *J. Am. Chem. Soc.*, 78 (1956) 1072.

8.3.1 *The exchange reaction between Mn(II) and Mn(I)*

1 D. S. Matteson and R. A. Bailey, *J. Am. Chem. Soc.*, 89 (1967) 6389.
2 D. S. Matteson and R. A. Bailey, *J. Am. Chem. Soc.*, 91 (1969) 1975.

8.3.2 *The exchange reaction between Mn(III) and Mn(II)*

1 M. Polissar, *J. Am. Chem. Soc.*, 58 (1936) 1372.
2 A. W. Adamson, *J. Phys. Coll. Chem.*, 55 (1951) 293.
3 H. Diebler and N. Sutin, *J. Phys. Chem.*, 68 (1964) 174.
4 A. W. Adamson, *J. Phys. Chem.*, 56 (1952) 858.
5 R. G. Wilkins and R. E. Yelin, *Inorg. Chem.*, 7 (1968) 2667.

8.3.3 *The exchange reaction between Mn(VII) and Mn(VI)*

1 W. F. Libby, *J. Am. Chem. Soc.*, 62 (1940) 1930.
2 H. C. Hornig, G. L. Zimmerman and W. F. Libby, *J. Am. Chem. Soc.*, 72 (1950) 3808.
3 A. W. Adamson, *J. Phys. Chem.*, 55 (1951) 293.
4 N. A. Bonner and H. A. Potratz, *J. Am. Chem. Soc.*, 73 (1951) 1845.
5 J. C. Sheppard and A. C. Wahl, *J. Am. Chem. Soc.*, 75 (1953) 5133.
6 J. C. Sheppard and A. C. Wahl, *J. Am. Chem. Soc.*, 79 (1957) 1020.
7 L. Gjertsen and A. C. Wahl, *J. Am. Chem. Soc.*, 81 (1959) 1572.
8 O. E. Meyers and J. C. Sheppard, *J. Am. Chem. Soc.*, 83 (1961) 4730.
9 A. D. Britt and W. M. Yen, *J. Am. Chem. Soc.*, 83 (1961) 4516.

8.3.4 *The exchange reaction between Mn(VII) and Mn(III)*

1 M. Polissar, *J. Am. Chem. Soc.*, 58 (1936) 1372.

8.3.5 *The exchange reaction between Mn(VII) and Mn(II)*

1 M. Polissar, *J. Am. Chem. Soc.*, 58 (1936) 1372.
2 A. W. Adamson, *J. Phys. Chem.*, 55 (1951) 293.
3 J. A. Happe and D. S. Martin, *J. Am. Chem. Soc.*, 77 (1955) 4212.

8.3.6 *The reaction between Mn(VII) and Mn(II)*

1 F. C. Tompkins, *Trans. Faraday Soc.*, 38 (1942) 128.
2 M. J. Polissar, *J. Phys. Chem.*, 39 (1935) 1057.
3 M. A. Guyard, *Bull. Soc. Chim.*, 1 (1864) 89.
4 H. F. Launer and D. M. Yost, *J. Am. Chem. Soc.*, 56 (1934) 2571.
5 G. R. Waterbury, A. M. Hayes and D. S. Martin, *J. Am. Chem. Soc.*, 74 (1952) 15.
6 D. R. Rosseinsky and M. J. Nicol, *Trans. Faraday Soc.*, 61 (1965) 2718.

8.4.1 *The exchange reaction between Fe(III) and Fe(II) in aqueous media*

1 L. van Alten and C. N. Rice, *J. Am. Chem. Soc.*, 70 (1948) 883.
2 V. J. Linnenbom and A. C. Wahl, *J. Am. Chem. Soc.*, 71 (1949) 2589.
3 H. A. Kierstead, *J. Chem. Phys.*, 18 (1950) 856.
4 R. H. Betts, H. S. A. Gilmour and R. K. Leigh, *J. Am. Chem. Soc.*, 72 (1950) 4978.
5 J. Silverman and R. W. Dodson, *J. Phys. Chem.*, 56 (1952) 846.
6 R. W. Dodson, *J. Am. Chem. Soc.*, 72 (1950) 3315.
7 R. W. Dodson, *J. Phys. Chem.*, 56 (1952) 852.
8 L. Eimer, A. I. Medalia and R. W. Dodson, *J. Chem. Phys.*, 20 (1952) 743.
9 J. Hudis and R. W. Dodson, *J. Am. Chem. Soc.*, 78 (1956) 912.
10 S. Fukushima and W. L. Reynolds, *Talanta*, 11 (1964) 283.

11 H. L. Reynolds and R. W. Lumry, *J. Chem. Phys.*, 23 (1955) 2460.
12 R. A. Horne, *J. Inorg. Nucl. Chem.*, 25 (1963) 1139.
13 R. A. Horne and E. H. Axelrod, *J. Chem. Phys.*, 40 (1964) 1518.

8.4.2 *The effect of inorganic ions on the exchange between Fe(III) and Fe(II)*

1 J. Silverman and R. W. Dodson, *J. Phys. Chem.*, 56 (1952) 846.
2 N. Sutin, J. K. Rowley and R. W. Dodson, *J. Phys. Chem.*, 65 (1961) 1248.
3 R. J. Campion, J. J. Conocchioli and N. Sutin, *J. Am. Chem. Soc.*, 86 (1964) 4591.
4 R. A. Horne, *J. Inorg. Nucl. Chem.*, 25 (1963) 1139.
5 J. Hudis and A. C. Wahl, *J. Am. Chem. Soc.*, 75 (1953) 4153.
6 J. Menashi, S. Fukushima, C. Foxx and W. L. Reynolds, *Inorg. Chem.*, 3 (1964) 1242.
7 G. S. Laurence, *Trans. Faraday Soc.*, 53 (1957) 1326.
8 T. J. Conocchioli and N. Sutin, *J. Am. Chem. Soc.*, 89 (1967) 282.
9 R. A. Horne and E. H. Axelrod, *J. Chem. Phys.*, 40 (1964) 1518.
10 D. Bunn, F. S. Dainton and S. Duckworth, *Trans. Faraday Soc.*, 57 (1961) 1131.
11 D. Bunn, F. S. Dainton and S. Duckworth, *Trans. Faraday Soc.*, 55 (1959) 1267.
12 J. C. Sheppard and L. C. Brown, *J. Phys. Chem.*, 67 (1963) 1025.
13 R. A. Horne, *Ph. D. Thesis*, Columbia (1955).
14 W. L. Reynolds and S. Fukushima, *Inorg. Chem.*, 2 (1963) 176.
15 R. L. S. Willix, *Trans. Faraday Soc.*, 59 (1963) 1315.
16 K. Bächmann and K. H. Lieser, *Z. Physik. Chem.* N.F., 36 (1963) 3.

8.4.3 *The effect of organic ligands on the exchange between Fe(III) and Fe(II)*

1 R. A. Horne, *J. Phys. Chem.*, 64 (1960) 1512.
2 J. C. Sheppard and L. C. Brown, *J. Phys. Chem.*, 67 (1963) 1025.
3 A. McAuley and C. H. Brubaker, *Inorg. Chem.*, 3 (1964) 273.
4 M. R. Chakrabarty, J. F. Stephens and E. S. Hanrahan, *Inorg. Chem.*, 5 (1966) 1617.
5 A. W. Adamson and K. S. Vorres, *J. Inorg. Nucl. Chem.*, 3 (1956) 206.
6 W. L. Reynolds, N. Liu and J. Mickus, *J. Am. Chem. Soc.*, 83 (1961) 1078.
7 L. Eimer and A. I. Medalia, *J. Am. Chem. Soc.*, 74 (1952) 1592.
8 E. Eichler and A. C. Wahl, *J. Am. Chem. Soc.*, 80 (1958) 4145.
9 M. W. Dietrich and A. C. Wahl, *J. Chem. Phys.*, 38 (1963) 1591.
10 D. W. Larsen and A. C. Wahl, *J. Chem. Phys.*, 41 (1964) 908.
11 D. W. Larsen and A. C. Wahl, *J. Chem. Phys.*, 43 (1964) 3765.
12 D. R. Stranks, *Discussions Faraday Soc.*, 29 (1960) 73.

8.4.4 *The exchange reaction between Fe(III) and Fe(II) in non-aqueous and mixed solvents*

1 D. Peschanski, *J. Chim. Phys.*, 50 (1953) 634.
2 R. A. Horne, *Exchange Reactions*, I.A.E.A., Vienna, 1965, p. 67.
3 N. Sutin, *J. Phys. Chem.*, 64 (1960) 1766.
4 A. G. Maddock, *Trans. Faraday Soc.*, 55 (1959) 1267.
5 J. Menashi, W. R. Reynolds and G. van Auken, *Inorg. Chem.*, 4 (1965) 299.
6 G. Wada and W. L. Reynolds, *Exchange Reactions*, I.A.E.A., Vienna, 1965, p. 59.

8.4.5 *The exchange reaction between hexacyanoferrate (III) and hexacyanoferrate (II)*

1 A. C. Thomson, *J. Am. Chem. Soc.*, 70 (1948) 1045.
2 C. Haenny and E. Wickler, *Helv. Chim. Acta*, 32 (1949) 2444.
3 C. Haenny and G. Rochat, *Helv. Chim. Acta*, 32 (1949) 2441.
4 J. W. Copple and A. W. Adamson, *J. Am. Chem. Soc.*, 72 (1950) 2276.
5 A. C. Wahl and C. F. Deck, *J. Am. Chem. Soc.*, 76 (1954) 4054.
6 A. C. Wahl, *Z. Electrochem.*, 64 (1964) 90.
7 A. Loewenstein, M. Shporer and G. Navon, *J. Am. Chem. Soc.*, 85 (1963) 2855.

8 M. Shporer, G. Ron, A. Loewenstein and G. Navon, *Inorg. Chem.*, 4 (1965) 361.
9 P. King, C. F. Deck and A. C. Wahl, *Abstr. Paper* 139th *Meeting Am. Chem. Soc.*, Missouri, 1961.
10 R. J. Campion, C. F. Deck, P. King and A. C. Wahl, *Inorg. Chem.*, 6 (1967) 672.
11 R. Stasiw and R. G. Wilkins, *Inorg. Chem.*, 8 (1969) 156.
12 A. Loewenstein and G. Ron, *Inorg. Chem.*, 6 (1967) 1604.

8.4.6 *Reactions of Fe(III) with Fe(II)*

1 N. Sutin and B. Gordon, *J. Am. Chem. Soc.*, 83 (1961) 70.
2 M. H. Ford-Smith and N. Sutin, *J. Am. Chem. Soc.*, 83 (1961) 1830.
3 B. M. Gordon, L. L. Williams and N. Sutin, *J. Am. Chem. Soc.*, 83 (1961) 2061.
4 N. Sutin, *Nature*, 190 (1961) 438.
5 N. Sutin and D. R. Christman, *J. Am. Chem. Soc.*, 83 (1961) 1773.
6 R. Stasiw and R. G. Wilkins, *Inorg. Chem.*, 8 (1969) 156.
7 R. G. Wilkins and R. E. Yelin, *Inorg. Chem.*, 7 (1968) 2667.

8.4.7 *The Fe(II)-catalysed aquation of Fe(III)*

1 R. J. Campion, T. J. Conocchioli and N. Sutin, *J. Am. Chem. Soc.*, 86 (1964) 4591.
2 T. J. Conocchioli and N. Sutin, *J. Am. Chem. Soc.*, 89 (1967) 282.
3 E. G. Moorhead and N. Sutin, *Inorg. Chem.*, 6 (1967) 428.

8.4.8 *The reaction between Fe(IV) and Fe(II)*

1 T. J. Conocchioli, E. J. Hamilton and N. Sutin, *J. Am. Chem. Soc.*, 87 (1965) 926.

8.4.9 *The exchange reaction of Ru(VII) and Ru(VI)*

1 E. V. Luoma and C. H. Brubaker, *Inorg. Chem.*, 5 (1966) 1618.

8.4.10 *Ru(II)-catalysed substitution reactions of Ru(III)*

1 J. F. Endicott and H. Taube, *J. Am. Chem. Soc.*, 84 (1962) 4985.

8.4.11 *The exchange reaction between Os(III) and Os(II)*

1 F. P. Dwyer and E. C. Garfas, *Nature*, 166 (1950) 481.
2 E. Eichler and A. C. Wahl, *J. Am. Chem. Soc.*, 80 (1958) 4145.
3 M. W. Dietrich and A. C. Wahl, *J. Chem. Phys.*, 38 (1963) 1591.
4 R. Campion, N. Purdie and N. Sutin, *J. Am. Chem. Soc.*, 85 (1963) 3528.

8.5.1 *The exchange reaction between Co(III) and Co(II) in aqueous media*

1 S. A. Hoshowsky, O. G., Holmes and K. J. McCallum, *Can. J. Res. B*, 27 (1949) 258.
2 N. A. Bonner and J. P. Hunt, *J. Am. Chem. Soc.*, 74 (1952) 1866.
3 N. A. Bonner and J. P. Hunt, *J. Am. Chem. Soc.*, 82 (1960) 3826.
4 J. Shankar and B. C. de Souza, *J. Inorg. Nucl. Chem.*, 24 (1962) 187.
5 J. Shankar and B. C. de Souza, *J. Inorg. Nucl. Chem.*, 24 (1962) 693.
6 L. H. Sutcliffe and J. R. Weber, *Trans. Faraday Soc.*, 52 (1956) 1225.
7 H. S. Habib and J. P. Hunt, *J. Am. Chem. Soc.*, 88 (1966) 1668.
8 J. Shankar and B. C. de Souza, *J. Inorg. Nucl. Chem.*, 29 (1967) 1983.
9 T. J. Conocchioli, G. H. Nancollas and N. Sutin, *Inorg. Chem.*, 5 (1966) 1.
10 P. G. Rasmussen and C. H. Brubaker, *Inorg. Chem.*, 3 (1964) 977.
11 J. P. Birk and J. Halpern, *J. Am. Chem. Soc.*, 90 (1968) 305.
12 A. W. Adamson, *J. Am. Chem. Soc.*, 73 (1951) 5710.

8.5.2 *Exchange reactions involving complexes of Co(III) and Co(II) with ammonia and organic ligands*

1 J. H. FLAGG, *J. Am. Chem. Soc.*, 63 (1941) 557.
2 K. J. MCCALLUM AND S. A. HOSHOWSKY, *J. Chem. Phys.*, 16 (1948) 254.
3 S. A. HOSHOWSKY, O. G. HOLMES AND K. J. MCCALLUM, *Can. J. Res.* B, 27 (1949) 258.
4 W. B. LEWIS, C. D. CORYELL AND J. W. IRVINE, *J. Chem. Soc.* S, (1949) 386.
5 D. R. STRANKS, *Discussions Faraday Soc.*, 29 (1960) 73, 131.
6 E. APPLEMAN, M. ANBAR AND H. TAUBE, *J. Phys. Chem.*, 63 (1959) 126.
7 F. P. DWYER AND A. M. SARGESON, *J. Phys. Chem.*, 65 (1961) 1892.
8 D. R. STRANKS, *Advances in Chemistry of Coordination Compounds*, Macmillan, New York, 1961.
9 B. WEST, *J. Chem. Soc.*, (1952) 3115.
10 A. W. ADAMSON AND K. S. VORRES, *J. Inorg. Nucl. Chem.*, 3 (1956) 206.
11 Y. A. IM AND D. H. BUSCH, *J. Am. Chem. Soc.*, 83 (1961) 3357.
12 Y. A. IM AND D. H. BUSCH, *J. Am. Chem. Soc.*, 83 (1961) 3362.
13 B. R. BAKER, F. BASOLO AND H. M. NEUMANN, *J. Phys. Chem.*, 63 (1959) 371.
14 P. ELLIS, R. G. WILKINS AND M. J. G. WILLIAMS, *J. Chem. Soc.*, (1957) 4456.
15 N. S. BIRADAR, D. R. STRANKS AND M. S. VAIDYA, *Trans. Faraday Soc.*, 58 (1962) 2421.
16 T. J. WILLIAMS AND J. P. HUNT, *J. Am. Chem. Soc.*, 90 (1968) 7210.
17 S. BRÜCKNER, V. CRESCENZI AND F. QUADRIFOGLIO, *J. Chem. Soc.*, A (1970) 1168.

8.5.3 *The exchange reaction between Co(III) and Co(II) in non-aqueous media*

1 J. J. GROSSMAN AND C. S. GARNER, *J. Chem. Phys.*, 28 (1958) 268.
2 B. WEST, *J. Chem. Soc.*, (1952) 3115.
3 B. R. BAKER, F. BASOLO AND H. M. NEUMANN, *J. Phys. Chem.*, 63 (1959) 371.

8.5.4 *Co(II)-catalysed substitution reactions of Co(III)*

1 J. P. CANDLIN, J. HALPERN AND S. NAKAMURA, *J. Am. Chem. Soc.*, 85 (1963) 2517.
2 A. W. ADAMSON, *J. Am. Chem. Soc.*, 78 (1956) 4260.
3 J. HALPERN AND S. NAKAMURA, *J. Am. Chem. Soc.*, 87 (1965) 3002.
4 J. P. BIRK AND J. HALPERN, *J. Am. Chem. Soc.*, 90 (1968) 305.
5 R. FARINA AND R. G. WILKINS, *Inorg. Chem.*, 7 (1968) 514.

8.5.5 *The reaction of Co(IV) with Co(II)*

1 D. BENSON, P. J. PROLL, J. WALKLEY AND L. H. SUTCLIFFE, *Discussions Faraday Soc.*, 29 (1960) 60.
2 P. J. PROLL, *Ph. D. Thesis*, Liverpool University (1962).

8.5.6 *Rh(I)-catalysed substitution reactions of Rh(III)*

1 J. V. RUND, F. BASOLO AND R. G. PEARSON, *Inorg. Chem.*, 3 (1964) 659.
2 R. D. GILLARD, J. A. OSBORNE AND G. WILKINSON, *J. Chem. Soc.*, (1965) 1951.
3 R. D. GILLARD, J. A. OSBORNE AND G. WILKINSON, *J. Chem. Soc.*, (1965) 4107.

8.5.7 *The exchange reaction between Ir(IV) and Ir(III)*

1 E. N. SLOTH AND C. S. GARNER, *J. Am. Chem. Soc.*, 77 (1955) 1440.
2 P. HURWITZ AND K. KUSTIN, *Trans. Faraday Soc.*, 62 (1966) 427.
3 P. HURWITZ AND K. KUSTIN, *Inorg. Chem.*, 3 (1963) 823.

8.5.8 *The reaction between complexes of Ir(IV) and Ir(III)*

1 P. HURWITZ AND K. KUSTIN, *Trans. Faraday Soc.*, 62 (1966) 427.

8.6.1 *The exchange reaction between Pt(IV) and Pt(II); Pt(II)-catalysed substitution reactions of Pt(IV)*

1 R. L. Rich and H. Taube, *J. Am. Chem. Soc.*, 76 (1954) 2608.
2 R. E. McCarley, D. S. Martin and L. T. Cox, *J. Inorg. Nucl. Chem.*, 7 (1958) 113.
3 F. Basolo, P. H. Wilks, R. G. Pearson and R. G. Wilkins, *J. Inorg. Nucl. Chem.*, 6 (1958) 161.
4 F. Basolo, A. F. Messing, P. H. Wilks, R. G. Wilkins and R. G. Pearson, *J. Inorg. Nucl. Chem.*, 8 (1958) 201.
5 L. T. Cox, S. B. Collins and D. S. Martin, *J. Inorg. Nucl. Chem.*, 17 (1961) 383.
6 F. Basolo, M. L. Morris and R. G. Pearson, *Discussions Faraday Soc.*, 29 (1960) 80.
7 R. C. Johnson and F. Basolo, *J. Inorg. Nucl. Chem.*, 13 (1960) 36.
8 H. R. Ellison, F. Basolo and R. G. Pearson, *J. Am. Chem. Soc.*, 83 (1961) 3943.
9 R. R. Rettew and R. C. Johnson, *Inorg. Chem.*, 4 (1965) 1565.
10 R. C. Johnson and E. R. Berger, *Inorg. Chem.*, 4 (1965) 1262.
11 W. R. Mason and R. C. Johnson, *Inorg. Chem.*, 4 (1965) 1258.
12 A. Peloso and R. Ettore, *J. Chem. Soc.* A, (1968) 2253.
13 W. R. Mason, *Inorg. Chem.*, 8 (1969) 1756.
14 A. Syamal and R. C. Johnson, *Inorg. Chem.*, 9 (1970) 265.
15 S. G. Bailey and R. C. Johnson, *Inorg. Chem.*, 8 (1969) 2596.
16 R. C. Johnson and E. R. Berger, *Inorg. Chem.*, 7 (1968) 1656.

9.1 *The exchange reaction between Ce(IV) and Ce(III)*

1 V. J. Linnenbom and A. C. Wahl, *J. Am. Chem. Soc.*, 71 (1949) 2589.
2 J. W. Gryder and R. W. Dodson, *J. Am. Chem. Soc.*, 71 (1949) 1894.
3 J. W. Gryder and R. W. Dodson, *J. Am. Chem. Soc.*, 73 (1951) 2890.
4 F. R. Parchen and F. R. Duke, *J. Am. Chem. Soc.*, 78 (1956) 1540.
5 S. Fronaeus and C. O. Ostman, *Acta Chem. Scand.*, 10 (1956) 769.
6 P. B. Sigler and B. J. Masters, *J. Am. Chem. Soc.*, 79 (1957) 6353.
7 G. E. Challenger and B. J. Masters, *J. Am. Chem. Soc.*, 77 (1955) 1063.
8 H. C. Hornig and W. F. Libby, *J. Phys. Chem.*, 56 (1952) 869.

9.2 *The exchange reaction between Eu(III) and Eu(II)*

1 D. J. Meier and C. S. Garner, *J. Am. Chem. Soc.*, 73 (1951) 1894.
2 D. J. Meier and C. S. Garner, *J. Phys. Chem.*, 56 (1952) 853.

10.1 *Exchange reactions between uranium ions*

1 R. H. Betts, *Can. J. Res.*, 26B (1948) 702.
2 E. Rona, *J. Am. Chem. Soc.*, 72 (1950) 4339.
3 B. J. Masters and L. L. Schwartz, *J. Am. Chem. Soc.*, 83 (1961) 2620.
4 D. M. Mathews, J. D. Hefley and E. S. Amis, *J. Phys. Chem.*, 63 (1959) 1236.
5 A. Indelli and E. S. Amis, *J. Am. Chem. Soc.*, 81 (1959) 4180.
6 S. L. Milton, A. Indelli and E. S. Amis, *J. Inorg. Nucl. Chem.*, 17 (1961) 325.
7 S. L. Milton, J. O. Wear and E. S. Amis, *J. Inorg. Nucl. Chem.*, 17 (1961) 317.
8 G. Gordon and H. Taube, *J. Inorg. Nucl. Chem.*, 16 (1961) 272.

10.2 *Reactions between uranium ions*

1 D. M. H. Kern and E. F. Orleman, *J. Am. Chem. Soc.*, 71 (1949) 2102.
2 D. M. H. Kern and E. F. Orleman, *J. Am. Chem. Soc.*, 75 (1953) 3059.
3 K. A. Kraus, F. Nelson and G. L. Johnson, *J. Am. Chem. Soc.*, 71 (1949) 2510.
4 F. R. Duke and R. C. Pinkerton, *J. Am. Chem. Soc.*, 73 (1951) 2361.
5 J. Koryta and J. Koutecky, *Coll. Czech. Chem. Commun.*, 20 (1955) 423.

6 H. Imai, *Bull. Chem. Soc. Japan*, 30 (1957) 873.
7 D. T. Pence and G. L. Booman, *Anal. Chem.*, 38 (1966) 1112.
8 H. G. Heal and J. G. N. Thomas, *Trans. Faraday Soc.*, 45 (1949) 11.
9 O. Fisher and O. Draka, *Coll. Czech. Chem. Commun.*, 24 (1959) 3046.
10 L. J. Heidt and K. A. Moon, *J. Am. Chem. Soc.*, 75 (1953) 5803.
11 L. J. Heidt, *J. Am. Chem. Soc.*, 76 (1954) 5962.
12 T. W. Newton and F. B. Baker, *Inorg. Chem.*, 4 (1965) 1166.
13 B. J. Masters and L. L. Schwartz, *J. Am. Chem. Soc.*, 83 (1961) 2620

10.3 *Exchange reactions between neptunium ions*

1 D. Cohen, J. C. Sullivan and J. C. Hindman, *J. Am. Chem. Soc.*, 76 (1954) 352.
2 D. Cohen, J. C. Sullivan and J. C. Hindman, *J. Am. Chem. Soc.*, 77 (1955) 4964.
3 J. C. Sullivan, D. Cohen and J. C. Hindman, *J. Am. Chem. Soc.*, 79 (1957) 3672.
4 D. Cohen, J. C. Sullivan, E. S. Amis and J. C. Hindman, *J. Am. Chem. Soc.*, 78 (1956) 1543.
5 J. C. Sullivan, D. Cohen and J. C. Hindman, *J. Am. Chem. Soc.*, 76 (1954) 4275.

10.4 *Reactions between neptunium ions*

1 J. C. Hindman, J. C. Sullivan and D. Cohen, *J. Am. Chem. Soc.*, 80 (1958) 1812.
2 J. C. Hindman, J. C. Sullivan and D. Cohen, *J. Am. Chem. Soc.*, 76 (1954) 3278.
3 J. C. Hindman, J. C. Sullivan and D. Cohen, *J. Am. Chem. Soc.*, 81 (1959) 2316.
4 J. C. Sullivan, D. Cohen and J. C. Hindman, *J. Am. Chem. Soc.*, 79 (1957) 4029.
5 D. Cohen, E. S. Amis, J. C. Sullivan and J. C. Hindman, *J. Phys. Chem.*, 60 (1956) 701.

10.5 *Exchange reactions between plutonium ions*

1 T. K. Keenan, *J. Am. Chem. Soc.*, 78 (1956) 2339.
2 T. K. Keenan, *J. Phys. Chem.*, 61 (1957) 1117.

10.6 *Reactions between plutonium ions*

1 O. E. Ogard and S. W. Rabideau, *J. Phys. Chem.*, 60 (1956) 812.
2 S. W. Rabideau and R. J. Kline, *J. Phys. Chem.*, 62 (1958) 617.
3 M. Alexi, Q. C. Johnson, H. D. Cowan and J. F. Lemons, *J. Inorg. Nucl. Chem.*, 29 (1967) 2327.
4 R. E. Connick, *J. Am. Chem. Soc.*, 71 (1949) 1528.
5 S. W. Rabideau, *J. Am. Chem. Soc.*, 79 (1957) 6350.
6 S. W. Rabideau, *J. Am. Chem. Soc.*, 75 (1953) 798.
7 R. E. Connick and W. H. McVey, *J. Am. Chem. Soc.*, 75 (1952) 474.
8 S. W. Rabideau and H. D. Cowan, *J. Am. Chem. Soc.*, 77 (1955) 6145.

10.7 *Exchange reactions between americium ions*

1 T. K. Keenan, R. A. Penneman and J. F. Suttle, *J. Phys. Chem.*, 59 (1955) 381.

10.8 *Reactions between americium ions*

1 G. R. Hall and P. D. Herniman, *J. Chem. Soc.*, (1954) 2214.
2 G. R. Hall and T. L. Markin, *J. Inorg. Nucl. Chem.*, 4 (1957) 296.
3 S. R. Gunn and B. B. Cunningham, *J. Am. Chem. Soc.*, 79 (1957) 1563.
4 R. A. Penneman and L. B. Asprey, *Intern. Conf. Peaceful Uses Atomic Energy*, Vol. 7, 1955, p. 355.
5 A. A. Zaitsev, V. N. Kosyakov, A. G. Rykov, Yu. P. Sobelev and G. N. Yakovlev, *Radiokhimiya*, 2 (1960) 339.
6 J. S. Coleman, *Inorg. Chem.*, 2 (1963) 53.
7 R. A. Penneman, J. S. Coleman and T. K. Keenan, *J. Inorg. Nucl. Chem.*, 17 (1961) 138.

Chapter 3

Oxidation–Reduction Reactions Between Complexes of Different Metals

D. BENSON

1. Introduction

The last two decades have seen a growing interest in the mechanism of inorganic reactions in solution. Nowhere is this activity more evident than in the topic covered by this review: the oxidation–reduction processes of metal complexes. This subject has been reviewed a number of times previously, notably by Taube[1] (1959), Halpern[2] (1961), Sutin[3] (1966), and Sykes[4] (1967). Other articles and books concerned, wholly or partly, with the topic include those by Stranks[5], Fraser[6], Strehlow[7], Reynolds and Lumry[8], Basolo and Pearson[9], and Candlin *et al.*[10]†. Important recent articles on the theoretical aspects are those by Marcus[11] and Ruff[12]. Elementary accounts of redox reactions are included in the books by Edwards[13], Sykes[14] and Benson[15]. The object of the present review is to provide a more detailed survey of the experimental work than has hitherto been available.

The material included is organised according to the periodic table as follows. Classification, in the first place, is on the basis of *oxidants*, arranged in order of their position across the periodic table, *i.e.*, from vanadium to lead. The reactions of the lanthanide cerium(IV) and the actinides are treated last. Within each section the order for each oxidant is of decreasing oxidation number. In general, reductants within each section and sub-section are arranged, again according to the periodic table, but in order of increasing oxidation number. However, in sections (6.1) and (6.2), dealing with oxidations by cobalt(III), the subject matter is such that no classification on the basis of reductant has been attempted.

In terms of gross features of mechanism, a redox reaction between transition metal complexes, having adjacent stable oxidation states, generally takes place in a simple one-equivalent change. For the post-transition and actinide elements, where there is usually a difference of two between the stable oxidation states, both single two-equivalent and consecutive one-equivalent changes are possible.

As regards intimate mechanism, electron transfer reactions of metal complexes are of two basic types. These have become known as *outer-sphere* and *inner-sphere* (see Chapter 4, Volume 2). In principle, an outer-sphere process occurs with substitution-inert reactants whose coordination shells remain intact in

† See, also, Sutin[274].

forming the activated complex. In other words, electron transfer proceeds more rapidly than substitution in the coordination shell of one of the reactants. Conversely, in inner-sphere processes substitution takes place prior to electron transfer. The direct and indirect criteria used in distinguishing between inner- and outer-sphere mechanisms have been discussed in detail by Sutin[3].

2. Oxidations by vanadium

2.1 OXIDATIONS BY VANADIUM(V)

The oxidation by vanadium(V) of iron(II), a reaction in which a metal–oxygen bond is broken, takes place according to the stoichiometric equation

$$VO_2^+ + Fe(II) + 2\,H^+ = VO^{2+} + Fe(III) + H_2O$$

Oxidation potentials lead to a value of 7.9×10^3 for the equilibrium constant. Kinetic data for the reaction (from 0 to 55.6 °C) in acid perchlorate solutions (over the range 0.047–1.0 *M*) have been obtained spectrophotometrically by following the disappearance of V(V) (which absorbs strongly between 305 and 350 mμ) as a function of time[16]. The second-order nature of the rate law

$$-d[V(V)]/dt = -d[Fe(II)]/dt = k'[V(V)][Fe(II)]$$

is rigidly adhered to over a wide range of reactant concentrations. The observed rate coefficient, k', shows a dependence on $[H^+]$, *viz.*

$$k' = a/[H^+] + b + c[H^+]$$

The three terms of this expression indicate three competitive activation processes, respectively

$$VO_2^+ + Fe^{2+} + H_2O \longrightarrow (VO_2FeOH^{2+})^{\ddagger} + H^+ \quad (2.1)$$

$$VO_2^+ + Fe^{2+} \longrightarrow (VO_2Fe^{3+})^{\ddagger} \quad (2.2)$$

$$VO_2^+ + Fe^{2+} + H^+ \longrightarrow (HVO_2Fe^{4+})^{\ddagger} \quad (2.3)$$

assuming inner-sphere structures for convenience. The term in a is barely significant, and the c term strongly predominates over the b term. At 25 °C and $\mu = 1$ *M* the value of c [the rate coefficient for the most important path (2.3)] is 3400 $l^2.mole^{-2}.sec^{-1}$. Under the same conditions a value of 60 $l.mole^{-1}.sec^{-1}$ can be assigned tentatively to b. The activation parameters corresponding to step (2.3), $\Delta H^{\ddagger}$ and $\Delta S^{\ddagger}$, are 1.52 kcal.mole^{-1} and -37.3 cal.deg^{-1}.mole^{-1}, respectively. It is noteworthy that the following reduction reactions of the analogous MO_2^+ actinide ions show a similar first-order hydrogen-ion dependence: $(NpO_2^+ + Fe^{2+})$[17], $(NpO_2^+ + Np^{3+})$[18], $(PuO_2^+ + PuO_2^+)$[19], and $(UO_2^+ + UO_2^+)$[20]. In Table 1 the activation parameters of the V(V)+Fe(II) reaction are

TABLE 1

ACTIVATION PARAMETERS FOR REACTIONS HAVING +4 ACTIVATED COMPLEXES[a]

Net activation process	$\Delta H^{\ddagger}$*(kcal. mole*$^{-1}$*)*	$\Delta S^{\ddagger}$*(cal.deg*$^{-1}$*. mole*$^{-1}$*)*	$S^{\ddagger}$ [b] *complex*	*Ref.*
$VO_2^+ + Fe^{2+} + H^+ \rightarrow (HVO_2Fe^{4+})^{\ddagger}$	1.52	-37.3 ± 0.6	-70 ± 1	16
$NpO_2^+ + Fe^{2+} + H^+ \rightarrow (HNpO_2Fe^{4+})^{\ddagger}$	8.6	-38	-69	17
$Fe^{2+} + Fe^{3+} + H_2O \rightarrow (FeOHFe^{4+})^{\ddagger} + H^+$	19.4	$+9.9 \pm 1.4$	-70 ± 1	22
$Cr^{2+} + Cr^{3+} + H_2O \rightarrow (CrOHCr^{4+})^{\ddagger} + H^+$	22	-2 ± 5	-79 ± 5	23
$Fe^{2+} + Co^{3+} + H_2O \rightarrow (FeOHCo^{4+})^{\ddagger} + H^+$	18.8	$+16 \pm 4$	-66 ± 4	24
$V^{3+} + VO^{2+} + H_2O \rightarrow (VOHVO^{4+})^{\ddagger} + H^+$	20.1	-3 ± 5	-77 ± 5	25
$V^{3+} + VO_2^+ \rightarrow (VO_2V^{4+})^{\ddagger}$	16.6	$+5 \pm 6$	-65 ± 6	26

[a] From Daugherty and Newton[16].
[b] $S^{\ddagger}$ complex is the formal ionic entropy of the activated complex: $S^{\ddagger}_{complex} = \Delta S^{\ddagger} + \Sigma S^0_{reactants}$[21]. Some values used for $S^0_{reactants}$ are estimated ones.

contrasted with those of other redox reactions for which the net charge on the activated complex is +4. The V(V)+Fe(II) reaction is remarkable in having a very low enthalpy of activation, considerably smaller than $\Delta H^{\ddagger}$ for the Np(V)+Fe(II) system (which also involves the breaking of a metal–oxygen bond), and rather surprisingly, lower than $\Delta H^{\ddagger}$ for the exchange between $Fe(CN)_6^{3-}$ and $Fe(CN)_6^{4-}$. The $\Delta S^{\ddagger}$ values in Table 1 are seen to vary widely from -38 to $+16$ cal.deg^{-1}.mole^{-1}. However, the (formal) ionic entropies ($S^{\ddagger}_{complex}$) of the +4 activated complexes are similar (-65 to -79 cal.deg^{-1}.mole^{-1}) and comparable to those for reaction between two actinide ions[21]. Thus it is apparent that the prime factor in deciding the entropy of these types of activated complexes is the charge and not the size of the complex.

The reduction of V(V) to V(IV) by Fe(II) has been also studied by Nicol and Rosseinsky[27,28]. These authors used a polarographic method employing a rotating platinum electrode–calomel electrode pair. Under certain conditions of applied voltage, the diffusion current depends only upon the concentration of Fe(II). Nicol and Rosseinsky conclude, after studying the reaction over wider ranges of conditions than those of Daugherty and Newton[16], that the complex dependence on hydrogen-ion concentration is better described by

$$k' = b + c[H^+] + d[H^+]^2$$

They attribute this to the existence of three reaction paths, given by eqns. (2.2) and (2.3) together with

$$VO_2^+ + Fe^{2+} + 2\,H^+ \longrightarrow (H_2VO_2Fe^{5+})^\ddagger$$

At 25 °C and $\mu = 3.0$ *M*, *b*, *c* and *d* values are quoted of 203 l.mole^{-1}.sec^{-1}, 6730 l^2.mole^{-2}.sec^{-1}, and 1830 l^3.mole^{-3}.sec^{-1}, respectively.

When an excess of vanadium(V) is reacted with tin(II) in dilute hydrochloric acid media, V(IV) is produced in amounts equivalent to Sn(II) oxidised, and no V(III) is obtained[29]†. When vanadium(V) is reduced by an excess of tin(II) in the same media, both V(IV) and V(III) are formed as products: the ratio of V(V) reduced to V(III) formed is constant except when the ratio $[V(V)]_0/[Sn(II)]_0$ is close to unity. V(IV) and V(III) were determined spectrophotometrically at 755 mμ and 400 mμ, respectively. There is considerable evidence against the scheme

$$V(V) + Sn(II) \longrightarrow V(IV) + Sn(III) \tag{2.4}$$

$$V(V) + Sn(III) \longrightarrow V(IV) + Sn(IV) \tag{2.5}$$

$$V(IV) + Sn(III) \longrightarrow V(III) + Sn(IV) \tag{2.6}$$

$$V(V) + V(III) \longrightarrow 2\,V(IV) \tag{2.7}$$

in which the reactant species undergo one-equivalent changes. Instead, an alternative scheme, made up of a primary two-equivalent change

$$V(V) + Sn(II) \longrightarrow V(III) + Sn(IV) \tag{2.8}$$

coupled with steps (2.4), (2.5) and (2.7), is more in accord with the experimental results on stoichiometry. From a detailed analysis of the kinetic implications of this reaction sequence it is concluded that about 90 % of effective encounters between V(V) and Sn(II) result in the direct formation of V(III) and Sn(IV), whilst the remaining 10 % yield V(IV) and Sn(III). It is conceivable that the reaction involves an association complex represented as $(V+Sn)^{VII}$ which, in addition to decaying to V(III)+Sn(IV), or V(IV)+Sn(III) (steps (2.8) and (2.4), respectively), can persist for a sufficient time in solution to react further with V(V) to generate V(IV) and Sn(IV), *viz.*

$$(V+Sn)^{VII} + V(V) \longrightarrow 2\,V(IV) + Sn(IV) \tag{2.9}$$

Evidence has been cited for a similar 1 : 1 complex formed between Sn(II) and U(VI) in hydrochloric acid[30]. Path (2.9) is expected to become more important at relatively high initial concentrations of V(V). For conditions in which the

† Although the rate was found to be too great to measure in hydrochloric acid media, the V(V)+Sn(II) reaction has been investigated more recently in perchloric acid[275].

initial concentration of V(V) is in excess of Sn(II), the principal mechanism for the generation of V(IV) is step (2.8) in combination with step (2.7).

2.2 OXIDATIONS BY VANADIUM(IV)

A V(III)–Cr(III) dimer has been detected as an intermediate in the V(IV)+ Cr(II) reaction[31] and its formation and subsequent decomposition studied[32]. Since V(IV) was arranged to be in excess of Cr^{2+}, no complications ensued from the latter ion bringing about further reduction of V(III) to V(II). Under these conditions the overall reaction corresponds accurately to

$$VO^{2+}+Cr^{2+}+2\,H^{+} = V^{3+}+Cr^{3+}+H_2O$$

Kinetic data were obtained over wavelengths ranging from 260 to 760 mμ. The presence of an intermediate is readily apparent: on mixing pale-blue solutions of V(IV) and Cr(II) a bright green colour develops immediately, and then fades slowly to the blue-purple colour characteristic of V^{3+} and Cr^{3+}. From the response of the rate to variations in acidity, it is deduced that the reaction proceeds mainly *via* a rapid "direct" route and that only a small portion occurs *via* the intensely-coloured intermediate at a measurable rate. It is clear that the direct path cannot be a two-equivalent change, involving V(II) and Cr(IV) as intermediates, since the product Cr(III) species is the aquo ion, and not the dimer which would be produced from the interaction of Cr(IV) and Cr(II)[33]. Also, as the dimeric species VOV^{4+} (formed from V(IV)+V(II)[34]) is not detectable at 425 mμ, appreciable concentrations of V(II) cannot be present. The rate of disappearance of the intermediate (which follows first-order kinetics) increases as the hydrogen-ion concentration increases although the dependence is complex: over ranges of 0.010 to 0.99 M $HClO_4$ and 5° to 25 °C at μ = 1.00 M the following expression applies

$$k' = (AK+B[H^+])/(K+[H^+]) \qquad (2.10)$$

where A and B are constants, and K is the equilibrium constant for equilibrium between acid and base forms of the intermediate. Formulating the intermediate as $VO(OH)_nCr^{(4-n)+}$ and the equilibrium as

$$VO(OH)_nCr^{(4-n)+} \rightleftharpoons VO(OH)_{n+1}Cr^{(3-n)+}+H^+$$

then a comparable expression to (2.10), *viz.*

$$k' = (k_0K+k_1[H^+])/(K+[H^+])$$

can be obtained by assuming that the intermediate decays by the two competitive processes

$$VO(OH)_{n+1}Cr^{(3-n)+} \xrightarrow{k_0} \text{products}$$

$$VO(OH)_nCr^{(4-n)+} \text{ (and/or } VO(OH)_{n+1}Cr^{(3-n)+} + H^+) \xrightarrow{k_1} \text{products}$$

On this basis $k_0 = 0.0170$ sec^{-1}, $k_1 = 0.645$ sec^{-1}, and $K = 0.739$ mole.l^{-1} at 25 °C. The corresponding activation parameters were determined also by Espenson[32]. By a method involving extrapolation of the first-order rate plots at various wavelengths to zero time, the absorption spectrum of the intermediate was revealed (Fig. 1). Furthermore, the value of K obtained from the kinetics was compatible with that derived from measurements on the acid dependence of the spectrum of the intermediate. Rate data for a number of binuclear intermediates are collected in Table 2. Espenson[32] shows there to be a correlation between the rate of decomposition of the dimer and the substitution lability of the more labile metal ion component. The latter is assessed in terms of the rate of substitution of SCN^- in the hydration sphere of the more labile hydrated metal ion.

The one-equivalent oxidation of Eu(II) by V(IV) takes place rapidly: at 0 °C, 0.2 M [H^+] and ~ 10^{-4} M reactant concentrations, the second-order rate coefficient[39] is ~ 10^3 l.mole^{-1}.sec^{-1} (2.64×10^3 l.mole^{-1}.sec^{-1} at 25 °C and μ = 1.0 M[276]).

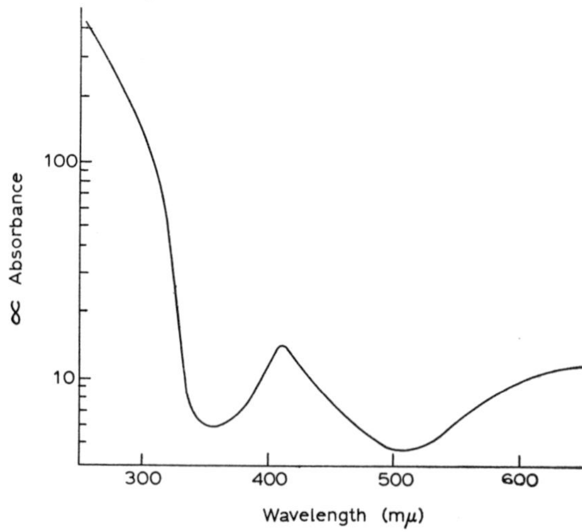

Fig. 1. Absorption spectrum of V(III)–Cr(III) dimer formed as an intermediate in the V(IV)+Cr(II) reaction (the ordinate is directly proportional to the absorbance); 0.200 M $HClO_4$; temp., 15.0 °C. (*From Espenson*[32], *by courtesy of The American Chemical Society.*)

TABLE 2

DECOMPOSITION RATES OF BINUCLEAR METAL COMPLEXES[a]

Mode of formation	*Binuclear species*	*Specific decomposition rate at 25.0 °C (sec^{-1})*	*log k for formation of SCN$^-$ complex*[b]	*Ref.*
Cr(IV)+Cr(II) and Cr(III)+OH$^-$	$CrOHCr^{5+}$	1.5×10^{-6}	Cr(III), −5.7	33b, 36
Np(VI)+Cr(III) and Np(V)+Cr(II)	NpO_2Cr^{4+}	2.3×10^{-6}	Cr(III), −5.7	37
V(IV)+V(II) and V(III)+OH$^-$	VOV^{4+}	$0.03 + 1.54[H^+]$	V(III), +1.8	34
V(IV)+Cr(II)	$VO(OH)_nCr^{(4-n)+}$	$0.017 + 0.645[H^+]$	V(III), +1.8	32
Fe(IV)+Fe(II) and Fe(III)+OH$^-$	$FeOFe^{4+}$	$0.35 + 3.5[H^+]$	Fe(III), +2.1	38

[a] From Espenson[32]. [b] From ref. 35 (k in l.mole^{-1}.sec^{-1}).

2.3 OXIDATIONS BY VANADIUM(III)

The kinetics of the oxidation of chromium(II) by vanadium(III) in acid perchlorate media have been studied spectrophotometrically between 0.2° and 35.0 °C over a range of 0.027–0.500 M $HClO_4$[40]. The oxygen-sensitivity of both reactants meant that the air had to be excluded in all kinetic runs. Also, since V(III) slowly reduces perchlorate ion, fresh solutions of V(III) were required for each experiment. In terms of stoichiometry the reaction conforms accurately to

$$V^{3+} + Cr^{2+} = V^{2+} + Cr^{3+}$$

and the rate law is given by

$$-d[V^{3+}]/dt = k'[V^{3+}][Cr^{2+}]$$

over wide ranges of wavelengths and reactant concentrations. No direct evidence could be adduced for the existence of intermediates. The rate response to variations in acidity demonstrates that the observed rate coefficient, k', can be considered in terms of two empirical parameters q and r

$$k' = q/(r + [H^+]) \tag{2.11}$$

To account for the inverse dependence of k' on $[H^+]$, Espenson[40] has proposed the following general mechanism

$$V^{3+} + Cr^{2+} + n\,H_2O \underset{k_{-1}}{\overset{k_1}{\rightleftharpoons}} V(OH)_nCr^{(5-n)+} + n\,H^+ \tag{2.12}$$

$$V(OH)_nCr^{(5-n)+} + (n-1)\,H^+ \xrightarrow{k_2} (e.g.,\ V^{2+} + CrOH^{2+}) \tag{2.13}$$

$$(CrOH^{2+} + H^+ \rightleftharpoons Cr^{3+} + H_2O,\ \text{rapid equilibrium}) \tag{2.14}$$

It will be noted that conversion of the intermediate $V(OH)_nCr^{(5-n)+}$ to products involves a different number of H^+ ions than its conversion back to reactants. It is considered likely that the binuclear intermediate has an inner-sphere structure. On applying the steady-state approximation to the concentration of this intermediate, it follows that

$$k' = (k_1 k_2/k_{-1})/((k_2/k_{-1}) + [H^+]) \tag{2.15}$$

Comparison of equations (2.11) and (2.15) reveals q and r to be k_1k_2/k_{-1} and k_2/k_{-1}, respectively. This enables k_1 to be calculated from q/r. In its simplest forms the structure of the reactive intermediate can be viewed as $V(OH)Cr^{4+}$ (when n is 1) or as $VOCr^{3+}$ (when n is 2). Similar species which have been characterized or implied kinetically are $CrOCr^{4+}$ (ref. 33), NpO_2Cr^{4+} (ref. 37), UO_2Cr^{4+} (ref. 31), VOV^{4+} (ref. 34), $UOPuO_2{}^{4+}$ (ref. 41), PuO_2Fe^{4+} (ref. 42) and $FeOFe^{4+}$ (ref. 38). Predictions on the rate of the V(III)+Cr(II) system, based upon Marcus theory[43], have been made by Dulz and Sutin[44] on the assumption that an outer-sphere process applies. The value arrived at by these authors is ~ 60 times lower than the experimental value.

Haim[45] has commented on the ambiguities inherent in the interpretation of rate laws, using the V(III)+Cr(II) reaction as an example. The experimental rate law found by Espenson[40]

$$\text{rate} = \frac{q[V^{3+}][Cr^{2+}]}{r + [H^+]}$$

has the limiting forms

$$\text{rate} = q[V^{3+}][Cr^{2+}]/r \text{ at low } [H^+]$$

and

$$\text{rate} = q[V^{3+}][Cr^{2+}]/[H^+] \text{ at high } [H^+]$$

It follows that the reaction proceeds *via* two consecutive activated complexes:

$(VCr^{5+})^{\ddagger}$ and $(V(OH)Cr^{4+})^{\ddagger}$. In Espenson's mechanism, equations (2.12) to (2.14), $(VCr^{5+})^{\ddagger}$ is formed directly from the interaction of $V^{3+}+Cr^{2+}$, an event which is followed by the spontaneous aquation of $(V(OH)Cr^{4+})^{\ddagger}$. However, as Haim points out, $(V(OH)Cr^{4+})^{\ddagger}$, formed from $VOH^{2+}+Cr^{2+}$, may precede $(VCr^{5+})^{\ddagger}$, formed from $V(OH)Cr^{4+}+H^{+}$, as shown by

$$V^{3+}+H_2O \rightleftharpoons VOH^{2+}+H^{+} \text{ rapid equilibrium, } K \tag{2.16}$$

$$VOH^{2+}+Cr^{2+} \underset{k_{-3}}{\overset{k_3}{\rightleftharpoons}} V(OH)Cr^{4+} \tag{2.17}$$

$$V(OH)Cr^{4+}+H^{+} \xrightarrow{k_4} V^{2+}+Cr^{3+}+H_2O \tag{2.18}$$

On this mechanism

$$\text{rate} = \frac{k_3\, k_4\, K[V^{3+}][Cr^{2+}]}{k_{-3}+k_4[H^{+}]}$$

whence $k_3 = 3.12\times10^2$ l.mole^{-1}.sec^{-1} and $k_{-3}/k_4 = 0.108$ mole.l^{-1} (at 25 °C), as compared with $k_1 = 5.76$ l.mole^{-1}.sec^{-1} and $k_2/k_{-1} = 0.108$ mole.l^{-1}. The experimental rate law, whilst defining the compositions of the activated complexes, is incapable of defining the sequence in which they are formed.

Adin and Sykes[46] have re-examined the hydrogen-ion dependence of the V(III)+Cr(II) system over a broad range of H^{+} concentrations from 0.45 *M* down to 0.016 *M*. They confirm the type of acid dependence quoted by Espenson and support the interpretation given by Haim, equations (2.16) to (2.18)†. At 25 °C and $\mu = 0.5$ *M*, the experimental parameters q and r are 0.50 and 0.10, respectively, so that $k_3 = 3.57\times10^2$ l.mole^{-1}.sec^{-1}, and $k_{-3}/k_4 = 0.1$ mole.l^{-1}. Adin and Sykes[46] find no evidence for the existence of the more complex hydrogen-ion dependence originally suggested by Sykes[47] in a re-analysis of Espenson's data[40].

The reduction of V(III) by Eu(II)

$$V(III)+Eu(II) = V(II)+Eu(III)$$

has been examined in perchlorate media[48]. Contrary to an earlier report[49], Eu(II) is not oxidised by perchlorate ions. A spectrophotometric method was employed in the kinetic work, the formation of V(II) being monitored at its peak of 850 mμ. Oxygen was rigorously excluded in all kinetic runs. The rate law is

† In a later paper, Espenson and Parker[277], reporting on an extensive study of the V(III)+Cr(II) system in chloride and perchlorate media, favour Haim's mechanism on grounds of reactivity patterns.

$$d[(V(II)]/dt = (k_5 + k'[H^+]^{-1})[V(III)][Eu(II)]$$

which is indicative of the steps

$$\left.\begin{array}{l} V^{3+} + Eu^{2+} \xrightarrow{k_5} \\ VOH^{2+} + Eu^{2+} \xrightarrow{k_6} \end{array}\right\} \text{products} \qquad \begin{array}{r}(2.19)\\(2.20)\end{array}$$

At 25 °C, k_5 is 9.0×10^{-3} l.mole^{-1}.sec^{-1} and $k_6 = k'/K = 2.0$ l.mole^{-1}.sec^{-1} (K is the hydrolysis constant of V^{3+}). Steps (2.19) and (2.20) have associated $\Delta H^{\ddagger}$ values of 11.4 and 6.2 kcal.mole^{-1}, and $\Delta S^{\ddagger}$ values of -30.1 ± 5 and $\sim -35 \pm 5$ cal.deg^{-1}.mole^{-1}, respectively. The much slower rate of the V(III)+Eu(II) reaction, as compared to the V(III)+Cr(II) system, is ascribed by Adin and Sykes to the relative difficulty in transferring electrons from *f*-orbitals than from *d*-orbitals. The sensitivity of the Eu(II) reaction to chloride ions is noted but not reported in detail.

3. Oxidations by chromium and molybdenum

3.1 OXIDATIONS BY CHROMIUM(VI)

The stoichiometric equation for oxidation of vanadium(IV) by chromium(VI) in acid perchlorate solutions is essentially

$$HCrO_4^- + 3\,VO^{2+} + H^+ = Cr^{3+} + 3\,VO_2^+ + H_2O$$

That $HCrO_4^-$, VO^{2+} and VO_2^+ are the predominant species under the conditions of Espenson's kinetic study[50] originates from evidence cited by Tong and King[51], and Rossotti and Rossotti[52], respectively. No binuclear species are detectable in the Cr(III) product. Vanadium(V) retards the reaction and the full form of the rate law is

$$\frac{-d[HCrO_4^-]}{dt} = \frac{[VO^{2+}]^2}{[VO_2^+]}(k'[HCrO_4^-] + k''[H^+][HCrO_4^-]^2) \qquad (3.1)$$

where $k' = 0.563$ l.mole^{-1}.sec^{-1} and $k'' = 5.4 \times 10^4$ l^3.mole^{-3}.sec^{-1} at 25 °C and $\mu = 1\ M$. The first term of the rate law predominates at low concentrations of Cr(VI) ($\leqq 4 \times 10^{-5}\ M$) and at low acidities ($\leqq 0.03\ M$). At higher concentrations of Cr(VI) and/or H^+ (such that $[HCrO_4^-][H^+]$ exceeds $2 \times 10^{-6}\ M^2$) the second term becomes important. Restricting attention to the first term, the form suggests a transition state of $(HCrO_4V^{2+} \pm n\,H_2O)^{\ddagger}$ with an average oxidation number for Cr and V of 4.5, *i.e.*, 0.5 $(2\times4+6-5)$. Such a transition state could stem from a V(III)+Cr(VI) or V(IV)+Cr(V) combination. In terms of oxidation numbers the possible mechanisms involve either disproportionation of V(IV) followed by a two-equivalent oxidation of V(III) (*A*), or a sequence of three one-equivalent steps (*B*); *viz.*

$$2\,\mathrm{V(IV)} \underset{k_{-1}}{\overset{k_1}{\rightleftharpoons}} \mathrm{V(V)+V(III)} \qquad \text{rapid, } K$$

$$(A)\quad \mathrm{V(III)+Cr(VI)} \xrightarrow{k_2} \mathrm{V(V)+Cr(IV)} \qquad \text{slow}$$

$$\mathrm{V(IV)+Cr(IV)} \rightleftharpoons \mathrm{V(V)+Cr(III)} \qquad \text{rapid}$$

or

$$\mathrm{V(IV)+Cr(VI)} \rightleftharpoons \mathrm{V(V)+Cr(V)} \qquad \text{rapid, unfavourable} \qquad (3.2)$$

$$(B)\quad \mathrm{V(IV)+Cr(V)} \longrightarrow \mathrm{V(V)+Cr(IV)} \qquad \text{slow} \qquad (3.3)$$

$$\mathrm{V(IV)+Cr(IV)} \rightleftharpoons \mathrm{V(V)+Cr(III)} \qquad \text{rapid} \qquad (3.4)$$

If mechanism (*A*) applied the Cr(VI)+V(IV) system would be anomalous when compared with the Cr(VI)+Fe(II) and Ce(IV)+Cr(III) reactions which have similar rate laws and Cr(V) → Cr(IV) transformations as rate-controlling steps. Apart from this there are other good reasons for rejecting mechanism (*A*). At 25 °C, K is 10^{-10} and k' is 0.56 $\mathrm{l.mole^{-1}.sec^{-1}}$, allowing k_2 to be calculated as 0.56×10^{10} $\mathrm{l.mole^{-1}.sec^{-1}}$ (since $k' = k_2K$). Furthermore, V(V) can only retard the reaction if the intermediate V(III) reacts with V(V) in preference to Cr(VI), *i.e.*, the conditions $k_{-1}[\mathrm{V(V)}] \gg k_2[\mathrm{Cr(VI)}]$ must apply. A value of $k_{-1} = 1.4\times10^4$ $\mathrm{l.mole^{-1}.sec^{-1}}$ at 25 °C has been obtained by Daugherty and Newton[53] from a study of the V(III)+V(V) system. Using this value would mean that the ratio [V(V)]/[Cr(VI)] should be very much larger than $4\times10^5(k_2/k_{-1})$. Mechanism (*A*) can be discarded on the grounds that this result is much higher than that achieved experimentally (the highest value of [V(V)]/[Cr(VI)] is $\sim 3\times10^2$). Mechanism (*B*) can be made more explicit by making use of the result that the rate is independent of $[\mathrm{H^+}]$. Thus, steps (3.2) and (3.3) can be rewritten as

$$\mathrm{VO^{2+}+HCrO_4^-+H_2O} \rightleftharpoons \mathrm{VO_2^++H_3CrO_4} \qquad \text{rapid}$$

$$\mathrm{VO^{2+}+H_3CrO_4} \rightarrow \mathrm{(H_3CrO_4VO^{2+})^\ddagger} \qquad \text{slow}$$

on the assumption that Cr(V) exhibits a coordination number of 4. The second term of the rate law (3.1), in which there is a second-order dependence on $[\mathrm{HCrO_4^-}]$, signifies the participation of $\mathrm{Cr_2O_7^{2-}}$ as a reactant†, and a transition

† The species $\mathrm{HCrO_4^-}$ and $\mathrm{Cr_2O_7^{2-}}$ are in equilibrium by

$$2\,\mathrm{HCrO_4^-} \rightleftharpoons \mathrm{Cr_2O_7^{2-}+H_2O}$$

The equilibrium constant has been measured spectrophotometrically[51] as 98 $\mathrm{l.mole^{-1}}$ at 25 °C and $\mu = 1\,M$.

state, consisting of two Cr atoms and one V atom, of composition $(HCr_2O_7V^{2+} \pm m\,H_2O)^{\ddagger}$. A reasonable mechanism, similar to (*B*), would have $Cr_2O_7^{2-}$ as a reactant in the first step and a dimer of Cr(V) and Cr(VI) as the intermediate in the slow stage.

The kinetics of the oxidation of iron(II) by chromium(VI)

$$HCrO_4^- + 3\,Fe^{2+} + 7\,H^+ = Cr^{3+} + 3\,Fe^{3+} + 4\,H_2O$$

were studied first by Benson[54] and subsequently by Gortner[55] and by Wagner and Preiss[56]. The early literature is discussed in detail by Westheimer[57]. More recently the reaction has been the subject of a comprehensive investigation by Espenson and King[58] under conditions where the dominant species of Cr(VI) and Fe(III) are $HCrO_4^-$, $FeCrO_4^+$, and $Fe(H_2O)_6^{3+}$. There is no evidence to suggest that the product Cr(III) is other than the simple hexaaquo ion; the formation of a dimeric species of Cr(III) is considered unlikely. The extent of complex formation between Fe^{3+} and $HCrO_4^-$, *viz.*

$$Fe^{3+} + HCrO_4^- \rightleftharpoons FeCrO_4^+ + H^+ \qquad K_1$$

received detailed study and a value of $K_1 = 1.4$ was assigned at 0 °C and $\mu = 0.0839\,M^{\dagger}$. Effectively, this result means that under these conditions ~ 8 to ~ 40 % of Cr(VI) is present as Fe(III)–chromate complex. Rate measurements on the Cr(VI)+Fe(II) reaction were made both by a direct spectrophotometric approach, and by the titrimetric method first used by Benson[54] (utilising the induced oxidation of iodide ion). Iron(III) retards the reaction and the corresponding rate law takes the form

$$\frac{-d[Fe^{2+}]}{dt} = \frac{[Fe^{2+}]^2[H^+]^3}{[Fe^{3+}]}(k'[HCrO_4^-] + k''[HCrO_4^-]^2) \tag{3.5}$$

with $k' = 6.2 \times 10^8$ $l^4.mole^{-4}.sec^{-1}$ and $k'' = 2.2 \times 10^{12}$ $l^5.mole^{-5}.sec^{-1}$ at 0 °C. The gross features of the mechanism are

$$Cr(VI) + Fe(II) \rightleftharpoons Cr(V) + Fe(III)$$

$$Cr(V) + Fe(II) \rightarrow Cr(IV) + Fe(III) \qquad \text{rate-determining}$$

$$Cr(IV) + Fe(II) \rightarrow Cr(III) + Fe(III) \qquad \text{rapid}$$

The first term of (3.5) implies a corresponding net activation process[21] represented

† See, also, ref. 278.

by

$$HCrO_4^- + 2\,Fe^{2+} + 3\,H^+ + (n-2)\,H_2O \rightarrow [FeOCrO(OH_2)_n{}^{3+}]^{\ddagger} + Fe^{3+}$$

That is, the activated complex contains one Cr(V) atom and one Fe(II) atom. Espenson[50b] has shown, from a consideration of the induced oxidation of iodide ion, that the reaction between Cr(VI) and Fe(II) requires one added proton. Consequently, the $[H^+]^3$-dependence of the rate can be viewed as the addition of two protons in a pre-equilibrium followed by the addition of a further proton in the slow step, *viz.*

$$HCrO_4^- + Fe^{2+} + 2\,H^+ \rightleftharpoons H_3CrO_4 + Fe^{3+} \qquad \text{rapid}$$

$$H_3CrO_4 + Fe^{2+} + H^+ \rightarrow (H_4CrO_4Fe^{3+})^{\ddagger} \qquad \text{slow}$$

The second term of (3.5) is indicative of an activated complex containing two Cr atoms with an average oxidation number of +5.5 together with one Fe(II) atom. This result suggests that the reactive entity in the rate-controlling step may well involve one Cr(V) atom and one Cr(VI) atom, *e.g.*, $Cr_2O_7H_n{}^{(n-3)+}$. Rosseinsky and Nicol[279], while finding no evidence for this pathway, have detected an additional term first-order in Fe(II).

The kinetics of the reduction of Cr(VI) by tris(1,10-phenanthroline)iron(II), *viz.*

$$HCrO_4^- + 3\,Fe(phen)_3{}^{2+} + 7\,H^+ = Cr^{3+} + 3\,Fe(phen)_3{}^{3+} + 4\,H_2O$$

have been studied by Espenson and King[58] over the temperature range 0–40 °C by a spectrophotometric method. No unequivocal conclusions can be drawn about the mechanism from the rate data except the lack of effect of $Fe(phen)_3{}^{3+}$ concentration on the rate suggests that reduction of a Cr(V) species is not rate-determining. In addition, the rate of the $Cr(VI) + Fe(phen)_3{}^{2+}$ reaction is much smaller than the $Cr(VI) + Fe^{2+}$ reaction. This may be due to different thermodynamic tendencies in the primary and rate-determining step

$$Cr(VI) + Fe(II) \rightarrow Cr(V) + Fe(III)$$

as reflected by differences in equilibrium constants of $\sim 10^6$. A further factor contributing to the greater rate of the Cr(VI)+Fe(II) reaction may be the availability of an inner-sphere transition state with an oxygen-bridged structure, Fe–O–Cr. In contrast, reaction of $Fe(phen)_3{}^{2+}$ must needs occur by an outer-sphere path[1]. These interpretations are supported by the results of Birk[280] on reactions of Cr(VI) with other Fe(II) complexes.

In the multi-equivalent oxidation of Sn(II) by Cr(VI)

$$2\,\mathrm{Cr(VI)}+3\,\mathrm{Sn(II)} = 2\,\mathrm{Cr(III)}+3\,\mathrm{Sn(IV)}$$

there is some evidence for the presence of Sn(III) or a related reducing species[59]. In the Mo(V)+Sn(II) system, however, formation of intermediates seems insignificant[281].

Mason and Kowalak[60] have examined the oxidation of As(III) by Cr(VI) in 0.2 *M* acetic acid–0.2 *M* potassium acetate buffers. When As(III) is in large excess over Cr(VI) the rate law is

$$\frac{-\mathrm{d[Cr(VI)]}}{\mathrm{d}t} = \frac{k_3 K_2[\mathrm{Cr(VI)}][\mathrm{As(III)}]}{1+K_2[\mathrm{As(III)}]}$$

in accord with the scheme

$$\mathrm{As(III)}+\mathrm{HCrO_4^-} \rightleftharpoons \mathrm{As(III)\cdot HCrO_4^-} \qquad \text{pre-equilibrium, } K_2$$

$$\mathrm{As(III)\cdot HCrO_4^-} \xrightarrow{k_3} \text{products} \qquad \text{rate-determining}$$

At 25 °C, and with the ionic strength adjusted to 1.5 *M* with potassium nitrate, the values of k_3 and K_2 are 3.76×10^{-4} $\mathrm{sec^{-1}}$ and 22.4 $\mathrm{l.mole^{-1}}$, respectively. At higher concentrations of Cr(VI) an analogous mechanism applies, $HCrO_4^-$ being replaced by $Cr_2O_7^{2-}$. No evidence is cited for the formation of As(IV); it seems likely that As(V) and Cr(IV) are formed in the slow step.

Kinetic data for the oxidation of neptunium(V) by chromium(VI), *viz.*

$$\mathrm{Cr(VI)}+3\,\mathrm{Np(V)} = \mathrm{Cr(III)}+3\,\mathrm{Np(VI)}$$

have been obtained by spectrophotometric means at wavelengths of 350 mμ [where the absorbance is largely due to Cr(VI)] and 980 mμ [where the absorbance is entirely due to Np(V)][61]. Formally, the process may be described as the simple transfer of one of the two available 5*f* electrons from Np(V) to the oxidant since both Np(V) and Np(VI) have the same linear O–Np–O structure. Fresh solutions of neptunium were prepared directly before each kinetic experiment to reduce the possibility of peroxide formation by radiolysis of the solutions by α-particles from the decay of ^{237}Np. The empirical rate law is not in line with the rate laws of the Cr(VI)+Fe(II) (p. 164) and Cr(VI)+V(IV) (p. 162) reactions and was derived by a computer programme designed to approximate, from a set of experimental data, the first derivative of an unknown function. At constant acidity the derived law takes the form

$$\frac{-\mathrm{d[NpO_2^+]}}{\mathrm{d}t} = \frac{k'[\mathrm{NpO_2^+}][\mathrm{Cr(VI)}]}{1+(k''[\mathrm{NpO_2^{2+}}]/[\mathrm{NpO_2^+}])} \qquad (3.6)$$

and has the following indefinite integral

$$t = A \log \left\{ \frac{[NpO_2^+]}{[Cr(VI)]_0 - [NpO_2^+]_0/3 + [NpO_2^+]/3} \right\} + \frac{B}{[NpO_2^+]} + C \quad (3.7)$$

in which the subscripts refer to initial concentrations of reactants, and the parameters A and B are defined in terms of initial concentrations and rate coefficients k' and k''. Values for A, B and C, computed from rate data (concentration of Np(V) *versus* time) using equation (3.7), were used to evaluate k' and k''. Agreement was attained between these results and those calculated directly from equation (3.6). The rate law was checked by observing the constancy of k' and k'' (at constant perchloric acid concentration and ionic strength) over a range of initial concentrations of Cr(VI), Np(V) and Np(VI). The average values are $k' = 8.26$ l.mole^{-1}.sec^{-1} and $k'' = 0.42$ at $[H^+] = 1.51$ M, $\mu = 2.00$ M and 25 °C. As a mechanism, Sullivan[61] has proposed a sequence of one-equivalent steps

$$Cr(VI) + NpO_2^+ \underset{k_{-4}}{\overset{k_4}{\rightleftharpoons}} Cr(V) + NpO_2^{2+}$$

$$Cr(V) + NpO_2^+ \overset{k_5}{\rightleftharpoons} Cr(IV) + NpO_2^{2+}$$

$$Cr(IV) + NpO_2^+ \rightarrow Cr(III) + NpO_2^{2+} \qquad \text{rapid}$$

On this basis Cr(V), not Cr(IV), is the kinetically important intermediate such that $k' = 3\,k_4$ and $k'' = k_{-4}/k_5$. The hydrogen-ion dependence of the reaction rate has been discussed. Furthermore, comparisons are drawn with the rate of the $Cr(VI) + Fe(phen)_3^{2+}$ reaction[58], and Sullivan has speculated on the intimate nature of both mechanisms in the light of Marcus theory[43].

3.2 OXIDATIONS BY CHROMIUM(III)

Oxidation of hexaaquovanadium(II) by $Cr(H_2O)_5SCN^{2+}$ occurs in two stages[62,282]

$$CrSCN^{2+} + V^{2+} \xrightarrow{k_1} Cr^{2+} + VNCS^{2+} \quad (3.8)$$

$$VNCS^{2+} \xrightarrow{k_2} V^{3+} + SCN^-$$

A value for k_1 of 9.8 l.mole^{-1}.sec^{-1}, at $[H^+] = 1.0$ M, $\mu = 1.0$ M and 25 °C, was obtained by following the rate of disappearance of $CrSCN^{2+}$ at 262 mμ[282]. By contrast, the rate coefficient for the reaction of $CrNCS^{2+}$ with V^{2+} is 1.7×10^{-4} l.mole^{-1}.sec^{-1} under the same conditions[62]. The formation and subsequent disappearance of $VNCS^{2+}$ was followed at 350 mμ and 400 mμ: k_2 is quoted as 0.99 sec^{-1} at 25 °C and $\mu = 1.0$ *M*. Reaction of $CrSCN^{2+}$ with Cr^{2+}, *viz.*[63,64]

$$CrSCN^{2+} + Cr^{2+} \longrightarrow Cr^{2+} + CrNCS^{2+}$$

is negligible in comparison to reaction (3.8). Baker *et al.*[62] suggest that the reaction of $CrSCN^{2+}$ with V^{2+} takes place by a thiocyanato-bridged transition state, in which the anion is bonded to both metal atoms.

The equilibrium constant of the system

$$\text{Cr(III)} + \text{Eu(II)} \underset{k_{-3}}{\overset{k_3}{\rightleftharpoons}} \text{Cr(II)} + \text{Eu(III)}$$

is estimated[39,48] (from redox potentials) to be 2.2. Spectrophotometric measurements (at 575 mμ, the absorption maximum for Cr^{3+}) reveal both forward and backward reactions to be extremely slow (5 % reaction in one day). Approximate values (± 20 %) are $k_3 \sim 1.7 \times 10^{-5}$ $l.mole^{-1}.sec^{-1}$ and $k_{-3} \sim 1.4 \times 10^{-5}$ $l.mole^{-1}.sec^{-1}$ at 25 °C and 0.5 *M* $HClO_4$. The addition of chloride ion, whilst having no influence on k_3, catalyses the back reaction appreciably, and

$$d[\text{Cr(III)}]/dt = k_{-4}[\text{Cr(II)}][\text{Eu(III)}][\text{Cl}^-]$$

k_{-4} being 9.8×10^{-4} $l^2.mole^{-2}.sec^{-1}$ under the same conditions. The substitution-inert $CrCl^{2+}$ species is formed as a product. With $CrCl^{2+}$ as the Cr(III) species, the reduction

$$CrCl^{2+} + Eu^{2+} \xrightarrow{k_4} Cr^{2+} + Eu^{3+} + Cl^-$$

has a rate coefficient k_4 of 2.23×10^{-3} $l.mole^{-1}.sec^{-1}$ at 25 °C and activation parameters $\Delta H^{\ddagger}$ and $\Delta S^{\ddagger}$ of 17.1 $kcal.mole^{-1}$ and 14.0 ± 5 $cal.deg^{-1}.mole^{-1}$, respectively. The marked increase in reactivity of $CrCl^{2+}$ as compared with Cr^{3+} indicates that an inner-sphere activated complex is involved. Although $(EuClCr^{4+})^{\ddagger}$ seems the likely structure, the other possibilities, $(EuH_2OClCr^{4+})^{\ddagger}$ and $(EuH_2OCrCl^{4+})^{\ddagger}$, cannot be excluded. It is of interest that, in the reactions of Eu(II) with Cr(III), V(III), V(IV) and Fe(III), there is an approximately linear relationship between the logarithms of the rate coefficients and the standard free-energy changes.

Adin and Sykes[65] have reported on the reduction of monohalogenochromium(III) complexes by Eu(II)

$$CrX^{2+} + Eu^{2+} \longrightarrow Cr^{2+} + Eu^{3+} + X^-$$

in perchloric acid solutions at $\mu = 1.0$ *M*. Corrections were applied for the simultaneous uncatalysed and Cr(II)-catalysed aquations of CrX^{2+}, *viz.*

$$CrX^{2+} \longrightarrow Cr^{3+} + X^-$$

$$Cr^{2+} + CrX^{2+} \longrightarrow Cr^{3+} + Cr^{2+} + X^-$$

At 25 °C the rate coefficients ($l.mole^{-1}.sec^{-1}$) for the reductions are given as CrI^{2+} (4.1×10^{-2}) > $CrBr^{2+}$ (3.4×10^{-3}) > $CrCl^{2+}$ (1.4×10^{-3}) > CrF^{2+} (6.0×10^{-4}). The reactivity sequence follows the "normal" order of bridging efficiency (see p. 194) and is the reverse to that found in the reactions of Eu(II) and Fe(II) with $Co(NH_3)_5X^{2+}$ (p. 190).

3.3 OXIDATIONS BY MOLYBDENUM(V)

The rate of oxidation of tris(2,2'-bipyridine)osmium(II) by octacyanomolybdate(V)

$$Mo(CN)_8{}^{3-} + Os(bipy)_3{}^{2+} \underset{k_r}{\overset{k_f}{\rightleftharpoons}} Mo(CN)_8{}^{4-} + Os(bipy)_3{}^{3+}$$

has been measured by the temperature-jump method[66]. In practice, application of this method to a rapid bimolecular electron-transfer can only be successful if the equilibrium constant of the system is close to unity. Consequently, for the perturbation to effect a measurable shift in the equilibrium position the system must show a considerable $\Delta S°$ value, as occurs, for example, in reactions between oppositely-charged species. The change in equilibrium between $Mo(CN)_8{}^{3-}$ and $Os(bipy)_3{}^{2+}$ followed by recording the absorbance of the latter at 480 mμ as a function of time after a temperature difference of ~ 10° had been induced. Calculation revealed k_f and k_r to be 2.0×10^9 $l.mole^{-1}.sec^{-1}$ and 4.0×10^9 $l.mole^{-1}.sec^{-1}$, respectively, at 10 °C and $\mu = 0.50$ *M*. These values are close to the diffusion-controlled limits as given by the Debye equation[67]. Application of Marcus theory[43] (see p. 247) allows rate coefficients to be estimated for the $Os(bipy)_3{}^{2+} + Os(bipy)_3{}^{3+}$ and $Mo(CN)_8{}^{4-} + Mo(CN)_8{}^{3-}$ exchange reactions: the calculated values are 1×10^7 $l.mole^{-1}.sec^{-1}$ and 3×10^4 $l.mole^{-1}.sec^{-1}$ whereas those found experimentally[68] are $> 1 \times 10^5$ $l.mole^{-1}.sec^{-1}$ and $\sim 3 \times 10^4$ $l.mole^{-1}.sec^{-1}$, respectively, at 10 °C and $\mu = 0$.

4. Oxidations by manganese and rhenium

4.1 OXIDATIONS BY MANGANESE(VII)

The kinetics of the oxidation of $Fe(CN)_6{}^{4-}$ by permanganate have been examined in phosphate buffers over the pH range[69] 1.6–6.3. The stoichiometry of the reaction was confirmed by measurements on the product $Fe(CN)_6^{3-}$ at 420 mμ. The technique employed to obtain kinetic data incorporated a rapid-mixing device; the rate of disappearance of MnO_4^-($Fe(CN)_6^{4-}$ in excess) was followed at 520 mμ using photographic recording of oscilloscope traces[70]. The reaction is first-order in both reactants; second-order rate coefficients were deduced from the equation

$$\log\left[\frac{A_t}{A_t+\varepsilon' c}\right] = -\frac{5ck't}{2.303} + \text{constant}$$

where A_t is the absorbance of MnO_4^-, $c = ([Fe(CN)_6^{4-}]_0/5)-[MnO_4^-]_0$ and ε' is the molar absorptivity of MnO_4^-. At 15.1 °C in phosphate buffer of pH 6.30 and $\mu = 0.0933$ M, the rate coefficient k' is 2.64×10^4 l.mole^{-1}.sec^{-1}. This value was verified by monitoring the rate of formation of $Fe(CN)_6^{3-}$ at 420 mμ, by which method $k' = 2.56\times10^4$ l.mole^{-1}.sec^{-1}. Increase in acid concentration increases the rate of reaction; activation parameters are quoted at low and high acidities. In the range pH 5–6 the rate-controlling step may involve an ion-pair formed with K^+ (or Na^+) from the buffer

$$K^+ + Fe(CN)_6^{4-} \rightleftharpoons KFe(CN)_6^{3-} \quad \text{rapid,} \quad K_1 = 350 \text{ l.mole}^{-1}$$

$$MnO_4^- + KFe(CN)_6^{3-} \xrightarrow{k_0} MnO_4^{2-} + KFe(CN)_6^{2-} \quad \text{slow}$$

As the pH is reduced to pH 3 the rate increases, and eventually achieves a limiting value at ~ pH 2.5. The following scheme, with protonated species, is suggested as likely at pH 1–2.

$$\left.\begin{array}{l} H^+ + Fe(CN)_6^{4-} \rightleftharpoons HFe(CN)_6^{3-} \\ H^+ + HFe(CN)_6^{3-} \rightleftharpoons H_2Fe(CN)_6^{2-} \end{array}\right\} \text{rapid equilibria}$$

$$MnO_4^- + H_2Fe(CN)_6^{2-} \xrightarrow{k_2} MnO_4^{2-} + H_2Fe(CN)_6^- \quad \text{slow}$$

The reaction between MnO_4^- and $HFe(CN)_6^{3-}$ (rate coefficient, k_1) is important in the intermediate pH region. Detailed treatment of the data yields a value of 17.4×10^4 l.mole^{-1}.sec^{-1} for k_1. The resemblance of k_1 to k_2 (52.0×10^4 l. mole^{-1}.sec^{-1}) suggests that, in terms of intimate mechanism, a single proton is necessary as a bridge between the reactants, electrostatic effects being unimportant. Comparison of the rate parameters with those calculated on the basis of Marcus theory[43] suggests that an outer-sphere process obtains. Replacement of K^+ by Na^+ ion in the buffer solution is shown not to produce specific cation effects.

Permanganate oxidises formatopentaamminecobalt(III) in a complicated fashion[71], the over-all reaction being compounded of a mixture of

$$MnO_4^- + 3\ Co(NH_3)_5{\cdot}OCHO^{2+} \longrightarrow 3\ CO_2 + 3\ Co^{2+} + MnO_2$$

and

$$2\ MnO_4^- + 3\ Co(NH_3)_5{\cdot}OCHO^{2+} \longrightarrow 3\ CO_2 + 3\ (NH_3)_5Co{\cdot}OH_2^{3+} + 2\ MnO_2$$

The ratio of unreduced to reduced cobalt is variable and is dependent on the concentration of MnO_4^-, *viz.*

$$[(NH_3)_5Co{\cdot}OH_2^{3+}]/[Co^{2+}] = 3 \times 10^2[MnO_4^-]$$

By way of contrast, oxidation of the organic ligand in oxalatopentaamminecobalt(III)[72] and *p*-aldehydobenzoatopentaamminecobalt(III)[73] is accompanied by reduction of the cobalt(III) centre in the case of one-equivalent oxidants, *e.g.* Ce(IV), but not in the case of two-equivalent oxidants (*e.g.* Cl_2). The rate law is simple

$$-d[Co(NH_3)_5{\cdot}OCHO^{2+}]/dt = k_3[Co(NH_3)_5{\cdot}OCHO^{2+}][MnO_4^-]$$

k_3, independent of $[H^+]$, is $\sim 2.5 \times 10^{-2}$ l.mole^{-1}.sec^{-1} at 0.1 °C, and 0.21 l.mole^{-1}.sec^{-1} at 25 °C, $[HClO_4] = 0.1\ M$ and $\mu = 1.0\ M$. Replacement of hydrogen by deuterium in the formato group, whilst bringing about no change in stoichiometry, leads to a $\sim$ 10-fold reduction in rate. Candlin and Halpern[71] consider that the first stage in the reaction is the formation of an intermediate

$$MnO_4^- + (NH_3)_5Co^{III}(OCHO^-)^{2+} \longrightarrow HMnO_4^- + (NH_3)_5Co^{III}(CO_2^-)^{2+}$$

in a one-equivalent step by abstraction of a hydrogen atom. The fate of the intermediate is to produce either $Co^{2+} + CO_2$ or $(NH_3)_5Co^{III}{\cdot}OH_2^{3+} + CO_2$. It seems likely also that the oxidation of free $HCOO^-$, in the oxidation of formic acid by MnO_4^-, occurs by hydrogen-atom and not hydrogen-ion transfer[74, 75]. The oxidation of $(NH_3)_5Co{\cdot}OCHO^{2+}$ by Co^{3+} and $S_2O_8^{2-}$ (catalysed by Ag^+) generates $(NH_3)_5Co{\cdot}OH_2^{3+}$ as the major product, whereas the oxalato and *p*-aldehydobenzoato complexes yield almost exclusively Co^{2+}.

In the multi-equivalent reduction of permanganate by Sn(II) there is evidence to suggest the presence of Sn(III) in the reacting system[59]. In this respect the reaction

$$Mn(V) + Sn(II) \longrightarrow Mn(IV) + Sn(III)$$

is considered the likely source of Sn(III).

References pp. 267–273

4.2 OXIDATIONS BY MANGANESE(III)

There is a paucity of information on reactions of the strongly-oxidising manganese(III) with inorganic substrates. The main reason for this neglect lies in the tendency of Mn(III) to disproportionate, *viz.*

$$2\,Mn(III) \rightleftharpoons Mn(II)+Mn(IV)(as\ MnO_2)$$

This probably takes place through the self-condensation of hydrolysed species and thus disproportionation is reduced at high acidities. This tendency can be suppressed also by the use of strong sulphuric acid ($\geqq 5\ M$) as the medium, or by employing the oxalato or pyrophosphato complex. In both these cases strong complexing occurs. Perchlorate possesses no such stabilising ability but Rosseinsky[76] reports that reasonably stable solutions of Mn(III) perchlorate ($\leqq \sim 10^{-3}\ M$ in $\sim 4\ M$ acid) can be prepared from acid permanganate and Mn(II) solutions, a large excess ($\sim$ 25-fold) of the latter preventing disproportionation of Mn(III).

Rosseinsky and Nicol[77] have made a kinetic study of the Mn(III)+V(IV) reaction

$$Mn(III)+V(IV) = Mn(II)+V(V)$$

by following the rate of appearance of V(V) at 325 mμ (ε = 284 l.mole^{-1}.cm^{-1}). A simple second-order rate law

$$d[V(V)]/dt = k_{obs}[Mn(III)][V(IV)]$$

obtains over the range 0.8 to 3.0 M perchloric acid at temperatures from 5 °C to 20 °C. With $\sim 10^{-4}\ M$ reactant concentrations, half-lives were of the order of 10–20 sec. At 20 °C and 3 M $HClO_4$, k_{obs} is 1.53×10^2 l.mole^{-1}.sec^{-1}. The rate increase with decrease in hydrogen-ion concentration is in keeping with a mechanism in which a hydrolysed species of Mn(III) takes part, *viz.*

$$Mn^{3+}+VO^{2+}+H_2O \xrightarrow{k_0} Mn^{2+}+VO_2^{\ +}+2\,H^+$$

$$Mn^{3+}+H_2O \rightleftharpoons MnOH^{2+}+H^+ \qquad \text{rapid equilibrium, } K_h$$

$$MnOH^{2+}+VO^{2+} \xrightarrow{k_1} Mn^{2+}+VO_2^{\ +}+H^+$$

The rate data were fitted to the relationship

$$k_{obs}([H^+]+K_h) = k_0[H^+]+k_1 K_h$$

Although the activation parameters corresponding to steps k_0 and k_1 are given, there is some doubt over the correct value for K_h (and the accompanying ΔH_h) under the conditions employed. There seems to be no special kinetic feature about the oxygen ligand in VO^{2+}, since the rates of reaction of the latter oxocation with various oxidants follow the same sequence (V(V)[78] > Mn(III)[79] > Co(III)[80] > Fe(III)[28] > Tl(III)[26]) as the corresponding reactions of Fe(II)[22,24,28,79,81].

An important study of Mn(III) reactions is that by Diebler and Sutin[82]. Mn(III) was prepared by anodic oxidation of up to 0.1 *M* Mn(II) in 1 to 6 *M* $HClO_4$. The nature of Mn(III) in perchloric acid is discussed[83]. The oxidation of Fe(II) by Mn(III) is kinetically second-order with corresponding rate coefficients of 1.46×10^4 and 1.67×10^4 $l.mole^{-1}.sec^{-1}$ in 3.0 and 1.0 *M* perchloric acid, respectively, at $\mu = 3.1$ *M* and 25 °C. The latter rate coefficient is consistent with the value of 6.92×10^3 $l.mole^{-1}.sec^{-1}$ at 15 °C (1.04 *M* $HClO_4$, $\mu = 3.04$ *M*) recorded by Rosseinsky[84]. The oxidation of various substituted 1,10-phenanthroline complexes of Fe(II) was investigated also; a linear relationship exists between log $k_{1,2}$ and log $K_{1,2}$ (Fig. 2). The gradient of the plot is 0.45, a value close to the theoretical value of 0.5 given by the Marcus equation [eqn. (12.6), p. 247]. Rate coefficients for the oxidation of Fe(II)–phenanthroline complexes in pyrophosphate–sulphate media are presented. A linear relationship is observed between log $k_{1,2}$ and the formal oxidation potentials of the Fe(II) complexes (Fig. 3). A rapid-mixing and flow apparatus[44] was employed for all the above reactions. In a recent paper, Rosseinsky and Nicol[79] confirm that the Mn(III)+ Fe(II) reaction is second order from measurements on the changes in the diffusion current of iron(II) at a rotating platinum electrode. In the ranges 0.54–3 *M* perchloric acid and 0.3–15 °C the observed rate coefficient has an acid dependence

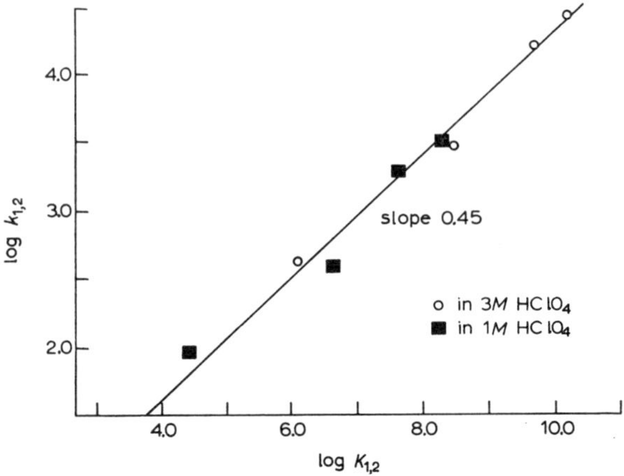

Fig. 2. Relationship between the logarithms of the rate coefficients ($k_{1,2}$) and the logarithms of the equilibrium constants ($K_{1,2}$) for the oxidation of various substituted Fe(II)–phenanthroline complexes by Mn(III) in 1 *M* and 3 *M* $HClO_4$ at 25.0 °C. (*From Diebler and Sutin*[82], *by courtesy of The American Chemical Society.*)

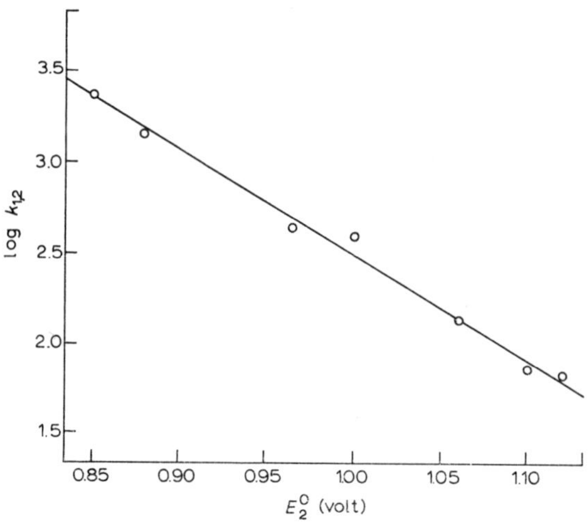

Fig. 3. Relationship between the logarithms of the rate coefficients ($k_{1,2}$) for the oxidation of various substituted Fe(II)–phenanthroline complexes by Mn(III) pyrophosphate (pH 1.0, ionic strength ~ 0.5 M) at 25.0 °C and the formal oxidation potentials (E_2^0) of the Fe(II) complexes. (*From Diebler and Sutin*[82], *by courtesy of The American Chemical Society.*)

similar to that encountered in the Mn(III)+V(IV) reaction[77] (p. 172). Accordingly, Rosseinsky and Nicol suggest the analogous scheme

$$Mn^{3+} + Fe^{2+} \rightarrow Mn^{2+} + Fe^{3+}$$

$$Mn^{3+} + H_2O \rightleftharpoons MnOH^{2+} + H^+ \qquad \text{rapid equilibrium, } K_h$$

$$MnOH^{2+} + Fe^{2+} \rightarrow Mn^{2+} + Fe^{3+} + OH^-$$

As noted above (p. 173), a lack of certainty over the correct value of K_h under the prevailing conditions gives rise to ambiguities in the values of individual rate coefficients.

Rosseinsky[76] has examined the oxidation of Hg(I) and Mn(III) in aqueous perchloric acid at 50 °C by a method involving titration of Mn(III) with Fe(II). At constant acidity ($[H^+] = 3.0\ M$) his rate data can be expressed as

$$-d[Mn(III)]/dt =$$
$$= k'[Mn(III)][Hg(I)_2]/[Hg(II)] + k''[Hg(I)_2][Mn(III)]^2/[Mn(II)]$$

where $Hg(I)_2$ represents $Hg_2^{2+} + Hg_2ClO_4^+$. Thus the reaction is retarded by both Hg(II) and Mn(II). This result suggests two rapid pre-equilibria

$$Hg(I)_2 \rightleftharpoons Hg(O) + Hg(II) \qquad K_1 \sim 10^{-9}\ \text{mole.l}^{-1}$$

and

$$2\,Mn(III) \rightleftharpoons Mn(IV) + Mn(II) \qquad K_2 \sim 10^{-3}$$

which are followed by two slow and rate-controlling steps

$$Mn(III) + Hg(O) \xrightarrow{k_2} Mn(II) + Hg(I) \qquad \text{slow} \tag{4.1}$$

$$Mn(IV) + Hg(I)_2 \xrightarrow{k_3} Mn(II) + 2Hg(II) \qquad \text{slow} \tag{4.2}$$

together with

$$Mn(III) + Hg(I) \longrightarrow Mn(II) + Hg(II) \qquad \text{rapid}$$

In this scheme Hg(I) is present[85] as Hg^+ ions (possibly ion-paired with ClO_4^-), and Mn(IV) is likely to be $Mn^{4+} + MnO^{2+}$. Step (4.2) may proceed either as written (two-equivalent change) or *via* two consecutive one-equivalent stages with Hg_2^{3+}, or Hg(I), +Hg(II), as intermediates. On this basis k' and k'' are identified as k_2K_1 and k_3K_2, respectively. Very approximate values of k_2 and k_3 were estimated (on the basis of a number of rather drastic assumptions) as 5×10^3 $l.mole^{-1}.sec^{-1}$ and $\geqq 50$ $l.mole^{-1}.sec^{-1}$, respectively, at 50 °C. Other possible mechanisms tested and found inadequate in explaining retardation of the reaction by products include (*a*) reactions $Mn(III) + Hg(I)_2$ and $Mn(IV) + Hg(O)$, and (*b*) a back reaction in step (4.1) and omission of (4.2). It is of interest to note that Mn(III) is capable of oxidising both Hg(O) *and* Hg(I) whereas Tl(III) oxidises only the former[86] and Co(III) the latter[80]. Also, of the three oxidants, only Co(III) is capable of oxidising Hg_2^{2+}, a result which has led Rosseinsky[76] to speculate on the ability of Co(III) to react via a high-spin species which can t_{2g}-overlap with Hg_2^{2+}.

4.3 OXIDATIONS BY RHENIUM(VII)

Rhenium(VII), as the perrhenate ion ReO_4^-, is reduced rapidly to rhenium(V) by tin(II) in hydrochloric acid solutions. Further reduction to rhenium(IV) takes place slowly[87]. Banerjea and Mohan[88] have followed the rate of formation of Re(IV) at 550 mμ. At 30 °C, with $[Re(V)] = 3 \times 10^{-3}$ *M* and $[Sn(II)] = 3 \times 10^{-2}$ *M*, the observed rate coefficient is 6.2×10^{-4} sec^{-1} in 3 *M* hydrochloric acid solution. The following mechanism is considered appropriate

$$ReOCl_5^{2-} + H_2SnCl_4 \longrightarrow Re(OH)Cl_5^{2-} + HSnCl_4$$

$$ReOCl_5^{2-} + HSnCl_4 \longrightarrow Re(OH)Cl_5^{2-} + SnCl_4$$

5. Oxidations by iron(III) and ruthenium(III)

5.1 OXIDATIONS BY IRON(III)

The reaction between complexes of the type $Fe(H_2O)_5X^{2+}$ and $V(H_2O)_6{}^{2+}$ has been investigated by a flow method[44] by measurements at the absorption maximum of the former species[62]. Rate coefficients are recorded in Table 3 when

TABLE 3

RATE COEFFICIENTS FOR THE REACTION BETWEEN $Fe(H_2O)_5X^{2+}$ AND $V(H_2O)_6{}^{2+}$ AT 25°C AND $\mu = 1.0$ *M* (ref. 62)

X	k(*l.mole*$^{-1}$.*sec*$^{-1}$)
H_2O	1.8×10^4
OH^-	$<4 \times 10^5$
Cl^-	4.6×10^5
NCS^-	6.6×10^5
N_3^-	5.2×10^5

X = H_2O, OH^-, Cl^-, NCS^-, and N_3^-; in all these cases the rates of reduction are rapid compared with the rates of substitution of X^- in $Fe(H_2O)_6{}^{3+}$. No evidence is evinced for formation of VX^{2+} complexes. It is noteworthy that OH^- exerts little effect on the rate of this reaction although it has a pronounced influence on the rate of other reactions of Fe(III)[89]. This result, together with the observation that the other anions have similar effects on the rate, would seem to indicate that the $FeX^{2+}+V^{2+}$ reactions proceed *via* an outer sphere path. The suggestion is made, however, that, although the $Fe^{3+}+V^{2+}$ reaction is undoubtedly of the outer-sphere type (since the rate of water replacement in V^{2+} is much slower than the electron-transfer), the $FeX^{2+}+V^{2+}$ reactions occur through inner-sphere, water-bridged activated complexes resulting from the loss of a water molecule coordinated to Fe(III).

Fe(III) oxidises V(III) to V(IV). Superficially the reaction is unexceptional: the stoichiometry is simple

$$\mathrm{Fe(III)+V(III) = Fe(II)+V(IV)} \tag{5.1}$$

and the process is first-order in each reactant (Fig. 4; plot A). However, a number of interesting features have been revealed by Higginson and Sykes in the course of a detailed kinetic study[90]. Measurements were made at 755 mμ (where VO^{2+} absorbs strongly). Precautions were taken against atmospheric oxidation. Reduction of perchlorate to chloride by V(III)[91] was slight under the conditions prevailing. On the assumption that the reaction proceeded as a single-stage process V(IV) was added (in concentrations similar to those of the reactants) in an effort

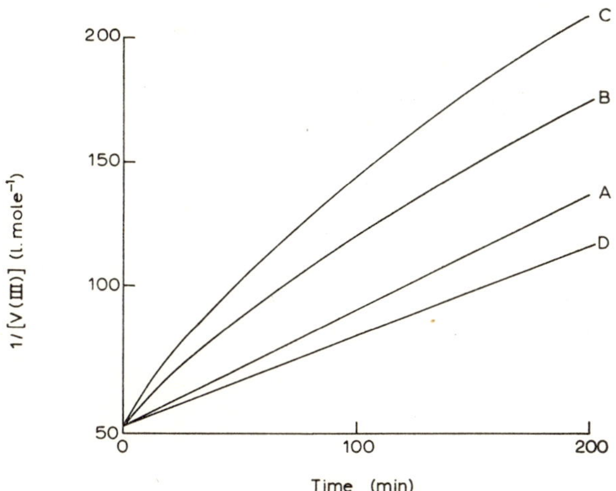

Fig. 4. Effect of initial presence of products upon the rate of the Fe(III)+V(III) reaction. Initial concentrations: [Fe(III)] = 0.0226 M; [V(III)] = 0.022 M; [Fe(II)] $\sim 10^{-4}$ M (A, B, C), 0.07–0.12 M (D); [V(IV)] $\sim 10^{-3}$ M (A, D), 0.057 M (B), 0.094 M (C); $[H^+] = 1.00$ M; ionic strength, 3.0 M; temp., 20.0 °C. (*From Higginson and Sykes*[90], *by courtesy of The Chemical Society.*)

to disclose a back-reaction. Instead V(IV) has a pronounced accelerating and not a retarding influence (Fig. 4; plots B and C). On the other hand, the presence of Fe(II) leads to a decrease in rate which then achieves a limiting value after the addition of a certain amount of Fe(II) (Fig. 4; plot D). A quantitative analysis of these and similar results allows the empirical rate law to be deduced as

$$-\mathrm{d}[\mathrm{V(III)}]/\mathrm{d}t = k_1[\mathrm{Fe(III)}][\mathrm{V(III)}]+k'[\mathrm{Fe(III)}][\mathrm{V(III)}][\mathrm{V(IV)}]/[\mathrm{Fe(II)}] \quad (5.2)$$

The first term in (5.2) corresponds to the simple path represented by (5.1), rate coefficient k_1. The form of the second term can be understood on the basis of the sequence

$$\mathrm{Fe(III)}+\mathrm{V(IV)} \underset{k_{-2}}{\overset{k_2}{\rightleftharpoons}} \mathrm{Fe(II)}+\mathrm{V(V)} \quad (5.3)$$

$$\mathrm{V(V)}+\mathrm{V(III)} \xrightarrow{k_3} 2\,\mathrm{V(IV)} \quad (5.4)$$

Although k_2/k_{-2} is very small ([V(V)] $\leqq 0.1\,\%$ of [V(III)] and [Fe(III)]), the forward reaction in (5.3) is sufficiently rapid for (5.3) and (5.4) to make a significant contribution in the absence of large amounts of Fe(II). The presence of added Fe(II) so reduces the amount of V(V) that (5.3) and (5.4) are no longer effective. Applying the steady-state approximation to V(V) in (5.3) and (5.4), it follows that

$$\frac{-\mathrm{d}[\mathrm{V(III)}]}{\mathrm{d}t} = k_1[\mathrm{Fe(III)}][\mathrm{V(III)}] + \frac{k_2 k_3[\mathrm{Fe(III)}][\mathrm{V(III)}][\mathrm{V(IV)}]}{k_{-2}[\mathrm{Fe(II)}]+k_3[\mathrm{V(III)}]} \quad (5.5)$$

Equation (5.5) reduces to the form of (5.2) if k_{-2} [Fe(II)] $\gg k_3$ [V(III)], whereupon $k' = k_3 k_2/k_{-2}$. Equation (5.5) can be written as

$$\frac{[\mathrm{Fe(III)}][\mathrm{V(IV)}]}{(-\mathrm{d}[\mathrm{V(III)}]/\mathrm{d}t) - k_1[\mathrm{Fe(III)}][\mathrm{V(III)}]} = \frac{k_{-2}[\mathrm{Fe(II)}]}{k_2 k_3[\mathrm{V(III)}]} + \frac{1}{k_2}$$

Plots of the left-hand side of this equation *versus* corresponding ratios of [Fe(II)]/[V(III)] are linear. Such plots appear to pass through the origin so that k_2 must be large. The gradients of the plots $(k_{-2}/k_2 k_3)$ yield a value for k_3 since k_{-2}/k_2 was determined separately by measurements on the EMF of a cell composed of Fe(III)+Fe(II) and V(V)+V(IV) half-cells. At 25 °C, $\mu = 3.0$ *M* and $[\mathrm{H}^+] = 1.0$ *M*, k_3 is calculated to be 260 $\mathrm{l.mole^{-1}.sec^{-1}}$. Step (5.1) was found to be acid dependent

$$k_1 = b + (c/[\mathrm{H}^+]) + (d/[\mathrm{H}^+]^2)$$

At 25 °C, approximate values of the constants *b*, *c* and *d* are 2.2×10^{-3} $\mathrm{l.mole^{-1}.sec^{-1}}$, 3.0×10^{-3} $\mathrm{sec^{-1}}$ and 3.4×10^{-3} $\mathrm{mole.l^{-1}.sec^{-1}}$, respectively. Three parallel routes (with activation complexes differing in the number of bound protons) are indicated. Activation parameters corresponding to k_1 and k_3 are reported.

Besides being catalysed by V(IV), the reaction is also subject to catalysis by Cu(II) due to

$$\mathrm{Cu(II)} + \mathrm{V(III)} \xrightarrow{k_4} \mathrm{Cu(I)} + \mathrm{V(IV)} \quad (5.6)$$

$$\mathrm{Fe(III)} + \mathrm{Cu(I)} \longrightarrow \mathrm{Fe(II)} + \mathrm{Cu(II)} \quad \text{rapid} \quad (5.7)$$

With Cu(II) present in similar concentrations to Fe(III) the contribution of the uncatalysed path is small. The rate coefficient k_4 varies with acidity according to $k_4 = g + h/[\mathrm{H}^+]$†. From Arrhenius plots of *g* and *h*, $\Delta H^\ddagger$ and $\Delta S^\ddagger$ were calculated and compared with those corresponding to path *b*. At 25 °C in 1.0 *M* perchloric acid, $k_4 = 0.30$ $\mathrm{l.mole^{-1}.sec^{-1}}$, $\Delta H_g^\ddagger = 21.3$ $\mathrm{kcal.mole^{-1}}$ and $\Delta S_g^\ddagger = 5 \pm 5$ $\mathrm{cal.deg^{-1}.mole^{-1}}$. Under the same conditions $k_1 = 0.025$ $\mathrm{l.mole^{-1}.sec^{-1}}$,

† See, also, ref. 283.

$\Delta H_b^{\ddagger} = 17.3 \pm 4.1$ kcal.mole^{-1} and $\Delta S_b^{\ddagger} = -15 \pm 13$ cal.deg^{-1}.mole^{-1}. Thus the efficiency of Cu(II) as a catalyst is due to a more favourable entropy of activation. In fact the enthalpy of activation for the catalysed reaction is greater than that for the uncatalysed reaction[26]. Recently, Espenson *et al.*[92,284] have measured the rate of the reverse of reaction (5.6) as

$$d[V^{3+}]/dt = e[VO^{2+}][Cu^{+}][H^{+}]$$

where $e = 3.84 \times 10^2$ l^2.mole^{-2}.sec^{-1} at 25 °C and $\mu = 3.0$ *M*[284]. These authors question the reality of the term involving g in the expression for k_4, as included by Higginson and Sykes[90]. Consequently, since $h = 0.384$ sec^{-1} then the equilibrium constant for reaction (5.6) is calculated as 1.1×10^3 l^2.mole^{-2} (e/h) at 25 °C[284]. Espenson *et al.*[92,285] have been able to measure also the rate coefficient of reaction (5.7) as 1.6×10^5 l.mole^{-1}.sec^{-1} at 25 °C in 1 *M* $HClO_4$[285].

Fe(III) oxidises the tantalum cluster-ion $Ta_6Cl_{12}{}^{2+}$ by way of two consecutive one-electron transfer steps[93] (see, also, refs. 286 and 287).

$$Fe^{3+} + Ta_6Cl_{12}{}^{2+} \overset{k_5}{\rightleftharpoons} Fe^{2+} + Ta_6Cl_{12}{}^{3+}$$

$$Fe^{3+} + Ta_6Cl_{12}{}^{3+} \underset{k_{-6}}{\overset{k_6}{\rightleftharpoons}} Fe^{2+} + Ta_6Cl_{12}{}^{4+}$$

At 15 °C and 0.020 *M* perchloric acid $k_5 = 620$ l.mole^{-1}.sec^{-1}, $k_6 = 3.72$ l. mole^{-1}.sec^{-1}, $k_{-6} = 201$ l.mole^{-1}.sec^{-1} and $K = k_6/k_{-6} = 0.018$ (*cf.* spectrophotometric value of 0.016 at 22 °C). As might be expected for redox reactions in which little molecular rearrangement is required, these steps are rapid.

The oxidation of Cr(II) by Fe(III) in perchloric acid is markedly catalysed by chloride ion. Taube and Myers[94] found that $Cr(H_2O)_5Cl^{2+}$ is formed along with $Cr(H_2O)_6{}^{3+}$ as products of the oxidation, the relative proportions of the two species depending on concentrations of H^+ and Cl^-. The suggestion was made that $Cr(H_2O)_5Cl^{2+}$ is produced by reaction of $Cr(H_2O)_6{}^{2+}$ with a chloro complex of Fe(III), *viz.*

$$Fe(H_2O)_5Cl^{2+} + Cr(H_2O)_6{}^{2+} \longrightarrow Fe(H_2O)_6{}^{2+} + Cr(H_2O)_5Cl^{2+}$$

and that the mechanism can be described in terms of an inner-sphere activated complex of composition $[(H_2O)_5FeClCr(H_2O)_6{}^{4+}]^{\ddagger}$. Ardon *et al.*[95] investigated the catalysed oxidation in 5.27 *M* perchloric acid at −50 °C. Under these conditions the process was described as instantaneous. The green product solution was analysed by ion-exchange, revealing $Cr(H_2O)_5Cl^{2+}$ to be the chief product of reaction. Since at −50 °C the formation of $Fe(H_2O)_5Cl^{2+}$ is relatively slow, it is deduced that Fe^{3+} may be incorporated into the activated complex without requiring the prior formation of $Fe(H_2O)_5Cl^{2+}$. Dulz and Sutin[96] have published a detailed kinetic study of the Cl^--catalysed oxidation (at 25 °C) based upon the use of flow techniques[44]†. The reaction is first-order in both Fe(III) and Cr(II), the second-order rate coefficient being 7.7×10^3 l.mole^{-1}.sec^{-1} for 1.00 M $HClO_4$. The acid dependence of the rate can be expressed as

$$\text{rate} = k_7[Fe^{3+}][Cr^{2+}] + k_8[FeOH^{2+}][Cr^{2+}]$$

Since K_h as defined by

$$Fe^{3+} + H_2O \rightleftharpoons FeOH^{2+} + H^+$$

is much less than $[H^+]$ then

$$k' = \text{rate}/[Fe(III)][Cr^{2+}] = k_7 + k_8 K_h/[H^+]$$

A plot of k' *versus* $1/[H^+]$ is linear. Using a value[97] for K_h of 1.69×10^{-3} enables k_7 and k_8 to be calculated from the slope and intercept: $k_7 = 2.3 \times 10^3$ l.mole^{-1}.sec^{-1} and $k_8 = 3.3 \times 10^6$ l.mole^{-1}.sec^{-1}, at 25 °C and $\mu = 1.00$ M. The effect of chloride was examined systematically as follows††. The formation and dissociation of $Fe(H_2O)_5Cl^{2+}$ by

$$Fe(H_2O)_6{}^{3+} + Cl^- \underset{k_{-9}}{\overset{k_9}{\rightleftharpoons}} Fe(H_2O)_5Cl^{2+}$$

has rate coefficients k_9 and k_{-9} of 19.4 l.mole^{-1}.sec^{-1} and 6.7 sec^{-1}, respectively, at 25 °C and $\mu = 1.00$ M[98]. With Cr(II) present, the rate of disappearance of $Fe(H_2O)_5Cl^{2+}$ is given by

$$-d[FeCl^{2+}]/dt = k_{-9}[FeCl^{2+}] - k_9[Fe^{3+}][Cl^-] + k_{10}[FeCl^{2+}][Cr^{2+}] \quad (5.8)$$

where k_{10} is defined by

$$Fe(H_2O)_5Cl^{2+} + Cr^{2+} \xrightarrow{k_{10}} \text{products}$$

It was shown that k_{10} is much larger than both k_9 and k_{-9} whereupon equation (5.8) becomes

$$-d[FeCl^{2+}]/dt = k_{10}[FeCl^{2+}][Cr^{2+}]$$

$Fe(H_2O)_5Cl^{2+}$ was followed spectrophotometrically at 336 mμ and k_{10} was

† The comparable reaction between $FeBr^{2+}$ and Cr^{2+} has been examined by Carlyle and Espenson[288].

†† Catalysis by chloride has been examined for the comparable Fe(III)+Np(III) reaction[289].

obtained as 2×10^7 l.mole^{-1}.sec^{-1} at 25 °C and $\mu = 1.00$ *M*. On the assumption that there exists a Cl^--catalysed path for the reaction, not requiring the participation of $Fe(H_2O)_5Cl^{2+}$ as a reactant, then the expression

$$-d[Fe(III)]/dt = k_7[Fe^{3+}][Cr^{2+}]+k_8[FeOH^{2+}][Cr^{2+}] \\ +k_9[Fe^{3+}][Cl^-]+k_{11}[Fe^{3+}][Cl^-][Cr^{2+}]$$

holds for experiments in which Cl^- is added only to the Cr(II) solution. For conditions where $[Cr(II)] \gg [Fe(III)]$ and $[Cl^-] \gg [Fe(III)]$, the apparent first-order rate coefficient is given by

$$k' = \{k_7+(k_8 K_h/[H^+])+k_{11}[Cl^-]\}[Cr^{2+}]+k_9[Cl^-]$$

Plots of k' *versus* $[Cr^{2+}]$ at fixed $[Cl^-]$ are linear and allow k_{11} to be calculated from the slopes as 2.2×10^4 l^2.mole^{-2}.sec^{-1}, at 25 °C and $\mu = 1.00$ *M*. Product yields of $Cr(H_2O)_5Cl^{2+}$ and $Cr(H_2O)_6{}^{3+}$ obtained experimentally were in excellent agreement with those calculated on the basis of the kinetic scheme. Dulz and Sutin[96] conclude that two routes exist for the chloride-catalysed oxidation, $Cr(H_2O)_5Cl^{2+}$ being formed in both paths, *viz.* inner-sphere process

$$Fe(H_2O)_5Cl^{2+}+Cr(H_2O)_6{}^{2+} \longrightarrow [(H_2O)_5Fe\text{–}Cl\text{–}Cr(H_2O)_5{}^{4+}]^{\ddagger} \\ \longrightarrow Fe(H_2O)_6{}^{2+}+Cr(H_2O)_5Cl^{2+}$$

and outer-sphere process, either

$$Fe(H_2O)_6{}^{3+}+Cr(H_2O)_5Cl^{+} \longrightarrow [(H_2O)_5FeH_2O\text{–}Cl\text{–}Cr(H_2O)_5{}^{4+}]^{\ddagger} \\ (\text{or } [(H_2O)_5FeH_2O\text{–}CrCl(H_2O)_4{}^{4+}]^{\ddagger}) \\ \longrightarrow Fe(H_2O)_6{}^{2+}+Cr(H_2O)_5Cl^{2+}$$

or

$$(H_2O)_6Fe^{3+}Cl^-+Cr(H_2O)_6{}^{2+} \longrightarrow [(H_2O)_5FeH_2O\text{–}Cl\text{–}Cr(H_2O)_5{}^{4+}]^{\ddagger} \\ \longrightarrow Fe(H_2O)_6{}^{2+}+Cr(H_2O)_5Cl^{2+}$$

The rate of the path involving $Fe(H_2O)_5Cl^{2+}$ as a reactant is ~ 3000 times faster than the alternative (outer-sphere) path. Table 4 summarises results on this reaction and the corresponding electron exchange reactions in terms of relative rate coefficients.

When a solution of Fe(III) and thiocyanate is mixed with a solution containing excess Cr^{2+} in a flow apparatus[44] the absorbance change can be seen to involve three distinct stages, corresponding to (*a*) the very rapid reaction between $FeNCS^{2+}$ and Cr^{2+} ($k \geqq 2 \times 10^7$ l.mole^{-1}.sec^{-1}); (*b*) the reactions $Fe^{3+}+Cr^{2+}$,

TABLE 4

RELATIVE RATE COEFFICIENTS FOR THE Fe(III)+Cr(II) AND RELATED REACTIONS[96]

	Relative rate coefficients		
X	$Cr(H_2O)_5X+Cr^{2+}$	$Fe(H_2O)_5X+Cr^{2+}$	$Fe(H_2O)_5X+Fe^{2+}$
H_2O	1	1	1
OH^-	$>8\times10^4$ [a]	1.4×10^3	1.1×10^3 [c]
Cl^-	$>2\times10^6$ [b]	$1\ \times10^4$	6.2[d]

[a] From ref. 23. [b] From ref. 99. [c] From ref. 22. [d] From ref. 100.

$FeOH^{2+}+Cr^{2+}$, and $Fe^{3+}+Cr^{2+}+SCN^-$ ($k = 2.3\times10^3$ l.mole^{-1}.sec^{-1}, 3.3×10^6 l.mole^{-1}.sec^{-1}, and $\sim 2\times10^5$ l^2.mole^{-2}.sec^{-1}, respectively); (*c*) the Cr(II)-catalysed isomerisation of $CrSCN^{2+}$ produced in (*a*) ($k = 42$ l.mole^{-1}. sec^{-1})[63]. Rate coefficients pertain to 1 *M* $HClO_4$ solutions at 25 °C. Thus an inner-sphere mechanism is demonstrated. The S-bonded thiocyanato complex, $CrSCN^{2+}$, is *not* produced when a solution of $Cr^{2+}+SCN^-$ is oxidised by Fe(III). $CrSCN^{2+}$ can be prepared by the gradual addition of a 5×10^{-3} *M* Cr^{2+} solution to an equal volume of a well-stirred solution of 5.5×10^{-3} *M* Fe(III) and 4.5×10^{-3} *M* SCN^-. The product solution is green whereas $CrNCS^{2+}$ solutions are purple.

Oxidation of the pentacyano complex of Co(II), $Co(CN)_5{}^{3-}$, by $Fe(CN)_6{}^{3-}$ results in the oxidant being retained in the coordination sphere of cobalt, and the inert binuclear ion $(NC)_5F^{II}CN\ Co^{III}(CN)_5{}^{6-}$ is formed[101]. This constitutes a piece of direct evidence for a bridged mechanism. Similarly, reaction of $Co(CN)_5{}^{3-}$ with oxygen results in the production of $(NC)_5Co^{III}OO\ Co^{III}(CN)_5{}^{6-}$.

The reduction of $Fe(CN)_6{}^{3-}$ by cobalt(II) ethylenediaminetetraacetate, $Co(EDTA)^{2-}$, yields $Fe(CN)_6{}^{4-}$ and $Co(EDTA)^-$ indirectly in two stages[102]

$$Co(EDTA)^{2-}+Fe(CN)_6{}^{3-} \rightleftharpoons Co(EDTA)\cdot Fe(CN)_6{}^{5-} \qquad \text{rapid equilibrium, } K_1$$

$$Co(EDTA)\cdot Fe(CN)_6{}^{5-} \xrightarrow{k_{12}} Co(EDTA)^- + Fe(CN)_6{}^{4-} \quad \text{rate-determining}$$

Under conditions of a large excess of $Co(EDTA)^{2-}$ the back-reaction is negligible and the rate is given by

$$\frac{d[B_{total}]}{dt} = \frac{k_{12}K_1[Co(EDTA)^{2-}]}{1+K_1[Co(EDTA)^{2-}]}[B_{total}] = k_{obs}[B_{total}]$$

where $[B_{total}]$ represents the sum of the concentrations of $Fe(CN)_6{}^{3-}$ and $Co(EDTA)\cdot Fe(CN)_6{}^{5-}$, and k_{obs} is the observed (first-order) rate coefficient. In accord with the rate law, plots of log $(A_t - A_\infty)$ *versus* time are linear, where A

and A_∞ are the absorbance values at 420 mμ, the absorption maximum of $Fe(CN)_6^{3-}$, at time t and after complete reaction. Furthermore, the form of the rate law is verified from the linearity of plots of $1/k_{obs}$ *versus* $1/[Co(EDTA)^{2-}]$. Slope and intercept values of such plots yield values for k_{12} and K_1 at 25 °C of 6.2×10^{-3} sec^{-1} and 670 l.mole^{-1}, respectively. An independent value of K_1 (710 l. mole^{-1}) was obtained spectrophotometrically. The overall activation energy and entropy are estimated to be 26 kcal.mole^{-1} and 16 cal.deg^{-1}.mole^{-1}, respectively. Adamson and Gonick[102] are of the opinion that the reaction sequence is more correctly a three-stage process involving the formation of the species $(EDTA)Co^{II}$–NC–$Fe^{III}(CN)_5^{5-}$ which first undergoes charge-transfer to $(EDTA)Co^{III}$–NC–$Fe^{II}(CN)_5^{5-}$ and then breaks down to $Co^{III}(EDTA)^-$ and $Fe^{II}(CN)_6^{4-}$. The second of these binuclear species is taken to be the intermediate, since magnetic susceptibility measurements reveal the intermediate to be diamagnetic. Structurally the intermediate resembles $(NC)_5Co^{III}$–NC–$Fe^{II}(CN)_5^{6-}$ formed between $Co(CN)_5^{3-}$ and $Fe(CN)_6^{3-}$[101]. The same system has been investigated in greater detail by Huchital and Wilkins[103], using rapid reaction techniques to characterise the intermediates formed. The kinetics of formation and decomposition of the bridged cyanide intermediate, $(EDTA)Co^{III}$–NC–$Fe^{II}(CN)_5^{5-}$

$$Co(EDTA)^{2-} + Fe(CN)_6^{3-} \underset{k_{-13}}{\overset{k_{13}}{\rightleftharpoons}} (EDTA)Co^{III}\text{–}NC\text{–}Fe^{II}(CN)_5^{5-} \quad K_1 \qquad (5.9)$$

have been studied with the aid of temperature-jump and stopped-flow equipment. It is suggested that the intermediate is formed by replacement of water from the Co(II) complex, which may react as $Co(EDTA)(H_2O)^{2-}$. In addition, evidence from stopped-flow measurements is presented for the inclusion of a second intermediate, *viz.*

$$(EDTA)Co^{III}\text{–}NC\text{–}Fe^{II}(CN)_5^{5-} + Fe(CN)_6^{3-} \underset{k_{-14}}{\overset{k_{14}}{\rightleftharpoons}} (EDTA)Co^{III}\text{–}NC\text{–}Fe^{III}(CN)_5^{4-} + Fe(CN)_6^{4-} \quad K_2 \qquad (5.10)$$

TABLE 5

RATE PARAMETERS FOR $Fe(CN)_6^{3-} + Co(EDTA)^{2-}$ REACTION[103]

Reaction	k_{13}(*l.mole*$^{-1}$.*sec*$^{-1}$)	k_{-13}(*sec*$^{-1}$)	K_1(*l.mole*$^{-1}$)
(5.9)	1.3×10^5 [a] 0.9×10^5 [b]	86[a]	1.5×10^3 [a] 1.6×10^3 [b] 1.6×10^3 [c]
	k_{14}(*l.mole*$^{-1}$.*sec*$^{-1}$)	k_{-14}(*l.mole*$^{-1}$.*sec*$^{-1}$)	K_2
(5.10)	2×10^3 [b]	3×10^4 [b]	0.07[b] 0.06[c]

[a] Temperature jump method. [b] Stopped flow method. [c] Direct spectrophotometry.

Rate parameters, at 25 °C, for reactions (5.9) and (5.10) are collected in Table 5. Reaction (5.10) is analogous to[101]

$$(CN)_5Co^{III}\text{–}NC\text{–}Fe^{II}(CN)_5{}^{6-} + Fe(CN)_6{}^{3-} \underset{k_{-15}}{\overset{k_{15}}{\rightleftharpoons}} (CN)_5Co^{III}\text{–}NC\text{–}Fe^{III}(CN)_5{}^{5-} + Fe(CN)_6{}^{4-} \quad K_3$$

Huchital and Wilkins[103] report $k_{15} = 1.1 \times 10^3$ l.mole^{-1}.sec^{-1}, $k_{-15} = 1.6 \times 10^4$ l.mole^{-1}.sec^{-1} and $K_3 = 0.07$, at $\mu = 0.1$ M and 25 °C. These results are close to those for reaction (5.10).

The reaction between Fe(III) and Sn(II) in dilute perchloric acid in the presence of chloride ions is first-order in Fe(III) concentration[104]. The order is maintained when bromide or iodide is present. The kinetic data seem to point to a fourth-order dependence on chloride ion. A minimum of three Cl^- ions in the activated complex seems necessary for the reaction to proceed at a measurable rate. Bromide and iodide show third-order dependences. The reaction is retarded by Sn(II) (first-order dependence) due to removal of halide ions from solution by complex formation. Estimates are given for the formation constants of the monochloro and monobromo Sn(II) complexes. In terms of catalytic power $I^- > Br^- > Cl^-$ and this is also the order of decreasing ease of oxidation of the halide ion by Fe(III). However, the state of complexing of Sn(II) and Fe(III) is given by $Cl^- > Br^- > I^-$. Apparently, electrostatic effects are not effective in deciding the rate. For the case of chloride ions, the chief activated complex is likely to have the composition $(FeSnCl_4{}^+)^\ddagger$. The kinetic data cannot resolve the way in which the Cl^- ions are distributed between Fe(III) and Sn(II).

The problem has been partially resolved in a later note by Peterson and Duke[105] describing their investigation of the reaction between Sn(II) and the ferricinium ion. Ferricinium perchlorate was prepared by oxidation of ferrocene with $AgClO_4$ in aqueous perchloric acid; from the nature of the ferricinium structure, Fe(III) is unlikely to complex with more than one chloride ion. The reaction, followed by absorbance measurements on the ferricinium ion at 615 mμ, is first-order in both reactants. The chloride-ion dependence indicates a total of five Cl^- ions in the activated complex, four of which are deduced to be associated with Sn(II) as $SnCl_4{}^-$.

Wetton and Higginson[59] have briefly investigated the reaction between Fe(III) and Sn(II). Spectrophotometric data for the rate of disappearance of Fe(III) were obtained at 335 mμ. In the absence of substantial amounts of Fe(II) and Sn(IV) the kinetics in 1.0 M hydrochloric acid are of simple second order, indicating that Sn(II) is present as the monomeric species. The addition of Fe(II) to the reacting system produces a pronounced retardation, the second-order plots showing curvature after 75 % reaction. The appropriate mechanism is

$$\mathrm{Fe(III)+Sn(II)} \underset{k_{-16}}{\overset{k_{16}}{\rightleftharpoons}} \mathrm{Fe(II)+Sn(III)}$$

$$\mathrm{Fe(III)+Sn(III)} \xrightarrow{k_{17}} \mathrm{Fe(II)+Sn(IV)}$$

This mechanism gives a quantitative fit with the observed kinetics: at 25 °C, k_{17}/k_{-16} is 1280 and k_{16} is 4.4 l.mole^{-1}.sec^{-1}. Furthermore, experiments in which the Co(III) complex, Co(YOH)H_2O, is included in the reacting system provide evidence for the presence of Sn(III) intermediates (H_4Y = EDTA).

By means of a stopped-flow technique, Carlyle and Espenson[105a,290] have subjected the reaction between europium(II) and iron(III) to a detailed examination. In perchloric acid solution two processes are discerned, *viz.*

$$\mathrm{Fe(H_2O)_6{}^{3+} + Eu^{2+}_{aq} \longrightarrow [FeEu(H_2O)^{5+}_m]^{\ddagger}}$$

and

$$\mathrm{Fe(H_2O)^{3+}_6 + Eu^{2+}_{aq} + H_2O \longrightarrow [FeEu(H_2O)_nOH^{4+}]^{\ddagger} + H^+}$$

In the presence of chloride ions there exists both an anion-catalysed reaction

$$\mathrm{Fe(H_2O)^{3+}_6 + Eu^{2+}_{aq} + Cl^- \longrightarrow Fe(H_2O)^{2+}_6 + Eu^{3+}_{aq} + Cl^-}$$

and an inner-sphere reaction (see, also, ref. 288)

$$\mathrm{(H_2O)_5FeCl^{2+} + Eu^{2+}_{aq} \longrightarrow Eu^{3+}_{aq} + Fe(H_2O)^{2+}_6 + Cl^-}$$

U(IV) is oxidised quantitatively to U(VI) by Fe(III) in dilute perchloric acid solutions, *viz.*

$$\mathrm{2\,Fe(III)+U(IV) = 2\,Fe(II)+U(VI)}$$

The reaction, as studied by Betts[106], was followed by measuring the amount of Fe(II) formed as a function of time, aliquots of the reaction mixture being quenched by a solution of *o*-phenanthroline at pH 4. At constant acidity and ionic strength the reaction is first-order in both Fe(III) and U(IV), *viz.*

$$\mathrm{d[Fe(II)]/d}t = -\mathrm{d[Fe(III)]/d}t = -2\,\mathrm{d[U(IV)]/d}t = 2\,k'\mathrm{[Fe(III)][U(IV)]}$$

where k' is 12.4 l.mole^{-1}.sec^{-1} when [$HClO_4$] = 1.02 M and μ = 1.02 M. Increase in acidity has a retarding influence: plots of log [H^+] *versus* log k' indicate a −1.8-order dependence on hydrogen-ion concentration. Increase of ionic strength (by addition of $NaClO_4$) is shown to produce a small increase in rate.

References pp. 267–273

At constant acidity both of following schemes are plausible

$$(A)\quad \begin{aligned} &\mathrm{Fe(III)+U(IV) \longrightarrow Fe(II)+U(V)} && \text{slow} \\ &\mathrm{Fe(III)+U(V) \longrightarrow Fe(II)+U(VI)} && \text{rapid} \end{aligned}$$

or

$$(B)\quad \begin{aligned} &\mathrm{Fe(III)+U(IV) \longrightarrow Fe(II)+U(V)} && \text{slow} \\ &\mathrm{2\,U(V) \longrightarrow U(VI)+U(IV)} && \text{rapid} \end{aligned}$$

Scheme (*B*) is considered unlikely on the grounds that, at the low concentrations of U(V) involved, the latter would disappear by oxidation with Fe(III) rather by dismutation. The hydrogen-ion dependence suggests that the rate-controlling step between Fe(III) and U(IV) can be visualised in terms of a series of competitive reactions of hydrolysed species of both reactants *viz.*

$$\mathrm{Fe^{3+} + H_2O \rightleftharpoons FeOH^{2+} + H^+} \qquad K_h$$

$$\mathrm{U^{4+} + H_2O \rightleftharpoons UOH^{3+} + H^+} \qquad K_4$$

$$\mathrm{Fe^{3+} + 2\,H_2O \rightleftharpoons Fe(OH)_2{}^+ + 2\,H^+} \qquad K_5$$

$$\mathrm{U^{4+} + 2\,H_2O \rightleftharpoons U(OH)_2{}^{2+} + 2\,H^+} \qquad K_6$$

$$\mathrm{Fe^{3+} + UOH^{3+}} \xrightarrow{k_{18}} \mathrm{Fe^{2+} + U(V)}$$

$$\mathrm{FeOH^{2+} + U^{4+}} \xrightarrow{k_{19}} \mathrm{Fe^{2+} + U(V)}$$

$$\mathrm{FeOH^{2+} + UOH^{3+}} \xrightarrow{k_{20}} \mathrm{Fe^{2+} + U(V)}$$

$$\mathrm{Fe(OH)_2{}^+ + U^{4+}} \xrightarrow{k_{21}} \mathrm{Fe^{2+} + U(V)} \tag{5.11}$$

$$\mathrm{Fe^{3+} + U(OH)_2{}^{2+}} \xrightarrow{k_{22}} \mathrm{Fe^{2+} + U(V)} \tag{5.12}$$

On the assumption that the concentration of $U(OH)_2{}^{2+}$ and $Fe(OH)_2{}^+$ are negligible when compared with the total concentration of Fe(III) and U(IV), the derived rate law is

$$\frac{\mathrm{d[Fe(II)]}}{\mathrm{d}t} = 2\mathrm{[Fe(III)][U(IV)]}\left(\frac{K'[\mathrm{H^+}]+K''}{[\mathrm{H^+}]^2+(K_h+K_4)[\mathrm{H^+}]+K_h K_4}\right)$$

so that

$$k' = \frac{K'[\mathrm{H^+}]+K''}{[\mathrm{H^+}]^2+(K_h+K_4)[\mathrm{H^+}]+K_h K_4} \tag{5.13}$$

where $K' = k_{18}K_4+k_{19}K_h$ and $K'' = k_{20}K_hK_4+k_{21}K_5+k_{22}K_6$. Writing D for the denominator of the right-hand side of (5.13)

$$k'D = K'[H^+]+K''$$

and plots of $k'D$ (calculated from known values of K_h and K_4) against $[H^+]$ are linear at temperature between 3.1 °C and 24.8 °C. The constants K' and K'', evaluated from the slopes and intercepts of such plots, are 2.98 sec^{-1} and 20.6 $mole.l^{-1}.sec^{-1}$, respectively, at 24.8 °C; the corresponding apparent activation energies are 22.5 and 24.5 $kcal.mole^{-1}$. If $Fe(OH)_2^+$ and $U(OH)_2^{2+}$, known to be present in extremely low concentrations, are insignificant kinetically ($k_{21} = k_{22} = 0$) then

$$\Delta H'' = E_{20}+\Delta H_h+\Delta H_4$$

where ΔH_h and ΔH_4 are the enthalpies of hydrolysis of Fe(III) and U(IV), respectively, and E_{20} is the true activation energy for $FeOH^{2+}+UOH^{3+}$. On this basis $E_{20} = 24.2-12.3-10.6 = 1.3$ $kcal.mole^{-1}$, a most unrealistic figure for electron-transfer processes. It is concluded, therefore, that the sequence involving two OH groups is, in fact, made up of (5.11) and (5.12). Maximum rates and activation parameters of the paths involving one OH group are given in Table 6. The Fe(III)+U(IV) reaction is greatly accelerated by HSO_4^- ions[107]. This is ascribed to the participation of sulphate complexes of both reactants, *e.g.*

$$U^{4+}+HSO_4^- \rightleftharpoons USO_4^{2+}+H^+$$

A -1.2-order dependence on hydrogen-ion concentration is noted. In sulphuric acid media the possibility that U(V) is consumed by disproportionation cannot be ruled out.

The rate law for the reduction of Fe(III) by Np(IV), *viz.*

$$Fe(III)+Np(IV) \rightleftharpoons Fe(II)+Np(V)$$

TABLE 6

MAXIMUM VALUES OF k_{18} AND k_{19} IN Fe(III)+U(IV) REACTION[106]

Condition	*Rate coefficient(l.mole^{-1}. sec^{-1}, 25° C)*	*E(kcal.mole^{-1})*	*$\Delta S^\ddagger$(cal.deg^{-1}. mole^{-1})*
$k_{19} = 0$, $k_{18} = K'/K_4$	107	11.9	−11.2
$k_{18} = 0$, $k_{19} = K'/K_h$	1860	10.2	−11.3

References pp. 267–273

is reported to be

$$d[Np(IV)]/dt = k''[Np(V)][Fe(II)][H^+]-k'[Np(IV)][Fe(III)]/[H^+]^3$$

The values given by Huizenga and Magnusson[17] for k' and k'' are 3.4 mole^2.l^{-2}.min^{-1} and 4.7 l^2.mole^{-2}.min^{-1}, respectively, at 25 °C in 1.0 M perchlorate solutions. The apparent activation energy corresponding to k' is 35 kcal.mole^{-1}. Nitrate complexing of Np(IV) reduces the forward rate.

5.2 OXIDATIONS BY RUTHENIUM(III)

$Ru(NH_3)_6^{3+}$, $Ru(NH_3)_5Cl^{2+}$, $Ru(H_2O)_5Cl^{2+}$, and *cis*-$Ru(NH_3)_4Cl_2^+$ are reduced by Cr^{2+} in perchlorate media[108,291–293] or *p*-toluenesulphonic acid media[293]. Chloride ion strongly catalyses the former reaction[108] according to

$$d[Ru(II)]/dt = (k+k'[Cl^-])[Ru(NH_3)_6^{3+}][Cr^{2+}]$$

At $\mu = 0.022\ M$, $k = 28$ l.mole^{-1}.sec^{-1}, $k' = 6\times10^2$ l^2.mole^{-2}.sec^{-1}. Ion-exchange experiments indicate that $CrCl^{2+}$ is the major product species.

6. Oxidations by cobalt(III)

A great deal of attention has been given to the oxidation–reduction reactions of cobalt(III). For convenience this section is subdivided into three parts: reactions involving inorganic bridging ligands (some outer-sphere systems are discussed also for comparison), reactions involving organic bridging ligands, and reactions of aquo complexes.

6.1 INORGANIC BRIDGING LIGANDS IN OXIDATIONS BY COBALT(III) COMPLEXES

The oxidation of Cr^{2+} with substitution-inert complexes of the type $Co(NH_3)_5X^{2+}$ has been the subject of a pioneering study by Taube *et al.*[94,109]. The range of complexes studied are those for which X = Cl^-, Br^-, I^-, F^-, SO_4^{2-}, H_2O and NH_3. The most significant fact to emerge from such reactions is that they are accompanied by quantitative transfer of X to the reductant. When X is Cl^-, no exchange occurs with chloride ion in solution during the course of reaction and it is surmised that an inner-sphere activated complex containing a Cr–Cl–Co bridge is involved, *i.e.*, $[(NH_3)_5Co\text{–}X\text{–}Cr^{4+}]^{\ddagger}$. Earley and Gorbitz[110], confirming the earlier results of Taube[111], have shown that the principal path of reduction of $Co(NH_3)_5Cl^{2+}$ by Cr^{2+} in the presence of pyrophosphate yields product Cr(III) incorporating both chloride and pyrophosphate.

The rates of oxidation of V^{2+} by complexes of the type $Co(NH_3)_5X$, where X = H_2O, NH_3 and Cl^-, have been examined by Zwickel and Taube[112] in H_2O and in D_2O solution. These workers have compared the results on such systems with data on the $Cr^{2+}+Co(NH_3)_6{}^{3+}$ reaction in H_2O and D_2O. The product V^{3+}, unlike Cr^{3+}, is substitution-labile. Consequently, a different approach is necessary for V^{2+} oxidations than is customary for Cr^{2+} oxidations. The specific rate coefficients for $Co(NH_3)_6{}^{3+}+V^{2+}$ and $Co(NH_3)_6{}^{3+}+Cr^{2+}$ vary with Cl^- concentration according to

$$k_{obs} = k+k'[Cl^-]$$

The rate coefficients, k and k', and the corresponding activation parameters are given in Tables 7 and 8. $Cr(H_2O)_5Cl^{2+}$ is the primary product in the chloride-dependent path for the $Co(NH_3)_6{}^{3+}+Cr^{2+}$ reaction. The specific rate coefficient for the $Co(NH_3)_5Cl^{2+}+V^{2+}$ reaction was determined as 342 l.mole^{-1}.min^{-1} at μ = 1.00 M and 25 °C. In general, reactions of V^{2+} contrast sharply with those of Cr^{2+} (except for $Co(NH_3)_6{}^{3+}+Cr^{2+}$) and $Cr(bipy)_3{}^{2+}$ by exhibiting chloride

TABLE 7

RATE DATA FOR $Co(NH_3)_6{}^{3+}$ REACTIONS[112]

Reductant	*Temp.(°C)*	k(l.mole^{-1}.min^{-1})	k'(l^2.mole^{-2}.min^{-1})
Cr^{2+}	25.0	0.0053	0.74
	37.0	0.014	1.70
	37.0	0.011[a]	1.35[a]
V^{2+}	25.0	0.22	1.27
	37.0	0.41	3.27
	37.0	0.24[a]	1.95[a]

Ionic strength 0.40 M. [a] In 100 % D_2O.

TABLE 8

ACTIVATION PARAMETERS FOR $Co(NH_3)_6{}^{3+}$ REACTIONS[112]

Reductant	*Path*	$\Delta H^‡$(kcal.mole^{-1})	$\Delta S^‡$(cal.deg^{-1}.mole^{-1})
Cr^{2+}	k	14.7	−30
Cr^{2+}	k'	12.4	−25
V^{2+}	k	9.1	−40
V^{2+}	k'	14.1	−20

TABLE 9

SECOND-ORDER RATE COEFFICIENTS ($l.mole^{-1}.sec^{-1}$ at 25 °C) FOR THE REDUCTION OF VARIOUS Co(III) COMPLEXES

Oxidant	Cr^{2+} [a]	V^{2+} [a]	Eu^{2+} [a]	$Cr(bipy)_3^{2+}$ [b]	$Ru(NH_3)_6^{2+}$ [c]	Fe^{2+} [a]
$Co(NH_3)_6^{3+}$	8.9×10^{-5} [d,e]	3.7×10^{-3} [d,e]	$2\ \times10^{-2}$ [b]	6.9×10^{2} [f]	1.1×10^{-2}	—
$Co(NH_3)_5OH_2^{3+}$	0.5	$\sim$ 0.5[e]	0.15	$5\ \times10^{4}$	3.0	—
$Co(NH_3)_5OH^{2+}$	1.5×10^{6} [e]	—	—	—	$4\ \times10^{-2}$	—
$Co(NH_3)_5F^{2+}$	$9\ \times10^{5}$ [g]	2.6	2.6×10^{4}	1.8×10^{3}	—	6.6×10^{-3}
$Co(NH_3)_5Cl^{2+}$	2.6×10^{6} [g]	$\sim$5	3.9×10^{2}	$8\ \times10^{5}$	2.6×10^{2}	1.3×10^{-3}
$Co(NH_3)_5Br^{2+}$	$> 2\ \times10^{6}$	$\sim$25	2.5×10^{2}	$5\ \times10^{6}$	1.6×10^{3}	7.3×10^{-4}
$Co(NH_3)_5I^{2+}$	$> 2\ \times10^{6}$	1.2×10^{2}	1.2×10^{2}	—	6.7×10^{3}	—
$Co(NH_3)_5N_3^{2+}$	$\sim 3\ \times10^{5}$	13	1.9×10^{2}	4.1×10^{4}	1.2	8.8×10^{-3} [h]
$Co(NH_3)_5NCS^{2+}$	19	0.3	$\sim$0.7	1.0×10^{4}	—	$<3\ \times10^{-6}$
$Co(NH_3)_5SO_4^{+}$	18	7.8[i]	1.4×10^{2}	4.5×10^{4}	—	—
$Co(NH_3)_5OAc^{2+}$	0.18[j]	0.43[k]	0.18[k]	1.2×10^{3}	—	$<5\ \times10^{-5}$
$Co(NH_3)_5NO_3^{2+}$	$\sim$90	—	$\sim1\times10^{2}$	—	34	—
$Co(NH_3)_5S_2O_3^{+}$	—	—	—	$8\ \times10^{4}$	—	—
$Co(NH_3)_5(maleate)^{+}$	—	—	—	1.0×10^{3}	—	—
$Co(NH_3)_5PO_4$	4.8×10^{9}	1.4×10^{7}	—	—	—	—
$Co(NH_3)_5PO_4H^{+}$	8.3×10^{3}	1.6×10^{2}	$5\ \times10^{2}$	—	—	—
$Co(NH_3)_5PO_4H_2^{2+}$	0.3	2.3	6	—	—	—
$Co(NH_3)_5PO_4H_3^{3+}$	0.3	4.5	$\sim$3	—	—	—
$Co(en)_3^{3+}$	$\sim2\ \times10^{-5}$ [d]	$\sim2\ \times10^{-4}$	$\sim5\ \times10^{-3}$	1.8×10^{3}	—	—

Data from refs. 113, 125 and 126, unless otherwise specified.
[a] $\mu = 1.0$ *M*, ClO_4^- medium, unless otherwise noted. [b] $\mu = 0.1$ *M*, ClO_4^- medium. [c] $\mu \sim 0.2$ *M*. [d] $\mu = 0.4$ *M*. [e] from refs. 112 and 121. [f] from ref. 114. [g] from ref. 127. [h] from ref. 128. [i] from ref. 122. [j] from ref. 129. [k] from ref. 130.

catalysis†. The rate data on the $Co(NH_3)_5(H_2O)^{3+}+V^{2+}$ reaction are marred by pronounced scatter of results. Previous arguments[1] used to explain the differences in the $Co(NH_3)_5(H_2O)^{3+}+Cr^{2+}$ and $Fe^{3+}+Cr^{2+}$ reactions, based upon variations in the lability of Co(III) and Fe(III), are considered doubtful. Zwickel and Taube[112] conclude that oxidations of V^{2+} by Co(III) complexes are more likely to proceed by outer-sphere routes††. These authors speculate on the intimate nature of the processes involved in the transfer of an electron from the reductant to the Co(III) centre.

Candlin *et al.*[113] have reported the results of kinetic investigations on the reduction of various pentaamminecobalt(III) complexes by Cr^{2+}, V^{2+}, Eu^{2+} and $Cr(bipy)_3{}^{2+}$. In most cases a stopped-flow apparatus[44] was used, the rate of disappearance of the Co(III) complex being measured at ~ 500 mμ. Data on $Cr(bipy)_3{}^{2+}$ reduction were obtained by following its decrease in absorbance at 562 mμ. Oxygen was eliminated during all kinetic experiments. The flow technique could measure half-lives within the range 5×10^{-3} sec to 30 sec. Second-order rate coefficients and activation parameters are given in Tables 9 and 10. Rate coefficients for reduction of phosphatopentaamminecobalt(III) complexes are recorded also; the order of reactivity with Cr^{2+}, V^{2+} and Eu^{2+} is $Co(NH_3)_5PO_4 >$ $Co(NH_3)_5PO_4H^+ > Co(NH_3)_5PO_4H_2{}^{2+} \sim Co(NH_3)_5PO_4H_3{}^{3+}$. The inner-sphere character of Cr^{2+} reductions is recognisable from the substantial effects caused by variation of the nature of ligand X in the oxidant. All the oxidants studied, excluding $Co(NH_3)_6{}^{3+}$ and $Co(en)_3{}^{3+}$ which react by outer-sphere processes, react with Cr^{2+} through the intervention of a bridging group. Reductions by $Cr(bipy)_3{}^{2+}$ are relatively insensitive to the nature of X and are design-

TABLE 10

ACTIVATION PARAMETERS FOR REDUCTION OF VARIOUS Co(III) COMPLEXES[113]

	Cr^{2+}		V^{2+}		Eu^{2+}	
Oxidant	$\Delta H^‡$	$\Delta S^‡$	$\Delta H^‡$	$\Delta S^‡$	$\Delta H^‡$	$\Delta S^‡$
$Co(NH_3)_5NH_3{}^{3+}$	14.7[a]	−30[a]	9.1[a]	−40[a]	—	—
$Co(NH_3)_5OH_2{}^{3+}$	2.9[b]	−52[b]	—	—	—	—
$Co(NH_3)_5Cl^{2+}$	—	—	—	—	5.0	−30
$Co(NH_3)_5Br^{2+}$	—	—	9.1	−22	4.7	−32
$Co(NH_3)_5I^{2+}$	—	—	(4.6)	(−32)	—	—
$Co(NH_3)_5N_3{}^{2+}$	—	—	11.7	−14	5.5	−30
$Co(NH_3)_5SO_4{}^{+}$	6.2	−32	6.1[c]	−34[c]	6.7	−26
$Co(NH_3)_5NCS^{2+}$	6.9	−29	—	—	—	—

$\Delta H^‡$ in kcal.mole^{-1}; $\Delta S^‡$ in cal.deg^{-1}.mole^{-1}. [a] From ref. 112. [b] From ref. 121. [c] From ref. 122.

† See, also, ref. 294.
†† See, however, ref. 295.

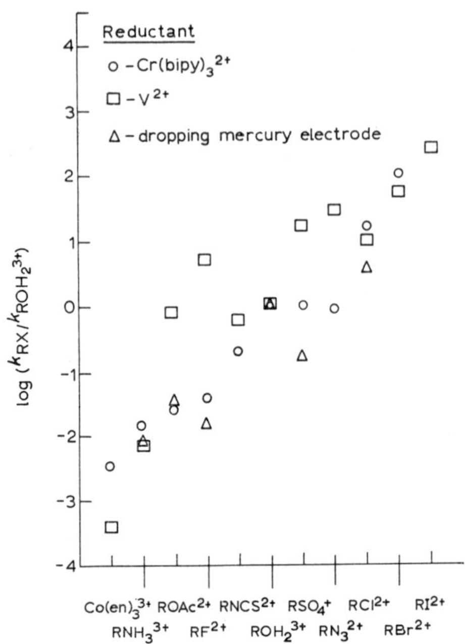

Fig. 5. Relative rate coefficients ($k_{RX}/k_{ROH_2{}^{3+}}$) at 25 °C, referred to the rate for $ROH_2{}^{3+}$ [R = $Co(NH_3)_5$], for the reduction of various Co(III) complexes arranged in approximate order of increasing reactivity. Ionic strengths: $Cr(bipy)_3{}^{2+}$, 0.1 *M*; V^{2+}, 1.0 *M*; dropping mercury electrode (from data of Vlček[116]), 0.4 *M*. (*From Candlin et al.*[113], *by courtesy of The American Chemical Society.*)

ated as outer-sphere reactions (as originally suggested by Zwickel and Taube[114]). Similarly, the kinetic data of Table 9 indicates that reactions of V^{2+} are probably outer-sphere. The close resemblance between V^{2+} and $Cr(bipy)_3{}^{2+}$, as regards patterns of reactivity, is clear from Fig. 5 and both show a correlation with the rates of electrochemical reduction of the Co(III) complexes as measured by Vlček using a dropping mercury electrode[115]. Vlček[115,116] has drawn attention to the fact that the order of increasing reactivity of the Co(III) complexes is very nearly the same as the order of decreasing excitation energy, $t_{2g}{}^6 \rightarrow t_{2g}{}^5 e_g$. The latter is evident from the shift in λ_{max} values with changes in field strength of ligand X. Reductions by Eu(II) are difficult to categorise: the pattern of reactivity towards halogeno complexes is $Co(NH_3)_5F^{2+}$ > $Co(NH_3)_5Cl^{2+}$ > $Co(NH_3)_5Br^{2+}$ > $Co(NH_3)_5I^{2+}$, the opposite of that shown by V^{2+}, $Cr(bipy)_3{}^{2+}$ and $Co(CN)_5{}^{3-}$. Candlin *et al.*[113] conclude that Eu(II) reductions proceed predominantly by an inner-sphere route. These authors comment also on the dangers inherent in assuming that reactions of even closely-related oxidants with a common reductant proceed by a common mechanism, particularly in the light of evidence to show that reduction of $Co(NH_3)_5X$ complexes by $Co(CN)_5{}^{3-}$ may occur through both inner- and outer-sphere paths[117,118] and that reduction by V^{2+} of complexes containing certain conjugated organic ligands takes place by remote attack on the ligand[119,120].

Diebler and Taube[123] have quoted rate parameters for the reduction of halopentaamminecobalt(III) complexes by Fe(II) and V(II). Rate coeffients for V^{2+} reductions are in reasonable agreement with those reported by Candlin *et al.*[113] (except for $Co(NH_3)_5Br^{2+}$). So also are Fe^{2+} reductions, taking into account differences in conditions. At 25.5 °C and $\mu = 1.7$ *M*, the second-order rate coefficients for Fe^{2+} reduction of $Co(NH_3)_5F^{2+}$, $Co(NH_3)_5Cl^{2+}$ and $Co(NH_3)_5Br^{2+}$ are 7.6×10^{-3}, 1.6×10^{-3}, and 0.92×10^{-3} l.mole^{-1}.sec^{-1}, respectively (compare results in Table 9). The respective values of $\Delta H^{\ddagger}$ are 13.4, 14.5 and 15.6 kcal.mole^{-1}, and of $\Delta S^{\ddagger}$ are -23, -23 and -20 cal.deg^{-1}.mole^{-1}. The observed reactivity pattern $F^- > Cl^- > Br^-$ is the reverse of the pattern encountered in the oxidation of Cr^{2+} with CrX^{2+} (ref. 99) and $Cr(NH_3)_5X^{2+}$ (ref. 124) and in the reduction of $Co(NH_3)_5X^{2+}$ by $Cr(bipy)_3{}^{2+}$ (ref. 114), $Ru(NH_3)_6{}^{2+}$ (ref. 125) and V^{2+} (ref. 113)†. However, the pattern $F^- > Cl^- > Br^-$ applies also in the $Co(NH_3)_5X^{2+} + Eu^{2+}$ system[113]. In the case of the $Co(NH_3)_5X^{2+} + Fe^{2+}$ systems there seems to be a correlation between rate and the thermodynamic stability of the complexes of Fe^{3+} with X^-: the relevant stability constants are $\sim 1.5 \times 10^5$, ~ 4 and < 0.1 for FeF^{2+}, $FeCl^{2+}$ and $FeBr^{2+}$, respectively. The reactions are assisted by free halide ions, as shown by

$$\text{rate} = k'[Fe^{2+}][Co(NH_3)_5X^{2+}][X^-]$$

When $Co(NH_3)_5F^{2+}$ is the oxidant, values of k' are 3.5×10^{-3}, 2.1×10^{-2} and 13 l^2.mole^{-2}.sec^{-1} for $X^- = Br^-$, Cl^- and F^-, respectively. With $Co(NH_3)_5Br^{2+}$, values of k' are $\leqq 1 \times 10^{-3}$ and 4 l^2.mole^{-2}.sec^{-1} for $X^- = Cl^-$ and F^-, respectively. These results apply for $\mu = 1.7$ *M* and 25.8 °C except for $X^- = F^-$ where the temperature is 25.1 °C and $\mu = 2.0$ *M*. Since the ratio of k' for F^- compared to k' for Cl^- is considerably different for $Co(NH_3)_5F^{2+}$ and $Co(NH_3)_5Br^{2+}$ (6×10^2 as opposed to $> 4 \times 10^3$) the stabilization order of Fe^{3+} by the halide cannot be invoked to explain the greater effect of F^- as compared to Cl^-.

Espenson[126] has made a study of reductions of acidopentaamminecobalt(III) ions by Fe^{2+}, *viz.*

$$Co(NH_3)_5X^{2+} + Fe^{2+} + 5\,H^+ = Co^{2+} + Fe^{3+} + X^- + 5\,NH_4{}^+$$

It proved necessary to correct the rate data in the case of $X = Cl^-$ and Br^- for the simultaneous aquation of the complex. Rate coefficients are essentially in agreement with those given in Table 9. Activation parameters are quoted when $X = F^-, Cl^-, Br^-$, and also for $Co(C_2O_4)_3{}^{3-}$.

† Parker and Espenson[296] find the pattern $Br^- > Cl^- > F^-$ in similar reductions by Cu^+

Candlin and Halpern[127] comment that the sequence of rapid rates observed for Cr^{2+} as a reductant (*i.e.* $Co(NH_3)_5I^{2+} > Co(NH_3)_5Br^{2+} > Co(NH_3)_5Cl^{2+} > Co(NH_3)_5F^{2+}$) is contrary to that found for the slow reactions of Fe^{2+} (ref. 126) and Eu^{2+} (ref. 113). All three reductants would appear to favour inner-sphere mechanisms, but in the case of Fe^{2+} and Eu^{2+} the order of reactivity seems to be connected with the stability of the product halide complex (FeX^{2+} or EuX^{2+}) which increases in the order $X = I^-$ to $X = F^-$. Or in other words, as pointed out by Halpern and Rabani[131], in the generalised inner-sphere reaction

$$\text{Co–X} + \text{red} \longrightarrow \text{Co} + \text{X–red}$$

it is reasonable to suppose that the order of reactivity, as X is varied, will depend on variations in the strength of the bond being broken (Co–X) *and* the bond being formed (X–red). The strength of both bonds increases in the order $I^- < Br^- < Cl^- < F^-$. The order of reactivity of various reductants (when $X = Cl^-$) is $H > Co(CN)_5^{3-} > Cr^{2+} > Eu^{2+} > Fe^{2+}$. In the case of mild reductants (for example, Eu^{2+} and Fe^{2+}) the formation of a bond to the reductant in the transition state is considered more important than bond breaking. Thus the reactivity order will be $Co(NH_3)_5F^{2+} > Co(NH_3)_5Cl^{2+} > Co(NH_3)_5Br^{2+} > Co(NH_3)_5I^{2+}$ (see, however, Adin and Sykes[65] on $CrX^{2+} + Eu^{2+}$, p. 169). For more reactive reductants (*e.g.* hydrogen atoms, Cr^{2+} and $Co(CN)_5^{3-}$) bond-breaking is more important in the transition state than bond-making so that the reactivity order is decided by the order of bond strengths, *i.e.*, $Co(NH_3)_5I^{2+} > Co(NH_3)_5Br^{2+} > Co(NH_3)_5Cl^{2+} > Co(NH_3)_5F^{2+}$.

The oxygen-(I) and sulphur-bonded(II) isomers of thiosulphatopentaamminecobalt(III)

```
            O                              O
            |                              |
(NH3)5Co–O–S–S              (NH3)5Co–S–S–O
            |                              |
            O                              O

      I                          II
```

are of comparable stability[132]. However, isomer I is reduced by Cr^{2+} 70 times more rapidly than isomer II. The rate of reduction of I resembles the rates of reduction of the sulphato[113] and sulphito complexes (Table 11).

Miller *et al.*[133] have examined in detail the relative efficiencies of oxoanions as bridging groups by means of a study of the rates of reaction of Cr(II), V(II), Eu(II) and Ti(III) with oxoanion complexes of pentaammine- and tetraammine-cobalt(III). The complexes used include metaborato, carbonato, nitro, nitrito, nitrato, sulphito, sulphato, aquosulphato, thiosulphato, selinito, selanato and phosphato. The rates of reduction correlate with the position in the periodic table of the central atom of the oxoanion group: the rate increases from Group

TABLE 11

REDUCTION OF $Co(NH_3)_5X$ COMPLEXES BY Cr^{2+} (from ref. 132)

X	*k(l.mole⁻¹.sec⁻¹, at 25 °C)*	*ΔH‡(kcal.mole⁻¹)*	*ΔS‡(cal.deg⁻¹.mole⁻¹)*	*Ref.*
SO_4^{2-}	18	8.3	−25	113
SO_3^{2-}	18.6	8.3	−26	132
O-bonded $S_2O_3^{2-}$	13.3	4.2	−39	132
S-bonded $S_2O_3^{2-}$	0.18	24.6	3	132

III to a maximum value at Group V, and then decreases. Furthermore, for a given ligand, the rate increases as the atomic weight of the central atom increases.

The reduction of cyanopentaamminecobalt(III) by Cr^{2+} in acidic perchlorate media proceeds in two distinct stages[134, 297]

$$Co(NH_3)_5CN^{2+}+Cr^{2+}+5\,H^+ \longrightarrow Co^{2+}+CrNC^{2+}+5\,NH_4^+ \tag{6.1}$$

$$CrNC^{2+} \longrightarrow CrCN^{2+} \tag{6.2}$$

The oxidation–reduction stage (6.1) has a second-order rate coefficient of 22.8 l. mole^{-1}.sec^{-1} ($\mu = 0.15\ M$) at 15 °C. The linkage isomerisation stage (6.2) is much slower with a first-order rate coefficient of 1.0×10^{-2} sec^{-1} in the absence of Cr^{2+} at 15 °C, $[H^+] = 0.4\ M$, $\mu = 1.0\ M$[297]. Since the rate of this stage is dependent on the concentration of Cr^{2+} [297], the isomerisation occurs also by the step

$$CrNC^{2+}+Cr^{2+} \longrightarrow Cr^{2+}+CrCN^{2+} \tag{6.3}$$

The species $CrCN^{2+}$ was characterised by means of its behaviour to ion-exchange, its absorption spectrum, and analysis of the CN^-/Cr ratio in the separated complex. The existence of the species $CrNC^{2+}$ was less clearly established from its absorption spectra.

The rates of reduction of *cis*- and *trans*-$Co(NH_3)_4(N_3)_2^+$ and $Co(NH_3)_5N_3^{2+}$ by Fe^{2+} in aqueous perchloric acid have been studied by Haim[128]. The *cis* isomer of the former complex reacts at an acid-independent rate whereas the rate for the *trans* isomer is acid-dependent. The respective rate laws at 25 °C are

$$\text{rate} = 11.1[Fe^{2+}][\textit{cis}\text{-}Co(NH_3)_4(N_3)_2^+]$$

and

$$\text{rate} = (4.4+82[H^+])[Fe^{2+}][\textit{trans}\text{-}Co(NH_3)_4(N_3)_2^+]$$

where time is expressed in min. $Co(NH_3)_5N_3^{2+}$ reacts at an acid-independent rate given by

$$\text{rate} = 0.52\ [Fe^{2+}][Co(NH_3)_5N_3^{2+}]$$

The order of reactivity towards Fe^{2+} is given by *cis*-$Co(NH_3)_4(N_3)_2^+$ > *trans*-$Co(NH_3)_4(N_3)_2^+ \gg Co(NH_3)_5N_3^{2+}$. In view of the lability of Fe(III) to substitution it is not possible to decide whether these reactions proceed by inner- or outer-sphere routes. But the findings can be explained readily on the basis of an inner-sphere mechanism. The suggestion is made[135] that *cis*-$Co(NH_3)_4$-$(N_3)_2^+$ reacts with Fe^{2+} through a double-bridged activated complex analogous to that formed between $Cr(N_3)_2^+$ and Cr^{2+}. Interpretation of the relatively high reactivity of *trans*-$Co(NH_3)_4(N_3)_2^+$ in terms of a *trans* effect[136] receives support from the observed acid catalysis which probably arises from the fact that removal of N_3^- *trans* to the bridging N_3^- is aided by attachment of a proton (see also ref. 137).

With a view to determining the equilibrium constant for the isomerisation, the rates of reduction of an equilibrium mixture of *cis*- and *trans*-$Co(NH_3)_4(OH_2)N_3^{2+}$ with Fe^{2+} have been measured by Haim[138]. At Fe^{2+} concentrations above 1.5×10^{-3} *M* the reaction with Fe^{2+} is too rapid for equilibrium to be established between *cis* and *trans* isomers, and two rates are observed. For Fe^{2+} concentrations below 1×10^{-4} *M*, however, equilibrium between *cis* and *trans* forms is maintained and only one rate is observed. Detailed analysis of the rate data yields the individual rate coefficients for the reduction of the *trans* and *cis* isomers by Fe^{2+} (24 $l.mole^{-1}.sec^{-1}$ and 0.355 $l.mole^{-1}.sec^{-1}$) as well as the rate coefficient and equilibrium constant for the *cis* to *trans* isomerisation (1.42×10^{-3} sec^{-1} and 0.22, respectively). All these results apply at perchlorate concentrations of 0.50 *M* and at 25 °C. Rate coefficients for the reduction of various azidoammine-cobalt(III) complexes are collected in Table 12. Haim[138] discusses the implications of these results on the basis that all these systems make use of azide bridges*. The effect of substitution in Co(III) by a non-bridging ligand is remarkable in terms of reactivity towards Fe^{2+}. The order of reactivity, *trans*-$Co(NH_3)_4(OH_2)N_3^{2+}$ > *trans*-$Co(NH_3)_4(N_3)_2^+$ > $Co(NH_3)_5N_3^{2+}$, is at va-

TABLE 12

RATE COEFFICIENTS FOR THE REDUCTION OF Co(III) AZIDE COMPLEXES BY Fe^{2+} (25 °C)[138]

Complex	*k*($l.mole^{-1}.sec^{-1}$)
$Co(NH_3)_5N_3^{2+}$	0.0087
trans-$Co(NH_3)_4(N_3)_2^+$	0.0733
cis-$Co(NH_3)_4(N_3)_2^+$	0.185
cis-$Co(NH_3)_4(OH_2)N_3^{2+}$	0.355
trans-$Co(NH_3)_4(N_3H)N_3^{2+}$	>1.37
trans-$Co(NH_3)_4(OH_2)N_3^{2+}$	24

* A direct test of this mechanism is awaited. Haim[138] notes that the primary Fe(III) product of reduction of *trans*-$Co(NH_3)_4(OH_2)N_3^{2+}$ should be detectable using a suitable flow apparatus.

riance with Orgel's suggestion[136] that the smaller the ligand field strength of the group *trans* to the bridging ligand ($N_3^- < H_2O < NH_3$ is accepted as the order of field strength) the higher the rate. Instead, this behaviour can be rationalised on the grounds that electron transfer necessitates the movement of both the bridging ligand and the group *trans* to it away from the Co(III) centre (see also ref. 128). Thus *trans*-$Co(NH_3)_4(OH_2)N_3^{2+}$ is more reactive than *trans*-$Co(NH_3)_4(N_3)_2^+$ because H_2O is more easily removed from Co(III) than is N_3^-.

Reductions of various Co(III) complexes by Fe(II) have been studied under high pressures[139]. The motivation for performing such experiments resides in the possibility that the volume of activation ($\Delta V^\ddagger$), like the entropy of activation, might be a criterion for distinguishing between inner- and outer-sphere reactions. For reactions of the type

$$(NH_3)_5CoX^{2+} + Fe^{2+} + 5\,H^+ = Co^{2+} + Fe^{3+} + X^- + 5\,NH_4^+$$

the formation of the activated complex on an inner-sphere mechanism, unlike that for an outer-sphere route, is accompanied by the release of a water molecule from the coordination shell of the reductant to the solvent, *viz.*

$$(NH_3)_5CoX^{2+} + Fe(H_2O)_6^{2+} \begin{array}{l} \nearrow [(NH_3)_5Co\text{–}X\text{–}Fe(H_2O)_5^{4+}]^\ddagger + H_2O \quad \text{inner-sphere} \\ \searrow [(NH_3)_5CoX\ H_2O\ Fe(H_2O)_5^{4+}]^\ddagger \quad \text{outer-sphere} \end{array}$$

On this basis $\Delta V^\ddagger$ should be more positive for an inner-sphere than for an outer-sphere reaction since a water molecule occupies a greater volume in the liquid phase than if it is coordinated. Second-order rate coefficients were determined at various pressures in the range 0.001 to 3.5 kbars, the rate decreasing with increase in pressure. The apparatus used was a modification of that first described by Osborn and Whalley[140]. Values of $\Delta V^\ddagger$ were calculated from the slopes of plots of log k *versus* pressure, since

$$-\mathrm{d}\log_e k/\mathrm{d}P = \Delta V^\ddagger/RT$$

The results obtained by Candlin and Halpern[139] are given in Table 13; in all cases, it is seen that $\Delta V^\ddagger$ is positive. These results strongly suggest that the Fe(II) reductions proceed by inner-sphere routes. However, to be convincing the method requires calibration by reactions of known mechanism.

Endicott and Taube[125] consider that there is cause for doubt over the generally-held views that $Cr(bipy)_3^{2+}$ is oxidised by an outer-sphere mechanism[114]. They suggest that, since the complex is very labile to substitution, coordination sites

TABLE 13

VOLUMES OF ACTIVATION ($\Delta V^{\ddagger}$) FOR REDUCTION OF VARIOUS Co(III) COMPLEXES BY Fe(II)[139]

Complex	$\Delta V^{\ddagger}$ ($cm^3.mole^{-1}$)
$Co(NH_3)_5F^{2+}$	+11
$Co(NH_3)_5Cl^{2+}$ [a]	+8
$Co(NH_3)_5Br^{2+}$ [a]	+8
$Co(NH_3)_5N_3^{2+}$	+14
cis-$Co(NH_3)_4(N_3)_2^+$	+14
trans-$Co(NH_3)_4(N_3)_2^+$ [b]	+2.2
trans-$Co(NH_3)_4(N_3)_2^+$ [c]	+2.8
$Co(HY)Cl^-$ [d]	+3

$HClO_4$ media; temp., 25 °C except when noted. [a] At 35 °C. [b] 0.001 *M* $HClO_4$. [c] 0.02 *M* $HClO_4$. [d] At 20 °C; Y^{4-} = ethylenediaminetetraacetate.

could well be exposed by the opening-up of a chelate ring. The complex $Ru(NH_3)_6^{2+}$ is a better choice for an outer-sphere reductant since it is substitution-inert and the oxidation product is $Ru(NH_3)_6^{3+}$. Endicott and Taube[125] have surveyed the reaction of this reagent with a variety of pentaamminecobalt(III) complexes†. The data given in Table 9 (p. 190) displays the parallelism between $Cr(bipy)_3^{2+}$ and $Ru(NH_3)_6^{2+}$ reductions as regards general patterns of reactivity, *e.g.* $Co(NH_3)_5OH_2^{3+}$ is more reactive than $Cr(NH_3)_5OH^{2+}$ towards both reagents. These results imply that the two reagents react by a common mechanism. Furthermore, there is a resemblance in the rates of reduction of the Co(III) complexes at the dropping mercury electrode[115] and by $Ru(NH_3)_6^{2+}$. Outer-sphere routes may operate for reduction of $Co(NH_3)_5NH_3^{3+}$, $Co(NH_3)_5OH_2^{3+}$ and $Co(NH_3)_5Cl^{2+}$ by V^{2+}. The pattern of reactivity for Cr^{2+} is quite different than that of the other reductants: with the exception of $Co(NH_3)_6^{3+}$, Cr^{2+} reacts *via* an inner-sphere mechanism. It is significant that $Co(NH_3)_5OH^{2+}$ is reduced much more rapidly than $Co(NH_3)_5OH_2^{3+}$. From a restricted study of Cu^+ reductions it appears likely that this reductant favours an outer-sphere process. Taube *et al.*[125,298] and Patel and Endicott[299] discuss the results from the standpoint of the Marcus theory of electron-transfer reactions[43].

Endicott and Taube[108] have investigated salt effects in the reactions of halogenopentaamminecobalt(III) complexes with $Ru(NH_3)_6^{2+}$, *i.e.*

$$Co(NH_3)_5Br^{2+} + Ru(NH_3)_6^{2+} + 5\,H^+ = Co^{2+} + Ru(NH_3)_6^{3+} + 5\,NH_4^+ + Br^- \qquad (k')$$

$$Co(en)_2Cl_2^+ + Ru(NH_3)_6^{2+} + 2\,H^+ = Co^{2+} + Ru(NH_3)_6^{3+} + 2\,enH^+ + 2\,Cl^- \qquad (k'')$$

† Meyer and Taube[298] have similarly investigated reduction of Fe(III) complexes by $Ru(NH_3)_6^{2+}$.

At ionic strengths less than 0.015 M

$$\log k' = 4.4\ \mu^{\frac{1}{2}} + \log k'_0$$

and

$$\log k'' = 2.5\ \mu^{\frac{1}{2}} + \log k''_0$$

In the case of the reduction of the complex $Co(NH_3)_5I^{2+}$ by $Ru(NH_3)_6{}^{2+}$ the kinetic concentration of $Ru(NH_3)_5I^{2+}$ is higher than the estimated equilibrium concentration, and thus the reaction would appear to occur by direct group transfer.

The reduction of a variety of *cis*- and *trans*-chlorobis(ethylenediamine)cobalt-(III) complexes (of the type $Co(en)_2XCl^{n+}$) by Fe^{2+} have been examined by Benson and Haim[141†]. These authors summarise the available information concerning the nature of the bridging ligand for complexes of this type. When $X = NH_3$, Cl^- must be the bridging ligand. When $X = SCN^-$, since $Co(NH_3)_5Cl^{2+}$ reacts very much faster[126] than $Co(NH_3)_5NCS^{2+}$, again Cl^- must act as the bridge. When $X = H_2O$, Cl^- bridging must occur because $Co(NH_3)_5Cl^{2+}$ is much more reactive than $Co(NH_3)_5OH_2{}^{3+}$. For the case where $X = N_3{}^-$, it appears that $N_3{}^-$ functions as the bridging ligand for the following reasons: (*a*) $Co(NH_3)_5N_3{}^{2+}$ is reduced by Fe^{2+} more rapidly than is $Co(NH_3)_5Cl^{2+}$, (refs. 126, 128, 138) *i.e.*, $N_3{}^-$ is more efficient than Cl^- when NH_3 is in a *trans* position to the bridging group; (*b*) replacement of NH_3 by $N_3{}^-$ in the *trans* position of $Co(NH_3)_5N_3{}^{2+}$ produces a much less marked effect in reactivity than does the replacement of NH_3 by Cl^- in the *trans* position of $Co(en)_2NH_3Cl^{2+}$, and therefore if Cl^- and $N_3{}^-$ function similarly as *trans* ligands in reduction of *trans*-$Co(en)_2N_3Cl^+$ then the presence of a *trans* Cl^- should help attack at $N_3{}^-$. When $X = Br^-$, it is likely that both Cl^- and Br^- are involved in bridging. The nature of the bridging group may be different for reduction of *cis* and *trans* isomers particularly when X and Cl^- have similar abilities for bridging. It is clear, however, that for those cases where $X = H_2O$, NH_3 or SCN^- the bridge is formed by Cl^- irrespective of whether X and Cl^- are *cis* or *trans* to one another. Benson and Haim[141] interpret their results (Table 14) for reductions of those $Co(en)_2XCl^{n+}$ complexes, which make use of the same bridge, on a model which assigns relative reactivities to two factors: (*a*) the ligand field strength of the group *trans* to the bridge[136], and (*b*) the energy required to stretch the metal–ligand bond along the z axis[1]. The latter factor is of importance since such stretching will bring about a lowering in the energy of the d_{z^2} orbital of the Co(III) centre, and therefore increase the availability of the orbital to an incoming electron. The general lack of information on force constants makes

† Compare ref. 300.

TABLE 14

COMPARISON OF THE RATE OF REDUCTION OF SOME $Co(en)_2XY^{n+}$ COMPLEXES BY CHROMIUM(II) AND IRON(II) AT 25 °C[64]

Oxidant	*Reductant*	*Relative rate coefficients*		
		trans-X/cis-X	*cis-X/cis-NH₃*	*trans-X/trans-NH₃*
$Co(en)_2NH_3(NCS)^{2+}$	Cr^{2+}	1.2	1.0	1.0
$Co(en)_2(NCS)_2^{+}$	Cr^{2+}	9×10^{-1}	4.8	3.7
$Co(en)_2OH_2(NCS)^{2+}$	Cr^{2+}	3×10^{1}	1.5×10^{1}	3.7×10^{2}
$Co(en)_2NH_3Cl^{2+}$ [a]	Fe^{2+}	3.7	1.0	1.0
$Co(en)_2(NCS)Cl^{+}$ [a]	Fe^{2+}	8×10^{-1}	10	2
$Co(en)_2OH_2Cl^{2+}$ [a]	Fe^{2+}	5.2×10^{2}	2.6×10^{1}	3.6×10^{3}

[a] Data from ref. 141.

direct comparisons difficult. It should be noted that measurements on nitrogen isotopic-fractionation factors for the reduction of various Co(III) ammine complexes with Cr^{2+} have not revealed stretching of Co–N bonds[142]. These results are to be expected in the light of the small difference in the bond distances in Co(II) and Co(III) ammines[143]. However, De Chant and Hunt[144] have shown that substantial distortion of the bond to the *trans* ligand occurs in the reaction of *trans*-$Cr(NH_3)_4(OH_2)Cl^{2+}$ with Cr^{2+}. For the *trans* series of $Co(en)_2XCl^{n+}$ the order of reactivity towards Fe^{2+} is $X = H_2O > Br^- > Cl^- > SCN^- > NH_3$ (overall change in reactivity of 3600). With the exception of $X = H_2O$, this coincides with the order expected in the light of relative ligand field strengths and ease of removal of the *trans* groups away from the Co(III) centre. For the *cis* series of complexes the order of reactivity is $Cl^- > H_2O > SCN^- > NH_3$ (overall change in reactivity of 90). Since, in all these cases, the group *trans* to the bridging ligand is ethylenediamine, then reactivity should be little affected by modification of the *cis* group. Any slight differences must arise from other factors than those considered to operate for *trans* isomers.

The rates of reaction of Cr(II) with a number of *cis*- and *trans*-$Co(en)_2(NCS)X^{n+}$ complexes (where $X = H_2O$, NH_3, Cl^- and SCN^-) have been examined by Haim and Sutin[64] using a flow apparatus[44]. Quantitative transfer of thiocyanate from cobalt to chromium is observed for all complexes except *cis*- and *trans*-$Co(en)_2(NCS)Cl^+$. These chloro complexes react *via* a chloride-bridged activated complex, the other complexes by thiocyanate-bridged activated complexes. The reduction of *trans*-$Co(en)_2OH_2(NCS)^{2+}$ by excess Cr^{2+} proceeds in two stages. Firstly, reduction of Co(III) takes place to give $CrSCN^{2+}$, and this is followed by the chromium(II)-catalysed isomerisation of the latter to give $CrNCS^{2+}$ ($k = 42$ l.mole^{-1}.sec^{-1} at 25 °C in 1 *M* $HClO_4$: see also ref. 63). The overall rate increases with decreasing acidity, the observed rate coefficient being given by

$$k' = k_1 + k_2 K_h/[H^+]$$

where k_1 and k_2 are the rate coefficients for *trans*-$Co(en)_2OH_2(NCS)^{2+}+Cr^{2+}$ and *trans*-$Co(en)_2(NCS)OH^{+}+Cr^{2+}$, respectively, and K_h is the equilibrium constant for

$$trans\text{-}Co(en)_2OH_2(NCS)^{2+} \rightleftharpoons trans\text{-}Co(en)_2(NCS)OH^{+}+H^{+}$$

From a linear plot of k' *versus* $1/[H^+]$, $k_1 = 1.4\times10^3$ l.mole^{-1}.sec^{-1} and $k_2K_h = 6.6$ sec^{-1}. Using a value of $\sim 5\times10^{-7}$ mole.l^{-1} for K_h, k_2 is estimated as $\sim 1\times10^7$ l.mole^{-1}.sec^{-1} at 25 °C. A decision as to whether SCN^- or OH^- acts as a bridging group is possible by considering the influence of acidity on thiocyanate transfer. On the assumption that the aquo and the hydroxo complexes use thiocyanate and hydroxide bridges, respectively, *viz.*

$$trans\text{-}Co(en)_2OH_2(NCS)^{2+}+Cr^{2+} \xrightarrow{k_1} (\text{Co–NCS–Cr})^{\ddagger} \longrightarrow CrSCN^{2+} + CrNCS^{2+}+Co(II)$$

$$trans\text{-}Co(en)_2(NCS)OH^{+}+Cr^{2+} \xrightarrow{k_2} (\text{Co–}\overset{\text{H}}{\overset{|}{\text{O}}}\text{–Cr})^{\ddagger} \longrightarrow Cr^{3+}+Co(II)$$

then the ratio $([CrSCN^{2+}]+[CrNCS^{2+}])/[Cr^{3+}]$ is equal to $k_1\ [H^+]/k_2K_h = 2.1\times10^2\ [H^+]$. This result is compatible with the value $1.4\times10^2\ [H^+]$ derived experimentally by analysis of the products. $CrSCN^{2+}$, formed in the reduction of *trans*-$Co(en)_2OH_2(NCS)^{2+}$, is believed to be formed from the attack of Cr^{2+} on the S atom of the thiocyanate group. On the other hand, $CrNCS^{2+}$ is probably formed by adjacent attack on the thiocyanate ligand, *i.e.* attack by Cr^{2+} on the N atom. In Table 14 a comparison is made between the rates of reduction of $Co(en)_2(NCS)X^{n+}$ by Cr(II) and $Co(en)_2ClX^{n+}$ by Fe(II). The following points are noteworthy: (*a*) when X = NH_3 or SCN^- there are only slight differences between the rates for reduction of *cis* and *trans* isomers, (*b*) substitution of SCN^- for NH_3 in either *cis* or *trans* positions produces only slight increases in rate, (*c*) substitution of H_2O for NH_3 produces a substantial rate increase, particularly for *trans* isomers. The overall similarities in the reactivity patterns for Fe(II) and Cr(II) are additional evidence for supposing that Fe(II) reductions use chloride bridges. Table 15 shows, for a number of related reactions, a comparison between the rate coefficients observed and those calculated from the Marcus equation (eqn. (12.6), p. 247) modified to

$$k_{1,2}/k_{1,3} = (k_{2,2}K_{1,2}/k_{3,3}K_{1,3})^{\frac{1}{2}} \tag{6.4}$$

assuming $f_{1,2}/f_{1,3} \sim 1$, where, for example, $k_{1,2}$ and $K_{1,2}$ are the rate coefficient and equilibrium constant for the $Co(NH_3)_5Cl^{2+}+Cr^{2+}$ reaction, $k_{1,3}$ and $K_{1,3}$ are the corresponding constants for the $Co(NH_3)_5Cl^{2+}+Fe^{2+}$ reaction, and $f_{1,2}$

TABLE 15

RATE COEFFICIENTS ($l.mole^{-1}.sec^{-1}$.) FOR SOME REACTIONS OF CHROMIUM(II) AND IRON(II) WITH COBALT(III) COMPLEXES AT 25.0 °C AND $\mu = 1.0\ M$ [a]

Reaction	*log k_{obs}*	*log k_{calc}* [b]
$Co(NH_3)_5Cl^{2+}+Fe^{2+}$	−2.9	—
$Co(NH_3)_5Cl^{2+}+Cr^{2+}$	6.4	6.6
$Co(NH_3)_5F^{2+}+Fe^{2+}$	−2.2	—
$Co(NH_3)_5F^{2+}+Cr^{2+}$	5.9	5.5
$Co(NH_3)_5N_3^{2+}+Fe^{2+}$	−2.0	—
$Co(NH_3)_5N_3^{2+}+Cr^{2+}$	5.5	5.7
cis-$Co(en)_2(NCS)Cl^{2+}+Fe^{2+}$	−3.8	—
cis-$Co(en)_2(NCS)Cl^{2+}+Cr^{2+}$	6.3	5.7
trans-$Co(en)_2(NCS)Cl^{2+}+Fe^{2+}$	−3.9	—
trans-$Co(en)_2(NCS)Cl^{2+}+Cr^{2+}$	6.4	5.6

[a] From Haim and Sutin[64], and refs. cited therein. [b] Calculated from eqn. (6.4).

and $f_{1,3}$ are as defined by Marcus[11]. Although equation (6.4) was first derived for outer-sphere processes, evidently it can be applied as well to inner-sphere reactions since observed and calculated rate coefficients are in good agreement (particularly among the pentaammine series).

Cannon and Earley[145] have measured the rates of reduction of *cis*- and *trans*-$Co(en)_2X(H_2O)^{3+}$ (where X = NH_3 or H_2O) by Cr^{2+}. From the observed inverse dependence on hydrogen-ion concentration it is concluded that the rate-determining step is the attack of the reductant on the conjugate base of the aquo ion. A bridged mechanism is supported by the work of Kruse and Taube[146] who showed that the Cr^{2+} reduction of *cis*-$Co(en)_2(H_2O)OH^{2+}$ results in the transfer of a single oxygen atom. Table 16 allows results of Cr^{2+} and Fe^{2+} reductions[141] to be compared; the data on the Fe^{2+} reactions refer to the corresponding chloro complexes. There is a vast difference in rates between these two systems: the rates of the $Co(en)_2(NH_3)OH^{2+}+Cr^{2+}$ reactions are about 10^{10} times greater than the rates of the $Co(en)_2(NH_3)Cl^{2+}+Fe^{2+}$ reactions. Assuming that this disparity in rate arises entirely from free energy differences in the Cr(III)–Cr(II) and Fe(III)–Fe(II) couples ($\Delta(\Delta G) = 27.6$ kcal.mole^{-1}), it is interesting that the Marcus treatment[43] gives

$$\log \frac{k_{Cr(II)}}{k_{Fe(II)}} \sim \frac{1}{2} \cdot \frac{\Delta(\Delta G)}{2.303\ \boldsymbol{RT}} = 10$$

although the equation was intended originally to describe the behaviour of outer-sphere reactions. For both Cr^{2+} and Fe^{2+} reductions the most reactive complex is the *trans*-aquo one. However, the effect of the *trans* ligand is less pronounced when Cr^{2+} is the reductant. Outward motion of the *trans* ligand in the formation of the activated complex is present in both Cr^{2+} and Fe^{2+} reductions but is much

TABLE 16

RATE COEFFICIENTS FOR REDUCTION OF $Co(en)_2XCl^{n+}$ AND $Co(en)_2XOH^{n+}$ COMPLEXES BY Fe^{2+} AND Cr^{2+}

	Fe^{2+} reductant [a]	*Cr^{2+} reductant* [b]
X	10^4k (*l.mole^{-1}.sec^{-1}*)	$10^{-6}k$(*l.mole^{-1}.sec^{-1}*)
trans-isomers		
NH_3	0.66	0.22
NCS	1.3	—
Cl	320	—
Br	360	—
N_3	620	—
H_2O	2400	2.6
cis-isomers		
NH_3	0.18	0.20
NCS	1.7	—
H_2O	4.5	0.79
Cl	16	—

[a] Chloro complexes[141]: $[ClO_4^-] = 1.0$ *M*, 25 °C. [b] Hydroxo complexes[145]: 25.5 °C.

more important in the latter case. This is consistent with the low isotopic fractionation factor found for *trans* nitrogen in Cr^{2+} reductions[142].

Espenson[147] has shown that the reaction of *cis*-$Co(en)_2(N_3)_2^+$ with V^{2+} takes place by an inner-sphere mechanism. This Co(III) complex was selected for investigation because it is particularly reactive towards V^{2+}, and also the dissociation of monoazido vanadium(III) is relatively slow. At low V^{2+} concentrations ($2\text{–}20 \times 10^{-4}$ *M*) the second-order rate coefficient is 32.9 $l.mole^{-1}.sec^{-1}$ at 25 °C, $[H^+] = 0.10$ *M* and $\mu = 1.0$ *M*. At higher V^{2+} concentrations (~ 0.1 *M*), using a stopped-flow apparatus, the kinetics are apparently first order at 520 mμ, a wavelength where VN_3^{2+} shows negligible absorbance. The rate coefficient under these conditions agrees with that obtained at low V^{2+} concentrations. However, the data obtained by monitoring the reaction at 350 mμ, an absorption maximum for VN_3^{2+}, deviates from first-order behaviour. Subtraction of the absorbances of stable reactants and products from the net absorption at 350 mμ gives the absorbance due to the VN_3^{2+} entity. Fig. 6 shows the formation and decay of the latter in a typical experiment. The variation of VN_3^{2+} concentration with ime was calculated on the basis of an inner-sphere sequence

$$cis\text{-}Co(en)_2(N_3)_2^+ + V^{2+} \xrightarrow{k_{ox}} VN_3^{2+} \xrightarrow{k_{aq}} V^{3+} + N_3^-$$

where k_{ox} and k_{aq} are the rate constants for the redox and aquation reactions, respectively. As shown in Fig. 6, the calculated and observed values agree nicely. That the VN_3^{2+} intermediate does not originate from interaction of the Co(III)

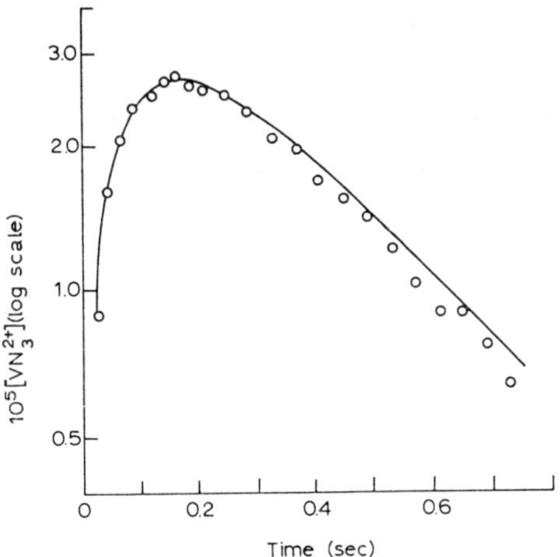

Fig. 6. The formation and decay (at 350 mμ) of the VN_3^{2+} intermediate formed in the *cis*-$Co(en)_2$-$(N_3)_2^+ + V^{2+}$ reaction. The points are the observed concentrations and the line is calculated from the known rate parameters. $[Co(III)]_0 = 1.5 \times 10^{-4}$ *M*, $[H^+] = 0.10$ *M*, $[V^{2+}] \sim 0.1$ *M*. The intermediate attains a maximum concentration of 2.7×10^{-5} *M* at 0.17 sec; calculated, 2.74×10^{-5} *M* at 0.173 sec. (*From Espenson*[147], *by courtesy of The American Chemical Society.*)

complex with VN_3^+ was proved by the observation that added HN_3 has no effect on the rate.

It is of incidental interest that a little work has been done on dicobalt systems. Doyle and Sykes[148] have made a study of the reduction of decammine-μ-amidodicobalt(III), $(NH_3)_5Co{\cdot}NH_2{\cdot}Co(NH_3)_5^{5+}$, by V(II). Since the rate is independent of hydrogen-ion concentration the mechanism cannot involve an amide bridge and must be outer-sphere, as it is in the case of the reduction of $Co(NH_3)_6^{3+}$ by V(II)[112,149]. Both the binuclear complex and $Co(NH_3)_6^{3+}$ are inert to substitution but the former is capable of functioning as a two-equivalent oxidant. Thus the two likely mechanisms are

$$(NH_3)_5Co{\cdot}NH_2{\cdot}Co(NH_3)_5^{5+} + V^{2+} \xrightarrow{H^+} Co^{2+} + Co(NH_3)_6^{3+} + V^{3+} + 5\,NH_4^+ \quad (6.5)$$

$$Co(NH_3)_6^{3+} + V^{2+} \xrightarrow{H^+} Co^{2+} + V^{3+} + 6\,NH_4^+ \quad (6.6)$$

and

$$(NH_3)_5Co{\cdot}NH_2{\cdot}Co(NH_3)_5^{5+} + V^{2+} \longrightarrow 2\,Co^{2+} + V(IV) + 11\,NH_4^+ \quad (6.7)$$

$$V(II) + V(IV) \longrightarrow 2\,V(III) \quad (6.8)$$

On the basis of the rate law alone

$$\text{rate} = k[V(II)][(NH_3)_5Co{\cdot}NH_2{\cdot}Co(NH_3)_5^{5+}]$$

the mechanisms are indistinguishable. For experiments in which V(II) was present in large excess over the complex, the formation and decay of an absorption peak at 425 mμ was noted during the initial stages of the reaction. Step (6.8) proceeds in part by means of a binuclear intermediate VOV^{4+}, characterised by its intense absorption ($\varepsilon \sim$ 6800 $l.mole^{-1}.cm^{-1}$) at the same wavelength[34]. Although this appears to be evidence for the two-equivalent mechanism, in fact VOV^{4+} is produced by a side reaction between V(II) and H^+ ions or water†. It is concluded that the reaction between V(II) and $(NH_3)_5Co{\cdot}NH_2{\cdot}Co(NH_3)_5{}^{5+}$ takes place predominantly through the one-equivalent route, (6.5) and (6.6), with the intermediate formation of $Co(NH_3)_6{}^{3+}$. Activation parameters for step (6.5) are $\Delta H^\ddagger$ = 9.5 $kcal.mole^{-1}$ and $\Delta S^\ddagger = -31 \pm 3$ $cal.deg^{-1}.mole^{-1}$ at $\mu = 0.4$ *M*. Activation parameters for step (6.6) are $\Delta H^\ddagger$ = 9.1 $kcal.mole^{-1}$ and $\Delta S^\ddagger$ = -40 $cal.deg^{-1}.mole^{-1}$ at $\mu = 1$ *M* (ref. 149). Step (6.5) is faster than (6.6) by a factor of ~ 30 at 25 °C; the differences in rate must arise from differences in activation entropies. Table 17 gives direct and relative kinetic data on the effect of anions on the reactions of V^{2+} with $(NH_3)_5Co{\cdot}NH_2{\cdot}Co(NH_3)_5{}^{5+}$ and $Co(NH_3)_6{}^{3+}$. The reactivity order is the same in both cases, *i.e.* $F^- > SO_4{}^{2-} > Cl^-$, and the relative rates are similar. The suggestion is made that the anion is brought into the activated complex by V^{2+} and is held there at the side of V(II) remote from the cobalt centre. Sykes[150-152] has examined other dicobalt systems: the reductions by Fe(II) of the peroxo complexes, $(NH_3)_4Co{\cdot}\mu(NH_2, O_2){\cdot}Co(NH_3)_4{}^{4+}$ and $(NH_3)_5Co{\cdot}O_2{\cdot}Co(NH_3)_5{}^{5+}$ (see, also, refs. 302 and 303, in which reduction of the latter complex has been studied using Cr(II), V(II) and Eu(II)).

The reductions by Fe(II) of chloro(ethylenediaminetriacetatoacetate)cobaltate (III), $Co(Y)Cl^{2-}$, and its conjugate acid, $Co(HY)Cl^-$, have been investigated by Pidcock and Higginson[153]. At hydrogen-ion concentrations $> 5 \times 10^{-3}$ *M*

TABLE 17

ANION EFFECTS IN THE REDUCTION OF $(NH_3)_5Co{\cdot}NH_2{\cdot}Co(NH_3)_5{}^{5+}$ AND $Co(NH_3)_6{}^{3+}$ BY V^{2+} AT 25 °C[148]

	$(NH_3)_5Co{\cdot}NH_2{\cdot}Co(NH_3)_5{}^{5+}$		$Co(NH_3)_6{}^{3+}$	
Anion	k' ($l^2.mole^{-2}.sec^{-1}$)	k'/k	k' ($l^2.mole^{-2}.sec^{-1}$)	k'/k
Uncatalysed	0.149 [a]	—	0.00441 [a]	—
$+Cl^-$	2.1	14	0.035	8
$+SO_4{}^{2-}$	1000	6700	8.50	1930
$+F^-$	8600	58000	91.7	21000

Ionic strength, 0.4 *M*. $k_{obs} = k + k'[X^-]$. [a] k values in $l.mole^{-1}.sec^{-1}$.

† The side reaction can be obviated by setting aside the V(II) before commencing a reaction, thus removing traces of oxygen[301].

the chief reaction is

$$Co(HY)Cl^- + Fe^{2+} = Co^{2+} + Fe(III)$$

It should be noted that the Co(II)–EDTA complex is unstable and dissociates to give the simple Co^{2+} ion. The product Fe(III) is largely Fe^{3+} along with some EDTA complex. The rate law obeyed is

$$-d[Co(HY)Cl^-]/dt = k_h[Co(HY)Cl^-][Fe^{2+}]$$

where $\log k_h = \log k_0 + A[H^+]$, and k_0 and A are constants. The complexes $Co(Y)Cl^{2-}$ and $Co(HY)Cl^-$ react with Fe^{2+} at comparable rates. Comparisons are instructive between this redox reaction and other cation-catalysed chloride-abstraction reactions[154] which occur without simultaneous oxidation–reduction. The intimate mechanisms are likely to be

$$Co^{III}(HY)Cl^- + Fe^{2+} \rightleftharpoons [Co^{III}(HY)Cl \cdots Fe^{II}] \longrightarrow [Co(HY) \cdot\cdot Cl \cdot\cdot Fe]^{\ddagger}$$
$$\longrightarrow Co^{II}(HY)^- + Fe^{III}Cl^{2+}$$

and

$$Co^{III}(HY)Cl^- + M^{2+} \rightleftharpoons [Co^{III}(HY)Cl \cdots M^{II}]$$
$$\longrightarrow [Co^{III}(HY) \cdot\cdot Cl \cdot\cdot M^{II}]^{\ddagger} \longrightarrow Co^{III}(Y) + H^+ + M^{II}Cl^+$$

where $M^{2+} = Mn^{2+}$, Co^{2+}, Ni^{2+}, Cd^{2+}, Pb^{2+} and Hg^{2+}. The fact that the reduction by Fe^{2+} has a lower activation energy may be a consequence of the weakening of the Co–Cl bond in the activated complex on reduction of the charge on the Co centre.

6.2 ORGANIC BRIDGING LIGANDS IN OXIDATIONS BY COBALT(III) COMPLEXES

A partial knowledge of the function of organic ligands as bridging groups has arisen out of an examination of the rates of reduction of over a hundred carboxylatopentaamminecobalt(III) complexes with Cr(II), V(II), Fe(II), and other reductants. It seems clear that the role of the organic ligand is to act as a mediator for electron transfer. For this to occur the ligand must contain groups capable of associating strongly with the reductant. Alternatively, it must contain a conjugated system of bonds. Three classes of interaction are recognised: *adjacent attack*, *adjacent attack with chelation*, and *remote attack*. The topic has been the subject of an extensive review by Taube[155] (see, also, Taube and Gould[304]).

Table 18 is a collection of rate parameters (taken, mainly, from Fraser[130])

TABLE 18

RATE PARAMETERS[a] FOR REDUCTION OF CARBOXYLATOPENTAAMMINE Co(III) COMPLEXES BY Cr^{2+}, V^{2+}, AND Eu^{2+}

Reductant	Cr^{2+}			V^{2+}			Eu^{2+}		
Ligand	k^b(*l.mole*$^{-1}$. *sec*$^{-1}$)	$\Delta H^\ddagger$(*kcal. mole*$^{-1}$)	$\Delta S^\ddagger$(*cal. deg*$^{-1}$.*mole*$^{-1}$)	k^b(*l.mole*$^{-1}$. *sec*$^{-1}$)	$\Delta H^\ddagger$(*kcal. mole*$^{-1}$)	$\Delta S^\ddagger$(*cal. deg*$^{-1}$.*mole*$^{-1}$)	k^b(*l.mole*$^{-1}$ *sec*$^{-1}$)	$\Delta H^\ddagger$(*kcal. mole*$^{-1}$)	$\Delta S^\ddagger$(*cal. deg*$^{-1}$.*mole*$^{-1}$)
Acetato	0.35[f]	8.2[f]	−33[f]	0.43	5.8	−41	0.18	4.4	−47
Chloroacetato	0.10	7.9	−37	1.25	9.4	−27	3.16	9.3	−25
Cyanoacetato	0.11	4.0	−49	1.13	9.4	−27	—	—	—
Dichloroacetato	0.074	2.5	−55	1.03	9.6	−26	2.08	6.9	−35
Benzoato	0.15[f]	9.0[f]	−33[f]	0.52	6.7	−37	0.24	6.2	−40
o-Chlorobenzoato	0.074	6.0	−43	0.57	10.5	−24	0.39	3.2	−50
o-Iodobenzoato	0.082	2.8	−54	0.90	15.8	− 5	0.28	3.3	−51
Salicylato	—	—	—	—	9.3	−27	—	—	—
o-Phthalato	0.075[c]	5.1[c]	−45[c]	1.01	10.2	−24	2.16	8.2	−30
m-Phthalato	0.093[c]	2.6[c]	−56[c]	0.60	9.0	−29	0.64	7.9	−33
p-Chlorobenzoato	0.21	10.0	−28	0.60	8.0	−33	—	—	—
p-Iodobenzoato	—	—	—	0.37	9.0	−30	0.27	2.1	—
p-Hydroxybenzoato	0.13	9.6	−30	0.53	9.3	−28	—	—	—
p-Cyanobenzoato	0.18	7.5	−37	0.88	10.2	−24	—	—	—
Formato	7[d] (7.2[f])	8.3[f]	−27[f]	—	—	—	—	—	—
Trifluoroacetato	0.052[e]	—	—	—	—	—	—	—	—

[a] From Fraser[130]. [b] For $\mu = 1.0$ *M*, 25 °C. [c] From ref. 156. [d] From ref. 155. [e] From ref. 157. [f] From ref. 304.

for the reduction by Cr^{2+}, V^{2+}, and Eu^{2+} of pentaamminecobalt(III) complexes containing those carboxylato groups which do not favour reaction by remote attack and have no tendency for chelation with the reductant (except for the salicylato and phthalato complexes). For these ligands, attack by the reducing agent occurs at the carboxyl group adjacent to the Co(III) centre (*A*)

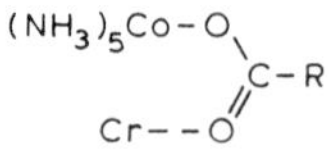

```
(NH3)5Co-O
          \
           C-R
          //
    Cr--O
     (A)
```

and the path for electron transfer is Co–O–C–O–red. The two oxygen atoms in the carboxylato complexes are not equivalent[158] (a result in conflict with a previous report[159]). In a systematic study, Fraser[130] has investigated the effect of varying the substituent R. The rates are, in general, independent of hydrogen-ion concentration, *viz.*

$$-\mathrm{d}[\mathrm{Co(NH_3)_5L^{2+}}]/\mathrm{d}t = k[\mathrm{Co(NH_3)_5L^{2+}}][\mathrm{red}]$$

where red = Cr^{2+}, V^{2+}, or Eu^{2+}, and the net reaction is, for example,

$$\mathrm{(NH_3)_5CoL^{2+}+Cr^{2+}+5\,H^+ = Co^{2+}+CrL^{2+}+5\,NH_4^+}$$

The results show that:

(*a*) variations in the nature of R have only a small effect on the rate coefficients although there is a 10^4–10^5-fold variation in the dissociation constants of the corresponding acids;

(*b*) there are, however, considerable variations in $\Delta H^\ddagger$ and $\Delta S^\ddagger$; furthermore, all $\Delta S^\ddagger$ values are strongly negative. A good linear relationship exists between the activation parameters, *viz.*

$$T\Delta S^\ddagger = \alpha\Delta H^\ddagger + \beta$$

For Cr^{2+}, $\alpha = 0.94$ and $\beta = 18.0$ kcal.mole^{-1}. For V^{2+}, $\alpha = 0.97$ and $\beta = 17.5$ kcal.mole^{-1};

(*c*) the rate decreases in the sequence $CH_3COO^- > ClCH_2COO^- > Cl_2CHCOO^-$, as might be expected since the electron-withdrawing power of the R group increases.

An explanation of the overall ~ 10-fold variation in the rates for the various complexes has been given by invoking steric factors[160]. Steric effects may account also for the fact that, although $HCOO^-$ is less basic than CH_3COO^-, the formato complex reacts much more rapidly than the acetato one[155].

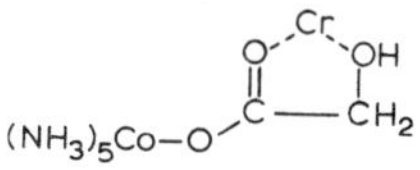

(B) 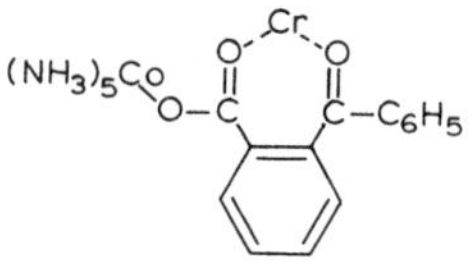 (C) 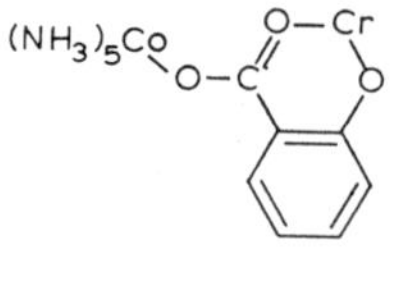 (D)

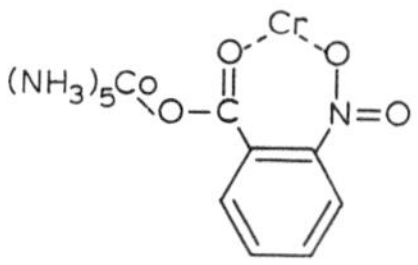

(E) 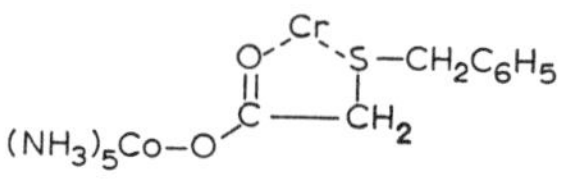 (F)

The rate of reaction is enhanced if the ligand is able to chelate with the reductant. Examples of chelating ligands are α-hydroxy acids, *e.g.*, glycolate (*B*) and lactate[161]; and those ligands containing a carbonyl group (*C*) or a hydroxy group (*D*) in an *ortho* position to the co-ordinated carboxyl group[155,157,161–164] (Table 19). *o*-Nitrobenzoato (*E*) (but not *o*-aminobenzoato) pentaammine complexes provide further examples. Chelation with Cr(II) can take place *via* a sulphur atom[157,164], as in the *S*-benzylthioglycolato complex (*F*). The efficiency of such ligands in promoting electron transfer from reductant to cobalt is imperfectly understood but is probably related to the provision of a firm route along which an electron may move. Good evidence for the existence of chelation is reported by Butler and Taube[161] in the case of reduction of glycolatopentaamminecobalt(III) by Cr^{2+}. The first product of reaction is a metastable Cr(III) species which has a higher absorptivity than the stable glycolatochromium(III) ion, and reverts to the latter with a half-life of ~ 22 h. Chelation of Cr^{2+} by the α-hydroxy group seems to explain the rate sequence glycolate < lactate < methyl-

TABLE 19

RATE PARAMETERS FOR REDUCTION OF SELECTED CARBOXYLATOPENTAAMMINE Co(III) COMPLEXES BY Cr^{2+}, SHOWING EFFECT OF CHELATION[155]

Ligand	$k(l.mole^{-1}.sec^{-1})$	$\Delta H^{\ddagger}(kcal.mole^{-1})$	$\Delta S^{\ddagger}(cal.deg^{-1}.mole^{-1})$
Glycolato	3.1	9.0	−26
Methoxyacetato	0.42	9.3	−23
Lactato	6.7	—	—
Methyllactato	11.8	9.1	−24
α-Malato	2.7	—	—
β-Malato	0.36	—	—
Malonato	0.29	—	—
Salicylato	0.15 [a]	—	—
Phthalato ion	2.7 [b]	—	—
Salicylato ion	$\sim 2\times10^{8}$ [c]	—	—

Ionic strength, 1.0 *M*; temp., 25 °C. [a] For $\mu = 3.0$ *M*, ref. 157. [b] From ref. 129.
[c] From ref. 181.

lactate since replacement of H by CH_3 on the α carbon should increase the basicity of the OH group. Chelation is an important factor in the reduction by Cr^{2+} of a variety of heterocyclic complexes (derived from pyridine, pyrazole and pyrazine) as is evidenced by marked spectral effects[163]. Huchital and Taube[165] have examined the rate of ring closure of malonatopentaamminecobalt(III). They find that a chelate ring is formed before oxidation of Cr^{2+} occurs. It is considered that the hydrogen-ion dependent path is not to be conceived in terms of a remote attack mechanism but that the activated complex is probably of the type (*G*), *viz.*

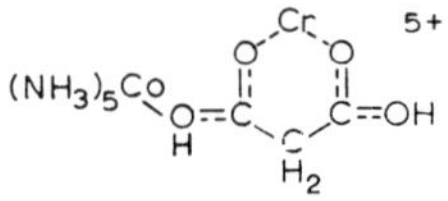

(G)

Ester hydrolysis accompanies electron transfer during the reduction by Cr(II) of pentaammine complexes containing half-esters of conjugated dibasic acids as ligands, for example*, methyl fumarate[119,166a], phenyl fumarate[119,167], methyl maleate[120], methyl terephthalate, and phenyl terephthalate[168]. Complete hydrolysis of the ester occurs in the reduction by V^{2+} and Eu^{2+} of the half-ester complex containing the methyl succinato group[169]. Huchital and Taube[170] have investigated the products formed in the reaction of methyl- and ethylmalonato pentaamminecobalt(III) complexes with Cr(II). In the case of the methyl complex, spectrophotometry and ion-exchange show that about 50 % of the ligand appears in the chelated form, $Cr(OOC)_2CH_2^+$, and for the ethyl complex, about 67 %. The corresponding amount of alcohol is found free in solution. The rest of the ligand is found as the monodentate ester malonato complex of Cr(III). V(II) and Eu(II) are ineffective in inducing hydrolysis. Contrary to an earlier report[169], no ester hydrolysis occurs with the succinato half-ester complex on reaction with Cr(II), V(II), or Eu(II). This latter observation can be rationalised on the basis that the chelate ring is stable for the succinato complex than it is for the malonato, the former complex reacting *via* simple adjacent attack[170]. Taube[155] suggests that in reductions by Cr(II) an intermediate of the type (*H*)

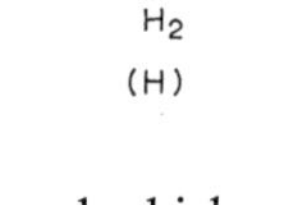

(H)

is formed which can decompose by rupture at the Cr–O bond (to form the half-

* On reinvestigation[166b], the reduction of (methylfumarato)pentaamminecobalt(III) by Cr(II), and by V(II) and Eu(II), has been found to produce little ester hydrolysis.

ester complex), or at the C–O bond (leading to hydrolysis and preservation of the chelate ring). However, this does not explain reductions by V(II) and Eu(II) where M^{3+}–O bond cleavage would predominate since V(III) and Eu(III) form complexes which are highly labile. Neither does it account for the oxygen-tracer observation that about 40 % alkyl–oxygen fission occurs in such systems[169]. The effects of chelation by non-bridging ligands have been covered by Fraser[122] who has shown that the rate of reduction of amminecobalt(III) acetato complexes by Cr(II) and V(II) is not altered when four NH_3 groups are replaced by two ethylenediamines, or when five ammonias are replaced by a tetraethylene-pentamine group. However, increasing chelation brings about a reduction in the rate for sulphato complexes.

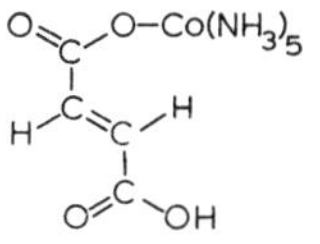

O=C(–O–Co(NH3)5)–CH=CH–C(=O)OH

(I)

When a dibasic ligand, containing a conjugated system, is attached to Co(III), *e.g.*, fumarate(*I*) or terephthalate, the Cr^{2+} reductant may react at either of the two carboxyl groups, *i.e.*, by both adjacent and remote attack. In these cases the rate law is compounded of two terms: a hydrogen-ion independent term and a hydrogen-ion dependent term[119,129,168]. The precise effects of conjugation on the rates of reduction of Co(III) complexes has given rise to much interest and speculation. Attempts have been made[171–173] to relate the rate of remote attack to the mobile bond order[174] between the terminal atoms of the bridging ligand. Another theoretical treatment is that given by Libby[175]. The remote attack mechanism has been suggested for a large number of other ligands, for example, oxalate, maleate[111], methyl monoesters of fumaric and maleic acid[111,166,167,176], *p*-aldehydobenzoate[119], 4-carboxylatopyridine[177], nicotinamide[178], substituted pyridines and pyrazole[179].

Fraser[180] has investigated the rates of reduction by Cr^{2+} of various cobalt(III) ammine complexes containing ligands coordinated by nitrogen, *e.g.*, urethane, methyl glycinate, benzocaine, ethyl nicotinate and isonicotinate and ethyl-4-aminobutyrate. These complexes react very much more rapidly than complexes with similar ligands attached to cobalt through oxygen. It seems likely that the first stage in the reduction of Co(III) complexes is the formation of a radical ion as a result of electron transfer to the organic ligand, the electron being subsequently transmitted to the Co(III) centre. Furthermore, it is possible that remote attack takes place only if the organic ligand is reducible. There is strong evidence to suggest the formation of an analogous radical-ion intermediate in the oxidation of formatopentaamminecobalt(III) by permanganate[71]. Although in principle ESR spectroscopy is capable of detecting such intermediates, there is at present no direct evidence of this kind. The subject of remote attack is discussed at length by Taube[155].

Halide effects have been reported for the Cr^{2+} reduction of $Co(NH_3)_5L^{n+}$ complexes (where L = NH_3, acetato or fumarato, FuH^-)[182]. Chloride exerts a stronger catalytic effect than bromide; in all cases the second-order rate coefficient takes the form $k_{obs} = k+k_X[X^-]$ where k is pH-independent for L = NH_3 or OAc^-, but not for L = FuH^-. When L = NH_3 or FuH^-, halide is captured into the inner coordination sphere of chromium, and products of the type XCr^{2+} and $XCrFuH^+$ result. $Co(NH_3)_5NH_3^{3+}$ is known to react *via* an outer-sphere mechanism[112], whereas $Co(NH_3)_5OAc^{2+}$ and $Co(NH_3)_5FuH^{2+}$ use bridge mechanisms[112,129]. It is surprising that Cl^- and Br^- have similar influences on both outer- and inner-sphere processes (Table 20). This similarity suggests that a Cr(II) halide complex, CrX^+, is formed rapidly and that this species takes part in the formation of the activated complex.

TABLE 20

KINETIC PARAMETERS IN THE HALIDE-CATALYSED REDUCTION OF $Co(NH_3)_5L^{n+}$ BY Cr^{2+} (AT 25 °C)[182]

L	$\mu(M)$	k(l.mole^{-1}.sec^{-1})	k_{Cl}(l^2.mole^{-2}.sec^{-1})	k_{Br}(l^2.mole^{-2}.sec^{-1})	k_{Cl}/k_{Br}	k_{Cl}/k(l.mole^{-1})	k_{Br}/k(l.mole^{-1})
NH_3	2.60	0.0072	0.60	0.31	1.9	83	43
FuH^-	3.4	1.00	0.69	0.30	2.3	0.69	0.30
OAc^-	1.65	0.34	0.20	0.064	3.1	0.59	0.10

TABLE 21

RATE PARAMETERS FOR THE OXIDATION OF Fe(II) BY VARIOUS EDTA AND HEDTA Co(III) COMPLEXES[184]

Complex	$\Delta H^\ddagger$(kcal.mole^{-1})	$\Delta S^\ddagger$(cal.deg^{-1}.mole^{-1})	k(25 °C)(l.mole^{-1}.min^{-1})
$Co(Y)^-$	11.8	-36 ± 5	0.0362
$Co(HY)H_2O$	11.1	-33 ± 3	0.510
$Co(Y)H_2O^-$	22.1	$+27 \pm 6$	53.9
$Co(YOH)H_2O$	14.2	-23.5 ± 1.4	0.277
$Co(YOH)OH^-$	13.7	$+6 \pm 5$	1.45×10^6
$Co(HY)Cl^-$	11.3	-21.9 ± 1.5	81.7

The kinetics for the reduction of glyoxalatopentaamminecobalt(III) by Cr^{2+} indicate that the carbonyl form is far more reactive ($k > 7 \times 10^3$ l.mole^{-1}.sec^{-1} at 25 °C) than the hydrate ($k \sim 1$ l.mole^{-1}.sec^{-1} at 25 °C)[183]. The hydrate form predominates and its rate of dehydration has been studied.

Wood and Higginson[184] have made a detailed study of the kinetics of oxidation of Fe(II) by a number of complexes of Co(III) with ethylenediaminetetraacetic acid (H_4Y = EDTA) and hydroxyethylethylenediaminetriacetic acid (H_3YOH = HEDTA). Rate data and activation parameters are quoted (Table 21) for the reduction of $Co(Y)^-$, $Co(HY)H_2O$, $Co(Y)H_2O^-$, $Co(YOH)H_2O$, $Co(YOH)OH^-$, and $Co(HY)Cl^-$ in perchloric acid media. The last two complexes are believed to react with Fe(II) through inner-sphere activated complexes involving hydroxo and chloride bridges, respectively. On the other hand, the reductions of $Co(Y)^-$ and $Co(YOH)H_2O$ are best interpreted as proceeding *via* carboxylato bridges. When the Co(III)–EDTA complex is reduced by Cr(II), three carboxylato groups are transferred which suggests that a triply-bridged transition state is involved in this reaction[185]. A number of redox reactions involving metal–EDTA and similar complexes have been studied briefly by Wilkins and Yelin[305].

6.3 OXIDATIONS BY AQUO COMPLEXES OF COBALT(III)

The course of the oxidation of V(III) by Co(III) has been the subject of an investigation by Rosseinsky and Higginson[80]. Fig. 7 illustrates the dependence of the concentrations of reactants and products upon time when Co(III) is in excess over V(III). Under the conditions applying in Fig. 7, V(III) is oxidised to V(IV) until 90 % of V(III) has been used up (*i.e.* at ~ 28 min). Then V(V) begins to be produced, and after V(III) has been totally consumed (~ 37 min) the reaction corresponds to the oxidation of V(IV) by Co(III). Co(III) reacts with V(III) and V(IV) at a similar rate under the same conditions*. It is clear that Co(III) is reduced by V(IV) as well as by V(III) during the reaction, and that reduction by

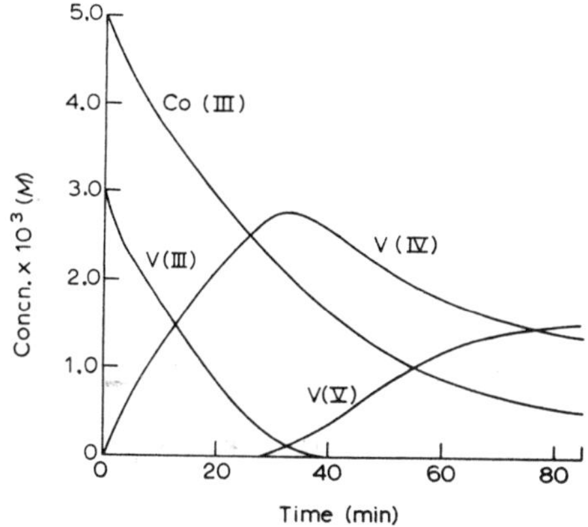

Fig. 7. Dependence of concentrations of reactants and products upon time in the Co(III)+V(III) reaction. Initial concentrations: [Co(III)] = 4.99×10^{-3} *M*; [V(III)] = 2.93×10^{-3} *M*; [V(IV)]= 3×10^{-5} *M*; [H^+] = 2.96 *M*; ionic strength = 3.0 *M*; temp., 5.0 °C. (*From Rosseinsky and Higginson*[80], *by courtesy of the Chemical Society.*)

* The redox potentials of the VO_2^+–VO^{2+} and VO^{2+}–V^{3+} couples are 1.00 and 0.36 V, respectively.

V(IV) becomes progressively more important as the reaction proceeds. The fact that V(V), the product of oxidation, is not detectable until all the V(III) is consumed is ascribed to the rapid reaction V(III)+V(V) → 2 V(IV) with a rate coefficient of $\sim 5\times10^3$ $l.mole^{-1}.min^{-1}$ at 5 °C, $[H^+] = 3$ M and $\mu = 3$ M. When the initial concentration of V(III) is greater than that of Co(III), the reaction between V(III) and V(V) is unimportant and

$$-d[Co(III)]/dt = k_c[Co(III)][V(III)]+k_b[Co(III)][V(IV)]$$

Values of k_c were derived from plots of this equation using known values of k_b (see below). At 0 °C, $[H^+] = 1$ M and $\mu = 3$ M, k_c is equal to 0.192 $l.mole^{-1}.sec^{-1}$. k_c varies inversely with hydrogen-ion concentration, the precise form of the function being uncertain but of the type $k_c = k_1+k'/[H^+]$.

The reaction between Co(III) and V(IV)

$$Co(III)+V(IV) = Co(II)+V(V)$$

in perchloric acid media (followed by monitoring Co(III) at 400 mμ) is second order and the rate is little affected by the presence of a large excess of V(V)[80]. However, high concentrations of Co(II) ($\sim$ 0.1 M) increase the rate; also chloride ions are found in the reaction mixture. Apparently, Co(II) catalyses the reduction of perchloric acid by V(IV). The second-order rate coefficient, k_b, can be expressed as $k_b = k_2+k''/[H^+]$ where k_2 and k'' are 0.245 $l.mole^{-1}.sec^{-1}$ and 2.08 sec^{-1}, respectively, at 20 °C and $\mu = 3.0$ M. The corresponding activation energies are 21.8 and 15.9 $kcal.mole^{-1}$, and the entropies of activation are 12 ± 9 and -5 ± 6 $cal.deg^{-1}.mole^{-1}$, respectively. At 0 °C, $[H^+] = 1$ M and $\mu = 3$ M, k_b is 0.260 $l.mole^{-1}.sec^{-1}$. Anion effects were not investigated.

Although the oxidation of Cr(III) by Ce(IV) is a rapid reaction in perchlorate solutions, the oxidation of Cr(III) by Co(III) in the same media occurs at a rate similar to or slower than that of the thermal decomposition of Co(III) in perchloric acid (3 M). However, the Co(III)+Cr(III) reaction is subject to catalysis by Ag(I) ion and Kirwin *et al.*[186] have made a kinetic study of this system, *viz.*

$$3\ Co(III)+Cr(III) \underset{Ag(I)}{=} 3\ Co(II)+Cr(VI)$$

Kinetic data were obtained by following the rate of appearance of Cr(VI) at 475 mμ, a wavelength where Cr(III), Co(II) and Co(III) absorb only slightly. Ag(II) is said to exhibit an absorption maximum at 475 mμ but does not interfere with the observations since it is present in low concentrations. The concentration of Cr(III) was maintained in excess over Co(III) and Ag(I). Plots of log $(A_\infty - A_t)$ *versus* time are linear where A_∞ and A_t are the absorbancies due to Cr(VI) after complete reaction, and after time t. At constant Cr(III), Co(II) and Co(III) concentrations the observed rate coefficient (k_{obs}) is proportional to Ag(I) concentration. In addition, plots of $1/k_{obs}$ *versus* 1/[Cr(III)] and $1/k_{obs}$ *versus* [Co(II)] are linear. Variations in acidity and perchlorate concentration have no effect on the rate. The kinetic results are discussed on the basis of the following mechanism

$$\mathrm{Co(III)+Ag(I)} \underset{k_{-3}}{\overset{k_3}{\rightleftharpoons}} \mathrm{Co(II)+Ag(II)}$$

$$\mathrm{Cr(III)+Ag(II)} \xrightarrow{k_4} \mathrm{Cr(IV)+Ag(I)}$$

$$\mathrm{Cr(IV)+Ag(II)} \longrightarrow \mathrm{Cr(V)+Ag(I)}$$

$$\mathrm{Cr(V)+Ag(II)} \longrightarrow \mathrm{Cr(VI)+Ag(I)}$$

Application of the steady-state treatment to the concentrations of Ag(II), Cr(IV) and Cr(V) leads to the rate law

$$\frac{-\mathrm{d[Co(III)]}}{3\,\mathrm{d}t} = \frac{\mathrm{d[Cr(VI)]}}{\mathrm{d}t} = \frac{k_3 k_4\mathrm{[Cr(III)][Ag(I)][Co(III)]}}{k_{-3}\mathrm{[Co(II)]}+3\,k_4\mathrm{[Cr(III)]}}$$

Since

$$\frac{\mathrm{d}t}{\mathrm{d[Cr(VI)]}} = \frac{-3\,\mathrm{d}t}{\mathrm{d[Co(III)]}} = \frac{3}{k_{obs}\mathrm{[Co(III)]}}$$

inversion of the rate law gives

$$\frac{1}{k_{obs}} = \frac{k_{-3}\mathrm{[Co(II)]}}{3\,k_3 k_4\mathrm{[Cr(III)][Ag(I)]}} + \frac{1}{k_3\mathrm{[Ag(I)]}}$$

in keeping with the linear relationships referred to above. Evaluation of the slopes and intercepts of these plots (along with the value of k_3/k_{-3} previously determined) allows the individual rate coefficients, k_3, k_{-3} and k_4, to be calculated. Table 22 contains these values together with the associated activation parameters.

The rate coefficient for the oxidation of Mn(II) by Co(III) has been determined

TABLE 22

RATE PARAMETERS FOR THE Ag(I)-CATALYSED Co(III)+Cr(III) REACTION AT 25 °C[a]

	Rate coefficient ($l.mole^{-1}.sec^{-1}$)	$\Delta H^‡$ (*kcal.mole*$^{-1}$)	$\Delta S^‡$ (*cal.deg*$^{-1}$*.mole*$^{-1}$)
k_3	37±7	18±2	7± 6
k_{-3}	50±7	6±3	−32±10
k_4	16±1	12±6	−14±20

[a] From Kirwin *et al.*[186] As noted by Sutin *et al.*[191], the original table has rate coefficients in error by a factor of 3.

by Diebler and Sutin[82] as 1.00×10^2 l.mole^{-1}.sec^{-1} in 3.0 M perchloric acid at 25 °C.

The reduction of Co(III) by Fe(II) in perchloric acid solution proceeds at a rate which is just accessible to conventional spectrophotometric measurements[24]. At 2 °C in 1 M acid with [Co(III)] = [Fe(II)] $\sim 5 \times 10^{-3}$ M the half-life is of the order of 4 sec. Kinetic data were obtained by sampling the reactant solution for unreacted Fe(II) at various times. To achieve this, aliquots of the reaction mixture were run into a quenching solution made up of ammoniacal 2,2′-bipyridine, and the absorbance of the $Fe(bipy)_3{}^{2+}$ complex measured at 522 mμ. Absorbancies of Fe(III) and Co(III) hydroxides and $Co(bipy)_3{}^{2+}$ are negligible at this wavelength. With the reactant concentrations equal, plots of 1/[Fe(II)] *versus* time are accurately linear (over a sixty-fold range of concentrations), showing the reaction to be second order, *viz.*

$$-\mathrm{d}[\mathrm{Fe(II)}]/\mathrm{d}t = -\mathrm{d}[\mathrm{Co(III)}]/\mathrm{d}t = k'[\mathrm{Co(III)}][\mathrm{Fe(II)}]$$

No indication is given of the reaction of Co(III) polymers although these are present in the reaction solutions[187]. It is noteworthy that the intercepts of the above plots do not coincide with the values obtained from the initial Fe(II) concentration. The "zero-time oxidation" is believed to arise from a finite quenching time together with a rapid reaction of hydrolysed species of the reactants. The rate of reaction is inversely proportional to the concentration of hydrogen ions. This result is taken as implying competitive reactions between $CoOH^{2+} + Fe^{2+}$ and $Co^{3+} + Fe^{2+}$, as described by the rate law

$$\text{rate} = k_5[\mathrm{Co^{3+}}][\mathrm{Fe^{2+}}] + k_6[\mathrm{CoOH^{2+}}][\mathrm{Fe^{2+}}]$$

Thus the observed rate coefficient, k', is given by

$$k' = k_5 + k_6 K_1/[\mathrm{H^+}]$$

where K_1 is the hydrolysis constant[188] for Co^{3+}. Slope and intercept values of plots of k' *versus* $1/[H^+]$ enable k_5 and k_6 to be calculated. At 0 °C and $\mu = 1.0$ M, k_5 and k_6 are 10 and ~ 6500 l.mole^{-1}.sec^{-1}, respectively. The corresponding activation energies and activation entropies are 9.1 and 7.9 kcal. mole^{-1} and -23 and -14 cal.deg^{-1}.mole^{-1}, respectively. It is apparent that the preferred path is that involving $CoOH^{2+}$. It is a common observation in metal-ion redox systems that the greater part of reaction proceeds *via* hydrolysed species as, for example, in the reactions Fe(III)+Fe(II)[22], Co(III)+Co(II)[189], and Fe(III)+Cr(II)[94]. Sulphate, unlike fluoride, increases the rate of reaction. Assuming this is due to the introduction of the step

$$\mathrm{CoSO_4{}^+ + Fe^{2+} \xrightarrow{k_7} Co^{2+} + FeSO_4{}^+}$$

then k_7 has a value of $\sim$ 4900 l.mole^{-1}.sec^{-1} at 0 °C. Anion effects were not investigated in detail.

Some key experiments have been performed by Sutin *et al.*[190] on the reaction between Co(III) and Fe(II) in the presence of chloride ions, and have established the process

$$CoCl^{2+} + Fe^{2+} \longrightarrow (CoClFe^{4+})^{\ddagger} \longrightarrow Co^{2+} + FeCl^{2+}$$

to be of the inner-sphere variety. A flow technique[44] was necessary since both reactants and both products are substitution-labile. When a solution containing 9.6×10^{-4} *M* Co(III), 2.9×10^{-2} *M* Co(II), 4.0×10^{-3} *M* Cl^- in 3 *M* perchloric acid is mixed with a solution of 9.2×10^{-2} *M* Fe(II) in 2.7 *M* perchloric acid, the reaction of $CoCl^{2+}$ with Fe^{2+} is complete within 2 or 3 msec ($k \geqq 5 \times 10^3$ l. mole^{-1}.sec^{-1}), and the $FeCl^{2+}$ produced decays with a half-life of 280 msec. $FeCl^{2+}$ has a λ_{max} of 336 mμ. When a solution of 9.6×10^{-4} *M* Co(III) and 5.0×10^{-3} *M* Co(II) in 3 *M* $HClO_4$ is mixed with one containing 9.2×10^{-2} *M* Fe(II) and 4×10^{-3} *M* Cl^- in 2.7 *M* $HClO_4$, $FeCl^{2+}$ is not detected at 336 mμ. However, at larger Co(III) concentrations (4.8×10^{-3} *M*) the slow formation of $FeCl^{2+}$ is observed due to

$$Fe^{3+} + Cl^- \longrightarrow FeCl^{2+}$$

The rate of oxidation of Fe(II) by Co(III) has been studied in the presence of Ag(I) by means of a stopped-flow apparatus[191]. Under conditions of excess Ag(I) and Fe(II), Ag(II) competes with Co(III) for Fe(II), and the reactions are

$$\text{Co(III)} + \text{Ag(I)} \underset{k_{-3}}{\overset{k_3}{\rightleftharpoons}} \text{Co(II)} + \text{Ag(II)}$$

$$\text{Ag(II)} + \text{Fe(II)} \xrightarrow{k_8} \text{Ag(I)} + \text{Fe(III)}$$

$$\text{Co(III)} + \text{Fe(II)} \xrightarrow{k_9} \text{Co(II)} + \text{Fe(III)}$$

At sufficiently high concentrations of Fe(II) the observed second-order rate coefficient for the disappearance of Co(III) can be expressed as

$$k_{obs} = k_9 + k_3[\text{Ag(I)}]/[\text{Fe(II)}]$$

thus allowing the determination of k_3 and k_9 from a plot of k_{obs} *versus* [Ag(I)]/[Fe(II)]: $k_3 = 41 \pm 3$ l.mole^{-1}.sec^{-1} and $k_9 = 330 \pm 7$ l.mole^{-1}.sec^{-1} at 25 °C. The oxidation of Co(II) by Ag(II) was also examined (at 470 mμ, the absorption maximum for Ag(II)) in the presence of excess Co(II) and Ag(I), when the first-order rate coefficient describing the approach to equilibrium is given by

$$k_{obs} = k_3[Ag(I)] + k_{-3}[Co(II)]$$

Values of k_{obs} vary linearly with [Co(II)], at constant [Ag(I)], and k_3 and k_{-3} were obtained as 37 ± 4 l.mole^{-1}sec^{-1} and 1.75×10^3 l.mole^{-1}.sec^{-1}, respectively (*cf.* Table 22). The two independently-determined values of k_3 are seen to be in close agreement. The equilibrium constant, k_3/k_{-3}, of the Co(III)+Ag(I) system is, thus, 2.1×10^{-2} (using average k_3) in 4 *M* perchloric acid at 25 °C. A value of $4 \pm 2 \times 10^{-2}$ is arrived at by considering the oxidation potentials of the Ag(I)–Ag(II) and Co(II)–Co(III) couples.

Despite the lability of Fe(III) complexes, Haim and Sutin[192] have been able to study the rapid reductions of some Co(III) complexes by Fe(II) and identify the primary Fe(III) products. A flow apparatus, described by Dulz and Sutin[44], enabled the rate of formation and decay of the spectra of the Fe(III) products

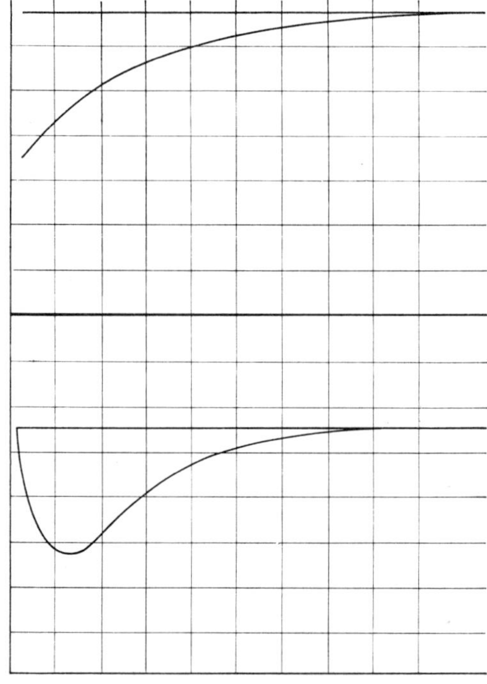

Fig. 8. Transmittance *versus* time curves for the $Co(C_2O_4)_3^{3-} + Fe^{2+}$ reaction. Upper curve shows disappearance of $Co(C_2O_4)_3^{3-}$ (wavelength, 600 mμ; abscissa scale, 500 msec per major division). Lower curve shows formation and decay of the intermediate $FeC_2O_4^+$ (wavelength, 310 mμ; abscissa scale, 2 sec per major division). $[Co(C_2O_4)_3^{3-}] = 1.0 \times 10^{-3}$ *M*; $[Fe^{2+}] = 2.5 \times 10^{-2}$ *M*; $[HClO_4] = 0.92$ *M*; ionic strength = 1.0 *M*; temp., 25 °C. (*From Haim and Sutin*[192], *by courtesy of The American Chemical Society.*)

to be observed and recorded (Fig. 8). A two-stage mechanism holds for the $Co(C_2O_4)_3^{3-}+Fe^{2+}$ reaction, *viz.*

$$Co(C_2O_4)_3^{3-}+Fe^{2+} \longrightarrow FeC_2O_4^{+}+Co^{2+}+2\,C_2O_4^{2-}$$

$$FeC_2O_4^{+}+2\,H^{+} \rightleftharpoons Fe^{3+}+H_2C_2O_4$$

The disappearance of $Co(C_2O_4)_3^{3-}$, followed at 600 mμ, gives a rate coefficient of 33 $l.mole^{-1}.sec^{-1}$ at 25 °C and $\mu = 1.0\ M$ (*cf.* the value of 1.15×10^3 $l.mole^{-1}.sec^{-1}$ as determined by Barrett and Baxendale[193] for 20 °C and $\mu = 0$). The rate of formation and disappearance of $FeC_2O_4^{+}$ was followed at 310 mμ. Under the conditions described in Fig. 8, the half-lives for the redox process and the decay of $FeC_2O_4^{+}$ are 0.85 sec and 2.4 sec, respectively. The latter value agrees well with the half-life as calculated from the data of Moorhead and Sutin[194]. The time required for $FeC_2O_4^{+}$ to achieve its maximum concentration is 2.1 sec (*cf.* 2.0 sec calculated on the basis of the kinetic scheme) and the ratio $[FeC_2O_4^{+}]_{max}/[Co(III)]_0$ is 0.51 (*cf.* 0.56 from the kinetic scheme). The results indicate that an oxalate-bridging mechanism applies. Similar conclusions are drawn for the Fe(II)-reduction of $Co(H{\cdot}EDTA)Cl^{-}$, $Co(en)_2(OH_2)Cl^{2+}$, $Co(NH_3)_3(OH_2)_2Cl^{2+}$, $Co(NH_3)_3(OH_2)_2N_3^{2+}$, and $Co(NH_3)_3(OH_2)C_2O_4^{+}$. In all cases a bridged transition state applies, and observations were made on the growth and decay of the corresponding Fe(III) complex.

The kinetics of the Co(III) oxidation of a number of substituted tris(1,10-phenanthroline) complexes of Fe(II) have been studied by Campion *et al.*[195], using a flow technique[44]. The second-order rate coefficients obtained for 5-methyl-1,10-phenanthroline, 1,10-phenanthroline, 5-chloro-1,10-phenanthroline and 5-nitro-1,10-phenanthroline in 3 M $HClO_4$ at 25 °C are given in Table 23 along with the formal oxidation potentials of the complexes. Table 24 gives observed and calculated values of slopes and intercepts of plots of $(\Delta G^{\ddagger}_{1,2}+1.15\,\boldsymbol{RT}\log f)$ *versus* $\Delta G^{\circ}_{1,2}$ (see eqn. (12.8), p. 247) for various redox reactions involving substituted Fe(II) phenanthroline complexes. The slopes, particularly, agree well with the calculated values. However, the intercepts for all the reactions are smaller than those calculated from the Marcus treatment. According to Campion

TABLE 23

SECOND-ORDER RATE COEFFICIENTS FOR THE OXIDATION OF Fe(II)-PHENANTHROLINE COMPLEXES BY Co(III) IN 3 M PERCHLORIC ACID AT 25.0 °C[195]

Ligand	$E^{\circ}(V)$	$10^{-3}k_{12}(l.mole^{-1}.sec^{-1})$
5-Methyl-1,10-phenanthroline	1.02	15.0
1,10-Phenanthroline	1.06	14.0
5-Chloro-1,10-phenanthroline	1.12	5.02
5-Nitro-1,10-phenanthroline	1.25	1.49

TABLE 24

SLOPES AND INTERCEPTS OF PLOTS OF $(\Delta G^{\ddagger}_{1,2}+1.15 \log f)$ *vs.* $\Delta G^{0}_{1,2}$ FOR OXIDATION–REDUCTION REACTIONS INVOLVING SUBSTITUTED Fe(II)–PHENANTHROLINE COMPLEXES AT 25.0 °C

Reaction	*Slope*		*Intercept(kcal.mole⁻¹)*		Medium
	Observed	*Calculated*	*Observed*	*Calculated*	
$Fe(phen)_3^{3+}+Fe^{2+}$ [a]	0.56	0.50	14.8	13.0	0.5 M H_2SO_4
$Ce(IV)+Fe(phen)_3^{2+}$ [b]	0.48	0.50	14.8	13.0	0.5 M H_2SO_4
$Mn(III)+Fe(phen)_3^{2+}$ [c]	0.49	0.50	17.6	15.8	1 and 3 M $HClO_4$
$Co(III)+Fe(phen)_3^{2+}$ [d]	0.51	0.50	19.5	12.9	3 M $HClO_4$

[a] From ref. 196. [b] From ref. 44. [c] From ref. 82. [d] From ref. 195.

et al.[195], this deviation may stem from non-cancellation of the non-electrostatic contributions to the work required to bring together the various pairs of reactants.

The reduction of Co(III) by Ag(I) in perchlorate solutions has been studied by Sutcliffe *et al.*[197]. Since the initial product of reaction is the very reactive Ag(II) species, all solutions were subject to preliminary ozonolysis to remove traces of reducible impurities. The final products of reaction are Co(II) and Ag(I). Kinetic data were obtained spectrophotometrically by following the disappearance of Co(III) at 605 mμ, a small correction being applied for the absorbance of Co(II). With Ag(I) in excess, the disappearance of Co(III) is second order, i.e., plots of the reciprocal of the corrected absorbance *versus* time are linear. The rate is directly proportional to the concentration of Ag(I), and inversely proportional to the square of the concentration of Co(II). These results can be understood in terms of the mechanism

$$\text{Co(III)}+\text{Ag(I)} \underset{k_{-3}}{\overset{k_3}{\rightleftharpoons}} \text{Co(II)}+\text{Ag(II)} \qquad K_2$$

$$\text{Ag(II)} \rightarrow \text{products}$$

Transient concentrations of Ag(II) were detected spectrophotometrically, and by electron spin resonance. The thermal decomposition of Ag(II) perchlorate, the subject of a separate study[198], takes place by

$$2\,Ag^{2+} \rightleftharpoons Ag^{+}+Ag^{3+} \qquad \text{rapid equilibrium, } K_3$$

$$Ag^{3+}+H_2O \rightleftharpoons AgO^{+}+2\,H^{+} \qquad \text{rapid equilibrium, } K_4$$

$$AgO^{+} \xrightarrow{k_{10}} Ag^{+}+\tfrac{1}{2}O_2 \qquad \text{slow}$$

and the rate law is

$$\frac{-\mathrm{d}[\mathrm{Ag}^{2+}]}{\mathrm{d}t} = \frac{k_{10}K_3K_4[\mathrm{Ag}^{2+}]^2}{[\mathrm{Ag}^+][\mathrm{H}^+]^2}$$

Reverting to the Co(III)+Ag(I) reaction, the concentration of Ag(II) is fixed by K_2 as

$$[\mathrm{Ag(II)}] = \frac{K_2[\mathrm{Co(III)}][\mathrm{Ag(I)}]}{[\mathrm{Co(II)}]}$$

and the derived rate law is

$$\frac{-\mathrm{d}[\mathrm{Co(III)}]}{\mathrm{d}t} = \frac{-\mathrm{d}[\mathrm{Ag}^{2+}]}{\mathrm{d}t} = \frac{k_{10}K_3K_4K_2{}^2[\mathrm{Co(III)}]^2[\mathrm{Ag(I)}]}{[\mathrm{Co(II)}]^2[\mathrm{H}^+]^2}$$

under conditions where [Ag(I)] > [Co(III)]. In keeping with this relationship, the rate is dependent on $[\mathrm{H}^+]^{-2}$ at constant ionic strength; the observed rate constant goes through a pronounced minimum at ~ 3.3 *M* perchloric acid. The overall $\Delta H^{\ddagger}$ values for the Ag(II) decomposition[198] and the Co(III)+Ag(I) reaction[197] are 11±2 and 34±4 kcal.mole^{-1}, respectively. Thus K_2 has an associated overall ΔH of 12 kcal.mole^{-1}. The value of 0.76 obtained for K_2 is in poor agreement with the value of 2.1×10^{-2} given in ref. 191 (see p. 218).

The slow reduction of cobalt(III) by mercury(I)

$$2\,\mathrm{Co(III)} + \mathrm{Hg(I)}_2 = 2\,\mathrm{Co(II)} + 2\,\mathrm{Hg(II)}$$

is stoichiometric under conditions of equivalent concentrations of the two reactants, or when mercury(I) is present in excess[80]. However, when Co(III) is in great excess the slow oxidation of water by Co(III) becomes important, and the reaction is then non-stoichiometric. The second-order nature of the rate law, *viz.*

$$-2\,\mathrm{d}[\mathrm{Hg(I)}_2]/\mathrm{d}t = -\mathrm{d}[\mathrm{Co(III)}]/\mathrm{d}t = 2\,k_{\mathrm{obs}}[\mathrm{Co(III)}][\mathrm{Hg(I)}_2]$$

is maintained in the presence of a large excess of Co(II) and Hg(II), the products of reaction. Kinetic data were obtained in most cases by monitoring the disappearance of Co(III) at 603 mμ. Variation in perchlorate-ion concentration does not affect the rate. This last observation contrasts with the behaviour shown by the Tl(III)+$\mathrm{Hg(I)_2}$ system, where the rate varies inversely with $\mathrm{ClO_4}^-$ concentration[86]. It follows from the observed acidity dependence that the second-order rate coefficient k_{obs} can be written as $k_{\mathrm{obs}} = k + k'/[\mathrm{H}^+]$ where $k = 0.02$ l.mole^{-1}.sec^{-1} and $k' = 0.36$ sec^{-1} at 19.9 °C and $\mu = 3.0$ *M*. The corresponding activation energies and activation entropies are 22.1 and 29.2 kcal.mole^{-1}, and 9±6 and 37±4 cal.deg^{-1}.mole^{-1}, respectively. Either of the following schemes is likely

$$\text{Co(III)} + \text{Hg(I)}_2 \longrightarrow \text{Co(II)} + \text{Hg(II)} + \text{Hg(I)} \qquad \text{slow}$$

$$\text{Co(III)} + \text{Hg(I)} \longrightarrow \text{Co(II)} + \text{Hg(II)} \qquad \text{rapid}$$

or

$$\text{Co(III)} + \text{Hg(I)}_2 \longrightarrow \text{Co(II)} + \text{Hg}_2{}^{3+} \qquad \text{slow}$$

$$\text{Co(III)} + \text{Hg}_2{}^{3+} \longrightarrow \text{Co(II)} + 2\ \text{Hg(II)} \qquad \text{rapid}$$

Since the rate does not display an inverse dependence on Hg(II) concentration, the oxidation of Hg atoms, in equilibrium with mercury(I) and mercury(II), can be discounted, although Hg atoms are kinetically important in the reduction of thallium(III) by mercury(I)[86]. It seems likely that the acid-dependent path (k') involves $CoOH^{2+}$. Anion effects were not investigated.

Co(III) oxidises Tl(I)[81] according to

$$2\ \text{Co(III)} + \text{Tl(I)} = 2\ \text{Co(II)} + \text{Tl(III)}$$

in a manner resembling the oxidation of Fe(II) by Tl(III)[199]. Owing to the instability of Co(III) in aqueous perchloric acid solutions, the kinetics of the reaction were examined under conditions where Tl(I) was in excess over Co(III), the disappearance of the latter being followed by titration. By employing low initial concentrations of Co(III) relative to Tl(I), and by avoiding low hydrogen-ion concentrations, the rate of decomposition was reduced to less than 10 % of the overall rate of disappearance. The presence initially of high concentrations of Co(III) causes a marked reduction in rate; on the other hand, Tl(III) has little effect. The simplest scheme is

$$\text{Co(III)} + \text{Tl(I)} \rightleftharpoons \text{Co(II)} + \text{Tl(II)}$$

$$\text{Co(III)} + \text{Tl(II)} \rightarrow \text{Co(II)} + \text{Tl(III)}$$

where Tl(II) is competed for by Co(II) and Co(III). At low Co(II) concentrations

$$-\mathrm{d}[\text{Co(III)}]/\mathrm{d}t = 2\,k_{\text{obs}}[\text{Co(III)}][\text{Tl(I)}]$$

and the pH-dependence of k_{obs} was investigated over the range 0.25–2.50 M $HClO_4$ between 0 °C and 25 °C at $\mu = 2.70\ M$. First-order plots (of log [Co(III)] *versus* time) deviated slightly, but significantly, from linearity at the beginning of reaction. This is thought to come about from the reaction of dimeric species of Co(III). The variation of k_{obs} with $[H^+]$ can be expressed as

$$k_{obs} = k_{11}+k_{12}K_1/[H^+]$$

Plots of k_{obs} *versus* $1/[H^+]$ allow k_{11} to be evaluated, but k_{12} cannot be defined exactly due to doubt over the value of K_1, the hydrolysis constant of Co^{3+}, and also to scatter in the data. k_{11} is 6.8×10^{-4} l.mole^{-1}.sec^{-1} at 15 °C, and 2.5×10^{-3} l.mole^{-1}.sec^{-1} at 25 °C. Using a value of 5×10^{-3} mole.l^{-1} for K_1 (ref. 200), $k_{12} \sim 1.7 \times 10^{-2}$ l.mole^{-1}.sec^{-1} at 15 °C. Unsuccessful attempts were made to correct for the reduction of Co(III) by water, and thus errors of up to ~ 10 % may be present in the quoted values of k_{11}. The latter has an associated activation energy and entropy of 26.4 kcal.mole^{-1} and 22 ± 7 cal.deg^{-1}.mole^{-1}, respectively. The rate of the Co(III)+Tl(I) reaction is increased by the addition of sulphate. It is deduced that the $SO_4{}^{2-}$ ion, and not the $HSO_4{}^-$ ion, is responsible for the catalysis, although the latter is present in much larger amounts. The back-reaction of Co(II) with Tl(II) is much less important in the presence of sulphate. Ashurst and Higginson[81] make a tentative suggestion that perchlorate ions participate in the Co(III)+Tl(I) system, *i.e.*, the activated complex contains one or more $ClO_4{}^-$ ions.

The binuclear complex of Co(III), bi-μ-hydroxobis(bioxalatocobaltate(III) $[(C_2O_4)_2Co^{III}(OH)_2Co^{III}(C_2O_4)_2{}^{4-}]$, is reduced by Sn(II) in dilute hydrochloric acid solutions[201]. The products of reaction are the mononuclear complexes, probably *cis*-bioxalatodiaquocobaltate(III). If trioxalatocobaltate(III) is included in the reaction mixture, then a considerable fraction of it is consumed during the course of the reaction of Sn(II) with the binuclear species. Since trioxalatocobaltate-(III) does not react with Sn(II) under the prevailing conditions, this result indicates that the reactive reducing species is Sn(III), generated by a one-equivalent process.

The Co(III) complexes $Co(NH_3)_6{}^{3+}$ and $Co(NH_3)_5OH^{2+}$ bring about oxidation of stannate(II) ion in strongly basic solution[202]. The rates were found to be independent of the concentration of the Co(III) complex. It is proposed that stannate(II) exists as a dimer, and that the monomer is the reactive species, the rate being close to half-order in stannate(II). Cyanide and thiosulphate catalyse the reaction but $Co(CN)_6{}^{3-}$ is immune to attack by stannate(II) ion. The experimental difficulties encountered in this study preclude a full analysis as regards mechanism.

The reaction between Co(III) and Ce(III) has been the subject of a detailed study in a series of papers by Sutcliffe and Weber[188,205,206]. Of particular value is the thorough investigation of the influence of anions on the rate in perchlorate media. The reaction was followed by measuring the disappearance of Co(III) at its absorption maximum of 650 mμ, a wavelength where both oxidation states of cerium are transparent and Co(II) absorbs only slightly. Changes in temperature and ionic strength affect the spectrum of Ce(III) at the 296 mμ maximum, but the spectrum is unaltered by variations in perchloric acid concentration[188]. From these results it is concluded that a complex, presumed to be $CeClO_4^{2+}$, exists between Ce(III) and perchlorate, and analysis of the spectral data allowed the equilibrium constant (K = 1.4 l.mole^{-1} at μ = 1.14 M and 25 °C) and associated thermodynamic parameters to be calculated. Using a different approach, Heidt and Berestecki[203] have obtained a value for K of 0.86 l.mole^{-1} at μ = 4.50 M and 25 °C. These authors suggest that the perchlorate complex is an extended (outer-sphere) one, namely, $Ce(H_2O)_6^{3+}ClO_4^-$. The oxidation of Ce(III) by Co(III) is stoichiometric, and first-order in both reactants. The back-reaction is not detectable in the presence of a large excess of Co(II) ($[Co(II)]_0/[Co(III)]_0 > 200$). The dependence on $HClO_4$ concentration is described by

$$k' = a/[HClO_4]$$

where k' is the observed second-order rate coefficient and a is a constant. At acidities below 0.2 M a colloidal Ce(IV) compound is formed, leading to interference in optical measurements. At constant acidity the rate increases with increase in ClO_4^- concentration (to show this effect the ionic strength was maintained constant, and lanthanum perchlorate substituted for sodium perchlorate). This dependence on ClO_4^- concentration can be expressed as

$$1/k' = b + c/[ClO_4^-]$$

where b and c are constants. Since previous work had precluded the possibility of a reactive hydrolysed species of cerium(III), the hydrogen-ion dependence was taken as evidence for the existence of $CoOH^{2+}$. The equilibrium constant for

$$Co^{3+} + H_2O \rightleftharpoons CoOH^{2+} + H^+ \qquad K_1$$

was determined successfully ($K_1 = 1.75 \times 10^{-2}$ mole.l^{-1} at 25 °C), despite experimental difficulties arising from the reduction of Co^{3+} by water at low acidities. It is deduced that the reaction

$$CoOH^{2+} + CeClO_4^{2+} \xrightarrow{k_{13}} Co(II) + Ce(IV)$$

is the rate-controlling step, as a result of the observed rate response to variations in acid and perchlorate concentrations. On the assumption that the total Co(III) and Ce(III) concentrations are $[Co^{3+}]+[CoOH^{2+}]$ and $[Ce^{3+}]+[CeClO_4^{2+}]$, respectively, it follows that

$$1/k' = [H^+]/k_{13}K_1 + [H^+]/k_{13}K_1K[ClO_4^-]$$

since K_1 is less than $[H^+]$. The empirical parameters, a, b and c, are then evaluated as follows

$$a = k_{13} K_1 K[ClO_4^-]/(1+K[ClO_4^-])$$

$$b = [H^+]/k_{13} K_1$$

$$c = [H^+]/k_{13} K_1 K$$

TABLE 25

RATE PARAMETERS FOR THE Co(III)+Ce(III) REACTION[188,205,206]

	k (l.mole^{-1}.sec^{-1}) at 25 °C	$\Delta H^\ddagger$ (kcal.mole^{-1})	$\Delta S^\ddagger$ (cal.deg^{-1}.mole^{-1})
$CoOH^{2+}+CeClO_4^{2+}$	95	19±2	14±7
$CoOH^{2+}+CeNO_3^{2+}$	93	14±2	−5±1
$CoOH^{2+}+CeF^{2+}$	8500	—	—
$CoSO_4^{+}+CeSO_4^{+}$	≦2000	—	—
$CoSO_4^{+}+CeClO_4^{2+}$	≦ 250	—	—
$CoSO_4^{+}+Ce^{3+}$	≦ 350	—	—
$Co^{3+}+CeSO_4^{+}$	≦ 200	—	—

The true rate coefficient, derived from these relationships, is given in Table 25, along with the corresponding $\Delta H^\ddagger$ and $\Delta S^\ddagger$ terms. A later spectrophotometric study[187] of Co(III) perchlorate solution has cast some doubt over the value of the formation constant (K_1) for $CoOH^{2+}$. It is suggested that $CoOH^{2+}$ undergoes slow dimerisation to $Co\text{–}O\text{–}Co^{4+}$ and/or $Co\text{–}O\text{–}CoOH^{3+}$, although further hydrolysis and/or polymerisation is possible. Furthermore, reaction between hydrolysed and dimeric species has been proposed as the rate-determining step in the reduction of Co(III) by water[200]. A kinetic study of the formation of the monochloro complex of Co(III) has given a considerably higher value for K_1 of 0.22 mole.l^{-1} at 25 °C[204]. Nitrate ion increases slightly the rate of the Co(III)+Ce(III) reaction in terms of a first-order dependence[205]. The effect is attributed to the introduction of a step involving $CoOH^{2+}$ and $CeNO_3^{2+}$. The formation constant of the latter ion received a spectrophotometric study. In contrast to the slight effect of NO_3^-, the addition of fluoride ion increases the rate markedly: the presence of 10^{-5} M F^- is sufficient to double the rate coefficient. The linear dependence of rate on F^- concentration is discussed on the basis of reaction between $CoOH^{2+}$ and CeF^{2+}. Rate parameters are collected in Table 25: experimental difficulties arising from the insolubility of cerium(III) fluoride did not allow calculation of the $\Delta H^\ddagger$ and $\Delta S^\ddagger$ values for the fluoride-catalysed path. That the product Ce(IV) is strongly complexed is shown by the observation that spent reaction mixtures were colourless instead of yellow. Sulphate complexes of both reactants participate in the Co(III)+Ce(III) reaction in bisulphate media; a linear relationship is found between the rate and the anion concentration[206].

The predominant reactive species of Co(III) are $CoSO_4^+$ at 20 °C, and $Co(SO_4)_2^-$ at higher temperatures than 30 °C. Rate data (in the form of maximum rate coefficients) are given in Table 25 for the various mechanistic steps.

The oxidation of Np(V) by Co(III) has the stoichiometry

$$\text{Co(III)} + \text{Np(V)} = \text{Co(II)} + \text{Np(VI)}$$

and a rate law

$$-\text{d}[\text{Np(V)}]/\text{d}t = k'[\text{Np(V)}][\text{Co(III)}][\text{H}^+]^{-0.11}$$

in perchlorate media at 25 °C[207]. Kinetic data were obtained by recording the absorbance of Np(V) (λ_{max} = 980 mμ) as a function of time. No catalysis was observed on the addition of SO_4^{2-}, Cl^-, HPO_4^{2-} or F^- ions. The form of the rate law indicates that the activated complex is composed of one Co(III) and one Np(V) ion (along with an undetermined number of water molecules). The small hydrogen-ion dependence is attributed to ionic strength effects arising from changes in solution composition. Application of the Marcus equation (p. 247) gives rise to a value of 2.8×10^6 l.mole^{-1}.sec^{-1} for the second-order rate coefficient at 25 °C. The observed value is 3.35×10^2 l.mole^{-1}.sec^{-1}. Similar discrepancies are noted for other Co(III) systems.

Np(V), present as a weak complex with Cr(III)†, is oxidised by Co(III) according to

$$\text{Co(III)} + \text{Np(V)}\cdot\text{Cr(III)} = \text{Co(II)} + \text{Np(VI)} + \text{Cr(III)}$$

Kinetic data were obtained by following the rate of disappearance of the complex at 993 mμ, and the Cr(III) product was identified[207] as $Cr(H_2O)_6^{3+}$. The rate is described by

$$-\text{d}[\text{Np(V)}\cdot\text{Cr(III)}]/\text{d}t = k''[\text{Co(III)}][\text{Np(V)}\cdot\text{Cr(III)}][\text{H}^+]^{-1.08}$$

at 25 °C. This result suggests a hydrolysis of Co(III) as a pre-equilibrium, followed by attack of the hydrolysed species at the Np(V) site in the complex as the rate-determining step. By this means the net formal charge of the activated complex is reduced. However, the stoichiometry and kinetics are consistent also with a scheme in which Co(III) attacks the complex at the Cr(III) centre, *viz.*

$$\text{Co(III)}\text{--}\text{Cr(III)}\cdot\text{Np(V)} \longrightarrow \text{Co(II)} + \text{Cr(IV)}\cdot\text{Np(V)}$$

† At 25 °C, the Np(V)·Cr(III) complex has an equilibrium constant of 2.62, and a rate coefficient for decomposition of 2.32×10^{-6} sec^{-1} (p. 259). The dissociation of this species during the course of the redox reaction is negligible[208].

whereupon Cr(IV) is destroyed either by direct electron exchange with Np(V), or by dissociation of the Cr(IV)·Np(V) complex, with subsequent reduction of Cr(IV) and oxidation of Np(V).

7. Oxidations by platinum(IV)

Beattie and Basolo[209] have investigated the reactions of the substitution-inert octahedral complexes of Pt(IV) with tris(bipyridine)chromium(II). A rapid-mixing, stopped-flow apparatus[210] was made use of in the majority of experiments. Kinetic data were obtained by following the disappearance of $Cr(bipy)_3^{2+}$ at 562 mμ, usually with Pt(IV) in excess. The stoichiometry corresponds to

$$Pt(IV)+2\,Cr(II) = \text{products}$$

and the rate law is simple second-order. The rate coefficients obtained are given in Table 26. The complex $Pt(en)_3^{4+}$ is reduced 20 times more rapidly than its conjugate base, $Pt(en)_2(en\text{-}H)^{3+}$ (where en-H represents ethylenediamine *minus* a hydrogen atom). A comparison of the rates of reduction of Pt(IV) and Co(III)[113] (see Table 9, p. 190) by the common reductant $Cr(bipy)_3^{2+}$ reveals: (*a*) the order of halogenopentaamminecobalt(III) complexes is $F^- < Cl^- < Br^-$, whereas that for the analogous Pt(IV) complexes is $Cl^- < I^- < Br^-$, (*b*) $Pt(en)_3^{4+}$ is reduced ten times faster than $Pt(NH_3)_6^{4+}$ whereas $Co(en)_3^{3+}$ reacts four times *slower* than $Co(NH_3)_6^{3+}$. The presence of a hydroxo ligand imparts inertness to Pt(IV), *trans*-$Pt(NH_3)_4ClOH^{2+}$ reacting nearly 10^4 times slower than *trans*-$Pt(NH_3)_4Cl_2^{2+}$. There is a general correlation between the rates of reduction of Pt(IV) complexes and their polarographic half-wave potentials. Beattie and Basolo[209] suggest that the rate-controlling step in the reductions is the formation of a Pt(III) intermediate by a one-electron, outer-sphere process, *cf.* the oxidation of Pt(II) complexes by hexachloroiridate (IV)[306].

TABLE 26

RATES OF REDUCTION OF Pt(IV) COMPLEXES BY $Cr(bipy)_3^{2+}$ AT 25 °C AND $\mu = 0.1\ M$[209]

Complex	*k (l.mole⁻¹.sec⁻¹)*
$Pt(NH_3)_5Cl^{3+}$	$(2.9 \pm 0.3) \times 10^5$
$Pt(NH_3)_5Br^{3+}$	$(8 \pm 4) \times 10^6$
$Pt(NH_3)_5I^{3+}$	$(3 \pm 1) \times 10^6$
$Pt(en)_3^{4+}$	$(7.7 \pm 1.0) \times 10^3$
$Pt(en)_2(en\text{-}H)^{3+}$	$(4 \pm 1) \times 10^2$
$Pt(NH_3)_6^{4+}$	$(7.7 \pm 1.0) \times 10^2$
trans-$Pt(NH_3)_4Cl_2^{2+}$	10^7
trans-$Pt(NH_3)_4ClOH^{2+}$	$(2 \pm 1) \times 10^3$

8. Oxidations by copper(II)

The rate of reduction of Cu(II) by Cr(II) has been measured in aqueous perchloric acid[92,307]. With Cu(II) in excess and in the absence of oxygen, the reaction corresponds to

$$Cu^{2+}+Cr^{2+} = Cu^{+}+Cr^{3+}$$

and is first-order in each reactant, the rate showing an acid dependence given by

$$d[Cu^{+}]/dt = (a+b/[H^{+}])[Cu^{2+}][Cr^{2+}]$$

At 24.6 °C and 1.00 M ClO_4^- concentration, a is 0.17 l.mole^{-1}.sec^{-1} ($\Delta H^{\ddagger}$ = 12.5 kcal.mole^{-1}, $\Delta S^{\ddagger}$ = −20.2 cal.deg^{-1}.mole^{-1}) and b is 0.587 sec^{-1} ($\Delta H^{\ddagger}$ = 17.1 kcal.mole^{-1}, $\Delta S^{\ddagger}$ = −2.1 cal.deg^{-1}.mole^{-1})[307]. The stoichiometry was checked by noting the decrease in absorbance of Cu^{2+} at 750 mμ. When Cr^{2+} is present initially in concentrations greater than Cu^{2+}, then metallic copper is produced (as a supersaturated solution) according to the step

$$Cu^{+}+Cr^{2+} \longrightarrow Cu^{0}+Cr^{3+}$$

Cu^{+} is generated also in the reduction of Cu^{2+} by V^{2+} in perchloric acid media[92,308]. Unlike the $Cu^{2+}+Cr^{2+}$ system, the rate is pH-independent and the second-order rate coefficient is 26.6 l.mole^{-1}.sec^{-1} at 25 °C in the range 0.04–1.0 M hydrogen-ion concentration. $\Delta H^{\ddagger}$ and $\Delta S^{\ddagger}$ are 11.4 kcal.mole^{-1} and −13.8 cal.deg^{-1}.mole^{-1}, respectively. By analogy with V^{2+} reductions the rapid rate and insensitivity to acid are taken as indicating an outer-sphere mechanism[62]. The greater reactivity of V^{2+} as compared to Cr^{2+} is interesting since Cr^{2+} is, in general, a much more powerful reducing agent. Cu^{+} cannot be prepared satisfactorily using Eu^{2+} as reductant since Cu metal is formed: the second-order rate coefficient for the $Cu^{2+}+Eu^{2+}$ reaction is estimated as ~ 3 l.mole^{-1}.sec^{-1} at 25 °C[92].

9. Oxidations by mercury(II)

The rate law for the oxidation of V(III) by Hg(II)

$$2\,Hg(II)+2\,V(III) = Hg(I)_2+2\,V(IV)$$

is complex[26], *viz.*

$$\frac{-d[V(III)]}{dt} = \frac{[Hg(II)][V(III)]^2}{p[V(IV)]+q[V(III)]} + \frac{[Hg(II)][V(III)]^2}{r[V(IV)]+s[Hg(II)]}$$

The second term contributes only 10–20 % of the overall rate and its precise form is uncertain. The kinetic parameters have been evaluated for the reaction at 15 °C. The first term is consistent with the sequence of steps

$$Hg(II)+V(III) \rightleftharpoons Hg(I)+V(IV)$$

$$Hg(I)+V(III) \rightarrow Hg(0)+V(IV)$$

$$Hg(II)+Hg(0) \rightarrow Hg(I)_2 \qquad \text{rapid}$$

The second term suggests the existence of an alternative path-way, possibly

$$2\,V(III) \rightleftharpoons V(IV)+V(II)$$

$$Hg(II)+V(II) \rightarrow Hg(0)+V(IV)$$

$$Hg(II)+Hg(0) \rightarrow Hg(I)_2 \qquad \text{rapid}$$

Mercury(II) oxidises Cr(II) in aqueous perchloric acid solutions according to the stoichiometric equation

$$2\,Hg(II)+2\,Cr(II) = Hg(I)_2+2\,Cr(III)$$

Spectrophotometric measurements on the rate of appearance of Cr(III) were made at 408 mμ over the temperature range 5–20 °C at $\mu = 2.0\ M$ (ref. 211). With Hg(II) in excess the rate law is

$$d[Cr(III)]/dt = 2\,k_{obs}[Hg(II)][Cr(II)]$$

where $k_{obs} = k_1+k_2'/[H^+]$. The hydrogen-ion dependent term predominates: at 20.0 °C, $k_1 = 0.040$ l.mole^{-1}.sec^{-1} and $k_2' = 0.895$ sec^{-1}. The k_1 term corresponds to the reaction between Cr^{2+} and Hg^{2+}, and the k_2' term to $Cr^{2+}+HgOH^+$ ($k_2' = k_2K_A$ where k_2 is the specific rate coefficient and K_A is the acid dissociation constant of Hg^{2+}, 2.8×10^{-4} mole.l^{-1}). At 25 °C, k_2 is 5.25×10^3 l.mole^{-1}.sec^{-1}. Added $Hg(I)_2$ or Cr(III) have no influence on the rate of the Hg(II)+Cr(II) system. However, when Cr(II) is added to $Hg(I)_2$ in the absence of Hg(II), then mercury metal precipitates since Cr(II) reduces the Hg(II) formed in the disproportionation

$$Hg(I)_2 \rightleftharpoons Hg(O)+Hg(II)$$

The following mechanism is likely

$$Hg(II)+Cr(II) \rightarrow Hg(I)+Cr(III) \qquad \text{rate-determining} \tag{9.1}$$

$$Hg(I)+Cr(II) \rightarrow Hg(O)+Cr(III) \qquad \text{rapid} \tag{9.2}$$

$$Hg(O)+Hg(II) \rightarrow Hg(I)_2 \qquad \text{rapid} \tag{9.3}$$

In this respect the final Cr(III) product is the monomeric species and not the green dimer. Furthermore, if Hg(II) is not in excess then reaction (9.3) cannot occur, and mercury precipitates. The reaction is appreciably catalysed by chloride ions. At 10 °C, the specific rate coefficients for the reaction of Cr^{2+} with $HgCl^+$, $HgOH^+$, and Hg^{2+} are 1.5, $\sim 2.3 \times 10^3$, and 1.75×10^{-2} $l.mole^{-1}.sec^{-1}$, respectively. The species $HgCl_2$ is much less reactive than $HgCl^+$.

A slow reaction takes place between Hg(II) and Fe(II) in perchloric acid solution *viz.*

$$2\,Hg(II)+2\,Fe(II) = Hg(I)_2+2\,Fe(III)$$

Preliminary work[212] records values for second-order rate coefficients of 2.03×10^{-6} $l.mole^{-1}.sec^{-1}$ at 80 °C and 3.75×10^{-6} $l.mole^{-1}.sec^{-1}$ at 90 °C. The activation energy is quoted as 15.5 $kcal.mole^{-1}$, and $\Delta S^{\ddagger}$ as -42 $cal.deg^{-1}.mole^{-1}$. The step

$$Hg^{2+}+Fe^{2+} \rightarrow Hg^{+}+Fe^{3+} \qquad \text{slow}$$

is rate-determining and is followed either by

$$2\,Hg^{+} \rightleftharpoons Hg_2{}^{2+} \qquad \text{rapid}$$

or by the two steps

$$Hg^{+}+Fe^{2+} \rightarrow Hg+Fe^{3+} \qquad \text{rapid}$$

$$Hg+Hg^{2+} \rightarrow Hg_2{}^{2+} \qquad \text{rapid}$$

In dilute perchloric acid solution the reaction between Hg(II) and Sn(II) is complete within 1 min for reactant concentrations of $\sim 10^{-3}$ *M*. After a detailed consideration, Wetton and Higginson[59] conclude that a two-equivalent primary reaction is likely.

10. Oxidations by thallium(III)

The development of fast reaction techniques has allowed a detailed kinetic study of the Tl(III)+V(III) system. Daugherty[213] followed the course of the reaction by monitoring the appearance of V(IV) at 760 mμ. 70–90 % completion of reaction corresponded to 25–30 sec. Spectrophotometric observations revealed no oxidation of V(IV) to V(V)*. The simple rate law

$$-\mathrm{d}[\mathrm{Tl(III)}]/\mathrm{d}t = k_{\mathrm{obs}}[\mathrm{Tl(III)}][\mathrm{V(III)}]$$

is uncomplicated by V(IV) or Tl(I) dependences: at reactant concentrations of $\sim 10^{-4}$ *M* the products of reaction do not affect the rate, even when present at concentrations of 10^{-2} *M*. Variations in perchloric acid concentrations over the range 0.30 to 1.90 *M* ($\mu = 2.0$ *M*) at temperatures 0.5, 14.0 and 25.1 °C show a -1.22-order dependence of the apparent second-order rate coefficient on H^+ concentration. This result strongly suggests that the mechanism can be visualised in terms of two parallel steps, of which the most important is $TlOH^{2+} + V^{3+}$. Activation parameters for this major path are given in Table 27, which includes

TABLE 27

ACTIVATION PARAMETERS FOR VARIOUS RELATED REACTIONS[213]

Net activation process	$\Delta H^{\ddagger}$.*(kcal. mole*$^{-1}$*)*	$\Delta S^{\ddagger}$*(cal.deg*$^{-1}$*. mole*$^{-1}$*)*	$S^{\ddagger}$ *complex* [a]	*Ref.*
$Tl^{3+} + V^{3+} + H_2O \rightarrow (TlOHV^{5+})^{\ddagger} + H^+$	13.9	4.1 ± 2.4	−86	213
$Fe^{3+} + V^{3+} + H_2O \rightarrow (FeOHV^{5+})^{\ddagger} + H^+$	17.6	-10.2 ± 5.3	−128	90
$Ti^{3+} + Fe^{3+} + H_2O \rightarrow (TiOHFe^{5+})^{\ddagger} + H^+$	13.4	−8.3	−124	[b]

[a] $S^{\ddagger}_{\mathrm{complex}} = \Delta S^{\ddagger} + \Sigma S^{0}_{\mathrm{reactants}}$[21]. [b] Higginson, unpublished results quoted in ref. 213.

also comparable information for the $Fe^{3+} + V^{3+}$ and $Fe^{3+} + Ti^{3+}$ reactions. Replacement of $HClO_4$ by HCl results in a pronounced decrease in rate, and the rate law deviates from pure second order; this chloride effect is attributed to the reduced reactivity of chloro complexes of Tl(III). The rate coefficient of the V(V)+V(III) reaction[53] is too small (by a factor of ~ 14) to account for the observed rate of formation of V(IV) in the mechanism

$$\left.\begin{array}{ll} \mathrm{Tl(III)} + \mathrm{V(III)} \rightarrow \mathrm{Tl(I)} + \mathrm{V(V)} & \text{rate-determining} \\ \mathrm{V(V)} + \mathrm{V(III)} \rightarrow 2\,\mathrm{V(IV)} & \text{rapid} \end{array}\right\} \quad (A)$$

and the alternative one-electron process is considered more appropriate, *viz.*

$$\left.\begin{array}{ll} \mathrm{Tl(III)} + \mathrm{V(III)} \rightarrow \mathrm{Tl(II)} + \mathrm{V(IV)} & \text{rate-determining} \\ \mathrm{Tl(II)} + \mathrm{V(III)} \rightarrow \mathrm{Tl(I)} + \mathrm{V(IV)} & \end{array}\right\} \quad (B)$$

† Higginson *et al.*[26] have observed the formation of V(V) in dilute sulphuric acid media in the presence of an excess of Tl(III). More V(V) is formed if V(IV) is present initially. In this media they suggest that mechanism (*B*) operates together with

$$\mathrm{Tl(II)} + \mathrm{V(IV)} \rightarrow \mathrm{Tl(I)} + \mathrm{V(V)}$$
$$\mathrm{V(V)} + \mathrm{V(III)} \rightarrow 2\,\mathrm{V(IV)}$$

This proposal is supported by the observation that the normally slow reaction between Tl(III) and Fe(II) can be induced by V(III) (p. 233). Consequently, Fe(II) must react with an active intermediate, which is likely to be Tl(II).

The rate of oxidation of V(IV) by Tl(III) is unaffected by the presence of Tl(I) but is considerably decreased by the addition of V(V)[26]. The rate law is

$$\frac{-\mathrm{d}[\mathrm{V(IV)}]}{\mathrm{d}t} = \frac{k'[\mathrm{Tl(III)}][\mathrm{V(IV)}]^2}{k''[\mathrm{V(V)}]+[\mathrm{V(IV)}]}$$

where $k' \sim 2.2\times10^{-2}$ l.mole^{-1}.sec^{-1} and $k'' \sim 42$ at 80 °C, $[\mathrm{H}^+] = 1.8$ *M* and $\mu = 3$ *M*. Consequently, the mechanism is similar to the Tl(III)+Fe(II) one[199], *i.e.*

$$\mathrm{Tl(III)+V(IV)} \rightleftharpoons \mathrm{Tl(II)+V(V)}$$

$$\mathrm{Tl(II)+V(IV)} \rightarrow \mathrm{Tl(I)+V(V)}$$

This scheme has a bearing[214] on the Tl(III)+Tl(I) exchange reaction where, on the basis of the observed rate law only (rate $\propto$[Tl(III)][Tl(I)]), it is impossible to discriminate between a single-stage and a two-stage two-equivalent process. Since the rate of the exchange reaction is comparable to the rate of oxidation of V(IV) by Tl(III), appreciable exchange occurs during the course of the latter reaction. It is argued[214] that the addition of Tl(I) should increase the rate of the Tl(III)+V(IV) reaction if Tl(II) were an intermediate in the exchange. However, as noted above, Tl(I) has no such effect, and it is concluded that the exchange takes place by a single two-equivalent step.

The rate of reduction of Tl(III) by Fe(II) was studied titrimetrically by Johnson[215] between 25 °C and 45 °C in aqueous perchloric acid (0.5 *M* to 2.0 *M*) at $\mu = 3.00$ *M*. At constant acidity the rate data in the initial stages of reaction conform to a second-order equation, the rate coefficient of which is not dependent on whether Tl(III) or Fe(II) is in excess. The second-order character of the reaction confirms early work on this system[216]. A non-linearity in the second-order plots in the last 30 % of reaction was noted, and proved to be particularly significant. Ashurst and Higginson[199] observed that Fe(III) retards the oxidation, thereby accounting for the curvature of the rate plots in the last stages of reaction. On the other hand, the addition of Tl(I) has no significant effect. On this basis, they proposed the scheme

$$\mathrm{Tl(III)+Fe(II)} \underset{k_{-1}}{\overset{k_1}{\rightleftharpoons}} \mathrm{Tl(II)+Fe(III)}$$

$$\mathrm{Tl(II)+Fe(II)} \xrightarrow{k_2} \mathrm{Tl(I)+Fe(III)}$$

in which Fe(III) and Fe(II) compete for the Tl(II) intermediate. The derived rate law, applicable to the whole course of the reaction, is then

$$\frac{-\mathrm{d}[\mathrm{Fe(II)}]}{\mathrm{d}t} = \frac{2\,k_1 k_2[\mathrm{Fe(II)}]^2[\mathrm{Tl(III)}]}{k_2[\mathrm{Fe(II)}]+k_{-1}[\mathrm{Fe(III)}]}$$

assuming a stationary-state concentration for Tl(II). The reaction was followed by the disappearance of Fe(II). From a detailed analysis of the kinetic data, values for k_2/k_{-1} and k_1 were obtained of 30.8 and 2.41×10^{-2} $\mathrm{l.mole^{-1}.sec^{-1}}$, respectively, at $[\mathrm{H^+}] = 1.00\ M$, $\mu = 3.00\ M$ and 25 °C. Contrary to earlier observations[217], oxygen has no detectable influence on the rate. Strict conformity of the data to the rate equation rules out the possibility that the alternative route involving Fe(IV) can make a significant contribution. Increase in acidity decreases the rate of the reaction, and the results are treated[215] in terms of a mechanism involving two pathways (with activated complexes made up of one or two $\mathrm{OH^-}$ ions) such that

$$k_1 = \frac{K_1}{(K_1+[\mathrm{H^+}])}\left(k_1' + \frac{k_1'' K_2}{[\mathrm{H^+}]}\right)$$

Plots of $k_1(K_1+[\mathrm{H^+}]/K_1)$ *versus* $1/[\mathrm{H^+}]$ are linear. The values of k_1' (intercept) and $k_1''K_2$ (slope) are dependent on the value assigned to K_1, the first hydrolysis constant of $\mathrm{Tl^{3+}}$. The latter quantity was estimated from previous data[218]. K_2 is the second hydrolysis constant of $\mathrm{Tl^{3+}}$. Rate parameters are collected in Table 28; k_1' has an associated activation energy[215] of 18.4 $\mathrm{kcal.mole^{-1}}$ and a $\Delta S^\ddagger$ value[215] of -7.3 $\mathrm{cal.deg^{-1}.mole^{-1}}$. The results of Ashurst and Higginson[199] at 25 °C are in excellent agreement with those of Johnson[215]. Sulphate ion is reported to catalyse the Tl(III)+Fe(II) reaction[215]. On the other hand, chloride ion inhibits the reaction in an unusual manner[217,219]. The rate exhibits a pronounced minimum at 0.1 to 0.2 M $\mathrm{Cl^-}$ concentration, and attains a limiting value at $\sim 2\ M$ $\mathrm{Cl^-}$ concentration. The conclusion is drawn that Tl(II) must complex with chloride.

The slow reaction between Tl(III) and Fe(II) is induced by V(III)[220]. The results indicate that an intermediate, probably Tl(II), is formed by the reaction

TABLE 28

RATE COEFFICIENTS AND EQUILIBRIUM CONSTANTS OF Tl(III)+Fe(II) REACTION[215]

Temp. (°C)	K_1(*mole.l*$^{-1}$)	$10^2k_1'$(*l.mole*$^{-1}$.*sec*$^{-1}$)	$10^2\ k_1''\ K_2$(*sec*$^{-1}$)
25.0	6.4	1.41	1.24
35.0	9.4	3.91	4.44
45.0	12.7	9.96	12.8

of Tl(III) and V(III), and that this intermediate is capable of reacting with Fe(II). The Tl(III)+U(IV) system is also subject to induction by V(III), although Tl(II) is not as reactive towards U(IV) as it is to Fe(II). Induction experiments on the effect of V(II) on the Tl(III)+U(IV) system reveal that Tl(II) reacts much more rapidly with V(II) than with V(III). Alternatively, they imply that Tl(II) is not formed in the Tl(III)+V(II) reaction. Spectrophotometric data for the rate of oxidation of V(II) by Tl(III) were obtained from measurements at 760 mμ, a wavelength where V(IV) absorbs more strongly than V(II), V(III), and Tl(III). As the reaction proceeds the observed second-order rate coefficient increases. This effect is attributable to the catalytic influence of V(III). The following sequence of reactions is an appropriate mechanistic framework

$$\text{Tl(III)} + \text{V(II)} \longrightarrow \text{Tl(I)} + \text{V(IV)} \quad (10.1)$$

$$\text{Tl(III)} + \text{V(II)} \longrightarrow \text{Tl(II)} + \text{V(III)} \quad (10.2)$$

$$\text{Tl(III)} + \text{V(III)} \longrightarrow \text{Tl(II)} + \text{V(IV)} \quad (10.3)$$

$$\text{Tl(II)} + \text{V(II)} \longrightarrow \text{Tl(I)} + \text{V(III)} \quad (10.4)$$

$$\text{Tl(II)} + \text{V(III)} \longrightarrow \text{Tl(I)} + \text{V(IV)} \quad (10.5)$$

$$\text{Tl(III)} + 2\,\text{V(II)} \longrightarrow \text{Tl(I)} + 2\,\text{V(III)} \quad (10.6)$$

$$2\,\text{Tl(II)} \longrightarrow \text{Tl(I)} + \text{Tl(III)} \quad (10.7)$$

$$\text{V(II)} + \text{V(IV)} \longrightarrow 2\,\text{V(III)} \quad (10.8)$$

Of these steps, the last three can be discounted: (10.6) on the grounds that there is no significant V(II) dependence, (10.7) is considered unimportant since Tl(II) is present only in minute concentrations, (10.8) is slow by comparison with the other steps in the set ($k' \sim 0.13$ l.mole^{-1}.sec^{-1} in 1 M $HClO_4$ at 0 °C)[221]. Both rate and stoichiometric data infer that the reaction between Tl(III) and V(II) occurs essentially by a two-electron oxidation (step (10.1)). In the presence of chloride, less V(IV) is produced. It is interesting to note that oxidation of V(II) by molecular oxygen or hydrogen peroxide generates V(IV)[222]. However, the oxidation of V(III) by Tl(III) does not occur as a two-electron step (see p. 231).

The kinetics of the oxidation of tris(bipyridyl)osmium(II) by Tl(III)[223]

$$\text{Tl(III)} + 2\,\text{Os(bipy)}_3{}^{2+} = \text{Tl(I)} + 2\,\text{Os(bipy)}_3{}^{3+}$$

show deviations from second-order behaviour similar to those encountered in the Tl(III)+Fe(II) reaction[199]. It is probable that an analogous mechanism operates, *viz.*

$$\text{Tl(III)} + \text{Os(bipy)}_3{}^{2+} \rightleftharpoons \text{Tl(II)} + \text{Os(bipy)}_3{}^{3+}$$

$$\text{Tl(II)} + \text{Os(bipy)}_3{}^{2+} \longrightarrow \text{Tl(I)} + \text{Os(bipy)}_3{}^{3+}$$

The reaction was followed by means of the strong absorption of the Os(II) complex at 480 mμ. Unlike the Tl(III)+Fe(II) system, there is a slight increase in rate as the hydrogen-ion concentration is increased. The kinetic data were interpreted on the basis that both Tl^{3+} and $TlOH^{2+}$ react with $Os(bipy)_3{}^{2+}$ (with rate coefficients k_3 and k_4, respectively). At 24.5 °C and $\mu = 2.99$ *M*, $k_3 = 36.0$ l.mole^{-1}.sec^{-1} and $k_4 = 14.7$ l.mole^{-1}.sec^{-1}; corresponding activation energies are 6.90 and 11.5 kcal.mole^{-1}. The latter values are considerably smaller than those for the Tl(III)+Tl(I) exchange[224] and for the Tl(III)+Fe(II) reaction[199]. On the other hand, all three reactions are subject to retardation by Cl^- ions.

Ce(IV), although a strong oxidant, reacts only very slowly with $Hg(I)_2$ whereas Tl(III), a two-equivalent oxidant, reacts relatively rapidly[86, 225] according to

$$\mathrm{Tl(III)+Hg(I)_2 = Tl(I)+2\,Hg(II)}$$

Rate data were obtained by following the rate of disappearance of $Hg(I)_2$ spectrophotometrically at 236 mμ[86]. The rate, unaltered by the addition of Tl(I) but reduced by Hg(II), is represented by

$$-\mathrm{d[Hg(I)_2]}/\mathrm{d}t = k'\mathrm{[Tl(III)][Hg(I)_2]/[Hg(II)]}$$

which becomes, on integration

$$\left(\frac{\mathrm{[Hg(II)]_0+2\,[Hg(I)_2]_0}}{\mathrm{[Tl(III)]_0-[Hg(I)_2]_0}}\right)\log\frac{\mathrm{[Hg(I)_2]_0}}{\mathrm{[Hg(I)_2]}} - \left(\frac{\mathrm{[Hg(II)]_0+2\,[Tl(III)]_0}}{\mathrm{[Tl(III)]_0-[Hg(I)_2]_0}}\right)\log\frac{\mathrm{[Tl(III)]_0}}{\mathrm{[Tl(III)]}} = \frac{k't}{2.303}$$

where the subscript o refers to initial concentrations. Plots of the left-hand side of this equation *versus* time are linear over a wide range of concentrations of reactants and products. The observed rate coefficient k', obtainable from the slopes of such plots, is found to be inversely proportional to $HClO_4$ concentration (2.4 to 6.0 *M* at $\mu = 6.0$ *M*), and also to $ClO_4{}^-$ concentration. The following scheme accounts very adequately for the observed kinetics

$$\left.\begin{array}{rl} \mathrm{Hg_2{}^{2+}+ClO_4{}^-} \overset{K_3}{\rightleftharpoons} \mathrm{Hg_2ClO_4{}^+} & (10.9)\\ \mathrm{Hg_2{}^{2+}} \overset{K_4}{\rightleftharpoons} \mathrm{Hg^{2+}+Hg} & (10.10)\\ \mathrm{Tl^{3+}+H_2O} \overset{K_1}{\rightleftharpoons} \mathrm{TlOH^{2+}+H^+} & (10.11) \end{array}\right\}\ \text{rapid equilibria}$$

$$\mathrm{TlOH^{2+}+Hg} \overset{k_5}{\rightarrow} \mathrm{Tl^++Hg^{2+}+OH^-}\quad \text{rate-determining} \qquad (10.12)$$

Accordingly, the derived rate law is

$$-\mathrm{d}[\mathrm{Hg(I)_2}]\mathrm{d}t = k_5[\mathrm{TlOH^{2+}}][\mathrm{Hg}]$$

or, alternatively

$$\frac{-\mathrm{d}[\mathrm{Hg(I)_2}]}{\mathrm{d}t} = \frac{k_5 K_4 K_1}{(1+K_3[\mathrm{ClO_4}^-][\mathrm{H}^+])} \cdot \frac{[\mathrm{Hg(I)_2}][\mathrm{Tl(III)}]}{[\mathrm{Hg(II)}]}$$

Thus k' is given by

$$k' = \frac{k_5 K_4 K_1}{(1+K_3[\mathrm{ClO_4}^-][\mathrm{H}^+])}$$

Using values of $K_3 = 0.91$ l.mole^{-1}, $K_4 = 5.5 \times 10^{-9}$ mole.l^{-1} and $K_1 = 0.073$ mole.l^{-1} in conjunction with a k' value of 4.2×10^{-5} sec^{-1} (at $[\mathrm{H}^+] = 3$ M and $[\mathrm{ClO_4}^-] = 3$ M), allows k_5 to be calculated as 1×10^6 l.mole^{-1}.sec^{-1} at 25 °C. Other metal ions known to complex with ClO_4^- ion include Fe^{3+} (ref. 226) and Ce^{3+} (ref. 188). Perchlorate complexing of Hg_2^{2+} has been reported independently[227]. Equation (10.13) gives the apparent activation energy E' in terms of the true activation energy E_5 [$\Delta H^‡$ of step (10.12)] and the enthalpies of reaction, ΔH_3, ΔH_4 and ΔH_1, corresponding to steps (10.9), (10.10) and (10.11) as

$$E' = E_5 + \Delta H_4 + \Delta H_1 - \Delta H_3 \tag{10.13}$$

Estimated values of ΔH_3, ΔH_4 and ΔH_1 are ~0, 10.6 and −0.5 kcal.mole^{-1}, respectively, and E' is 24.4 kcal.mole^{-1} and therefore E_5 is given as $\sim 14 \pm 3$ kcal.mole^{-1}. The corresponding $\Delta S^‡$ term is 13 ± 10 cal.deg^{-1}.mole^{-1}. It is noteworthy that the kinetics demonstrate the direct reaction

$$\mathrm{TlOH^{2+}} + \mathrm{Hg_2}^{2+} \rightarrow \mathrm{Tl}^+ + 2\,\mathrm{Hg}^{2+} + \mathrm{OH}^-$$

to be negligible, although the concentration of Hg_2^{2+} ions can be 10^5 times greater than that of Hg atoms. The slow step (10.12) bears a formal resemblance to the rate-determining step in the Tl(III)+Tl(I) exchange[218], *viz.*

$$\mathrm{TlOH^{2+}} + \mathrm{Tl}^+ \rightarrow \mathrm{Tl}^+ + \mathrm{Tl}^{3+} + \mathrm{OH}^-$$

particularly as the Tl^+ and Hg species are isoelectronic. However, this step is ~ 10^9 to 10^{10} times slower than (10.12) even though the activation energies are similar. The profound difference in rate strongly suggests that the principal factor responsible is the charge of the reactant ions: this is reflected in disparate $\Delta S^‡$

terms. Chloride and bromide ions catalyse the Tl(III)+$Hg(I)_2$ reaction (showing non-linear dependences) by complexing with Hg^{2+}, thus increasing the concentration of Hg atoms by displacing the dismutation step (10.10).

Wetton and Higginson[59] report that the reaction between Tl(III) and Sn(II) is complete in less than 10 sec at −6 °C in ~ 1.3 *M* HCl for ~ 10^{-2} *M* reactant concentrations. It appears that it takes place as a single-stage two-equivalent process without the intervention of Sn(III), since added $Co(NH_3)_4(H_2O)Cl^{2+}$ is not consumed during the course of the reaction.

The two-equivalent oxidation of U(IV) by Tl(III), *viz.*

$$\mathrm{Tl(III)+U(IV) = Tl(I)+U(VI)}$$

obeys a simple second-order rate law, addition of products having no effect on the rate[228]. Rate measurements were made at 650 mμ, the absorption maximum of U(VI). Increase of $HClO_4$ concentration leads to a decrease in rate: rate $\alpha[H^+]^{-n}$ where n ranges from 1.48 at 16 °C to 1.39 at 25 °C. This result is interpreted as arising from the availability of two simultaneous pathways involving hydrolysed species (UOH^{3+} and/or $TlOH^{2+}$). The corresponding rate law

$$-\mathrm{d[U(IV)]/d}t = [\mathrm{Tl}^{3+}][\mathrm{U}^{4+}](k_1'[\mathrm{H}^+]^{-1}+k_2'[\mathrm{H}^+]^{-2})$$

can be written in terms of K_5 and K_1, the hydrolysis constants of U^{4+} and Tl^{3+}, *viz.*

$$-\mathrm{d[U(IV)]/d}t = [\mathrm{Tl(III)}][\mathrm{U(IV)}](k_1'[\mathrm{H}^+]+k_2')/([\mathrm{H}^+]+K_5)([\mathrm{H}^+]+K_1)$$

Thus k_{obs}, the observed (second-order) rate coefficient, is given by

$$k_{\mathrm{obs}} = (k_1'[\mathrm{H}^+]+k_2')/([\mathrm{H}^+]+K_5)([\mathrm{H}^+]+K_1)$$

Use was made of known values of K_5 = 0.021 mole.l^{-1} and K_1 = 0.073 mole.l^{-1} at 25 °C and $\mu \sim 3$ *M*, along with estimated values of these constants at other temperatures (assuming, in each case, an enthalpy of hydrolysis of 11.0 kcal. mole^{-1}). Plots of $k_{obs}([H^+]+K_5)([H^+]+K_1)$ *versus* $[H^+]$ are linear, and k_1' and k_2' are obtained from the slopes and intercepts. At 25 °C and μ = 2.9 *M*, the rate parameters for the first path are: $k_1' = 2.11\times10^{-2}$ sec^{-1}, $\Delta H^\ddagger$ = 24.6 kcal. mole^{-1} and $\Delta S^\ddagger = 16\pm7$ cal.deg^{-1}.mole^{-1}. For the second path, the corresponding values are: $k_2' = 2.13\times10^{-2}$ mole.l^{-1}.sec^{-1}, $\Delta H^\ddagger$ = 21.7 kcal.mole^{-1} and $\Delta S^\ddagger = 7\pm7$ cal.deg^{-1}.mole^{-1}. Paths 1 and 2 involve the activated complexes $(UOHTl^{6+})^\ddagger$ and $(UOTl^{5+})^\ddagger$, respectively. The formation of U–O bonds is a necessary requirement since changes in coordination occur from U(IV) to U(V) (UO_2^+) to U(VI) (UO_2^{2+}). Addition of $NaClO_4$ increases the rate of the reac-

References pp. 267–273

tion. The effect of various anions and cations on the reaction is summarised as follows. Cl^- inhibits the reaction whereas SO_4^{2-} enhances the rate, results ascribed to the formation of complexes of Tl(III): parallel behaviour is noted in the Tl(III)+Tl(I) exchange[229], and in the Tl(III)+Fe(II) reaction[219]. Addition of small amounts of Cu(II), Ag(I) and Hg(II) leave the rate unaffected. These ions have a marked effect on the rate of oxidation of U(IV) by oxygen, a reaction known to proceed by a chain mechanism[230]. It is concluded that the general features of the Tl(III)+U(IV) reaction are in accord more with a single two-electron step than with successive one-electron steps for the following reasons:

(*a*) The reaction is considerably faster than the Tl(III)+Fe(II) system, generally accepted to occur *via* one-electron steps;

(*b*) The reaction shows no abnormality in $\Delta H^\ddagger$ and $\Delta S^\ddagger$ which seems to be characteristic of processes, *e.g.* Tl(III)+Fe(II), involving Tl(II).

(*c*) If the primary process generated Tl(II), then a chain process of the sort

$$\text{U(IV)}+\text{Tl(II)} \rightarrow \text{U(V)}+\text{Tl(I)}$$

$$\text{U(V)}+\text{Tl(III)} \rightarrow \text{U(VI)}+\text{Tl(II)}$$

would result. However, there is no such indication, particularly since the rate is not sensitive to metallic ions like Cu(II).

The reaction between Tl(III) and U(IV) is one of the few redox reactions which have been studied in a mixed solvent[231]. Solutions were kept under nitrogen. There are striking differences between the rate in aqueous perchloric acid and methanol–aqueous perchloric acid solutions. In the latter media the order with respect to Tl(III), U(IV), and H^+ alters as the solvent composition is changed (Table 29). For 25 % methanol–75 % water solvent the kinetic orders of 1.0, 1.5 and −1.33 with respect to U(IV), Tl(III), and H^+, respectively, are consistent with the existence of two competing paths whose net activation processes are

$$Tl^{3+}+U^{4+}+H_2O \xrightarrow{k'} (Tl{\cdot}HO{\cdot}U^{6+})^\ddagger+H^+$$

$$2\,Tl^{3+}+U^{4+}+2\,H_2O \xrightarrow{k''} (Tl{\cdot}HO{\cdot}U{\cdot}OH{\cdot}Tl^{8+})^\ddagger+2\,H^+$$

TABLE 29

Tl(III)+U(IV) REACTION: KINETIC ORDERS IN AQUEOUS METHANOL[231]

Methanol (%)	*Order with respect to*		
	U(IV)	*Tl(III)*	*H^+*
0	1.00	1.00	−1.39 to −1.48
25	1.00	1.50	−1.33
50	0.50	1.25	−0.25
75	0.33	0.67	−0.67

The corresponding rate law is

$$-\mathrm{d}[\mathrm{U(IV)}]/\mathrm{d}t = k'[\mathrm{Tl}^{3+}][\mathrm{U}^{4+}][\mathrm{H}^+]^{-1} + k''[\mathrm{Tl}^{3+}]^2[\mathrm{U}^{4+}][\mathrm{H}^+]^{-2}$$

and

$$\frac{-\mathrm{d}[\mathrm{U(IV)}]}{\mathrm{d}t}\left(\frac{([\mathrm{H}^+]+K_5)([\mathrm{H}^+]+K_1)^2}{[\mathrm{U(VI)}][\mathrm{Tl(III)}]^2[\mathrm{H}^+]}\right) = k'\left(\frac{([\mathrm{H}^+]+K_1)}{[\mathrm{Tl(III)}]}\right) + k'' \quad (10.14)$$

Using values of K_5 and K_1, identical with those made use of by Harkness and Halpern[228], the rate coefficients k' and k'' were evaluated from the slopes and intercepts of linear plots of the left-hand side of equation (10.14) *versus* ([H$^+$]+

TABLE 30

Tl(III)+U(IV) REACTION: KINETIC DATA IN 25 % METHANOL–75 % WATER MEDIA[231]

Path	*Rate coefficient(sec^{-1}) at 25 °C*	$\Delta H^\ddagger$*(kcal.mole^{-1})*	$\Delta S^\ddagger$*(cal.deg^{-1}.mole^{-1})*
k'	0.0041	3.11	−59.0
k''	0.98	4.47	−43.6

Ionic strength, 2.9 *M*.

K_1)/[Tl(III)]. Table 30 contains these results together with $\Delta H^\ddagger$ and $\Delta S^\ddagger$ values. The reaction in 25 % methanol is faster than in water, although $\Delta S^\ddagger$ is strongly negative. To some extent this may be rationalised in terms of the structure of the activated complex $(\mathrm{Tl{\cdot}HO{\cdot}U{\cdot}OH{\cdot}Tl}^{8+})^\ddagger$, since the formation of UO_2^{2+} would appear to be facilitated by the positions of the two oxygen atoms relative to the U atom. The role of the solvent in governing the progress of the reaction is discussed by Jones and Amis[231]. In contrast to the reaction in aqueous solution, chloride is ineffective as an inhibitor and sulphate is a poor catalyst in 75 % methanol media. Moreover, Cu(II) and Hg(II) exert a marked catalytic effect whereas Ag(I) strongly retards the reaction. These latter observations recall those recorded for the U(IV)+O_2 system[230]. A step-wise process, involving U(V) and Tl(II), is advanced, *viz.*

$$\mathrm{Tl(III)+U(IV) \rightarrow Tl(II)+U(V)}$$

$$\mathrm{U(V)+Cu(II) \rightarrow U(VI)+Cu(I)}$$

$$\mathrm{Tl(II)+Cu(I) \rightarrow Tl(I)+Cu(II)}$$

Ag(I) is believed to act as an inhibitor, as is likely in the U(IV)+O_2 reaction, by bringing about a chain-breaking step with U(V): the well-defined induction period is proportional to the initial Ag(I) concentration. During this time, a colloidal suspension of metallic Ag is formed, and after this is complete the reaction proceeds at its normal rate.

Wear[232] has subjected the Tl(III)+U(IV) reaction to a close scrutiny, the kinetics being studied over a 2000-fold range of reactant concentrations. Solutions were kept under nitrogen. The range of concentrations examined by Harkness and Halpern[228] were U(IV) = 3.5×10^{-3} to 11.0×10^{-3} *M*, Tl(III) = 5×10^{-3} to 21×10^{-3} *M*. Wear finds that the orders with respect to the reactant vary with

TABLE 31

Tl(III)+U(IV) REACTION: VARIATION OF KINETIC ORDERS WITH CONCENTRATION

Region	10^3[*U(IV)*] (*M*)	10^3[*Tl(III)*] (*M*)	[H^+] (*M*)	*U(IV) order*	*Tl(III) order*	H^+*order*
I	0.5–50	1–50	0.25–2.75	0.88	0.90	−1.4
II	0.03–0.5	5.0	0.90	1.2	0.90	−1.4
III	6.3	0.02–1.0	0.90	0.88	0.67	−1.4
IV	0.07–0.22	0.12–0.36	0.50–1.90	1.6	0.38	−1.5

Ionic strength, 2.9 *M*; temp., 25 °C. From Wear[232] (by courtesy of Sandia Laboratories and U.S. Atomic Energy Commission).

concentration[232]. His results are summarised in Table 31*. Qualitatively, the data are equally consistent with a chain mechanism

$$\mathrm{Tl(III)+U(IV) \rightarrow Tl(II)+U(V)}$$
$$\mathrm{Tl(II)+U(IV) \rightarrow Tl(I)+U(V)}$$
$$\mathrm{Tl(II)+U(V) \rightarrow Tl(I)+U(VI)}$$
$$\mathrm{2\,U(V) \rightarrow U(IV)+U(VI)}$$
$$\mathrm{2\,Tl(II) \rightarrow Tl(I)+Tl(III)}$$
$$\mathrm{Tl(III)+U(V) \rightarrow Tl(II)+U(VI)}$$

Love *et al.*[233] have examined the oxidation of U(IV) by Tl(III) in the presence of tartaric acid. The kinetics are complex: tartaric acid increases the initial rate but, in its presence, reaction ceases before all the Tl(III) is consumed. The disagreement between the rate coefficients of Love *et al.* and those of Harkness and Halpern[228] is ascribed to the presence of Fe(III) impurities in Tl(III) stock solutions. However, as pointed out by Newton and Baker[250], this explanation cannot be valid since the results of Harkness and Halpern are not dependent on the initial reactant concentrations.

---

* The rate laws reported by Wear[232] have been criticised by Newton and Baker[250].

11. Oxidations by lead(IV)

The reduction of Pb(IV) acetate by Co(II) acetate in acetic acid[234] exhibits more complex kinetics than the Pb(IV)+Ce(III) system (p. 242). The expected stoichiometry of 1 : 2, corresponding to

$$\text{Pb(IV)} + 2\,\text{Co(II)} = \text{Pb(II)} + 2\,\text{Co(III)}$$

is found in the anhydrous solvent but the presence of water causes a change in stoichiometry (Fig. 9). Alcohol also brings about a similar effect (the addition of 10 % by volume of methanol causes the stoichiometry to be reduced to 1 : 1.1). The fact that methanol and water do not effect the decomposition of Pb(IV) or of Co(III) acetates, under these conditions, strongly suggests that these additives must be reacting with a transient intermediate, thus causing the final concentration of Co(III) to decrease. Kinetic data, for the most part obtained for the appearance of Co(III) at 400 mμ, show that the order with respect to Pb(IV) is unity under all conditions, but that the order with respect to Co(II) is non-integral (1.5). At 25.0 °C the observed rate coefficient is 1.70 $\text{l}^{\frac{3}{2}}.\text{mole}^{-\frac{3}{2}}.\text{sec}^{-1}$; the apparent activation energy and entropy are 17.5 kcal.mole^{-1} and -8 ± 4 $\text{cal.deg}^{-1}.\text{mole}^{-1}$, respectively. Furthermore, the presence of Pb(II) causes the order in Co(II) to increase. Of the two reaction products, only Pb(II) in large excess causes a retardation. The kinetic observations are interpreted in terms of a scheme in-

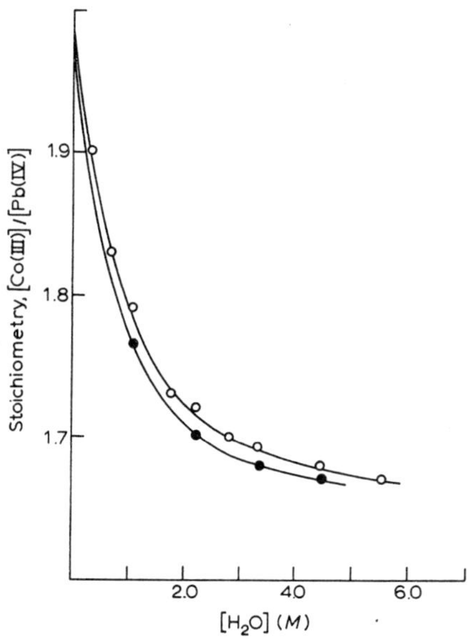

Fig. 9. Effect of water on the stoichiometry of the Pb(IV)+Co(II) reaction in acetic acid at temperatures of 23 °C (○) and 37 °C (●). (*From Benson et al.*[234], *by courtesy of The Faraday Society.*)

volving a Co(II) dimer, and Pb(III) and Co(IV) as intermediate species, *viz.*

$$\text{Co(II)}+\text{Co(II)} \rightleftharpoons [\text{Co(II)}]_2$$

$$\text{Pb(IV)}+[\text{Co(II)}]_2 \rightarrow \text{Pb(II)}+2\ \text{Co(III)}$$

$$\text{Pb(IV)}+[\text{Co(II)}]_2 \rightarrow \text{Pb(III)}+\text{Co(III)}+\text{Co(II)}$$

$$\text{Pb(IV)}+[\text{Co(II)}]_2 \rightarrow \text{Pb(II)}+\text{Co(IV)}+\text{Co(II)}$$

$$\text{Pb(IV)}+\text{Co(II)} \rightleftharpoons \text{Pb(II)}+\text{Co(IV)}$$

$$\text{Pb(IV)}+\text{Co(II)} \rightarrow \text{Pb(III)}+\text{Co(III)}$$

$$\text{Co(IV)}+\text{Co(II)} \rightarrow 2\ \text{Co(III)}$$

$$\text{Pb(III)}+\text{Co(II)} \rightarrow \text{Pb(II)}+\text{Co(III)}$$

The postulation of the +4 oxidation state of cobalt is necessary to account for the retarding influence of Pb(II). The existence of a dimeric species of Co(II) acetate is required by the rate law and is confirmed by spectrophotometric and solubility measurements[235]. The existence of ionic species of the reactants is inferred by the rate increase on addition of sodium acetate, an observation which cannot be attributed to a salt effect because sodium perchlorate produces a rate decrease. On this scheme an explanation of the effect of water on the stoichiometry is that the step

$$\text{Co(IV)}+\text{Co(II)} \rightarrow 2\ \text{Co(III)}$$

is eliminated, Co(IV) being reduced directly by water, possibly by

$$\text{Co(IV)}+\text{H}_2\text{O} = \text{Co(II)}+2\ \text{H}^+ +\tfrac{1}{2}\ \text{O}_2$$

The even greater influence of methanol may be due to its reaction with Pb(III).

Benson and Sutcliffe[236] have made a kinetic study of the reaction between lead(IV) acetate and cerium(III) acetate in anhydrous acetic acid, *viz.*

$$\text{Pb(IV)}+2\ \text{Ce(III)} = \text{Pb(II)}+2\ \text{Ce(IV)}$$

The reaction was followed by observing the appearance of the yellow colour of Ce(IV) at 400 mμ, with Pb(IV) present in excess concentration. Pb(IV) was varied in the region 8.6×10^{-3} *M* to 4.4×10^{-2} *M* while Ce(III) was kept at $\sim 4\times10^{-4}$ *M*. A practical difficulty encountered was the photochemical instability of Ce(IV) acetate. Under the above conditions and in the temperature range 30–47 °C, the reaction is strictly first-order in each reactant. The observed rate coefficient at 30.0 °C is 1.48×10^{-2} l.mole^{-1}.sec^{-1} and the apparent activation energy and entropy are 19.5 kcal.mole^{-1} and -2.5 ± 1.5 cal.deg^{-1}.mole^{-1}, respectively. The rate is unaltered by the presence of Ce(IV) at similar concentrations to that of the initial Ce(III) concentration. Similarly, Pb(II) has no influence on the rate, unless present in very large excess when a slight increase is observed. The presence of sodium perchlorate produces a marked retardation and the results were treated in terms of an ionic strength effect. On the other hand, added sodium acetate increases the rate, there being an initial first-order dependence on the concentration of this salt. The addition of ethanol or benzene leaves the order of reaction unaffected, but both these additives produce an increase in rate. In the case of ethanol there is a first-order dependence, whereas for benzene a linear relationship exists between log k_{obs} and the reciprocal of the bulk dielectric constant, suggesting a solvent effect. Since both Pb(II) and Ce(IV) fail to retard the reaction, the mechanism is probably

$$\mathrm{Pb(IV)+Ce(III)} \rightarrow \mathrm{Pb(III)+Ce(IV)} \qquad \text{slow}$$

$$\mathrm{Pb(III)+Ce(III)} \rightarrow \mathrm{Pb(II)+Ce(IV)} \qquad \text{rapid}$$

although no direct evidence was adduced for the participation of Pb(III). In the presence of sodium acetate, ionic species take part in the reaction[237], *e.g.*, $Pb(OAc)_6^{2-}$, $Pb(OAc)_5^{-}$ and $Ce(OAc)_4^{-}$. Confirmatory evidence for these comes from ion-migration and spectrophotometric studies. The effect of ethanol is attributed to the formation of a reactive complex between the alcohol and Pb(IV). In this respect, it is interesting that ethanol also increases the rate of oxidation of *t*-butyl hydroperoxide by Pb(IV)[238]. In the latter system a complex exists between the organic substrate and Pb(IV), the rate-controlling step is

$$\mathrm{Pb(IV)\cdot ROOH+ROOH} \rightarrow \text{products}$$

and the reaction is second-order in peroxide. Ethanol supplants the peroxide molecule from the complex so that the step becomes

$$\mathrm{Pb(IV)\cdot EtOH+ROOH} \rightarrow \text{products}$$

and the reaction order with respect to peroxide approaches unity.

12. Oxidations by cerium(IV)

In acidic sulphate media, the multi-equivalent oxidation of Cr(III) by Ce(IV)

$$3\ \mathrm{Ce(IV)+Cr(III)} = 3\ \mathrm{Ce(III)+Cr(VI)}$$

takes place at a convenient rate for spectrophotometric measurements[239]. However, the reaction was found to be inconveniently fast in perchloric acid solutions. Kinetic results were obtained at 492 mμ and 500 mμ, wavelengths where Ce(III) is transparent but Ce(IV), Cr(III) and Cr(VI) absorb to varying degrees. No evidence was obtained for the presence of dimeric species of Ce(IV); this reactant is presumed to be highly complexed with sulphate (or bisulphate) ions[240†]. Although dimers of Cr(VI) are possible, monomer and dimer have similar absorptivity values at the chosen wavelengths[51]. The reaction demonstrates an inverse dependence on the concentration of Ce(III), and a square dependence on Ce(IV) concentration, *viz.*

$$\frac{\mathrm{d[Cr(VI)]}}{\mathrm{d}t} = \frac{k_{\mathrm{obs}}\mathrm{[Ce(IV)]^2[Cr(III)]}}{\mathrm{[Ce(III)]}}$$

Consequently, the activated complex ($X^{\ddagger}$) of the rate-determining step is composed of one Ce atom and one Cr atom, the average oxidation state of each atom being +4 [$X^{\ddagger}$ = 2 Ce(IV)+Cr(III)−Ce(III)]. A sequence of one-equivalent steps are in accord with the rate law, *viz.*

$$\mathrm{Ce(IV)+Cr(III)} \underset{k_{-1}}{\overset{k_1}{\rightleftharpoons}} \mathrm{Ce(III)+Cr(IV)} \tag{12.1}$$

$$\mathrm{Ce(IV)+Cr(IV)} \underset{k_{-2}}{\overset{k_2}{\rightleftharpoons}} \mathrm{Ce(III)+Cr(V)} \qquad \text{rate-determining} \tag{12.2}$$

$$\mathrm{Ce(IV)+Cr(V)} \overset{k_3}{\rightleftharpoons} \mathrm{Ce(III)+Cr(VI)} \tag{12.3}$$

The concentration of Cr(IV) is maintained by equilibrium (12.1) as

$$\mathrm{[Cr(IV)]} = \frac{k_1\mathrm{[Cr(III)][Ce(IV)]}}{k_{-1}\mathrm{[Ce(III)]}}$$

and since the rate is controlled by (12.2)

$$\text{rate} = k_2\mathrm{[Ce(IV)][Cr(IV)]} = \frac{k_1 k_2\mathrm{[Ce(IV)]^2[Cr(III)]}}{k_{-1}\mathrm{[Ce(III)]}} \tag{12.4}$$

Thus the observed rate coefficient k_{obs} is equivalent to $k_1 k_2/k_{-1}$. Because of the limitations of the kinetic procedure, there is some small doubt over the inclusion of a term in Ce(IV) concentration in the denominator of (12.4). However, the relative insignificance of this term means that k_{-1}[Ce(III)] > k_2[Ce(IV)]. Also, the inequality k_3[Ce(IV)] ≫ k_{-2}[Ce(III)] is a consequence of the observed kinetics. As in the Cr(VI)+Fe(II) system[58], the slow stage involves the inter-

† See, also, ref. 309.

conversion of Cr(IV) and Cr(V). There is strong evidence that such a transformation requires a change in coordination number of the metal ion. The observed rate coefficient shows an approximate inverse dependence on the square of HSO_4^- concentration.

The cerium(IV) oxidation of Mn(II) has been briefly reported on by Aspray *et al.*[241a]. The reaction was followed from the rate of appearance of Mn(III) at 505 mμ, using solutions in 4.5 M sulphuric acid. The results indicate the occurrence of an equilibrium

$$\text{Ce(IV)} + \text{Mn(II)} \rightleftharpoons \text{Ce(III)} + \text{Mn(III)}$$

At 20.8 °C, $[\text{Mn(II)}] = 8.21 \times 10^{-2}\ M$, $[\text{Ce(IV)}] = 1.63 \times 10^{-3}\ M$, the forward rate coefficient is 0.213 $\text{l.mole}^{-1}.\text{sec}^{-1}$; the corresponding equilibrium constant is 0.015, in good agreement with redox potential data. Rechnitz *et al.*[241b] have examined the system in greater detail. They find the values for the apparent rate coefficients of the forward and backward steps to be 0.435 $\text{l.mole}^{-1}.\text{sec}^{-1}$ and 5.9 $\text{l.mole}^{-1}.\text{sec}^{-1}$, respectively, in 3.0 M sulphuric acid at 25 °C. Rates were measured by monitoring cerium(IV) at 400 mμ. The forward path has $\Delta H^{\ddagger}$ and $\Delta S^{\ddagger}$ values of 13.0 kcal.mole^{-1} and -17 $\text{cal.deg}^{-1}.\text{mole}^{-1}$; the corresponding values for the Ce(III)+Mn(III) reaction are 15.9 kcal.mole^{-1} and -1.7 cal. $\text{deg}^{-1}.\text{mole}^{-1}$. From the observed dependences on reaction media the kinetically important Ce(IV) and Mn(III) species are $Ce(SO_4)_2$ and $MnOH^{2+}$.

The kinetics of the oxidation of Fe(II) by Ce(IV) in aqueous perchloric acid have been studied, using reactant concentrations in the range 10^{-5} to 10^{-6} M (ref. 242). A quenching method was utilised to monitor the disappearance of Fe(II). The reaction conforms to a 1 : 1 stoichiometry and is of simple second order, *viz.*

$$-\text{d}[\text{Fe(II)}]/\text{d}t = k'[\text{Ce(IV)}][\text{Fe(II)}]$$

However, the intercepts of log [Fe(II)]/[Ce(IV)] *versus* time plots deviate from the values expected for the initial concentrations of the reactants. This "apparent zero-time oxidation", which is reproducible, is believed to result from a finite quenching time, and the reaction of Fe(II) with a very reactive Ce(IV) species. Added amounts of Ce(III) and Fe(III) leave the rate unaffected. At constant ionic strength, k' varies inversely with hydrogen-ion concentration in the range 0.05 to 1.00 M; for $[H^+] > 1.0\ M$, k' increases with increasing $[H^+]$. In general

$$k' = a[\text{H}^+] + b + c/[\text{H}^+]$$

which signifies reaction by three simultaneous routes. These are

$$Ce^{4+} + Fe(II) \xrightarrow{k_4} Ce(III) + Fe(III)$$

$$CeOH^{3+} + Fe(II) \xrightarrow{k_5} Ce(III) + Fe(III)$$

$$Ce(OH)_2{}^{2+} + Fe(II) \xrightarrow{k_6} Ce(III) + Fe(III)$$

Therefore

$$k' = \frac{(k_4[H^+]/K_1) + k_5 + (k_6 K_2/[H^+])}{([H^+]/K_1) + 1 + (K_2/[H^+])} \tag{12.5}$$

where K_1 and K_2 are the hydrolysis constants of Ce^{4+} and $CeOH^{3+}$, respectively. When $[H^+] < 1$ M, $K_1 \gg [H^+]$ and equation (12.5) reduces to

$$k' = k_5 + \frac{(k_6 - k_5)K_2}{([H^+] + K_2)}$$

In accordance with this relationship, plots of k' *versus* $K_2/(K_2 + [H^+])$ are fairly linear, using a value of 0.08 for K_2. Values of k_5 and k_6 are obtained from the slopes ($= k_6 - k_5$) and intercepts ($= k_5$) of such plots. At 0.3 °C and $\mu = 2.0$ M, $k_4 = 5865 \pm 1500$ l.mole^{-1}.sec^{-1}, $k_5 = 1000 \pm 200$ l.mole^{-1}.sec^{-1}, and $k_6 = 4830 \pm 500$ l.mole^{-1}.sec^{-1}. At unit ionic strength, k' has an associated $\Delta H^\ddagger$ of 9.4 kcal.mole^{-1}, and a $\Delta S^\ddagger$ of -6 ± 3 cal.deg^{-1}.mole^{-1}. Dainton *et al.*[242], suggest that the similarity between k_5 and k_6 may arise from common transition states of the type $(Ce^{IV}\text{–}OH^{3+} \cdots Fe^{II})$ and $(OHCe^{IV}\text{–}OH^{2+} \cdots Fe^{II})$. Electron transfer may take place by hydrogen-atom transfer, or electron conduction through a hydrogen-bonded intermediate. The possible mechanism of the direct acid-dependent path is discussed in terms of

(*a*) charge transfer through symmetrical bridges of protonated water molecules, *i.e.*

$$\begin{array}{ccccc} & H & H & H & \\ & | & | & | & \\ (H_2O)_nCe^{IV} & O \cdots & H\text{–}O^+\text{–}H & \cdots O & Fe^{II}(H_2O)_m \\ & | & & | & \\ & H & & H & \end{array}$$

and

(*b*) electron tunnelling through extended acid chains of the type $-(HClO_4-)_n$ which may exist in cold, concentrated acid solutions.

The Ce(IV)+Fe(II) system is catalysed by HSO_4^- and F^-, but not by Cl^- ions. Since there is a linear relationship between k' and $[HSO_4^-]$, the pathway

$$CeSO_4{}^{2+} + Fe^{2+} \xrightarrow{k_7} Ce(III) + Fe(III)$$

is likely. Assuming a value of 3500 for the equilibrium constant of

$$Ce^{4+} + HSO_4^- \rightleftharpoons CeSO_4^{2+} + H^+$$

then k_7 is $\sim 5 \times 10^3$ l.mole^{-1}.sec^{-1} at 0 °C and $\mu = 0.23$ M. This value agrees quite well with that calculated from the data of Dulz and Sutin[44] for the reaction in sulphuric acid solution.

Dulz and Sutin[44] have measured the rates of oxidation of Fe(II) and tris(1,10-phenanthroline) Fe(II) complexes by Ce(IV) with a view to test the linear free energy relationships† predicted by Marcus[243, 244]. These rapid reactions were studied in sulphuric acid media using a stopped-flow apparatus. In 0.50 M H_2SO_4 at 25 °C the Ce(IV)+Fe(II) and Ce(IV)+$Fe(phen)_3^{2+}$ systems have (second-order) rate coefficients of 1.3×10^6 l.mole^{-1}.sec^{-1} and 1.42×10^5 l.mole^{-1}.sec^{-1}, respectively. The corresponding activation energies are 9.5 kcal.mole^{-1} and 6.5 kcal.mole^{-1}. According to the Marcus theory of electron transfer reactions[43], if $k_{1,2}$ and $K_{1,2}$ are the rate and equilibrium constant, respectively, for the oxidation–reduction reaction, and $k_{1,1}$ and $k_{2,2}$ are the rate coefficients of the exchange reactions, then

$$k_{1,2} = (k_{1,1}\, k_{2,2}\, K_{1,2} f)^{\frac{1}{2}} \tag{12.6}$$

where f is defined by

$$\log f = \frac{(\log K_{1,2})^2}{4 \log (k_{1,1}\, k_{2,2}/Z^2)} \tag{12.7}$$

Z is the collision frequency of two uncharged molecules in solution, and has a value of 10^{11} l.mole^{-1}.sec^{-1}. In terms of free energies, equation (12.6) can be written as

$$\Delta G^{\ddagger}_{1,2} = 0.50\, \Delta G^{\ddagger}_{1,1} + 0.50\, \Delta G^{\ddagger}_{2,2} + 0.50\, \Delta G^{0}_{1,2} - 1.15 \boldsymbol{R}T \log f \tag{12.8}$$

where $\Delta G^{\ddagger}$ represents the free energies of activation of the various processes, and $\Delta G^{0}_{1,2}$ is the standard free-energy change of the redox reaction. Consequently, a linear relationship between $(\Delta G^{\ddagger}_{1,2} - 0.5\ \Delta G^{\ddagger}_{1,1})$ and $\Delta G^{0}_{1,2}$ is predicted. Fig. 10 contains such a plot for the Ce(IV) oxidation of a number of Fe(II) phenanthroline complexes. Similar data are included for the oxidation of Fe^{2+} by various Fe(III) phenanthroline complexes in 0.50 M $HClO_4$ at 25 °C[196]. Rate coefficients and (formal) oxidation potentials are set out in Table 32. A straight

* Other linear free-energy relationships are those for the oxidation of a series of Fe(II) phenanthroline complexes by Co(III)[195] and Mn(III)[82].

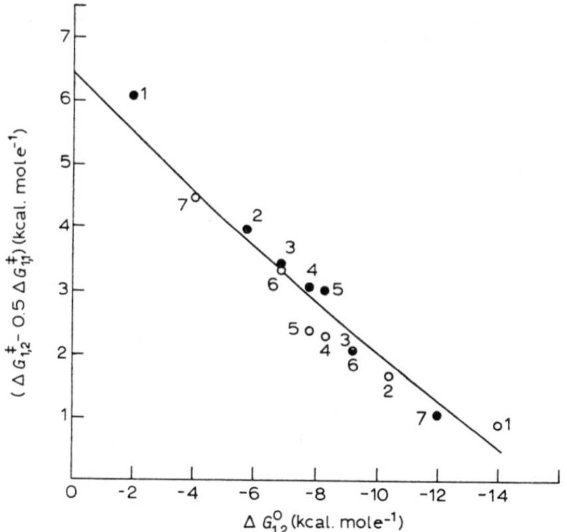

Fig. 10. Relationship between $(\Delta G^{\ddagger}_{1,2} - 0.5\,\Delta G^{\ddagger}_{1,1})$ and the standard free energy change (ΔG^{0}_{12}) of the redox reactions at 25 °C. Open circles, Ce(IV)+$Fe(phen)_3^{2+}$ reactions in 0.50 M H_2SO_4. Closed circles, Fe^{2+}+$Fe(phen)_3^{3+}$ reactions in 0.50 M $HClO_4$. Numbers refer to complexes in Table 32. (*From Dulz and Sutin*[44], *by courtesy of The American Chemical Society*.)

TABLE 32

RATE COEFFICIENTS FOR THE OXIDATION OF Fe(II) COMPLEXES BY Ce(IV) IN 0.50 M H_2SO_4 AT 25.0 °C[44]

Complex		E^0 *of complex* (V)[a]	$k(l.mole^{-1}.sec^{-1})$
Iron(II)		0.68	1.3×10^6
Tris(3,4,7,8-tetramethyl-1,10-phenanthroline)iron(II)	(1)	0.83	1.6×10^6
Tris(5,6-dimethyl-1,10-phenanthroline)iron(II)	(2)	0.99	4.3×10^5
Tris(5-methyl-1,10-phenanthroline)iron(II)	(3)	1.04	2.2×10^5
Tris(1,10-phenanthroline)iron(II)	(4)	1.08	1.42×10^5
Tris(5-phenyl-1,10-phenanthroline)iron(II)	(5)	1.10	1.2×10^5
Tris(5-chloro-1,10-phenanthroline)iron(II)	(6)	1.14	2.5×10^4
Tris(5-nitro-1,10-phenanthroline)iron(II)	(7)	1.26	3.9×10^3

[a] The formal oxidation potentials of the complexes.

line of slope 0.50 has been drawn through the data for reactions which have $\Delta G^{0}_{1,2}$ close to zero, when equation (12.8) simplifies to

$$\Delta G^{\ddagger}_{1,2} = 0.50\,\Delta G^{\ddagger}_{1,1} + 0.50\,\Delta G^{\ddagger}_{2,2} + 0.50\,\Delta G^{0}_{1,2} \tag{12.9}$$

The intercept, 6.5 kcal.mole^{-1}, leads to a value of 13.0 kcal.mole^{-1} for the (average) $\Delta G^{\ddagger}$ of the various phenanthroline Fe(II)–Fe(III) exchanges. Such a $\Delta G^{\ddagger}_{2,2}$ value corresponds to an average rate coefficient, $k_{2,2}$, of 2×10^3 l.mole^{-1}.

TABLE 33

COMPARISON OF OBSERVED AND CALCULATED RATE COEFFICIENTS AT 25.0 °C[44]

Reaction	$k_{1,1}$ *(l.mole^{-1}.sec^{-1})*	$k_{2,2}$ *(l.mole^{-1}.sec^{-1})*	$k_{1,2}$ *observed (l.mole^{-1}.sec^{-1})*	$k_{1,2}$ *calculated (l.mole^{-1}.sec^{-1})*
Fe(II)+Ce(IV)	4.0	4.4	1.3×10^6	6×10^5
Cr(II)+Fe(III)	$\leqq 2\times10^{-5}$	4.0	$\sim 8\times10^3$	$\leqq 6\times10^5$
V(II)+Fe(III)	1.0×10^{-2}	4.0	$>10^5$ [a]	9×10^5
Eu(II)+Fe(III)	$\leqq 1\times10^{-4}$	4.0	$>10^5$ [a]	$\leqq 2\times10^6$
Cr(II)+Co(III)	$\leqq 2\times10^{-5}$	~ 5	$> 3\times10^2$ [a]	$\leqq 1\times10^{10}$
V(II)+Co(III)	1.0×10^{-2}	~ 5	$> 3\times10^2$ [a]	$\sim 2\times10^{10}$
$Cr^{2+}+V^{3+}$	$\leqq 2\times10^{-5}$	1.0×10^{-2}	—	$\leqq 2\times10^{-2}$
$Fe^{2+}+Co^{3+}$	4.0	~ 5	42 [a]	$\sim 6\times10^6$

[a] From ref. 24.

sec^{-1}. The curve ($\Delta G^0_{1,2}$ more negative than 5 kcal.mole^{-1}) was calculated from equation (12.8) using this value for $k_{2,2}$ in equation (12.7). The general conclusion is that the data adequately comply with the predictions of the Marcus theory. Table 33 is a collection of observed and calculated rate coefficients for a variety of redox reactions. Adamson *et al.*[242] have criticised the value of $k_{1,2}$ for the Ce(IV)+Fe(II) system, as calculated by Dulz and Sutin[44], on the grounds that the rate of the Fe(III)+Fe(II) exchange ($k_{1,1}$) and the corresponding oxidation potential relate to $HClO_4$ media, whereas the rate ($k_{2,2}$) and oxidation potentia of the Ce(IV)+Ce(III) system are for H_2SO_4 media. Adamson *et al.*[242] arrive at a calculated value of 1.3×10^6 l.mole^{-1}.sec^{-1} for the rate coefficient ($k_{1,2}$) of the Ce(IV)+Fe(II) reaction in 0.5 *M* $HClO_4$ at 0 °C. Since this value is very much at variance with the observed value (700 l.mole^{-1}.sec^{-1}), they conclude that this oxidation takes place by an atom-transfer mechanism, to which the theoretical treatment of Marcus is not appropriate.

Rate coefficients (and oxidation potentials) are given in Table 34 for the Ce(IV)

TABLE 34

SECOND-ORDER RATE COEFFICIENTS FOR THE OXIDATION OF Fe(II) COMPLEXES BY Ce(IV) IN 0.5 *M* SULPHURIC ACID AT 25.0 °C[195]

Complex	*E°(V)*	$10^{-6}k$ *(l.mole^{-1}.sec^{-1})*
$Fe(phen)_3^{2+}$	1.07	0.142 [a]
$Fe(phen)_2(CN)_2$	0.79	7.11
$Fe(phen)(CN)_4^{2-}$	0.65	8.88
$Fe(bipy)_3^{2+}$	1.05	0.196
$Fe(bipy)_2(CN)_2$	0.81	8.40
$Fe(bipy)(CN)_4^{2-}$	0.67	12.5
$Fe(CN)_6^{4-}$	0.69	1.90

[a] From ref. 44.

TABLE 35

COMPARISON OF OBSERVED AND CALCULATED RATE COEFFICIENTS[195]

Reaction	$k_{1,2}$ *observed(l.mole^{-1}.sec^{-1})*	$k_{1,2}$ *calculated(l.mole^{-1}.sec^{-1})*
$Ce(IV)+W(CN)_8^{4-}$	>108	6.1×10^8
$Ce(IV)+Fe(CN)_6^{4-}$	1.9×10^6	6.0×10^6
$Ce(IV)+Mo(CN)_8^{4-}$	1.4×10^7	1.3×10^7
$IrCl_6^{2-}+W(CN)_8^{4-}$	6.1×10^7	8.1×10^7
$IrCl_6^{2-}+Fe(CN)_6^{4-}$	3.8×10^5	5.7×10^5
$IrCl_6^{2-}+Mo(CN)_8^{4-}$	1.9×10^6	1.0×10^6
$Mo(CN)_8^{3-}+W(CN)_8^{4-}$	5.0×10^6	1.7×10^7
$Mo(CN)_8^{3-}+Fe(CN)_6^{4-}$	3.0×10^4	2.7×10^4
$Fe(CN)_6^{3-}+W(CN)_8^{4-}$	4.3×10^4	5.1×10^4

oxidation of $Fe(bipy)_3^{2+}$, $Fe(bipy)_2(CN)_2$, $Fe(bipy)(CN)_4^{2-}$, $Fe(CN)_6^{4-}$, and corresponding phenanthroline complexes[195]. Mixed-ligand complexes react more rapidly than complexes containing identical ligands.

Campion *et al.*[195] have compared the rates of oxidation of $W(CN)_8^{4-}$, $Mo(CN)_8^{4-}$, and $Fe(CN)_6^{4-}$ by Ce(IV) with those calculated from the Marcus theory[43]. The results on these systems are given in Table 35 together with data on a number of related reactions. Agreement between observed and calculated rate coefficients is good.

In conforming to an expected linear free energy relationship, the Ce(IV) oxidation of various 1,10-phenanthroline and bipyridyl complexes of Ru(II) in 0.5 *M* sulphuric acid are consistent with the requirements of the Marcus treatment[245]. The results for the oxidation of the 3- and 5-sulphonic-substituted ferroin complexes by Ce(IV) suggest that the ligand does not function as an electron mediator, and that the mechanism is outer-sphere in type. Second-order rate coefficients for the oxidation of $Ru(phen)_3^{2+}$, $Ru(bipy)_3^{2+}$, and $Ru(terpy)_3^{2+}$ are 5.8×10^3, 8.8×10^3, and 7.0×10^3 l.mole^{-1}.sec^{-1}, respectively, in 0.5 *M* H_2SO_4 at 25 °C; a rapid-mixing device was employed.

The rate of oxidation of mercury(I) by Ce(IV) is slow in any medium but ~ 3.6 times faster in 2 *M* perchloric acid than in 1 *M* sulphuric acid, achieving a maximum in the former medium at ~ 4 *M*, and then decreasing[246]. Sulphate ion retards the reaction: the rate increase observed in $HClO_4$ solutions is ascribed to the formation of less complexed, more reactive species of Ce(IV). The kinetics of the reaction between Hg(I) perchlorate and Ce(IV) sulphate have been examined in 2.0 *M* perchloric acid at 50.0 °C, under which conditions the rate law

$$-d[Ce(IV)]/dt = k'[Ce(IV)][Hg(I)_2]$$

obtains, Hg(II) and Ce(III) having no effect on the rate. Mechanistically the reaction is best envisaged as

$$Ce(IV)+(Hg\text{–}Hg)^{2+} \rightarrow Ce(III)+Hg(I)+Hg(II) \quad \text{slow}$$

$$Ce(IV)+Hg(I) \rightarrow Ce(III)+Hg(II) \quad \text{rapid}$$

a scheme in which the Hg–Hg bond is broken simultaneously with electron transfer. At 50 °C in 2.0 M $HClO_4$, the rate coefficient of the slow step is given as 0.14 $l.mole^{-1}.sec^{-1}$; the corresponding apparent activation energy is 14.4 $kcal.mole^{-1}$.

Silver(I) is effective as a catalyst in the oxidation of mercury(I) by Ce(IV)[26]. In dilute $HClO_4$ solutions no direct reaction takes place between Ce(IV) and mercury(I), or between Ce(IV) and Ag(I). The catalysed reaction obeys, in the presence of excess $Hg(I)_2$, the expression

$$-d[Ce(IV)]/dt = -2\,d[Hg(I)_2]/dt$$
$$= 2\,k_8'[Ce(IV)] = 2\,k_8[Ce(IV)][Ag(I)]_0 \quad (12.10)$$

where $[Ag(I)]_0$ represents the initial concentration of Ag(I). When $Hg(I)_2$ is in small excess only and Ce(III) is present initially, first-order plots for the rate of disappearance of Ce(IV) show curvature, the observed rate coefficient (k_8') decreasing as Ce(III) increases. This result suggests a back-reaction involving Ce(III). The scheme proposed is

$$Ce(IV)+Ag(I) \underset{k_{-8}}{\overset{k_8}{\rightleftharpoons}} Ce(III)+Ag(II)$$

$$Ag(II)+Hg(I)_2 \overset{k_9}{\rightarrow} Ag(I)+Hg(I)+Hg(II)$$

$$Ce(IV)+Hg(I) \rightarrow Ce(III)+Hg(II) \quad \text{rapid}$$

Assuming the steady-state hypothesis to apply to Ag(II)

$$\frac{-d[Ce(IV)]}{dt} = \frac{-2\,d[Hg(I)_2]}{dt} = \frac{2\,k_8\,k_9[Ce(IV)][Ag(I)]_0[Hg(I)_2]}{k_{-8}[Ce(III)]+k_9[Hg(I)_2]} \quad (12.11)$$

If $Hg(I)_2$ is in large excess, equation (12.11) reduces to (12.10) since $k_{-8} < k_9$ ($k_{-8}/k_9 = 0.198$ at 1.5 M $[H^+]$, $\mu = 3.0$ M and 20 °C).

In mixed solutions, 2.0 M in $HClO_4$ and 0.1 M in H_2SO_4, the rates of oxidation of $Hg(I)_2$ by Ce(IV), as catalysed by Ag(I) and Mn(II), are [247]

$$-d[Ce(IV)]/dt = 0.304[Ce(IV)][Ag(I)]+0.0218[Ce(IV)][Ag(I)]/[Hg(I)_2]$$

and

$$-d[Ce(IV)/dt = 1.28[Ce(IV)][Mn(II)]+0.143[Ce(IV)][Mn(II)]/[Hg(I)_2]$$

respectively, at a temperature of 50.0 °C (time expressed in min); corresponding activation energies are 13.5 and 12.1 kcal.mole^{-1}. In the latter case the scheme suggested is

$$Ce(IV)+Mn(II) \rightleftharpoons Ce(III)+Mn(III)$$

$$Mn(III)+Hg(I)_2 \rightarrow Hg(I)+Hg(II)+Mn(II)$$

$$Ce(IV)+Hg(I) \rightarrow Ce(III)+Hg(II) \quad \text{rapid}$$

$$Ce(IV)+Mn(III) \rightarrow Ce(III)+Mn(IV)$$

$$Mn(IV)+Hg(I)_2 \rightarrow Mn(II)+2\,Hg(II)$$

although this does not explain the inverse dependence on $[Hg(I)_2]$ at low concentrations. Spectrophotometric evidence was adduced for the presence of Mn(III) as an intermediate. The $Ce(IV)+Hg(I)_2$ reaction, doubly catalysed by Ag(I)+Mn(II), was examined in both $HClO_4$ and H_2SO_4. In the former medium, the kinetics are complex, and the rate shows an enhanced catalytic effect; in the latter medium, the kinetics are analytically soluble, and the catalytic effect is additive. Tentative mechanisms are proposed.

The mechanism proposed by Dorfman and Gryder[248] to account for the reduction of Ce(IV) by Tl(I), in 6.18 *M* nitric acid at 54 °C, includes a dimeric species of Ce(IV) and a Ce(IV)–Ce(III) binuclear species, *viz.*

$$2\,Ce(IV) \rightleftharpoons [Ce(IV)]_2 \qquad K_3$$

$$Ce(III)+Ce(IV) \rightleftharpoons [Ce(III){\cdot}Ce(IV)] \quad K_4$$

$$Ce(IV)+OH^- \underset{k_{-10}}{\overset{k_{10}}{\rightleftharpoons}} Ce(III)+OH$$

$$Tl(I)+OH \underset{k_{-11}}{\overset{k_{11}}{\rightleftharpoons}} Tl(II)+OH^-$$

$$Ce(IV)+Tl(I) \underset{k_{-12}}{\overset{k_{12}}{\rightleftharpoons}} Ce(III)+Tl(II)$$

$$Ce(IV)+Tl(II) \xrightarrow{k_{13}} Ce(III)+Tl(III)$$

Using concentrations in mole.l^{-1} and time in sec, Dorfman and Gryder find $K_3 = 18$, $K_4 = 2$, $k_{10} = 1.33\times10^{-5}$, $k_{12} = 3.81\times10^{-4}$, $k_{-10}/k_{11} = 0.021$, $k_{-10}k_{-11}/k_{11}k_{13} = 5.4\times10^{-5}$, $k_{-11}/k_{13} = 2.6\times10^{-3}$, and $k_{-12}/k_{13} = 1.52\times10^{-3}$

The Ag(I)-catalysed oxidation of Tl(I) by Ce(IV) can be explained[26] by a series of reactions

$$Ce(IV)+Ag(I) \underset{k_{-8}}{\overset{k_8}{\rightleftharpoons}} Ce(III)+Ag(II)$$

$$Ag(II)+Tl(I) \xrightarrow{k_{14}} Ag(I)+Tl(II)$$

$$Ce(IV)+Tl(II) \rightarrow Ce(III)+Tl(III) \quad \text{rapid}$$

analogous to those for the Ag(I)-catalysed reaction of Ce(IV) and mercury(I)[26]. From these the derived rate law is

$$\frac{-\mathrm{d}[\mathrm{Ce(IV)}]}{\mathrm{d}t} = \frac{-2\,\mathrm{d}[\mathrm{Tl(I)}]}{\mathrm{d}t} = \frac{2\,k_8\,k_{14}[\mathrm{Ce(IV)}][\mathrm{Ag(I)}]_0[\mathrm{Tl(I)}]}{k_{-8}[\mathrm{Ce(III)}]+k_{14}[\mathrm{Tl(I)}]} \tag{12.12}$$

However, the term k_{-8}[Ce(III)] cannot be neglected since k_{-8}/k_{14} is 35.7. Neither can k_8 be evaluated from first-order plots for the disappearance of Ce(IV). The integrated form of equation (12.12) allows k_{-8}/k_{14} and k_8 to be obtained. The ratio k_9/k_{14}, obtained from k_{-8}/k_9 and k_{-8}/k_{14}, is 180 at 1.5 M [H^+] and 20 °C. This corresponds to the ratio of the rate coefficients for the oxidation of $Hg(I)_2$ and of Tl(I) by Ag(II). Higginson *et al.*[26] quote a value of 185 for the ratio of rate coefficients for oxidation of these two species by Co(III). The dependence of k_8 on hydrogen-ion concentration is described. Schenk and Bazzelle[310] have studied the Ce(IV)+Tl(I) system, both uncatalysed and catalysed by Ag(I) and Mn(III), in sulphuric acid media.

Cerium(IV) oxidises tin(II) in aqueous sulphuric acid probably by a two-step path involving Sn(III)[249]. At low Sn(IV) concentrations and low sulphate concentration the reaction is second order, and the suggestion is made that the reactant species are $Ce(SO_4)_3{}^{2-}$ and $SnSO_4$. In mixed chloride–sulphate media the Ce(IV)+Sn(II) reaction, in the presence of trioxalatocobaltate(III), produces an intermediate which consumes the Co(III) complex[59]. This result is interpreted as being evidence for the presence of Sn(III) in the reacting system.

13. Oxidations by uranium, neptunium and plutonium

The redox reactions of the actinide elements have been the subject of a recent and authoritative review by Newton and Baker[250]. The net activation process concept is used to interpret the experimental data. Empirical correlations shown to exist include those between the entropies of the activated complexes and their charges, and, for a set of similar reactions, between $\Delta G^{\ddagger}$ and ΔG^0, and $\Delta H^{\ddagger}$ and ΔH^0. The present state of the evidence for binuclear species is discussed.

13.1 OXIDATIONS BY URANIUM(VI)

The oxidation of vanadium(II) by uranium(VI)[251], *viz.*

$$\mathrm{U(VI)+V(II)} \rightarrow \mathrm{U(V)+V(III)} \tag{13.1}$$

$$\mathrm{U(V)+V(II)} \rightarrow \mathrm{U(IV)+V(III)} \tag{13.2}$$

References pp. 267–273

is complicated by the presence of the additional reactions

$$2\,\mathrm{U(V)} \rightarrow \mathrm{U(IV)} + \mathrm{U(VI)} \tag{13.3}$$

and

$$\mathrm{U(V)} + \mathrm{V(III)} \rightarrow \mathrm{U(IV)} + \mathrm{V(IV)} \tag{13.4}$$

However, in the presence of excess V(IV) the U(V) formed in reaction (13.1) can be reoxidised to U(VI) by

$$\mathrm{U(V)} + \mathrm{V(IV)} \rightarrow \mathrm{U(VI)} + \mathrm{V(III)} \tag{13.5}$$

Since reaction (13.5) is more rapid than (13.2), (13.3), or (13.4), reactions (13.1) and (13.5) are predominant under these conditions, and the overall process corresponds to the U(VI)-catalysed reaction of V(II) and V(IV). In the absence of U(VI) this reaction is relatively slow[221]. Kinetic data were obtained spectrophotometrically at 760 mμ (where V(IV) is the principal absorbing species) in 0.05 to 2.0 M perchloric acid between 0.6° and 36.8 °C. The rate law is

$$-\mathrm{d}[\mathrm{V(IV)}]/\mathrm{d}t = -\mathrm{d}[\mathrm{V(II)}]/\mathrm{d}t = k_0[\mathrm{V(II)}][\mathrm{V(IV)}] + k_1[\mathrm{U(VI)}][\mathrm{V(II)}]$$

where k_0 and k_1 refer to the uncatalysed and catalysed reactions, respectively. At 25 °C, $[H^+] = 1\ M$, $\mu = 2.0\ M$, $[\mathrm{V(II)}] = 2\times10^{-3}\ M$, $[\mathrm{V(IV)}] = 2.55\times10^{-3}\ M$, k_0 and k_1 have values of 1.6 and $\sim$ 71 l.mole^{-1}.sec^{-1}, respectively. No direct evidence was adduced for the presence of U(IV). The response of the rate to variations in hydrogen-ion concentration is only very slight. Accordingly, the principal net activation process for the catalysed reaction is

$$UO_2{}^{2+} + V^{2+} \rightarrow (UO_2V^{4+})^{\ddagger}$$

and the corresponding $\Delta S^{\ddagger}$ and $\Delta H^{\ddagger}$ values are quoted as -26.1 ± 0.4 cal.deg^{-1}.mole^{-1} and 7.1 kcal.mole^{-1} at $\mu = 2.0\ M$. It is not known whether the reaction is of the inner-sphere type or not, as no evidence was obtained for a binuclear intermediate. The reaction is catalysed by chloride and sulphate ions; the former effect receives a detailed discussion by Newton and Baker[251], as does the ionic strength dependence.

U(VI) oxidises V(III) slowly in acid perchlorate solutions, the first step being

$$\mathrm{U(VI)} + \mathrm{V(III)} \underset{k_{-2}}{\overset{k_2}{\rightleftharpoons}} \mathrm{U(V)} + \mathrm{V(IV)} \tag{13.6}$$

which is followed by disproportionation of U(V). Newton and Baker[252] have made use of the rapid reaction of V(V) with Fe(III)[106]

$$\mathrm{U(V)} + \mathrm{Fe(III)} \overset{k_3}{\rightarrow} \mathrm{U(VI)} + \mathrm{Fe(II)} \tag{13.7}$$

to obtain kinetic information on reaction (13.6). Reactions (13.6) and (13.7), taken in combination, represent the U(VI)-catalysed oxidation of V(III) by Fe(III). The uncatalysed reaction has been described by Higginson and Sykes[90] (p. 176), *viz.*

$$V(III)+Fe(III) \xrightarrow{k_4} V(IV)+Fe(II) \quad (13.8)$$

$$V(IV)+Fe(III) \underset{k_{-5}}{\overset{k_5}{\rightleftharpoons}} V(V)+Fe(II) \quad (13.9)$$

$$V(III)+V(V) \xrightarrow{k_6} 2\,V(IV) \quad (13.10)$$

The catalysed reaction was followed by measuring the rate of appearance of V(IV) at 760 mμ. Comprehensively, on the basis of reactions (13.6) to (13.10), the rate law is

$$\frac{d[V(IV)]}{dt} = \frac{k_2[U(VI)][V(III)]}{1+(k_{-2}/k_3)[V(IV)]/[Fe(III)]} + k_4[V(III)][Fe(III)] + \frac{k_5[V(IV)][Fe(III)]}{1+(k_{-5}/k_6)[Fe(II)]/[V(III)]} \quad (13.11)$$

In equation (13.11), the first term corresponds to the catalysed part of the reaction and the remaining terms, which make a relatively small contribution, apply to the uncatalysed part. Kinetic data at constant acidity were in good agreement with the integrated form of the calculated rate expression. The rate coefficients k_2, k_4, k_5, and the ratio k_{-5}/k_6 were evaluated. Almost linear plots of log k_2 *versus* log [H^+] were obtained at four temperatures with slopes close to -1.8. This result suggests that the dominant activated complex is that formed by loss of two H^+ ions, *viz.*

$$UO_2^{2+}+V^{3+}+H_2O \rightarrow (VO{\cdot}UO_2^{3+})^{\ddagger}+2\,H^+ \quad (13.12)$$

although the alternative path, involving the loss of one H^+ ion, *viz.*

$$UO_2^{2+}+V^{3+}+H_2O \rightarrow (VOH{\cdot}UO_2^{4+})^{\ddagger}+H^+ \quad (13.13)$$

must be present also. The rate data are shown to be more in accord with consecutive reactions and an inner-sphere binuclear intermediate, rather than with parallel reactions. A number of mechanisms are possible, *e.g.*

$$V^{3+}+H_2O \rightleftharpoons VOH^{2+}+H^+ \qquad \text{rapid equilibrium}$$

$$UO_2^{2+}+VOH^{2+} \rightleftharpoons VO{\cdot}UO_2^{3+}+H^+ \qquad \text{rate-determining}$$

$$VO{\cdot}UO_2^{3+} \rightleftharpoons VO^{2+}+UO_2^{+} \qquad \text{rate-determining}$$

TABLE 36

ACTIVATION PARAMETERS[252] FOR $MO_2^{2+}+V^{3+}+H_2O \rightarrow (VOHMO_2^{4+})^{\ddagger}+H^+$

M	$\Delta H^{\ddagger}$*(kcal.mole⁻¹)*	$\Delta S^{\ddagger}$*(cal.deg⁻¹.mole⁻¹)*	*Ref.*
U	17.7±0.3	3.8±0.9	252
Np	13 ±2	−9 ±6	a
Pu	16 ±1	−5 ±3	254

[a] Recalculated from data in ref. 253.

This contrasts with the Pu(VI)+V(III) reaction which occurs by parallel paths. The activation parameters corresponding to the net activation processes (13.12) and (13.13) are $\Delta H^{\ddagger} = 22.1$ and 17.7 kcal.mole^{-1}, and $\Delta S^{\ddagger} = 12.9 \pm 0.5$ and 3.8 ± 0.9 cal.deg^{-1}.mole^{-1}. respectively. Table 36 contains similar data for the analogous reactions of NpO_2^{2+} (ref. 253) and PuO_2^{2+} (ref. 254) with V^{3+}. The positive $\Delta S^{\ddagger}$ value for the $UO_2^{2+}+V^{3+}$ reaction has led Newton and Baker[252] to suppose that the U(VI) reaction is inner-sphere, whereas the Np(VI) and Pu(VI) reactions are outer-sphere.

On mixing acid solutions of U(VI) and Cr(II) at 0 °C, a rapid reaction takes place and a bright green solution is produced. A slower reaction then occurs to yield a solution having the darker green colour characteristic of U(IV) and Cr(III). The appearance of the absorption spectrum of the principal intermediate leads to the conclusion[31] that it is a complex formed between U(V) and Cr(III), *viz.*

$$U(VI)+Cr(II) \rightarrow U(V)\cdot Cr(III)$$

From the lack of pH dependence it is likely that the intermediate species is $CrOUO^{4+}$. Reduction of this complex by Cr(II)

$$U(V)\cdot Cr(III)+Cr(II) \rightarrow 2\,Cr(III)+U(IV)$$

proceeds rapidly with half-lives between 4 and 8 min at 0 °C. The intermediate species is reactive towards Tl(III) and V(IV), displaying rates which are proportional to the concentration of the intermediate, but independent of the concentration of oxidant. Efficient transfer of oxygen from UO_2^{2+} to $Cr(H_2O)_6^{3+}$ takes place[255]. Evidence was sought for an analogous intermediate in the Pu(VI)+Cr(II) system but the results were negative.

Baes[256] has briefly investigated the reduction of U(VI) by Fe(II) in the strongly complexing medium of phosphoric acid.

Moore[30] has made a brief study of the slow reduction of U(VI) by Sn(II) in hydrochloric acid media. Chloride ion has a pronounced effect on the rate. Spectrophotometric evidence is cited for the complexing interaction of U(VI) and Sn(II).

13.2 OXIDATIONS BY NEPTUNIUM

The reaction between Np(VI) and V(III) in perchlorate media proceeds *via* consecutive steps of comparable rate[253], *viz.*

$$Np(VI)+V(III) \xrightarrow{k_1} Np(V)+V(IV)$$

$$Np(VI)+V(IV) \xrightarrow{k_2} Np(V)+V(V)$$

At 25 °C and in 2.0 M $HClO_4$, k_1 is estimated as 17.5 ± 1.6 $l.mole^{-1}.sec^{-1}$ and k_2 as 30 ± 15 $l.mole^{-1}.sec^{-1}$. The hydrogen-ion dependence of the reaction is

$$\text{rate} = (k+k'/[H^+])[Np(VI)][V(III)]$$

where k and k' have values of 6.3 $l.mole^{-1}.sec^{-1}$ and 20.3 sec^{-1}, respectively. Likely steps are between $NpO_2^{2+}+V^{3+}$, together with $NpO_2^{2+}+VOH^{2+}$ and/or $NpO_2OH^{+}+V^{3+}$. The activation energies and entropies of the k and k' routes are, respectively, 32 $kcal.mole^{-1}$ and 52 ± 16 $cal.deg^{-1}.mole^{-1}$, and 13 $kcal.mole^{-1}$ and -9 ± 6 $cal.deg^{-1}.mole^{-1}$ (as recalculated by Newton and Baker [250]).

Np(VI) oxidises U(IV) in a two-equivalent process

$$2\,Np(VI)+U(IV) = 2\,Np(V)+U(VI)$$

In aqueous perchloric acid the rate law found by Sullivan *et al.*[257] is

$$d[NpO_2^{+}]/dt = 2\,k'[NpO_2^{2+}][U^{4+}]/[H^+]$$

which suggests the scheme

$$U^{4+}+H_2O \rightleftharpoons UOH^{3+}+H^+ \qquad K_1$$

$$NpO_2^{2+}+UOH^{3+} \xrightarrow{k_3} NpO_2^{+}+U(V) \qquad \text{rate-determining}$$

$$NpO_2^{2+}+U(V) \rightarrow NpO_2^{+}+UO_2^{2+} \qquad \text{rapid}$$

Assuming a steady-state concentration for U(V), k' is identified as K_1k_3. The parameters describing the net activation process

$$NpO_2^{2+}+U^{4+}+H_2O \rightarrow (UOHNpO_2^{5+})^{\ddagger}+H^+$$

are $\Delta H^{\ddagger} = 18.2$ $kcal.mole^{-1}$ and $\Delta S^{\ddagger} = 7.4 \pm 0.8$ $cal.deg^{-1}.mole^{-1}$. At 25 °C and $\mu = 2$ M, k' is 10.8 sec^{-1}. Spectrophotometric, potentiometric and proton relaxation evidence has been cited for the specific interaction of Np(V) and U(VI) in acid media[258]. A value of 0.690 l.mole^{-1} has been derived for the equilibrium constant of the complex $NpO_2{}^+ \cdot UO_2{}^{2+}$ at 25 °C.

Np(V) oxidises V(III) to V(IV) in perchloric acid according to

$$\mathrm{Np(V)+V(III) = Np(IV)+V(IV)}$$

that is

$$NpO_2{}^+ + V^{3+} + 2\,H^+ = Np^{4+} + VO^{2+} + H_2O$$

However, at acid concentrations less than 0.5 M, Np^{3+} is formed by

$$Np^{4+} + V^{3+} + H_2O \rightleftharpoons Np^{3+} + VO^{2+} + 2\,H^+ \qquad K_2$$

where $K_2 \sim 6\times10^{-4}$ at 25 °C and $[ClO_4{}^-] = 3.0$ M (ref. 259). The rate law of the main reaction is

$$\frac{-d[NpO_2{}^+]}{dt} = \left(k_4 + k' \frac{[Np^{4+}]}{[VO^{2+}]}\right)[NpO_2{}^+][V^{3+}]$$

Plots of $(-d[NpO_2{}^+]/dt)/[NpO_2{}^+][V^{3+}]$ *versus* $[Np^{4+}]/[VO^{2+}]$ are linear with intercepts k_4 and slopes k'. The following mechanism accounts fairly adequately for the kinetics

$$NpO_2{}^+ + V^{3+} \xrightarrow{k_4} NpO^{2+} + VO^{2+}$$

$$NpO^{2+} + 2\,H^+ \rightleftharpoons Np^{4+} + H_2O \qquad \text{rapid equilibrium}$$

$$Np^{4+} + V^{3+} + H_2O \rightleftharpoons Np^{3+} + VO^{2+} + 2\,H^+ \qquad \text{rapid equilibrium, } K_2$$

$$\mathrm{Np(V)+Np(III)} \xrightarrow{k_5} \mathrm{2\,Np(IV)}$$

Thus

$$\frac{-d[NpO_2{}^+]}{dt} = \left(k_4 + \frac{k_5 K_2 [Np^{4+}]}{[VO^{2+}][H^+]^2}\right)[NpO_2{}^+][V^{3+}]$$

whereupon $k' = k_5 K_2/[H^+]^2$. At 25 °C, $[ClO_4{}^-] = 3.0$ M, and $[H^+] = 0.13$ M, k_5 is 35 l.mole^{-1}.sec^{-1}, and k' is calculated as 1.3 l.mole^{-1}.sec^{-1} (the observed value is 3.6 l.mole^{-1}.sec^{-1}). The postulated mechanism has points in common with that for the Fe(III)+V(III) system[90].

Np(V) and Cr(III) interact to form a binuclear complex, *viz.*

$$\mathrm{O{-}Np{-}O^+} + Cr(H_2O)_6{}^{3+} = \mathrm{O{-}Np{-}O{\cdot}}Cr(H_2O)_5{}^{4+} + H_2O$$

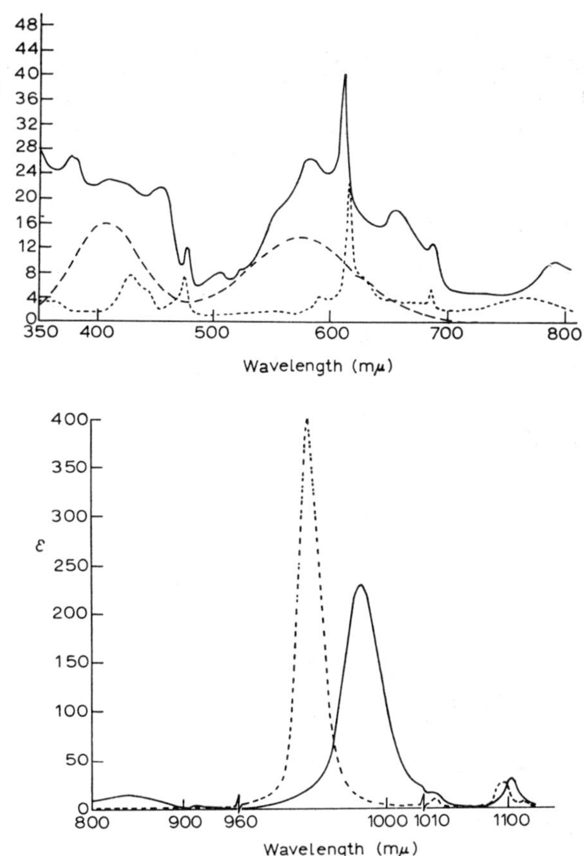

Fig. 11. Absorption spectrum of Np(V)–Cr(III) dimer (———), Cr(III) (– – –), and Np(V) (- - -); 1.0 *M* $HClO_4$; temp., 25 °C. (*From Sullivan*[208], *by courtesy of The American Chemical Society.*)

The complex has been separated by ion exchange and characterised by direct analysis[208]. The complex has a distinctive absorption spectrum (Fig. 11), quite unlike that of Np(V) and Cr(III). The rate coefficient for the first-order decomposition of the complex is 2.32×10^{-6} sec^{-1} at 25 °C in 1.0 *M* $HClO_4$. Sullivan[37] has obtained a value for the equilibrium constant of the complex, $K = [Np(V) \cdot Cr(III)]/[Np(V)][Cr(III)]$, of 2.62 ± 0.48 at 25 °C by spectrophotometric experiments. The associated thermodynamic functions are: $\Delta H = -3.3$ kcal. $mole^{-1}$ and $\Delta S = -9.0$ cal.deg^{-1}.$mole^{-1}$. The rates of decay and aquation of the complex, measured at 992 mμ, were investigated in detail. The same complex is formed when Np(VI) is reduced by Cr(II), and it is concluded that the latter reaction proceeds through both inner- and outer-sphere paths. It is noteworthy that the substitution-inert Rh(III), like Cr(III), forms a complex with Np(V)[260]. This bright-yellow Np(V)·Rh(III) dimer has been separated by ion-exchange and its absorption spectrum recorded. Using spectrophotometric means, the equilibrium constant, $K = [\mathrm{Np(V)\cdot Rh(III)}]/[\mathrm{Np(V)}][\mathrm{Rh(III)}]$, has been determined as 3.31 $\mathrm{l.mole^{-1}}$ at 25 °C. Associated values of ΔH and ΔS are -3.6 $\mathrm{kcal.mole^{-1}}$ and -10 ± 3 $\mathrm{cal.deg^{-1}.mole^{-1}}$. It may be significant that K for the Np(V)·Cr(III) analogue has a similar value at 25 °C (2.62 $\mathrm{l.mole^{-1}}$)[37]. The rate coefficient for the dissociation of the complex into Np(V) and Rh(III) is 4.38×10^{-4} $\mathrm{sec^{-1}}$ at 50 °C ($[\mathrm{H^+}] = 1.00$ M, $\mu = 1.00$ M); the activation energy is 27.3 kcal. $\mathrm{mole^{-1}}$ and $\Delta S^{\ddagger}$ is 8.6 ± 0.9 $\mathrm{cal.deg^{-1}.mole^{-1}}$. The rate of approach to equilibrium has also been measured. The dissociations of Np(V)·Rh(III) and Np(V)·Cr(III) are catalysed by HF in a first-order manner.

In 1 M perchloric acid solution, excess Np(VI) is reduced by Cr(II) to Np(IV) along with smaller amounts of Np(V) and Np(V)–Cr(III) complex[37]. Excess Np(V) is reduced by Cr(II) according to

$$\mathrm{Np(V)+Cr(II) = Np(IV)+Cr(III)}$$

and a Np(III) intermediate has been detected at $[\mathrm{H^+}] < 0.2$ M. Thompson and Sullivan[261] discuss their kinetic results on this reaction in the light of the following scheme

$$\mathrm{Np(V)+Cr(II)} \xrightarrow{k_6} \mathrm{Np(IV)+Cr(III)} \qquad (13.14)$$

$$\mathrm{Np(IV)+Cr(II)} \xrightarrow{k_7} \mathrm{Np(III)+Cr(III)} \qquad (13.15)$$

$$\mathrm{Np(V)+Np(III)} \xrightarrow{k_5} \mathrm{2\,Np(IV)} \qquad (13.16)$$

The alternative scheme

$$\mathrm{Np(V)+Cr(II)} \longrightarrow \mathrm{Np(III)+Cr(IV)}$$

$$\mathrm{Cr(II)+Cr(IV)} \longrightarrow \mathrm{2\,Cr(III)}$$

$$\mathrm{Np(V)+Np(III)} \longrightarrow \mathrm{2\,Np(IV)}$$

is rejected on the grounds that (*a*) no dimeric Cr(III) species was detected, and (*b*) for agreement with the empirical rate data, the Np(III)+Np(V) reaction would require to have a rate coefficient ~ 40 times greater than the observed value. The rate law for the one-equivalent reduction of Np(V) by Cr(II), reaction (13.14), is

$$-\mathrm{d}[\mathrm{Np(V)}]/\mathrm{d}t = k_6[\mathrm{Np(V)}][\mathrm{Cr(II)}][\mathrm{H^+}]^{0.78}$$

The term $k_6[\mathrm{H^+}]^{0.78}$ is re-expressed as $k_0[\mathrm{H^+}]e^{\beta[\mathrm{H^+}]}$ according to the Harned treatment with $k_0 = 1146\ l^2.mole^{-2}.sec^{-1}$ and $\beta = -1.142$ at 25 °C. Activation parameters corresponding to the k_0 term are $\Delta H^{\ddagger} = 1.85$ kcal.mole^{-1} and $\Delta S^{\ddagger} = -38.4$ cal.deg^{-1}.mole^{-1}. Oxygen-18 tracer experiments strongly suggest the participation of an inner-sphere activated complex. Kinetic data on the reaction between Np(V) and Np(III), reaction (13.16), conform to the expression

$$-d[Np(V)]/dt = k_5[Np(V)][Np(III)][H^+]^{0.56}$$

where $k_5 = 5.40$ (mole, l and sec units) at 25 °C and $\mu = 0.2\ M$ (see also refs. 18 and 259). The effect of Cl^- and HSO_4^- ions on the rate of the Np(V)+Np(III) and Np(V)+Cr(II) reactions are similar in that Cl^- is ineffective as a catalyst, whereas both reactions are sensitive to HSO_4^-.

Shastri *et al.*[262] have examined the kinetics of the reduction of Np(V) by U(IV). The rate of the reaction, zero order in Np(V) and first order in U(IV), shows an inverse dependence on the square of the hydrogen-ion concentration. The gross features of the mechanism are believed to be

$$Np(V)+U(IV) \rightarrow Np(IV)+U(V)$$

$$Np(IV)+U(IV) \rightarrow Np(III)+U(V)$$

$$Np(V)+Np(III) \rightarrow 2\ Np(IV)$$

$$Np(V)+U(V) \rightarrow Np(IV)+U(VI)$$

$$Np(IV)+U(V) \rightarrow Np(III)+U(VI)$$

At $[H^+] = 0.1\ M$, U(IV) reacts about six times more rapidly with Np(IV) than with Np(V). Unlike nitrate and sulphate, chloride ion accelerates the reaction.

The reaction between Np(IV) and Cr(II), reaction (13.15), has been the subject of a separate investigation by Thompson and Sullivan[263]. The rate law is

$$-d[Np(IV)]/dt = k'[Np(IV)][Cr(II)][H^+]^{-1.27}$$

in perchloric acid solutions at 25 °C†. At the same temperature, $\mu = 1.00\ M$ and $[HClO_4] = 1.00\ M$, k' is 4.29 l.mole^{-1}.sec^{-1}. In the presence of $1.0 \times 10^{-3}\ M$ Cl^- and $1.0 \times 10^{-3}\ M$ HSO_4^-, the k' values are 4.33 and 6.26 l.mole^{-1}.sec^{-1}, respectively.

13.3 OXIDATIONS BY PLUTONIUM

In contrast to the reactions of Pu(VI) and Pu(IV) with Ti(III), the reaction between Pu(V) and Ti(III) was found by Rabideau and Kline[264] to be immea-

† The Np(IV)+V(II) reaction shows a quite different hydrogen-ion dependence and is thought to be outer-sphere in type.

References pp. 267–273

surably fast, and therefore they consider the overall reduction of Pu(VI) to Pu(III) as being composed of two consecutive second-order reactions, *viz.*

$$\mathrm{Pu(VI)+2\,Ti(III) = Pu(IV)+2\,Ti(IV)} \tag{13.17}$$

$$\mathrm{Pu(IV)+Ti(III) = Pu(III)+Ti(IV)} \tag{13.18}$$

The interaction of Pu(VI) and Pu(III) is insignificant. From absorbance–time data for PuO_2^{2+} at 830 mμ

$$-\mathrm{d[Pu(VI)]/d}t = \mathrm{k_1'[Pu(VI)][Ti(III)]/[H^+]}$$

This form of rate law suggests

$$\mathrm{PuO_2^{2+}+TiOH^{2+} \rightarrow PuO_2^{+}+TiO^{2+}+H^+}$$

as the rate-controlling step. Using a value of 65.5 $\mathrm{sec^{-1}}$ for k_2', the observed rate coefficient of (13.18), the kinetic data were given an iterative treatment by a computer method. At 25 °C in 1 *M* perchloric acid, the average value of k_1' was shown to be 108 $\mathrm{sec^{-1}}$. For the net process

$$\mathrm{PuO_2^{2+}+Ti^{3+}+H_2O \rightarrow (PuO_2TiOH^{4+})^{\ddagger}+H^+}$$

$\Delta H^{\ddagger}$ and $\Delta S^{\ddagger}$ are 10.3 $\mathrm{kcal.mole^{-1}}$ and -14.7 ± 1.3 $\mathrm{cal.deg^{-1}.mole^{-1}}$. A practical difficulty is that Ti(III) slowly reduces perchlorate ions to chloride[265]. However, added chloride ion was demonstrated to have no effect on the rate of the Pu(VI)+Ti(III) system. It remains unclear why the reduction of PuO_2^{2+} by Ti(III) should be more difficult to bring about than is the reduction of PuO_2^{+}. It is interesting that, with V(III) as reductant, the reduction of PuO_2^{+} takes place slowly[254].

Pu(VI) oxidises V^{3+} by

$$\mathrm{PuO_2^{2+}+V^{3+}+H_2O = PuO_2^{+}+VO^{2+}+2\,H^+}$$

The progress of the reaction was followed at 830 mμ[254]. Kinetically, the oxidation is second order and it proceeds *via* parallel paths showing a dependence on both the inverse first and inverse second powers of $[H^+]$, the respective rate coefficients being 2.12 $\mathrm{sec^{-1}}$ and 0.228 $\mathrm{l.mole^{-1}.sec^{-1}}$ at 25 °C and $\mu = 2$ *M*. The route in which a single hydrogen ion is liberated predominates at 25 °C, and has $\Delta H^{\ddagger} =$ 15.5 $\mathrm{kcal.mole^{-1}}$ and $\Delta S^{\ddagger} = -5\pm2$ $\mathrm{cal.deg^{-1}.mole^{-1}}$. Rabideau[254] suggests the following scheme as likely

$$\mathrm{V^{3+}+H_2O \rightleftharpoons VOH^{2+}+H^+} \quad \text{rapid} \tag{13.19}$$

$$\mathrm{PuO_2^{2+}+VOH^{2+} \rightarrow PuO_2^{+}+VO^{2+}+H^+} \quad \text{slow} \tag{13.20}$$

$$\mathrm{PuO_2^{2+}+V(OH)_2^{+} \rightarrow PuO_2^{+}+VO^{2+}+H_2O} \quad \text{slow} \tag{13.21}$$

The kinetics are equally in accord with the replacement of reaction (13.20) by one between $PuO_2OH^+ + V^{3+}$, and reaction (13.21) by $PuO_2OH^+ + VOH^{2+}$.
At low concentrations ($< 10^{-4}$ M) Pu(VI) is reduced by Fe(II) to Pu(V), *viz.*

$$\mathrm{Pu(VI)+Fe(II) = Pu(V)+Fe(III)} \qquad (13.22)$$

At higher concentrations Pu(V) is reduced further

$$\mathrm{Pu(V)+Fe(II) = Pu(IV)+Fe(III)} \qquad (13.23)$$

The inverse hydrogen-ion dependence of the rate of reaction (13.22) is exceptionally complex and is summarised by

$$-\mathrm{d[Pu(VI)]/d}t = \mathrm{[Pu(VI)][Fe(II)]}\{A+(B+C[\mathrm{H}^+])^{-1}\}$$

from experiments in the range 0.05 to 2.0 M $HClO_4$ at $\mu = 2.0$ M (ref. 42). Rate data were obtained by spectrophotometric measurements of Pu(VI) at 830 mμ between 0 °C and 25 °C. The existence of a binuclear intermediate, of the type Pu(V)·Fe(III), is inferred from the form of the rate law. Thermodynamic quantities of activation are reported for the three activated complexes through which the reaction proceeds. Ionic strength effects were investigated, as also was the influence of chloride ions on the rate. The Pu(V)·Fe(III) dimer is analogous to the complexes formed between Np(V) and Fe(III)[208], and U(V) and Cr(III)[31].
Likewise, chloride ion has a marked accelerating effect on the two-equivalent reduction of Pu(VI) by Sn(II). On the assumption that the rate-determining step is

$$\mathrm{Pu(VI)+Sn(II) \rightarrow Pu(V)+Sn(III)}$$

two mechanisms are possible[266]: (*a*) a second Pu(VI) is reduced by Sn(III); the Pu(V) formed then disproportionates, *viz.*

$$\mathrm{Pu(VI)+Sn(III) \rightarrow Pu(V)+Sn(IV)}$$

$$\mathrm{2\,Pu(V) \rightarrow Pu(IV)+Pu(VI)}$$

(*b*) the reduction of a second Pu(VI) takes place, along with reduction of Pu(V) by Sn(II), *viz.*

$$\mathrm{Pu(VI)+Sn(III) \rightarrow Pu(V)+Sn(IV)}$$

$$\mathrm{Pu(V)+Sn(II) \rightarrow Pu(IV)+Sn(III)}$$

$$\mathrm{Pu(V)+Sn(III) \rightarrow Pu(IV)+Sn(IV)}$$

References pp. 267–273

Neither of these possibilities are likely: the disproportionation of Pu(V), and the Pu(V)+Sn(II) reaction are too slow to account for the rapidity and kinetics of the overall reaction. That the Pu(VI)+Sn(II) reaction is very much faster than the Pu(V)+Sn(II) reduction is taken as evidence for the occurrence of a single two-equivalent process, *viz.*

$$\text{Pu(VI)}+\text{Sn(II)} \rightarrow \text{Pu(IV)}+\text{Sn(IV)}$$

The chloride-ion dependence indicates the importance of two activated complexes, $(PuO_2SnCl_3{}^+)^\ddagger$ and $(PuO_2SnCl_4)^\ddagger$, with $\Delta H^\ddagger$ values of 14.0 and 14.6 kcal.mole^{-1}, and $\Delta S^\ddagger$ values of 4.4±7 and 8.0±5.5 cal.deg^{-1}.mole^{-1}, respectively (as re-calculated by Newton and Baker[250]). In terms of the apparent rate coefficient, k'

$$k' = a[\text{Cl}^-]^3 + b[\text{Cl}^-]^4$$

where a = 433 l^4.mole^{-4}.sec^{-1} and b = 768 l^5.mole^{-5}.sec^{-1} at 2.4 °C. As in the Pu(IV)+Sn(II) system, the rate is insensitive to variations in acidity.

When 2×10^{-3} *M* solutions of Pu(VI) and U(IV) are mixed in 1 *M* $HClO_4$ solution, the main plutonium product is Pu(V) along with smaller amounts of Pu(III)[41]. When 10^{-4} *M* concentrations are used, the stoichiometry does not deviate significantly from

$$2\,\text{Pu(VI)}+\text{U(IV)} = 2\,\text{Pu(V)}+\text{U(VI)}$$

Newton[41] has shown that no complications ensue from the reaction of the intermediate U(V) with oxygen, since the latter has no effect on the rate. A simple second-order rate equation applies, the disappearance of Pu(VI) being followed at 830 mμ, and the probable mechanism is

$$\text{Pu(VI)}+\text{U(IV)} \rightarrow \text{Pu(V)}+\text{U(V)} \quad \text{rate-determining} \tag{13.24}$$

followed by

$$\text{Pu(VI)}+\text{U(V)} \rightarrow \text{Pu(V)}+\text{U(VI)} \quad \text{rapid} \tag{13.25}$$

or

$$2\,\text{U(V)} \rightarrow \text{U(IV)}+\text{U(VI)} \quad \text{rapid} \tag{13.26}$$

To be consistent with the observed first-order dependences on Pu(VI) and U(IV), it is necessary that steps (13.24) and (13.25) do not occur simultaneously. The rate of reaction decreases with increasing hydrogen-ion concentration. It appears that two activated complexes (and a binuclear intermediate) are involved in the reaction, as represented by the net activation processes

$$PuO_2^{2+} + U^{4+} + 2\,H_2O \rightarrow (H_2OUOHPuO_2^{5+})^{\ddagger} + H^+ \quad (13.27)$$

and

$$PuO_2^{2+} + U^{4+} + 2\,H_2O \rightarrow (HOUOHPuO_2^{4+})^{\ddagger} + 2\,H^+ \quad (13.28)$$

The respective $\Delta H^{\ddagger}$ and $\Delta S^{\ddagger}$ values of (13.27) and (13.28) are 17.6 kcal.mole^{-1} and 3.4±1.5 cal.deg^{-1}.mole^{-1}, and 21.4 kcal.mole^{-1} and 18.1±1.0 cal.deg^{-1}.mole^{-1}. Newton[41] discusses the apparent similarities between the U(VI)+U(IV)[267] Np(VI)+Np(IV)[268], Pu(VI)+Pu(IV)[267], and Pu(VI)+U(IV) systems.

The reduction of Pu(IV) by Ti(III) to blue Pu(III)

$$\text{Pu(IV)} + \text{Ti(III)} = \text{Pu(III)} + \text{Ti(IV)}$$

has been investigated by Rabideau and Kline[269], using a spectrophotometric method to follow the disappearance of Pu(IV) at 469 mμ. Rate data can be expressed in terms of the principal species as

$$-\text{d[Pu(IV)]}/\text{d}t = k_2'[Pu^{4+}][Ti^{3+}]/[H^+]$$

At 2.4 °C and $\mu = 2.02\ M$, k_2' is 12.7 sec^{-1}; at 25 °C and $\mu = 1.02\ M$, k_2' is 65.5 sec^{-1}. In terms of the net activation process

$$Pu^{4+} + Ti^{3+} + H_2O \rightarrow (PuTiOH^{6+})^{\ddagger} + H^+$$

$\Delta H^{\ddagger}$ and $\Delta S^{\ddagger}$ are 16.7 kcal.mole^{-1} and 5.9±2 cal.deg^{-1}.mole^{-1}, respectively. Chloride ion has a slight accelerating influence on the reaction.

Pu(IV) oxidises V(III) stoichiometrically in perchloric acid solution, *viz.*

$$Pu^{4+} + V^{3+} + H_2O = Pu^{3+} + 2\,H^+ + VO^{2+}$$

and the rate law is given by [270]

$$\begin{aligned} -\text{d[Pu(IV)]}/\text{d}t &= k'[Pu^{4+}][V^{3+}]/[H^+] + k''[Pu^{4+}][V^{3+}]/[H^+]^2 \\ &= k_{obs}[\text{Pu(IV)}][V^{3+}] \end{aligned}$$

Spectrophotometric values for k' and k'' at 2.4 °C ($\mu = 2\ M$) are 1.70 sec^{-1} and 1.71 mole.l^{-1}.sec^{-1}, respectively. The corresponding activation parameters, $\Delta H^{\ddagger}$ and $\Delta S^{\ddagger}$, are 17.1 kcal.mole^{-1} and 4.8±1.7 cal.deg^{-1}.mole^{-1} for the k' path, and 21.5 kcal.mole^{-1} and 20.8±1.4 cal.deg^{-1}.mole^{-1} for the k'' path. The rate of reaction is not susceptible to the addition of chloride ions. This result is of importance because ClO_4^- is reduced to Cl^- by V(III).

References pp. 267–273

The Pu(IV)+Fe(II) reaction proceeds stoichiometrically in perchloric acid media[271], *viz.*

$$\text{Pu(IV)}+\text{Fe(II)} = \text{Pu(III)}+\text{Fe(III)}$$

The process is first order in each of the reactants and the rate is unaffected by the presence of Fe(III). However, there is some uncertainty about the inclusion of a term in Pu(III) in the rate law. From the observed hydrogen-ion dependence, Newton and Cowan[271] conclude that the principal reaction path has an activated complex formed from Pu^{4+}, Fe^{2+} and water with the prior loss of one hydrogen ion. The probable form of rate law is

$$-\text{d}[\text{Pu(IV)}]/\text{d}t = k_3[\text{Pu}^{4+}][\text{Fe}^{2+}]+k'[\text{Pu}^{4+}][\text{Fe}^{2+}]/[\text{H}^+]$$

where $k_3 = 0.177$ l.mole^{-1}.sec^{-1} and $k' = 3.12$ sec^{-1} at 2.5 °C. The activated complex $(PuOHFe^{5+})^{\ddagger}$ has $\Delta H^{\ddagger} = 19.1$ kcal.mole^{-1}, and $\Delta S^{\ddagger} = 13.3$ cal.deg^{-1}.mole^{-1}. The pronounced increase of rate on the addition of chloride ion is attributed to the provision of a new reaction path involving $(PuClFe^{5+})^{\ddagger}$, for which $\Delta H^{\ddagger} = 14.4$ kcal.mole^{-1} and $\Delta S^{\ddagger} = 0.6$ cal.deg^{-1}.mole^{-1}. Sulphate ion also increases the rate.

Sn(II) reduces Pu(IV) by

$$2\,\text{Pu(IV)}+\text{Sn(II)} = 2\,\text{Pu(III)}+\text{Sn(IV)}$$

In perchloric acid media the reaction is extremely slow and is complicated by the formation of polymeric species of tin, and by heterogeneity. Rabideau[272] has examined the kinetics in mixed perchlorate–chloride solutions, in which media no turbidity is apparent. The rate expression is complex, *viz.*

$$-\text{d}[\text{Sn(II)}]/\text{d}t = c[\text{Pu}^{4+}][\text{Sn}^{2+}][\text{Cl}^-]^4+d[\text{Pu}^{4+}][\text{Sn}^{2+}][\text{Cl}^-]^5$$

but shows a lack of dependence on hydrogen-ion concentration. At 25 °C, c and d are 720 l^5.mole^{-5}.sec^{-1} and 1636 l^6.mole^{-6}.sec^{-1}, respectively. The net activation process

$$\text{Pu}^{4+}+\text{Sn}^{2+}+4\,\text{Cl}^- \rightarrow (\text{PuCl}_4\text{Sn}^{2+})^{\ddagger}$$

has $\Delta H^{\ddagger}$ and $\Delta S^{\ddagger}$ equal to 26.9 kcal.mole^{-1} and 44.7 cal.deg^{-1}.mole^{-1}, whereas the process

$$\text{Pu}^{4+}+\text{Sn}^{2+}+5\,\text{Cl}^- \rightarrow (\text{PuCl}_5\text{Sn}^{+})^{\ddagger}$$

has a $\Delta H^{\ddagger}$ of 24.1 kcal.mole^{-1}, and a $\Delta S^{\ddagger}$ of 37.0 cal.deg^{-1}.mole^{-1}. A possible mechanism is

$$PuCl^{3+} + SnCl_3^- \rightarrow PuCl^{2+} + SnCl_3$$

$$PuCl_2^{2+} + SnCl_3^- \rightarrow PuCl_2^+ + SnCl_3$$

$$PuCl^{3+} + SnCl_3 \rightarrow PuCl^{2+} + SnCl_3^+ \qquad \text{rapid}$$

although there is no direct evidence for the participation of Sn(III).

The reduction of Pu(IV) by U(IV)

$$2\,Pu(IV) + U(IV) = 2\,Pu(III) + U(VI)$$

obeys the rate law[273]

$$-d[Pu(IV)]/dt = k[Pu^{4+}][U^{4+}]/[H^+]^2$$

in accordance with the net activation process

$$Pu^{4+} + U^{4+} + H_2O \rightarrow (PuOU^{6+})^{\ddagger} + 2\,H^+$$

The corresponding activation parameters are $\Delta H^{\ddagger} = 24.3$ kcal.mole^{-1} and $\Delta S^{\ddagger} = 30.1 \pm 1.9$ cal.deg^{-1}.mole^{-1}. Sulphate ion catalyses the reaction.

REFERENCES

1 H. Taube, in *Advances in Inorganic Chemistry and Radiochemistry*, ed. H. J. Emeléus and A. G. Sharpe, Vol. 1, Academic Press, New York, 1959, p. 1.
2 J. Halpern, *Quart. Rev.*, 15 (1961) 207.
3 N. Sutin, *Ann. Rev. Phys. Chem.*, 17 (1966) 119.
4 A. G. Sykes, in *Advances in Inorganic Chemistry and Radiochemistry*, ed. H. J. Emeléus and A. G. Sharpe, Vol. 10, Academic Press, New York, 1967, p. 153.
5 D. R. Stranks, in *Modern Coordination Chemistry*, ed. J. Lewis and R. G. Wilkins, Interscience, 1960, p. 78.
6 R. T. M. Fraser, *Rev. Pure Appl. Chem.*, 11 (1961) 64.
7 H. Strehlow, *Ann. Rev. Phys. Chem.*, 16 (1965) 167.
8 W. L. Reynolds and R. W. Lumry, *Mechanisms of Electron Transfer*, Ronald Press, New York, 1966.
9 F. Basolo and R. G. Pearson, *Mechanism of Inorganic Reactions*, 2nd Ed., Wiley, New York, 1967.
10 J. P. Candlin, K. A. Taylor and D. T. Thompson, *Reactions of Transition Metal Complexes*, Elsevier, Amsterdam, 1968.
11 R. A. Marcus, *Ann. Rev. Phys. Chem.*, 15 (1964) 155.
12 I. Ruff, *Quart. Rev.*, 22 (1968) 199.
13 J. O. Edwards, *Inorganic Reaction Mechanisms*, Benjamin, New York, 1964.
14 A. G. Sykes, *Kinetics of Inorganic Reactions*, Pergamon, London, 1966.
15 D. Benson, *Mechanisms of Inorganic Reactions in Solution*, McGraw-Hill, London, 1968.

16 N. A. Daugherty and T. W. Newton, *J. Phys. Chem.*, 67 (1963) 1090.
17 J. R. Huizenga and L. B. Magnusson, *J. Am. Chem. Soc.*, 73 (1951) 3202.
18 J. C. Hindman, J. C. Sullivan and D. Cohen, *J. Am. Chem. Soc.*, 80 (1958) 1812.
19 S. W. Rabideau, *J. Am. Chem. Soc.*, 79 (1957) 6350.
20 H. Imai, *Bull. Chem. Soc. Japan*, 30 (1957) 873.
21 T. W. Newton and S. W. Rabideau, *J. Phys. Chem.*, 63 (1959) 365.
22 J. Silverman and R. W. Dodson, *J. Phys. Chem.*, 56 (1952) 846.
23 A. Anderson and N. A. Bonner, *J. Am. Chem. Soc.*, 76 (1954) 3826.
24 L. E. Bennett and J. C. Sheppard, *J. Phys. Chem.*, 66 (1962) 1275.
25 S. C. Furman and C. S. Garner, *J. Am. Chem. Soc.*, 74 (1952) 2333.
26 W. C. E. Higginson, D. R. Rosseinsky, J. B. Stead and A. G. Sykes, *Discussions Faraday Soc.*, 29 (1960) 49.
27 M. J. Nicol and D. R. Rosseinsky, *Proc. Chem. Soc.*, (1963) 16.
28 D. R. Rosseinsky and M. J. Nicol, *Electrochim. Acta*, 11 (1966) 1069.
29 D. J. Drye, W. C. E. Higginson and P. Knowles, *J. Chem. Soc.*, (1962) 1137.
30 R. L. Moore, *J. Am. Chem. Soc.*, 77 (1955) 1504.
31 T. W. Newton and F. B. Baker, *Inorg. Chem.*, 1 (1962) 368.
32 J. H. Espenson, *Inorg. Chem.*, 4 (1965) 1533.
33a M. Ardon and R. A. Plane, *J. Am. Chem. Soc.*, 81 (1959) 3197.
33b R. E. Connick and M. G. Thompson, *148th National Meeting of American Chemical Society*, Inorg. Division, Papers 23 and 24, September, 1964.
34 T. W. Newton and F. B. Baker, *Inorg. Chem.*, 3 (1964) 569.
35 M. Eigen and R. G. Wilkins, *Advan. Chem. Ser.*, ed. R. F. Gould, No. 49, American Chemical Society, 1965, p. 55.
36 R. W. Kolaczkowski and R. A. Plane, *Inorg. Chem.*, 3 (1964) 322.
37 J. C. Sullivan, *Inorg. Chem.*, 3 (1964) 315.
38 T. J. Conocchioli, E. J. Hamilton and N. Sutin, *J. Am. Chem. Soc.*, 87 (1965) 926.
39 A. Adin and A. G. Sykes, *Nature*, 209 (1966) 804.
40 J. H. Espenson, *Inorg. Chem.*, 4 (1965) 1025.
41 T. W. Newton, *J. Phys. Chem.*, 62 (1958) 943.
42 T. W. Newton and F. B. Baker, *J. Phys. Chem.*, 67 (1963) 1425.
43 R. A. Marcus, *J. Phys. Chem.*, 67 (1963) 853.
44 G. Dulz and N. Sutin, *Inorg. Chem.*, 2 (1963) 917.
45 A. Haim, *Inorg. Chem.*, 5 (1966) 2081.
46 A. Adin and A. G. Sykes, *J. Chem. Soc. A*, (1968) 351.
47 A. G. Sykes, *Chem. Commun.*, (1965) 442.
48 A. Adin and A. G. Sykes, *J. Chem. Soc. A*, (1966), 1230.
49 D. J. Meier and C. S. Garner, *J. Phys. Chem.*, 56 (1952) 853.
50a J. H. Espenson, *J. Am. Chem. Soc.*, 86 (1964) 1883;
50b J. H. Espenson, *J. Am. Chem. Soc.*, 86 (1964) 5101.
51 J. Y. P. Tong and E. L. King, *J. Am. Chem. Soc.*, 75 (1953) 6180.
52 F. J. C. Rossotti and H. S. Rossotti, *Acta Chem. Scand.*, 9 (1955) 1177; 10 (1956) 957.
53 N. A. Daugherty and T. W. Newton, *J. Phys. Chem.*, 68 (1964) 612.
54 C. Benson, *J. Phys. Chem.*, 7 (1903) 1, 356.
55 R. A. Gortner, *J. Phys. Chem.*, 12 (1908) 632.
56 C. Wagner and W. Preiss, *Z. Anorg. Chem.*, 168 (1928) 265.
57 F. H. Westheimer, *Chem. Rev.*, 45 (1949) 419.
58 J. H. Espenson and E. L. King, *J. Am. Chem. Soc.*, 85 (1963) 3328.
59 E. A. M. Wetton and W. C. E. Higginson, *J. Chem. Soc.*, (1965) 5890.
60 J. G. Mason and A. D. Kowalak, *Inorg. Chem.*, 3 (1964) 1248.
61 J. C. Sullivan, *J. Am. Chem. Soc.*, 87 (1965) 1495.
62 B. R. Baker, M. Orhanovic and N. Sutin, *J. Am. Chem. Soc.*, 89 (1967) 722.
63 A. Haim and N. Sutin, *J. Am. Chem. Soc.*, 87 (1965) 4210.
64 A. Haim and N. Sutin, *J. Am. Chem. Soc.*, 88 (1966) 434.
65 A. Adin and A. G. Sykes, *J. Chem. Soc. A*, (1968) 354.
66 C. Czerlinski and M. Eigen, *Z. Elektrochem.*, 63 (1959) 652; G. G. Hammes and J. I.

STEINFELD, *J. Am. Chem. Soc.*, 84 (1962) 4639.
67 P. DEBYE, *Trans. Electrochem. Soc.*, 82 (1942) 265.
68 E. EICHLER AND A. C. WAHL, *J. Am. Chem. Soc.*, 80 (1958) 4145; M. W. DIETRICH AND A. C. WAHL, *J. Chem. Phys.*, 38 (1963) 1591.
69 M. A. RAWOOF AND J. R. SUTTER, *J. Phys. Chem.*, 71 (1967) 2767.
70 L. J. KIRSCHENBAUM AND J. R. SUTTER, *J. Phys. Chem.*, 70 (1966) 3863.
71 J. P. CANDLIN AND J. HALPERN, *J. Am. Chem. Soc.*, 85 (1963) 2518.
72 P. SAFFIR AND H. TAUBE, *J. Am. Chem. Soc.*, 82 (1960) 13.
73 R. T. M. FRASER AND H. TAUBE, *J. Am. Chem. Soc.*, 82 (1960) 4152.
74 S. M. TAYLOR AND J. HALPERN, *J. Am. Chem. Soc.*, 81 (1959) 2933.
75 K. B. WIBERG AND R. STEWART, *J. Am. Chem. Soc.*, 78 (1956) 1214.
76 D. R. ROSSEINSKY, *J. Chem. Soc.*, (1963) 1181.
77 D. R. ROSSEINSKY AND M. J. NICOL, *J. Chem. Soc. A*, (1968) 1022.
78 C. R. GIULIANO AND H. M. MCCONNELL, *J. Inorg. Nucl. Chem.*, 9 (1959) 171.
79 D. R. ROSSEINSKY AND M. J. NICOL, *Trans. Faraday Soc.*, 64 (1968) 2410.
80 D. R. ROSSEINSKY AND W. C. E. HIGGINSON, *J. Chem. Soc.*, (1960) 31.
81 K. G. ASHURST AND W. C. E. HIGGINSON, *J. Chem. Soc.*, (1956) 343.
82 H. DIEBLER AND N. SUTIN, *J. Phys. Chem.*, 68 (1964) 174.
83 see also, G. DAVIES, L. J. KIRSCHENBAUM AND K. KUSTIN, *Inorg. Chem.*, 7 (1968) 146.
84 M. J. NICOL AND D. R. ROSSEINSKY, *Chem. Ind. London*, (1963) 1166.
85 A. W. ADAMSON, *J. Phys. Colloid Chem.*, 55 (1951) 293.
86 A. M. ARMSTRONG AND J. HALPERN, *Can. J. Chem.*, 35 (1957) 1020.
87 E. K. MAUN AND N. DAVIDSON, *J. Am. Chem. Soc.*, 72 (1950) 2254.
88 D. BANERJEA AND M. S. MOHAN, *J. Indian Chem. Soc.*, 40 (1963) 188.
89 E. G. MOORHEAD AND N. SUTIN, *Inorg. Chem.*, 6 (1967) 428.
90 W. C. E. HIGGINSON AND A. G. SYKES, *J. Chem. Soc.*, (1962) 2841. See also ref. 45.
91 E. L. KING AND C. S. GARNER, *J. Phys. Chem.*, 58 (1954) 29.
92 J. H. ESPENSON, K. SHAW AND O. J. PARKER, *J. Am. Chem. Soc.*, 89 (1967) 5730.
93 J. H. ESPENSON AND R. E. MCCARLEY, *J. Am. Chem. Soc.*, 88 (1966) 1063.
94 H. TAUBE AND H. MYERS, *J. Am. Chem. Soc.*, 76 (1954) 2103.
95 A. ARDON, J. LEVITAN AND H. TAUBE, *J. Am. Chem. Soc.*, 84 (1962) 872.
96 G. DULZ AND N. SUTIN, *J. Am. Chem. Soc.*, 86 (1964) 829.
97 R. M. MILBURN AND W. C. VOSBURGH, *J. Am. Chem. Soc.*, 77 (1955) 1352.
98 From results of R. E. CONNICK AND C. P. COPPEL, *J. Am. Chem. Soc.*, 81 (1959) 6389, and M. J. M. WOODS, P. K. GALLAGHER AND E. L. KING, *Inorg. Chem.*, 1 (1962) 55.
99 D. L. BALL AND E. L. KING, *J. Am. Chem. Soc.*, 80 (1958) 1091.
100 N. SUTIN, J. K. ROWLEY AND R. W. DODSON, *J. Phys. Chem.*, 65 (1961) 1248.
101 A. HAIM AND W. K. WILMARTH, *J. Am. Chem. Soc.*, 83 (1961) 509.
102 A. W. ADAMSON AND E. GONICK, *Inorg. Chem.*, 2 (1963) 129.
103 D. H. HUCHITAL AND R. G. WILKINS, *Inorg. Chem.*, 6 (1967) 1022.
104 F. R. DUKE AND R. C. PINKERTON, *J. Am. Chem. Soc.*, 73 (1951) 3045.
105 N. C. PETERSON AND F. R. DUKE, *J. Phys. Chem.*, 67 (1963) 531.
105a D. W. CARLYLE AND J. H. ESPENSON, *J. Am. Chem. Soc.*, 90 (1968) 2272.
106 R. H. BETTS, *Can. J. Chem.*, 33 (1955) 1780.
107 S. MINC, J. SOBKOWSKI AND M. STOK, *Nukleonika*, 10 (1965) 747.
108 J. F. ENDICOTT AND H. TAUBE, *Inorg. Chem.*, 4 (1965) 437.
109 H. TAUBE, H. MYERS AND R. L. RICH, *J. Am. Chem. Soc.*, 75 (1953) 4118.
110 J. E. EARLEY AND J. H. GORBITZ, *J. Inorg. Nucl. Chem.*, 25 (1963) 306.
111 H. TAUBE, *J. Am. Chem. Soc.*, 77 (1955) 4481.
112 A. ZWICKEL AND H. TAUBE, *J. Am. Chem. Soc.*, 83 (1961) 793.
113 J. P. CANDLIN, J. HALPERN AND D. L. TRIMM, *J. Am. Chem. Soc.*, 86 (1964) 1019.
114 A. ZWICKEL AND H. TAUBE, *Discussions Faraday Soc.*, 29 (1960) 42.
115 A. A. VLČEK, *Advances in the Chemistry of the Coordination Compounds*, ed. S. KIRSCHNER, MacMillan, New York, 1961, p. 590.
116 A. A. VLČEK, *Discussions Faraday Soc.*, 26 (1958) 164.
117 J. P. CANDLIN, J. HALPERN AND S. NAKAMURA, *J. Am. Chem. Soc.*, 85 (1963) 2517.

118 J. HALPERN AND S. NAKAMURA, *J. Am. Chem. Soc.*, 87 (1965) 3002.
119 R. T. M. FRASER AND H. TAUBE, *J. Am. Chem. Soc.*, 83 (1961) 2239.
120 R. T. M. FRASER AND H. TAUBE, *J. Am. Chem. Soc.*, 83 (1961) 2242.
121 A. ZWICKEL AND H. TAUBE, *J. Am. Chem. Soc.*, 81 (1959) 1288.
122 R. T. M. FRASER, *Inorg. Chem.* 2 (1963) 954.
123 H. DIEBLER AND H. TAUBE, *Inorg. Chem.*, 4 (1965) 1029.
124 A. E. OGARD AND H. TAUBE, *J. Am. Chem. Soc.*, 80 (1958) 1084.
125 J. F. ENDICOTT AND H. TAUBE, *J. Am. Chem. Soc.*, 86 (1964) 1686.
126 J. H. ESPENSON, *Inorg. Chem.*, 4 (1965) 121.
127 J. P. CANDLIN AND J. HALPERN, *Inorg. Chem.*, 4 (1965) 766.
128 A. HAIM, *J. Am. Chem. Soc.*, 85 (1963) 1016.
129 D. K. SEBERA AND H. TAUBE, *J. Am. Chem. Soc.*, 83 (1961) 1785.
130 R. T. M. FRASER, *Advances in the Chemistry of Coordination Compounds*, ed. S. KIRSCHNER, MacMillan, New York, 1961, p. 287.
131 J. HALPERN AND J. RABANI, *J. Am. Chem. Soc.*, 88 (1966) 699.
132 D. E. PETERS AND R. T. M. FRASER, *J. Am. Chem. Soc.*, 87 (1965) 2758.
133 R. G. MILLER, D. E. PETERS AND R. T. M. FRASER, *Proc. Symp. Exchange Reactions*, Brookhaven National Laboratory, Intern. Atomic Energy Agency, Vienna, 1965, p. 203.
134 J. H. ESPENSON AND J. P. BIRK, *J. Am. Chem. Soc.*, 87 (1965) 3280.
135 R. SNELLGROVE AND E. L. KING, *J. Am. Chem. Soc.*, 84 (1962) 4610.
136 L. ORGEL, *Rept. 10th Solvay Conference*, Brussels, 1956, p. 289.
137 A. HAIM AND W. K. WILMARTH, *Inorg. Chem.*, 1 (1962) 583.
138 A. HAIM, *J. Am. Chem. Soc.*, 86 (1964) 2352.
139 J. P. CANDLIN AND J. HALPERN, *Inorg. Chem.*, 4 (1965) 1086.
140 A. R. OSBORN AND E. WHALLEY, *Can. J. Chem.*, 39 (1961) 1094.
141 P. BENSON AND A. HAIM, *J. Am. Chem. Soc.*, 87 (1965) 3826.
142 M. GREEN, K. SCHUG AND H. TAUBE, *Inorg. Chem.*, 4 (1965) 1184.
143 M. T. BARNET, B. M. CRAVEN, H. C. FREEMAN, N. E. KIME AND J. A. IBERS, *Chem. Commun.*, (1966) 307.
144 J. M. DE CHANT AND J. B. HUNT, *J. Am. Chem. Soc.*, 89 (1967) 5988.
145 R. D. CANNON AND J. E. EARLEY, *J. Am. Chem. Soc.*, 87 (1965) 5264.
146 W. KRUSE AND H. TAUBE, *J. Am. Chem. Soc.*, 82 (1960) 526.
147 J. H. ESPENSON, *J. Am. Chem. Soc.*, 89 (1967) 1276.
148 J. DOYLE AND A. G. SYKES, *J. Chem. Soc. A*, (1967) 795.
149 P. H. DODEL AND H. TAUBE, *Z. Phys. Chem.*, 44 (1965) 92.
150 A. G. SYKES, *Trans. Faraday Soc.*, 58 (1962) 543.
151 A. G. SYKES, *Trans. Faraday Soc.*, 59 (1963) 1325.
152 A. G. SYKES, *Trans. Faraday Soc.*, 59 (1963) 1334.
153 A. PIDCOCK AND W. C. E. HIGGINSON, *J. Chem. Soc.*, (1963) 2798.
154 R. DYKE AND W. C. E. HIGGINSON, *J. Chem. Soc.*, (1963) 2788.
155 H. TAUBE, in *Mechanisms of Inorganic Reactions, Advan. Chem. Ser.*, ed. R. F. GOULD, No. 49, Am. Chem. Soc., 1965, pp. 107, 117.
156 H. TAUBE, *Can. J. Chem.*, 37 (1959) 129.
157 E. S. GOULD AND H. TAUBE, *J. Am. Chem. Soc.*, 86 (1964) 1318.
158 E. B. FLEISCHER AND R. FROST, *J. Am. Chem. Soc.*, 87 (1965) 3998.
159 R. T. M. FRASER, *Nature*, 202 (1964) 691.
160 R. T. M. FRASER, *Nature*, 205 (1965) 1207.
161 R. D. BUTLER AND H. TAUBE, *J. Am. Chem. Soc.*, 87 (1965) 5597.
162 R. T. M. FRASER, *J. Am. Chem. Soc.*, 85 (1963) 1747.
163 E. S. GOULD, *J. Am. Chem. Soc.*, 87 (1965) 4730.
164 E. S. GOULD, *J. Am. Chem. Soc.*, 88 (1966) 2983.
165 D. H. HUCHITAL AND H. TAUBE, *Inorg. Chem.*, 4 (1965) 1660.
166a R. T. M. FRASER, D. K. SEBERA AND H. TAUBE, *J. Am. Chem. Soc.*, 81 (1959) 2906.
166b J. K. HURST AND H. TAUBE, *J. Am. Chem. Soc.*, 90 (1968) 1178.
167 R. T. M. FRASER AND H. TAUBE, *J. Am. Chem. Soc.*, 81 (1959) 5000.
168 R. T. M. FRASER, *J. Am. Chem. Soc.*, 83 (1961) 564.

169 R. T. M. Fraser, *J. Am. Chem. Soc.*, 84 (1962) 3436.
170 D. H. Huchital and H. Taube, *J. Am. Chem. Soc.*, 87 (1965) 5371.
171 J. Halpern and L. E. Orgel, *Discussions Faraday Soc.*, 29 (1960) 32.
172 R. T. M. Fraser, *J. Am. Chem. Soc.*, 83 (1961) 4920.
173 P. V. Manning, R. C. Jarnagin and M. Silver, *J. Phys. Chem.*, 68 (1964) 265.
174 C. A. Coulson and H. C. Longuet-Higgins, *Proc. Roy. Soc. London, Ser. A*, 191 (1947) 39.
175 W. F. Libby, *J. Chem. Phys.*, 38 (1963) 420.
176 R. T. M. Fraser and H. Taube, *J. Am. Chem. Soc.*, 81 (1959) 5514.
177 E. S. Gould and H. Taube, *J. Am. Chem. Soc.*, 85 (1963) 3706.
178 F. R. Nordmeyer and H. Taube, *J. Am. Chem. Soc.*, 88 (1966) 4295; 90 (1968) 1162.
179 E. S. Gould, *J. Am. Chem. Soc.*, 89 (1967) 5792.
180 R. T. M. Fraser, *Inorg. Chem.*, 3 (1964) 1561.
181 G. Svatos and H. Taube, *J. Am. Chem. Soc.*, 83 (1961) 4172.
182 P. V. Manning and R. C. Jarnagin, *J. Phys. Chem.*, 67 (1963) 2884.
183 H. J. Price and H. Taube, *J. Am. Chem. Soc.*, 89 (1967) 269.
184 P. B. Wood and W. C. E. Higginson, *J. Chem. Soc.*, (1965) 2116.
185 P. B. Wood and W. C. E. Higginson, *Proc. Chem. Soc.*, (1964) 109.
186 J. B. Kirwin, P. J. Proll and L. H. Sutcliffe, *Trans. Faraday Soc.*, 60 (1964) 119.
187 L. H. Sutcliffe and J. R. Weber, *J. Inorg. Nucl. Chem.*, 12 (1960) 281.
188 L. H. Sutcliffe and J. R. Weber, *Trans. Faraday Soc.*, 52 (1956) 1225.
189 N. A. Bonner and J. P. Hunt, *J. Am. Chem. Soc.*, 82 (1960) 3826.
190 T. J. Conocchioli, G. H. Nancollas and N. Sutin, *J. Am. Chem. Soc.*, 86 (1964) 1453.
191 D. H. Huchital, N. Sutin and B. Warnqvist, *Inorg. Chem.*, 6 (1967) 838.
192 A. Haim and N. Sutin, *J. Am. Chem. Soc.*, 88 (1966) 5343.
193 J. Barrett and J. H. Baxendale, *Trans. Faraday Soc.*, 52 (1956) 210.
194 E. G. Moorhead and N. Sutin, *Inorg. Chem.*, 5 (1966) 1866.
195 R. J. Campion, N. Purdie and N. Sutin, *Inorg. Chem.*, 3 (1964) 1091.
196 M. H. Ford-Smith and N. Sutin, *J. Am. Chem.*, 83 (1961) 1830.
197 J. B. Kirwin, F. D. Peat, P. J. Proll and L. H. Sutcliffe, *J. Phys. Chem.*, 67 (1963) 2288.
198 J. B. Kirwin, F. D. Peat, P. J. Proll and L. H. Sutcliffe, *J. Phys. Chem.*, 67 (1963) 1617.
199 K. G. Ashurst and W. C. E. Higginson, *J. Chem. Soc.*, (1953) 3044.
200 J. H. Baxendale and C. F. Wells, *Trans. Faraday Soc.*, 53 (1957) 800.
201 W. C. E. Higginson, R. T. Leigh and R. Nightingale, *J. Chem. Soc.*, (1962) 435.
202 R. C. Pinkerton and F. R. Duke, *J. Am. Chem. Soc.*, 74 (1952) 1535.
203 L. J. Heidt and J. Berestecki, *J. Am. Chem. Soc.*, 77 (1955) 2049.
204 T. J. Conocchioli, G. H. Nancollas and N. Sutin, *Inorg. Chem.*, 5 (1966) 1.
205 L. H. Sutcliffe and J. R. Weber, *Trans. Faraday Soc.*, 55 (1959) 1892.
206 L. H. Sutcliffe and J. R. Weber, *Trans. Faraday Soc.*, 57 (1961) 91.
207 J. C. Sullivan and R. C. Thompson, *Inorg. Chem.*, 6 (1967) 1795.
208 J. C. Sullivan, *J. Am. Chem. Soc.*, 84 (1962) 4256.
209 J. K. Beattie and F. Basolo, *Inorg. Chem.*, 6 (1967) 2069.
210 R. G. Pearson and J. W. Moore, *Inorg. Chem.*, 5 (1966) 1523.
211 J. Doyle and A. G. Sykes, *J. Chem. Soc. A*, (1968) 215.
212 A. W. Adamson, *Discussions Faraday Soc.*, 29 (1960) 125.
213 N. A. Daugherty, *J. Am. Chem. Soc.*, 87 (1965) 5026.
214 A. G. Sykes, *J. Chem. Soc.*, (1961) 5549.
215 C. E. Johnson, *J. Am. Chem. Soc.*, 74 (1952) 959.
216 A. J. Berry, *J. Chem. Soc.*, 121 (1922) 394.
217 O. L. Forchheimer and R. P. Epple, *J. Am. Chem. Soc.*, 74 (1952) 5772.
218 R. J. Prestwood and A. C. Wahl, *J. Am. Chem. Soc.*, 71 (1949) 3137.
219 F. R. Duke and B. Bornong, *J. Phys. Chem.*, 60 (1956) 1015.
220 F. B. Baker, W. D. Brewer and T. W. Newton, *Inorg. Chem.*, 5 (1966) 1294. See also ref. 213.
221 T. W. Newton and F. B. Baker, *J. Phys. Chem.*, 68 (1964) 228.
222 J. H. Swinehart, *Inorg. Chem.*, 4 (1965) 1069.
223 D. H. Irvine, *J. Chem. Soc.*, (1957) 1841.

224 G. HARBOTTLE AND R. W. DODSON, *J. Am. Chem. Soc.*, 73 (1951) 2442.
225 A. M. ARMSTRONG, J. HALPERN AND W. C. E. HIGGINSON, *J. Phys. Chem.*, 60 (1956) 1661.
226 K. W. SYKES, in *Kinetics and Mechanism of Inorganic Reactions in Solution*, Chemical Society, London, 1954, p. 64.
227 S. HIETANEN AND L. G. SILLÉN, *Arkiv Kemi*, 10 (1956) 103.
228 A. C. HARKNESS AND J. HALPERN, *J. Am. Chem. Soc.*, 81 (1959) 3526.
229 See, for example, C. H. BRUBAKER AND J. P. MICKEL, *J. Inorg. Nucl. Chem.*, 4 (1957) 55.
230 J. HALPERN AND J. G. SMITH, *Can. J. Chem.*, 34 (1956) 1419.
231 F. A. JONES AND E. S. AMIS, *J. Inorg. Nucl. Chem.*, 26 (1964) 1045.
232 J. O. WEAR, *J. Chem. Soc.*, (1965) 5596.
233 C. M. LOVE, L. P. QUINN AND C. H. BRUBAKER, *J. Inorg. Nucl. Chem.*, 27 (1965) 2183.
234 D. BENSON, P. J. PROLL, L. H. SUTCLIFFE AND J. WALKLEY, *Discussions Faraday Soc.*, 29 (1960) 60.
235 W. P. TAPPMEYER AND A. W. DAVIDSON, *Inorg. Chem.*, 2 (1963) 823.
236 D. BENSON AND L. H. SUTCLIFFE, *Trans. Faraday Soc.*, 56 (1960) 246.
237 See also, D. BENSON, L. H. SUTCLIFFE AND J. WALKLEY, *J. Am. Chem. Soc.*, 81 (1959) 4488.
238 D. BENSON AND L. H. SUTCLIFFE, *Trans. Faraday Soc.*, 55 (1959) 2107.
239 J. Y. P. TONG AND E. L. KING, *J. Am. Chem. Soc.*, 82 (1960) 3805.
240 See, for example, T. J. HARDWICKE AND E. ROBERTSON, *Can. J. Chem.*, 29 (1951) 828.
241a M. J. ASPRAY, D. R. ROSSEINSKY AND G. B. SHAW, *Chem. Ind. London*, (1963) 911.
241b G. A. RECHNITZ, G. N. RAO AND G. P. RAO, *Anal. Chem.*, 38 (1966) 1900.
242 M. G. ADAMSON, F. S. DAINTON AND P. GLENTWORTH, *Trans Faraday Soc.*, 61 (1965) 689.
243 R. A. MARCUS, *Discussions Faraday Soc.*, 29 (1960) 21.
244 R. A. MARCUS, *Can. J. Chem.*, 37 (1959) 155.
245 J. D. MILLER AND R. H. PRINCE, *J. Chem. Soc.*, (1965) 5749; *J. Chem. Soc. A*, (1966) 1370.
246 W. H. MCCURDY AND G. G. GUILBAULT, *J. Phys. Chem.*, 64 (1960) 1825.
247 G. G. GUILBAULT AND W. H. MCCURDY, *J. Phys. Chem.*, 70 (1966) 656.
248 M. K. DORFMAN AND J. W. GRYDER, *Inorg. Chem.*, 1 (1962) 799; J. W. GRYDER AND M. K. DORFMAN, *J. Am. Chem. Soc.*, 83 (1961) 1254.
249 C. H. BRUBAKER AND A. J. COURT, *J. Am. Chem. Soc.*, 78 (1956) 5530.
250 T. W. NEWTON AND F. B. BAKER, in *Advan. Chem. Ser.*, No. 71, Am. Chem. Soc., 1967, p. 268.
251 T. W. NEWTON AND F. B. BAKER, *J. Phys. Chem.*, 69 (1965) 176.
252 T. W. NEWTON AND F. B. BAKER, *J. Phys. Chem.*, 70 (1966) 1943.
253 J. C. SHEPPARD, *J. Phys. Chem.*, 68 (1964) 1190.
254 S. W. RABIDEAU, *J. Phys. Chem.*, 62 (1958) 414.
255 G. GORDON, *Inorg. Chem.*, 2 (1963) 1277.
256 C. F. BAES, *J. Phys. Chem.*, 60 (1956) 805.
257 J. C. SULLIVAN, A. J. ZIELEN AND J. C. HINDMAN, *J. Am. Chem. Soc.*, 82 (1960) 5288.
258 J. C. SULLIVAN, J. C. HINDMAN AND A. J. ZIELEN, *J. Am. Chem. Soc.*, 83 (1961) 3373.
259 E. H. APPELMAN AND J. C. SULLIVAN, *J. Phys. Chem.*, 66 (1962) 442.
260 R. K. MURMANN AND J. C. SULLIVAN, *Inorg. Chem.*, 6 (1967) 892.
261 R. C. THOMPSON AND J. C. SULLIVAN, *J. Am. Chem. Soc.*, 89 (1967) 1098.
262 N. K. SHASTRI, E. S. AMIS AND J. O. WEAR, *J. Inorg. Nucl. Chem.*, 27 (1965) 2413.
263 R. C. THOMPSON AND J. C. SULLIVAN, *J. Am. Chem. Soc.*, 89 (1967) 1096.
264 S. W. RABIDEAU AND R. J. KLINE, *J. Phys. Chem.*, 63 (1959) 1502.
265 F. R. DUKE AND P. R. QUINNEY, *J. Am. Chem. Soc.*, 76 (1954) 3800.
266 S. W. RABIDEAU AND B. J. MASTERS, *J. Phys. Chem.*, 65 (1961) 1256.
267 See for example, S. W. RABIDEAU, *J. Am. Chem. Soc.*, 79 (1957) 6350.
268 See for example, J. C. HINDMAN, J. C. SULLIVAN AND D. COHEN, *J. Am. Chem. Soc.*, 76 (1954) 3278.
269 S. W. RABIDEAU AND R. J. KLINE, *J. Phys. Chem.*, 64 (1960) 193.
270 S. W. RABIDEAU AND R. J. KLINE, *J. Inorg. Nucl. Chem.*, 14 (1960) 91.
271 T. W. NEWTON AND H. D. COWAN, *J. Phys. Chem.*, 64 (1960) 244.
272 S. W. RABIDEAU, *J. Phys. Chem.*, 64 (1960) 1491.
273 T. W. NEWTON, *J. Phys. Chem.*, 63 (1959) 1493.

SUPPLEMENTARY REFERENCES

274 N. Sutin, *Accounts Chem. Res.*, 1 (1968) 225.
275 N. A. Daugherty and B. Schiefelbein, *J. Am. Chem. Soc.*, 91 (1969) 4328.
276 J. H. Espenson and R. J. Christensen, *J. Am. Chem. Soc.*, 91 (1969) 7311.
277 J. H. Espenson and O. J. Parker, *J. Am. Chem. Soc.*, 90 (1968) 3689.
278 J. H. Espenson and S. R. Helzer, *Inorg. Chem.*, 8 (1969) 1051.
279 D. R. Rosseinsky and M. J. Nicol, *J. Chem. Soc.* A, (1969) 2887.
280 J. P. Birk, *J. Am. Chem. Soc.*, 91 (1969) 3189.
281 A. A. Bergh and G. P. Haight, *Inorg. Chem.*, 8 (1969) 189.
282 M. Orhanović, H. N. Po and N. Sutin, *J. Am. Chem. Soc.*, 90 (1968) 7224.
283 O. J. Parker and J. H. Espenson, *J. Am. Chem. Soc.*, 91 (1969) 1313.
284 K. Shaw and J. H. Espenson, *J. Am. Chem. Soc.*, 90 (1968) 6622.
285 O. J. Parker and J. H. Espenson, *Inorg. Chem.*, 8 (1969) 1523.
286 J. H. Espenson, *Inorg. Chem.*, 7 (1968) 631.
287 J. H. Espenson and D. J. Boone, *Inorg. Chem.*, 7 (1968) 636.
288 D. W. Carlyle and J. H. Espenson, *Inorg. Chem.*, 8 (1969) 575.
289 T. W. Newton, G. E. McCrary and W. G. Clark, *J. Phys. Chem.*, 72 (1968) 4333.
290 D. W. Carlyle and J. H. Espenson, *J. Am. Chem. Soc.*, 91 (1969) 599.
291 J. A. Stritar and H. Taube, *Inorg. Chem.*, 8 (1969) 2281.
292 D. Seewald, N. Sutin and K. D. Watkins, *J. Am. Chem. Soc.*, 91 (1969) 7307.
293 W. G. Movius and R. G. Linck, *J. Am. Chem. Soc.*, 91 (1969) 5394.
294 D. E. Pennington and A. Haim, *Inorg. Chem.*, 7 (1968) 1659.
295 K. M. Davies and J. H. Espenson, *Chem. Commun.*, (1969) 111; *J. Am. Chem. Soc.*, 91 (1969) 3093.
296 D. J. Parker and J. H. Espenson, *J. Am. Chem. Soc.*, 91 (1969) 1968.
297 J. P. Birk and J. H. Espenson, *J. Am. Chem. Soc.*, 90 (1968) 1153.
298 T. J. Meyer and H. Taube, *Inorg. Chem.*, 7 (1968) 2369.
299 R. C. Patel and J. F. Endicott, *J. Am. Chem. Soc.*, 90 (1968) 6364.
300 R. G. Linck, *Inorg. Chem.*, 7 (1968) 2394.
301 J. Doyle and A. G. Sykes, *J. Chem. Soc.* A, (1968) 2836.
302 R. Davies and A. G. Sykes, *J. Chem. Soc.* A, (1968) 2831.
303 A. B. Hoffman and H. Taube, *Inorg. Chem.*, 7 (1968) 1971.
304 H. Taube and E. S. Gould, *Accounts Chem. Res.*, 2 (1969) 321.
305 R. G. Wilkins and R. E. Yelin, *Inorg. Chem.*, 7 (1968) 2667.
306 J. Halpern and M. Pribanić, *J. Am. Chem. Soc.*, 90 (1968) 5942.
307 K. Shaw and J. H. Espenson, *Inorg. Chem.*, 7 (1968) 1619.
308 O. J. Parker and J. H. Espenson, *Inorg. Chem.*, 8 (1969) 185.
309 M. W. Hsu, H. G. Kruszyna and R. M. Milburn, *Inorg. Chem.*, 8 (1969) 2201.
310 G. H. Schenk and W. E. Bazzelle, *Anal. Chem.*, 40 (1968) 162.
311 M. J. Burkhart and T. W. Newton, *J. Phys. Chem.*, 73 (1969) 1741.

Chapter 4

Oxidation–Reduction Reactions between Covalent Compounds and Metal Ions

T. J. KEMP

1. Introduction

1.1 SCOPE AND PATTERN OF THIS CHAPTER

The vast number of thermodynamically possible reactions obtained by permuting oxidants and reductants within the scope of this review present major problems of classification and selection. To only a limited extent is the modernity or detail of a paper indicative of its relevance, some of the definitive papers having been published before 1950. Discussion has been concentrated, therefore, at points where a kinetic investigation of a reaction has resulted in a real advance in our understanding both of its mechanism and of those of related reactions, and work which has been more of a confirmatory nature will not receive comparable consideration. Detailed reference to products, spectra, *etc.* will be made only when the kinetics produce real ambiguities.

Presentation of existing data can be made in several ways. Wiberg[1] and Stewart[2] have taken a series of oxidising metal ions singly or in very small groups and have examined the reactions of each member with a range of reductants. An alternative approach, adopted by Waters[3], is to compare the reactions of a given reductant, *e.g.* iodide ion, hydrogen peroxide, aliphatic hydrocarbon, ketone, alcohol, olefin and acetylene, with a range of oxidising metal ions, noting the roles of complex-formation, hydrolysis, inner and outer-sphere processes and other mechanistic features.

Classification exclusively in terms of a few basic mechanisms is the ideal approach, but in a comprehensive review of this kind, one is presented with all reactions, and not merely the well-documented (and well-behaved) ones which are readily denoted as inner- or outer-sphere electron transfer, hydrogen atom transfer from coordinated solvent, ligand transfer, concerted electron transfer, *etc.* Such an approach has been made on a more limited scale[4]. Turney[5] has considered reactions in terms of the charges and complexing of oxidant and reductant but this approach leaves a large number to be coped with under further categories.

As regards oxidation by metal ions, we have chosen to select groups of from two to eleven metal ions, the members of which are known to display similar mechanisms in their reaction with several substrates. This basic similarity of mechanism implies certain properties common to a group, such as equal increments between stable valency states, comparable redox potentials, ability to form π-complexes, *etc.* This means that different oxidation states of a given transition metal may be represented in different groups. These groups consist of

Cr(VI) and Mn(VII)

Pb(IV), Tl(III), Hg(II), Hg(I), Bi(V), Au(III), Pt(IV), Pd(II), Rh(III), Ru(III) and Mo(VI)

Ag(II), Ag(III), Co(III), Ce(IV), Mn(III), V(V), Ir(IV), Np(VI) and Pu(VI)

Fe(III) (including ferricyanide), Ag(I), Cu(II), Cu(I), Np(V) and Mo(V)

Reductions by metal ions are covered in Section 6 in terms of (*i*) electron-acceptance and (*ii*) electron-acceptance concerted with homolytic fission. One group of reactions, which includes oxidations and reductions by metal ions, is that between a metal ion and a neutral free radical. These form a self-contained class which is treated separately in Section 7.

This procedure emphasises the overall division of oxidants into "one-equivalent" and "two-equivalent" types defined[6-8] in terms of how many equivalents of reducing species, *e.g.* electrons or hydrogen atoms, are taken up by the oxidant in the primary act, as deduced from the behaviour of the oxidant towards either hydrazine[6], sulphite ion[8] or captive ligand[9,10] according to the schemes (*a*)–(*c*) respectively.

One-equivalent oxidant

(*a*)

$$M^{n+} + N_2H_4 \rightarrow M^{(n-1)+} + N_2H_3\cdot + H^+ \quad (1)$$

$$2\,N_2H_3\cdot \rightarrow N_4H_6 \quad (2)$$

$$N_4H_6 \rightarrow N_2 + 2\,NH_3 \quad (3)$$

(*b*)

$$M^{n+} + SO_3^{2-} \rightarrow M^{(n-1)+} + \cdot SO_3^- \quad (4)$$

$$2\,\cdot SO_3^- \rightarrow S_2O_6^{2-} \text{ (dithionate)} \quad (5)$$

(*c*)

$$M^{n+} + [(NH_3)_5Co(III)\text{–}OCOCO_2H]^{2+} \rightarrow M^{(n-1)+} + [(NH_3)_5Co(III)\text{–}OCOCO_2\cdot]^{2+} + H^+$$

$$[(NH_3)_5Co(III)\text{–}OCOCO_2\cdot]^{2+} \rightarrow [(NH_3)_5Co(II)OH_2]^{2+} + 2\,CO_2 \text{ (very fast)}$$

References pp. 493–509

Two-equivalent oxidant

(*a*)

$$M^{n+} + N_2H_4 \rightarrow N_2H_2 + M^{(n-2)+} + 2\,H^+$$

$$2\,N_2H_2 \rightarrow N_4H_4$$

$$N_4H_4 \begin{cases} \nearrow N_2 + N_2H_4 \\ \searrow NH_3 + HN_3 \end{cases}$$

(*b*)

$$M^{n+} + SO_3^{2-} \rightarrow M^{(n-2)+} + SO_3$$

$$SO_3 + H_2O \rightarrow SO_4^{2-} + 2\,H^+$$

(*c*)

$$M^{n+} + [(NH_3)_5Co(III){-}OCOCO_2H]^{2+} \rightarrow M^{(n-2)+} + [(NH_3)_5Co(III)OH_2]^{3+} + 2\,CO_2 + H^+$$

It should be noted that although an oxidant may be categorised as two-equivalent from these reactions, it may, on occasion, function as a one-equivalent reagent. Three-equivalent changes are extremely rare.

The sub-classification of the oxidising metal ions derives from overall reactivity, which is only crudely related to redox potential and is gauged largely with hindsight.

The coverage of reducing substrates is in terms of increasing molecular complexity. The inorganic substrates are dealt with in terms of periodic group although azide and cyanide appear with halide, and sulphurous, phosphorous and arsenious acids are taken together, as are also the gases carbon monoxide and hydrogen. The organic compounds are considered in the order aliphatic and aromatic hydrocarbons, alcohols, aldehydes, carboxylic acids, ethers and amines followed by polyfunctional compounds (unless one of the functional groups is not in any way involved in the reaction).

The cases of oxidation and reduction of "bound" ligands are included in the section appropriate to the ligand involved.

1.2 CATEGORISATION OF OXIDANTS AS ONE- OR TWO-EQUIVALENT

While it is very convenient to consider separately (*i*) those oxidants which undergo changes of oxidation state by only one unit altogether or by a series of changes of which the first involves a single unit and (*ii*) those which undergo changes of oxidation state by two units, this scheme does depend on an initial correct classification. This is obvious in certain instances, thus cerium exists only in two oxidation states, +4 and +3, and can participate only in one-equivalent changes. In many cases, however, the stages of reduction are far from obvious and recourse to the modes of oxidation of sulphite, hydrazine and captive ligand must be made (*vide supra*).

The use of sulphite has, however, been called into question recently. According to the scheme all one-equivalent oxidants should produce at least some dithionate (eqns. 4 and 5). However, Veprek-Šiška *et al.*[12–14] have found that while substitution-labile one-equivalent oxidising metal complexes give sulphate and dithionate, substitution-inert one-equivalent oxidising complexes give sulphate only.

The following mechanisms for generation of both products were suggested

$$M^{n+} + SO_3^{2-} \rightarrow (MSO_3)^{n+IV} \quad (6)$$

$$M^{n+} + (MSO_3)^{n+IV} \rightarrow M^{(n-1)+} + (MSO_3)^{n+V} \quad (7)$$

$$(MSO_3)^{n+V} \rightarrow M^{(n-1)+} + SO_4^{2-} \quad (8)$$

$$2\,(MSO_3)^{n+IV} \rightarrow 2\,M^{(n-1)+} + S_2O_6^{2-} \quad (9)$$

The species $(MSO_3)^{n+IV}$ represents an inner-sphere complex between sulphite and the oxidant formed by ligand-displacement. $(MSO_3)^{n+V}$ is formed by abstracting an electron from $(MSO_3)^{n+IV}$.

Reactions (6), (7) and (9) occur with substitution-labile oxidants and (6), (7) and (8) with substitution-inert oxidants. It is postulated that the species $(MSO_3)^{n+IV}$ involves metal–oxygen coordination for labile M^{n+}, and metal–sulphur coordination for inert M^{n+}. The latter mode is thought to prevent sulphur–sulphur bond formation.

Brown and Higginson[15] have proposed an alternative interpretation of the data. The stationary-state concentration of $(MSO_3)^{n+IV}$ is likely to be very low for substitution into an inert oxidising complex and hence step (9), which depends on the *square* of this concentration, is rather unlikely, particularly as it involves rupture of two inert metal–sulphur bonds. They have also discussed the parallel case of hydrazine, for which a similar distinction between substitution-labile and -inert one-equivalent oxidising complexes is apparent. Substitution-inert ferricyanide ion[16] produces only nitrogen, whilst substitution-labile Mn(III) pyrophosphate[17] gives some nitrogen and ammonia *via* N_4H_6 (the analogue of $S_2O_6^{2-}$) and some nitrogen *via* N_2H_2 (the analogue of SO_4^{2-}). (The distinction between the separate origins of the nitrogen is based on isotopic studies[18]). Coordination isomerism of the type proposed for sulphite is not feasible for N_2H_4 and a more general explanation of the dichotomy is preferable. Brown and Higginson[15] suggest that whereas substitution-labile oxidants may release the sulphite radical-ion, which then dimerises or is oxidised to sulphate, substitution-inert oxidants retain the radical-anion until a second oxidation (7) occurs.

2. Oxidation by Cr(VI) and Mn(VII)

2.1 GENERAL FEATURES

Reactions of these oxidants share a variety of features. Both function as either one- or two-equivalent reagents, depending on the substrate involved. Complete reaction involves several stages of reduction of the metal ions and considerable evidence for the transient existence of unusual valency states has been obtained. In the form of chromic acid and acidic, neutral or alkaline permanganate, Cr(VI) and Mn(VII) have been very widely used in synthetic organic chemistry and in analytical chemistry and a correspondingly large body of non-kinetic information relevant to the mechanisms has accumulated. It is probable that more kinetic studies have been performed using Cr(VI) and Mn(VII) than any other metal-ion oxidant. One contrasting feature is that the oxidising power of Cr(VI) is normally confined to acidic solutions.

Some relevant oxidation potentials are[19]

$$Cr_2O_7^{2-} + 14\,H^+ + 6\,e^- \rightleftharpoons 2\,Cr^{3+} + 7\,H_2O \qquad +1.33\ V$$

$$MnO_4^- + 8\,H^+ + 5\,e^- \rightleftharpoons Mn^{2+} + 4\,H_2O \qquad +1.51\ V$$

$$MnO_4^- + e^- \rightleftharpoons MnO_4^{2-} \qquad +0.564\ V$$

$$MnO_4^- + 4\,H^+ + 3\,e^- \rightleftharpoons MnO_2 + 2\,H_2O \qquad +1.695\ V$$

2.1.1 Oxidation states involved in reduction of Cr(VI) and Mn(VII)

Cr(VI) is normally reduced to Cr(III) since Cr(V) and Cr(IV) are very unstable under ordinary reaction conditions. Westheimer[20] has discussed critically the roles of Cr(V) and Cr(IV) as reactive intermediates, (*vide infra*), and Wiberg[21] has summarised the inorganic chemistry of compounds containing Cr(V) and Cr(IV).

The situation with Mn(VII) is more complicated. The final stage of reduction depends both on the pH of the medium and upon the presence of some ion capable of stabilising an intermediate valency state by forming an insoluble precipitate or a complex ion. Oxidations by permanganate in alkaline solution normally produce a precipitate of MnO_2, but addition of barium ions[22] results in the precipitation of barium manganate, $BaMnO_4$, containing Mn(VI). This does not imply a one-equivalent reduction of Mn(VII) however, for the reaction

$$Mn(V) + Mn(VII) \rightarrow 2\,Mn(VI)$$

is rapid[23]. Mn(V) exists only in strong alkali and its role in the reduction of Mn(VII) is essentially that of a reactive intermediate.

In acidic solution MnO_2 is usually the end product, although particularly vigorous reductants, *e.g.* iodide and oxalate ions, convert permanganate to manganous ions. Mn(III) is stable only in acidic solution or in the form of a complex, *e.g.* with pyrophosphate ion, and it has seldom been reported as the end product of a permanganate oxidation, *e.g.* for that of Mn(II) in a phosphate buffer and for those of alcohols and ethers in the presence of fluoride ion[24].

2.1.2 *Solution equilibria of oxy-anions of Cr(VI) and Mn(VII)*

Equilibria relevant only to kinetic studies of oxidations will be mentioned. For Cr(VI) in aqueous solution these are at 25 °C[25,26]

$$H_2CrO_4 \rightleftharpoons H^+ + HCrO_4^- \qquad K_1 = 4.1 \text{ mole.l}^{-1}$$

$$HCrO_4^- \rightleftharpoons H^+ + CrO_4^{2-} \qquad K_2 = 1.3 \times 10^{-6} \text{ mole.l}^{-1}$$

$$2\,HCrO_4^- \rightleftharpoons Cr_2O_7{}^{2-} + H_2O \qquad K_3 = 155$$

Addition of ions such as chloride[27], sulphate and phosphate (denoted X^-)[28,29] produces a new equilibrium

$$H^+ + HO\text{-}\overset{\overset{\displaystyle O}{\|}}{\underset{\underset{\displaystyle O}{\|}}{Cr}}\text{-}O^- + X^- \rightleftharpoons X\text{-}\overset{\overset{\displaystyle O}{\|}}{\underset{\underset{\displaystyle O}{\|}}{Cr}}\text{-}O^- + H_2O$$

Addition of acetic acid[27] or trifluoroacetic acid[30] also influences the acid chromate ion (*vide infra*).

It transpires[20] that $Cr_2O_7{}^{2-}$ plays a minor part in oxidation mechanisms and that H_2CrO_4 is the reactive form of Cr(VI) and accordingly most reactions are acid-catalysed.

Permanganic acid has a pK of -2.25 in perchloric acid[31] and one of -4.6 in sulphuric acid[32] when the Hammett acidity function, H_0, is used. Accordingly $HMnO_4$ is present to a significant extent only in strongly acidic solutions and comparatively few of the reactions which have been examined involve anything other than MnO_4^-.

2.2 OXIDATION OF INORGANIC COVALENT SPECIES

2.2.1 *Halide ions*

The oxidation of iodide ion by aqueous chromic acid at low acidity is very slow but is subject to marked enhancement in rate on addition of ferrous ion[33].

Iodide ion is very slowly oxidised by Fe(III) and the iodide must therefore be oxidised by some species other than Cr(VI) or Fe(III). For a large excess of iodide, the stoichiometry has been reported to be[34]

$$\mathrm{Cr(VI)+2\,I^- + Fe(II) = Cr(III) + I_2 + Fe(III)}$$

from which it is clear that the Fe(II) is not acting as a true catalyst, but that it is probably involved in a scheme[20]

$$\mathrm{Cr(VI)+Fe(II) \rightarrow Cr(V)+Fe(III)} \qquad (10)$$

$$\mathrm{Cr(V)+I^- \rightarrow Cr(III)+IO^-} \qquad (11)$$

$$\mathrm{IO^- + I^- + 2\,H^+ \rightarrow I_2 + H_2O} \qquad (12)$$

This is an example of an *induced* reaction, which is the subject of the next chapter of this book. That step (11) is involved rather than

$$\mathrm{Cr(V)+I^- \rightarrow Cr(IV)+I\cdot}$$

is argued by Westheimer[20] from the relative effects of arsenious acid upon the chromic acid oxidations of manganous and iodide ions. The latter has an induction factor, defined as the number of equivalents of substrate oxidised to the number of equivalents of inductor oxidised, of 2 but the former has one of 0.5. The reactions involved are:

Manganous ion

$$\mathrm{H_3AsO_3 + Cr(VI) \rightarrow H_3AsO_4 + Cr(IV)}$$

$$\mathrm{Cr(IV) + Mn(II) \rightarrow Cr(III) + Mn(III)}$$

$$\overline{\mathrm{H_3AsO_3 + Cr(VI) + Mn(II) = H_3AsO_4 + Cr(III) + Mn(III)}}$$

Iodide ion

$$\mathrm{H_3AsO_3 + Cr(VI) \rightarrow H_3AsO_4 + Cr(IV)}$$

$$\mathrm{Cr(IV) + Cr(VI) \rightarrow 2\,Cr(V)}$$

$$\mathrm{2\,[Cr(V) + I^- \rightarrow Cr(III) + IO^-]}$$

$$\mathrm{2\,[IO^- + I^- + 2\,H^+ \rightarrow I_2 + H_2O]}$$

$$\overline{\mathrm{H_3AsO_3 + 2\,Cr(VI) + 4\,I^- = H_3AsO_4 + 2\,Cr(III) + 2\,I_2}}$$

It is assumed that Cr(IV) reacts quickly with Mn(II) but slowly with iodide ion for otherwise the induction factors would not be as found.

Westheimer[20] has reviewed other inductions of the chromic acid oxidation of iodide, indicating how these reactions afford insight into the mechanism of the simple oxidation.

An early study[35] of the kinetics of the simple iodide–chromic acid reaction yielded the rate law

$$-\mathrm{d}[\mathrm{Cr(VI)}]/\mathrm{d}t = k[\mathrm{Cr(VI)}]\{[\mathrm{H}^+][\mathrm{I}^-]+k'[\mathrm{H}^+]^2[\mathrm{I}^-]^2\}$$

A more recent examination[36] confirmed the existence of an unusual ionic-strength dependence of the reaction rate[35], which features a minimum at $\mu \sim 0.83$, but it was noted that the initial kinetics differ from those occurring later in the reaction. Consideration of the equilibria prevailing in aqueous solutions of Cr(VI) produced a simplified rate law

$$-\mathrm{d}[\mathrm{I}^-]/\mathrm{d}t = k[\mathrm{HCrO_4^-}][\mathrm{I}^-]^2[\mathrm{H_3O^+}]^2$$

where k at 25 °C ($\mu = 0.1$) is $(1.25 \pm 0.08) \times 10^3$ $\mathrm{l^4.mole^{-4}.sec^{-1}}$ and E is 5.6 $\mathrm{kcal.mole^{-1}}$. The autoretardation is connected with the production of molecular iodine since addition of I_2 reduces the initial reaction rate. An ion-pair or a complex between Cr(VI) and I^- is viewed as the active oxidant giving Cr(IV) and iodine which can, however, back-react, *viz.*

$$\mathrm{Complex} + \mathrm{I}^- \underset{k_{-13}}{\overset{k_{13}}{\rightleftharpoons}} \mathrm{I_2} + \mathrm{Cr(IV)} \tag{13}$$

$$\mathrm{Cr(IV)} + \mathrm{Cr(VI)} \underset{k_{-14}}{\overset{k_{14}}{\rightleftharpoons}} 2\,\mathrm{Cr(V)} \tag{14}$$

$$\mathrm{Cr(V)} + \mathrm{I}^- \rightarrow \mathrm{IO}^- + \mathrm{Cr(III)}$$

$$2\,\mathrm{H}^+ + \mathrm{IO}^- + \mathrm{I}^- \rightarrow \mathrm{H_2O} + \mathrm{I_2}$$

Further analysis of the kinetics over the course of reaction afforded values for k_{14}/k_{-13}, *e.g.* 0.10 at 20 °C ($\mu = 0.1$).

Measurements over a wider range of reactant concentrations[36a] favour a more complex rate law

$$-\mathrm{d}[\mathrm{HCrO_4^-}]/\mathrm{d}t = [\mathrm{HCrO_4}^-]\{k_1[\mathrm{H_3O^+}][\mathrm{I}^-]+k_2[\mathrm{H_3O^+}]^2[\mathrm{I}^-]^2 + k_3[\mathrm{H_3O^+}]^3[\mathrm{I}^-]\}$$

where, at 20.3 °C ($\mu = 0.130$ M $\mathrm{NaClO_4}$), $k_1 = 0.206 \pm 0.009$ $\mathrm{l^2.mole^{-2}.sec^{-1}}$, $k_2 = 111 \pm 7$ $\mathrm{l^4.mole^{-4}.sec^{-1}}$, $k_3 = 154 \pm 3$ $\mathrm{l^4.mole^{-4}.sec^{-1}}$, $E_2 = 10.3 \pm 1.5$ $\mathrm{kcal.mole^{-1}}$, $\Delta S_2^\ddagger = -16 \pm 6$ eu, $E_3 = 6.9 \pm 1.0$ $\mathrm{kcal.mole^{-1}}$ and $\Delta S_3^\ddagger = -27 \pm 3$ eu. The second order term in $[\mathrm{I}^-]$ is considered to correspond to attack of $\mathrm{ICrO_3}^-$ upon I^-.

The permanganate oxidation of iodide has been the subject of a recent detailed study by a rapid mixing technique[37]. At 35 °C and at an ionic strength of 0.9 M over the pH range 3–6 the rate expression is

$$-\mathrm{d}[\mathrm{MnO_4^-}]/\mathrm{d}t = [\mathrm{MnO_4^-}][\mathrm{I^-}]\{k_2+k_3\, a_{H}+\}$$

In the region of pH 5, $k_2 = 51.2$ l.mole^{-1}.sec^{-1} and $k_3 = 1.70\times10^7$ l.^{2}mole^{-2}.sec^{-1}. Activation parameters for the second-order path are $E = 1.90$ kcal.mole^{-1} and $\Delta S^{\ddagger} = -45.8$ eu while those for the third order path are 4.37 kcal.mole^{-1} and -14.4 eu, respectively. The authors accommodate both the low energies of activation and the observed orders in the mechanisms:

pH-independent path

$$\mathrm{MnO_4^-} + \mathrm{I^-} \rightleftharpoons (\mathrm{O_3MnOI})^{2-} \qquad \text{(rapid)}$$

$$(\mathrm{O_3MnOI})^{2-} + \mathrm{H_2O} \rightarrow \mathrm{HOI} + \mathrm{HMnO_4^{2-}} \qquad \text{(slow)}$$

pH-dependent path

$$(\mathrm{O_3MnOI})^{2-} + \mathrm{H_3O^+} \rightarrow \mathrm{HOI} + \mathrm{H_2MnO_4^-} \qquad \text{(slow)}$$

The HOI would be rapidly reduced by iodide and the Mn(V) species would be expected either to disproportionate or to oxidise further iodide. This reaction scheme has features in common with the analogous reaction with cyanide ion discussed below.

The oxidation of bromide ion by Cr(VI) has been examined at length by Bobtelsky *et al.*[38–40]. The rate equation is

$$\mathrm{d}[\mathrm{Br_2}]/\mathrm{d}t = k[\mathrm{Cr(VI)}][\mathrm{Br^-}]f[\mathrm{H_2SO_4}]$$

The apparent first-order rate coefficient obtained using excess oxidant increased exponentially with increase in acidity in the range $5\ N < [\mathrm{H_3O^+}] < 12\ N$. The reaction is first-order with respect to added manganous ions (k increasing sharply), but the activation energy (11.0 kcal.mole^{-1}) remains unchanged. At appreciable catalyst concentrations the reaction becomes almost zero-order with respect to bromide ion. The mechanism appears to be a slow oxidation of Mn(II) to Mn(III) followed by a rapid reduction of the latter by bromide. This reaction is considered further in the section on Mn(II)-catalysis of chromic acid oxidations (p. 327).

2.2.2 *Cyanide ion*

Only the permanganate oxidation of this ion has been studied kinetically. At pH 12–14.6 the reaction has the stoichiometry

$$2\,MnO_4^- + CN^- + 2\,OH^- = 2\,MnO_4^{2-} + OCN^- + H_2O$$

but at pH 6–12, the reaction is complex and non-stoichiometric, yielding cyanate, cyanide and carbon dioxide, whilst cyanogen is formed between pH 6 and 9.

The reaction is very slow in acid solution, and has a maximum rate near pH 9. It is quite rapid, however, in the high pH region in which Stewart and Van der Linden[41] found evidence for two reaction paths. At pH > 13 and at low reactant concentrations, the kinetics are

$$-d[MnO_4^-]/dt = k_2[MnO_4^-][CN^-]$$

but at lower basicity and at higher reactant concentrations a more complex and pH-dependent path dominates, the rate law being

$$\begin{aligned} -d[MnO_4^-]/dt &= k_3[MnO_4^-][CN^-]^2[OH^-]^{-1} \\ &= k_3'[MnO_4^-][CN^-]^2[H^+] \\ &= k_3''[MnO_4^-][CN^-][HCN] \end{aligned}$$

Freund[42] has reported similar kinetics for the "simple" reaction and gives k_2 as $6.4\times10^7 \exp(-9.0\times10^3/\boldsymbol{RT})$ l.mole^{-1}.sec^{-1}.

Stewart and Van der Linden[41] also examined the incorporation of ^{18}O into the cyanate from labelled permanganate. The percentage of transfer varied with alkalinity and the authors believe that significant oxygen-transfer occurs in the second-order reaction, but not in the complex reaction. Accordingly the mechanism for the second-order reaction is proposed to be

$$MnO_4^- + CN^- \rightarrow [O_3Mn \ldots O \ldots CN]^{2-} \rightarrow MnO_3^- + OCN^- \quad \text{(slow)}$$

$$MnO_3^- + MnO_4^- + 2\,OH^- \rightarrow 2\,MnO_4^{2-} + H_2O \quad \text{(fast)}$$

Conversely, a mechanism involving oxygen-transfer cannot be written for the complex reaction, and the authors propose a sequence

$$\begin{aligned} HCN + CN^- &\rightleftharpoons H(CN)_2^- && \text{(fast)} \\ MnO_4^- + H(CN)_2^- &\rightarrow (CN)_2 + H^+ + MnO_4^{3-} && \text{(slow)} \\ (CN)_2 + 2\,OH^- &\rightarrow CN^- + OCN^- + H_2O && \text{(fast)} \\ 3\,MnO_4^{3-} + 4\,H_2O &= 2\,MnO_2 + MnO_4^- + 8\,OH^- && \text{(fast)} \end{aligned}$$

The slow step of this reaction corresponds to removal of hydride from an anion and finds several counterparts in oxidations of organic compounds by MnO_4^-. The anion may have the structure

$$\begin{array}{c} H\text{–}C\text{=}N^- \\ | \\ C\equiv N \end{array}$$

2.2.3 *Oxides of hydrogen*

Vepřek-Šiška *et al.*[43–45] have recently shown that the oxidation of hydroxide ion by Mn(VII) is very much slower than has been suggested, being profoundly influenced by trace quantities of transition metal cations.

The reaction between chromic acid and hydrogen peroxide gives CrO_5 and involves no immediate change in the oxidation state of the metal atom. However, CrO_5 is decomposed by acid to $Cr(H_2O)_6^{3+}$ although there is some evidence for the intermediacy of other species, and a brief reference to the overall reaction merits inclusion in this review.

A stopped-flow study[46] of the formation of CrO_5 yielded the following rate law for acidities of 0.01 to 0.05 *M*

$$d[CrO_5]/dt = k_3[H^+][H_2O_2][HCrO_4^-]$$

where $k_3 = (4.4 \pm 2.0) \times 10^7 \exp[-(4.5 \pm 0.2) \times 10^3/RT]$, 2.0×10^4 $l^2.mole^{-2}.sec^{-1}$ at 25 °C. At higher acidities (up to 6 *M* HNO_3) and at 4 °C, k_{obs} was reasonably well represented[47] by

$$k_{obs} = 5.0 \times 10^3 \frac{[H_2O_2][H^+]}{(1 + 0.1[H^+])}$$

These data are considered to favour the mechanism

$$HCrO_4^- + H^+ \rightleftharpoons H_2CrO_4$$

$$H_2CrO_4 + H_2O_2 \rightarrow H_2CrO_5 + H_2O \quad \text{(slow)}$$

$$H_2CrO_5 + H_2O_2 \rightarrow CrO_5 + 2\,H_2O \quad \text{(fast)}$$

The disappearance of CrO_5 in perchloric acid follows the law

$$-d[CrO_5]/dt = k_3'[CrO_5][H^+]^2$$

with $k_3' = 6.8 \times 10^9 (\exp -12.8 \times 10^3/RT)\, l^2.mole^{-2}.sec^{-1}$, 2.7 at 25 °C.

Permanganate does not form a stable peroxy compound with hydrogen peroxide but oxidises it rapidly to oxygen and water. The stoichiometric equation depends on pH, with Mn(II), Mn(IV) and Mn(VI) being formed in acidic, neutral and alkaline solutions respectively. Chang[48] found D_2O_2 to react with MnO_4^- in D_2O with a rate only 15 % of that for H_2O_2 in H_2O, the reaction being auto-catalysed by Mn^{2+} ion. At 0.1 *M* acidity k_2 equals 2.9×10^3 l.mole^{-1}.sec^{-1} (18 °C)[49].

2.2.4 *Oxy-acids of sulphur*

Haight *et al.*[50] have published a detailed account of the kinetics and stoichiometry of the oxidation of buffered bisulphite ion by chromic acid. The reaction is fast and its study requires a rapid mixing technique. The stoichiometry varies from a Cr(VI)/S(IV) molar ratio of 1 : 2 to 2 : 3 as the initial concentrations are changed in the range $0.12 \leqslant [\text{Cr(VI)}]/[\text{S(IV)}] \leqslant 1.4$ and this was explained in terms of competition between two overall reactions

$$2\,HCrO_4^- + 4\,HSO_3^- + 6\,H^+ = 2\,Cr^{3+} + 2\,SO_4^{2-} + S_2O_6^{2-} + 6\,H_2O \tag{15}$$

$$2\,HCrO_4^- + 3\,HSO_3^- + 5\,H^+ = 2\,Cr^{3+} + 3\,SO_4^{2-} + 5\,H_2O \tag{16}$$

The observed rate law, which applies only to (15) in 0.5 *M* sodium acetate buffer in the pH range 4.18–5.05 is

$$-\mathrm{d}[\text{Cr(VI)}]/\mathrm{d}t = k_{\text{obs}} \frac{[\text{Cr(VI)}][\text{S(IV)}]^2[\text{H}^+]}{1 + K_1[\text{S(IV)}]} \tag{17}$$

K_1, which has a value of 36 l.mole^{-1} at 25 °C, is interpreted as being the equilibrium constant for the reaction

$$HCrO_4^- + HSO_3^- \rightleftharpoons CrSO_6^{2-} + H_2O \tag{18}$$

k_{obs} is 1.37×10^8 l^3.mole^{-3}.sec^{-1} at 25.0 °C and the activation energy and entropy are 4.5 kcal.mole^{-1} and -13 eu, respectively. A further point is that the sulphate formed during the reaction is bound to Cr(III) at its conclusion.

The presence of the denominator term in the rate equation (17) suggests that the equilibrium (18) precedes the oxidation step. Two sequences of reactions are proposed (see below), depending on whether the sulphite radical ion dimerises (20) or attacks further acid chromate ion (21). It should be noted that of the species prevalent in dilute aqueous chromic acid, namely CrO_4^{2-}, $Cr_2O_7^{2-}$, $HCrO_4^-$ and H_2CrO_4, only the last is regarded as possessing oxidising powers. This fact, noted by Westheimer[20], is tacitly assumed in all recent discussion of mechanisms of oxidations by Cr(VI)

$$HSO_3^- + H^+ \rightleftharpoons SO_2 + H_2O \qquad K$$

$$SO_2 + CrSO_6^{2-} \rightleftharpoons \{[O_2SOCrO_2OSO_2]^{2-}\}^{\ddagger} \quad \text{(slow)}$$

$$\{[O_2SOCrO_2OSO_2]^{2-}\}^{\ddagger} + 4\,H_2O + 2\,H^+ \rightarrow [SO_4Cr(H_2O)_5]^+ + \cdot SO_3^- \quad (19)$$

$$2\cdot SO_3^- \rightarrow S_2O_6^{2-} \quad (20)$$

$$\cdot SO_3^- + HCrO_4^- \rightarrow SO_4^{2-} + Cr(V) \quad (21)$$

$$H^+ + HCrO_4^- + CrSO_6^{2-} \rightarrow O_3CrOCrO_2OSO_2^{2-} + H_2O \quad (22)$$

$$O_3CrOCrO_2OSO_2^{2-} \rightarrow 2\,Cr(V) + SO_4^{2-} \quad (23)$$

$$Cr(V) + S(IV) \rightarrow Cr(III) + S(VI) \quad (24)$$

The entity marked with a double dagger is regarded by the authors as an activated complex. Its breakdown (19) may well consist of a sequence of rapid steps rather than the single step implied, which involves a three-electron transfer and double protonation of a transition state subsequent to its formation. Steps (22)–(24) were invoked to explain the complete oxidation of S(IV) to S(VI) at higher Cr(VI) concentrations.

The authors contrast this oxidation with those of As(III) and of alcohols, both of which involve analogous pre-equilibria (*vide infra*).

At high ratios of reductant to oxidant, conditions which favour tetrathionate formation at the expense of sulphate, the Cr(VI) oxidation of thiosulphate follows kinetics[695]

$$\frac{-d[CrS_2O_6^{2-}]}{dt} = \frac{K(k_2 + k_3[H_3O^+])[HSO_3^-]^2[CrS_2O_6^{2-}]}{3(1 + K[HS_2O_3^-])}$$

and at 20 °C $k_2 = (1.42 \pm 0.21) \times 10^2$ l.mole^{-1}.sec^{-1}, $k_3 = (1.55 \pm 0.22) \times 10^4$ l^2.mole^{-2}.sec^{-1}. $E_2 = 8.8 \pm 0.7$ kcal.mole^{-1}, $\Delta S_2^{\ddagger} = -20.8 \pm 2.5$ eu, $E_3 = 6.8 \pm 2.4$ kcal.mole^{-1} and $\Delta S_3^{\ddagger} = -17.7 \pm 8.0$ eu. K, the formation constant of $CrS_2O_6^{2-}$,

$$HCrO_4^- + HS_2O_3^- \rightleftharpoons H_2O + CrS_2O_6^{2-}$$

is unusually high (1.24×10^4 l.mole^{-1}) and essentially all Cr(VI) is in the form of the complex under the reaction conditions of high [$S_2O_3^{2-}$]. The activated complex contains one Cr(VI) and two $S_2O_3^{2-}$ species and two-electron reduction of Cr(VI) occurs with concomitant one-electron oxidation of *both* $S_2O_3^{2-}$ species to yield $S_4O_6^{2-}$. No evidence for Cr(V) or $S_2O_3\cdot$ was found from attempts to initiate induction and polymerisation reactions.

A similar investigation[715] produced a simpler rate law

$$-d[Cr(VI)]/dt = k[Cr(VI)][S_2O_3^{2-}]^2(H_3O^+]$$

with $k = 5.5 \times 10^4$ l^3.mole^{-3}.sec^{-1} (25 °C, pH 4, $\mu = 0.23$ M), $E = 7.4$ kcal. mole^{-1} and $k_{D_2O}/k_{H_2O} = 1.4$. Spectrophotometric evidence for $CrS_2O_6^-$ was obtained, yielding $K = 2.2 \times 10^5$ l.mole^{-1} (18.4 °C).

The chromic acid oxidation of dithionic acid is independent of oxidant concentration and its rate is equal to that of the acid-catalysed hydrolysis to sulphite and sulphate, which must therefore constitute the rate-determining process[51].

2.2.5 *Nitrite ion*

The rate law for the MnO_4^- oxidation of NO_2^- to nitrate is reported to be[51a]

$$-d[MnO_4^-]/dt = k[MnO_4^-][NO_2^-](1+k'/[H_3O^+])$$

with $E = 11.5$ kcal.mole^{-1} and $\Delta S^{\ddagger} = -9.9$ eu.

2.2.6 *Trivalent phosphorus compounds*

Haight *et al.*[52–52b] have made a comparative study of the Cr(VI) oxidations of phosphorous and hypophosphorous acids and several organophosphorus analogues. Neither $H_2PO_3^-$ nor $H_2PO_2^-$ ions are oxidised at an appreciable rate; the former ion yields an anhydride, $HO_2POCrO_3^{2-}$, with a formation constant of 7 ± 1 l.mole^{-1} (24 °C, acetic acid–sodium acetate buffer at pH 4) but $H_2PO_2^-$ does not form an anhydride. In acidic media both neutral acids are oxidised according to the rate law[52, 52a]

$$\frac{-d[Cr(VI)]}{dt} = \frac{[Cr(VI)][P(III)](k_1[H_3O^+]+k_2[P(III)])}{1+K[P(III)]}$$

where, at 25 °C, for H_3PO_3 $k_1 = 2.0 \times 10^{-3}$, $k_2 = 6.0 \times 10^{-3}$ (both l^2.mole^{-2} .sec^{-1}) and $K = 16$ l.mole^{-1} and for H_3PO_2 $k_1 = 6.5 \times 10^{-3}$, $k_2 = 13.5 \times 10^{-3}$ and $K = 11$ (same units, respectively). The slower oxidation of $DP(OH)_2{=}O$ indicates a primary kinetic isotope effect of *ca.* 4 and chloride ion inhibits reaction.

The form of the rate law suggests significant complex formation, *cf.* $H_2PO_3^-$, and the isotope effect demonstrates slow P–H cleavage, *viz.*

$$HCrO_4^- + HP(OH)_2{=}O \rightleftharpoons {}^-OCrO_2{-}O{-}PH(OH){=}O + H_2O \qquad K$$

$$-H^+ \upharpoonleft\!\downharpoonright +H^+$$

$$HOCrO_2{-}O{-}PH(OH){=}O$$

$$HO{-}CrO_2{-}O{-}\overset{\overset{\displaystyle O}{\|}}{\underset{\underset{\displaystyle H:B}{|}}{P}}{-}OH \rightarrow HO{-}CrO_2^- + O{=}\overset{\overset{\displaystyle O}{|}}{P}{-}OH + BH^+$$

$$2\,Cr(IV) \rightarrow Cr(V) + Cr(III)$$

$$P(III) + Cr(V) \rightarrow P(V) + Cr(III)$$

(B is a base, *e.g.* H_2O or $H_2PO_3^-$)

This is entirely analogous to the Westheimer mechanism for isopropanol oxidation (Section 2.2.3).

Clearly for P(III) to be oxidised, the P atom must be bonded simultaneously to (*i*) a hydrogen atom, (*ii*) an oxygen atom linked to the oxidant and (*iii*) a free, unionised hydroxyl group [for otherwise $H_2PO_3^-$ would reduce Cr(VI)][52a]. In conformity with these conclusions replacement of the hydroxylic protons in H_3PO_3 by ethyl groups to give diethyl phosphite results in a reduction of Cr(VI) characterised by rate-determining ester-hydrolysis to ethanol and monoethyl phosphite, both of which are oxidised normally[52b]. Also $DP(OC_2H_5)_2{=}O$ is oxidised at virtually the same rate as $HP(OC_2H_5)_2{=}O$ but oxidation of $C_2H_5P(OC_2H_5)_2{=}O$ proceeds very slowly.

A reinvestigation[663] of the hypophosphite oxidation produced rather slower rates and a differing rate law including a term $k_0\,[HCrO_4^-][H_3PO_2]$ corresponding to a complex $H_2PCrO_5^-$ for which optical measurements provided support.

2.2.7 Arsenious acid

The chromic acid oxidation of As(III) was first studied by De Lury[54], whose results have been recalculated by Westheimer[20] to afford a rate law

$$-d[Cr(VI)]/dt = k[HCrO_4^-][H_3AsO_3][H^+]^2$$

Westheimer[20] has also reviewed the induced oxidations by the Cr(VI)–As(III) couple of iodide, bromide and manganous ions (*vide supra*). The induction factor of 0.5 for Mn(II) implies an intermediate tetravalent chromium species; however, the factor of 2 for iodide points to a pentavalent chromium intermediate. Both observations are best accommodated in the following scheme, which does not involve the unstable As(IV)

$$HCrO_4^- + As(III) \rightarrow Cr(IV) + AsO_4^-$$

$$Cr(IV) + Cr(VI) \rightarrow 2\ Cr(V)$$

$$Cr(V) + As(III) \rightarrow Cr(III) + AsO_4^-$$

Edwards[55] has also re-examined De Lury's data and finds evidence for an extra term in the rate law, which becomes

$$-d[Cr(VI)]/dt = k[HCrO_4^-][H_3AsO_3][H^+] + k'[HCrO_4^-][H_3AsO_3][H^+]^2$$

The reaction in 0.2 M acetic acid–0.2 M sodium acetate buffers at an ionic strength of 1.5 M (KNO_3) and a temperature of 25 °C, has been reinvestigated recently by Mason and Kowalak[56]. They report the following rate law for a high [As(III)]/[Cr(VI)] ratio

$$-d[Cr(VI)]/dt = \frac{2\,kK[As(III)][HCrO_4^-]}{1 + K[As(III)]}$$

For the conditions quoted, $k = 3.76 \times 10^{-4}$ sec^{-1} and $K = 22.4$ l.mole^{-1}. The law is consistent with the mechanism

$$HCrO_4^- + As(III) \rightleftharpoons As(III){\cdot}HCrO_4^- \qquad K$$

$$As(III){\cdot}HCrO_4^- \rightarrow \text{products} \qquad \text{(slow)}$$

A second series of experiments was conducted with [Cr(VI)] > [As(III)] and a modified rate law was obtained, *viz.*

$$-d[Cr(VI)]/dt = \frac{1.68 \times 10^{-2}[As(III)][HCrO_4^-]}{1 + 22.4\,[HCrO_4^-]} + 2.47 \times 10^{-2}[Cr_2O_7^{2-}][As(III)]$$

The authors believe that a similar mechanism to the above operates for the dichromate-ion dependent path, but that the equilibrium constant K is much smaller. This oxidation presents two novel features, namely the lack of an acidity dependence of the rate and the participation of a term involving dichromatic ion.

Kolthoff and Fineman[57] have examined the reaction in *alkaline* media in the presence and absence of oxygen, obtaining the rate law

$$-d[Cr(VI)]/dt = k_2[CrO_4^{2-}][As(III)]$$

References pp. 493–509

with $k_2 = 1.61 \times 10^{-3}$ l.mole^{-1}.sec^{-1} at 30 °C, pH 9.1 and an ionic strength of 1.75. k_2 was independent of acidity above this pH, but the reaction was acid catalysed at lower pH. This conclusion was based on a single measurement, however, and the disagreement with the results of Mason and Kowalak[56] is not serious. The Cr(VI)–As(III) reaction was found to induce the reduction of molecular oxygen, thereby confirming the early work of Kessler[58]. For further details, the reader is referred to the chapter on induced reactions.

Wiberg[59] has reviewed this oxidation and suggests that the complex of Mason and Kowalak is a mixed anhydride, *cf.* the oxidations of sulphur dioxide (p. 285) and formic acid (p. 316). The anhydride cannot be of neutral charge, however, as no acidity dependence is observed and a possibility is

$$H_3AsO_3 + HCrO_4^- \rightleftharpoons \underset{\displaystyle OH}{HO\text{–}\underset{|}{As}\text{–}O\text{–}CrO_2\text{–}O^-} + H_2O$$

$$\underset{\displaystyle OH}{HO\text{–}\underset{|}{As}\text{–}O\text{–}CrO_2\text{–}O^-} \rightarrow As(V) + Cr(IV) \qquad \text{(slow)}$$

Further complication is apparent when the reaction is investigated in $H_2PO_4^-$–HPO_4^{2-} buffers[664]; while the rate shows a first-order dependence on total [Cr(VI)] and [As(III)], a complex dependence on the buffer composition is found, indicating two activated complexes of composition $H_2PO_4^- \cdot HCrO_4^- \cdot$ As(III) and $HPO_4^{2-} \cdot HCrO_4^- \cdot$As(III), which correspond to attack of $HCrPO_7^{2-}$ upon neutral As(III) and $As(OH)_2O^-$ respectively.

2.2.8 Carbon monoxide

The oxidation by permanganate proceeds readily in acid and neutral solutions (to give MnO_2) and in basic solution (to give MnO_4^{2-}). The rate law is[60, 61]

$$-d[CO]/dt = k[CO][MnO_4^-]$$

with $E = 13.6$ kcal.mole^{-1} and $\Delta S^\ddagger = 17$ eu over the pH range 1 to 13. Catalysis by Ag^+ and Hg^{2+} ions was observed (*cf.* oxidation of hydrogen by MnO_4^-) following a general rate law[61]

$$-d[CO]/dt = k[CO][MnO_4^-][cat]$$

For cat = Ag^+, $k(0\ °C) = 1.10 \times 10^5$ l^2.mole^{-2}.sec^{-1}, $E = 1.80$ kcal.mole^{-1} and $\Delta S^\ddagger = -31$ eu; for cat = Hg^{2+}, $k(0\ °C) = 1.06 \times 10^3$ l^2.mole^{-2}.sec^{-1}, $E = 7.0$ kcal.mole^{-1} and $\Delta S^\ddagger = 21$ eu. It was suggested that intermediates of the type

$$[\text{-Hg-CO-OMnO}_3]$$

are involved. It is pertinent[62] that CO is readily oxidised by solid $AgMnO_4$.

2.2.9 *Molecular hydrogen*

Unlike Mn(VII) chromic acid oxidises hydrogen only in the presence of a catalyst, namely, argentous ion[63]. The stoichiometry is

$$Cr_2O_7^{2-} + 3\,H_2 + 8\,H^+ = 2\,Cr^{3+} + 7\,H_2O$$

and no change in the concentration of Ag(I) was found. No evidence of complex formation between Cr(VI) and Ag(I) was obtained, and the rate law at 0.50 *M* ionic strength is

$$-d[H_2]/dt = k[Ag^+]^2[H_2]$$

with $k = 4 \times 10^8 \exp(-15.8 \times 10^3/RT)$ $l^2.mole^{-2}.sec^{-1}$.

When ceric ions were substituted for chromic acid, the reaction was still zero-order with respect to metal ion, the rate of reduction of which was unchanged. The mechanism favoured by the authors depends on formation of a complex of silver ions and hydrogen, *viz.*

$$Ag^+ + H_2 \rightleftharpoons AgH_2^+ \qquad \text{(fast)}$$

$$AgH_2^+ + Ag^+ \rightleftharpoons (Ag_2 \ldots H_2)^{2+} \qquad \text{(slow)}$$

$$(Ag_2 \ldots H_2)^{2+} + \tfrac{2}{3}\,Cr(VI) \rightarrow 2\,Ag^+ + \tfrac{2}{3}\,Cr^{3+} + 2\,H^+ \qquad \text{(fast)}$$

Other possible pre-equilibria were also considered.

Webster and Halpern[64] found the stoichiometry of the permanganate oxidation in acidic solution to be

$$2\,MnO_4^- + 3\,H_2 + 2\,H^+ = 2\,MnO_2 + 4\,H_2O$$

and in alkaline solution (0.3–0.6 *M* sodium hydroxide) to be

$$2\,MnO_4^- + H_2 + 2\,OH^- = 2\,MnO_4^{2-} + 2\,H_2O$$

In acidic solution the kinetics are

$$-d[H_2]/dt = k_2[H_2][MnO_4^-]$$

with $k_2 = 4.2\times10^9 \exp[-(14.7\pm0.5)\times10^3/RT]$ l.mole^{-1}.sec^{-1} ($\mu = 0.3$ M). The reaction is subject to slight acid catalysis. In neutral solution the rate expression has the same form[64, 65], but k is lower by 15 %. In basic solution also no change in the law was observed and no significant salt effect was found at any pH. The reaction has the same velocity in deuterium oxide[66].

The authors favour a two-equivalent oxidation in order to avoid the necessity to postulate free hydrogen atoms as intermediates in the reaction, *viz.*

$$MnO_4^- + H_2 \rightarrow MnO_4^{3-} + 2\,H^+ \text{ or } HMnO_4^{2-} + H^+$$

or

$$MnO_4^- + H_2 \rightarrow MnO_3^- + H_2O$$

The reaction was also found to be catalysed by silver ion[64], being of the third order, with $k_3 = 7.5\times10^7 \exp[(-9.3\pm0.5)\times10^3/RT]$ l^2.mole^{-2}.sec^{-1} ($\mu = 0.3$ M). Ag^+ is thought to form a complex with MnO_4^- (a process for which there is spectroscopic support), which attacks the hydrogen molecule, *viz.*

$$Ag^+ + MnO_4^- \rightleftharpoons AgMnO_4 \quad \text{(fast)}$$

$$AgMnO_4 + H_2 \rightarrow AgH^+ + MnO_4^{2-} + H^+ \quad \text{(slow)}$$

$$AgH^+ + MnO_4^- \rightarrow Ag^+ + H^+ + MnO_4^{2-} \quad \text{(fast)}$$

2.3 OXIDATION OF MONOFUNCTIONAL ORGANIC MOLECULES

It is not within the scope of this review to deal with products of oxidation of organic molecules in cases other than those for which a kinetic analysis has also been attempted. However, much insight into the modes of reactivity of Cr(VI) and Mn(VII) is to be gained from such information and the reader is referred to the recent excellent reviews of Wiberg[67] on Cr(VI) and Stewart[68] on Mn(VIII).

One major difference between the published accounts of oxidation of inorganic and organic species is the more prevalent use of non-aqueous solvents for the latter, which gives rise to several complicating features.

2.3.1 Aliphatic hydrocarbons

Work has been concentrated on hydrocarbons containing activating groups, such as phenyl or *tert*-butyl, in order to achieve convenient rates of reaction. The relative rates of oxidation of two series of hydrocarbons with chromic acid in 95 % acetic acid are[69, 70]

(*a*)	cyclohexane	methylcyclohexane	toluene	ethylbenzene	diphenylmethane
	0.01	0.08	0.16	0.50	1.00

(*b*)	toluene	ethylbenzene	isopropylbenzene	*tert*-butylbenzene
	1	7.2	71.1	0.019

By comparing a number of aliphatic hydrocarbons of different structure, Mareš and Roček[70] compute relative rates of oxidation of methyl, methylene and methine groups to be 1 : 114 : 7000–18,000.

The simplest compounds which have been examined are the series $CH_3(CH_2)_nCH_3$[71]. The observed rate law is[72]

$$-\mathrm{d[Cr(VI)]/d}t = k\mathrm{[Cr(VI)][alkane]}h_0$$

k was found to increase regularly with n and at a temperature of 50 °C, an acidity of 0.2 M (H_2SO_4) and a solvent composition of 99 % acetic acid, k is in good agreement with values calculated from

$$k = nk_{CH_2} \tag{25}$$

where $k_{CH_2} = 5.73 \times 10^{-2}$ l.mole^{-1}.sec^{-1}. The same rate law was found for cyclohexane[72] and the oxidations of both cyclohexane and n-heptane were greatly retarded on addition of 4 % of water to the acetic acid medium. This effect has been found subsequently for many oxidations of organic substances by chromic acid in acetic acid (*vide infra*).

Roček *et al.*[73] have also measured rate coefficients for a series of cyclo-alkanes, $(CH_2)_n (n = 4$ to $14)$, and find the analogue of k_{CH_2} in equation (25) to fluctuate with ring size in a manner corresponding exactly to the enthalpy of combustion of the cycloalkane concerned per methylene group, provided n is greater than five, *i.e.* there exists a direct correlation between reactivity and thermochemical strain.

For straight chain and cycloalkanes, Roček *et al.*[73] prefer a mechanism involving hydride ion abstraction to give a partly-developed carbonium ion which suffers further reaction with the Cr(IV) portion before it can become free to give acetate or olefin

$$\mathrm{{>}C(H){-}H \cdots O(H){-}\overset{+}{Cr}(=O)(OH)_2} \xrightarrow{-\mathrm{H_2O}} \left[\mathrm{{>}\overset{+}{C}{-}H \quad O{=}Cr(OH)_2}\right] \longrightarrow \mathrm{{>}C(H){-}O{-}\overset{+}{Cr}(OH)_2}$$

The oxidation by Cr(VI) of aliphatic hydrocarbons containing a tertiary carbon atom has been studied by several groups of workers. Sager and Bradley[74] showed that oxidation of triethylmethane yields triethylcarbinol as the primary product with a primary kinetic isotope effect of about 1.6 (later corrected by Wiberg and Foster[75] to 3.1) for deuterium substitution at the tertiary C–H bond. Oxidations of norbornane and bornane proceed readily[74], but the production of β-camphor from the latter indicates the *methylene* group to be oxidised in preference to the bridgehead tertiary C–H bond. Mareš and Roček[70] sought evidence for steric hindrance in their examination of the relative rates of oxidation by Cr(VI) of the following series

	$(CH_3)_3CH$	$(CH_3)_2CH(neo\text{-}C_5H_{11})$	$CH_3CH(neo\text{-}C_5H_{11})_2$
(relative) k_2	1	0.674	0.104

They also compared compounds which might be expected[76] to give strong neighbouring group effects should a carbonium ion intermediate be formed, *viz.*

	cyclohexane	bornane	isobornane
k_2 (relative)	1	2.0	8.0

These effects are very much smaller than those found for the first-order solvolyses of bornyl and isobornyl chlorides, which differ in relative rate by several orders of magnitude. However, the authors argue that this does not necessarily disprove participation by a carbonium ion in the mechanism which they proposed for the oxidation of straight-chain alkanes (see above).

Wiberg and Foster[75] oxidised (+)-3-methylheptane in 91 % acetic acid with chromic acid and obtained (+)-3-methyl-3-heptanol with 70–85 % retention of configuration. This implies that should the initial product of oxidation be either $MeEtPrC^+$ or $MeEtPrC\cdot$, then neither diffuses away from the solvent cage before reacting further. Slow initial formation of $[R_3\dot{C}\cdot Cr(V)]$ would be expected to be followed by rapid transfer of a second electron to give $[R_3C^+\cdot Cr(IV)]$. The latter could be an ester of Cr(IV) which suffers Cr–O fission to give R_3COH with retention of configuration.

Closely related to this work is that of Wiberg and Evans[69] on the Cr(VI) oxidation of diphenylmethane in 95 % acetic acid. This has the rate law

$$-d[Cr(VI)]/dt = k[\text{substrate}][CrO_3]h_0$$

and activation parameters; $E = 15.8$ kcal.mole^{-1}, $\Delta S^{\ddagger} = 21.2$ eu. A primary kinetic isotope effect of 6.4 was obtained at 30 °C and electron-releasing groups enhanced the reaction rate ($\rho^+ = -1.17$). The appearance of the *total* concentration of hexavalent chromium in the rate equation is significant. In many oxidations, of both organic and inorganic substrates, reactivity is confined to H_2CrO_4 even in an aqueous acetic acid medium, *e.g.* aromatic aldehydes (p. 310). Evidently dichromate ion has comparable reactivity in this particular oxidation.

Writing the oxidant as H_2CrO_4 while noting the participation of $HCr_2O_7^-$, Wiberg and Evans[69] present five mechanisms which accommodate both the kinetics and the C–H cleavage implied by the isotope effect, *viz.*

A. $Ph_2CH_2 + H_2CrO_4 \longrightarrow Ph_2\dot{C}H + Cr(V)$

B. $Ph_2CH_2 + H_2CrO_4 \longrightarrow Ph_2\overset{+}{C}H + Cr(IV)$

C. $Ph_2C{-}H$ (H … $O{=}CrO_3H_2$) $\longrightarrow Ph_2C(OH){-}H + Cr(IV)$

D. $Ph_2C{-}H$ (H … O, $O{=}Cr(OH)(OH)$) $\longrightarrow Ph_2C(H){-}O{-}Cr(OH)_3$ (HO–Cr–OH, OH)

E. $O{=}CrO_3H_2$ / $Ph{-}CH(H){-}Ph \longrightarrow Ph{-}CH(O{-}Cr(OH)_2O^-){-}Ph$

The lack of steric effects in oxidations of hydrocarbons by Cr(VI) renders D and E unacceptable. The activated complex of scheme C is non-linear and hence does not comply with the magnitude of the observed isotope effect. Two pieces of evidence are quoted which indicate A to be the more probable of the remaining two. Firstly, the ρ^+ constant of -1.17 is more in agreement with that obtained for bromine atom abstraction from toluenes (-1.369 to -1.806)[77] than those found for solvolyses involving electron-deficient carbon (-2.57 to -4.67)[78]. Secondly, the correlation between the relative rates of oxidation of the series

$$PhCH_3, \quad PhC_2H_5, \quad Ph_2CH_2, \quad Ph_3CH$$

and the corresponding relative rates of hydrogen abstraction by $CCl_3\cdot$ is much superior to that between oxidation rates and relative rates of solvolysis of the corresponding chlorides. It should be noted, however, that this type of argument has received criticism[79].

Wiberg prefers mechanism A to the carbonium-ion mechanism with the proviso that the radical is oxidised before inversion occurs. The carbonium ion formed must rapidly acquire an oxygen atom to prevent inversion and the two processes may be synchronous. The minor role which free carbonium ions may play in the reaction has been discussed[80].

A quantity of earlier work exists on chromic acid oxidation of hydrocarbons. It was noted that diphenylmethane and other hydrocarbons in glacial acetic acid solution are oxidised rapidly during the initial stages but that reaction is autoretarded[81]. The autoretardation is eliminated on adding 2.5 % of sulphuric acid. The orders of the reaction with respect to diphenylmethane and Cr(VI) are one and two respectively[82], the latter differing from that found by Wiberg and

Evans[69]. The more recent result is to be preferred in view of the wider range of Cr(VI) concentration employed.

Electron-releasing groups facilitate the oxidation of a series of *para*-substituted toluenes, although determination of the rate coefficients again depended on a second-order decay plot of the concentration of Cr(VI)[83].

Chromyl chloride combines with arylalkanes in inert solvents to deposit an Étard complex of the general formula $ArH \cdot 2\,CrO_2Cl_2$. On hydrolysis this yields the expected oxidation product. Study of the reaction has been from three standpoints, namely the formation and the spectroscopy of the complex and the products of its decomposition. The latter aspect has been covered by Wiberg[67], who gives details of the reaction of *o*-nitrotoluene to give an apparently normal complex which, on hydrolysis, reverts mainly to the starting compound[84]. This implies that oxidation occurs only during hydrolysis. However, both the magnetic susceptibility[85, 85a] (3.2 ± 0.06 Bohr magnetons per Cr atom) and the ESR spectrum[86] of the complex with toluene, indicate the presence of Cr(IV)[87].

The oxidations of toluene, diphenylmethane and triphenylmethane are second order[88], relative rates at 22 °C being

	$PhCH_3$	Ph_2CH_2	Ph_3CH
k_2(l.mole^{-1}.sec^{-1})	1.28×10^{-4}	1.67×10^{-2}	0.16

The effect of solvent upon k_2 has been reported[89], and it was concluded that the activated complex is not sufficiently polar to be called "ionic". The oxidations of toluene[90] and triphenylmethane[91] exhibit primary kinetic deuterium isotope effects of 2.4 and *ca.* 4 respectively. No isotopic mixing occurred during formation of the Étard complex from a mixture of normal and deuterated *o*-nitrotoluene[91]. The chromyl chloride oxidation of a series of substituted diphenylmethanes revealed that electron-withdrawing substituents slow reaction while electron-releasing groups have the opposite effect, the values of ρ and ρ^+ being -2.28 ± 0.08 and -2.20 ± 0.07 respectively[92].

Stairs[93] comments that this ρ value is strongly dependent on the temperature but his data have been criticised by Duffin and Tucker[94], who prefer their method of observing the rate of formation of the adduct to that of estimating total residual oxidising power employed by Stairs, and they find ρ^+ (25°) to be -2.32 ± 0.10 as compared to a value of -2.20 ± 0.07 at 40 °C. These values are considerably more negative than those found for chromic acid oxidation of diphenylmethanes (-1.17) and toluenes (-1.12).

The relative rates of oxidation of phenylmethanes cover too small a range to be compatible with carbonium ion formation (*cf.* the discussion on chromic acid oxidation of diphenylmethane, p. 295), and an initial reaction to give a radical *plus* Cr(V) followed by rapid transfer of a second electron to form Cr(IV) is more likely[67, 87], *viz.*

$$Ar_3CH + CrO_2Cl_2 \rightarrow [Ar_3\dot{C}.OCr(OH)Cl_2]$$
$$[Ar_3\dot{C}.OCr(OH)Cl_2] \rightarrow Ar_3C\text{–}O\text{–}Cr(OH)Cl_2$$

Ar_2CH_2 would react further to give $Ar_2C[OCr(OH)Cl_2]_2$, hydrolysis of which would give the observed products.

Necşoiu *et al.*[79, 95] dispute these conclusions and prefer an S_E2 process, *viz.*

$$\gt C\text{–}H \longrightarrow \gt C\cdots(O\text{–}CrCl_2 / H\cdots O) \longrightarrow \gt C\text{–}O\text{–}Cr(Cl)(OH)\text{–}Cl$$

The transition state does not involve a large degree of charge separation and hence the relative rates of oxidation of toluene, diphenylmethane and triphenylmethane may be explained in terms of an ionic mechanism.

The aqueous permanganate oxidation of hydrocarbons has been relatively neglected for reasons of low solubility, and discussion of the oxidations of C–H groups has derived largely from studies of certain carboxylic acids or alcohols when the functional group is not directly involved. The oxidations of 4-methylhexanoic acid, 5-methylheptane sulphonic acid and *p-sec*-butylbenzoic acids are of a total kinetic order of two[96] at pH 13. Oxidation of 4-methylhexanoic acid proceeds with 35 % retention of configuration at pH 13, but that of *p-sec*-butylbenzoic acid displays only 6 % retention. About 25 % of ^{18}O was transferred to the product, 4-hydroxy-4-methylhexanoate lactone, from labelled MnO_4^-, but only 0.6 % to α-hydroxy-*p-sec*-butylbenzoic acid. As in the case of chromic acid, three modes of electron-transfer deserve consideration

$$R_3CH + MnO_4^- \longrightarrow R_3C^+ + Mn(V)$$
$$R_3CH + MnO_4^- \longrightarrow R_3C\cdot + Mn(VI)$$
$$R_3C\text{–}H + O_2Mn(O)O^- \longrightarrow R_3C\text{–}O\text{–}Mn(OH)(O)O^-$$

The factor of seven variation between k_2 for the ordinary and benzylic tertiary hydrogen is too small to be associated with carbonium ion formation. The observed degrees of retention and ^{18}O transfer imply a caged radical rapidly reducing Mn(VI) to Mn(V) by accepting oxygen

$$[R_3\dot{C}.MnO_4H^-] \rightarrow R_3C\text{–}O\text{–}MnO_3H$$

Solvolysis follows, with carboxyl participation where this is feasible, to yield R_3COH, although several modes are involved. Other stereochemical work has been summarised by Stewart[68].

A subsequent detailed analysis[665] of the permanganate oxidation of the tertiary hydrogen atom of 4-phenylvaleric acid in 2.5 M potassium hydroxide solution supports the caged radical mechanism. The reaction order is two overall, k_H/k_D is *ca.* 11.5, ring substitution has little effect on the rate ($\rho \sim 0$) and the oxidation proceeds with a net 30–40 % retention of optical configuration.

In an earlier series of experiments, Cullis and Ladbury[97] examined the kinetics of the permanganate oxidation of hydrocarbons in acetic acid solution. Initial attack on toluene occurs at the methyl group and a total order of two was found. Electron-withdrawing agents reduced the rate of oxidation. However, the effects of added salts were complex and the authors believe that lower oxidation states of manganese are responsible for the oxidation. The oxidation of ethylbenzene produced acetophenone, the process being second-order with

$$k_2 = 1.9 \times 10^{10} \exp(-14.8 \times 10^3/\boldsymbol{R}T) \text{ l.mole}^{-1}.\text{sec}^{-1}.$$

However, added pyrosphosphate reduced the rate by 40 %.

2.3.2 *Olefins and acetylenes*

The oxidation of olefins by Cr(VI) follows kinetics[97a]

$$-\mathrm{d}[\mathrm{Cr(VI)}]/\mathrm{d}t = k_3[\mathrm{Cr(VI)}][\text{olefin}][\mathrm{H_3O^+}]$$

The reaction is insensitive to steric effects, but k_3 increases with the *number* of alkyl substituents although it is not influenced by position of the alkyl groups (unlike oxidation by Tl(III))[97b]. In these respects Cr(VI) oxidation resembles bromination, chlorination and epoxidation and a symmetrical transition state of the type depicted is favoured.

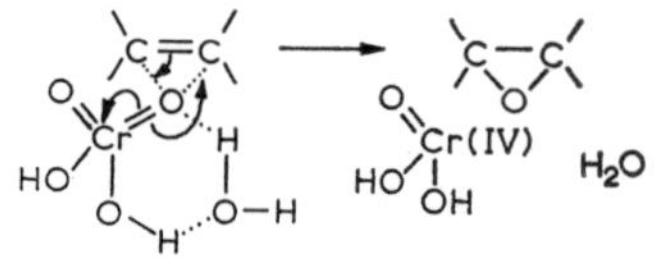

Chromyl chloride oxidation of alkenes proceeds *via* the formation of adducts at a rate necessitating stopped-flow techniques. At 15 °C the formation of 1 : 1 adduct from styrene and oxidant in CCl_4 solution[707] is simple second-order with $k_2 = 37.0$ l.mole^{-1}.sec^{-1}. Measurements with substituted styrenes yielded $\rho^+ = -1.99$. $E = 9.0$ kcal.mole^{-1} and $\Delta S^‡ = -23.8$ eu for styrene itself. Hydrolysis of the styrene adduct yields mostly phenylacetaldehyde (76.5 %) and benzaldehyde (21.1 %). Essentially similar results were obtained with a set of 15 alkenes[708] and some cycloalkenes[709] with E and $\Delta S^{\ddagger}$ values in the ranges 6.0–7.6 kcal.mole^{-1} and 27–41 eu, respectively. Formation of rearranged products

$$R_1R_2C{=}CR_3R_4 \rightarrow R_1{-}\overset{\displaystyle R_3}{\underset{\displaystyle R_2}{C}}{-}COR_4$$

suggests alkyl migration on hydrolysis of the adduct.

The oxidation by permanganate has been investigated both at pH 6.5 and 13 using salts of unsaturated carboxylic acids to facilitate dissolution[98]. At low substrate concentrations ($\sim 10^{-3}$ M) and pH 13, the reaction is first-order with respect to both oxidant and substrate but at 2–3 $\times 10^{-3}$ M substrate concentration significant deviation is apparent, especially at 0 °C. The *initial* rate is independent of base concentration but the rate at later stages is strongly affected. This suggests that the reduction of a second permanganate ion is influenced by hydroxide ion as in the following scheme

$$MnO_4^- + \text{crotonate ion} \xrightarrow{k_1} B \tag{26}$$

$$MnO_4^- + B \xrightarrow{k_2} C \tag{27}$$

Thus $d[MnO_4^-]/dt = -k_1[MnO_4^-][\text{crotonate}] - k_2[MnO_4^-][B]$ and $d[B]/dt = k_1[MnO_4^-][\text{crotonate}] - k_2[MnO_4^-][B]$. Elimination of time as a variable followed by integration and further manipulation leads to an expression for k_2.

At 25 °C and pH 13, $k_1 = 220$ l.mole^{-1}.sec^{-1}; at pH 6.5, $k_1 = 232$ l.mole^{-1}.sec^{-1}. The reaction at pH 6.5 showed no deviation from simple second-order behaviour even at 0° C and the individual rate plots were straight. Evidently hydroxide ion does not affect the initial stage of reaction, which has a low activation energy (5.4±0.7 kcal.mole^{-1}) and an activation entropy of −32±2 eu.

At pH 13 the final products are manganate and *cis*-diol. The observation of a large degree of transfer of ^{18}O from labelled MnO_4^- into oleate ion[99] at pH 12 suggests formation of a cyclic ester, *viz.*

$$>C{=}C< \; + \; MnO_4^- \rightleftharpoons \; >\!C(-O-)\!-\!C(-O-)\!< \text{(cyclic ester with } Mn(=O)O^-\text{)}$$

The oxidation of a series of olefins reveals the reaction to be very insensitive to electronic effects[98]. Phenyl and methyl substitution of the olefin mildly accelerate reaction. In all cases k_1 is pH-independent. Data are collected in Table 1.

The rate of permanganate oxidation of acetylenedicarboxylic acid to CO_2 at pH 0.25–5.0 requires the use of the stopped-flow method, which yielded simple

TABLE 1

RATE PARAMETERS FOR FIRST STAGE OF THE OXIDATION OF OLEFINS BY PERMANGANATE

Temperature = 25 °C; pH 13.

Olefin	k *(l.mole^{-1}.sec^{-1})*	E *(kcal.mole^{-1})*	$\Delta S^{\ddagger}$*(eu)*
Allyl alcohol	137±7	8.1±0.7	−24±2
Acrylate ion	330±17	5.9±0.7	−30±2
Crotonate ion	220±11	5.4±0.7	−32±2
Vinylacetate ion	111±6	7.0±0.7	−28±2
4-Bromocrotonate ion	325±16	—	—
2-Pentenoate ion	242±12	—	—
3-Pentenoate ion	455±23	—	—
4-Pentenoate ion	167±8	7.2±0.7	−27±2
Cinnamate ion	503±25	3.9±0.7	−36±2
p-Methylcinnamate ion	278±14	—	—
p-Methoxycinnamate ion	264±14	—	—
p-Chlorocinnamate ion	302±15	—	—
Oleate ion	70	—	—

second-order kinetics[710]. Although the removal of MnO_4^- (λ_{max} 540 nm) under conditions of excess substrate showed clean first-order behaviour, a transient built up and then decayed at *ca.* 250 nm, the region of absorption by Mn(III). Quenching the reaction at this point in time enabled identification of oxalic acid as another intermediate. At pH > 3, when the dianion of the substrate is the only significant species, E = 6.7 kcal.mole^{-1} and $\Delta S^{\ddagger}$ = −32 eu.

2.3.3 *Alcohols*

The oxidation of alcohols to carbonyl compounds by hexavalent chromium is a reaction of the greatest synthetic importance. Wiberg's recent review[67] covers the range of compounds to which the method has been applied together with the resulting yields and products. Very many examples of the reaction have been studied from the mechanistic point of view, but the work of Westheimer and his group has contributed most to our understanding of the reaction, providing, indeed, a model for many subsequent investigations. In the late nineteen-fifties Roček[100, 101] proposed a mechanism alternative to Westheimer's so-called "ester mechanism"[20] but more recently the two proponents have achieved a common viewpoint[102]. The general reaction has been reviewed several times[1–3, 5], and it is sufficient to enumerate the main points, which relate to isopropanol, unless stated to the contrary.

(a) Products

In general, the substrate R_1R_2CHOH, where R_1 or R_1 and R_2 may be H, is oxidised to $R_1R_2C{=}O$ and the oxidant is reduced to chromic ion. Complications may occur when R_1 is not a simple alkyl group (see later).

(*b*) *Kinetics*

In aqueous solutions of moderate acidity ($< 1\,M$) the rate expression is[103, 104]

$$-\mathrm{d}[\mathrm{Cr(VI)}]/\mathrm{d}t = k_1[\mathrm{HCrO_4^-}][\mathrm{R_2CHOH}][\mathrm{H_3O^+}] + k_2[\mathrm{HCrO_4^-}][\mathrm{R_2CHOH}][\mathrm{H_3O^+}]^2$$

The dependence on $[HCrO_4^-]$ is a feature of many other oxidations by chromic acid and is characterised by a falling-off of the apparent first-order rate coefficient for disappearance of Cr(VI) at higher concentrations of oxidant. In more strongly acidic solutions the kinetics are[100, 101]

$$-\mathrm{d}[\mathrm{Cr(VI)}/\mathrm{d}t = k[\mathrm{Cr(VI)}][\mathrm{R_2CHOH}]h_0$$

At an h_0 value of 1, $k = 1.4\times10^7\ \exp(-13.1\times10^3/\boldsymbol{R}T)$ l.mole^{-1}.sec^{-1}. At acidities $H_0 = -3$ to -9, the rates of oxidation of 2-propanol, 1,1,1-trifluoro-2-propanol and 1,1,1,3,3,3-hexafluoro-2-propanol reach maxima[105] at different values of h_0.

(*c*) *Primary isotope effect*

At low acidity the relative rates of oxidation of isopropanol and $(CH_3)_2CDOH$ (contaminated with "light" isopropanol) were found to be 6.4 ± 1.9 at 40 °C after correction for incomplete deuteration[106]. k_H/k_T has also been determined and the results are in agreement[107]. Under extremely acid conditions k_H/k_D decreases to 1.3[105].

(*d*) *Solvent isotope effect*

Results on the relative rates of oxidation in light and heavy water were considered[108] in terms of the two components of the rate expression. $k_1(D_2O/H_2O)$ is 2.44 and $k_2(D_2O/H_2O)$ is 6.26.

(*e*) *Replacement of hydroxylic hydrogen by an alkyl group*

This replacement (by an isopropyl group to give di-isopropyl ether) reduces the oxidation rate[108] by a factor of about 1500.

(*f*) *Effects of substituents in the group R_1*

The oxidation of a series of *meta*- and *para*-substituted α-phenylethanols shows that electron-donating substituents facilitate reaction ($\rho = -1.01$)[109]. A similar study[110] of primary aliphatic alcohols confirmed this trend ($\rho^* = -1.06\pm0.06$).

(*g*) *Base catalysis*

An observation[104] that the oxidation is subject to marked catalysis by pyridine was later refuted[101,111].

(*h*) *Induced oxidation*

The oxidation induces the oxidation of manganous ion to MnO_2 with an induction factor (see p. 280) of 0.5[112]. The rate of reduction of Cr(VI) is reduced simultaneously to a maximum extent of 50 %[103].

(*i*) *Steric effect*

The alcohol 3β, 28-diacetoxy-6β-hydroxy-18β-12-oleanene was oxidised in an aqueous acetic acid medium[102]. Deuteration at the six position had no effect on the rate in solvents of high (> 80 %) acetic acid content, but the isotope effect reached 2 in 60 % acetic acid. Increasing the acetic acid content of the medium produced a much larger effect on the oxidation rate of cyclohexanol than of the polycyclic alcohol.

(*j*) *Stopped flow investigation*

The resolution of the overall reaction into steps implied by the steric effect (above) has been achieved[113,113b] for the oxidation of isopropanol. In 97 % aqueous acetic acid a rapid reaction, $k_2 \approx 1.25 \times 10^4$ l.mole^{-1}.sec^{-1} (15 °C, $\mu = 0.183\ M$ $NaClO_4$), which is unaffected by deuteration, precedes the oxidation. Evidence for an intermediate has been reported for the oxidation of 1,1,1-trifluoro-2-propanol at very high acidities[105].

Numerous further investigations of the reaction have been reported, but those outlined above bear most directly on the problem.

Westheimer's "ester mechanism" is

$$HCrO_4^- + H_3O^+ \rightleftharpoons H_2CrO_4 + H_2O \qquad (28)\ (\text{fast})$$

$$R_2CHOH + H_2CrO_4 \rightleftharpoons R_2\underset{H}{\underset{|}{C}}-O-\overset{O}{\overset{\|}{\underset{O}{\underset{\|}{Cr}}}}-OH \qquad (29)\ (\text{fast})$$

$$H_2O + R_2\underset{H}{\underset{|}{C}}-O-\overset{O}{\overset{\|}{\underset{O}{\underset{\|}{Cr}}}}-OH \longrightarrow R_2C{=}O + H_3O^+ + \overset{O}{\overset{\|}{\underset{O^-}{\underset{|}{Cr}}}}-OH \qquad (30)\ (\text{slow})$$

$$Cr(IV) + Cr(VI) \rightleftharpoons 2\,Cr(V) \qquad (\text{fast})$$

$$Cr(V) + R_2CHOH \longrightarrow R_2C{=}O + Cr(III) \qquad (\text{fast})$$

The importance of Cr(V) as an intermediate has been confirmed by a stopped-flow investigation[113,113b]. A new species is formed absorbing at 510 nm and displaying a narrow ESR signal centred on $g = 1.9805$. This species is reduced by the alcohol with a primary kinetic isotope effect of 9.1 ± 0.8[113b]. The term in $[H_3O^+]^2$ in the rate expression corresponds to a protonated ester.

Oxidation of cyclobutanol by a Cr(VI)–V(IV) couple appears to involve attack of Cr(IV) upon the substrate to yield a free radical, (Section 2.5). This implies the following possible variation in the Westheimer scheme for a secondary alcohol[20]

$$\text{Cr(IV)} + R_2CHOH \rightarrow \text{Cr(III)} + R_2\dot{C}OH + H^+$$

$$R_2\dot{C}OH + \text{Cr(VI)} \rightarrow R_2C{=}O + \text{Cr(V)} + H^+$$

$$\text{Cr(V)} + R_2CHOH \rightarrow R_2C{=}O + \text{Cr(III)}$$

The rival mechanism advanced by Roček and Krupička[101] has, as the first step

$$R_2C(H)(O{-}H) + HO(O{=})Cr(O)(OH) \longrightarrow R_2C{=}O + H_2O + O{=}Cr(OH)_2$$

The two major differences are: (*i*) the absence of any complex or ester in the second mechanism, and (*ii*) the direction of the movement of the electron pair of the C–H bond, the hydrogen becoming proton-like in the transition state of the Westheimer scheme but hydride-like in that of the Roček mechanism.

Both schemes accommodate the kinetics, the primary isotope effect and the induction factor, which indicates that Cr(IV) is the initial stage of reduction of the oxidant. Roček's mechanism does not accord with the solvent isotope effect of 2.5 per proton, which has just the value to be expected for acid-catalysis, for the O–H bond cleavage should be subject to a primary isotope effect of about 7. The ester mechanism is not open to this criticim.

The main objections to the ester mechanism are based on the substituent effects and, to a lesser degree, the kinetics in strongly acidic solutions. The departure of the hydrogen as a proton in the slow step should be facilitated by electron-withdrawing groups, but the converse is found. Substituents are, of course, bound to affect both the esterification equilibrium (29) and the oxidation step (30) and the observed ρ is the sum of the individual values. Spectroscopic evidence has been obtained[114,115], however, for the existence of 1 : 1 complexes in solution between alcohols and acid chromate ion, *e.g.* K_{25° for isopropanol is 0.083 $l.mole^{-1}$ ($\mu = 0.001\ M$). These esters can be extracted into benzene[104] and are rapidly decomposed therein on addition of pyridine to give the oxidation product, although hydrolysis is a competing reaction[116]. The pyridine-catalysed oxidative breakdown in benzene has a primary kinetic isotope effect of about 5 (ref. 116).

These results do not prove that the ester is an essential intermediate in aqueous solution even though it is present, but the result with the hindred triterpene is convincing[102]. In this case the esterification step, which is normally fast, has become rate-determining and the disappearance of the isotope effect must mean that C–H cleavage occurs after the formation of the ester and not independently of it. The generality of this result is apparent from the stopped-flow investigation of isopropanol oxidation [113].

The h_0 dependence of the oxidation in strongly acidic media was taken by Roček as implying that no water molecule is involved in the transition state and that consequently the ester mechanism as portrayed above cannot hold. However, the Zucker–Hammett hypothesis upon which this argument is based, *i.e.* that a reaction forming a transition state containing a water molecule will follow a $[H_3O^+]$ dependence, but that otherwise an h_0 dependence will be followed, may not be valid, and in any case the ester can be depicted as breaking down as follows[117, 118]

$$R_2C(H)\text{–O–}Cr(=O)_2(OH) \longrightarrow R_2C{=}O + H\text{–O–}Cr(=O)(OH)$$

The maximum rate reported at very high acidity accords with protonation of the ester[105]. To summarise, the ester mechanism has gained general acceptance, although the substituent effects have yet to be explained wholly satisfactorily and the exact nature of the transition state, *i.e.* whether it is of considerable or only slight carbonyl character, remains a contentious issue[67, 119].

Many interesting observations not bearing directly on the general mechanism have been reported. Certain secondary alcohols undergo cleavage[120], *e.g.*

$$C_6H_5\text{–}CH(OH)\text{–}C(CH_3)_3 \xrightarrow{Cr(VI)} C_6H_5C(=O)\text{–}C(CH_3)_3,\ C_6H_5CHO \text{ and } (CH_3)_3COH$$

which is up to 70 % of the total reaction in the case of phenyl-*tert*-butylcarbinol. Addition of manganous or cerous ions, which react with Cr(IV), greatly reduces the degree of cleavage. $C_6H_5CD(OH)C(CH_3)_3$ is oxidised at only one-twelfth of the rate of the "light" alcohol at 0 °C. Cleavage does not break the C–H(D) bond and hence the isotope effect implies that cleavage does not occur by attack of Cr(VI) upon the alcohol, for otherwise the oxidation route would be completely usurped for the "heavy" alcohol. Presumably, therefore, cleavage is entirely due to attack of Cr(IV) or Cr(V).

The following mechanism for cleavage was proposed[121]

$$C_6H_5CH(OH)C(CH_3)_3 + Cr(V) \rightleftharpoons C_6H_5\text{-}CH(O\text{-}Cr(V))\text{-}C(CH_3)_3$$

$$C_6H_5\text{-}CH(O\text{-}Cr(V))\text{-}C(CH_3)_3 \rightarrow C_6H_5CHO + Cr(III) + (CH_3)_3C^+$$

$$(CH_3)_3C^+ + H_2O \rightarrow (CH_3)_3COH + H^+$$

This is supported by the observation[120] that ^{18}O-labelled alcohol is cleaved to unlabelled *t*-butanol. Wiberg[67] has given other examples of this type of cleavage, and has dealt with other complications such as further oxidation of products, formation of hemiacetals, *etc.*

The reactivity of Cr(IV) towards cyclobutanol to give 4-hydroxybutyraldehyde (ref. [186b]) (Section 2.5) suggests the possibility of participation of this species in cleavage processes, and Nave and Trahanovsky[666] have demonstrated *radical* processes in the Cr(VI) oxidation of 2-aryl-1-phenylethanols by trapping $C_6H_5CH_2\cdot$ with molecular oxygen (to give benzyl alcohol). Ring-substitution of the 2-aryl group produced good Hammett plots both for ketone formation ($\rho = -0.10 \pm 0.02$) and for cleavage ($\rho = 1.06 \pm 0.04$). The Westheimer cleavage mechanism is accordingly modified to

$$Cr(VI) + RCHOHR' \rightarrow Cr(IV) + RCOR'$$

$$Cr(IV) + RCHOHR' \rightarrow Cr(III) + RCHO + R'\cdot$$

$$Cr(VI) + R'\cdot \rightarrow Cr(V) + R'OH$$

$$2Cr(V) \rightarrow Cr(VI) + Cr(IV)$$

Interest has been shown by several groups on the effect of solvent and of added anions upon the oxidation of alcohols. The oxidation of isopropanol proceeds 2500 times faster in 86.5 % acetic acid than in water at the same hydrogen ion concentration[27]. The kinetics and primary kinetic isotope effect are essentially the same as in water. Addition of chloride ion strongly inhibits the oxidation and the spectrum of chromic acid is modified. The effect of chloride ion was rationalised[27] in terms of the equilibrium,

$$H^+ + Cl^- + HCrO_4^- \rightleftharpoons ClCrO_3^- + H_2O$$

where $K \simeq 10^5$ l^2.mole^{-2}, and the chlorochromate ion is presumed to be of negligible oxidising power. This type of equilibrium has been detected spectrophotometrically for phosphate[28], sulphate[122, 123] and other inorganic anions[29] and its influence on the rate of oxidation of isopropanol has been examined systematically[29]. The indications from optical measurements are that the anion A^- does not enter $HCrO_4^-$ to any great extent but that it does enter the neutral acid, *viz.*

$$H_2CrO_4 + HA \rightleftharpoons H_2O + HCrO_3A$$

where A = OPO_3H_2, –Cl, $-OSO_3H$, $-OClO_3$ and $-ONO_2$. A correlation exists[29] (with the exception of HNO_3) between the pK values of chromic acid in aqueous solution of the mineral acids and pK values of the mineral acids themselves.

These conclusions receive support from the kinetic data[29]. The rate of oxidation by the acid chromate ion is independent of the nature of the mineral acid but at higher acidity, when the oxidation by chromic acid becomes dominant, the rate depends not only upon the acidity but also upon the acid concerned, the oxidising ability of the species, $HCrO_3A$, increasing in the order

$$H_3CrPO_7 < HCrClO_3 < H_2CrSO_7 < HCrClO_7 < HCrNO_6$$

for a given value of h_0. From the mechanistic point of view, however, this merely means the replacement of an OH group on the chromium atoms in reaction (30) by A.

The effect of adding large quantities of acetic acid to the medium is more complicated. The acceleration of the oxidation rate of isopropanol was ascribed initially to a shift of the esterification equilibrium to the right (reaction 29). However, Roček found that acceleration by acetic acid occurs for oxidations which cannot involve a pre-equilibrium esterification, *e.g.* those of aliphatic[72] and alicyclic[72, 124] hydrocarbons. The obvious alternative, *i.e.* that acetic acid combines with chromic acid, *viz.*

$$H_2CrO_4 + CH_3CO_2H \rightleftharpoons CH_3COOCrO_2OH + H_2O$$

has been proposed[124, 125], and is supported by a very detailed examination of the oxidation of isopropanol in trifluoroacetic acid[30]. A plot of log k_2 for the latter medium against H_0 revealed a kink at $H_0 = 0.30$. A similar kink appears on graphs of the acidity dependences of the oxidation in various aqueous mineral acids (*vide supra*) and corresponds to a changeover of the oxidant from $HCrO_4^-$ to $HCrO_3A$. The reaction is independent of H_0 for $H_0 < -1.9$. The thermodynamic parameters and primary kinetic isotope effect for the reaction in CF_3CO_2H are approximately the same as in water. At high acidities chloride ion exhibited a very strong specific retarding effect ascribed to the process

$$HOCrO_2OCOCF_3 + Cl^- \rightleftharpoons HOCrO_2Cl^- + CF_3CO_2^-$$

Of relevance also is the effect on chromic acid oxidation of aqueous isopropanol of systematically replacing water by acetone[126]. Most parameters, including the primary kinetic isotope effect, the Hammett ρ value and salt effects are largely unaltered, but the rate is enhanced by a factor of 7×10^2 in 93.3 % acetone compared with that in water and depends on the h_0 function. These observations cannot be accommodated by invoking substitution of an anion into $HCrO_4^-$ and the enhancement may originate in a shift of the equilibrium (29) to the right.

A number of effects of conformation have been reported[118]. "Crowding" of the hydroxyl group results in faster oxidation and alcohols with axial hydroxyl groups are oxidised more readily than the equatorial isomers, *e.g.* *cis*-4-*tert*-butylcyclohexanol is oxidised about three times faster[127] than the *trans* isomer at 25 °C.

TABLE 2

OXIDATION OF SUBSTITUTED PHENYLTRIFLUOROMETHYLCARBINOLS BY CHROMIC ACID AT 25 °C

Substituent	k_2 *(averaged) for "light" alcohol (l.mole^{-1}.sec^{-1})*	k_H/k_D
CH_3	0.544	7.40±0.20
H	0.256	8.53±0.15
m-Br	0.111	9.80±0.20
m-NO_2	0.0494	12.20±0.20
3,5-Dinitro	0.0117	12.93±0.78

One further striking kinetic feature is the magnitude of the primary kinetic isotope effect for the oxidation of certain aryltrifluoromethylcarbinols[128] (Table 2). It is clear that a correlation exists between the isotope effect and the oxidation rate, slow oxidations exhibiting large isotopes effects. The magnitude of the isotope effects may be due to one of several effects including loss of bending frequencies in the transition state or quantum-mechanical tunnelling[129]. That large isotope effects are often associated with neighbouring groups of fluorine atoms is apparent from the analogous results with permanganate (p. 308).

The oxidation of tertiary alcohols by chromic acid is comparatively slow and shows a zero-order dependence of the rate upon oxidant concentration[130, 131]. For 1-methylcyclohexanol[131] the kinetics are

$$-\mathrm{d[Cr(VI)]/d}t = k[\text{alcohol}]h_0$$

Clearly the rate-determining step is the elimination of water from the protonated alcohol to form a carbonium ion, which loses a proton to give the readily oxidised olefin.

The Cr(VI) oxidation of di- and triarylcarbinols in acetic acid–sulphuric acid mixtures shows normal kinetics (first-order each in alcohol and Cr(VI) and an h_0 dependence up to $H_0 \sim -1$)[131a]. Electron-donating substituents appear in the phenolic component and the rate-determining step is believed to involve a 1,2-

aryl shift, *viz.*

$$Ar_3C\text{–}O\text{–}CrO_3H_2^+ \rightarrow Ar_2\underset{\underset{Ar}{\vdots\delta+\vdots}}{C\text{–}O}^{\delta+}\ldots CrO_3H_2 \rightarrow Ar_2C\text{=}\overset{+}{O}\text{–}Ar$$

$$Ar_2C\text{=}\overset{+}{O}\text{–}Ar + H_2O \rightarrow Ar_2C\text{=}O + ArOH + H^+$$

The reaction parameters for tri- and diarylcarbinol oxidation respectively are $\rho^+ = -0.879$ and $\rho = -0.54$. For migration in triarylcarbinols $\rho^+ = -1.44$. For triphenylcarbinol at an H_R of -3.95, $E = 12.4$ kcal.mole^{-1} and $\Delta S^\ddagger = -25.1$ eu whilst for benzhydrol at $H_R = -2.62$, $E = 6.5$ kcal.mole^{-1} and $\Delta S^\ddagger = -37.1$ eu.

The oxidation of alcohols to carbonyl compounds by permanganate proceeds most rapidly in basic solution and it is with this medium that the majority of kinetic studies have been performed.

As with chromic acid, tertiary alcohols are oxidised only very slowly with degradation. The rate expression for oxidation of secondary alcohols is[132, 133]

$$-d[MnO_4^-]/dt = k[R_2CHOH][MnO_4^-][OH^-]$$

and in basic solution the stoichiometry for benzhydrol oxidation is[134]

$$(C_6H_5)_2CHOH + 2\,MnO_4^- + 2\,OH^- = (C_6H_5)_2CO + 2\,MnO_4^{2-} + 2\,H_2O$$

The kinetic parameters are $E = 6.3$ kcal.mole^{-1} and $\Delta S^\ddagger = -38.4$ eu, and at 25 °C the reaction exhibits a primary kinetic isotope effect of 6.6. When ^{18}O-labelled MnO_4^- was employed, no labelled oxygen appeared in the benzophenone. The mechanism[132] involves abstraction of hydrogen, either as a hydride ion or a hydrogen atom, from the anion of the alcohol

$$R_2CHOH \rightleftharpoons R_2CHO^- + H^+$$

$$R_2CHO^- + MnO_4^- \begin{cases} \rightarrow R_2C\text{=}O + HMnO_4^{2-} \\ \rightarrow R_2\dot{C}\text{–}O^- + HMnO_4^- \end{cases}$$

A final decision between these alternatives has yet to be made.

The oxidation by basic permanganate of phenyltrifluoromethyl carbinols[134, 134a], which has the same rate law as benzydrol, is characterised by very large primary kinetic isotope effects (*cf.* Cr(VI) p. 307).

Substituent	H	p-CH_3	m-Br
k_H/k_D	16.0	16.1	16.2
k_H/k_T	57.1 ± 2.4	–	–

That reaction rates are barely affected by the nuclear substitution ill-accords with the hydride abstraction mechanism. However, nuclear substitution affects both the ionisation of the alcohol as well as the oxidation step. The value of k_H/k_T agrees well with that of 55.5 calculated from k_H/k_D by means of the Swain relationship,[134b] *i.e.* $(k_H/k_D)^{1.442} = k_H/k_T$.

A few results have been reported on the oxidation of cyclohexanol by acidic permanganate[24, 135]. In the absence of added fluoride ions the reaction is first-order in both alcohol and oxidant[135], the apparent first-order rate coefficient (for excess alcohol) at 25 °C following an acidity dependence $k = 3.5+16.0$ $[H_3O^+]sec^{-1}$. k_H/k_D depends on acidity (3.2 in dilute acid, 2.4 in 1 *M* acid) and k_{D_2O}/k_{H_2O} is 1.74. Addition of fluoride[24] permitted observation of the reaction for longer periods (before precipitation) and under these conditions methanol is attacked at about the same rates as di-isopropyl ether, although dioxan is oxidised over twenty times more slowly. The lack of specificity and the isotope effect indicates that a hydride-ion abstraction mechanism operates under these conditions. (The reactivity of di-isopropyl ether towards two-equivalent oxidants is illustrated by its reaction with Hg(II).) Similar results were obtained with buffered permanganate.

An analogous study[668] of the oxidation of benzyl alcohol by permanganate in aqueous perchloric acid yielded a rate law

$$-d[MnO_4^-]/dt = k_3[MnO_4^-][C_6H_5CH_2OH][H_3O^+]$$

with $k_3 = 0.32\ l^2.mole^{-2}.sec^{-1}$ at 35 °C ($\mu = 1.92\ M$). Mn^{2+} and F^- ions are without effect on the rate and $E = 11.4$ kcal.mole^{-1}, $\Delta S^\ddagger = -27.2$ eu. Hydride-ion transfer to $HMnO_4$ is proposed as the mechanism of reaction.

The MnO_4^- oxidation of di- and triarylcarbinols in moderately strong sulphuric acid ($H_0 < -0.50$) is first-order in alcohol and zero-order in oxidant; plots of $\log k$ *versus* H_0 give good straight lines of slope approximately -1 for both series of substrates[135a]. k_H/k_D for oxidation of $(C_6H_5)_2CDOH$ is only 1.08 and the Hammett ρ values are -1.02 for secondary alcohols and -1.39 for tertiary alcohols. These results indicate rate-determining carbonium-ion formation, *viz.*

$$Ar_2CHOH+H^+ \rightleftharpoons Ar_2CH^+ + H_2O$$

2.3.4 *Aldehydes*

Aldehydes exist in aqueous solution in one or both of the forms

$$RCH{=}O+H_2O \rightleftharpoons RCH(OH)_2 \tag{31}$$

and in any example the solute may act either as a *gem*-diol or a carbonyl compound.

References pp. 493–509

The chromic acid oxidation of benzaldehyde to benzoic acid in aqueous solution was examined by Graham and Westheimer[117], and in acetic acid solution by Wiberg and Mill[136]. With water as solvent the kinetics are

$$-\mathrm{d}[\mathrm{Cr(VI)}]/\mathrm{d}t = k[\mathrm{RCHO}][\mathrm{HCrO_4^-}]\{[\mathrm{H_3O^+}]+k'[\mathrm{H_3O^+}]^2\}$$

and in acetic acid solution

$$-\mathrm{d}[\mathrm{Cr(VI)}]/\mathrm{d}t = k[\mathrm{RCHO}][\mathrm{HCrO_4^-}]h_0$$

For the latter reaction $E = 13.2$ kcal.mole^{-1} and $\Delta S^{\ddagger} = -28$ eu.

Both groups reported a sharp decrease (40–65 %) in the rate on addition of Mn(II) ions. Wiberg and Mill[136] reported a deuterium isotope effect of 4.3 (30 °C) and that electron-withdrawing groups facilitate reaction ($\rho = +1.02$), the latter result confirming the data of Lucchi[137, 138] ($\rho = +1.06$). The opposite effect of electron-withdrawing groups for aliphatic aldehydes has been reported by Roček[139] ($\rho^* = -1.2$), who explains the difference in terms of the hydration pre-equilibrium (31) which favours the *gem*-diol form for aliphatic aldehydes and the carbonyl form for aromatic aldehydes. Otherwise the oxidation of aliphatic aldehydes strictly parallels that of benzaldehyde as regards kinetics[140, 141] (formaldehyde), Arrhenius parameters (formaldehyde[141], acetaldehyde, *n*-butyraldehyde, isobutyraldehyde[142]), primary kinetic and solvent isotope effects (formaldehyde[141], acetaldehyde[143]), and retardation by Mn(II) with induced oxidation (formaldehyde[140]).

Accordingly, the most reasonable mechanism is the exact analogue of that proposed for alcohol oxidation

$$\mathrm{RCHO + H^+ + HCrO_4^-} \rightleftharpoons \mathrm{R{-}C(OH)(H){-}OCrO_3H}$$

$$\mathrm{R{-}C(OH)(H){-}O{-}Cr(=O)_2OH} \longrightarrow \mathrm{R{-}C(OH){=}O + Cr(IV)}$$

Although the fate of Cr(IV) is uncertain, (*cf.* the alcohol oxidation), some characteristics of the intermediate chromium species have been obtained by Wiberg and Richardson[144] from a study of competitions between benzaldehyde and each of several substituted benzaldehydes. The competition between the two aldehydes for Cr(VI) is measured simply by their separate reactivities; that for the Cr(V) or Cr(IV) is obtained from estimation of residual aldehyde by a ^{14}C-labelling technique. If Cr(V) is involved then ρ values for oxidation by Cr(VI) and Cr(V) are 0.77 and 0.45, respectively. An isotope effect of 4.1 for oxidation of benzaldehyde by Cr(V) was obtained likewise.

Wiberg and Lepse[145] have examined the oxidation of benzaldehyde by chromyl acetate in an acetic anhydride medium. The orange colour of the oxidant changes to dark brown on addition of the aldehyde. Analysis showed that each chromium atom retains one equivalent of oxidising power and the spectrum rules out the possibility of a Cr(VI): Cr(III) complex. Isotope dilution analysis indicated that almost 25 % of the oxidant was consumed by the solvent under some conditions and this effect was minimised by employing excess aldehyde. The oxidation proceeds in two stages, both steps displaying a total kinetic order of two. Some substituted benzaldehydes show detailed differences of order in one step or the other. The ρ values for the two steps were estimated to be -0.2 and -0.9 respectively, and both display a moderate primary kinetic isotope effect. ^{18}O is transferred from labelled oxidant to the aldehyde and an ESR spectrum obtained from the reaction mixture was assigned to Cr(V). The authors prefer a one-equivalent oxidation mechanism to give an acyl radical (corresponding to the first step); the resulting Cr(V) effects a similar one-equivalent oxidation (corresponding to the second step), *viz.*

$$RCHO + (CH_3CO)_2CrO_2 \rightarrow R\dot{C}{=}O + Cr(V)$$
$$R\dot{C}O + (CH_3CO)_2CrO_2 \rightarrow R\overset{+}{C}O + Cr(V)$$
$$RCHO + Cr(V) \rightarrow R\dot{C}O + Cr(IV)$$
$$R\dot{C}O + Cr(V) \rightarrow R\overset{+}{C}O + Cr(IV)$$

The permanganate oxidation of aromatic aldehydes has been studied by several groups[146, 147]. The production of benzoic acid from benzaldehyde necessitates the use of buffers in all save the most basic solution (pH 12). Wiberg and Stewart[146] confined their measurements to a pH range of 5 to 13; at pH < 5 an autocatalytic reaction appeared. Under these conditions the end product is Mn(IV) or Mn(VI) (depending on pH) and the reaction is first-order in both aldehyde and oxidant. The acidity dependence is complex; the reaction showing a small negative ρ value (-0.248) in neutral solution and a large positive value ($+1.83$) in an alkaline medium. It is clear that two paths exist, one predominating at about pH 7 and the other at pH 11. The primary kinetic isotope effect is much larger at pH 6.5 than at pH 11–12. Again, 75 % oxygen transfer (determined with ^{18}O) occurs from labelled MnO_4^- to aldehyde at pH 5.5 but only about 26 % at pH 12.4. Rate parameters were obtained at pH 6.5 ($E = 10.3$ kcal.mole^{-1} and $\Delta S^\ddagger = -26.2$ eu for benzaldehyde). Detailed examination of the acidity dependence for the two paths indicated that in acidic solution general acid catalysis is operative, but that in alkaline solution catalysis is specific to hydroxide ion, the order with respect to OH^- being one-half†. A comparison with the manganate-ion oxidation of benzaldehydes was made and it was found that Mn(VI) attacks a series of

† This result has been questioned recently[706].

substituted benzaldehydes at similar rates as alkaline MnO_4^- and an identical ρ value is obtained.

Two mechanisms are proposed by the authors. For the acidic–neutral solutions

$$ArCHO + MnO_4^- + H_2O \rightleftharpoons Ar\text{–}\underset{\displaystyle H}{\overset{\displaystyle OH}{C}}\text{–}O\text{–}\underset{\displaystyle O}{\overset{\displaystyle O}{\overset{\|}{\underset{\|}{Mn}}}}{=}O + OH^-$$

$$Ar\text{–}\underset{\displaystyle H}{\overset{\displaystyle OH}{C}}\text{–}O\text{–}MnO_3 + B: \rightarrow Ar\text{–}\overset{\displaystyle OH}{C}{=}O + Mn(V)O_3^- \quad HB$$

$$3\,MnO_3^- + H_2O = 2\,MnO_2 + MnO_4^- + 2\,OH^-$$

H_2O may be replaced by any acid, HA, and a cyclic mechanism for the breakdown of the ester is quite feasible. For oxidation in alkali the fractional order in hydroxide ion, the low k_H/k_D and low degree of oxygen-transfer from oxidant are taken as symptomatic of a free-radical chain reaction of the type

$$MnO_4^- + OH^- \rightarrow MnO_4^{2-} + OH\cdot$$

$$OH\cdot + RCHO \rightarrow R\text{–}CH(OH)\text{–}O\cdot$$

$$R\text{–}CH(OH)\text{–}O\cdot + MnO_4^- \rightarrow RCO_2H + MnO_3^- + OH\cdot$$

with termination by dimerisation of OH·. This particular scheme does not fit the observed kinetics and it should operate for *any* organic substrate readily oxidised by hydroxyl radical, of which there are a large number. The kinetics are, however, unique to aromatic aldehydes.

Investigation of the oxidation of aliphatic aldehydes[32] has been confined to *gem*-diols which behave as secondary alcohols, being most easily oxidised in the anionic forms, *e.g.*

$$\underset{\displaystyle O^-}{CF_3CH}(OH) \quad \text{and} \quad \underset{\displaystyle O^-}{CF_3CH}\text{–}O^-$$

although in very strong acid the permanganic acid itself effects rapid oxidation. All these routes for fluoral hydrate display a substantial primary kinetic isotope effect and Arrhenius parameters have been obtained (Table 3).

As has frequently been the case, the evidence at present available does not permit discrimination between the possibilities of hydride ion- or hydrogen atom-abstraction by Mn(VII).

TABLE 3

KINETIC ISOTOPE EFFECTS AND ARRHENIUS PARAMETERS FOR THE OXIDATION OF FLUORAL HYDRATE BY Mn(VII) AND Mn(VI)

Reaction		k_H/k_D	*E (kcal. mole^{-1})*	$\Delta S^‡$*(eu)*
MnO_4^{2-}–dianion		4.8	—	—
MnO_4^{-}–dianion		5.1	7.2	−31.4
MnO_4^{-}–monoanion	pH 10.2	10.1	13.0	−18.4
	pH 7.6	13.6	—	—
$HMnO_4$–neutral aldehyde		6.3	13.5	−17.8

2.3.5 Phenols

In a perchlorate medium the oxidation of hydroquinone to *p*-benzoquinone by chromic acid obeys the rate expression[148]

$$-d[Cr(VI)]/dt = k[Cr(VI)][hydroquinone][H_3O^+]^{1.25}[quinone]^0$$

however, a power series of the type

$$a+b[H_3O^+]+c[H_3O^+]^2$$

also represents the acidity dependence adequately. At pH 2 the *second*-order coefficient is 53.4 ± 1.2 l.mole^{-1}.sec^{-1} ($\mu = 5 \times 10^{-2}$ *M*). The activation energies for the terms involving *b* and *c* are 4.6 ± 1.0 and 5.2 ± 0.9 kcal.mole^{-1}, respectively. Use of a stopped-flow apparatus and high reactant concentrations in an effort to identify an ester intermediate was unsuccessful.

The permanganate oxidation of phenols is complicated by the intervention of lower oxidation states of manganese, (*cf.* the oxidation of toluene, p. 298). For example, the oxidation of 2,6-dinitrophenol[149, 150] in weakly acidic solution displays an induction period, following second-order kinetics thereafter. However, addition of potassium fluoride inhibits reaction almost completely, but manganous ions strongly accelerate it.

Free-radical intermediates in the oxidation of a series of quinol phosphates by permanganate at pH 11.7 have been characterised by an ESR rapid-mixing technique[151]. Clearly a one-equivalent oxidation step to give a semiquinone phosphate radical operates in this case.

2.3.6 Ketones

The chromic acid oxidation of cyclohexanone to adipic acid *via* 2-hydroxy-cyclohexanone and cyclohexane-1,2-dione[152] is third order[153], *viz.*

$$\frac{-d[Cr(VI)]}{dt} = k[ketone][HCrO_4^-][H_3O^+]$$

where $k = 1.05 \times 10^{-2}$ $l^2.mole^{-2}.sec^{-1}$ at 50 °C ($\mu = 0.646$ M). The rate of oxidation is considerably lower than that of enolisation at the same temperature and the question arises at to whether oxidation proceeds *via* attack of Cr(VI) upon the ketone or the enol, and the observation of a primary kinetic isotope effect of 4.0 (50 °C) on changing the substrate to 2,2,6,6-tetradeuterocyclohexanone does not clarify matters, for the enolisation itself is subject to an isotope effect of the same order of magnitude[154]. However, the reaction proceeds four times faster in heavy than in light water and this increase, which is considerably larger than the factor of 2.0 to 2.5 for proton-catalysed reactions, was taken as evidence by the authors that enolisation must be taking place as a necessary pre-requisite for reaction, for this equilibrium is favoured by a factor of 2.0–2.5 on changing to heavy water. The mechanism therefore becomes

$$\text{cyclohexanone} \underset{k_K}{\overset{k_E}{\rightleftharpoons}} \text{cyclohexenol (OH)} \qquad \text{(fast)}$$

$$\text{cyclohexenol (OH)} + H^+ + HCrO_4^- \rightleftharpoons \text{cyclohexenyl–O–CrO}_2\text{–OH} \qquad \text{(fast)}$$

$$\text{cyclohexenyl–O–CrO}_2\text{–OH} \xrightarrow{k} \text{cyclohexanone cation (+)} + CrO_2 + OH^- \qquad \text{(slow)}$$

$$\text{cyclohexanone cation (+)} \xrightarrow{H_2O} \text{2-hydroxycyclohexanone (OH)} \qquad \text{(fast)}$$

Roček and Riehl[155] have pointed out that the enolisation scheme leads to kinetics

$$\frac{-d[\mathrm{Cr(VI)}]}{dt} = \frac{k_E k[\text{ketone}][\mathrm{Cr(VI)}]}{k_K + k[\mathrm{Cr(VI)}]}$$

This forecasts that should k[Cr(VI)] considerably exceed k_K then a change of reaction order with respect to Cr(VI) from one to zero is to be expected. This they observed for isobutyrophenone in 99 % acetic acid and 2-chlorocyclohexanone in water at moderate concentrations of oxidant.

Permanganate attacks ketones at both high and low pH. Alkaline permanganate oxidises acetone[156], the rate law being

$$-d[MnO_4^-]/dt = k_{obs}[\text{acetone}][OH^-][MnO_4^-]$$

where $k_{obs} = 107$ $l^2.mole^{-2}.sec^{-1}$ at 25 °C. These kinetics suggest the mechanism

$$\text{acetone} + OH^- \underset{k_{-1}}{\overset{k_1}{\rightleftharpoons}} \text{enolate} + H_2O$$

$$\text{enolate} + MnO_4^- \overset{k_2}{\rightarrow} \text{product} + \mathrm{Mn(V)}$$

$$\mathrm{Mn(V)} + \mathrm{Mn(VII)} \xrightarrow{\text{fast}} 2\ \mathrm{Mn(VI)}$$

A steady-state approximation for enolate ion yields

$$-\frac{d[MnO_4^-]}{dt} = \frac{2\,k_1\,k_2[MnO_4^-][\text{acetone}][OH^-]}{k_{-1}[H_2O]+k[MnO_4^-]}$$

Inclusion of an oxidation of neutral enol is incompatible with the observed rate law.

Oxidation of [2-^{14}C]acetone produced the following yields of products identified by isotope dilution analysis

acetol	4 %	pyruvaldehyde	0 %	lactic acid	6 %
pyruvic acid	27 %	oxalic acid	19 %	acetic acid	47 %

These figures suggest an oxidation sequence

$$CH_3\overset{\overset{O}{\|}}{C}CH_3 \rightarrow CH_3\overset{\overset{O}{\|}}{C}CH_2OH \rightarrow CH_3\overset{\overset{O}{\|}}{C}CHO \rightarrow CH_3COCO_2H \begin{cases} \nearrow CH_3CO_2H \\ \searrow (CO_2H)_2 \end{cases}$$

(with a direct route from CH_3COCH_3 to CH_3COCHO, and $CH_3COCHO \xrightarrow{OH^-} CH_3CH(OH)CO_2^-$)

Rate laws and coefficients were determined for the oxidation of all intermediate compounds by MnO_4^- and were compatible with the scheme as presented, *i.e.* including a route for *direct* oxidation to pyruvaldehyde. An estimate of k_2 ($\sim 5\times10^7$ l.mole^{-1}.sec^{-1}) suggests an initial electron-transfer to give $CH_3COCH_2\cdot$. This can then be rapidly oxidised in two ways

$$CH_3\overset{\overset{O}{\|}}{C}CH_2\cdot \begin{cases} \nearrow CH_3\overset{\overset{O}{\|}}{C}CH_2^+ + Mn(VI) \rightarrow CH_3COCH_2OH \\ \searrow CH_3C\text{–}\underset{\underset{H}{|}}{CH}\text{–}OMn(VI)O_3^- \end{cases}$$

$$CH_3C\text{–}\underset{\underset{H}{|}}{CH}\text{–}OMn(VI)O_3^- \xrightarrow{OH^-} CH_3\overset{\overset{O}{\|}}{C}\text{–}CHO + H_2O + Mn(IV)$$

A few kinetic measurements on the *acid* permanganate oxidation of cyclohexanone have been reported by Littler[157]. The reaction is first order in ketone but the order in oxidant was not clearly established.

2.3.7 Monocarboxylic acids

Formic acid is rather atypical and should be considered separately. The chromic acid oxidation has been examined in some detail by Kemp and Waters[141], following the earlier work of Snethlage[158]. At acidities of up to 3.6 M ($HClO_4$) the reaction follows kinetics

$$\frac{-\mathrm{d}[\mathrm{Cr(VI)}]}{\mathrm{d}t} = k[\text{formic acid}][\mathrm{HCrO_4^-}][\mathrm{H_3O^+}]^2$$

with $E = 11.6$ kcal.mole^{-1} and $\Delta S^{\ddagger} = -38$ eu. However, at an ionic strength of 6.0 M (ClO_4^-), the $[H_3O^+]^2$ term reverts to h_0. Several effects associated also with the Cr(VI) oxidation of alcohols were noted, including a primary kinetic isotope effect of 7.15 at 25 °C, a solvent isotope effect of 5.74 and a retardation of rate by Mn(II), the latter in a sulphate-ion medium only. It is probable, therefore that a Westheimer-type mechanism operates through an intermediate formylchromic anhydride, $HCOOCrO_3H$, which undergoes protonation before breaking down to CO_2 and Cr(IV), *cf.* reactions (28)–(30).

The permanganate oxidation of formic acid has attracted much attention. The reaction is pH-independent above pH 5 and involves formate ion. At lower pH's the rate is much lower until permanganic acids begins to be formed at very low pH[159].

Above pH 5 the reaction presumably involves two anions and, indeed, a positive salt effect is found[160, 161] and the rate expression is

$$-\mathrm{d}[\mathrm{MnO_4^-}]/\mathrm{d}t = k_2[\mathrm{HCO_2^-}][\mathrm{MnO_4^-}]$$

A large primary kinetic isotope effect has been reported by several groups[160,162–164], the most comprehensive study having been made by Bell and Onwood[160] in an attempt to assess the role of quantum-mechanical tunnelling. This was found to be unimportant and the rate coefficients at infinite dilution were expressed as

$$k_{\mathrm{H}} = (4.72 \pm 0.60) \times 10^8 \exp(-11{,}849 \pm 75/\boldsymbol{RT})\mathrm{l.mole^{-1}.sec^{-1}}$$
$$k_{\mathrm{D}} = (3.94 \pm 0.06) \times 10^8 \exp(-13{,}056 \pm 13/\boldsymbol{RT})\mathrm{l.mole^{-1}.sec^{-1}}.$$

A small solvent isotope effect was found by Bell and Onwood[160] ($k_{H_2O}/k_{D_2O} = 1.08$) in contradiction to that of only 0.38 reported by Taylor and Halpern[163]. Over one-third of the oxygen present in the carbonate originated from the oxidant when ^{18}O-labelled permanganate was used[163]. The reaction is subject to pronounced catalysis by ferric ions[163].

The available data do not permit a conclusive mechanism to be formulated. C–H cleavage is clearly rate-determining but no choice between a one- or two-equivalent process can yet be made. A simple hydride-ion or hydrogen-atom transfer is not compatible with the observed ^{18}O exchange.

The reaction between permangante ion and neutral formic acid follows similar bimolecular kinetics[163] with $k_2 = 1.1 \times 10^9 \exp(-16.4 \times 10^3/\boldsymbol{RT})$l.mole^{-1}.sec^{-1}. No primary kinetic isotope effect was found for this path either in light or heavy water. However, Mocek and Stewart[159] have reported that in very strong sulphuric acid the oxidations of neutral substrate by both $HMnO_4$ and MnO_3^+ display substantial isotope effects.

The oxidation of formate bound to cobalt(III) ("captive formate") has been examined by Candlin and Halpern[11]. The kinetics are

$$-\mathrm{d[complex]}/\mathrm{d}t = k[(NH_3)_5Co{\cdot}O_2CH^{2+}][MnO_4^-]$$

and the stoichiometry follows two paths

$$(NH_3)_5Co{\cdot}O_2CH^{2+} + \tfrac{1}{3}\,MnO_4^- = CO_2 + Co^{2+} + \tfrac{1}{3}\,MnO_2$$
$$(NH_3)_5Co{\cdot}O_2CH^{2+} + \tfrac{2}{3}\,MnO_4^- = CO_2 + (NH_3)_5Co{\cdot}OH_2^{3+} + \tfrac{2}{3}\,MnO_2$$

C-deuteration of the captive formate reduces the rate by a factor of 10.5 without affecting the stoichiometry. The ratio of unreduced to reduced cobalt was found to depend on oxidant concentration

$$\frac{[(NH_3)_5Co{\cdot}OH_2^{3+}]}{[Co^{2+}]} = 3 \times 10^2[MnO_4^-]$$

The authors propose that an initial abstraction of a hydrogen atom produces a complex containing bound carbon dioxide radical-anion. This complex could break down in two ways

$$[(NH_3)_5Co{\cdot}({}^-O_2CH)]^{2+} + MnO_4^- \rightarrow [(NH_3)_5Co(III)({\cdot}CO_2^-)]^{2+} + HMnO_4^-$$

$$[(NH_3)_5Co(III)({\cdot}CO_2^-)]^{2+} \begin{array}{l} \nearrow\ Co^{2+} + CO_2 \\ \underset{MnO_4^-}{\searrow}\ [(NH_3)_5Co(III)OH_2]^{3+} + CO_2 \end{array}$$

The ten-fold reduction in rate on deuteration without changing the stoichiometry provides strong evidence that both sets of products originate in a common step. The Arrhenius parameters for this reaction, (E = 13.3 kcal.mole^{-1}, $\Delta S^\ddagger = -17$ eu) invite comparison with those for the oxidation of free formate ion (E = 12.4 kcal.mole^{-1}, $\Delta S^\ddagger = -15$ eu).

The oxidation of other monocarboxylic acids by both Cr(VI) and Mn(VII) is slow. Mareš and Roček[165] examined the effect of CO_2H groups on the oxidation rates of methine and methylene groups. With a series of dicarboxylic acids

$HO_2C(CH_2)_nCO_2H$ it was found that beyond a certain point the increment in the rate coefficient for an increases of one in n is constant at 5.2×10^{-3} l.mole^{-1}.sec^{-1}, which compares with a value of 5.73×10^{-3} l.mole^{-1}.sec^{-1} for n-paraffin oxidation (p. 293). It is clear that carboxylic acids behave as paraffins except that a slight retardation due to the inductive effect of $-CO_2H$ is apparent. Permanganate behaves in much the same way and some examples of carboxylic acid oxidation have been cited in the section on hydrocarbons.

2.3.8 *Amines*

A comprehensive account of the oxidation of benzylamine by both neutral and alkaline permanganate has been published recently[166]. The neutral amine is the active reductant at all pH values and at pH 9.9 good second-order kinetics were obtained. Under these conditions the stoichiometry is

$$3\ C_6H_5CH_2NH_2 + 2\ MnO_4^- = 3\ C_6H_5CH{=}NH + 2\ MnO_2 + 2\ OH^- + 2\ H_2O$$

The pH-dependence of the rate is complicated. The rate increases very rapidly between pH values of 8 and 10 and the authors correlate this with the literature value of pK_{BH^+} of benzylamine of 9.34. The rate levels off at pH 11–12, but at pH 13–14 a reaction the rate of which is linearly dependent on hydroxyl ion concentration occurs.

In the pH range 8–10 the reaction shows a deuterium kinetic isotope effect of 7.0 for oxidation of $C_6H_5CD_2NH_2$. $C_6H_5CH_2ND_2$ (in D_2O) is oxidised at the same rate as the light compound. A series of ring-substituted analogues were examined and the data afforded a ρ^+ value of -0.28. Activation parameters were determined for a series of benzylamines and all fell in a range of 11.0 to 12.3 kcal.mole^{-1} (E) and -21.6 to -24.8 eu ($\Delta S^{\ddagger}$). These large negative values are of interest as they resemble those found for permanganate oxidation of both neutral molecules (H_2, HCO_2H) and of anions (alkoxides). For this path the authors propose a transition state formed thus

$$C_6H_5CH_2NH_2 + MnO_4^- \longrightarrow \left[\begin{matrix} (\delta+)C_6H_5\text{–}C(NH_2)\cdots H\cdots \overset{\delta-}{O}MnO_3^- \\ \updownarrow \\ C_6H_5\text{–}C(\overset{\delta+}{N}H_2)\cdots H\cdots \overset{\delta-}{O}MnO_3^- \end{matrix} \right]^-$$

I

This could decompose in one or both of two ways

$$[\text{I}]^- \begin{cases} \rightarrow ArCH{=}\overset{+}{N}H_2 + HMnO_4^{2-} \\ \rightarrow Ar\dot{C}HNH_2 + HMnO_4^- \end{cases}$$

The present information does not allow discrimination between these two routes. The high-pH path probably involves concerted attack of OH^- and MnO_4^- upon neutral amine because the possibility of attack of oxidant upon $C_6H_5CH_2NH^-$ or $C_6H_5\bar{C}HNH_2$ is too remote in view of known acidities of amines.

In the same paper the authors list the responses of a number of amines and amides towards permanganate and record the effect of freezing the medium upon the oxidation of benzylamine.

TABLE 4

RATE DATA FOR THE PERMANGANATE OXIDATION OF SOME ALKYLAMINES

Temperature unspecified, probably about 20 °C.

Amine	k_2(*l.mole*$^{-1}$.*sec*$^{-1}$)	pK_a
$(C_2H_5)_3N$	3.08	10.65
$(C_2H_5)_2NH$	0.944	10.98
$C_2H_5NH_2$	0.0828	10.63
$(CH_3)_3N$	3.36	9.92
$(CD_3)_3N$	1.82	10.155

Rosenblatt *et al.*[167] have examined the effect of structure and isotopic substitution upon the permanganate oxidation of some alkylamines (Table 4). The isotope effect of 1.84 is considered to be sufficiently low to be compatible with aminium radical-cation formation, and it is felt that, while C–H cleavage is significant for oxidation of primary amines, the dominant mode of oxidation of tertiary amines is electron-transfer, *e.g.*

$$MnO_4^- + N(CH_3)_3 \rightleftharpoons \cdot\overset{+}{N}(CH_3)_3 + MnO_4^{2-} \quad \text{(slow)}$$

$$\cdot\overset{+}{N}(CH_3)_3 \rightarrow H^+ + H_2C\cdots N(CH_3)_2 \quad \text{(fast)}$$

$$H_2C\cdots N(CH_3)_2 + MnO_4^- \rightarrow H_2C{=}N(CH_3)_2{}^+ \quad \text{(fast)}$$

$$H_2C{=}N(CH_3)_2{}^+ + H_2O \rightarrow H_2CO + HN(CH_3)_2$$

2.3.9 Nitroalkanes

A stopped-flow examination[167a] of the alkaline permanganate oxidation of phenylnitromethane

$$3\,C_6H_5CH{=}NO_2^- + 2\,MnO_4^- + H_2O = 2\,C_6H_5CHO + 2\,MnO_2 + 3\,NO_2^- + 2\,OH^-$$

indicates a rate law

$$-d[MnO_4^-]/dt = k_2[MnO_4^-][C_6H_5CH{=}NO_2^-]$$

which suggests a transition state

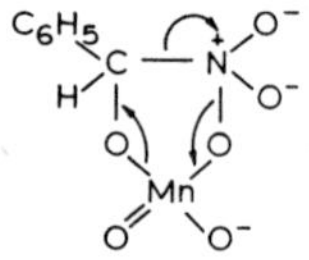

The kinetics and mechanisms of the MnO_4^- oxidations of nitrocyclohexane and nitrocyclopentane are entirely similar[167b]. The combined rate data for solutions at 0.5 *M* ionic strength are

Substrate	k_2*(274°K)(l.mole*$^{-1}$*.sec*$^{-1}$*)*	*E(kcal.mole*$^{-1}$*)*	$\Delta S^\ddagger$*(eu)*
Phenylnitromethane	180±20	8.1±0.2	−20±1
Nitrocyclohexane	310±20	8.05	−27.8
Nitrocyclopentane	52±10	11.1	−20.0

2.4 OXIDATION OF POLYFUNCTIONAL ORGANIC MOLECULES

2.4.1 Glycols

Oxidation of glycols can proceed by two routes: (*i*) formation of an α-hydroxy-carbonyl compound and (*ii*) carbon–carbon fission. In the case of chromic acid, successive C-methylation of ethylene glycol increases the degree of cleavage[168] until, with pinacol, cleavage is at least 70 % quantitative[169]. Chatterji and Mukherjee[169–171] have examined four glycols obtaining rate laws and Arrhenius parameters. In all cases the oxidation rate depends on the first powers of the glycol and acid chromate ion concentrations. The remaining data are summarised in Table 5. Manganous ion strongly retards the oxidation of all the glycols except pinacol. Despite its much larger activation energy, the oxidation of pinacol is *ca.* 400 times faster than that of ethylene glycol[171]. It is clear that an ordinary

TABLE 5

ACIDITY-DEPENDENCES AND RATE PARAMETERS FOR THE CHROMIC ACID OXIDATION OF α-GLYCOLS

Glycol	*Dependence on* $[H_3O^+]$	*A* (l^3*mole*$^{-3}$*.sec*$^{-1}$)	*E (kcal.mole*$^{-1}$*)*	*Ref.*
Ethylene glycol[a]	$[H_3O^+]+k[H_3O^+]^2$	2.2×10^5	10.7	170
Propylene glycol	$[H_3O^+]^2$	8.1×10^5	10.4	169, 171
2,3-Butylene glycol	$[H_3O^+]^2$	4.5×10^7	12.2	169, 171
Pinacol	$[H_3O^+]$	[b]1.07×10^{12}	17.2	169, 171

[a] Arrhenius data refer to $[H_3O^+]^2$ dependence term.
[b] Units l^2.mole^{-2}.sec^{-1}.

alcohol type oxidation mechanism operates almost exclusively for ethylene glycol, but that increasing C-methylation favours the second (cleavage) mechanism. This view is supported: (*i*) by the fitting of the result for ethylene glycol to the Taft plot for oxidation of primary alcohols and (*ii*) by its proximity to the point for 2-methoxyethanol on this plot[110].

The oxidation of pinacol was studied in further depth by Chang and Westheimer[172]. The total kinetic order of three was confirmed and the reaction was found to proceed 2.7 times faster in heavy than light water. In contrast with ethylene glycol, O-methylation of one group reduces the oxidation rate by a factor of the order of 1500, producing complex kinetics containing an essentially zero-order term in chromic acid. The reaction is not sensitive to oxygen, but induces the oxidation of Mn(II) ion to MnO_2 with a factor just under 0.3. The oxidation rate is reduced by Mn(II) by a factor of about 2.5 at chromic acid concentrations of 6×10^{-4} *M*. At high oxidant concentrations (0.145 *M*) no retardation is found. This probably accounts for the failure of Chatterji and Mukherjee to observe retardation[169].

The solvent isotope effect suggests that no O–H cleavage is involved in the slow step and the effect of O-methylation indicates that a cyclic complex is involved. The induction factor is probably obscured by the reaction of Mn(III) and MnO_2 with pinacol itself. The typical glycol-cleavage mechanism advocated for oxidations by Pb(IV) and I(VII) (p. 349) may well operate, *viz.*

$$H_2CrO_4 + \begin{array}{c}R_2C-OH\\|\\R_2C-OH\end{array} \rightleftharpoons \begin{array}{c}R_2C-O\\|\\R_2C-O\end{array}\!\!>CrO_2 + 2H_2O \quad \text{(fast)}$$

$$\begin{array}{c}R_2C-O\\|\\R_2C-O\end{array}\!\!>CrO_2 \longrightarrow \begin{array}{c}R_2C=O\\ \\R_2C=O\end{array} \quad CrO_2 \quad \text{(slow)}$$

Strong support for the cyclic ester intermediate comes from the measurement of the relative rates of oxidation by chromic acid of *cis*- and *trans*-1,2-dimethyl-1,2-cyclopentanediol[173]. In water and in 90 % acetic acid k_{cis}/k_{trans} is 17,000 and 800, respectively. Both oxidations are first order both in glycol and $HCrO_4^-$. The rate with the *cis*-isomer involves a hydrogen ion concentration dependence; the reaction with the *trans*-isomer was examined in more acidic solutions and the rate followed an h_0 dependence. The solvent isotope effect of k_{D_2O}/k_{H_2O} of 2.0 is also normal. Oxidation of some diastereoisomeric 1,2-diols has revealed lower $k(dl)/k(meso)$ ratios than expected on the basis of decomposition of a cyclic ester. This has been interpreted[174] in terms of rate-determining ester formation (*cf.* the hindered triterpene study of Westheimer *et al.*[102], p. 302). A similar proposal has been made by Kwart and Bretzger[175] to explain the relative rates of oxidation of cyclohexyl and cyclopentyl pinacols.

2.4.2 Allylic alcohols

The equatorial allylic alcohol 3β-hydroxyandrost-4-ene is oxidised by Cr(VI) 310 times faster than the saturated 3β-hydroxy-5α-androstan-17-one, 5.7 times faster than the axial 3α-hydroxyandrost-4-ene and 6.9 times faster than its 3-deuterated analogue[176]. The greater speed of oxidation of the equatorial isomer is in contrast to the pattern observed for saturated alcohols and probably arises from resonance between the double bond and the incipient carbonyl group.

Allyl alcohols behave essentially as olefins towards MnO_4^- (p. 300).

2.4.3 Ketols, keto-aldehydes and keto-acids

The rapid oxidations of certain of these polyfunctional compounds by alkaline permanganate were examined[156] as a supplement to the study of acetone oxidation (p. 314). The oxidations of acetol and pyruvaldehyde show identical rate laws of the form

$$-\mathrm{d}[\mathrm{MnO_4^-}]/\mathrm{d}t = k_2[\text{substrate}][\mathrm{MnO_4^-}]+k_3[\text{substrate}][\mathrm{MnO_4^-}][\mathrm{OH^-}]$$

with the following values of the rate coefficients at 25 °C

	acetol	*pyruvaldehyde*
k_2(l.mole^{-1}.sec^{-1})	50	5.6×10^2
k_3(l^2.mole^{-2}.sec^{-1})	1.2×10^3	9.3×10^2

Pyruvate ion is oxidised according to the kinetics

$$-\mathrm{d}[\mathrm{MnO_4^-}]/\mathrm{d}t = k_3[\text{substrate}][\mathrm{MnO_4^-}][\mathrm{OH^-}]+k_0[\text{substrate}][\mathrm{OH^-}]$$

with $k_3 = 4.0\times10^2$ l^2.mole^{-2}.sec^{-1} and $k_0 = 0.9$ l.mole^{-1}.sec^{-1}. These various results were regarded by the authors[156] as of too preliminary a nature to justify any mechanistic proposals.

2.4.4 Dicarboxylic acids

The examples of oxalic and malonic acids are atypical and are considered separately. Other dicarboxylic acids react essentially as alkanes (p. 297).

Early work of Dhar[177] established that oxidation of oxalic acid by chromic acid occurs readily, but some of his kinetic data are unreliable as the substrate itself acted as the source of hydrogen ions. The reaction is first-order in oxidant and is subject to strong manganous ion catalysis (as opposed to the customary retardation), the catalysed reaction being zero-order in chromic acid. This observation is related to those found in the manganous-ion catalysed oxidations of several organic compounds discussed at the end of this section.

Chandra *et al.*[178] and also Bakore and Jain[178a] determined the rate law for the uncatalysed reaction to be

$$-\mathrm{d}[\mathrm{Cr(VI)}]/\mathrm{d}t = k_3[\mathrm{HCrO_4^-}][\text{oxalic acid}]^2[\mathrm{H_3O^+}]^0$$

with $k_3 = 3.7\times10^7 \exp(12.0\times10^3/\boldsymbol{RT})$ $\mathrm{l^2.mole^{-2}.sec^{-1}}$. Rao and Ayyar[179] obtain a different rate law

$$-\mathrm{d}[\mathrm{Cr(VI)}]/\mathrm{d}t = k[\mathrm{Cr(VI)}][\text{oxalic acid}]^{2.5}$$

A third set of results, those of Durham[179a], indicates two paths, *viz.*

$$-d[\mathrm{Cr(VI)}]/\mathrm{d}t = (k_1[\mathrm{C_2O_4}^{2-}]+k_2[\mathrm{C_2O_4}^{2-}]^2)[\mathrm{H_3O^+}]^3[\mathrm{HCrO_4^-}]$$

Of the various results, these appear to the reviewer to be the most meticulously obtained and they do not fit the other two rate laws suggested. At 25 °C (μ = 1.0 *M* perchlorate), $k_1 = 5.10\times10^5$ $\mathrm{l^4.mole^{-4}.sec^{-1}}$ and $k_2 = 4.78\times10^{10}$ $\mathrm{l^5.mole^{-5}.sec^{-1}}$. The corresponding reaction intermediates are a neutral chelate mono-oxalato complex and a (non-chelate) bis-oxalato complex of Cr(VI).

The permanganate oxidation of oxalic acid has been studied exhaustively and has been reviewed by Ladbury and Cullis[180]. It is characterised by an induction period and a sigmoid dependence of rate upon time. Addition of manganous ions eliminates the induction period and produces first-order decay kinetics[181, 182]. Addition of fluoride ions, however, practically eliminates reaction[182].

It is clear that some slow reduction of permanganate occurs during the induction period to give lower valency states of manganese which form oxalato complexes of varying stability. Oxidative break-down of these gives manganous ions which react rapidly with permanganate unless removed by complexing, *e.g.* with fluoride.

Oxidation of malonic acid by Cr(VI) has been noted briefly by Snethlage[183] and later by Kemp and Waters[184]. The kinetics are simple second-order but the acidity dependence is complex. Heckner *et al.*[134c] find the alkaline permanganate oxidation of malonic acid (and also of *o*- and *p*-toluic acids and of *p*-toluene-sulphonic acid) to be retarded by added Mn(VI), *viz.*

$$\frac{-\mathrm{d}[\mathrm{MnO_4^-}]}{\mathrm{d}t} = \frac{k[\mathrm{MnO_4^-}][\mathrm{OH^-}][\text{substrate}]^{0.6-0.7}}{[\mathrm{MnO_4}^{2-}]}$$

These authors invoke the pre-equilibrium

$$OH^- + MnO_4^- \rightleftharpoons MnO_4^{2-} + OH\cdot$$

the validity of which rests on the value of π^0 for the couple OH·/OH⁻. Latimer's value[19] of +2.0 V, compared with that of +0.56 V for the Mn(VII)/Mn(VI) couple, would seem to exclude it but Stein's value of +1.25 V (pH 14)[134d] increases its plausibility. However, the extreme sensitivity of alkaline permanganate to trace metal ions[45a] renders Heckner's conclusions suspect as no special precautions to remove these ions appear to have been taken.

2.4.5 Hydroxy-acids

Bakore and Narain[185] obtained the following kinetics for the oxidations by chromic acid of lactic, malic and mandelic acids

$$-d[Cr(VI)]/dt = k[HCrO_4^-][\text{hydroxy-acid}][H_3O^+]$$

The rate of oxidation is reduced by one half on addition of manganous ions and the following Arrhenius parameters were recorded

	A ($l^2.mole^{-2}.sec^{-1}$)	E ($kcal.mole^{-1}$)
Lactic acid	5.2×10^4	9.04
Malic acid	6.6×10^4	8.96
Mandelic acid	2.2×10^4	7.90

Kemp and Waters[186] found a primary kinetic isotope effect of 8.7 for oxidation of C-deuterated mandelic acid and noted a large difference in rate between the oxidations of mandelic acid (k_2 at 24.4 °C = 1.7 $l.mole^{-1}.sec^{-1}$) and α-hydroxy-isobutyric acid (k_2 at 24.4 °C = 5.6×10^{-3} $l.mole^{-1}.sec^{-1}$) – a difference not reproduced for the oxidation of these compounds by the one-equivalent reagent, manganic sulphate. The various data are fully in accord with a Westheimer-type mechanism, *viz.*

$$RCH(OH)CO_2H + HCrO_4^- + H_3O^+ \rightleftharpoons R(HO_2C)CH{-}O{-}CrO_2{-}OH + 2\,H_2O$$

$$R(HO_2C)CH{-}O{-}CrO_2{-}OH \longrightarrow R(HO_2C)C{=}O + HO{-}CrO{-}OH \quad \text{(slow)}$$

Decarboxylation of the keto-acid would then ensue.

The oxidation of lactate ion by alkaline permanganate (0.1 *M* base) was examined as part of the general study of the oxidation of acetone and its possible oxidation products[156]. The reaction is first-order in lactate ion, and, if it is assumed that it is also first-order in oxidant, then the second-order rate coefficient is 2 $l.mole^{-1}.sec^{-1}$ at 25 °C.

No published work exists on the kinetics of acid permanganate oxidation of hydroxy-acids although Pink and Stewart[187] have noted that benzilic acid is oxidised by acid permanganate to benzophenone in an autocatalytic process with k_{H_2O}/k_{D_2O} of unity.

In unpublished work[188] Kemp has found that although the sigmoid character of the $[MnO_4^-]$ *versus* time plots obtained with acidic solutions makes kinetic analysis difficult, several features are apparent from comparison of collected plots (Figs. 1 and 2). Firstly, the "slow" or induction period is longer for mandelic than for C-deuterated mandelic acid and even the "fast" stage involving lower valency states of Mn is more prolonged for the deuterated acid. Secondly, although the induction periods for oxidation of several α-hydroxyacids follow the sequence, $(CH_3)_2C(OH)CO_2H > CH_2(OH)CO_2H > CH_3CH(OH)CO_2H > C_6H_5CH(OH)CO_2H$ the rates of the "fast" stages follow a different sequence from the reverse of the above, *viz.* $C_6H_5CH(OH)CO_2H > (CH_3)_2C(OH)CO_2H > CH_3CH(OH)CO_2H > CH_2(OH)CO_2H$. Experience with both two- and one-equivalent reagents with respect to this series suggests that the induction stage

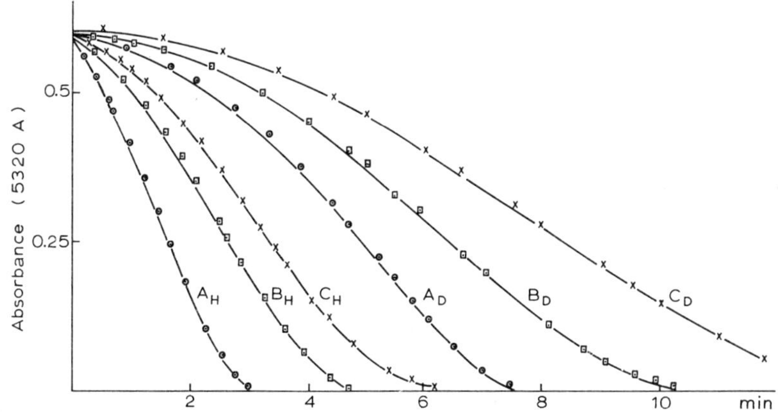

Fig. 1. Isotope effect for acid permanganate oxidation of mandelic acid. Temperature = 26.2 °C; $[MnO_4^-] = 1.4 \times 10^{-3}$ *M*; $[H_2SO_4] = 1.69$ *M*.

	A	*B*	*C*
10^3 $[C_6H_5CH(OH)CO_2H]$	10.4	7.3	5.2
10^3 $[C_6H_5CD(OH)CO_2H]$	9.9	6.9	4.9

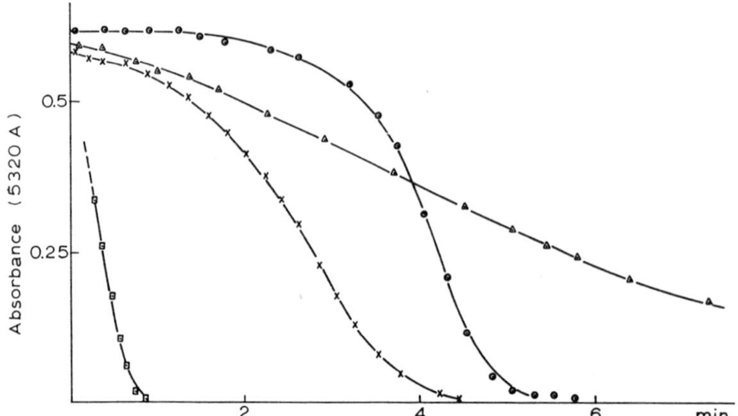

Fig. 2. Relative rates of oxidation of some α-hydroxy acids by acid permangante. Temperature = 25.6 °C; $[MnO_4^-] = 1.4 \times 10^{-3}$ *M*; $[H_2SO_4] = 1.69$ *M*. ⊡, $C_6H_5(OH)CO_2H$ (0.0693 *M*); ⊙, $(CH_3)_2C(OH)CO_2H$ (0.0630 *M*); ×, $CH_3CH(OH)CO_2H$ (0.0704 *M*); △, $CH_2(OH)CO_2H$ (0.142 *M*).

probably involves two-equivalent oxidation of the hydrogen atom bound to the hydroxylated carbon atom (to give the primary kinetic isotope effect and the sluggish oxidation of α-hydroxyisobutyric acid) and that the second stage involves one-equivalent attack by Mn(III) to give a rate sequence identical with that for direct attack by manganic sulphate[186] (Fig. 4, p. 394).

2.4.6 *Boronic acids*

The oxidation of *tert*-butylboronic acid by chromic acid has the stoichiometry

$$CrO_3 + t\text{-}C_4H_9B(OH)_2 \overset{H_2O}{=} t\text{-}C_4H_9OH + B(OH)_3 + Cr^{3+}$$

and follows simple second-order kinetics[189]

$$-d[Cr(VI)]/dt = k[HCrO_4^-][t\text{-}C_4H_9B(OH)_2]$$

The acidity dependence is complex and indicates that no extra proton to give H_2CrO_4 is required, but that $H_3CrO_4^+$ is an active oxidant in this reaction. The rate is very sensitive to the nature of the alkyl group, *viz*.

R	CH_3	C_2H_5	$t\text{-}C_4H_9$
$10^7\, k_2$(l.mole^{-1}.sec^{-1}) at 30 °C in 0.114 *M* $HClO_4$	2.4	6600	750,000

The authors propose a mechanism

$$HCrO_4^- + RB(OH)_2 \rightleftharpoons \left[R{-}B(OH)_2{-}OCrO_3H \right]^-$$

$$\left[R{-}B(OH)_2{-}OCrO_3H \right]^- \rightarrow \left[\text{cyclic transition state: } (HO)_2B \cdots R \cdots O{=}Cr(O)(OH){-}O \; (\delta-) \right]^- \quad \text{(slow)}$$

$$\rightarrow B(OH)_3 + ROH + Cr(V)$$

2.4.7 *Furfurals*

Examination of the permanganate oxidation of furfural and 5-substituted furfurals, at pH 11.5–13.3 in a stopped-flow apparatus, *viz.*

$$3\ X{-}C_4H_2O{-}CHO + 2\ MnO_4^- + OH^- = 3\ X{-}C_4H_2O{-}CO_2^- + 2\ MnO_2 + 2\ H_2O$$

indicates a rate law[705]

$$-d[MnO_4^-]/dt = k_0[\text{furfural}][MnO_4^-]+k_1[\text{furfural}][MnO_4^-][OH^-]$$

The second term dominates and is characterised by $\rho = +1.30$ (using σ-*meta* constants), $E = 10.8$ kcal.mole^{-1}, $\Delta S^{\ddagger} = -22.8$ eu and $k_H/k_D > 1.8$. ^{18}O experiments show the major source of oxygen in the acid produced is the solvent, which suggests the hydrate anion is the reactive form, *viz.*

$$C_4H_3O{-}C(O^-)(OH){-}H + MnO_4^- \xrightarrow{\text{slow}} C_4H_3O{-}C(=O){-}OH + HMnO_4^{2-}$$

2.5 SOME METAL-ION CATALYSED REACTIONS OF CHROMIC ACID

Dhar[177] noted that the oxidation of oxalic acid by chromic acid is markedly accelerated on adding manganous ions, the reaction order in Cr(VI) changing from one to zero. Bobtelsky and Glasner[40] found the oxidation of bromide ions by chromic acid in aqueous sulphuric acid to follow kinetics

$$-d[Cr(VI)]/dt = k[Mn(II)][Cr(VI)]f[H_3O^+]$$

i.e. to be independent of the concentration of bromide (p. 282). Their results were confirmed by Kemp and Waters[184] using aqueous sulphuric acid (2.4 M) and are readily rationalised in terms of the scheme

$$\mathrm{Cr(VI)+Mn(II) \rightleftharpoons Cr(V)+Mn(III)} \tag{32}$$

$$\mathrm{Cr(V)+Mn(II) \rightleftharpoons Cr(IV)+Mn(III)} \tag{33}$$

$$\mathrm{Cr(IV)+Mn(II) \rightleftharpoons Cr(III)+Mn(III)} \tag{34}$$

$$\mathrm{Cr(VI)+3\ Mn(II) \rightleftharpoons Cr(III)+3\ Mn(III)} \tag{35}$$

$$\mathrm{Mn(III)+Br^- \rightarrow Mn(II)+Br\cdot}$$

$$\mathrm{2\ Br\cdot \rightarrow Br_2}$$

The slow step is (32), which excludes any participation by bromide in determining the overall rate.

Kemp and Waters[184] also found the oxidations of cyclohexanone and of mandelic, malonic and α-hydroxyisobutyric acids by Cr(VI) to be Mn(II)-catalysed. In these cases, as with oxalic acid, the [Cr(VI)] *versus* time plots are almost linear and the reaction becomes first order in substrate (or involves Michaelis–Menten kinetics), and, except at lowest catalyst concentrations, approximately first order in [Mn(II)]. Detailed examination of the initial rate of oxidation of α-hydroxyrobutyric acid as a function of oxidant concentration revealed, however, that the dependence is

$$-\mathrm{d[Cr(VI)]}/\mathrm{d}t = k\mathrm{[substrate][Mn(II)][HCrO_4^-]^{\frac{1}{3}}}$$

This can be explained in terms of the pre-equilibria (32)–(34) followed by a *slow* oxidation of the substrate by Mn(III). It is probable that the substrates are chelated to Mn(II) and Mn(III) throughout the process. The rate of oxidation of the substrate is given by

$$\text{rate} = k\mathrm{[Mn(III)][substrate]}$$

However, [Mn(III)] is related to [Cr(VI)] by equilibrium (35), and

$$\mathrm{[Mn(III)]} = K^{\frac{1}{3}}\mathrm{[HCrO_4^-]^{\frac{1}{3}}[Mn(II)]/[Cr(III)]^{\frac{1}{3}}}$$

hence

$$-\mathrm{d[Cr(VI)]}/\mathrm{d}t = kK^{\frac{1}{3}}\mathrm{[substrate][Mn(II)][HCrO_4^-]^{\frac{1}{3}}/[Cr(III)]^{\frac{1}{3}}}$$

In confirmation of the suggestion that the active reagent of the Cr(VI)–Mn(II) couple is Mn(III) it was found[186] that, while the rates of oxidation of mandelic and α-hydroxyisobutyric acids by Cr(VI) are in the ratio 300 : 1, the ratios for oxidation by the couple and by Mn(III) are, respectively, 2.9 and 2.3. Furthermore, the values of k_H/k_D for oxidation of C-deuteromandelic acid by Cr(VI), Mn(III) and Cr(VI)–Mn(II) respectively are 8.7, 1.1 and approximately 1.0.

More detailed results on the less strongly-catalysed oxidation of mandelic acid[186a] shows that here the catalytic effect depends on the initial [Mn(II)]/[Cr(VI)] ratio; at ratios less than 10^2 Mn(II) *retards* reaction. At high ratios the catalysed reaction is *not* retarded by a hundred-fold excess of Cr(III) ions, which is evidence against the multiple equilibria (32)–(34), and for this substrate the reaction order in substrate is fractional. Step (33) is discounted in a modified reaction scheme which attributes part of the oxidation to attack by Cr(IV) and Mn(III) upon the substrate, the Cr(IV) being formed during initial attack on substrate by Cr(VI). At low [Mn(II)], the more strongly oxidising Cr(IV) is replaced by Mn(III) *via* reaction (34); at high [Mn(II)], Cr(VI) is reduced by Mn(II) as well as by mandelic acid.

Preliminary results have been reported[186b] of oxidation of cyclobutanol by the Cr(VI)–V(IV) couple to 4-hydroxybutyraldehyde. This proceeds at the same rate as the oxidation of V(IV) by Cr(VI)[186c] and cannot involve attack of Cr(V) upon the alcohol, for this oxidation state is formed in a rapid pre-equilibrium, but rather attack by Cr(IV), *viz.*

$$\mathrm{Cr(VI)+V(IV)} \rightleftharpoons \mathrm{Cr(V)+V(V)} \qquad \text{(fast)}$$

$$\mathrm{Cr(V)+V(IV)} \rightleftharpoons \mathrm{Cr(IV)+V(V)} \qquad \text{(slow)}$$

$$\mathrm{Cr(IV)+C_4H_7OH} \rightarrow \mathrm{Cr(III)+R\cdot} \qquad \text{(fast)}$$

$$\mathrm{R\cdot+V(V)} \rightarrow \mathrm{HOCH_2CH_2CH_2CHO+V(IV)} \qquad \text{(fast)}$$

The oxidation by chromic acid alone leads to a mixture of cyclobutanone and 4-hydroxybutyraldehyde; the existence of an isotope effect for the oxidation of 1-deuteriocyclohexanol suggests that Cr(VI) produces the ketone and lower oxidation states of chromium produce the cleavage product.

3. Oxidation by Pb(IV), Tl(III), Hg(II), Hg(I), Bi(V), Au(III), Pt(IV), Pd(II), Rh(III), Ru(III) and Mo(VI)

3.1 GENERAL FEATURES

The first three members of this series appear at the bottom of the B subgroups of the periodic groups 4, 3 and 2. They exhibit the so-called "inert-pair" effect and normally assume oxidation states of +4, +2; +3, +1 and +2, 0 respectively, *i.e.* differing by two units. The species Hg^+, Tl^{2+} and Pb^{3+} are of high energy and appear only as short-lived intermediates, *e.g.* in the oxidation of Tl^+ by OH· during pulse radiolysis[190]. The other oxidants are also mainly two-equivalent in their action.

All these oxidants form π-complexes by accepting electrons from olefinic bonds, a property which has been widely discussed[191]. Oxidations by these species are not, however, restricted to olefinic compounds and there is considerable evidence that they are not totally restricted to two-equivalent action. Kinetic data on their oxidations, once rare, has become profuse in the last decade both for aqueous and non-aqueous media.

The redox potentials are given in Table 6.

TABLE 6

OXIDATION POTENTIALS OF SOME TWO-EQUIVALENT REAGENTS[19]

Reaction	π^0 (*Volt*)
$Pb^{4+}+2e^- = Pb^{2+}$	*ca.* +1.7
$Tl^{3+}+2e^- = Tl^+$	+1.25
$2\,Hg^{2+}+2e^- = Hg_2^{2+}$	+0.920
$Hg_2^{2+}+2e^- = 2\,Hg$	+0.789
$Au^{3+}+2e^- = Au^+$	*ca.* +1.41
$PtCl_6^{2-}+2e^- = PtCl_4^{2-}+2\,Cl^-$	+0.68
$Pd^{2+}+2e^- = Pd$	+0.987

Pb(IV) is most usually employed as the tetraacetate and the action of this compound is complex in that it can function either as a two-equivalent oxidant giving Pb(II) or as a source of acetoxy radicals, *viz.*

$$Pb(O.COCH_3)_4 \rightarrow Pb(O.COCH_3)_3\cdot + CH_3CO_2\cdot$$

It can also act as a source of $\cdot CH_2CO_2H$ radicals (*vide infra*).

A recent authoritative review on the numerous reactions which have been studied without determination of kinetics is available[192] and discussion here will be restricted to the relatively few reactions which have been examined kinetically.

3.2 OXIDATION OF INORGANIC SPECIES

Although Pb(IV) is sufficiently strong an oxidant to oxidise halides, no kinetic data are available. Complexes of Pt(IV) and Au(III) oxidise iodide and thiocyanate ions but the other oxidants are weaker and form stable halo-complexes. However, some simple molecules such as hypophosphorous acid, carbon monoxide and molecular hydrogen are oxidised by the weaker members.

3.2.1 Halide and pseudohalide ions

The Pt(IV) oxidation of iodide ion was studied[193] using *bis*-triethylarsine and *bis*-triethylphosphine complexes to prevent interference from substitution reactions. The observation of isosbestic points with *trans*-$[Pt(AsEt_3)_2Cl_4]$ and *trans*-$[Pt(AsEt_3)_2Br_4]$ indicates a single step reaction and quantitative production of I_3^- per mole of oxidant was recorded. Simple second-order kinetics were obtained with these oxidants and also with *trans*-$[Pt(PEt_3)_2Cl_4]$ and rate data are summarised in Table 7. The stoichiometry is

$$trans\text{-}[Pt(AsEt_3)_2Cl_4]+5\,I^- = trans\text{-}[Pt(AsEt_3)I_2]+4\,Cl^-+I_3^-$$

TABLE 7

RATE DATA FOR THE OXIDATION OF IODIDE ION BY COMPLEXES OF Pt(IV)[193]

Complex	k_2 (*l.mole*$^{-1}$.*sec*$^{-1}$)	*Temperature* (°*C*)	*E* (*kcal.mole*$^{-1}$)	$\Delta S^{\ddagger}$(*eu*)
trans-$[Pt(AsEt_3)_2Cl_4]$	$(0.5\ \pm0.03)\times10^{-3}$	35	20.1 ± 0.6	$-11\ \pm3$
trans-$[Pt(AsEt_3)_2Br_4]$	1.56 ± 0.04	25	10.6 ± 0.4	$-25\ \pm2$
trans-$[Pt(PEt_3)_2Cl_4]$	$(3.5\ \pm0.15)\times10^{-3}$	35	16.2 ± 0.4	-19.5 ± 2

The authors acknowledge that the present data are insufficient to allow a detailed mechanism to be given but they reject the possibility of slow S_N2 substitution followed by a rapid redox process.

The oxidation of potassium thiocyanate by $AuBr_4^-$ is first-order in oxidant and the pH dependence indicates that it is also first-order in thiocyanate ion, which is oxidised much faster than HCNS[194]. The activation parameters are $E = 6.4\pm0.4$ kcal.mole^{-1} and $\Delta S^{\ddagger} = 26\pm2$ eu.

Bi(V) in aqueous perchloric acid is very strongly oxidising but kinetic studies have been confined to a few stopped-flow measurements on oxidation of iodide, bromide and chloride ions[667]. The appearance of Bi(III)–halide complexes was first-order with respect to Bi(III) and in all cases the first-order rate coefficient, k_1, was the same, *i.e.* 161 ± 8 sec^{-1} at 25 °C ($[H_3O^+] = 0.5\ M$, $\mu = 2.0\ M$), irrespective of the nature or concentration of the halide. A preliminary attack on solvent is compatible with these interesting results, *viz.*

$$Bi(V)+H_2O \rightleftharpoons Bi(IV)+H^++OH\cdot$$

$$Bi(IV)+X^- \rightarrow Bi(III)+X\cdot$$

$$OH\cdot+X^- \rightarrow OH^-+X\cdot$$

$$2X\cdot \rightarrow X_2$$

These authors also cite unpublished work on Sb(V) oxidation of I^- which follows a rate law involving the term $[Sb(V)][I^-]^2$.

3.2.2 *Oxy-acids of sulphur*

The work with iodide (preceding sub-section) was extended to thiosulphate[193], and isosbestic points and second-order kinetics were again obtained with the various Pt(IV) complexes (Table 8). Two $S_2O_3^{2-}$ ions are consumed per mole of Pt(IV) reduced, suggesting tetrathionate to be the product of oxidation, *viz.*

$$trans\text{-}[Pt(AsEt_3)_2Cl_4]+2\,S_2O_3^{2-} = trans\text{-}[Pt(AsEt_3)_2Cl_2]+S_4O_6^{2-}+2\,Cl^-$$

TABLE 8

RATE DATA FOR THE OXIDATION OF THIOSULPHATE ION BY COMPLEXES OF Pt(IV)[193]

Complex	k_2 (*l.mole^{-1}.sec^{-1}*)	*Temperature* (°C)	*E* (*kcal.mole^{-1}*)	$\Delta S^‡$(*eu*)
trans-$[Pt(AsEt_3)_2Cl_4]$	0.86 ± 0.03	35	15.5 ± 0.4	-10.5 ± 2
trans-$[Pt(AsEt_3)_2Br_4]$	$(2.6\pm0.1)\times10^3$	25	9.8 ± 0.5	-12 ± 3
trans-$[Pt(PEt_3)_2Cl_4]$	2.06 ± 0.08	35	14.5 ± 0.4	-12 ± 2

The rate of oxidation of sulphite ion by tetrabromoaurate[194] is pH-dependent and analysis of the kinetic data leads to the rate law

$$-d[AuBr_4^-]/dt = k_2[HSO_3^-][AuBr_4^-]$$

i.e. SO_2 and H_2SO_3 are unreactive. At 16 °C k_2 is $(4.0\pm0.4)\times10^4$ l.mole^{-1}.sec^{-1} and E is 19.4 ± 1 kcal.mole^{-1}.

3.2.3 *Hydrazine*

Oxidation by Mo(VI) as MoO_4^{2-} at low pH (1.2 to 3.2) gives N_2 stoichiometrically, *viz.*

$$4\,Mo(VI)+N_2H_5^+ = N_2+2\,[Mo(V)]_2+5\,H^+$$

The kinetics are[194a]

$$-d[N_2H_5^+]/dt = k_2'[Mo(VI)][N_2H_5^+] = +d[Mo(V)_2]/dt$$

k_2' is fractionally acid-inverse, which probably reflects equilibria involving the oxidising ion. E = 14.3 kcal.mole^{-1} and $\Delta S^‡ = -8.4$ eu at pH 1.60, $\mu = 0.22$ *M*. Di-imide was detected as an intermediate both mass spectrometrically and by trapping with *cis*-1,2-cyclohexane dicarboxylic acid. The products indicate purely two-equivalent behaviour, *viz.* the steps

$$N_2H_5^+ + Mo(VI) \rightarrow N_2H_2 + 3\,H^+ + Mo(IV) \qquad \text{(slow)}$$

$$H^+ + 2\,N_2H_2 \rightarrow N_2 + N_2H_5^+ \qquad \text{(fast)}$$

$$Mo(VI) + Mo(IV) \rightarrow 2\,Mo(V) \rightleftharpoons [Mo(V)]_2 \qquad \text{(fast)}$$

The catalysis by Mo(VI) of the oxidation of $N_2H_5^+$ to N_2 by methylene blue depends on the steps given above, Mo(VI) being regenerated by methylene blue oxidation of the Mo(V) dimer[194b]. The latter reaction was studied independently and

$$-\mathrm{d}[\text{methylene blue}]/\mathrm{d}t = k[Mo(V)_2]$$

where k (25 °C) = $(8.6 \pm 0.4) \times 10^{-5}$ sec^{-1} and E = 23.4 ± 0.2 kcal.mole^{-1}. These figures are very similar to those reported for oxidation of Mo(V) dimer by iodine and oxygen (pp. 468 and 450) and suggest a rate-determining first-order dissociation of the dimer to yield a reactive, monomeric Mo(V).

3.2.4 Nitrite ion

Ellison *et al.*[195] propose that the substitution of nitrite ion into chloroammineplatinum(IV) complexes involves oxidation of some nitrite by Pt(IV) to give Pt(II) which then functions as a catalyst. Prior addition of Pt(II) removes an induction period and produces a new rate law

$$\text{rate} = k_3[Pt(IV)][Pt(II)][NO_2^-]$$

Substitution into *trans*-$[Pt(en)_2Cl_2]^{2+}$ catalysed by $[Pt(en)_2]^{2+}$ (en = ethylenediamine) would proceed as follows

$$Pt(en)_2^{2+} + NO_2^- \rightleftharpoons Pt(en)_2NO_2^+ \qquad \text{(fast)}$$

$$Pt(en)_2Cl_2^{2+} + Pt(en)_2NO_2^+ \underset{\text{fast}}{\overset{\text{slow}}{\rightleftharpoons}} [ClPt(en)_2 \ldots Cl \ldots Pt(en)_2NO_2]^{3+}$$

$$\text{slow} \upharpoonleft\downharpoonright \text{fast}$$

$$ClPt(en)_2^+ + ClPt(en)_2NO_2^{2+}$$

$$Pt(en)_2Cl^+ \rightleftharpoons Pt(en)_2^{2+} + Cl^- \qquad \text{(fast)}$$

At 50 °C (μ = 0.22 M) with these reactants k_3 equals 10.1 l^2.mole^{-2}.sec^{-1}.

Redox behaviour of this type is considered to influence many substitution reactions of Pt(IV)[194].

3.2.5 Hypophosphorous acid

Mitchell[196] has shown that mercuric chloride oxidises this substrate to phosphorous acid

$$2\,HgCl_2+H_3PO_2+H_2O = Hg_2Cl_2+2\,HCl+H_3PO_3$$

The reaction shows a first-order dependence on substrate concentration but, except at very low concentration, is zero-order with respect to oxidant; moreover, the zero-order rate coefficient is the same as that observed with oxidations by iodine, cupric chloride and silver nitrate. The reaction is acid-catalysed. The oxidation is completely analogous to the halogenation of ketones and involves a slow tautomeric equilibrium followed by rapid oxidation, *viz.*

$$\mathrm{O{=}\underset{\displaystyle H}{\overset{\displaystyle H}{\overset{|}{\underset{|}{P}}}}{-}OH} \underset{}{\overset{\text{slow}}{\rightleftharpoons}} \mathrm{HO{-}\underset{\displaystyle H}{\underset{|}{P}}{-}OH} \xrightarrow[\text{fast}]{M^{n+}} P(OH)_3+M^{(n-2)+}$$

In contrast, thallic perchlorate oxidation follows a rate law

$$-\frac{d[Tl(III)]}{dt} = \frac{kK[Tl(III)][H_3PO_2]}{1+K[H_3PO_2]}$$

suggesting breakdown of a complex formed in a pre-equilibrium step[669]. Addition of chloride ion increased the rate by a factor of up to ten, presumably as a result of complexing. For the pre-equilibrium, $\Delta H = 7.8$ kcal.mole^{-1}, $\Delta S = 31$ eu and for the redox breakdown of the complex, $E = 13.7$ kcal.mole^{-1}, $\Delta S^{\ddagger} = -14.2$ eu.

3.2.6 Carbon monoxide

Tl(III) and Hg(I) are inert to this substrate but Hg(II) oxidises it to CO_2 at moderate temperatures[197]

$$2\,Hg^{2+}+CO+H_2O = Hg_2^{2+}+CO_2+2\,H^+$$

The rate law is

$$-d[CO]/dt = k_2[CO][Hg^{2+}]$$

with $E = 15.2$ kcal.mole^{-1} and $\Delta S^{\ddagger} = -13$ eu.

Methanolic mercuric acetate absorbs carbon monoxide to give a stable derivative. $AcO.HgCO.OCH_3$[198], *i.e.* CO has inserted into the Hg–OCH_3 bond. This implies the possibility of a similar insertion between Hg and a water ligand, *viz.*

$$[H_2O\text{–}Hg\text{–}OH_2]^{2+} + CO \rightarrow [H_2O\text{–}Hg\text{–}CO\text{–}OH]^+ + H^+ \quad \text{(slow)}$$

$$[H_2O\text{–}Hg\text{–}CO\text{–}OH]^+ \rightarrow Hg + CO_2 + H^+ \quad \text{(fast)}$$

$$Hg + Hg^{2+} \rightarrow Hg_2^{2+} \quad \text{(fast)}$$

Pd(II) in the form of K_2PdBr_4 oxidises CO to CO_2 in aqueous solution[199]

$$PdBr_4^{2-} + CO + H_2O = Pd + CO_2 + 2\,H^+ + 4\,Br^-$$

The appearance of CO_2 in the gas phase before precipitation of metal implies the production of a soluble Pd(0) complex such as $[Pd(CO)_2Br_2]^{2-}$. This receives support from the observation that three times as much CO is consumed as CO_2 liberated *before* deposition of Pd. The following reactions may be involved

$$PdBr_4^{2-} + CO \rightleftharpoons [PdBr_3CO]^- + Br^-$$

$$[PdBr_3CO]^- + CO \rightleftharpoons [PdBr_2(CO)_2]^0 + Br^-$$

$$[Pd(II)Br_2(CO)_2]^0 + 2\,H_2O \xrightarrow{k_3} Pd(0)\text{ complex} \rightarrow Pd(0) + 2\,CO_2 + 4\,H^+ + 2\,Br^-$$

The authors believe the Pd(0) complex to be dimeric, containing Pd(0) and Pd(II). At 5 °C, k_3 is of the order of 1.0 l.mole^{-1}.sec^{-1}, but it is influenced in a curious manner by addition of bromide ion, being increased initially and then sharply decreased. The effect itself is very temperature-dependent.

A few data have been reported on the analogous oxidation by $PdCl_4^{2-}$ which proceeds by a similar mechanism[200].

3.2.7 Molecular hydrogen

Along with Cu^{2+}, MnO_4^- and Ag^+, the two oxidation states of mercury are reduced by molecular hydrogen[201, 202]. Halpern[201, 203] considers that hydrogen is oxidised by two general mechanisms corresponding to rate equations of the type

$$\text{Class I} \quad -d[\text{oxidant}]/dt = k_2[\text{oxidant}][H_2]$$

$$\text{Class II} \quad -d[\text{oxidant}]/dt = k_3[\text{oxidant}]^2[H_2]$$

(Hg(I) should fall into Class I on this basis but it is regarded as a source of *two* oxidising ions per molecule.)

The first kinetic class probably corresponds to heterolysis of the H–H bond, the second class to homolysis. Hg(II) falls, with Cu(II), Ag(I) and MnO_4^-, into Class I and Hg(I), with Ag(I) and the Ag(I)–MnO_4^- couple, into Class II. The activation energies for Hg(I) and Hg(II) are 20.4 ± 0.6 and 18.1 ± 0.6 kcal.mole^{-1}, respectively, and the activation entropies are, respectively, -10.2 ± 2 and -12.2 ± 2 eu. The rate-determining steps are believed to be

$$Hg^{2+} + H_2 \rightarrow HgH^+ + H^+$$

$$Hg_2^{2+} + H_2 \rightarrow 2\ HgH^+$$

The configuration of the transition state for Class I reactions is seen as[203]

$$\begin{matrix} L-M^{n+}-OH_2 \\ H \text{---} H \end{matrix} \longrightarrow \begin{matrix} L-M^{(n-1)+} & \overset{+}{O}H_2 \\ | & | \\ H & H \end{matrix} \qquad (36)$$

(L = any other ligand). Strong complexing of the mercury atom, *e.g.* by Cl^- or ethylenediamine, reduces the rate by 2 or 3 powers of ten[204], but this is in any case a general feature of Hg(II) oxidations.

$PdCl_4^{2-}$, Rh(III) and Ru(III) act as homogeneous catalysts for reduction of $FeCl_3$ by molecular hydrogen[205–207]. The kinetics of all three activation reactions fall into Class I. The Arrhenius parameters are

	A(*l.mole*$^{-1}$.*sec*$^{-1}$)	*E*(*kcal.mole*$^{-1}$)
$PdCl_4^{2-}$	6.6×10^{11}	20
Rh(III)	2.3×10^{15}	25.2
Ru(III)	4.0×10^{14}	23.8

(36) represents the mechanism. Ru(III) is singular in not oxidising hydrogen in the absence of $FeCl_3$[207]. Oxidation by Ru(IV) proceeds *via* Ru(III) catalysis[207].

3.3 OXIDATION OF MONOFUNCTIONAL ORGANIC MOLECULES

3.3.1 Olefins

These ions, with the exception of Pb(IV), form complexes with olefins and this process is a preliminary to oxidation when this occurs. A recent review[208] of the action of Pd(II) covers aspects of structure and bonding as well as kinetics, and a similar but older, review exists for Hg(II)[209]. The oxidation of olefins by thallic species was discovered only recently and no previous review has included discussion of it.

Acetoxylation at an allylic position[210] is the typical reaction of lead tetraacetate but no kinetic data are available. Product studies favour a heterolytic mechanism[211].

Henry has examined in detail the oxidation of several olefins both by thallic perchlorate in an aqueous perchloric acid medium[212, 213] and by thallic acetate in aqueous acetic acid[214]. The reaction displays mixed stoichiometry

$$CR_1R_2{=}CR_3R_4 + Tl(III) + H_2O = Tl(I) + CR_1R_2R_3COR_4 + 2\,H^+$$
$$CR_1R_2{=}CR_3R_4 + Tl(III) + 2\,H_2O = Tl(I) + HOCR_1R_2CR_3R_4OH + 2\,H^+$$

the relative importance of the paths depending on the nature of the alkyl groups R_1, *etc.* The kinetics in aqueous mineral acid are

$$-d[Tl(III)]/dt = k_2[Tl(III)][\text{olefin}]$$

k_2 increases with increase of acidity but this is a salt effect and the dependence is really an inverse one with respect to the activity of water. The relative rates in the two media (but not *between* the media) are

	Ethylene	*Propene*	*1-Butene*	*cis-2-Butene*	*trans-2-Butene*	*Isobutene*
Perchloric acid	1	167	162	58	13.6	~200,000
Acetic acid	1	153	157	60	35	~230,000

The product distribution is also largely unaffected by change of medium. The basic kinetics in acetic acid are also unchanged but the effect of addition of acetate ions indicates a linear relationship between k_2 and $[Tl(OAc)_2^+]/[Tl(III)]$. $Tl(OAc)_2^+$ is regarded as the most significant oxidising entity over a wide range of acetate concentration although $Tl(OAc)^{2+}$ and Tl^{3+} become important at low acetate concentrations. The mechanism, which is basically the same for both sets of reaction conditions, is

$$Tl(III) + R_1R_2C{=}CR_3R_4 \underset{k_{-1}}{\overset{k_1}{\rightleftharpoons}} \left[\begin{matrix} R_1R_2C \\ | \\ R_3R_4C \end{matrix} \cdots Tl\right]^{3+} \quad \text{(I)} \tag{37}$$

$$I + H_2O \underset{k_{-2}}{\overset{k_2}{\rightleftharpoons}} {}^{2+}Tl{-}CR_1R_2{-}CR_3R_4OH + H^+ \tag{38}$$

$${}^{2+}Tl{-}CR_1R_2{-}CR_3R_4OH \xrightarrow{k_3} \xrightarrow{H_2O} Tl^+ + CR_1ROH{-}CR_3R_4OH + H^+ \tag{39}$$

$${}^{2+}Tl{-}CR_1R_2{-}CR_3R_4OH \xrightarrow{k_3} Tl^+ + CR_1R_2R_3COR_4 + H^+ \tag{40}$$

k_1 is very large and (37) is unlikely to be rate-determining. (39) or (40) cannot be the slow step because the acid-retardation implied by step (38) is not found. Reaction (38) is probably, therefore, the slow step. Several schemes for the breakdown of the oxythallation adduct are considered to account for the shift of the group R_3, and it is probable that more than one mechanism operates[213].

The reaction between Hg(II) and olefins has been examined from several angles and work prior to 1950 has been summarised by Chatt[209]. Several types of complex and product are formed, depending on the olefin, which involve no change in the oxidation state of the mercury atom. Propenyl ethers have long been known to produce the corresponding glycol plus metallic mercury but no kinetics are available[215].

It has been shown recently[216] that the 1 : 1 complexes between Hg(II) and a number of olefins are themselves subject to allylic oxidation by Hg(II) at 80 °C. Typical stoichiometry is

$$CH_3CH(OH)CH_2.Hg^+ + 3\ Hg^{2+} = CH_2{=}CH.CHO + 2\ Hg_2^{2+} + 3\ H^+$$

The kinetics, which involve examination of *two* species containing Hg(II) are of the form

$$\frac{+\mathrm{d}[\text{acraldehyde}]}{\mathrm{d}t} = -\frac{1}{4}\mathrm{d}\frac{[\text{Hg(II)}]}{\mathrm{d}t} = k_2[\text{adduct}][\text{Hg}^{2+}]$$

where Hg^{2+} refers to mercuric trifluoroacetate and $k_2 = 4.1 \times 10^{14} \exp(-28.8 \times 10^3/RT)$ l.mole^{-1}.sec^{-1}. Further alkylation of the olefin results in large increases in rate; a $-CH_2R$ group is oxidized to –COR in preference to a methyl group; the double bond always remains intact.

Oxidation of propene labelled with $^{13}C(CH_3{-}CH{=}{}^*CH_2)$ yields acraldehyde (88 %), acetone (10 %) and propanal (2 %)[217]. The labelled acraldehyde consists of equal amounts of $\overset{*}{C}H_2{=}CH{-}CHO$ and $CH_2{=}CH{-}\overset{*}{C}HO$. The scrambling of the ^{13}C is seen as a result of the formation of a symmetrical intermediate

CH
H_2C CH_2
$\overset{+}{Hg}$

This could be oxidised by a second Hg(II) species in the slow step, which must involve two molecules containing Hg(II).

The stoichiometry of the oxidation of ethylene by palladous chloride is

$$C_2H_4 + PdCl_2 + H_2O = CH_3CHO + Pd + 2\ HCl$$

Products from very many other olefins have been detailed by Smidt *et al.*[218]. All monoolefins with at least one hydrogen atom on each carbon atom of the double bond yield the corresponding ketone, with the carbonyl group appearing at the carbon atom initially subject to the higher inductive effect. Reaction in acetic acid produces complex mixtures of compounds including glycol acetates. Considerable kinetic and other data are available, and are summarised below for the reaction in aqueous solution.

(*i*) The kinetics for the oxidation of ethylene, propene and the three butenes are of the form[219–223]

$$-\frac{\mathrm{d[olefin]}}{\mathrm{d}t} = k'K_1 \frac{[\mathrm{PdCl_4}^{2-}][\mathrm{olefin}]}{[\mathrm{Cl^-}]^2[\mathrm{H_3O^+}]}$$

where K_1 is the equilibrium constant for the reaction

$$\mathrm{PdCl_4}^{2-} + \mathrm{RCH{=}CHR} \rightleftharpoons \left[\begin{matrix}\mathrm{RCH} \\ \| \\ \mathrm{RCH}\end{matrix} \longrightarrow \mathrm{PdCl_3}\right]^- + \mathrm{Cl^-} \qquad (41)$$

k' varies according to an Arrhenius expression with $E = 20.4$ kcal.mole^{-1} and $\Delta S^\ddagger = -8.7$ eu. K_1 is 17.4 ± 0.4 at 25 °C with $\mu = 1\ M$ (ClO_4^-) and is pH-independent.

(*ii*) The absorption of olefin by the solution is initially very rapid and to an extent greater than required for simple saturation. The degree of absorption diminishes as the chloride ion concentration is increased but is unaffected by change of pH[219, 220].

(*iii*) Oxidation of C_2D_4 produces only a secondary isotope effect[219].

(*iv*) Oxidation of C_2H_4 in D_2O proceeds four times more slowly[224] than in H_2O and no introduction of deuterium into the acetaldehyde occurs[225].

(*v*) The dependence of rate upon ionic strength (added $NaClO_4$) involves a maximum rate at $\mu = 0.4\ M$[219, 221].

(*vi*) A dimeric species $[PdCl_2C_2H_4]_2$ can be prepared in benzene and isolated[226]. It is decomposed by water to acetaldehyde, palladium metal and hydrochloric acid.

A mechanism accommodating these data has been proposed[219]; the pre-equilibrium (41) is rapidly attained and is followed by

$$[\mathrm{PdCl_3C_2H_4}]^- + \mathrm{H_2O} \underset{k_{-2}}{\overset{k_2}{\rightleftharpoons}} [\mathrm{PdCl_2(H_2O)C_2H_4}] + \mathrm{Cl^-} \qquad (K_2 = k_2/k_{-2}) \quad (42)$$

$$[\mathrm{PdCl_2(H_2O)C_2H_4}] + \mathrm{H_2O} \underset{k_{-3}}{\overset{k_3}{\rightleftharpoons}} [\mathrm{PdCl_2(OH)C_2H_4}]^- + \mathrm{H_3O^+} \qquad (K_3 = k_3/k_{-3}) \quad (43)$$

$$[\mathrm{PdCl_2(OH)C_2H_4}]^- \underset{k_{-4}}{\overset{k_4}{\rightleftharpoons}} \mathrm{CH_2OH{-}CH_2{-}PdCl} + \mathrm{Cl^-} \qquad (\mathrm{slow}) \quad (44)$$

$$\mathrm{CH_2OH{-}CH_2{-}PdCl} \overset{k_5}{\rightarrow} \mathrm{CH_3CHO} + \mathrm{Pd(0)} + \mathrm{HCl} \qquad (45)$$

This gives the rate expression

$$\frac{-\mathrm{d}[C_2H_4]}{\mathrm{d}t} = k_4 K_1 K_2 K_3 \frac{[PdCl_4{}^{2-}][C_2H_4]}{[Cl^-]^2[H_3O^+]}$$

in agreement with experiment. The slow step (44) corresponds to the transformation of a π-complex into a σ-complex, *viz.*

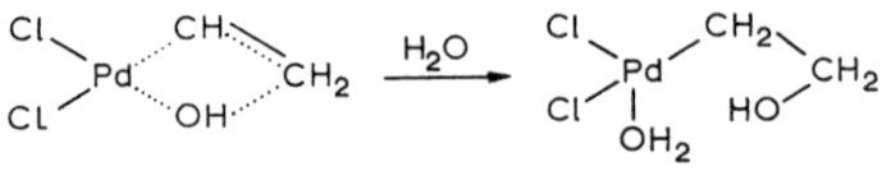

$$\mathrm{Cl_2Pd(\cdots CH_2{=}CH_2\cdots OH)} \xrightarrow{H_2O} \mathrm{Cl_2Pd(OH_2)\text{–}CH_2CH_2OH}$$

The breakdown of the σ-complex is envisaged by Henry[219] as

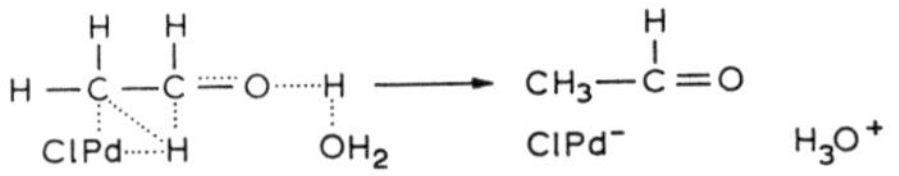

$$\mathrm{H{-}CH(ClPd)\text{–}CH{=}O\cdots H \cdots OH_2} \longrightarrow \mathrm{CH_3{-}CH{=}O} \quad \mathrm{ClPd^-} \quad \mathrm{H_3O^+}$$

Cu(II) compounds are frequently used in conjunction with Pd(II) in the oxidation of olefins in the Wacker process. Their role has been viewed as that of catalyst for autoxidation of Pd metal back to Pd(II). Dozono and Shiba[227] report the rate of oxidation of ethylene by a $PdCl_2$–$CuCl_2$ couple to be given by

$$\text{Rate} = k \frac{[PdCl_2]}{[H_2O^+][Cl^-]^2} p_{C_2H_4}$$

with $E = 10.4$ kcal.mole^{-1}. The agreement with the uncatalysed reactions and the non-involvement of Cu(II) concentration supports the catalytic role. The situation may, however, be more complicated (*vide infra*).

The oxidation of ethylene by palladous acetate in acetic acid has been examined by Moiseev *et al.*[228–230]. This reaction shows mixed stoichiometry[228]

$$C_2H_4 + PdCl_2 + 2\,CH_3CO_2^- = CH_3CO_2CH{=}CH_2 + Pd(0) + CH_3CO_2H + 2\,Cl^-$$

$$C_2H_4 + PdCl_2 + 2\,CH_3CO_2^- = (CH_3CO_2)_2CHCH_3 + Pd(0) + 2\,Cl^-$$

The kinetics are[229]

$$\text{rate} = k_2[Pd(II)][C_2H_4]$$

with $E = 17.2$ kcal.mole^{-1} and $\Delta S^\ddagger = -10.7$ eu. No incorporation of deuterium into the ethylidene diacetate occurs when the reaction is carried out in CH_3CO_2D[230].

One interesting difference between Pd(II) in aqueous and acetic acid solutions is that whilst oxidation of C_2D_4 by aqueous Pd(II) displays no primary kinetic isotope effect, the oxidation of CH_3–CD=CH_2 by $PdCl_2$ in acetic acid–isooctane affords a[231] k_H/k_D value of 2.8. The products are a mixture of propenyl and isopropenyl acetates in a 64 : 36 ratio unaffected by deuteration and with isotopic *retention.* A hydride-ion shift in the slow step is proposed to account for the isotope effect. This would be expected to result in a lower yield for labelled isopropenyl acetate compared with propenyl acetate. Aguiló[208] has commented that Pd(II)-catalysed isomerisation of the olefin could account for the isotopic pattern of the products.

Moiseev *et al.*, who proposed initially[232] that ethylidene diacetate was produced from addition of acetic acid to vinyl acetate, later showed this to be impossible from the result of reaction in CH_3CO_2D, preferring the following mechanism[230]

$$\left[Cl_3Pd \leftarrow \begin{matrix} CH_2 \\ \| \\ CH_2 \end{matrix}\right]^- + CH_3CO_2H \longrightarrow [Cl_3Pd-CH_2-CH_2O_2CCH_3]^{2-} + H^+$$

$$\downarrow$$

$$3\,Cl^- + Pd^0 + CH_3-\overset{+}{C}H-O_2CCH_3$$

$$CH_3\overset{+}{C}HO_2CCH_3 \longrightarrow CH_2=CH-O_2CCH_3 + H^+ \quad (46)$$

$$CH_3\overset{+}{C}HO_2CCH_3 \xrightarrow{CH_3CO_2H} (CH_3CO_2)_2\,CHCH_3 + H^+ \quad (47)$$

Under certain conditions a combination of Pd(II) and Cu(II) in acetic acid oxidises olefins to saturated products which neither reagent produces alone[233]. Although Cu(II) continues to catalyse the production of vinyl acetate through step (46) by a redox mechanism, the following new reaction can be effected

$$C_2H_4 + CuCl_2 + CH_3CO_2^- \xrightarrow[CH_3CO_2H+H_2O]{PdCl_2} \left\{\begin{matrix} CH_3CO_2CH_2CH_2Cl \\ CH_3CO_2CH_2CH_2O_2CCH_3 \\ CH_3CO_2CH_2CH_2OH \end{matrix}\right\} + CuCl$$

Palladium metal is not produced in the new reaction and the substitution of a twenty-fold excess of lithium chloride for cupric chloride prevented reaction; kinetic data revealed first-order dependences upon both Pd(II) and Cu(II). The distribution of products varied in an unpredictable way with reactant concentrations. The following mechanism was proposed by Henry[233] ($X = Cl^-$ or $CH_3CO_2^-$)

$$C_2H_4 + PdX_2 + CH_3CO_2^- \rightarrow XPdCH_2CH_2O_2CCH_3 + X^-$$

$$XPdCH_2CH_2O_2CCH_3 \rightarrow HPdX + CH_2{=}CHO_2CCH_3$$

$$XPdCH_2CH_2O_2CCH_3 + 2\,CuX_2 = PdX_2 + XCH_2CH_2O_2CCH_3 + 2\,CuX$$

The nature of this final step is ill-defined.

Rhodium trichloride oxidises ethylene in dimethylacetamide solution to a

References pp. 493–509

mixture of acetaldehyde, but-1-ene and but-2-ene. In general, 1 mole of CH_3CHO is produced per mole of Rh(III). Initial rate measurements indicate the law[234]

$$-\mathrm{d}[C_2H_4]/\mathrm{d}t = \frac{k[\mathrm{Rh(III)}][C_2H_4]^{1>n\geqslant 0}}{[\mathrm{Cl}^-]}$$

The following scheme is consistent with the available results

$$\mathrm{Rh(III)Cl}_n \underset{k_{-1}}{\overset{k_1}{\rightleftharpoons}} \mathrm{Rh(III)Cl}_{n-1} + \mathrm{Cl}^-$$

$$\mathrm{Rh(III)Cl}_{n-1} + C_2H_4 \underset{k_{-2}}{\overset{k_2}{\rightleftharpoons}} \mathrm{Rh(III)Cl}_{n-1}(C_2H_4)$$

$$\mathrm{Rh(III)Cl}_{n-1}(C_2H_4) + H_2O \rightarrow \mathrm{Rh(I)Cl}_{n-1} + CH_3CHO + 2\,H^+ \qquad \text{(fast)}$$

$$\mathrm{Rh(I)Cl}_{n-1} + C_2H_4 \rightarrow \mathrm{Rh(I)\ complex} \qquad \text{(fast)}$$

The H_2O is considered to come from coordinated water in $\mathrm{Rh(III)Cl_3.3\,H_2O}$. Assuming a steady state for Rh(III) Cl_{n-1} yields

$$\frac{-\mathrm{d}[C_2H_4]}{\mathrm{d}t} = \frac{k_1 k_2[\mathrm{Rh(III)Cl}_n][C_2H_4]}{k_{-1}[\mathrm{Cl}^-]+k_2[C_2H_4]}$$

The observed results fit this equation reasonably well, and at 80 °C $k_1 \approx 10^{-3}$ $\sec^{-1}$, $k_{-1}[\mathrm{Cl}^-]/k_2 \approx 0.025$ mole.l^{-1} with (for k_1) $E = 10.1$ kcal.mole^{-1} and $\Delta S^\ddagger = -46$ eu.

3.3.2 Arylcyclopropanes

Thallium(III) triacetate oxidatively cleaves phenylcyclopropane and its ring-substituted analogues in anhydrous acetic acid[234a] to yield mainly (> 90 %) the corresponding 1-aryl-1, 3-diacetoxypropane together with a little of the cinnamyl acetate. The reactions are of the first order in each component provided account is taken of the double salt formation between Tl(III) and Tl(I) which renders an *additional* Tl(III) molecule inactive following the reduction of each Tl(III) in a two-equivalent process. For phenylcyclopropane $E = 13.0 \pm 0.2$ kcal.mole^{-1} and $\Delta S^\ddagger = -29.2 \pm 0.6$ eu. Electron-releasing groups facilitate reaction ($\rho = -4.3$). By analogy with Hg(II) acetate which cleaves the ring to yield an organomercurial compound, the proposed mechanism is

C_6H_5–(cyclopropyl) + $\mathrm{Tl(OAc)_3}$ $\xrightarrow{\text{slow}}$ C_6H_5–CH(OAc)–CH$_2$–CH$_2$–$\mathrm{Tl(OAc)_3}$

↓ fast

C_6H_5–CH(OAc)–CH$_2$–CH$_2$–OAc + C_6H_5–CH=CH–CH$_2$–OAc + TlOAc

An entirely analogous study[234b] using $Hg(OAc)_2$ showed that for this reagent reaction ceases at the organometallic compound unless the temperature is increased to 135 °C ($k_{dec} = 8.6 \times 10^{-7}$ sec^{-1}).

Lead tetraacetate reacts more slowly with arylcyclopropanes then either thallic or mercuric acetates although the distribution of products is similar[234c]. The reactions are of the first order in each reactant; for phenylcyclopropane $E = 20.2 \pm 0.3$ kcal.mole^{-1} and $\Delta S^{\ddagger} = -17.8 \pm 1.0$ eu and a good correlation between rate of cleavage and σ^+ is obtained, giving $\rho^+ = -1.75$. Addition of perchloric acids catalyses reaction through equilibrium formation of the more highly electrophilic $Pb(OAc)_3^+ ClO_4^-$ ($K_{298°K} = 23.7$ l.mole^{-1}); under these conditions ρ^+ is -1.3 which argues against an ion-pair description of the reactive species in the tetraacetate oxidation.

3.3.3 Alcohols

Kinetic studies have been reported only for the oxidations by Hg(II) and Tl(III), although the oxidation by Pb(IV) is receiving mechanistic study at the present time.

1-Propanol reacts with lead tetraacetate in boiling benzene solution to give a complex mixture of products including 1-propylacetate (35 %) and 1,1-dipropoxypropane (10 %)[235]. The intermediate $RCH_2CH_2\text{–O–}Pb(OAc)_3$ is considered to decompose both homolytically and heterolytically on the basis of observed products, *e.g.* 2-propanol gives 2 % of isopropyl phenyl ether in benzene solution and the reaction is subject to catalysis by pyridine. The production of substituted tetrahydrofurans from long-chain alcohols[236] has been reviewed[237]. Triphenyl methanol gives high yields of hemiketal ester[238, 239]; in general the course of reaction is

$$X\text{–}C_6H_4\text{–}C(C_6H_5)_2\text{–OH} \xrightarrow{Pb(IV)} X\text{–}C_6H_4\text{–}C(C_6H_5)(OAc)\text{–O–}C_6H_4\text{–}Y$$

e.g. X = NO$_2$ → X = NO$_2$, Y = H; X = H, Y = NO$_2$

The migratory aptitude for *p*-nitrophenyl relative to phenyl is 4.4 ± 0.3 which was interpreted as indicating an exclusively homolytic mechanism. The following chain scheme was proposed

$$Ar_3COH + Pb(OAc)_4 \rightleftharpoons Ar_3C\text{–O–}Pb(OAc)_3 + HOAc$$

$$Ar_3C\text{–O–}Pb(OAc)_3 \rightarrow Ar_3C\text{–O}\cdot + Pb(OAc)_3\cdot \quad \text{(slow)}$$

$$Ar_3C\text{–O}\cdot \rightarrow Ar_2\dot{C}OAr$$

$$Ar_2\dot{C}OAr + Ar_3CO\text{–}Pb(OAc)_3 \rightarrow \text{product} + Ar_3CO\dot{P}b(OAc)_2$$

$$Ar_3CO\dot{P}b(OAc)_2 \rightarrow Pb(OAc)_2 + Ar_3CO\cdot$$

$$Ar_2\dot{C}OAr + (AcO)_3Pb\cdot \rightarrow \text{product} + Pb(OAc)_2$$

A brief study of the oxidations of cyclohexanol and cyclohexanol-1-*d* by Tl(III) indicated the rate expression to be[135]

$$-\mathrm{d}[\mathrm{Tl(III)}]/\mathrm{d}t = k[\text{alcohol}][\mathrm{Tl(III)}]/f[\mathrm{H_3O^+}]$$

The values of k_H/k_D and k_{D_2O}/k_{H_2O} are, respectively, 5.5 and 1.7. The acidity dependence was explained in terms of the partial hydrolysis[240]

$$\mathrm{Tl^{3+} + 2\,H_2O \rightleftharpoons Tl(OH)^{2+} + H_3O^+}$$

where Tl^{3+} is the only significant oxidant. This does not, however, explain the solvent isotope effect, for *K* (hydrolysis) is the same in H_2O/D_2O mixtures as in H_2O, and its origin may lie in differential solvation of the transition states.

Two studies have been performed by Littler[24, 135] on the oxidation of cyclohexanol by Hg(II), the second leading to more detailed and reliable data. The reaction is first-order in both oxidant and substrate but the rate is independent of acidity. *E* is 24.8 kcal.mole^{-1}, $\Delta S^{\ddagger}$ is 1 eu, k_H/k_D is 3.0 and k_{D_2O}/k_{H_2O} is 1.30. At 50 °C di-isopropyl ether is attacked at about one-half the rate of isopropanol, which implies that hydride ion abstraction is occurring in both cases. This is supported in the case of cyclohexanol by the isotope effects.

The oxidation of alcohols by Pd(II) and Rh(III) has been noted but no kinetic data are yet available[241–243].

3.3.4 *Hydroperoxides*

Only *tert*-hydroperoxides have been examined kinetically and discussion will be restricted to these. The stoichiometry of the lead tetraacetate oxidation is not straightforward, but the main product is the corresponding *tert*-alcohol[244] and one mole of *tert*-butyl hydroperoxide consumes just over two moles of oxidant[245]. The kinetics with this substrate are[245]

$$-\mathrm{d}[\mathrm{Pb(IV)}]/\mathrm{d}t = k_3[\mathrm{Pb(IV)}][t\text{-}\mathrm{BuOOH}]^2$$

with an activation energy of 20.0±0.5 kcal.mole^{-1}. Pb(II) fails to retard reaction, but added sodium acetate accelerates reaction to an extent directly proportional to its concentration. Sodium acetate is very weakly dissociated in acetic acid and this observation implies a *second*-order dependence on acetate ion, as in the auto-decomposition of lead tetraacetate (p. 346). Addition of benzene and also of ethanol increases the rate, the latter to a degree proportional to its concentration. In pure methanol the reaction order in hydroperoxide changes to unity.

The kinetics are in keeping with the scheme

$$Pb(OAc)_4 + OAc^- \rightleftharpoons Pb(OAc)_5^-$$

$$Pb(OAc)_5^- + OAc^- \rightleftharpoons Pb(OAc)_6^{2-}$$

$$Pb(OAc)_6^{2-} + ROOH \rightleftharpoons Pb(OAc)_6^{2-}.ROOH$$

$$Pb(OAc)_5^- + ROOH \rightleftharpoons [Pb(OAc)_5.ROOH]^-$$

$$[Pb(OAc)_5.ROOH]^- + ROOH \rightarrow RO_2\cdot$$

$$Pb(OAc)_6^{2-}.ROOH + ROOH \rightarrow RO_2\cdot$$

Added alcohol can replace ROOH in its purely coordinative capacity.

A non-kinetic study of the oxidation of cumyl hydroperoxide[246] by Pb(IV) to acetophenone and dimethylphenylcarbinol gives useful complementary data.

3.3.5 Formic acid

Although formic acid is oxidised to CO_2 by Pb(IV), kinetic data exist only for the oxidations with Hg(II), Hg(I) and Tl(III) in aqueous mineral acids[247–249] and with Pd(II) in acetic acid[249a]. Rate equations are of the form

$$-d[HCO_2H]/dt = k\frac{[HCO_2H][Ox]}{[H_3O^+]}$$

Kinetics are obtained directly for Hg(II), Hg(I) and Tl(III). With Pd(II) a catalytic system was used[249a], depending on the continuous reoxidation of Pd(0) by a Cu(II)–O_2 couple. The inverse acidity dependence could result from a hydrolysis of the type

$$Tl^{3+} + H_2O \rightleftharpoons TlOH^{2+} + H^+$$

where only $TlOH^{2+}$ is strongly oxidising, but is more probably due to ionisation of the substrate to the readily oxidised HCO_2^-; thus

$$-d[HCO_2H]/dt = k_2[HCO_2^-][Ox]$$

The data are summarised in Table 9. At high concentrations of formic acid the reaction becomes less than first-order in substrate[248]; this indicates the possibility of complex-formation, but a medium effect may also be influential in the vicinity of 1 *M* formic acid. Complex-formation affects the kinetics of the Tl(III) oxidation at all but the lowest reactant concentrations[249].

TABLE 9

SUMMARY OF DATA FOR OXIDATION OF FORMATE ION[247, 249a]

Oxidant	*Temp. range* (°C)	*E* (*kcal.mole*$^{-1}$)	$\Delta S^{\ddagger}$(*eu*)	k_H/k_D
Hg^{2+}	36–61	20.6	3	3.4 (46.5 °C)
Hg^{2+}	60–80	21.6	0	3.9 (65.1 °C)
Tl^{3+}	65–85	26.6	21	3.4 (75.2 °C)
Pd(II)	100–125	22.7	−6	2.0 (112 °C)

A two-equivalent oxidation mechanism is favoured for all oxidants[247, 249], *viz.*

$$HCO_2^- + M^{n+} \overset{K}{\rightleftharpoons} HCO_2.M^{(n-1)+} \qquad \text{(fast)}$$

$$HCO_2.M^{(n-1)+} \overset{k}{\rightarrow} H^+ + CO_2 + M^{(n-2)+} \qquad \text{(slow)}$$

In the case of Tl(III) the overall rate coefficient has been resolved into a product kK for the two steps[247, 249]. The large positive $\Delta S^{\ddagger}$ is due almost entirely to the initial association, which was also studied spectroscopically. An alternative rate determining step in the Pd(II) oxidation is hydride ion transfer to Pd(II)[249a].

Oxidation of formic acid by mercuric chloride is the subject of several early kinetic studies. Dhar[177] showed the reaction to be first-order in oxidant and substrate and to be subject to strong retardation by added chloride ions in agreement with earlier work. The reaction is also subject to retardation by added acid and presumably involves formate ion as the principal reactant.

3.3.6 Higher carboxylic acids

The auto-decomposition of lead tetraacetate in acetic acid, which normally occurs at reflux temperature[250], can be studied at 50 °C in the presence of sodium acetate[251, 251a]. The principal products of both the uncatalysed and catalysed decompositions are acetoxyacetic acid and carbon dioxide. The kinetic order of the "normal" decay of Pb(IV) is complex and evidence was obtained that oxidation of products is significant after the earliest stages. The evidence indicates that slow, simple homolytic breakage of lead tetraacetate to give $Pb(OAc)_3\cdot$ and $AcO\cdot$ does not occur but that the solvent plays an integral part, *e.g.*

$$Pb(OAc)_4 + HOAc \rightarrow (AcO)_3Pb\cdot + HOAc + \cdot CH_2CO_2H$$

The radical $\cdot CH_2CO_2H$ is regarded as originating from the solvent. This primary step is followed by a multitude of secondary reactions. An entirely analogous scheme, producing $(CH_3)_2CH\text{–O–}\dot{C}(CH_3)_2$, was proposed for decomposition in di-isopropylether[250].

In the presence of acetate ions (1.0 M) the rate of decomposition decreases as the oxidant concentration is increased[251, 251a], although the reaction tends towards first-order behaviour both in Pb(IV) and $CH_3CO_2^-$ as the oxidant concentration is lowered towards 2×10^{-4} M[251a]. Under these simplyfying conditions $E = 12.2 \pm 0.2$ kcal.mole^{-1} and $\Delta S^{\ddagger} = -38 \pm 0.7$ eu. Norman and Poustie[251a] favour a non-radical mechanism, *viz.*

$$\text{B:} \curvearrowright \; \begin{matrix} \text{AcO—Pb(OAc)}_2 \\ \text{H} \quad\quad \text{O} \\ \text{CH}_2\text{—CO} \end{matrix} \longrightarrow \text{AcO.CH}_2\text{CO}_2^- + \text{BH}^+ + \text{Pb(OAc)}_2$$

which is supported by the abilities of other bases to catalyse the decomposition. Benson *et al.*[251] prefer a second-order dependence on acetate ion and produce evidence that the complexity of the kinetic decay of Pb(IV) at higher concentrations stems from product-retardation. They write a radical mechanism with $Pb(OAc)_6^{2-}$ as the active oxidant. Norman and Poustie[251a] propose a base-catalysed dimerisation in competition with the oxidation to account for the kinetic deviation, and were able to isolate a new lead compound from reaction mixtures exhibiting 4 equivalents of oxidising power per mole.

3.3.7 Ketones

Pb(IV), Tl(III) and Hg(II) attack ketones at rates approximating to the corresponding rate of enolisation

$$-\overset{\overset{\text{O}}{\|}}{\text{C}}-\text{CHR}- \rightleftharpoons -\overset{\overset{\text{OH}}{|}}{\text{C}}=\text{CR}-$$

Lead tetraacetate attacks several ketones, the rates being those of enolisation[252], *viz.*

$$\text{rate} = k[\text{ketone}][\text{Pb(IV)}]^0$$

Clearly a two-equivalent oxidation of the enol is taking place. Product work has been summarised by Criegee[253] and in general α-acetoxylation occurs.

Oxidation of cyclohexanone by thallic perchlorate[157] has similar kinetics but includes an acidity dependence of the form (at 25 °C, $\mu = 1.3$ M)

$$k = k'\{4.8 \times 10^{-5} + 11.1 \times 10^{-5}[H_3O^+]\}$$

The rate of oxidation is, however, considerably slower than the rate of enolisation as measured by iodination.

Mercuric perchlorate has been shown to attack cyclohexanone, the reaction being zero-order in the salt and the rate being that of enolisation with a primary kinetic isotope effect of the same magnitude, to give a mercurated ketone[154]

$$-CH{=}C(-O{-}H)- \; + \; Hg(II) \longrightarrow -CH(Hg^{+})-C(=O)-$$

distinguished by its ultraviolet spectrum (λ_{max} 2260 A, $\varepsilon = 8.53 \times 10^3$). Similar data were reported for acetone[154]. These are *not* oxidation processes but are necessary preliminaries.

The ultimate product of Hg(II) oxidation is cyclohexane-1,2-dione which is probably produced from 2-hydroxycyclohexanone. The oxidation by thallic perchlorate produces cyclopentanecarboxylic acid in 75 % yield[254]; this contrasts with the production of 2-acetoxycyclohexanone by thallic acetate[255]. The mechanism favoured for the production of the carboxylic acid is

$$\text{cyclohexanone} \rightarrow \text{2-}TlX_2\text{-cyclohexanone} \rightleftharpoons \text{1,1-(H-O)(OH)-2-}TlX_2\text{-cyclohexane} \rightarrow \text{cyclopentane-}CO_2H$$

2-hydroxycyclohexanone does not produce the acid on treatment with thallic perchlorate and the reaction with 2,2′,6,6′-tetradeutero-cyclohexanone produces a tri-deuterated acid. Other possibilities were eliminated.

3.3.8 *Ethers*

The ready oxidation of di-isopropyl ether by Hg(II) perchlorate[24] is a good indication that this oxidant can function as a hydride-ion acceptor, *viz.*

$$Hg^{2+} + H{-}C(CH_3)_2{-}O{-}CH(CH_3)_2 \rightarrow HgH^{+} + (CH_3)_2C{=}\overset{+}{O}{-}CH(CH_3)_2 \qquad \text{(slow)}$$

$$HgH^{+} + Hg^{2+} \rightarrow Hg_2{}^{2+} + H^{+} \qquad \text{(fast)}$$

$$(CH_3)_2C{=}\overset{+}{O}{-}CH(CH_3)_2 + H_2O \rightarrow (CH_3)_2C{=}O + (CH_3)_2CHOH \qquad \text{(fast)}$$

The kinetics are simple second-order. At 40 °C ($\mu = [H^+] = 0.188$ M) k_2 is 5.1×10^{-6} l.mole^{-1}.sec^{-1} and E is found to be 16.6 kcal.mole^{-1}. These data are, of course, relevant to the problem of Hg(II) oxidation of secondary alcohols.

3.4 OXIDATION OF POLYFUNCTIONAL MOLECULES

3.4.1 Glycols

Lead tetraacetate is an important reagent for glycol cleavage, which has the stoichiometry

$$\begin{matrix} R_2C\text{–}OH \\ | \\ R_2C\text{–}OH \end{matrix} + Pb(OAc)_4 = 2\ R_2C{=}O + Pb(OAc)_2 + 2\ HOAc$$

It can be used both in polar and non-polar solvents and complements the Malaprade oxidation with periodate, with which it has been compared and contrasted[256]. Much has been published on the mechanism of its action, particularly on the stereochemical aspects of its reactivity, and only a summary of the main features can be presented.

The rate equation for the oxidation of ethylene glycol in acetic acid is[257]

$$-d[Pb(IV)]/dt = k_2[Pb(IV)][glycol]$$

with $k_2 = 1.95 \times 10^{14} \exp(-20.9 \times 10^3/\boldsymbol{RT})$ l.mole^{-1}.sec^{-1}. This has been confirmed for a series of glycols by Cordner and Pausacker[258], who also showed that electron-releasing substituents accelerate oxidation of benzpinacols while electron-withdrawing groups retard it.

Acidification with trichloracetic acid catalyses oxidation[259], the fractional increase in the rate coefficient per mole of acid added, *viz.* $(\Delta k/k_0)/[\text{acid}]$, being of the order of two. Strong catalysis by alkali metal acetates has been observed for several oxidations, *e.g.* of *cis*-cyclohexane-1,2-diol[260], formic acid[261], methyl mannoside and galactoside[261] and several α-hydroxycarboxylic acids[260].

Retardation by acetic acid itself, which reduces the oxidation rate of *trans*-cyclohexane-1:2-diol by three orders of magnitude[262], suggests the existence of a pre-equilibrium

$$\begin{matrix} R_2C\text{–}OH \\ | \\ R_2C\text{–}OH \end{matrix} + Pb(OAc)_4 \rightleftharpoons \begin{matrix} R_2C\text{–}O\text{–}Pb(OAc)_3 \\ | \\ R_2C\text{–}OH \end{matrix} + HOAc$$

Changing the reaction medium from acetic acid to water does not reduce the efficiency of the oxidant[263] and, indeed, gradual dilution of an acetic acid medium with water, methanol or benzene increases the rates of oxidation of several glycols of factors of 500 to 1000 (ref. 260). This effect raises the question of whether the catalysis by trichloracetic acid (*vide supra*) is solely an effect of acidity.

The main discussion of mechanism has been centred around the stereochemical specificity of the fission, particularly with respect to the role of cyclic intermediates formed by the reaction[264]

$$\begin{array}{l} R_2C-OH \\ | \\ R_2C-OH \end{array} + Pb(OAc)_4 \rightleftharpoons \begin{array}{l} R_2C-O-Pb(OAc)_3 \\ | \\ R_2C-OH \end{array} + HOAc$$

$$\updownarrow$$

$$\begin{array}{l} R_2C-O \\ | \\ R_2C-O \end{array} \!\!> Pb(OAc)_2 + HOAc \qquad (48)$$

Criegee *et al.*[265] have measured the oxidation rates of a series of cyclic 1:2-diols. For up to seven-membered rings the *cis* form is more reactive than the *trans*, k_{cis}/k_{trans} reaching 10^3 for cyclopentane-1:2-diol but falling to 22 for cyclohexane-1:2-diol and becoming less than one for nine-membered and larger rings. Racemic forms of the diols R–CH(OH)–CH(OH)–R are oxidised 20–40 times faster than the meso forms[265]. However, dihydrophenanthrene-9,10-diols(I)

(I) (II)

are sometimes far more readily oxidised in their *trans*-configurations (Table 10)[265, 266].

Trans-decalin-9,10-diol(II) is of interest because of its inability to form cyclic esters. It is oxidised by lead tetraacetate with a second-order rate coefficient of 0.148 $l.mole^{-1}.sec^{-1}$ 20 °C which is 10^2 times less than that of the *cis*-isomer[267]. The rate coefficients for the *cis* and *trans* forms of the hydrindane analogue are 5400 and 1.5 $l.mole^{-1}.sec^{-1}$ under comparable conditions[268].

It seems, however, that di-tertiary glycols are especially reactive; Angyal and Young[269] have found that reaction between lead tetraacetate and the di-secondary

TABLE 10

SECOND-ORDER RATE COEFFICIENTS (20 °C) FOR THE OXIDATION BY Pb(IV) OF SYMMETRICAL 9,10-DISUBSTITUTED DIHYDROPHENANTHRENE-9,10-DIOLS[265]

R	k_{cis}	k_{trans}
	($l.mole^{-1}.sec^{-1}$)	
H	9.7	130
CH_3	7.5	192
C_2H_5	304	1310
C_6H_6	286	24.7
p-$CH_3 \cdot C_6H_4$	407	32.4

diol cholestane-3β,6β,7α-triol(III) is immeasurably slow at 50 °C.

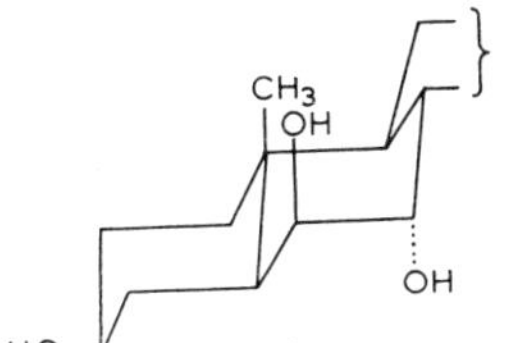

(III)

In a related study the same workers reported[270] that the camphanediols(IV) and (V)

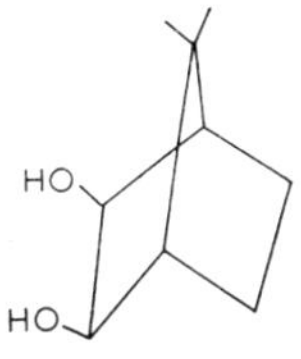

(IV) and (V)

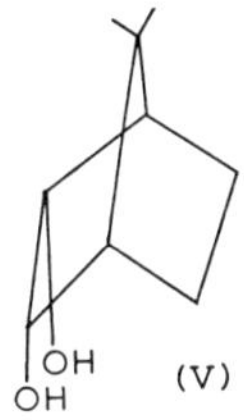

are oxidised at least 6×10^4 times faster than the pair (VI) and (VII)

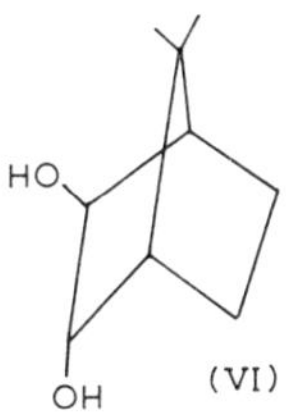

(VI) and (VII)

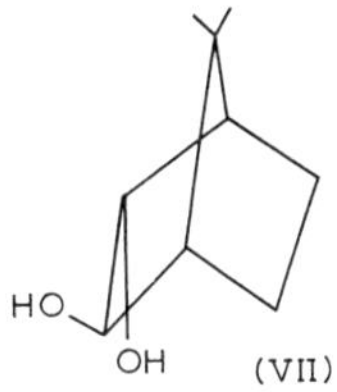

It is clear then that more than one mechanism is operative for glycol fission. In the case of *cis*-cyclopentanediols and camphanediols a cyclic ester is a necessary intermediate. For *trans*-decalin-9,10-diol a non-cyclic mechanism must operate which cannot function for cholestane-3β,6β,7α-triol and is inefficient for *trans*-camphanediols. It is pertinent that while the fission of glycols capable of forming cyclic esters proceeds several hundred times faster in benzene than in acetic acid, the reactions of *trans*-decalin-9,10-diol and *trans*-hydrindane-1,6-diol are 4–5-fold slower in benzene[265].

A solution of lead tetraacetate in pyridine rapidly oxidises the most recalcitrant *trans*-diols, especially if a considerable excess of oxidant (3–4 moles) is used, implying yet a further mechanism for the action of this versatile oxidant[271, 272].

The "normal" oxidation, *i.e.* that of *cis*-diols, occurs by an oxidative breakdown of the cyclic complex, *viz.*

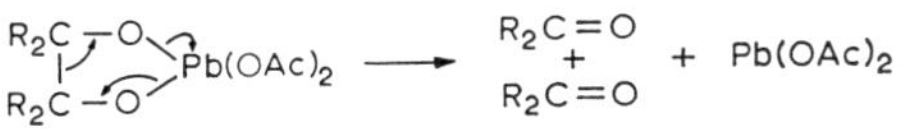

$$\begin{matrix} R_2C-O \\ | \\ R_2C-O \end{matrix} \!\!> Pb(OAc)_2 \longrightarrow \begin{matrix} R_2C=O \\ + \\ R_2C=O \end{matrix} + Pb(OAc)_2$$

The "slow" or "*trans*" oxidation presumably involves an intermediate of the kind

$$H-O-C-C-O-Pb(OAc)_3$$

the breakdown of which is readily subject to the observed base catalysis.

3.4.2 *α-Hydroxycarboxylic acids*

A few rate coefficients for oxidation by lead tetraacetate have been reported. These, together with those for some carbohydrates, are referred to in the section on glycols (p. 349).

3.4.3 *Dicarboxylic acids*

The oxidation of oxalic acid by mercuric chloride to give CO_2 and mercurous chloride is a classic example of an induced reaction. This reaction is extremely slow unless small quantities of chromic acid and manganous ions are added, whereon facile reduction takes place[177, 273]. Addition of permanganate or persulphate and some *reducing* agents is also effective and the oxidation proceeds readily under photo- or X-irradiation (Eder's reaction). The large quantum yield points to a chain mechanism[274], which could also function with an "inducing" oxidant, *viz.*

$$\text{Oxidant} + (CO_2H)_2 \rightarrow CO_2 + \cdot CO_2^- + 2\,H^+ + \text{ox}^-$$

$$\cdot CO_2^- + HgCl_2 \rightarrow CO_2 + HgCl\cdot + Cl^-$$

$$HgCl\cdot + (CO_2H)_2 \rightarrow Hg + Cl^- + \cdot CO_2^- + CO_2 + 2\,H^+$$

The oxidation of oxalic acid by $AuCl_4^-$ represents one of the few examples of a kinetic study[275] of Au(III) oxidation falling within the present category, *viz.*

$$AuCl_4^- + H_2C_2O_4 = AuCl_2^- + 2\,CO_2 + 2\,HCl$$

A slow, second stage leads to metallic gold. The rate law was determined, under conditions unfavourable to hydrolysis of chloraurate ion, to be

$$-d[\text{Au(III)}]/dt = k[H_2C_2O_4][AuCl_4^-]/\{([H_3O^+]/K)+1\}$$

which may be written

$$-d[\text{Au(III)}]/dt = k_2[HC_2O_4^-][AuCl_4^-]$$

where $k_2 = 2.5 \times 10^7 \exp(-13.0 \times 10^3/RT)$ l.mole^{-1}.sec^{-1} and at 20 °C is $(4.98 \pm 0.18) \times 10^{-3}$ l.mole^{-1}.sec^{-1}. Under conditions favouring the (rapid) hydrolysis of chloraurate ion

$$AuCl_4^- + H_2O \rightleftharpoons [AuCl_3OH]^- + Cl^- + H^+$$

the reaction becomes faster. k_2 for the reaction between the hydroxotrichlorogold-(III) ion and $HC_2O_4^-$ is 0.9 ± 0.1 l.mole^{-1}.sec^{-1} at 20 °C ($\mu = 0.15\ M$), the Arrhenius expression being $k_2 = 1 \times 10^{10} \exp(-13.5 \times 10^3/RT)$ l.mole^{-1}.sec^{-1}. This represents a further example of the greater reactivity of a hydroxo-ion as compared with its fully aquated or chlorinated counterpart.

4. Oxidation by Ag(II), Ag(III), Co(III), Ce(IV), Mn(III), V(V), Ir(IV), Np(VI) and Pu(VI)

4.1 INORGANIC CHEMISTRY OF THESE OXIDATION STATES

These share tendencies to behave as one-equivalent oxidants with redox potentials in the range 1.0 to 2.0 V and to undergo hydrolysis in aqueous solution above pH 2. They form complexes with anions, *e.g.* sulphate, which make their reactivity dependent on the nature of the medium both from the point of redox potential and of ligand displacement (Table 11).

The segregation of these oxidants from Cu(II) and Fe(III) is based upon their much greater reactivity and their restriction to low pH.

TABLE 11

REDOX POTENTIALS OF STRONG ONE-EQUIVALENT OXIDANTS[19]

Complex	*Medium*	π^0 (*Volt*)
$Ag^{2+} + e^- = Ag^+$	$HClO_4$	1.970
	HNO_3	1.927
$Co^{3+} + e^- = Co^{2+}$	3 M HNO_3	1.842
$Ce^{4+} + e^- = Ce^{3+}$	1 M $HClO_4$	1.70
	0.5–2.0 M HNO_3	1.61
	1 M H_2SO_4	1.44
	1 M HCl	1.28
$Mn^{3+} + e^- = Mn^{2+}$	3 M $HClO_4$	1.56
	H_2SO_4	1.51
	$H_4P_2O_7$	1.15[421]
$IrCl_6^{2+} + e^-\ IrCl_6^{3-}$	—	1.017
$V^{5+} + e^- = V^{4+}$	HCl	1.00
$NpO_2^{2+} + e^- = NpO_2^+$	1 M HCl	1.15
$PuO_2^{2+} + e^- = PuO_2^+$	—	0.93

4.1.1 Silver species

McMillan[276] has reviewed the chemistry of Ag(II) and Ag(III). Paramagnetism and electron spin resonance studies confirm the presence of Ag(II) (as opposed to equimolar Ag(I)+Ag(III)). The colours of Ag(II) solutions in various mineral acids indicate the existence of complexes, the oxidising power of which is apparent from their decomposition even at 0 °C, although high acidity promotes stability. Rapid isotope exchange[277] between Ag(I) and Ag(II) is considered to result from the equilibrium

$$2\,\text{Ag(II)} \rightleftharpoons \text{Ag(I)} + \text{Ag(III)}$$

Both Ag(II) and Ag(III) have been considered to be the active species in the Ag(I)-catalysed oxidation of many compounds by persulphate ion. Salts of Ag(III) have been prepared but only a single kinetic study (of the decomposition of water by the ethylene dibiguanide nitrate) has been reported (p. 366).

Oxidations by persulphate ion have been reviewed by House[278]. Silver-ion catalysed reactions normally obey the rate expression

$$-\text{d}[S_2O_8^{2-}]/\text{d}t = k_2[S_2O_8^{2-}][\text{Ag(I)}][\text{reductant}]^0$$

[Ag(I)] is, of course, unchanged during reaction. For reducing agents which are positive ions, all values of k_2 fall in a narrow range, particularly after correction for ionic strength[278]. k_2 for reductants bearing zero or negative charge is up to 10^2 times greater and varies over a much wider range. Within a given class of organic substrate, *e.g.* alcohol, α-hydroxy acid, variations in k_2 are small[278a].

Several mechanisms have been put forward which are consistent with the kinetics

$$(A)\quad Ag^+ + S_2O_8^{2-} \rightarrow Ag^{3+} + 2\,SO_4^{2-} \qquad \text{(slow)}$$

$$(B)\quad S_2O_8^{2-} \rightleftharpoons 2\,\cdot SO_4^- \qquad \text{(fast)}$$

$$2\,\cdot SO_4^- + Ag^+ \rightarrow Ag^{3+} + 2\,SO_4^{2-} \qquad \text{(slow)}$$

$$(C)\quad Ag^+ + S_2O_8^{2-} \rightarrow AgS_2O_8^- \qquad \text{(slow)}$$

$$AgS_2O_8^- \rightarrow Ag^{3+} + 2\,SO_4^{2-} \qquad \text{(fast)}$$

or

$$AgS_2O_8^- \rightarrow Ag^{2+} + SO_4^{2-} + \cdot SO_4^- \qquad \text{(fast)}$$

The species $AgS_2O_8^-$ may represent only a transition state and not a reactive intermediate.

References to the numerous substrates which have been oxidised by Ag^+–$S_2O_8^{2-}$ are given by House. Additional data have been reported more recently[278b–d]. Since all the oxidations follow the same basic kinetics and involve an oxidising metal ion in trace quantities, the details are not reiterated here. The simple oxidation of Ag(I) by persulphate is discussed on p. 475.

4.1.2 Cobalt (III)

The chemistry of Co(III) in dilute aqueous acidic solution is complicated by (*i*) oxidation of the solvent, (*ii*) complex formation with counter-ions, (*iii*) hydrolysis and (*iv*) apparently extensive dimerisation. These phenomena are discussed further in the section on oxidation of water.

4.1.3 Cerium(IV)

The redox potential and the reactivity of this oxidation state depend strongly upon the anion (Table 11). Strong complexes are formed[279e] with SO_4^{2-}. Even in perchloric acid, hydrolysis and polymerisation greatly complicate kinetics. The co-ordination number of Ce(IV) in solution is not established[279].

4.1.4 Manganese(III)

This oxidation state is stable only when complexed by anions, *e.g.* PO_4^{3-}, or in moderately acidic solutions. Perchloric acid solutions of Mn(III) perchlorate have been prepared[280]; these are fairly stable, although the omnipresent equilibrium

$$2\ \mathrm{Mn(III)} \rightleftharpoons \mathrm{Mn(II)} + \mathrm{Mn(IV)}$$

leads to slow deposition of MnO_2. This equilibrium has a constant[281] of *ca.* 10^{-3}, but its effect is apparent in several $[\mathrm{Mn(III)}]^2/[\mathrm{Mn(II)}]$ dependences which have been reported for reactions of Mn(III) (*vide infra*), *e.g.* with mercurous ion[280]. The hydrolysis of Mn^{3+} in aqueous perchloric acid has been examined

$$\mathrm{Mn^{3+}\ aq} \rightleftharpoons \mathrm{MnOH^{2+}} + \mathrm{H_3O^+}$$

and at 25 °C the equilibrium constant is about unity ($\mu = 4\ M$)[282]. Other aspects of Mn(III) chemistry have been reviewed recently[282a].

4.1.5 Vanadium(V)

This is stable in acid solution as a monomeric oxy-cation of the type VO_2^+ (ref. 283). Most oxidations by V(V) are acid-catalysed and this generality indicates protonation of the cation to VO_2H^{2+} or $V(OH)_3^{2+}$ in a rapid pre-equilibrium. Waters and Littler[283] have summarised the evidence for regarding V(V) as acting chiefly in a one-equivalent, rather than in a two-equivalent, capacity.

4.1.6 *Iridium(IV) as* $IrCl_6^{2-}$

This oxidant, which has been utilised only recently[284], is almost substitution-inert, and any reaction proceeding *via* substitution would be expected to have an activation energy of the order of 30 $kcal.mole^{-1}$. Those oxidations which have been examined have much lower activation energies. The reduction product is $IrCl_6^{3-}$ which is also solvolysed slowly.

4.2 OXIDATION OF INORGANIC SPECIES

4.2.1 *Chloride ion*

Ceric perchlorate oxidises Cl^- to $\frac{1}{2}$ Cl_2 according to the rate expression[285]

$$-d[Ce(IV)]/dt = k[Ce(IV)]f[H_3O^+]f'[Cl^-]$$

The chloride ion dependence indicates the participation of several chloro-complexes of Ce(IV); the acid catalysis was ascribed to the suppression of the hydrolysis

$$Ce(H_2O)_n{}^{4+} \rightleftharpoons Ce(H_2O)_{n-1}OH^{3+} + H^+$$

Reaction may involve formation of $\cdot Cl_2^-$ rather than Cl atoms, *viz.*

$$CeCl_2{}^{2+} \rightarrow Ce^{3+} + \cdot Cl_2^-$$

Study of the Co(III) oxidation of Cl^- has been restricted to observation of the $CoCl^{2+}$ intermediate complex and measurement of its formation rate[703, 704].

4.2.2 *Bromide ion*

In a sulphate ion medium, ceric sulphate oxidises Br^- to $\frac{1}{2}$ Br_2 according to the rate expression[286]

$$-d[Ce(IV)]/dt = \frac{[Ce(IV)]}{[SO_4^{2-}]}\{k_3[Br^-]^2 + k_2[Br^-]\}$$

The dominant species of Ce(IV) existing under the reaction conditions is $Ce(SO_4)_3^{2-}$ and the activated complexes for the two paths must have compositions $Ce(SO_4)_2Br_2^{2-}$ and $Ce(SO_4)_2Br^-$. The latter path is subject to chloride-ion catalysis of the form $k_{total} = k_0 + k'[Cl^-]$ which suggests an activated complex $Ce(SO_4)_2ClBr_2{}^{2-}$. Slow oxidative breakdown of the complexes containing bromide gives Ce(III) and Br atoms or $\cdot Br_2^-$. The latter go on to form molecular bromine; however, their presence has been detected in this reaction from their ability to add to butadiene to form dibromooctadienes[287].

A stopped-flow examination of the Co(III) perchlorate oxidation produced kinetics[673]

$$-\frac{d[Co(III)]}{dt} = [Co(III)][Br^-]\left(k_1 + \frac{k'_2}{[H_3O^+]}\right)$$

indicating both Co^{3+} and $CoOH^{2+}$ to be active oxidants. k'_2 may be written k_2K_h where K_h represents the hydrolysis constant of Co^{3+}; k_1 and k_2 refer to the reactions

$$Co^{3+} + Br^- \rightarrow Co(II) + Br\cdot$$

$$CoOH^{2+} + Br^- \rightarrow Co(II) + Br\cdot$$

respectively. At 15 °C, $k_1 = 11.2 \pm 1.0$ l.mole^{-1}.sec^{-1}, $k_2K_h = 10.6 \pm 0.8$ sec^{-1}, $k_2 = 1410 \pm 100$ l.mole^{-1}.sec^{-1}, $E_1 = 22 \pm 4$ kcal.mole^{-1}, $\Delta S^\ddagger = 23 \pm 10$ eu, $E_2 = 10 \pm 3$ kcal.mole^{-1} and $\Delta S^\ddagger = -14 \pm 8$ eu.

The oxidation by Mn(III) in a perchlorate medium[288] follows two kinetic paths depending on the bromide concentration, *viz.*

region of "high" $[Br^-]$ ($[Br^-] > 6 \times 10^{-4}$ *M*): $-d[Mn(III)]/dt = k_3[Mn(III)][Br^-]^2$

region of "low" $[Br^-]$ ($[Br^-] < 5 \times 10^{-4}$ *M*): $-d[Mn(III)]/dt = k_2[Mn(III)][Br^-]$

The stoichiometry $\Delta[Mn(III)] : \Delta[Br^-] : \Delta[Br_2]$ is 2 : 2 : 1 and induced polymerisation indicates the intermediacy of bromine atoms or radical ions, *viz.*

Route (1)

$$Mn(III) + Br^- \rightleftharpoons Mn(III)Br^- \quad \text{(fast)}$$

$$Mn(III)Br^- + Br^- \rightarrow Mn(II) + \cdot Br_2^- \quad \text{(slow)}$$

$$Mn(III) + \cdot Br_2^- \rightarrow Mn(II) + Br_2 \quad \text{(fast)}$$

Route (2)

$$Mn(III)Br^- \rightarrow Mn(II) + Br\cdot \quad \text{(slow)}$$

$$2\,Br\cdot \rightarrow Br_2$$

Both reaction paths are acid-catalysed and are subject to retardation by specific ions probably by removal of free Br^-. The second-order dependence with respect to reductant has several precedents, *e.g.* Fe(III) oxidation of I^- and Mn(III) oxidation of HN_3. The acid catalysis results from suppression of the hydrolysis to $MnOH^{2+}$ which is ineffective in this oxidation.

Strongly acidic vanadium(V) oxidises bromide in a sulphate ion medium[289]. The reaction is first-order in both oxidant and sulphuric acid. The dependence of the rate on bromide ion concentration is complex and a maximum is exhibited at certain acidities. A more satisfactory examination is that of Julian and Waters[290] who employed a perchlorate ion medium and controlled the ionic strength. They used several organic substrates which acted as captors for bromine radical species. The rate of reduction of V(V) is independent of the substrate employed and almost independent of substrate concentration. At a given acidity the kinetics are

$$-\mathrm{d}[\mathrm{V(V)}]/\mathrm{d}t = k[\mathrm{V(V)}][\mathrm{Br^-}]^3[\mathrm{substrate}]^0$$

The rate depends on the *square* of the acidity function h_0 at constant ionic strength (5 M).

A mechanism based on multiple pre-equilibria was proposed

$$\mathrm{VO_2^+ + H_3O^+ \rightleftharpoons V(OH)_3^{2+}} \qquad K_1$$

$$\mathrm{V(OH)_3^{2+} + Br^- \rightleftharpoons V(OH)_3Br^+} \qquad K_2$$

$$\mathrm{V(OH)_3Br^+ + H^+ \rightleftharpoons V(OH)_2Br^{2+} + H_2O} \qquad K_3$$

$$\mathrm{V(OH)_2Br^{2+} + Br^- \rightleftharpoons V(OH)_2Br_2^+} \qquad K_4$$

i.e. $[\mathrm{V(OH)_2Br_2^+}] = K_1 K_2 K_3 K_4[\mathrm{VO_2^+}][\mathrm{Br^-}]^2[\mathrm{H^+}]^2$

followed by a slow oxidation step

$$\mathrm{V(OH)_2Br_2^+ + Br^- \rightarrow V(IV)(OH)_2Br^+ + \cdot Br_2^-}$$

4.2.3 *Iodide ion*

10^{-5} M aqueous solutions of iodopentaminecobalt(III) decompose[291] with first-order kinetics at 45 °C with $k_1 = 6.0 \times 10^{-5}$ sec^{-1}. 10^{-3} M solutions decompose faster after an initial induction period at the "normal" rate. Product analysis shows the "fast" decomposition to be a mixture of a redox process leading to iodine and substitution leading to aquopentaminecobalt(III) and iodide. Addition of sodium iodide (to ~ 10^{-3} M) accelerates the decomposition and eliminates the induction period, thus

$$k_{\text{observed}} = k_1 + k_2[\text{I}^-](\text{sec}^{-1})$$

where $k_1 = (6.7 \pm 0.3) \times 10^{-5}$ sec^{-1} and $k_2 = 0.195 \pm 0.002$ l.mole^{-1}.sec^{-1} at 45 °C. k_1 is considered to refer to the aquation process and k_2 to the oxidation of I^- by the complex. The latter reaction is responsible for the autoacceleration effect in initially iodide-free solution when the concentration of free iodide ion liberated after aquation becomes significant.

The oxidation of I^- by Ce(IV) sulphate proceeds in two stages, *i.e.* an extremely rapid step ($k_2 = 3.3 \times 10^4$ l.mole^{-1}.sec^{-1} at 18 °C)[49] followed by a much slower reaction[286, 292]. This behaviour is also found in the oxidation of leucomalachite green (p. 406) and is attributed to two forms of Ce(IV) in slow equilibrium.

The V(V) oxidation of I^- has been studied extensively[294–296]. It is found to be greatly affected by dissolved oxygen. The early data[239] for the anaerobic oxidation have been supplanted by those of Ramsey *et al.*[294], who determined the rate law in a perchlorate ion medium to be

$$-\text{d}[\text{V(V)}]/\text{d}t = k[\text{V(V)}][\text{I}^-][\text{H}_3\text{O}^+]^2$$

The reactive species might be $V(OH)_2I^{2+}$, produced in a similar manner to the analogous species in the oxidation of bromide ion, which could undergo one-equivalent breakdown to V(IV) and atomic iodine. Ramsey *et al.*[294] postulate transfer of OH^+ to iodide ion, but the intermediacy of $\cdot I_2^-$ is referred to in a later study of the oxygen effect[295] to account for the relation[296]

$$\text{induction factor} = \frac{\text{equivalents } O_2 \text{ reduced } (l^{-1})}{\text{equivalents V(V) reduced} (l^{-1})} = \frac{k}{[\text{V(V)}]_0^{\frac{1}{2}}}$$

where $[\text{V(V)}]_0$ refers to the initial concentration. This is an example of an induced reaction and for further details reference should be made to the original paper.

For an anaerobic aqueous sulphuric acid medium the kinetics are[295a]

$$\text{d}[I_2]/\text{d}t = (k_3[\text{H}_3\text{O}^+] + k_4[\text{H}_3\text{O}^+]^2)[\text{V(V)}][\text{I}^-]$$

where $k_3 = (1.6 \pm 0.1) \times 10^{-2}$ l^2.mole^{-2}.sec^{-1} and $k_4 = (8.31 \pm 0.05) \times 10^{-2}$ l^3.mole^{-3}.sec^{-1} at 20 °C ($\mu = 1$ *M* $NaHSO_4$); $E_4 = 11.7$ kcal.mole^{-1} and $\Delta S_4^\ddagger = -12.5$ eu. From Ramsey's data[294] a figure for k_4 (with $HClO_4$) of $(6.5 \pm 1.2) \times 10^{-2}$ l^3.mole^{-3}.sec^{-1} at 25 °C can be obtained. The reactive complex is again considered to be $V(OH)_2I^{2+}$ or its protonated form and the authors propose the dominant species of V(V) to be $VO(OH)^{2+}$ (and not VO_2^+ as most other workers).

4.2.4 Hydrazoic acid

The quantitive oxidation to N_2 by Co(III) in perchloric acid shows a kinetic dependence on the initial [Co(III)]/[HN_3] ratio. When [Co(III)] ~ [HN_3] the rate expression is[296b]

$$-\mathrm{d}[\mathrm{Co(III)}]/\mathrm{d}t = k'[\mathrm{Co(III)}][\mathrm{HN_3}]$$

where $\log k' = n \log [\mathrm{H_3O^+}]$+constant and at 25 °C, $n = -0.973 \pm 0.026$ and $k' = 17.5 \pm 2$ l.mole^{-1}.sec^{-1} (2 *M* $HClO_4$). *E* is 25.4±0.3 kcal.mole^{-1} and ^{15}N tracer studies reveal the pathway

$$\mathrm{Co(III)+H^{15}N\text{–}N\text{–}^{15}N \rightarrow {}^{15}N\text{–}N+0.5\ {}^{15}N\text{–}^{15}N+Co(II)+H^+}$$

An inner-sphere oxidation of HN_3 by $CoOH^{2+}$ to $N_3\cdot$ is proposed, the azide radicals yielding nitrogen in a bimolecular process.

When [Co(III)] ≪ [HN_3] the kinetics become[296c]

$$-\mathrm{d}[\mathrm{Co(III)}]/\mathrm{d}t = k_2[\mathrm{Co(III)}][\mathrm{HN_3}]^2[\mathrm{H_3O^+}]^0$$

where k_3 (299.6 °K) = $(3.4 \pm 0.2) \times 10^3$ l^2.mole^{-2}.sec^{-1} ($\mu = 5.0$ *M* ClO_4^-), $E_3 = 12.6$ kcal.mole^{-1} and $\Delta S_3^\ddagger = -3.6$ eu. The reactants are deduced to be Co(III) (unhydrolysed) and HN_3 in the reaction scheme

$$\mathrm{Co(III)+HN_3 \rightleftharpoons Co(III)HN_3}$$

$$\mathrm{Co(III)HN_3+HN_3 \xrightarrow{k_1} Co(II)+\cdot H_2N_6^+} \quad \text{(slow)}$$

$$\mathrm{Co(III)+\cdot H_2N_6^+ \xrightarrow{k_2} Co(II)+3\,N_2+2\,H^+} \quad \text{(fast)}$$

The kinetics and mechanism bear some resemblance to those reported by the same workers for the Mn(III) oxidation of HN_3; Co(III) may be in a dimeric form.

This apparent duality of mechanism has been reinvestigated carefully by one of the groups involved[697], using experimental conditions very similar to those employed by the other (Wells and Mays[296c]), and a sharp discrepancy is revealed both as regards the rate law and activation energy. A *further* stopped-flow investigation[698] supports the results of Sullivan *et al.*[296b, 697].

Mn(III) in perchloric acid forms a pink complex with hydrazoic acid, which decomposes thermally[297] to give nitrogen with stoichiometry Δ[Mn(III)]/Δ[HN_3] = 1.0 and with kinetics

$$-\mathrm{d}[\mathrm{Mn(III)}]/\mathrm{d}t = k_{\mathrm{obs}}[\mathrm{Mn(III)}][\mathrm{HN_3}]^2$$

k_{obs} increases with acidity. The following mechanism was suggested

$$Mn^{3+} + H_2O \underset{k_{-1}}{\overset{k_1}{\rightleftharpoons}} MnOH^{2+} + H^+ \quad \text{(fast) } K_1 \qquad (49)$$

$$Mn^{3+} + HN_3 \underset{k_{-2}}{\overset{k_2}{\rightleftharpoons}} MnHN_3{}^{3+} \quad \text{(fast) } K_2 \qquad (50)$$

$$MnHN_3{}^{3+} + HN_3 \underset{k_4}{\overset{k_3}{\rightleftharpoons}} Mn(II) + \cdot H_2N_6^+ \quad \text{(slow)} \qquad (51)$$

$$Mn^{3+} + \cdot H_2N_6{}^+ \rightarrow Mn(II) + 3\,N_2 + 2\,H^+ \quad \text{(fast)} \qquad (52)$$

The rate of disappearance of total Mn(III) is given by

$$\frac{-d[Mn(III)]}{dt} = \frac{2\,k_3\,K_2[Mn(III)][HN_3]^2}{1+(k_1/[H^+])+k_2[HN_3]}$$

K_1 is 0.93 at 25 °C (μ = 4.0 M) and if $K_2[HN_3] \ll 1+(K_1/[H^+])$ then

$$k_{obs} = 2\,k_3\,K_2/\{1+(K_1/[H^+])\}$$

and $k_3 K_2 = 2.70 \times 10^3$ l^2.mole^{-2}.sec^{-1} at 25 °C (μ = 4.0 M). Composite Arrhenius parameters for $k_3 K_2$ are E = 19.6 kcal.mole^{-1} and $\Delta S^\ddagger$ = 23 eu[297a]. Step (51) is seen as an oxidation of one molecule of HN_3 in the inner coordination sphere concerted with that of a second HN_3 in the outer sphere.

A subsequent investigation[297b] has produced values of K for the equilibrium

$$Mn^{3+} + HN_3 \rightleftharpoons MnN_3{}^{2+} + H^+$$

of 89±18 (spectrophotometric) and 70±14 (kinetic) at 25 °C (μ = 3.8 M $NaClO_4$). In contrast to the previous study a *second*-order dependence on [Mn(III)] was found

$$\frac{-d[Mn(III)]}{dt} = \frac{k[Mn(III)]^2[HN_3]^2 f'[H_3O^+]}{f[Mn(II)]}$$

A diazido-Mn(IV) species is postulated to be the reactive complex to explain the retardation by Mn(II).

Ethylenediaminetetraacetomanganate(III) and azide form a complex[298]

$$[Mn(III).EDTA.OH_2]^- + N_3^- \rightleftharpoons [Mn(III).EDTA.N_3]^{2-} + H_2O$$

with K = 32.1 at 25 °C (μ = 0.25 M). This decomposes to give the corresponding Mn(II) complex *plus* nitrogen gas in first-order fashion with E = 17.7±1.0 kcal.mole^{-1} and $\Delta S^\ddagger$ = −20.6±1.08 eu, presumably *via* $N_3\cdot$ radicals.

References pp. 493–509

The oscilloscope traces recorded during stopped-flow reaction of Ce(IV) perchlorate with hydrazoic acid show an initial increase of absorption complete within a few msec followed by a slower, first-order decay extending over *ca.* 20 msec which is independent of both substrate concentration and acidity and also of temperature, implying $E = 0^{674}$; k_1 (averaged) $= (2.5 \pm 0.3) \times 10^2$ sec^{-1}. A hydrazidocerium(IV) complex is envisaged, breaking down to Ce(III) and $N_3\cdot$; alternatively a higher complex may yield $H_2N_6{}^{+}\cdot$

4.2.5 Bromine

The oxidation to bromate by Co(III) perchlorate[296a], *viz.*

$$Br_2 + 10\ Co(III) + 6\ H_2O = 2\ BrO_3^- + 10\ Co(II) + 12\ H^+$$

shows complex kinetics. At 0 °C mixing of the reactants results in a rapid drop in the absorption of Co(III) followed by a "slow" reaction. At 20 °C the "slow" reaction suddenly gives way to a much faster reaction provided the acidity is high (> 1 *M*). The "slow" reaction follows the rate law

$$-d[Co(III)]/dt = k[Co(III)][Br_2]/[H_3O^+]^2$$

which is thought to correspond to reaction between Br_2 and hydrolysed Co(III) dimer. For this reaction $E = 29.6$ kcal.mole^{-1} and $\Delta S^{\ddagger} = 47$ eu.

The other features are considered to correspond to (at 0 °C) formation of a Co(III)–Br_2 complex with different spectroscopic properties and (at 20 °C) to incursion of a path involving OH·. The "slow" reaction must consist of several steps, commencing possibly with a two-equivalent oxidation of Br_2 to 2 HOBr by dimeric Co(III).

4.2.6 Chlorine dioxide

The oxidation by Co(III) to chlorate in an acidic perchlorate medium has a simple stoichiometry

$$Co(III) + Cl(IV) = Co(II) + Cl(V)$$

and the rate law is[298a]

$$\frac{-d[ClO_2]}{dt} = \frac{k_2[ClO_2][Co(III)]}{[H_3O^+]} \cdot e^{\beta[H^+]}$$

where the (small) exponential term is of little mechanistic significance. Arrhenius parameters for k_2 are $E = 22.8 \pm 0.3$ kcal.mole^{-1} and $\Delta S^{\ddagger} = 24.9 \pm 0.9$ eu. ^{18}O from enriched water was incorporated into product chlorate to an extent of one atom per chlorate molecule. The oxidising species is $CoOH^{2+}$ but no physical evidence for an inner-sphere complex with ClO_2 was obtained.

4.2.7 *Hydrazine and methylhydrazines*

The kinetics and stoichiometries for the reactions

$$2\,Mn(III) + 2\,N_2H_5^+ = N_4H_8{}^{2+}\ (\text{or } N_3H_5^{2+} + NH_4^+) + 2\,Mn(II) + 2\,H^+\ (\text{or } H^+)$$

$$4\,Mn(III) + CH_3N_2H_4^+ = CH_3NH_2N{=}NNH_2CH_3^{2+} + 4\,Mn(II) + 4\,H^+$$

$$2\,Mn(III) + 2\,N,N\text{-}(CH_3)_2N_2H_3^+ = (CH_3)_2NHNHNHNH(CH_3)_2^{2+} + 2\,Mn(II) + 2\,H^+$$

$$4\,H_2O + 8\,Mn(III) + 2\,N,N'\text{-}(CH_3)_2N_2H_3^+ = 2\,CH_3NH_2NOH^{2+} + 8\,Mn(II) + 6\,H^+ + 2\,CH_3OH$$

$$2\,Mn(III) + 2\,(CH_3)_3N_2H_2^+ = (CH_3)_2NHN(CH_3)N(CH_3)NH(CH_3)_2^{2+} + 2\,Mn(II) + 2\,H^+$$

$$2\,H_2O + 2\,Mn(III) + 2\,(CH_3)_4N_2H_2^+ = (CH_3)_2NHN(CH_3)N(CH_3)NH(CH_3)_2^{2+} + 2\,Mn(II) + 2\,H^+ + 2\,CH_3OH$$

have been determined[298b] in acid perchlorate media. All reactions are simple first-order in both oxidant and hydrazine but the second order rate coefficients are acid-dependent, normally decreasing with increasing $[HClO_4]$ except for tetramethylhydrazine when an *increase* is found. All the hydrazines are almost entirely monoprotonated under the reaction conditions and the reactions are viewed as the combined attacks of Mn^{3+} and $MnOH^{2+}$ upon the protonated base, BH_2^+, *viz.*

$$Mn^{3+} + BH_2^+ \xrightarrow{k_1} Mn^{2+} + BH^{+}\cdot + H^+$$

$$MnOH^{2+} + BH_2^+ \xrightarrow{k'_1} Mn^{2+} + BH^{+}\cdot + H_2O$$

Substitution of a value of 0.93 for the hydrolysis constant of Mn^{3+} (aq) yields the individual values for k_1 and k'_1 at 25 °C (accuracy ±10 %) below.

Substrate	$10^{-3}\ k_1$ (*l.mole*$^{-1}$.*sec*$^{-1}$)	$10^{-3}\ k'_1$ (*l.mole*$^{-1}$.*sec*$^{-1}$)
N_2H_4	0.52	10.3
$CH_3N_2H_3$	0.08	3.0
N,N-$(CH_3)_2N_2H_2$	0.13	4.0
N,N'-$(CH_3)_2N_2H_2$	0.04	0.7_3
$(CH_3)_3N_2H$	0.02	0.4_3
$(CH_3)_4N_2$	0.28	0.0_3

As is quite often the case, the hydrolysed form of the oxidant is more reactive than the hexa-aquo species. The stoichiometries suggest dimerisation, rather than secondary oxidation, is the normal fate of the hydrazoyl radical. The variations in k_1 and k_1' suggest oxidation occurs at the non-protonated nitrogen atom.

4.2.8 *Hydroxylamine, O-methyhydroxylamine and nitrous acid*

Oxidations by Mn(III) perchlorate have stoichiometries[298c]

$$6\ \text{Mn(III)} + NH_3OH^+ + 2\ H_2O = 6\ \text{Mn(II)} + NO_3^- + 8\ H^+$$

$$2\ \text{Mn(III)} + 2\ NH_3OCH_3{}^+ = 2\ \text{Mn(II)} + N_2H_3(OCH_3)_2{}^+ + 3\ H^+$$

$$2\ \text{Mn(III)} + HNO_2 + H_2O = 2\ \text{Mn(II)} + NO_3^- + 3\ H^+$$

These three reactions require stopped-flow technique for kinetic study. All have a reaction order of two (*i.e.* one in each reactant) with k_2 independent of Mn^{2+} and ionic strength but pH-dependent, indicating both Mn^{3+} and $MnOH^{2+}$ to be reactive. A matrix of second order-rate-coefficients (l.mole^{-1}.sec^{-1}) at 25±1 °C is given below

	Mn^{3+}	$MnOH^{2+}$
NH_3OH^+	$(1.4 \pm 0.1) \times 10^3$	$(3.1 \pm 0.3) \times 10^3$
$NH_3OCH_3{}^+$	< 0.5	6.1 ± 0.4
HNO_2	$(2.2 \pm 0.2) \times 10^4$	$(4.9 \pm 0.4) \times 10^4$

The mechanisms of all three oxidations involve initial slow one-equivalent oxidation to a radical ($NH_2O\cdot$, $\cdot NHOCH_3$ and $NO_2\cdot$) followed either by rapid secondary oxidation (of $NH_2O\cdot$ or $NO_2\cdot$) or by radical dimerisation ($\cdot NHOCH_3$).

Ce(IV) sulphate oxidation of NH_3OH^+ shows a stoichiometry depending on the relative initial concentrations, but at low [Ce(IV)]/[NH_3OH^+], the conditions for kinetic experiments, a value of 1.0 is realised for Δ[Ce(IV)]/Δ[NH_3OH^+][298d]. The reaction orders with respect to oxidant and substrate are both one but k_2 is reduced by addition of sulphate ions, indicating $Ce(SO_4)_3{}^{2-}$, the dominant species present, to be less reactive than $Ce(SO_4)_2$. An inner-sphere oxidation to $NH_2O\cdot$ is proposed, the radical then dimerises (at low [Ce(IV)]) or is oxidised by Ce(IV) to HNO which decomposes to N_2O. $E_2 = 14.8$ kcal.mole^{-1} and $\Delta S^{\ddagger} = -7$ eu.

4.2.9 Oxides of hydrogen

The rate expression for the oxidation of water by silver(II) perchlorate is[299, 300]

$$-\mathrm{d}[\mathrm{Ag(II)}]/\mathrm{d}t = k\frac{[\mathrm{Ag(II)}]^2[\mathrm{ClO_4^-}]^2}{[\mathrm{Ag(I)}][\mathrm{H_3O^+}]^2}$$

with $E = 11.0 \pm 2.0$ kcal.mole^{-1}. The inverse dependence upon Ag(I) concentration indicates a rapid pre-equilibrium

$$2\,\mathrm{Ag(II)} \rightleftharpoons \mathrm{Ag(I)} + \mathrm{Ag(III)}$$

for which there exists radio-chemical evidence[277], and the acidity dependence indicates a hydrolysis

$$\mathrm{Ag^{3+}} + \mathrm{H_2O} \rightleftharpoons \mathrm{AgO^+} + 2\,\mathrm{H^+}$$

The oxidation step

$$\mathrm{AgO^+} \rightarrow \mathrm{Ag^+} + \tfrac{1}{2}\,\mathrm{O_2}$$

is rate-determining. The dependence on perchlorate ion combines ion-pairing and general ionic strength effects.

The kinetics are the same in sulphuric[301] and phosphoric acid[302] media as in perchloric acid. The main kinetic features are reproduced in a nitric acid medium[303] although a term, $k[\mathrm{Ag(II)}]^4/[\mathrm{Ag(I)}]$, is also involved. Both terms include a strong inverse acidity dependence; however, the role of the nitrate ions[304] obscures this.

A more recent examination[305] produced two rate laws corresponding to different concentrations of oxidant, *viz.*

(A) at $[\mathrm{Ag(II)}] > 10^{-4}\,M$

$$-\mathrm{d}[\mathrm{Ag(II)}]/\mathrm{d}t = \frac{k_1[\mathrm{Ag(II)}]^2[\mathrm{NO_3^-}]^0}{k_2 + k_3[\mathrm{Ag(I)}][\mathrm{H^+}]^2} = k_{\mathrm{A}}[\mathrm{Ag(II)}]^2$$

with $E_{\mathrm{A}} = 23.6 \pm 0.4$ kcal.mole^{-1},

(B) at $[Ag(II)] < 10^{-4}\ M$

$$-d[Ag(II)]/dt = \frac{(k_4+k_5[Ag(I)])[Ag(II)]}{(k_6+k_7[Ag(I)])(k_8+k_9[H^+]^2)} = k_B[Ag(II)]$$

with $E_B = 22.9 \pm 0.9$ kcal.mole^{-1}.
The authors propose that AgO^+ oxidises water to H_2O_2 which is rapidly oxidised in turn by Ag(II) to $HO_2\cdot$, itself oxidised by further Ag(II) to O_2. Application of the steady-state approximation to [Ag(III)], $[AgO^+]$, $[H_2O_2]$ and $[HO_2\cdot]$ produces the observed rate law (A).

A few data exist on the oxidation of water by Ag(III) ethylenedibiguanide nitrate[306]. The decomposition is only approximately first-order in oxidant in both water and dilute nitric acid. The activation energies are very different for the two media, (H_2O, 27.8 kcal.mole^{-1}; HNO_3, 18.65 kcal.mole^{-1}). A is about 10^{15} for water and 10^9 for nitric acid solutions. These data are inadequate for mechanistic interpretation.

The oxidation of water by Co(III) has attracted several groups. The rate law

$$-d[Co(III)]/dt = k_1[Co(III)]+k_2[Co(III)]^2$$

was proposed by Bawn and White[307], who also noted strong acid retardation of the rate thus confirming the earlier work of Noyes and Deahl[308]. The rate equation applies to both perchlorate and sulphate ion media, in which the decompositions have activation energies of 34.0 to 40.0 kcal.mole^{-1}, respectively. Later workers[309, 310] have, however, preferred the rate law

$$-d[Co(III)]/dt = k[Co(III)]^{\frac{3}{2}}[Co(II)]^0$$

which holds for both perchlorate and sulphate ion media. In a perchlorate medium k varies inversely as $[H_3O^+]^2$. The order in oxidant implies[309] that Co(III) is largely dimeric

$$2\ CoOH^{2+} \rightleftharpoons (Co\text{–}O\text{–}Co)^{4+} + H_2O$$

and that the slow oxidation step involves both dimer and monomer, *viz.*

$$CoOH^{2+} + (Co\text{–}O\text{–}Co)^{4+} \rightarrow 3\ Co^{2+} + HO_2\cdot$$

followed by a fast reduction by $HO_2\cdot$, *viz.*

$$CoOH^{2+} + HO_2\cdot \rightarrow Co^{2+} + H_2O + O_2$$

The acid retardation originates from suppression of hydrolyses of the type[309]

$$Co^{3+} + H_2O \rightleftharpoons CoOH^{2+} + H^+$$
$$(Co\text{–}O\text{–}Co)^{4+} + 2\,H_2O \rightleftharpoons (HOCo\text{–}O\text{–}CoOH)^{2+} + 2\,H^+$$

Spectroscopic support exists both for hydrolytic equilibria of this kind[309,311] and for the presence, especially at high oxidant concentrations, of a further species, probably of a polynuclear type[312]. The most convincing support for the dimer theory, however, comes from the elegant study of Anbar and Pecht[313] who used ^{18}O-labelled Co(III) to oxidise $H_2{}^{16}O$, making the due correction for ligand exchange in this substitution-inert complex. Anbar and Pecht conclude that nearly all the oxygen evolved originates from the inner sphere of Co(III) and not from the solvent. It is concluded that the oxygen is formed from binuclear complexes

$$\left[(H_2O)_4Co\begin{matrix} H \\ O \\ \\ O \\ H \end{matrix}Co(OH_2)_4\right]^{4+}$$

Very little hydrogen peroxide was found as a product (by isotope dilution analysis), and it was ruled out as an intermediate by addition of ^{18}O-labelled H_2O_2 before reaction.

A preliminary account of the kinetics of the oxidation of hydrogen peroxide by Ag(II) nitrate in 3 M HNO_3 has appeared[314], the rate law being

$$-\mathrm{d}[Ag(II)]/\mathrm{d}t = k_2[Ag(II)][H_2O_2]$$

E is 10.6 ± 0.8 kcal.mole^{-1}. Use of ^{18}O-enriched water failed to enhance the ^{18}O content of the evolved oxygen, which must therefore originate exclusively in the peroxide. For $HClO_4$ media, however, decay in the region of Ag(II) absorption is always first-order[314a], the rate coefficient being independent of [Ag(II)], $[H_2O_2]$, $[H^+]$ and Ag(I) (*cf.* the results of Wells and Mays on the Mn(III)–H_2O_2 system). Accordingly reaction is considered to involve breakdown of a complex, $Ag^{2+}.\,HO_2^-$ with $k = 11$ sec^{-1} (25 °C, $\mu = 4.0\ M$) and $E \simeq 0$ kcal.mole^{-1}.

The Co(III) perchlorate oxidation of hydrogen peroxide[309] affords a rate law

$$-\mathrm{d}[Co(III)]/\mathrm{d}t = \frac{k[Co(III)][H_2O_2]}{[H_3O^+]},$$

where $k = 1.0 \times 10^{21} \exp(-26.0 \times 10^3/\boldsymbol{RT})$ l.mole^{-1}.sec^{-1}. Spectrophotometric examination revealed rapid reversible complex formation, *e.g.*

$$Co(III) + H_2O_2 \rightleftharpoons [Co(III) \cdot HO_2^-]^{2+} + H^+$$

If a dimer of Co(III) is present then two one-equivalent oxidations of HO_2^- can occur synchronously to give H^+ and O_2; otherwise $HO_2\cdot$ is an intermediate.

The reaction between isothiocyanatopentaminecobalt(III) and hydrogen peroxide has been studied kinetically[315]. Although Co^{2+} (aq) is formed as one of numerous products, the overall reaction is essentially an oxidation of ligand by the peroxide. The reaction is first-order in both oxidant and reductant although there is a complex acidity dependence.

The Ce(IV) sulphate oxidation has a stoichiometry[316]

$$2\,\text{Ce(IV)}+H_2O_2 = 2\,\text{Ce(III)}+O_2+2\,H^+$$

which implies a clean reaction. A stopped-flow study[317] involved a critical test of the mechanism

$$\text{Ce(IV)}+H_2O_2 \underset{k_{-1}}{\overset{k_1}{\rightleftharpoons}} \text{Ce(III)}+HO_2\cdot+H^+$$

$$\text{Ce(IV)}+HO_2\cdot \xrightarrow{k_2} \text{Ce(III)}+O_2+H^+$$

Applying a steady-state approximation to $[HO_2\cdot]$, the mechanism leads to the rate law

$$\frac{-\text{d[Ce(IV)]}}{\text{d}t} = \frac{2\,k_1\,k_2[\text{Ce(IV)}]^2[H_2O_2]}{k_{-1}[\text{Ce(III)}]+k_2[\text{Ce(IV)}]}$$

When $[\text{Ce(III)}] \gg [\text{Ce(IV)}]$ or $[\text{Ce(IV)}] > H_2O_2$ the expression is simplified and tests using various concentrations of oxidant and peroxide confirmed both the simple laws. k_1 was found to be 10^6 l.mole^{-1}.sec^{-1} at 25.0 °C. ESR flow studies confirm the presence of $HO_2\cdot$ in the reaction[318, 319].

The rate of oxidation with Ce(IV) perchlorate depends on the method of preparation[320]. The material from certain preparations gives a deep red complex, containing two equivalents of Ce(IV) to one molecule of H_2O_2, which decomposes in second order fashion–presumably by means of two concerted one-equivalent oxidations of the substrate. Other preparations give no complex and decompose peroxide much faster. The difference is thought to lie in the degree of association of the oxidant (*cf.* the Ce(IV) oxidation of iodide ion, p. 359).

Oscilloscope traces obtained on mixing Ce(IV) perchlorate and H_2O_2 in a stopped-flow apparatus reveal an initial build-up of absorption at 350 nm complete within a few msec, suggesting formation of a complex, followed by a first-order decay almost complete within 20 msec and independent of initial [Ce(IV)], $[H_2O_2]$ and $[H_3O^+]$ and of added Ce(III); at 25 °C $k_1 = (2.8\pm0.2)\times10^2$ sec^{-1} and a similar value is found over a temperature range of 18–43 °C implying $E = 0$. Breakdown of a Ce(IV)–peroxide complex to $HO_2\cdot$ followed by oxidation of $HO_2\cdot$ to O_2 is proposed[675].

Mn(III) perchlorate in $HClO_4$ oxidises H_2O_2 with a stoichiometry of 2 Mn(III): 1 H_2O_2[321]. Stopped-flow studies show the reaction to be approximately first-order in both reactants but that the second-order rate coefficient k_2 varies as follows

$$\frac{1}{k_2} = A+B\frac{[\mathrm{Mn(II)}]}{[\mathrm{Mn(III)}]_0}+C\frac{[\mathrm{H_2O_2}]_0}{[\mathrm{Mn(III)}]_0} \tag{53}$$

A radical mechanism is proposed, *viz.*

$$\mathrm{Mn^{3+}+H_2O_2} \underset{k_{-1}}{\overset{k_1}{\rightleftharpoons}} \mathrm{Mn^{2+}+{\cdot}H_2O_2{}^+}$$

$$\mathrm{MnOH^{2+}+H_2O_2} \underset{k'_{-1}}{\overset{k'_1}{\rightleftharpoons}} \mathrm{Mn^{2+}+HO_2{\cdot}+H_2O}$$

$$\mathrm{Mn^{3+}+{\cdot}H_2O_2{}^+} \overset{k_2}{\rightarrow} \mathrm{Mn^{2+}+2\,H^++O_2}$$

$$\mathrm{MnOH^{2+}+{\cdot}H_2O_2{}^+} \overset{k'_2}{\rightarrow} \mathrm{Mn^{2+}+H_3O^++O_2}$$

$$\mathrm{H_2O_2+{\cdot}H_2O_2{}^+} \overset{k_3}{\rightarrow} \mathrm{OH{\cdot}+H_3O^++O_2}$$

$$\mathrm{Mn^{2+}+OH{\cdot}} \overset{k_4}{\rightarrow} \mathrm{Mn(III)+OH^-}$$

A steady state treatment for $[{\cdot}H_2O_2{}^+]$ and $[OH{\cdot}]$ leads to an expression identical with equation (53). At 25 °C $k_1 = 7.3\times10^4$ l.mole^{-1}.sec^{-1} and $k'_1 = 3.2\times10^4$ l.mole^{-1}.sec^{-1}.

A second recent stopped-flow study[322, 323] indicated some enhancement of the initial absorption of Mn(III) at 470 nm on mixing with H_2O_2, and a complex of formula $MnHO_2{}^{2+}$ is incorporated into the reaction scheme. The decomposition of this constitutes the rate-determining step ($k_1 = 80$ sec^{-1} at 25 °C with $\mu = 4.0$ M). k_1 is independent of initial $[\mathrm{Mn(III)}]/[\mathrm{H_2O_2}]$, of acidity (between 0.5 and 3.7 M) and of temperature. No dependence on [Mn(II)] was looked for.

There appear to be discrepancies between the two reports as regards the existence of a complex, the order of reaction and the role of Mn(II). Until this has been resolved experimentally the results of Davies *et al.*[321] which cover a wider range of experimental conditions are to be preferred.

Vanadium(V) does not oxidise hydrogen peroxide but forms peroxy complexes $VO(O_2)^+$ and $VO(O_2)_2^-$. Kinetic data are available[47].

4.2.10 Sulphur compounds

The silver(II) oxidation of dithionate[324] is of interest because this reductant is rather inert and oxidation is often preceded by rate-determining disproportionation (see Cr(VI)–dithionate, p. 287). The stoichiometry is

$$2\,\mathrm{Ag(II)} + \mathrm{S_2O_6}^{2-} + 2\,\mathrm{H_2O} = 2\,\mathrm{Ag}^+ + 2\,\mathrm{HSO_4}^- + 2\,\mathrm{H}^+$$

and the kinetics are

$$-\mathrm{d}[\mathrm{Ag(II)}]/\mathrm{d}t = k[\mathrm{H_3O^+}][\mathrm{Ag(II)}][\mathrm{S_2O_6}^{2-}][\mathrm{Ag(I)}]^0$$

where $k = 5.5 \times 10^2$ $\mathrm{l^2.mole^{-2}.sec^{-1}}$ at $\mu = 3.5$ M and 25.0 °C. This provides a contrast to other Ag(II) oxidations mentioned in this review which involve Ag(III). The simplest mechanism consistent with the kinetics is

$$\begin{aligned} \mathrm{H}^+ + \mathrm{S_2O_6}^{2-} &\rightleftharpoons \mathrm{HS_2O_6}^- && \text{(fast)} \\ \mathrm{Ag(II)} + \mathrm{HS_2O_6}^- + \mathrm{H_2O} &\rightarrow \mathrm{Ag(I)} + \mathrm{HSO_3}\cdot + \mathrm{HSO_4}^- + \mathrm{H}^+ && \text{(slow)} \\ \mathrm{Ag(II)} + \mathrm{HSO_3}\cdot + \mathrm{H_2O} &\rightarrow \mathrm{Ag(I)} + \mathrm{HSO_4}^- + 2\,\mathrm{H}^+ && \text{(fast)} \end{aligned}$$

4.2.11 Hypophosphorous acid

Carrol[325] has shown the oxidation by Co(III) perchlorate to have complex kinetics, *viz.*

$$\begin{aligned} -\mathrm{d}[\mathrm{Co(III)}]/\mathrm{d}t = {} & k_1[\mathrm{Co(III)}][\mathrm{H_2PO_2}^-][\mathrm{Co(II)}]^{\frac{1}{3}}[\mathrm{H_3O^+}]^{-\frac{1}{3}} \\ & + k_2[\mathrm{Co(III)}]^{\frac{3}{2}}[\mathrm{H_3PO_2}]^{\frac{1}{2}} + k_3[\mathrm{H_3PO_2}][\mathrm{Co(III)}] \end{aligned}$$

k_3 is considered to relate to the active form of the neutral acid (p. 334) and k_2 to the predominant form. k_3 is much smaller than the value reported for halogenation of the active form. Further details are available in the original thesis.

The analogous work with Ce(IV) perchlorate has been reported by Carroll and Thomas[326]. The rate expression is complex, *viz.*

$$-\mathrm{d}[\mathrm{Ce(IV)}]/\mathrm{d}t = k[\mathrm{H_3O^+}]\sum_{n=1}^{n=3}[\mathrm{Ce(H_2PO_2)}_n{}^{(4-n)+}]$$

and spectrophotometric data confirm the existence of hypophosphite complexes of Ce(IV). No reaction involving free hypoposphorous acid is evident and in this sense the oxidation is unique, (*e.g.* compare the Hg(II) oxidation on p. 334). The kinetics show that three complexes are undergoing oxidative breakdown to give radicals. Similar complexes play a role in the Ce(IV) sulphate oxidation[327] which follows a rate law

$$-\mathrm{d}[\mathrm{Ce(IV)}]/\mathrm{d}t = \frac{kK[\mathrm{H_3O^+}][\mathrm{H_3PO_2}][\mathrm{Ce(IV)}]}{K'[\mathrm{HSO_4}^-]\{1 + K[\mathrm{H_3PO_2}]\}}$$

where the equilibrium constants refer to

$$Ce(SO_4)_2 + H_3PO_2 \rightleftharpoons Ce(SO_4)_2(H_3PO_2) \qquad K$$
$$Ce(SO_4)_2 + HSO_4^- \rightleftharpoons Ce(SO_4)_3^{2-} + H_3O^+ \qquad K'$$

and for the overall reaction $E = 17.2$ kcal.mole^{-1}. The mechanism is presumably the same as for Ce(IV) perchlorate, involving

$$Ce(SO_4)_2(H_3PO_2) \rightarrow Ce(III) + \text{radical} \qquad \text{(slow)}$$

Again, therefore, the Mitchell type of mechanism (p. 334) is not operating.

4.2.12 *Arsenious acid*

The Ce(IV) oxidation of arsenite has been examined in various acids for which the sequence of rates is $HClO_4 > HNO_3 > H_2SO_4$[328]. The kinetics are simple second order. E in $HClO_4$ is 9.55 kcal.mole^{-1}. The chief kinetic interest in this reaction is, however, centred on its remarkable acceleration on addition of minute quantities of iodine[329], Ru(IV)[330, 331] Ru(VI) or Os(VIII)[332, 333]. The kinetics are complicated and although catalysis is not the subject of this review, the above references have been included.

The V(V) sulphate oxidation of arsenite[334] shows an interesting second-order dependence on oxidant concentration, the rate expression being

$$+\mathrm{d}[V(IV)]/\mathrm{d}t = k[As(III)][V(V)]^2 f[H_3O^+]$$

The reaction is catalysed by adding phosphoric acid. Presumably *two* V(V) entities are bound to the arsenite in the transition state and effect concerted one-equivalent oxidations to avoid formation of the energetic As(IV).

4.2.13 *Antimony(III)*

Oxidation by Ce(IV) sulphate of antimony(III) chloride in ~ 1–2 M H_2SO_4 follows a kinetic law[334a]

$$\frac{-\mathrm{d}[Ce(IV)]}{\mathrm{d}t} = \frac{k'[Ce(IV)][Sb(III)][Cl^-][H_3O^+]}{[HSO_4^-]^2(1+K'[Cl^-])}$$

which is interpreted in terms of the equilibria

$$Ce^{4+} + HSO_4^- \rightleftharpoons CeSO_4^{2+} + H^+ \qquad (K_1)$$
$$CeSO_4^{2+} + HSO_4^- \rightleftharpoons Ce(SO_4)_2 + H^+ \qquad (K_2)$$
$$Ce(SO_4)_2 + HSO_4^- \rightleftharpoons Ce(SO_4)_3^{2-} + H^+ \qquad (K_3)$$
$$CeSO_4^{2+} + Cl^- \rightleftharpoons CeSO_4Cl^+ \qquad (K_4)$$

where at 25 °C, $K_1 = 3.5 \times 10^3$, $K_2 = 2.0 \times 10^2$ and $K_3 = 20$ (ref. 278e). $CeSO_4Cl^+$ is regarded as the active species in the oxidation but $Ce(SO_4)_3^{2-}$ is predominant under these conditions, hence

$$[CeSO_4{}^{2+}] \simeq [Ce(IV)][H^+]/K_2 K_3[HSO_4^-]^2$$

and

$$\frac{-d[Ce(IV)]}{dt} = \frac{kK_4[Ce(IV)][H^+][Cl^-]}{K_2 K_3[HSO_4^-]^2(1+K_4[Cl^-])}$$

i.e. $k' = kK_4/K_2K_3$ and $K' = K_4$. A chloride ion bridge mechanism is favoured for the electron transfer to give Ce(III) and labile Sb(IV). Composite Arrhenius parameters were found to be $E = 16.0$ kcal.mole^{-1}, $\Delta S^\ddagger \sim 0$.

In a perchlorate medium the kinetics are found to be[676]

$$-\frac{d[Ce(IV)]}{dt} = \frac{2[Ce(IV)][Sb(III)](k_2[H_3O^+]+k_3K_2)}{[H_3O^+]+K_2}$$

where k_2 and k_3 refer to oxidation by $CeOH^{3+}$ and $Ce(OH)_2{}^{2+}$ and are 35 and 300 l.mole^{-1}.sec^{-1} respectively at 25 °C ($\mu = 2\ M$) and K_2 is the second hydrolysis constant of Ce^{4+}, *i.e.* to $Ce(OH)_2{}^{2+}$. Also $E_2 = 19.7$ kcal.mole^{-1}, $\Delta S_2^\ddagger = 7.2$ eu, $E_3 = 9.6$ kcal.mole^{-1} and $\Delta S_3^\ddagger = -27.6$ eu. The rate–acidity profile shows a maximum at 0.15 M acid, a phenomenon related to hydrolysis of $HSbO_2$ to SbO^+. Strong catalysis by halide ions is considered to favour a mechanism of atom-transfer rather than of electron-transfer.

4.3 OXIDATION OF MONOFUNCTIONAL ORGANIC MOLECULES

The reactivity of these oxidants towards organic substrates depends in a rough manner upon their redox potentials. Ag(II) and Co(III) attack unactivated and only slightly activated C–H bonds in cyclohexane, toluene and benzene and Ce(IV) perchlorate attacks saturated alcohols much faster than do Ce(IV) sulphate, V(V) or Mn(III). The last three are sluggish in action towards all but the active C–H and C–C bonds in polyfunctional compounds such as glycols and hydroxy-acids. They are, however, more reactive towards ketones than the two-equivalent reagents Cr(VI) and Mn(VIII) and in some cases oxidise them at a rate exceeding that of enolisation.

Complex formation frequently occurs to modify the kinetics and in some cases becomes measurable spectroscopically. Most oxidations are accordingly of an inner-sphere nature although that of hydrocarbons is more probably outer-sphere.

The one-equivalent oxidation of an organic molecule should yield a free radical or radical-cation. Direct or indirect evidence for the intermediacy of these species has been obtained in certain cases.

4.3.1 Alkanes

The slow oxidation of cyclohexane by Co(III) is mentioned in the following section.

4.3.2 Activated alkyl groups and polynuclear aromatics

No data are available on Ag(II) oxidation but a number of alkyl substituted aromatic hydrocarbons are oxidised by Co(III) perchlorate in aqueous acetonitrile[335]. Cyclohexane is attacked more slowly[335]. The reaction is generally first-order with respect to both oxidant and substrate and the reactivity sequence is phenanthrene, anthracene > napthalene > biphenyl > p-$Bu^t.C_6H_4CH_3$ > $(C_6H_5)_2CH_2 > C_6H_5C_2H_5 \sim C_6H_5CH_2CH_2C_6H_5 > C_6H_{12} > p\text{-}NO_2C_6H_4CH_3$. The second-order rate coefficient follows an acidity dependence of the kind $k_2 = a+b/[H^+]$ which is almost universal for oxidations by Co(III), having been reported for oxidations of Ce(III)[311], Hg(I), V(IV), V(III)[336] and many organic substrates. This has been ascribed to a hydrolysis of the type

$$Co(H_2O)_6{}^{3+} \rightleftharpoons Co(OH)(H_2O)_5{}^{2+} + H^+$$

with the hydroxopentaaquo complex deemed the reactive species.

The following outer-sphere oxidation mechanism is favoured over the radical-cation formation which is a feature of similar oxidations by Mn(III) acetate (p. 375). With toluene

$$C_6H_5{-}CH_2{-}H \quad HO{-}Co(III)\,(H_2O)_5 \longrightarrow C_6H_5CH_2\cdot + Co(H_2O)_6{}^{2+}$$

This is consistent with the observed products of oxidation, *i.e.* benzyl alcohol, benzaldehyde and benzoic acid and with the observed oxidation of cyclohexane. Radical-cations are, however, probably formed in oxidation of napthalene and anthracene. The increase of oxidation rate with acetonitrile concentration was intepreted[336] in terms of a more reactive complex between Co(III) and CH_3CN. The production of substituted benzophenones at high CH_3CN concentration indicates the participation of a second route of oxidation.

At 15 °C (70 % CH_3CN, 0.77 M H^+, μ = 0.9 M) k_2 for the Co(III) oxidation of toluene is 4×10^{-3} l.mole^{-1}.sec^{-1}, E = 28.0 kcal.mole^{-1} and $\Delta S^{\ddagger}$ = 24 eu.

This compares with a k_2 of 8.4×10^{-3} l.mole^{-1}.sec^{-1} at 40 °C ([H^+] = 1.0 *M*, 50 % acetic acid) for the analogous oxidation by Ce(IV) perchlorate in aqueous acetic–perchloric acid mixture[337] and one of 5.7×10^{-3} l.mole^{-1}.sec^{-1} for oxidation by V(V) sulphate[338] at 60 °C in 70 % aqueous acetic–sulphuric acid mixture ([H^+] ~ 4.5 *M*), both reactions giving benzaldehyde.

The oxidations by Ce(IV) and V(V) are first-order in oxidant and substrate. The V(V) oxidation rate depends on h_0 and a ρ value of -3.75 is obtained; the Ce(IV) oxidation rate depends on the square of h_0 and the ρ value is $+1.7$. Both reactions probably involve a mechanism similar to that of the oxidation by Co(III).

A few data on the Mn(III) sulphate oxidation of toluene have been reported[339]. $k_2 = 1.58 \times 10^{-4}$ l.mole^{-1}.sec^{-1} at 35 °C and E = 32.40 kcal.mole^{-1}.The main products are benzyl alcohol and benzaldehyde.

Quite different kinetics are exhibited by the anaerobic oxidation of alkylbenzenes by cobaltic acetate in a 95 % acetic acid medium[339a], *viz.*

$$-\mathrm{d}[\mathrm{Co(III)}]/\mathrm{d}t = k[\mathrm{ArH}][\mathrm{Co(III)}]^2/[\mathrm{Co(II)}]$$

This suggests a fast pre-equilibrium involving electron transfer followed by slow oxidation of a radical-cation, *viz.*

$$\mathrm{C_6H_5CH_3 + Co(III) \rightleftharpoons \cdot C_6H_5CH_3{}^+ + Co(II)}$$
$$\mathrm{\cdot C_6H_5CH_3{}^+ + Co(OAc)_3 \rightarrow C_6H_5CH_2OAc + Co(OAc)_2 + H^+}$$

Such a pre-equilibrium closely parallels that suggested by Dewar *et al.*[427] for the manganic acetate oxidations of several aromatic ethers and amines (p. 405). Other features of the reaction are a ρ^+ value of -0.7 and identical activation energies of 25.3 kcal.mole^{-1} for oxidation of toluene, ethylbenzene, cumene, diphenylmethane and triphenylmethane.

Direct ESR evidence for the intermediacy of radical-cations was obtained on flowing solutions of Co(III) acetate and a variety of substituted benzenes and polynuclear aromatics together in glacial acetic acid or trifluoroacetic acid solution[677]. A ρ^+ value of -2.4 was reported for a series of toluenes but addition of chloride ions, which greatly accelerated the reaction rate, resulted in ρ^+ falling to -1.35. Only trace quantities of $-CH_2OAC$ adducts were obtained and benzyl acetate is the chief product from toluene, in conformity with the equation given above.

4.3.3 Olefins

Co(III) sulphate oxidises several olefins in aqueous solution[340] with simple second-order kinetics and with k_2 values from 0.4 to 8.0 l.mole^{-1}.sec^{-1} at 25.0 °C

(H_2SO_4 = 0.09 *M*, μ = 2.0 *M*). Arrhenius parameters are: $A \sim 10^{20}$ l.mole^{-1}.sec^{-1} and $E \sim 28.5$ kcal.mole^{-1}. Successive alkylation of the olefinic bond increases the rate of reaction. One unusual feature is the lack of any acidity dependence. This implies that $Co(H_2O)_6^{3+}$ is the active oxidant and that a radical cation is formed initially; the lack of any retardation by added Co(II) means that the initial step is irreversible, *viz.*

$$RCH{=}CH_2 + Co(H_2O)_6^{3+} \rightarrow R\dot{C}H{-}\overset{+}{C}H_2 + Co(II)$$

The products formed are a complex mixture, but it appears that initially the double bond is split to give aldehydes which suffer subsequent oxidation. Glycol formation may precede C–C fission.

Addition of a Co(III) sulphate solution in sulphuric acid to an olefin dissolved in acetic acid results in reduction of Co(III) at a rate commensurate with that observed for aqueous solution and with identical kinetics[341]. Prior treatment of the Co(III) solution with acetic acid, however, causes the rate of reduction of Co(III) to become almost independent of olefin concentration. Evidently a Co(III)–acetate complex is formed in the mixture of acids which oxidises only after a rate-determining dissociation. However, this complex cannot be formed instantly, and uncomplexed Co(III) can attack olefins in acetic acid in a manner similar to that in water.

4.3.4 Benzene

The oxidation by Co(III) is first-order in oxidant and in substrate[342, 343] and the acidity dependence has the form

$$k_2 = k' + k''/[H^+]$$

where $k' = 1.8 \times 10^{11} \exp(-19.0 \times 10^3/\boldsymbol{RT})$ l.mole^{-1}.sec^{-1} and $k'' = 8.7 \times 10^{18} \exp(-29.0 \times 10^3/\boldsymbol{RT})$ l^2.mole^{-2}.sec^{-1}. The products were identified spectroscopically to be *p*-benzoquinone and muconic acid. The oxidation presumably involves the production of phenyl radicals, *viz.*

$$Co(H_2O)_5OH^{2+} + C_6H_6 \rightarrow Co(H_2O)_6^{2+} + C_6H_5\cdot$$

which are rapidly oxidised to phenol.

Oxidation of benzene (and also chlorobenzene and toluene) by Mn(III) acetate in glacial acetic acid gives a mixture of products including benzyl acetate (from benzene) indicating an initial attack on the aromatic by $\cdot CH_2CO_2H$[343a]. The kinetics and actual rate of disappearance of Mn(III) are the same for C_6H_6 and

C_6H_5Cl as in the absence of aromatic, suggesting auto-decomposition of the oxidant as the slow step (p. 386). The oxidation of toluene was somewhat faster, implying an additional electron-transfer pathway (*cf.* the oxidation of aromatic ethers and amines, p. 405).

Essentially the same conclusions have been reached by Heiba *et al.*[343b] who found the principal products of oxidation of aromatics to be $-CH_2OAc$ adducts although incursion of an electron-transfer mechanism to give ring-acetylation is apparent when the ionisation potential of the aromatic is less than 8.0 eV. This ionic path is strongly quenched by addition of excess acetate ion. Substituted toluenes are attacked by $\cdot CH_2CO_2H$ at rates showing a good correlation with σ^+ ($\rho^+ = -0.63$) and for toluene $k_H/k_D = 5.46$ at 130 °C.

4.3.5 Alcohols

All the oxidants convert primary and secondary alcohols to aldehydes and ketones respectively, albeit with a great range of velocities. Co(III) attacks even tertiary alcohols readily but the other oxidants generally require the presence of a hydrogen atom on the hydroxylated carbon atom. Spectroscopic evidence indicates the formation of complexes between oxidant and substrate in some instances and this is supported by the frequence occurrence of Michaelis–Menten kinetics. Carbon–carbon bond fission occurs in certain cases.

It has been demonstrated spectroscopically that Ce(IV)[344, 345] and V(V)[346] perchlorates and Ce(IV) nitrate[346a] form complexes with alcohols of composition $[ROH{\cdot}Ce(IV)]^{4+}$ and $[ROH{\cdot}V(OH)_3]^{2+}$. The agreement between the determined formation constant and the Michaelis–Menten constant for Ce(IV) oxidation is good evidence for the role of these complexes in the oxidation process. The oxidations by Co(III)[347] and V(V)[346] perchlorates have kinetics

$$-\mathrm{d}[\mathrm{Co(III)}]/\mathrm{d}t = k[\mathrm{ROH}][\mathrm{Co(III)}]\{a+b/h_0\}$$

$$-\mathrm{d}[\mathrm{V(V)}]/\mathrm{d}t = k[\mathrm{ROH}][\mathrm{V(V)}][\mathrm{H_3O^+}]$$

and the rate expressions for oxidation by Ce(IV) sulphate[348] and perchlorate[344, 345, 349], and Mn(III) sulphate[135] are of the form (Ox signifies oxidant)

$$-\mathrm{d}[\mathrm{Ox}]/\mathrm{d}t = \frac{kK[\mathrm{Ox}][\mathrm{ROH}]}{1+K[\mathrm{ROH}]}$$

where k is independent of acidity. The kinetics of Co(III) sulphate oxidation are more complex[350], but the oxidation by Mn(III) perchlorate[351, 351a] displays simple second-order kinetics with k_2 independent of acidity.

The other main piece of kinetic evidence concerns the primary kinetic isotope effect. 1-Deutero alcohols are oxidised more slowly than the corresponding protio compounds in all cases, the values of k_H/k_D being: Co(III)[347], 1.72 (10 °C); Ce(IV) sulphate[348], 1.9 (50 °C); Mn(III) perchlorate[351], 3.0 (25 °C); Mn(III) sulphate[135], 1.6 (50 °C); V(V) sulphate[346], 4.5 (50 °C); V(V) perchlorate[346], 3.6 (50 °C). These imply rate-determining C–H cleavage, although the significantly low values of k_H/k_D probably originate in the geometry of the transition state.

These data have been interpreted[352] as favouring the following general mechanism

$$M(H_2O)_nL_m^{z+} + ROH \rightleftharpoons [M(H_2O)_{n-1}L_mROH]^{z+} + H_2O \qquad \text{(fast)}$$

$$R_1R_2C(H)-O(H)\rightarrow ML_{m-1}^{z+}\,(L) \longrightarrow R_1R_2\dot{C}-O\rightarrow ML_{m-1}^{(z-1)+} + HL \qquad \text{(slow)}$$

$$R_1R_2\dot{C}-O\rightarrow ML_{m-1}^{(z-1)+} \xrightarrow{Ox} R_1R_2\overset{+}{C}-O(H)\rightarrow ML_{m-1}^{(z-1)+} \xrightarrow{H_2O} R_1R_2C{=}O + H^+ + ML_{m-1}^{(z-1)+} \qquad \text{(fast)}$$

L is a ligand which may be further H_2O and the oxidant will have gained a proton in the case of V(V), or lost one in the case of Co(III).

The very fast oxidation of the radical precludes its detection and identification by ESR; however, reacting mixtures are capable of initiating polymerisation of acrylonitrile[346]. The oxidations of allylic alcohols by V(V) perchlorate are *ca.* thirty times faster than those of saturated alcohols[353]. This is supporting evidence for radical intermediates in view of the expected delocalisation of the free electron

$$RCH{=}CH{-}\dot{C}HOH \leftrightarrow R\dot{C}H{-}CH{=}CHOH$$

Oxidation of unsaturated alcohols by Mn(III) pyrophosphate is also faster than that of saturated alcohols[354].

The presence of radical-stabilising groups can lead to C–C fission, for example in the V(V) oxidations of 2-phenylethanol and α-*tert*-butylbenzyl alcohol[355], *viz.*

$$Bu^t(C_6H_5)CHOH \xrightarrow{V(V)} Bu^t(C_6H_5)CH{-}O(H)\rightarrow V(OH)_3^{2+} \longrightarrow Bu^{t}\cdot + C_6H_5CH{=}\overset{+}{O}H + V^{IV}(OH)_3^{+}$$

This type of fission has been observed in a detailed examination of the oxidation of tertiary alcohols by Co(III)[347]. The kinetics are similar to those reported for cyclohexanol (*vide supra*) although the rate is about 40 times less. The possibility of alkoxyl radical formation seems attractive, for Co(III) is known to oxidise water at an appreciable rate, unlike the other oxidants which are also inert towards tertiary alcohols (although it has recently been found that Ce(IV) perchlorate oxidises *tert*-butanol[356]). The radical $R_1R_2R_3C–O·$ would then undergo rapid fission to give the mixture of ketones found experimentally[357]. Replacement of one or more methyl groups of *tert*-butanol by higher alkyl groups results in a 30–40 fold increase in the rate coefficient[357], which is more readily explained in terms of a synchronous decomposition of the incipient alkoxyl radical (*cf.* the V(V) oxidation of α-*tert*-butylbenzyl alcohol) than a subsequent breakdown.

Product studies show (*i*) that the ease of elimination of $R_1·$ from $R_1R_2R_3COH$ depends on the relative stabilities of $R_1·$, $R_2·$ and $R_3·$ and (*ii*) that $R_1·$ can be eliminated from R_1R_2HCOH, *i.e.* that secondary alcohols can undergo a measure of C–C fission *via* an alkoxyl radical. The extent of C–C fission increases when R_1R_2DCOH is employed, but the k_H/k_D values for several alcohols do not accord with the observed products[358]. This paradox has been discussed at length[359].

4.3.7 *Hydroperoxides*

Cobaltic acetate oxidises *tert*-butyl hydroperoxide to a mixture of *tert*-butanol, di-*tert*-butyl peroxide and oxygen with essentially second-order kinetics[360]. The reaction does not involve O–O fission, the mechanism suggested being

$$\mathrm{Co(III)+ROOH \rightarrow Co(II)+RO_2\cdot+H^+}$$
$$\mathrm{2\,RO_2\cdot \rightarrow ROOR+O_2}$$
$$\mathrm{RO_2\cdot+ROOH \rightarrow RO\cdot+ROH+O_2}$$
$$\mathrm{RO\cdot+ROOH \rightarrow ROH+RO_2\cdot}$$
$$\mathrm{RO\cdot+Co(II) \rightarrow RO^-+Co(III)}$$

Simple second-order kinetics have also been found for the Co(III) sulphate oxidation of the hydroperoxide of 2-methyl-but-2-ene in aqueous solution, although the reaction also shows both acid-inverse and acid-independent routes[361]. Co(III) in *aqueous* sulphuric acid oxidises *tert*-butyl hydroperoxide with kinetics

$$-\mathrm{d[Co(III)]/d}t = k\mathrm{[ROOH][Co(III)]/[H_3O^+]}$$

It is not clear whether the inverse acidity dependence stems from ionisation of ROOH or from hydrolysis of Co^{3+} aq.

4.3.7 *Aldehydes*

These comprise two classes, namely enolising and non-enolising. The former can undergo oxidation in either or both tautomeric forms; the latter exist partly or, in some cases, wholly as *gem*-diols.

Typical non-enolising aldehydes are formaldehyde and benzaldehyde, which are oxidised by Co(III)[141, 350, 362], Ce(IV) perchlorate[362, 364] and sulphate[365], and Mn(III)[366]. The main kinetic features and the primary kinetic isotope effects are the same as for the analogous cyclohexanol oxidations (section 4.3.5) and it is highly probable that the same general mechanism operates. k_{H_2O}/k_{D_2O} for Co(III) oxidation of formaldehyde is 1.81 (ref. 141), a value in agreement with the observed acid-retardation, *i.e. not* in accordance with abstraction of a hydroxylic hydrogen atom from $H_2C(OH)_2$.

The V(V) perchlorate oxidations of formaldehyde[141] and chloral hydrate[188] display an unusual rate expression, *viz.*

$$-\mathrm{d}[\mathrm{V(V)}]/\mathrm{d}t = k_1[\mathrm{substrate}]\{[\mathrm{V(V)}][\mathrm{H_3O^+}]+k_2[\mathrm{V(V)}]^2(\mathrm{H_3O^+}]^2\}$$

The second term relates to the simultaneous reduction of two V(V) species by the aldehyde, possibly *via* the mechanism

$$\mathrm{RCH(OH)_2} + \mathrm{V(OH)_3^{2+}} \rightleftharpoons \mathrm{RCHOH(HO \rightarrow V(OH)_3^{2+})} \quad \text{(fast)}$$

$$\mathrm{RCHOH(HO \rightarrow V(OH)_3^{2+})} + \mathrm{V(OH)_3^{2+}} \rightleftharpoons \mathrm{R\text{-}CH(OH \rightarrow V(OH)_3^{2+})_2} \quad \text{(fast)}$$

$$\mathrm{R\text{-}CH(O\text{-}H \cdots V^{2+}(OH)_3)_2} \longrightarrow \mathrm{RC(=O)OH} + 2\,\mathrm{V(IV)} \quad \text{(slow)}$$

The oxidation of formaldehyde by V(V) has an isotope effect of *ca* 4.5 which accords with the depicted C–H cleavage[141].

The Co(III) perchlorate oxidation of substituted and unsubstituted benzaldehydes has kinetics and a low isotope effect (2.3 at 10 °C) in complete analogy with cyclohexanol and formaldehyde[367]. Ring-substitution by electronegative groups accelerates reaction.

Two studies of the Ce(IV) perchlorate oxidation of benzaldehyde in aqueous acetic acid have been reported[363, 364, 364a]. The rate law is of the form

$$-\mathrm{d}[\mathrm{Ce(IV)}]/\mathrm{d}t = k[\mathrm{Ce(IV)}]f[\mathrm{H^+}]([\mathrm{ArCHO}]+k'[\mathrm{ArCHO}]^2)$$

which indicates two transition state complexes differing in composition by one molecule of aldehyde. Both paths display a primary kinetic isotope effect (3.78 for k)[364]. Substituent effects afford a ρ value of -0.72 for the first term which suggests hydrogen atom abstraction. The second term is characterised by a ρ value of 0.74.

The rate of the V(V) perchlorate oxidation of isobutyraldehyde is given by[368]

$$-\mathrm{d}[\mathrm{V(V)}]/\mathrm{d}t = k[\text{isobutyraldehyde}][\mathrm{H_3O^+}][\mathrm{V(V)}]^0$$

and is the same as the rate of bromination and iodination under the same conditions. Evidently acid-catalysed enolisation is rate-determining, *viz.*

$$(\mathrm{CH_3})_2\mathrm{CH{-}CH{=}O} \overset{\mathrm{H_3O^+}}{\rightleftharpoons} (\mathrm{CH_3})_2\mathrm{C{=}CH{-}OH} \qquad \text{(slow)}$$

Propionaldehyde and *n*-butyraldehyde are oxidised more slowly then they enolise and the rate expressions are[368]

$$-\mathrm{d}[\mathrm{V(V)}]/\mathrm{d}t = \frac{k[\mathrm{V(V)}][\text{aldehyde}]}{1+K[\text{aldehyde}]}$$

The acidity dependences are not simple. V(V) is thought to form a complex with the enol which undergoes slow oxidative breakdown. Propionaldehyde and *n*-butyraldehyde are, however, oxidised by Mn(III) pyrophosphate with a zero-order dependence on oxidant concentration[369] but first-order dependences on substrate and H_3O^+ concentrations. Here oxidation immediately follows enol formation. Ce(IV) sulphate oxidises acetaldehyde at a rate much faster than enolisation[370].

4.3.8 Ketones

These exist in several tautomeric forms, if the carbonyl group is flanked by two different α-CH bonds, *viz.*

$$\mathrm{R_1CH_2C({=}O)CH_2R_2} \rightleftharpoons \mathrm{R_1CH{=}C(OH){-}CH_2R_2} \rightleftharpoons \mathrm{R_1CH_2C(OH){=}CHR_2}$$

Discussion of ketone oxidation has centred around the identity of the molecule undergoing oxidation. This has been clearly resolved in some, but not all, cases, the evidence resting on (*i*), the relative rates of enolisation and oxidation, (*ii*) kinetic orders and (*iii*) isotope effects. A general feature of the oxidations of ketones by one-equivalent reagents is that the rate for a given oxidant exceeds that for oxidation of a secondary alcohol by the same oxidant. The most attractive explanation is that the radical formed from a ketone is stabilised by delocalisation, *viz.*

$$\mathrm{R\dot{C}H{-}C({=}O){-}R} \leftrightarrow \mathrm{RCH{=}C(O\cdot){-}R}$$

TABLE 12

KINETIC DATA FOR ONE-EQUIVALENT OXIDATION OF KETONES

Oxidant	*Rate expression*	k_H/k_D	k_{D_2O}/k_{H_2O}	*Ref.*
$Co(III)(HClO_4$ and $H_2SO_4)$	$k[Co(III)][ketone](a+b/[H^+])$	1.0 (10 °C)	0.5	371, 372
$Ce(IV)(HClO_4)$	$\frac{kK[Ce(IV)][ketone](a+b[H^+])}{(1+K[ketone])}$	2.0 (25 °C)	—	188, 373
$Ce(IV)(HNO_3)$	$k[Ce(IV)][ketone]f[H^+]$	—	—	374
$Ce(IV)(H_2SO_4)$	$k[Ce(IV)][ketone](a+b[H^+])$; also as for Ce(IV) in $HClO_4$ and HNO_3	6.0 (11 °C)	1.3	370, 372, 375
$Mn(III)(H_2SO_4)$	$k[Mn(III)][ketone][H^+]^0$	4.1 (11 °C)	—	372, 714
$Mn(III)(H_3PO_4)$	$k[ketone]f[Mn(III)]f'[H^+]$	—	—	376
$V(V)(HClO_4)$	$k[V(V)][ketone][H^+]$	—	—	377
$V(V)(H_2SO_4)$	complex	4.2 (50 °C)	2.4	372, 377
$IrCl_6^{2-}(HClO_4)$	$k[ketone][H^+][IrCl_6^{2-}]^0$ (low acidity)	—	—	378

more effectively than the radical from an alcohol. This situation is reversed for two-equivalent oxidants, *e.g.* chromic acid.

Kinetic data exist for all these oxidants[371–378] and some are given in Table 12. The important features are: (*i*) Ce(IV) perchlorate forms 1 : 1 complexes with ketones with spectroscopically determined formation constants in good agreement with kinetic values[373]; (*ii*) only Co(III) fails to give an appreciable primary kinetic isotope effect[371, 372] (Ir(IV) has yet to be examined in this respect); (*iii*) the acidity dependence for Co(III) oxidation is characteristic of the oxidant[371, 372] and (*iv*) in some cases [Co(III)[371, 372], Ce(IV) perchlorate[373], Mn(III) sulphate[372]] the rate of disappearance of ketone considerably exceeds the corresponding rate of enolisation; however, with Mn(III) pyrophosphate[376] and Ir(IV)[378] the rates of the two processes are identical and with Ce(IV) sulphate and V(V)[372] the rate of enolisation of ketone exceeds its rate of oxidation. (The opposite has been stated for Ce(IV) sulphate[372], but this was based on an erroneous value for k(enolisation) for cyclohexanone[154].) The oxidation of acetophenone by Mn(III) acetate in acetic acid is a crucial step in the Mn(II)-catalysed autoxidation of this substrate. The rate of autoxidation equals that of enolisation, determined by isotopic exchange[379], under these conditions, and evidently Mn(III) attacks the enolic form.

Products detected or isolated from these oxidations include the corresponding α-hydroxy ketone and α-diketone and also adipic acid (from cyclohexanone) in up to 95 % yield. However, $IrCl_6^{2-}$ gives α-chloroketone in quantitative yield[378]. Evidently when the rate of oxidation exceeds enolisation attack is on the keto form, probably *via* a complex, although this is definite only for Ce(IV) perchlorate, to give a radical, *e.g.*

+ Ce(III)

This is in accordance with the primary kinetic isotope effect for Mn(III) sulphate[312]. With Co(III) electron abstraction may occur to give a radical-cation which suffers further oxidation. The alternative explanation of the lack of an isotope effect is that formation of the Co(III)–ketone complex is rate-determining; this lacks, however, other kinetic support[371].

The mode of oxidation by Mn(III) pyrophosphate[376] also seems clear cut from (*i*) the agreement between oxidation and enolisation rates for cyclohexanone and (*ii*) the tendency for the rate to become independent of Mn(III) concentration at high concentrations. Several other ketones, however, were oxidised rather more slowly than they enolised[376].

Preliminary results on the anaerobic oxidation of cyclohexanone by $IrCl_6^{2-}$ suggest that, although enolisation is rate-determining at pH 3 (when the oxidation rate is equivalent to the enolisation rate and the reaction is zero-order in oxidant but first order in acidity), the subsequent reaction of the radical is one of ligand-capture[378], *viz.*

$$IrCl_6^{2-} + \text{enol} \rightarrow IrCl_6^{3-} + \cdot(\text{enol})^+ \quad (54)$$

$$\cdot(\text{enol})^+ \rightarrow \text{2-oxocyclohexyl radical (}\dot{C}\text{H)} \quad (55)$$

$$\text{2-oxocyclohexyl radical} + IrCl_6^{2-} + H_2O \rightarrow \text{2-chlorocyclohexanone} + IrCl_5OH_2^{2-} \quad (56)$$

Evidence for (56) includes the almost quantitative formation of chlorocyclohexanone and the production of 50 % of the Ir(III) in the form of $[IrCl_5OH_2]^{2-}$. At pH 1 an acid-independent reaction predominates with the rate parameters, $E = 16.4$ kcal.mole^{-1} and $\Delta S^{\ddagger} = -12.6$ eu. $[IrCl_5OH_2]^-$ also oxidises cyclohexanone by an acid-independent path, with $E = 16.6$ kcal.mole^{-1} and $\Delta S^{\ddagger} = -7.3$ eu.

The oxidations of 2-chloro- and 2-hydroxycyclohexanone by $IrCl_6^{2-}$ show the same general kinetic behaviour, indicating prior enolisation, and analogous products, *i.e.* dichlorocyclohexanone and $IrCl_5OH^{2-}$ from chlorocyclohexanone (ref. [699]).

The main mechanistic difficulty involves Ce(IV) sulphate and V(V), which attack ketones at rates less than those of enolisation. The kinetics and relative oxidation rates are of little assistance in discriminating between attack on keto and enol forms. However, the solvent isotope effects in these oxidations are in strict accord with the measured acidity-dependences: if attack had occurred on the enol then the change of solvent should have enhanced the expected k_{D_2O}/k_{H_2O} by a factor of 2.5, because K_{enol} is favoured to such an extent in D_2O as compared with H_2O. No such enhancement is found[372] and attack is, therefore, on the keto form in both cases, in contrast to the oxidation by Cr(VI) for which similar considerations lead to the opposite conclusion. (p. 314).

4.3.9 Ethers

The remarkable inertness[24] of dialkyl ethers to one-equivalent oxidants is good evidence that the readier oxidation of alcohols involves more than simple electron abstraction. Di-isopropyl ether is oxidised by Co(III) in CH_3CN–H_2O mixtures with complicated kinetics[380]; individual runs show first-order decay of Co(III) but the rate coefficients increase with increasing [Co(III)], and the order with respect to substrate is less than one but is neither fractional nor of a Michaelis–Menten type. The main product is acetone and the following reaction sequence is proposed

$$\mathrm{Pr^iOPr^i + Co(III) \rightarrow (CH_3)_2\dot{C}OPr^i + Co(II) + H^+} \qquad \text{(slow)}$$

$$\mathrm{(CH_3)_2\dot{C}OPr^i \rightarrow (CH_3)_2CO + Pr^i\cdot}$$

$$\mathrm{Pr^i\cdot + Co(III)aq \rightarrow Pr^iOH + Co(II) + H^+}$$

$$\mathrm{(CH_3)_2\dot{C}OPr^i + Co(III)aq \rightarrow (CH_3)_2CO + Pr^iOH + Co(II) + H^+}$$

The oxidation step may be of an inner sphere type, and the first two steps may be concerted. Some acetone is produced in a path not involving consumption of Co(III) and a short chain reaction may participate, including the step

$$\mathrm{Pr^i\cdot + Pr^iOPr^i \rightarrow Me_2\dot{C}OPr^i + C_3H_8}$$

The oxidation of di-2-chloroethyl ether[380] is first-order with respect to ether, but is autocatalytic and chloride ion is liberated. A hydrogen atom abstraction process, similar to that above, probably takes place, *viz.*

$$\mathrm{ClCH_2CH_2OCH_2CH_2Cl + Co(III) \rightarrow ClCH_2\dot{C}HOCH_2CH_2Cl + Co(II) + H^+}$$

followed by breakdown of the radical

$$\mathrm{ClCH_2\dot{C}HOCH_2CH_2Cl \rightarrow ClCH_2CHO + ClCH_2CH_2\cdot}$$

$$\mathrm{ClCH_2CH_2\cdot + Co(III)aq \rightarrow ClCH_2CH_2OH + Co(II) + H^+}$$

and chloride ion could be released by hydrolysis of the chlorhydrin.

Benzyl ethers are covered on p. 404.

4.3.10 *Carboxylic acids*

Only Co(III) has sufficient reactivity to oxidise RCO_2H at an appreciable rate; however, all these oxidants attack the atypical formic acid which can function like a secondary alcohol.

The kinetics of oxidation of propionic acid in aqueous perchloric acid by Co(III) perchlorate are[381]

$$-d[Co(III)]/dt = \frac{k_2K[Co(III)][RCO_2H]}{([H^+]+K[RCO_2H])}$$

Similar kinetics are given by phenylacetic acid, but with isobutyric and pivalic acids the rates are given by the simple expression $k_2[Co(III)][RCO_2H]/[H^+]$. The oxidation of $C_6H_5CD_2CO_2H$ proceeds at 80 % of the rate of the protio compound. The relative rates of oxidation of a series of acids of formula RCO_2H at 10 C° are

R	C_2H_5	$(CH_3)_2CH$	$(CH_3)_3C$	$C_6H_5CH_2$
$Kk_2(HClO_4)$	0.0034	0.33	0.68	1.25

Detailed product analyses indicate the major route of oxidation to be one of oxidative decarboxylation. The detailed mechanism is

$$RCO_2H+Co(H_2O)_6{}^{3+} \rightleftharpoons [RCO_2.Co(H_2O)_5]^{2+}+H^++H_2O \quad \text{(fast)}$$

$$[R\text{–}CO\text{–}O\text{–}Co(III)]^{2+} \rightarrow R\cdot+CO_2+Co^{2+}+5\,H_2O \quad \text{(slow)}$$

$$R\cdot+Co^{3+} \rightarrow R^++Co^{2+} \quad \text{(fast)}$$

$$R^++H_2O \rightarrow ROH+H^+ \quad \text{(fast)}$$

The synchronous departure of R· (as opposed to a two-step process) is supported by (*i*) the considerable effect on the rate of varying R and (*ii*) the isotope effect. R· can also be scavenged by bromoform giving RBr [381a]. As with nearly all oxidations by Co(III), these are characterised by large E values (19 to 23 kcal. mole^{-1}) and large A values (10^{14}–10^{18} l.mole^{-1}.sec^{-1}).

Complementary to the work with aqueous acidic media is the study of the homolytic decompositions of Co(III) carboxylates in carboxylic acid media by Lande and Kochi[381b]. For example, Co(III) is reduced in pivalic acid media with first-order kinetics with E = 30.6 kcal.mole^{-1}, $\Delta S^\ddagger$ = 8 eu and k_H/k_D = 1.28±0.10 (69 °C). The main oxidation products were found to be isobutylene and *tert*-butyl pivalate, which suggests that $(CH_3)_3C\cdot$ is an intermediate. Oxidative decarboxylation is the probable course in the analogous oxidations of *n*-butyric and isobutyric acids, in view of the production of propane and CO_2 under normal conditions but the almost quantitative formation of propene in the presence of catalytic amounts of Cu(II). The various steps are

$$CH_3CH_2CH_2CO_2Co(III) \rightarrow CH_3CH_2CH_2\cdot + CO_2 + Co(II)$$

$$CH_3CH_2CH_2\cdot + \text{solvent} \rightarrow CH_3CH_2CH_3 + \text{solvent radical}$$

$$CH_3CH_2CH_2\cdot + Cu(II) \rightarrow CH_2{=}CHCH_3 + H^+ + Cu(I)$$

$$Cu(I) + Co(III) \rightarrow Cu(II) + Co(II)$$

The rates of oxidation of α-aminoacids in aqueous acidic media by Co(III) perchlorate are comparable with those of carboxylic acids, whilst those of amines are much slower[712]. Accordingly, the $R\dot{C}H\overset{+}{N}H_3$ fragment is considered to be formed from $RCH(CO_2H)\overset{+}{N}H_3$, which then undergoes further oxidation to NH_3 and RCHO.

Although Ce(IV) oxidation of carboxylic acids is slow and incomplete under similar reaction conditions[343b], the rate is greatly enhanced on addition of perchloric acid. No kinetics were obtained but product analysis of the oxidations of *n*-butyric, isobutyric, pivalic and acetic acids indicates an identical oxidative decarboxylation to take place. Photochemical decomposition of Ce(IV) carboxylates is highly efficient ($\phi \sim$ unity) and Cu(II) diverts the course of reaction in the same way as in the thermal oxidation by Co(III). Direct spectroscopic evidence for the intermediate formation of alkyl radicals was obtained by Greatorex and Kemp[381c] who photoirradiated several Ce(IV) carboxylates in a degassed perchloric acid glass at 77 °K in the cavity of an electron spin resonance spectro-

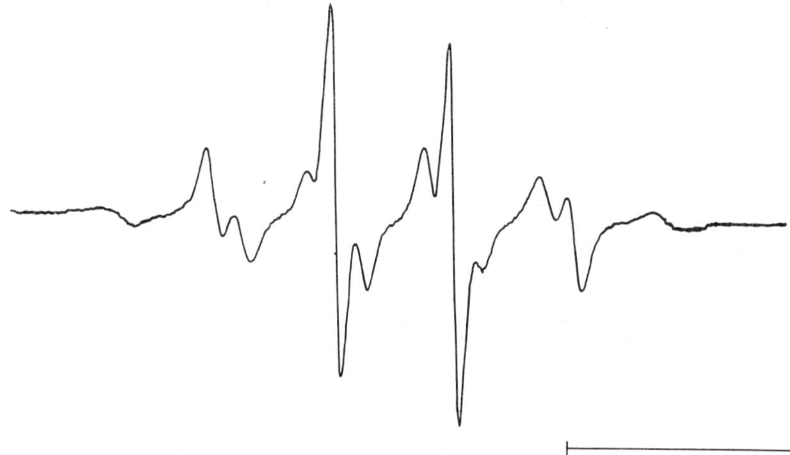

Fig. 3. Electron spin resonance spectrum of ethyl radical generated during photolysis of an aqueous acidic solution of Ce(IV) and propionic acid at 77 °C. The scale at lower right-hand-side = 50 G.

meter. The appropriate alkyl radical was obtained in each case; the production of $C_2H_5\cdot$ from propionic acid is illustrated (Fig. 3).

Kinetics for Ce(IV) perchlorate oxidation of acetic acid in $HClO_4$ media at 50–60 °C approximate to first-order in both oxidant and substrate and a plot of k_2^{-1} *versus* $[H_3O^+]^{-1}$ is linear[711]. A complex was identified (λ_{max} 286 nm) and more detailed examination of the substrate dependence revealed Michaelis–Menten kinetics. At 50 °C and unit acidity, k_2 for oxidation of several acids was determined,

Substrate acid	$ClCH_2CO_2H$	CH_3CO_2H	$C_2H_5CO_2H$	$(CH_3)_2CHCO_2H$
10^4k_2 (*l.mole*$^{-1}$.*sec*$^{-1}$) (50 °*C*)	0.71	1.26	3.26	17.5

A mechanism of oxidative decarboxylation was proposed[711].

A slow oxidation of acetic acid by Mn(III) acetate occurs at 100 °C to give mainly acetoxyacetic acid and CO_2 with an activation energy of 28 kcal.mole^{-1}. In the presence of excess Mn(II) a first-order disappearance of oxidant is found[343a]. The low yield of methane is incompatible with an initial homolysis of the type

$$Mn(O_2CCH_3)_3 \rightarrow Mn(O_2CCH_3)_2 + CH_3CO_2\cdot$$

and it seems that the primary radical is $\cdot CH_2CO_2H$ produced by an attack on the solvent. In this respect autodecomposition of manganic acetate resembles that of lead acetate in glacial acetic acid (p. 346) although a one-equivalent process seems more likely in the present instance, *viz.*

$$CH_3CO{-}O{-}Mn(OAc)_2 \;\;(H{-}CH_2{-}C(=O)OH) \longrightarrow CH_3CO{-}O(H) \;\; Mn(OAc)_2 \;\; \cdot CH{-}C(=O)OH$$

The oxidations of formic acid by Co(III)[141, 382] and V(V)[383] are straightforward, being first-order with respect to both oxidant and substrate and acid-inverse and slightly acid-catalysed respectively. The primary kinetic isotope effects are 1.5_2 (25 °C) for Co(III) and 4.1 (61.5 C°) for V(V). The low value for Co(III) is analogous to those for Co(III) oxidations of secondary alcohols, formaldehyde and *m*-nitrobenzaldehyde (*vide supra*). k_{D_2O}/k_{H_2O} for the Co(III) oxidation is about 1.0, which is curiously high for an acid-inverse reaction[383]. The mechanisms clearly parallel those for oxidation of alcohols (p. 376) where R_1 and R_2 become doubly bonded oxygen.

The oxidation by Mn(III) sulphate shows remarkable kinetics[366], *viz.*

$$\frac{-\mathrm{d}[\mathrm{Mn(III)}]}{\mathrm{d}t} = \frac{kK[\mathrm{Mn(III)}][\mathrm{HCO_2H}]}{1+K[\mathrm{HCO_2H}]}\left(a+\frac{b}{[\mathrm{H^+}]}\right) + \frac{k'K'[\mathrm{Mn(III)}]^3[\mathrm{HCO_2H}]}{(1+K'[\mathrm{HCO_2H}])[\mathrm{Mn(II)}]^{1.25}}\left(a'+\frac{b'}{[\mathrm{H^+}]}\right)$$

The first term is analogous to the rate expression for the Mn(III) oxidation of cyclohexanol (*vide supra*) and displays a primary isotope effect of similar magnitude (2.2 at 50 °C). The second term shows an isotope effect of 4.3 for replacement of HCO_2H by DCO_2H. The oxidations of malonic acid[384, 385] and Hg(I)[280] involve $[\mathrm{Mn(III)}]^2/[\mathrm{Mn(II)}]$ terms and these are readily explained by the equilibrium

$$2\,\mathrm{Mn(III)} \rightleftharpoons \mathrm{Mn(IV)} + \mathrm{Mn(II)}$$

and attack of the two-equivalent oxidant Mn(IV) upon the substrate. The formic acid oxidation probably involves concerted attack of Mn(III) and Mn(IV), *viz.*

$$\mathrm{HCO_2H} + \mathrm{Mn(IV)}^{n+} \rightleftharpoons \mathrm{HCO_2-Mn(IV)}^{(n-1)+} + \mathrm{H^+}$$

$$\mathrm{Mn(III)}\cdots\mathrm{O{=}C(H){-}O{-}Mn(IV)} \longrightarrow \mathrm{Mn(II)} + \mathrm{H^+} + \mathrm{CO_2} + \mathrm{Mn(III)}$$

Coordinated attack of two molecules of V(V) upon formaldehyde, chloral hydrate and malonic acid is mentioned elsewhere in this section.

The oxidation of formic acid by Ce(IV) sulphate which is reported as being very slow, is accelerated by X-irradiation[386]; OH· is the active oxidant.

4.4 OXIDATION OF POLYFUNCTIONAL ORGANIC MOLECULES

The main differences between these oxidations and those of monofunctional compounds are (*i*) the greater number of possible sites of attack, (*ii*) the more frequent modification of kinetics by complex formation and (*iii*) the almost inevitable greater reactivity.

4.4.1 Unsaturated and benzylic alcohols

These are generally oxidised with the same kinetics as saturated alcohols but more rapidly. In addition to the example of allyl alcohol[353, 354] discussed previously (p. 377), several reports exist of oxidations of benzylic alcohols. The importance of a 1 : 1 complex ($K = 0.8 \pm 0.2$ at 25 °C, $[\mathrm{HClO_4}] = 0.525\ M$) in the Ce(IV) oxidation is clear from agreement of spectroscopic and kinetic data[386a]. For a

References pp. 493–509

series of 2-aryl-1-phenylethanols, which are cleaved by Ce(IV) to benzaldehyde and a substituted benzyl radical, K is relatively insensitive to ring-substitution[386b] but the overall oxidation rate gives a good Hammett plot when σ^+ parameters are used ($\rho = -2.0$)[386c]. By contrast, oxidation of ring-substituted benzyl alcohols by V(V) in perchlorate media[386d] shows a concave upwards curve when $\log k/k_0$ is plotted against σ or σ^+, which the authors adduce as evidence for competing reactions, *i.e.* one- and two-equivalent paths.

4.4.2 *Glycols*

The oxidations of these by Ce(IV), Mn(III) and V(V) have been studied extensively[387–394]. Kinetic data are summarised in Table 13. The main tasks are those of (*i*) discriminating between C–C cleavage and oxidation of >CHOH groups to >C=O and (*ii*) assessing the role of cyclic complexes.

1 : 1 complexing affects the kinetics in several cases and with Ce(IV) perchlorate and nitrate oxidations, K(kinetic) equals K (spectroscopic). Some evidence exists for higher complexes, which break down less easily than the 1 : 1 complex[390]. The cleavage of pinacol to acetone in good yield in all cases is clear evidence for C–C fission, and polyhydric alcohols yield formic acid as the end product. Isolation of 2-hydroxycarbonyl compounds from reaction mixtures is difficult because of their much greater case of oxidation once formed, and evidence for the oxidation of >CHOH to >C=O is mainly kinetic. k_H/k_D for the Ce(IV) sulphate oxidation of butane-2 : 3-diol to (mainly) acetaldehyde is too low (1.17)[390] to be of diagnostic value. However, the V(V) oxidation of this substrate has a higher isotope effect (2.7) which indicates considerable, if not, preponderant, C–H as opposed to C–C, cleavage[390].

The acidity dependences of V(V) oxidations are significant. That of pinacol[394], which undergoes 100 % C–C cleavage, is $(a+bh_0)$. The first (acid-independent) term is rare in V(V) oxidations and implies that VO_2^+ is the active oxidant; the second term implies, on the basis of the Zucker–Hammett hypothesis, that the transition state has the structure (*B*), the mechanism being

$$\begin{matrix} R_2C{-}OH \\ | \\ R_2C{-}OH \end{matrix} + VO_2^+ \rightleftharpoons \begin{matrix} R_2C{-}O \\ R_2C{-}O \end{matrix} V^+ \begin{matrix} OH \\ OH \end{matrix} \quad (A)$$

$$-H_3O^+ \updownarrow +H_3O^+$$

$$\begin{matrix} R_2\dot{C}{-}O \\ R_2C{=}O \end{matrix} V^{2+} \begin{matrix} OH \\ OH_2 \end{matrix} \xleftarrow{\text{slow}} \begin{matrix} R_2C{-}O \\ R_2C{-}O \end{matrix} V^{2+} \begin{matrix} OH \\ OH_2 \end{matrix} \quad (B) + H_2O$$

(Structure (*A*) also breaks down in a similar manner.)

This contrasts with the V(V) oxidation of cyclohexanol which exhibits an $[H_3O^+]$ dependence and involves an extra water molecule in the transition

TABLE 13

KINETICS OF THE OXIDATIONS OF GLYCOLS

Oxidant	*Glycol (G)*	*Kinetic rate law*	k_H/k_D	k_{D_2O}/k_{H_2O}	*Ref*
Ce(IV)($HClO_4$)	2,3-butanediol	$[Ce(IV)]\left\{\frac{k_1K_1[G]+k_2K_1K_2[G]^2}{1+K_1[G]+K_1K_2[G]^2}\right\}$	—	—	387
Ce(IV)($HClO_4$)	glycerol, *cis*- and *trans*-1,2-cyclohexanediol, *trans*-2-methoxycyclohexanol	$\frac{kK[Ce(IV)][G]}{1+K[G]}$	—	—	388 349
Ce(IV)($HClO_4$ +H_2SO_4 mixture)	*cis*- and *trans*-1,2- -cyclopentanediols, *trans*-2-methoxycyclopentanol	$k_2[Ce(IV)][G]$	—	—	349
Ce(IV)(HNO_3)	2,3-butanediol	$\frac{kK[Ce(IV)][G]f[H^+]}{1+K[G]}$	—	—	389
Ce(IV)(H_2SO_4)	ethylene glycol	$k[Ce(IV)]f[G]$	—	—	390
Ce(IV)(H_2SO_4)	glycerol	$k[Ce(IV)][G][H^+]$	—	—	388
Ce(IV)(H_2SO_4)	2,3-butanediol	$k[Ce(IV)][G]$	1.17	—	390
Ce(IV)(H_2SO_4)	pinacol	$k[Ce(IV)][G]$	—	1.02	390,391
Mn(III)(H_3PO_4)	pinacol	$\frac{kK[Mn(III)][G]f[H^+]}{K[G]+[H_2P_2O_7^{2-}]}$			392
Mn(III)(H_3PO_4)	*trans*-1 : 2-dimethyl-cyclopentane-1 : 2-diol	$k[Mn(III)][G]$	—	—	393
Mn(III)(H_3PO_4)	*trans*-1 : 2-dimethyl-cyclohexane-1 : 2 diol	$k[Mn(III)][G]$	—	—	393
Mn(III)(H_3PO_4)	*cis*-cyclohexane-1 : 2-diol	$k[Mn(III)]^2[G]$	—	—	393
Mn(III)(H_3PO_4)	methyl α-d-glucopyranoside and mannopyranoside	$\frac{kK[Mn(III)][G]}{1+K[G]}$	—	—	393
V(V)($HClO_4$)	pinacol	$k[V(V)][G][a+bh_0]$	—	—	394
V(V)($HClO_4$)	pinacol monomethyl ether	$k[V(V)][G][H^+]$	—	—	394
V(V)(H_2SO_4)	pinacol	—	—	1.18	390
V(V)(H_2SO_4)	ethylene glycol	—	—	1.35	390
V(V)(H_2SO_4)	butane-2 : 3-diol	$k[V(V)][G]$	2.7	—	390
V(V)(H_2SO_4)	pinacol monomethyl ether	$k[(V)]f[H^+]$	—	—	390

state, *viz.*

$$R_2CHOH + VO_2^+ + H_3O^+ \rightleftharpoons R_2C(H)-O(H)\rightarrow V^{2+}(OH)_2-OH$$

and with that of pinacol monomethyl ether which cannot form a cyclic ester like (*A*) and also shows an $[H_3O^+]$ dependence[394], the transition state being

$$CH_3O-C(CH_3)_2-C(CH_3)_2-O(H)\rightarrow V^{2+}(OH)_2-OH$$

Thus the acidity dependence of V(V) oxidations gives some guidance as to the mode of fission; chelation followed by C–C cleavage is associated either with zero-order or an h_0-dependence and C–H cleavage with a $[H_3O^+]$ dependence. Ditertiary glycols fall into the former category and disecondary into the latter[395]. Steric effects in ditertiary glycol oxidation indicate that a cyclic ester is a necessary intermediate, for example 1,2-dimethylcyclohexane-1 : 2-diols are oxidised at one-hundredth the rate of pinacol[395].

Evidence concerning the relative extents of C–C and C–H fission is less well defined for Ce(IV) and Mn(III) as compared with V(V). Pinacol is cleaved to acetone in all cases, but while Mn(III) pyrophosphate [like V(V)] oxidises pinacol much faster than butane-2 : 3-diol, the rate ratio with Ce(IV) is only approximately 3 and the production of acetaldehyde from butane-2 : 3-diol by Ce(IV) oxidation demonstrates C–C cleavage[390]. It is probable, therefore, that Mn(III) oxidises the disecondary glycol by C–H fission.

Relative rate studies[395] of cyclic *cis*- and *trans*-1 : 2-diols give no clear pattern of results. *Cis*-cyclohexane-1 : 2-diol is oxidised by V(V) faster than the *trans* isomer, but the reverse is found for the isomers of 1 : 2-dimethylcyclohexane-1 : 2-diol.

4.4.3 Unsaturated aldehydes

Except at very low Mn(III) concentrations the oxidation of acraldehyde by Mn(III) pyrophosphate obeys the rate expression[396]

$$-\mathrm{d}[\mathrm{Mn(III)}]/\mathrm{d}t = k[\text{acraldehyde}][\mathrm{H_3O^+}][\mathrm{Mn(III)}]^0[\mathrm{Mn(II)}]^0$$

Similar kinetics are obtained with α-methylacraldehyde[396], but with crotonaldehyde the reaction is essentially first order in Mn(III), tending to zero-order only at relatively high [Mn(III)]. The oxidation of acraldehyde is viewed as a rapid reaction preceded by a slow acid-catalysed hydration, *viz.*

$$\mathrm{CH_2{=}CH{-}CH{=}O + H^+ \rightleftharpoons \left\{ \begin{array}{l} CH_2{=}CH{-}CH{=}\overset{+}{O}H \\ \updownarrow \\ CH_2{=}CH{-}\overset{+}{C}H{-}OH \\ \updownarrow \\ {}^+CH_2{-}CH{=}CH{-}OH \end{array} \right.}$$

$$\mathrm{H^+ + HO{-}CH_2{-}CH{=}CH{-}OH \xleftarrow[\text{slow}]{H_2O} {}^+CH_2{-}CH{=}CH{-}OH}$$

$$\mathrm{HO{-}CH_2{-}CH{=}CH{-}OH \xrightarrow[\text{fast}]{Mn(III)} \underset{\text{glyceraldehyde hydrate}}{HOCH_2.CHOH.CH(OH)_2}}$$

The immediate product of oxidation is degraded further. Ordinary enolisation of the type

$$\mathrm{CH_2{=}CH{-}CH{=}O \rightleftharpoons CH_2{=}C{=}CHOH}$$

is unlikely *per se* and cannot account for the comparable oxidation of α-methyl-acraldehyde.

Benzaldehyde and related compounds are dealt with under aldehydes (p. 379).

4.4.4 *Unsaturated carboxylic acids*

Oxidation by Co(III) perchlorate shows the same kinetics as that of saturated carboxylic acids (p. 384) namely

$$-\mathrm{d}[\mathrm{Co(III)}]/\mathrm{d}t = k[\mathrm{Co(III)}][\mathrm{substrate}](a+b/[\mathrm{H_3O^+}])$$

For cinnamic acid at 9.6 °C, $a = 0.107$, $b = 1.25$ and $k = 0.69$ l.mole^{-1}.sec^{-1}, $E = 26.7 \pm 0.5$ kcal.mole^{-1} and $\Delta S^{\ddagger} = 34.5$ eu. Identification of products of oxidation of a number of acids indicates two concurrent mechanisms. Predominating is direct attack on the double bond to give, ultimately, cleavage products, *e.g.* benzaldehyde from cinnamic acid (some phenylacetaldehyde is also found, indicating oxidative decarboxylation to occur) and also acetophenone from 3-phenylcrotonic acid.

These relatively facile oxidations may involve hydroxyl transfer, in preference to radical-cation formation, *viz.*

$$\mathrm{>C{=}C<} \;\; \mathrm{H_2O{-}Co(III)} \longrightarrow \mathrm{>C(HO)\text{-}C<} \;\; \mathrm{H^+} \;\; \mathrm{Co(II)}$$

4.4.5 *Hydroxy ketones*

A comparison[397] of the V(V) oxidations of acetoin, $CH_3CH(OH)COCH_3$, and 3-hydroxy-3-methylbutan-2-one, $(CH_3)_2C(OH)COCH_3$, shows that whilst both rate laws include first-order terms in substrate and oxidant, the acidity dependence for the former compound is purely h_0 but that for the latter is $(a+bh_0)$. The C-methyl compound consumes only 2 equivalents of V(V) to give acetone and a mechanism similar to that for the oxidation of pinacol is proposed[397], *viz.*

$$(CH_3)_2C(OH)COCH_3 + VO_2^+ \rightleftharpoons [(CH_3)_2C{-}O{-}V^+(OH)(=O){\leftarrow}O{=}C{-}CH_3]$$

$$[(CH_3)_2C{-}O{-}V^+(OH)(=O){\leftarrow}O{=}C{-}CH_3] \underset{}{\overset{H^+}{\rightleftharpoons}} [(CH_3)_2C{-}O{-}V^{2+}(OH)(OH){\leftarrow}O{=}C{-}CH_3]$$

$$\longrightarrow (CH_3)_2CO + CH_3\dot{C}O + V(IV)$$

Acetoin consumes 4 equivalents of V(V) to produce some biacetyl *via* C–H fission; however, this cleavage is not accompanied by a hydronium-ion concentration dependence of the rate thereby differing from a secondary alcohol oxidation. The mechanism of breakdown of the complex is depicted as follows

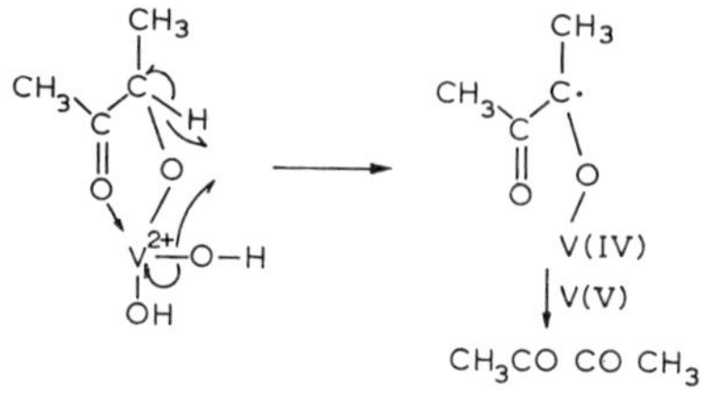

$CH_3CO\,CO\,CH_3$

It is noteworthy that while V(V) attacks α-hydroxy ketones faster than unsubstituted ketones by two orders of magnitude[377], Mn(III) pyrophosphate oxidises α-hydroxycyclohexanone more slowly than cyclohexanone[376].

4.4.6 *Hydroxy acids*

These oxidations have attracted wide interest[188, 398–404] and both specialised and comparative studies have been published. The rate laws which are summarised in Table 14 are not distinctive although the acidity dependence of the V(V) oxidations of some substrates suggests by analogy that a pinacol-type of oxidation may occur (*cf.* V(V)–pinacol complexes, p. 388), *viz.*

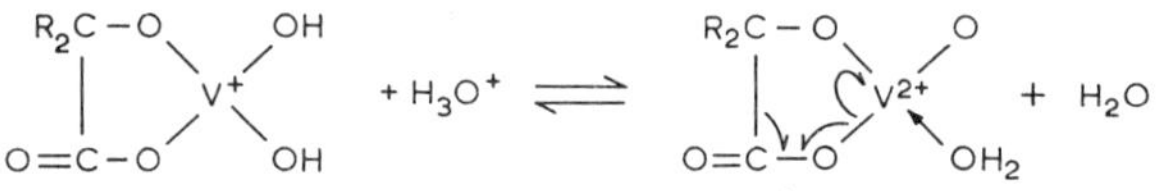

The existence of a transient complex between malic acid (HA) and Co(III) has been demonstrated optically ($\lambda_{max} \sim 275$ nm)[398],

$$Co^{3+} + HA \underset{k_{-1}}{\overset{k_1}{\rightleftharpoons}} Co.HA^{3+} \qquad K_1$$

$$CoOH^{2+} + HA \underset{k_{-2}}{\overset{k_2}{\rightleftharpoons}} CoOH.HA^{2+} \qquad K_2$$

where, at 7 °C ($\mu = 0.25$ *M*), $k_1 = 5.4 \pm 1.2$ l.mole^{-1}.sec^{-1}, $k_2 = 70 \pm 10$ l.mole^{-1}.sec^{-1} and $K_1 = 34.4 \pm 4$ mole.l^{-1}.

The main question is the site of initial oxidation of $RCH(OH)CO_2H$ (to give the possible radicals I–IV)

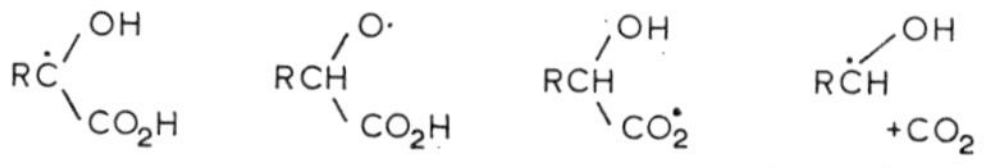

I II III IV

TABLE 14

KINETIC DATA FOR THE OXIDATION OF α-HYDROXY ACIDS

Oxidant	*Substrate*	*Rate law*	*E (kcal.mole⁻¹)*	*ΔS‡(eu)*	*Ref.*
Co(III)	mandelic acid	k[Co(III)][substrate] (at low substrate concentration only)	—	—	188
Co(III)	malic acid	$k\{[Co^{3+}\text{substrate}] + k'[CoOH^{2+}\text{substrate}]\{(a+b/[H^+])$	15.3±1.0 (k) 25.1±1.0 (k')	15.7±2 (k)	398
Ce(IV) (sulphate)	glycollic acid	$\dfrac{k[\text{Ce(IV)}][\text{substrate}]}{\{1+K[HSO_4^-]+K'[HSO_4^-]^2\}}$	10.1±0.8	−42.1±2	399
Ce(IV) (sulphate)	mandelic acid dl-malic acid lactic acid	$\dfrac{k[\text{Ce(IV)}][\text{substrate}]}{[H_2SO_4]^2}$	—	—	400
Ce(IV) (sulphate)	lactic acid	$\dfrac{kK[\text{Ce(IV)}][\text{substrate}]}{(1+K[\text{substrate}])[H^+]^2}$	22.8	3.1	401
Ce(IV) (sulphate)	benzilic acid	$\dfrac{k[\text{Ce(IV)}][\text{substrate}]}{[H_2SO_4]^2}$	22.3	−8.14	402
Mn(III) (H_3PO_4)	(+)-tartaric acid dl-malic acid *meso*-tartaric acid	$\dfrac{k[\text{Mn(III)}][\text{substrate}]}{(1+K[\text{substrate}])f[\text{Mn(II)}]}$	—	—	403
V(V) ($HClO_4$)	mandelic acid dl-malic acid latic acid	$k[\text{V(V)}][\text{substrate}](a+bh_0)$	—	—	394
V(V) ($HClO_4$)	α-hydroxy-isobutyric acid	$k[\text{V(V)}][\text{substrate}][H_3O^+]$ $k(\text{V(V)}][\text{substrate}]f[H_3O^+]$ (Plot of k^{-1} *versus* $[H_3O^+]^{-1}$ linear with intercept)	21.8 22.0	−2.8 −1.0	404 404a
V(V) (H_2SO_4)	α-hydroxy-isobutyric acid	As for V(V)($HClO_4$)	23.0	1.5	404a

Radical I can be ruled out because it would be oxidised to a α-keto acid which would be rapidly further oxidised to RCO_2H; in fact the stoichiometry for V(V) oxidation is 2 V(V) : 1 molecule substrate in all cases and the major product is always RCHO (or R_1R_2CO from $R_1R_2C(OH)CO_2H$). These data, are, however, compatible with the production of radicals II–IV and discrimination can be made only with the aid of kinetics.

Krishna and Tewari[400] favour II in Ce(IV) oxidation; Waters *et al.*[186, 394] prefer C–C fission to give IV for both V(V) and Mn(III). Kemp and Waters[186] have established two main features of these oxidations, namely, (*i*) k_H/k_D for the oxidations of $C_6H_5CD(OH)CO_2H$ and the "light" compound are: V(V), 2.0; Mn(III) sulphate, 1.2; Ce(IV) sulphate, 1.1; and (*ii*), that the trend of rates of oxidation of mandelic, α-hydroxyisobutyric, lactic and glycollic acids is as expected for Ce(IV) and Mn(III) if stabilisation of the radical $R\dot{C}HOH$ is important, but is altered for V(V) (Fig. 4). It appears from the latter observations that the presence of α-hydrogen atoms causes a drop in rate by a factor of almost 10 per α-hydrogen

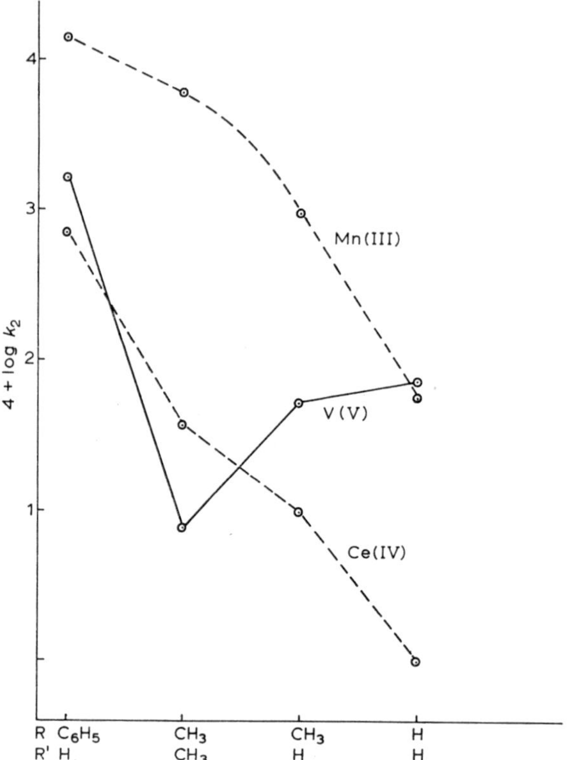

Fig. 4. Relative rates of oxidation of α-hydroxyacids $RR'C(OH)CO_2H$ by Mn(III), Ce(IV) and V(V) in dilute sulphuric acid medium. Temperature = 24.4 °C (Mn(III)) or 26.6 °C (V(V) and Ce(IV)) (other conditions given in Ref. 186).

for Mn(III) oxidation, the drop is less marked with Ce(IV) and is reversed for V(V). The isotope effects reveal that C–H breakage is of some importance in the V(V) oxidation, but much less significant for the other oxidants; indeed, at 35 °C the oxidation of benzilic acid by Ce(IV) sulphate is about 9 times faster than that of mandelic acid[402].

One anomaly is the $[H_3O^+]$ dependence found for the V(V) oxidation of α-hydroxyisobutyric acid to acetone[404], because the C–H fission normally associated with such an acidity dependence cannot occur. This dependence has been disputed in a recent re-examination[404a].

4.4.7 *α-Mercaptocarboxylic acids*

A stopped-flow study[405] of the Ce(IV) sulphate oxidations of several of these substances to the corresponding disulphide produced a general rate law

$$-\mathrm{d}[\mathrm{Ce(IV)}]/\mathrm{d}t = k_2[\mathrm{Ce(IV)}][\alpha\text{-mercapto acid}]$$

with the kinetic parameters, $[H_3O^+] = 0.512\ M$ given below (k_2 at 23.4 °C)

	thiomalic acid	*thioglycollic acid*	*thiolactic acid*
k_2 (l.mole^{-1}.sec^{-1})	1.15×10^{-3}	7.04×10^{-2}	4.12×10^{-2}
E (kcal.mole^{-1})	7.6	7.9	6.8
$\Delta S^{\ddagger}$(eu)	-30 ± 4	-27 ± 4	-31 ± 3

The stoichiometry of 1 Ce(IV) : 1 molecule substrate accords with the isolation of disulphides and implies that the rate of dimerisation of $R_1R_2C(CO_2H)S\cdot$ far exceeds that of oxidation by further Ce(IV).

The oxidation of 2-mercaptosuccinic acid by V(V) proceeds *via* a coloured intermediate (λ_{max} 460, 520 nm) to give the disulphide. The following scheme was put forward[406]

$$VO_2^+ + RSH \underset{k_{-1}}{\overset{k_1}{\rightleftharpoons}} VO_2^+\cdot RSH \qquad K_1$$

$$VO_2^+\cdot RSH \xrightarrow{k_2} VO^{2+} + RS\cdot + OH^-$$

$$VO_2^+\cdot RSH + RSH \xrightarrow{k_3} V(III) + RSSR + 2\ OH^-$$

where, at 15.3 °C, $k_1 = 465\pm50$ l.mole^{-1}.sec^{-1}, $k_{-1} = 44.5\pm3$ sec^{-1}, $K_1 = 10.5$ l.mole^{-1}, $k_2 = 5\pm0.2$ sec^{-1} and $k_3 = 40\pm3$ l.mole^{-1}.sec^{-1}.

Intermediate complexes $Co(RSH)^{3+}$ and $Co(OH)(RSH)^{2+}$ were also apparent in the rapid oxidation of thiomalic acid (RSH) by Co(III) in a perchlorate medium [406a]. These were formed in second order processes with $k_2 = 8.2\pm2.1$ and 320 ± 50 l.mole^{-1}.sec^{-1}, respectively, at 7 °C ($\mu = 0.25\ M$). Breakdown of the two complexes gave a combined rate law at 12.7 °C ($\mu = 1.0\ M$)

$$k_{oxid} = 0.05 + \frac{0.27}{[H_3O^+]}\ (\text{sec}^{-1})$$

The two terms correspond to breakdown of $Co(RSH)^{3+}$ and $Co(OH)(RSH)^{2+}$ respectively with total activation energies of 37.1 ± 1.5 and 15.2 ± 1.0 kcal.mole^{-1}.

4.4.8 *α-Keto acids*

Mn(III) pyrophosphate readily oxidises pyruvic acid, *viz.*

$$CH_3COCO_2H + 2\ Mn(III) + H_2O = CH_3CO_2H + 2\ Mn(II) + CO_2 + 2\ H^+$$

the rate law being[376]

$$\frac{-\mathrm{d}[\mathrm{Mn(III)}]}{\mathrm{d}t} =$$

$$\frac{kK[\mathrm{Mn(III)}][\text{pyruvic acid}]}{(1+K[\text{pyruvic acid}])} \cdot \frac{1}{1+K'[\text{pyrophosphate}]} \cdot \frac{f'[\mathrm{H^+}]}{f[\mathrm{Mn(II)}]}$$

The retardation by Mn(II) is only 18 % at most. The mechanism proposed involves acetyl radicals and does not incorporate enolisation, *viz.*

$$\begin{matrix} CH_3-C=O \\ | \quad\quad Mn(III) \\ O=C-O \end{matrix} \longrightarrow CH_3-\dot{C}=O + Mn(II) + CO_2$$

4.4.9 *Oxalic acid*

This is an typical example of a dicarboxylic acid in that C–C cleavage is the only route for oxidation. No study of the Co(III) oxidation has been made although it is highly probable that reaction would proceed through an oxalate complex. The thermal decomposition of $Co(Ox)_3{}^{3-}$ has been shown[407, 408] to be a first-order process and probably involves an internal redox reaction, *viz.*

$$Co(C_2O_4)_3{}^{3-} + H_2O \rightleftharpoons [Co(C_2O_4)_2.H_2O.CO_2\text{–}CO_2^-]^{3-}$$
$$\downarrow \text{(slow)}$$
$$Co(C_2O_4)_2^{2-} + \cdot CO_2^- + CO_2$$

The lack of exchange between ^{14}C-labelled oxalate in solution and bound oxalate rules out the existence of free $\cdot C_2O_4^-$ in this reaction. The same intermediate is thought to participate in isotopic Co(II)–Co(III) exchange in oxalate solution[407].

The most recent study[716] indicates that both $Co(C_2O_4)_2(H_2O)^-$ and $Co(C_2O_4)_3^{2-}$ decompose by the dissociation of one end of the chelate to give an intermediate (I) capable of attacking a second molecule of complex to give Co^{2+} and CO_2 *without* participation of free radical intermediates, *viz.*

$$Co(C_2O_4)_2(H_2O)_2^- \underset{k_{-1}}{\overset{k_1}{\rightleftharpoons}} I$$

$$I + Co(C_2O_4)_2(H_2O)_2^- \xrightarrow{k_2} 2\,Co^{2+} + 2CO_2 + 3C_2O_4^{2-}$$

Such a scheme yields a rate law

$$\mathrm{d}[\mathrm{Co(III)}]/\mathrm{d}t = -2k_1k_2[\mathrm{Co(III)}]^2/(k_{-1}+k_2[\mathrm{Co(III)}])$$

which on integration gives

$$\ln[\mathrm{Co(III)}]-(k_{-1}/k_2)/[\mathrm{Co(III)}] = -2k_1t+\text{constant}$$

identical with the experimental rate law.

The kinetics of the Ce(IV) sulphate oxidation of oxalic acid are simple second order[409, 410], although the rate coefficient is inversely proportional to both hydrogen and bisulphate-ion concentrations[411], and it is also reduced at very high oxalic acid concentrations[412]. Values of the activation energy from 13.4±1.5 (ref. 411) to 16.5±0.4 (ref. 409) kcal.mole^{-1} have been reported. An intermediate has been detected spectroscopically[411]; this decays in first-order fashion with E = 10.5±0.5 kcal.mole^{-1} and with a rate independent of acidity. However, the extent of formation of this complex is reduced as the acidity is increased[411], and it appears that a less reactive dioxalato complex is formed at higher substrate concentrations[412].

The oxidation by Mn(III) chloride involves three complexes and the kinetic data of Taube[413, 414] are summarised in Table 15. The greater thermal stability of the *tris*-complex is considered to result from the lowering of the free energy relative to the transition state as compared with *bis*- and *mono*-complexes. The study of $MnC_2O_4^+$ was based on the Mn(III)-catalysed chlorine oxidation of oxalic acid[414].

One study has been made with a chelate complex, *trans*-1,2-diaminocyclohexanetetraaceatomanganate(III)[415], which is either pentadentate or hexadentate, a water molecule occupying the sixth or seventh coordination position respectively, and hence chelation of the oxalate is very unlikely. The reaction is first-order both in oxidant and oxalate and is retarded by increase of acidity. The $HC_2O_4^-$ ion is, therefore, attacked more slowly than the $C_2O_4^{2-}$ ion but both forms are oxidised as follows

$$\mathrm{Mn(chelate)^- + HC_2O_4^- \rightarrow Mn(chelate)^{2-} + H^+ + \cdot C_2O_4^-}$$
$$\mathrm{Mn(chelate)^- + \cdot C_2O_4^- \rightarrow Mn(chelate)^{2-} + 2\ CO_2}$$

TABLE 15

DECOMPOSITION OF MANGANIOXALATES

	$Mn(C_2O_4)^+$	$Mn(C_2O_4)_2^-$	$Mn(C_2O_4)_3^{3-}$
k_1 (sec^{-1}) at 25.2 °C	0.197	7.7×10^{-4}	3.4×10^{-4}
E (kcal.mole^{-1})	18.3±0.5	—	22.2±0.5
A (sec^{-1})	2×10^{14}	—	3×10^{14}

The activation parameters for the oxidations of $HC_2O_4^-$ and $C_2O_4^{2-}$ are

	E (kcal.mole^{-1})	$\Delta S^\ddagger$ (*eu*)
$HC_2O_4^-$	21.5 ± 0.2	$+ 3.5 \pm 0.7$
$C_2O_4^{2-}$	16.6 ± 0.4	-13.0 ± 1.3

The V(V) oxidation has some interesting kinetic features. The dependence of rate upon acidity shows a sharp minimum in the region of 3 M[334, 416], and the early determinations of Bobtelsky and Glasner[334] showing second- and first-order dependence upon oxalic acid and V(V) concentration, respectively, have been confirmed[416] for conditions of low acidity (1 M). At high acidity (> 4 M) the kinetics revert to a "normal" form, *viz.*

$$-\mathrm{d}[\mathrm{V(V)}]/\mathrm{d}t = k[\mathrm{V(V)}][\text{oxalic acid}][\mathrm{H_3O^+}]$$

At low acidity a complex is formed[417] between oxalic acid and the feeble oxidant VO_2^+, *viz.*

$$VO_2^+ + H_2C_2O_4 \rightleftharpoons (HO)_2\overset{+}{V}(O\text{-}CO)_2$$

This may take up a second molecule of oxalic acid, *viz.*

$$(HO)_2\overset{+}{V}(O\text{-}CO)_2 + H_2C_2O_4 \rightleftharpoons (HO)(HO_2C.COO)\overset{+}{V}(O\text{-}CO)_2 + H_2O$$

and slow decomposition of the $V(V)Ox_2$ complex would produce the unusual kinetics. At lower acidities ionisation of this complex produces an uncharged complex in which the oxidising power of V(V) would be greatly reduced. However, combination with a third molecule of oxalic acid produces the complex

$$(HO_2C.CO.O)(^-O_2C.COO)\overset{+}{V}(O\text{-}CO)_2$$

which retains a positively charged V(V) centre and can undergo internal oxidation–reduction.

The oxidation by V(V) is also subject to strong catalysis by Mn(II)[418]. The kinetics are changed to

$$\frac{-\mathrm{d}[\mathrm{V(V)}]}{\mathrm{d}t} = \frac{kK_1[\mathrm{V(V)}][\text{oxalic acid}]}{1+K_1[\text{oxalic acid}]} \cdot \frac{[\mathrm{Mn(II)}]}{1+K_2[\mathrm{V(IV)}]}$$

although at higher [Mn(II)] the reaction becomes zero-order in Mn(II). These data are consistent with a slow oxidation by V(V) of a mono-oxalato complex of Mn(II), *viz.*

$$\mathrm{V(V)+Mn(II)C_2O_4.aq \rightleftharpoons V(IV)+Mn(III)C_2O_4{}^+.aq}$$

followed by a relatively rapid breakdown of Mn(III) $C_2O_4{}^+$·aq. The activation energy of the catalysed reaction is only 4.7 kcal.mole^{-1}, whilst that for the direct oxidations by V(V) and Mn(III) are 19.2 (ref. 419) and 18.3 (ref. 418) kcal.mole^{-1}, respectively. An entirely analogous catalysis has been observed for the V(V) oxidation of malonic acid (see below).

The $NpO_2{}^{2+}$ oxidation of oxalic acid in aqueous perchloric acid provides one of the few examples of redox kinetic studies of Np(VI)[670]. The rate law is

$$+\frac{\mathrm{d}[\mathrm{Np(V)}]}{\mathrm{d}t} = k\,\frac{[\mathrm{Np(VI)}][\text{total oxalate}]}{[\mathrm{H_3O^+}]}$$

Admixture of the reactants caused the pink colour of NpO_2^{2+} to change to yellow, suggesting complex formation, and one mole of oxalic acid consumed *four* moles of oxidant to give Np(V) as the final product, identified optically. k is independent of ionic strength and is 0.012±0.001 sec^{-1} at 25 °C ($\mu = 1\ M$) with $E = 15.5$ kcal.mole^{-1}. Breakdown of an oxalatoneptunium(VI) complex of low formation constant is presumably the mechanism.

4.4.10 *Malonic acid*

Kinetics of the reactions with all the oxidants have been reported (Table 16). The usual product is formic acid, which is the first molecule formed resistant to very rapid secondary oxidation. Six equivalents of Ce(IV) are destroyed in oxidising one molecule of substrate to one of HCO_2H[420], *viz.*

$$\mathrm{CH_2(CO_2H)_2+6\ Ce(IV)+2\ H_2O = 2\ CO_2+HCO_2H+6\ Ce(III)+6\ H^+}$$

The readily oxidised intermediates are probably tartronic and glyoxylic acids[384, 420] in all the oxidations, *viz.*

$$\mathrm{CH_2(CO_2H)_2 \rightarrow HO.CH(CO_2H)_2 \rightarrow OHC.CO_2H \rightarrow HCO_2H}$$

TABLE 16

KINETICS OF OXIDATION OF MALONIC ACID

Oxidant	*Kinetic rate law*	*E (kcal.mole^{-1})*	*$\Delta S^{\ddagger}$(eu)*	*Ref.*
Co(III)	$\frac{kK[\text{Co(III)}][\text{substrate}]}{1+K[\text{substrate}]}$	—	—	419
Ce(IV) perchlorate	$\frac{k[\text{Ce(IV)}][\text{substrate}]^0}{f[\text{H}_3\text{O}^+]}$	9.2±0.4	−38.6±1.2	420
Ce(IV) nitrate		11.5±0.5	−31.4±1.0	420
Ce(IV) sulphate		16.1±0.5	−16.5±1.0	420
Ce(IV) sulphate	$k[\text{Ce(IV)}]^n[\text{substrate}]$, $1 < n < 2$	—	—	188
Mn(III) sulphate	$\frac{kK[\text{Mn(III)}]^2[\text{substrate}]}{(1+K[\text{substrate}])} \cdot \frac{(1+k''[\text{H}_3\text{O}^+])}{(1+k'[\text{Mn(II)}])}$	13.2±1.0	—	385
Mn(III) pyrophosphate	$\frac{kK[\text{Mn(III)}]^2[\text{substrate}]}{(1+K[\text{substrate}])} \cdot \frac{1}{(1+k'[\text{Mn(II)}])}$	—	—	384
V(V) sulphate	$k([\text{V(V)}]+k'[\text{V(V)}]^2)[\text{substrate}]/f[\text{H}_3\text{O}^+]$	19.7	—	383

The kinetics are straightforward only in the case of Co(III) oxidation, which involves slow breakdown of a malonato complex[419], *viz.*

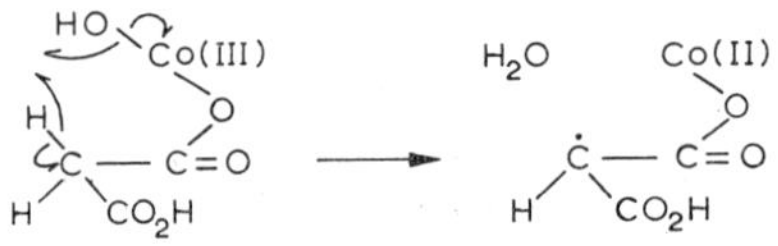

Oxidative decarboxylation, after the manner of RCO_2H (*vide supra*), is ruled out because the product, glycollic acid, is only slowly oxidised by Mn(III) pyrophosphate yet all equivalents of Mn(III) are consumed rapidly[384].

Sengupta and Aditya[420] find the usual reactivity sequence for Ce(IV) salts, *viz.* $ClO_4^- > NO_3^- \gg SO_4^{2-}$, but note that plots of log (Ce(IV)] *versus* time are *linear* even when equal molar concentrations of Ce(IV) and malonic acid are taken. This implies a first-order dependence on [Ce(IV)] and zero-order dependence on malonic acid concentration. Kemp[188], however, has found a clear first-order dependence on malonic acid concentration for the Ce(IV) sulphate oxidation, using an excess of reductant and making a four-fold variation in reductant concentration. Moreover, consumption of Ce(IV) was intermediate between first- and second-order. Further work is needed to resolve this discrepancy.

Waters *et al.* have found the remarkable kinetics for the Mn(III) oxidation in both H_3PO_4 (ref. 384) and H_2SO_4 (ref. 385) media, which implies a mechanism dependent upon the nature of Mn(III) rather than a mere one-equivalent oxidation, for Mn(III) pyrophosphate and sulphate have very different redox potentials of 1.15 (ref. 421) and 1.51 (ref. 19), respectively. An original mechanism[384] for the pyrophosphate oxidation depends on a reversible equilibrium (ignoring inorganic ligands), *viz.*

$$Mn(III)+CH_2(CO_2H)_2 \rightleftharpoons CH_2(CO_2H)_2.Mn(III) \quad \text{(fast)} \quad (57)$$

$$CH_2(CO_2H)_2.Mn(III) \underset{\text{slow}}{\overset{\text{slow}}{\rightleftharpoons}} \cdot CH(CO_2H)_2 + Mn(II) + H^+ \quad (58)$$

$$Mn(III)+\cdot CH(CO_2H)_2+H_2O \rightarrow Mn(II)+HO.CH(CO_2H)_2+H^+ \quad \text{(slow)} \quad (59)$$

This has the attraction of explaining other characteristic features of this reaction, namely: (*i*) the sensitivity to oxygen of the rate and course of reaction (because of capture of O_2 by $\cdot CH(CO_2H)_2$), (*ii*) acceleration by added acrylonitrile to produce a carboxyl group-containing polymer, and (*iii*) powerful induced oxidations by the reaction mixture of molecules inert to Mn(III) pyrophosphate, *e.g.* ethanol, diethyl ether.

The latter effect is presumed to result from *oxidation* by $\cdot CH(CO_2H)_2$. Of relevance also are the kinetics of the Mn(III) pyrophosphate oxidation of ethyl- and benzylmalonic acids[384]; the order in Mn(III) is approximately one, no retardation by Mn(II) is found and induced oxidations do not occur. This has been explained in terms of a lowering by R of the redox potential of $R\dot{C}(CO_2H)$ to a value insufficient to reoxidise Mn(II) pyrophosphate in reaction (58).

The free-radical scheme, however, fails to account for the following: (*i*) It cannot be easily generalised to cover the identical kinetics of the Mn(III) sulphate oxidation; if $\cdot CH(CO_2H)$ has an oxidation potential comparable with Mn(III)/Mn(II) pyrophosphate then it cannot appreciably reoxidise Mn(II) sulphate, (*ii*) If $\cdot CH(CO_2H)$ reoxidises Mn(II) sulphate then it should be capable of reoxidising both V(IV) sulphate (of the V(V)/V(IV) pair, potential 1.0 V) and Mn(II) sulphate in the V(V) oxidation of malonic acid; that it does neither can be seen from the rate laws of these oxidations which show no Mn(II)-retardation (*vide infra*). (*iii*) The not dissimilar kinetics of the Mn(III) sulphate oxidation of formic acid (*vide supra*) and mercurous ion[280].

Waters *et al.*[385, 422] have overcome these objections, whilst preserving the principal feature of the "oxidising radical" mechanism, by proposing formation of a malonato Mn(IV) complex stabilised by extensive delocalisation, *viz.*

$$Mn(III) + CH_2(CO_2H)_2 \rightleftharpoons CH_2(CO_2H)_2.Mn(III) \quad \text{(fast)}$$

$$CH_2(CO_2H)_2.Mn(III) + Mn(III) \rightleftharpoons [CH_2(CO_2H)_2Mn(IV)] + Mn(II) \quad \text{(fast)}$$

Mn(IV)$^{2+}$ [cyclic malonato complex: O—C(=O), CH, O—C, O$^-$] ↔ Mn(III)$^+$ [O—C(=O), ĊH, O—C(=O)] ↔ Mn(III)$^+$ [O—C(=O), CH, O—C, O·]

$$[CH_2(CO_2H)_2.Mn(IV)] + H_2O \longrightarrow Mn(II) + HO.CH(CO_2H)_2 + 2H^+ \quad \text{(slow)}$$

The second of the resonance structures is the source of the radical reactivity displayed during oxidation and the Mn(III) in this structure must be low-spin to preserve multiplicity. Substitution at the *meso*-position could provide steric hindrance to analogous decompositions and reactions of the Mn(IV) complexes of ethyl- and benzylmalonic acids, and a conventional one-equivalent oxidation step becomes dominant.

The V(V) oxidation[383] is about 10^7 and 10^3 times slower than those by Mn(III) sulphate and pyrophosphate respectively. The $[V(V)]^2$ dependence indicates two concerted one-equivalent oxidations; one possible mechanism involves oxygen-transfer, *viz.*

```
O   OH                              O   OH
 \C/  V2+(OH)2                       \C/
  |    OH           slow              |
  C                ----->          H-C-OH        + 2V(IV)
H  C-OH                               C-OH
    O                                 ||
(HO)2V2+                              O
```

The absence of retardation by V(IV) rules out a mechanism analogous to that of Mn(III) oxidation. Mn(II) ions strongly catalyse reaction[418], altering the kinetics to those observed[418] for the Mn(II)-catalysed oxidation of oxalic acid by V(V) (preceding sub-section) except that the [V(V)] dependence has a Michaelis–Menten, form rather than being first-order. E is reduced from 19.7 to 6.9 kcal.mole^{-1}, and a similar mechanism is believed to operate.

4.4.11 *Other dicarboxylic acids*

Co(III) perchlorate oxidations of succinic, aspartic, maleic and fumaric acids all obey the rate expression[422a]

$$-\mathrm{d}[\mathrm{Co(III)}]/\mathrm{d}t = k[\mathrm{Co(III)}][\mathrm{substrate}]/[\mathrm{H_3O^+}]$$

although at succinic acid concentrations greater than 0.12 M Michaelis–Menten kinetics were observed.

The oxidation of ethylenediaminetetraacetic acid (EDTA) by $PuO_2{}^{2+}$ and $NpO_2{}^{2+}$ to give the quinquevalent metal ions[671, 672] in perchlorate media is first-order in both oxidant and substrate and the stoichiometry, Δ[M(VI)]/Δ[EDTA], is 6 in both cases. The Np(VI) oxidation[671] shows a fractional dependence on acidity and has parameters E = 23.0 kcal.mole^{-1}, $\Delta S^{\ddagger}$ = 12.3 eu.

4.4.12 *Phenols and hydroquinone*

Rapid oxidation by acidic hexachloriridate of phenol and 2,6-dimethylphenol takes place to give the corresponding phenoxyl radical[423]. At low Ir(III) concentrations this step is rate-determining, with $k_1 = 40$ l.mole^{-1}.sec^{-1} at 20 °C ($E = 10.5$ kcal.mole^{-1} and $\Delta S^{\ddagger} = -18$ eu), for reaction with 2,6-dimethylphenol, and with corresponding values of 8.3×10^8 l.mole^{-1}.sec^{-1}, 3.1 kcal.mole^{-1} and -11 eu for reaction with 2,6-dimethylphenoxide anion. At higher Ir(III) concentrations the reverse of the electron-transfer step becomes significant and the following kinetics are obtained

$$\frac{-\mathrm{d}[\mathrm{Ir(IV)}]}{\mathrm{d}t} = \frac{4\,K_1{}^2k_2[\mathrm{ArOH}]^2[\mathrm{Ir(IV)}]^2}{[\mathrm{H_3O^+}]^2[\mathrm{Ir(III)}]^2} + \frac{4\,K_1{}^2k_3[\mathrm{ArOH}][\mathrm{Ir(IV)}]^2}{[\mathrm{H_3O^+}][\mathrm{Ir(III)}]} + \frac{4\,K_1\,K_4\,k_5[\mathrm{ArOH}]^2[\mathrm{Ir(IV)}]^2}{[\mathrm{H_3O}]^+[\mathrm{Ir(III)}]}$$

where the coefficients refer to the following reactions

$$\mathrm{ArOH + Ir(IV)} \underset{k_{-1}}{\overset{k_1}{\rightleftharpoons}} \mathrm{ArO\cdot + H^+ + Ir(III)} \qquad K_1$$

$$2\,\mathrm{ArO\cdot} \xrightarrow{k_2} \mathrm{dimer} \xrightarrow[\text{fast}]{2\,\mathrm{Ir(IV)}} \text{3,3',5,5'-tetramethyldiphenoquinone}$$

$$\mathrm{ArO\cdot + Ir(IV)} \xrightarrow{k_3} \mathrm{Ir(III) + ArO^+} \xrightarrow[\text{fast}]{2\,\mathrm{Ir(IV)}} \text{2,6-dimethylbenzoquinone}$$

$$\mathrm{ArO\cdot + ArOH} \underset{k_{-4}}{\overset{k_4}{\rightleftharpoons}} \text{dimer radical} \qquad k_4/k_{-4} = K_4$$

$$\text{dimer radical} + \mathrm{Ir(IV)} \xrightarrow{k_5} \mathrm{dimer} \xrightarrow[\text{fast}]{2\,\mathrm{Ir(IV)}} \text{3,3',5,5'-tetramethyldiphenoquinone}$$

Also significant is the reverse of the first step but involving phenoxide ion, *viz.*

$$\mathrm{ArO\cdot + Ir(III)} \xrightarrow{k'_{-1}} \mathrm{ArO^- + Ir(IV)}$$

The following values of the coefficients at 20 °C have been deduced[423]: $k_2 = 5.6 \times 10^9$ l.mole^{-1}.sec^{-1} (ref. 424), $k_3 = 6 \times 10^4$ l.mole^{-1}.sec^{-1}, $k_5K_4 = 6 \times 10^8$ l^2.mole^{-2}.sec^{-1}, $k_{-1} = 4.8 \times 10^6$ l^2.mole^{-2}.sec^{-1}, $k'_{-1} = 2.2 \times 10^3$ l.mole^{-1}.sec^{-1}.

Trisacetylacetonecobalt(III) oxidises phenols to phenoxyl radicals at 71–120 °C in chlorobenzene with simple second-order kinetics, ($E = 33$ kcal.mole^{-1} and $\Delta S^{\ddagger} = 13.6$ eu)[425]. When 2,4,6-*tert*-butylphenol was employed, the characteristic ESR spectrum of the phenoxyl radical was obtained with an intensity corresponding to almost quantitative conversion, *viz.*

$$\mathrm{ArOH + Co(acac)_3 = ArO\cdot + Hacac + Co(acac)_2}$$

An increase in absorption at 605 nm occurred on mixing Co(III) perchlorate and hydroquinone solutions in a stopped-flow apparatus, suggesting complex formation, *cf.* Ce(IV) oxidation. The decay was first-order with k_1 proportional to substrate concentration, but insensitive to acidity, suggesting Co^{3+} rather than $CoOH^{2+}$ to be the active oxidant, and $E_1 = 9.7 \pm 0.5$ kcal.mole^{-1}, $\Delta S^{\ddagger} = 14 \pm 2$ eu[678a].

Oxidation of hydroquinone[425a] by Mn(III) in acid perchlorate media is simple second-order. The acidity dependence indicates both Mn^{3+} and $MnOH^{2+}$ to be effective oxidants with second-order rate coefficients at 25 °C of $(0.48 \pm 0.12) \times 10^4$ l.mole^{-1}.sec^{-1} for Mn^{3+} and $(3.28 \pm 0.43) \times 10^4$ l.mole^{-1}.sec^{-1} for $MnOH^{2+}$. A second investigation[678] included an observation of an increase in absorption at 470 nm on mixing the reactants suggesting fast complex formation; this absorption then decayed with first-order kinetics and with a rate linearly dependent upon hydroquinone concentration. k_2 decreases with increasing acidity in a manner indicating the complex between $MnOH^{2+}$ and hydroquinone to be the exclusive reactant. $E_2 = 14.0 \pm 0.7$ kcal.mole^{-1} and $\Delta S^{\ddagger} = 7.5 \pm 2.1$ eu.

Ce(IV) oxidation of hydroquinone[425b] is first-order in Ce(IV) and zero-order in hydroquinone and the rate is acid-independent in the range 0.2 to 2.0 *M* $HClO_4$. A rapid initial rise in absorption following mixing in a stopped-flow apparatus suggests a high degree of formation of an intermediate complex, which then undergoes first-order decay with $k_1 = 105 \pm 6$ sec^{-1} over the temperature range 18–45 °C, *i.e.* $E = 0$.

4.4.13 Aromatic ethers and amines

Benzyl and phenyl ethers are oxidised by Co(III) with essentially second-order kinetics, but with k_2 inversely dependent upon acidity[426]. The reactivity sequence is

$$C_6H_5OC_6H_5 > C_6H_5CH_2OC_6H_{11} > C_6H_5CH_2OCH_2C_6H_5 > C_6H_5CH_2OCH_3$$

The main oxidation product from dibenzyl ether is benzaldehyde (up to 80 % yield) with smaller amounts of benzyl alcohol and benzoic acid. The rates of oxidation are only slightly affected by major stereochemical changes, and it is considered that an outer-sphere oxidation of the ether is followed by radical breakdown, *viz.*

$$C_6H_5CH_2OCH_2C_6H_5 + Co(III) \rightarrow C_6H_5\dot{C}HOCH_2C_6H_5 + Co(II) + H^+ \quad \text{(slow)}$$

$$C_6H_5\dot{C}HOCH_2C_6H_5 \rightarrow C_6H_5CHO + C_6H_5CH_2\cdot \quad \text{(fast)}$$

$$C_6H_5CH_2\cdot + Co(III) + H_2O \rightarrow C_6H_5CH_2OH + Co(II) + H^+ \quad \text{(fast)}$$

$$C_6H_5CH_2OH + 2\,Co(III) \rightarrow C_6H_5CHO + 2\,Co(II) + 2\,H^+ \quad \text{(fast)}$$

$$C_6H_5CHO + 2\,Co(III) + H_2O \rightarrow C_6H_5CO_2H + 2\,Co(II) + 2\,H^+ \quad \text{(slow)}$$

Benzyl methyl ether produces methyl benzoate, benzaldehyde and benzoic acid but essentially no dibenzyl or benzyl alcohol. The initial radical is, therefore $C_6H_5\dot{C}HOCH_3$ and methyl benzoate is produced *via* the paths

$$C_6H_5\dot{C}HOCH_3 + Co(III).aq \rightarrow C_6H_5CH(OH)OCH_3 + Co(II) + H^+$$

$$C_6H_5CH(OH)OCH_3 + Co(III) \rightarrow C_6H_5\dot{C}(OH)OCH_3 + Co(II) + H^+$$

$$C_6H_5\dot{C}(OH)OCH_3 + Co(III) \rightarrow C_6H_5COOCH_3 + Co(II) + H^+$$

Production of considerable amounts of cyclohexanol and cyclohexanone as well as benzaldehyde and benzoic acid in the oxidation of benzyl cyclohexyl ether shows the primary radical to be $C_6H_5\dot{C}HOC_6H_{11}$. Abstraction from aliphatic C–H bonds cannot occur in the case of diphenyl ether which is oxidised rapidly, and removal of a π-electron is likely.

Dewar *et al.*[427–429] have determined the kinetics for the oxidations in acetic acid of a number of aromatic ethers and amines by manganic acetate. With *p*-methoxytoluene (PMT)[427], for example, the reaction has the following stoichiometry for the main route of oxidation

$$CH_3OC_6H_4CH_3 + 2\,Mn(OCOCH_3)_3 = CH_3OC_6H_4CH_2OCOCH_3 + CH_3CO_2H + 2\,Mn(OCOCH_3)_2$$

and the rate law is

$$\frac{-d[Mn(III)]}{dt} = \frac{k[Mn(III)][PMT]}{[Mn(II)]}$$

k is unaffected by added sodium acetate but deuteration of the CH_3 group reduces k by a factor of 5 at 70 °C. The retardation by Mn(II) is rationalised as follows[427]

$$PMT + Mn(III) \rightleftharpoons \cdot PMT^+ + Mn(II) \quad \text{(fast)}$$

$$\cdot PMT^+ \rightarrow CH_3OC_6H_4CH_2\cdot + H^+ \quad \text{(slow)}$$

$$CH_3OC_6H_4CH_2\cdot + Mn(III) \rightarrow CH_3OC_6H_4CH_2^+ + Mn(II) \quad \text{(fast)}$$

$$CH_3OC_6H_4CH_2^+ + CH_3CO_2H \rightarrow CH_3OC_6H_4CH_2OCOCH_3 + H^+ \quad \text{(fast)}$$

The acetate ion relinquished by the metal atom in the first step probably remains in the vicinity of the radical-cation to promote the removal of the proton in the slow step.

Identical kinetics are exhibited in the analogous oxidations of 1- and 2-methoxynaphthalene to 4-methoxyl-1-naphthyl acetate and 2-methoxy-1,4-naphthoquinone respectively[428]. In these cases the radical-cations may react with acetate ion thus

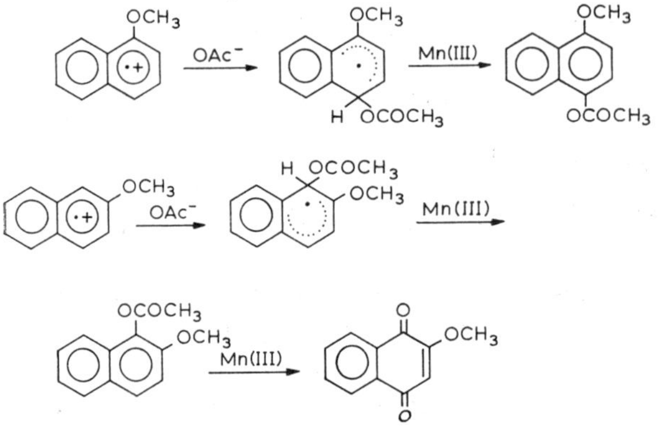

Similar acetoxylations have been reported for several other aromatic ethers, amines and polynuclear hydrocarbons[429] and more complex reaction products have also been identified, although kinetic data are lacking.

The Ce(IV) oxidation of the leuco-base of Malachite Green yields a dye, *viz.*

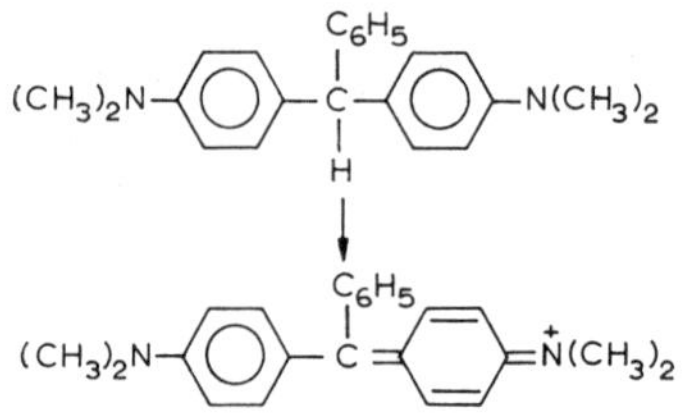

Swain and Hedberg[292] have shown that the tertiary alcohol is not an intermediate, for the dehydration process is slower than the rate of formation of dye. Instead it is proposed that the tertiary hydrogen is removed to give a radical-cation which is further oxidised to the carbonium ion. The oxidation involves two steps; one fast and one very much slower. This parallels the Ce(IV) oxidation of iodide ion and is therefore probably a function of the oxidant.

4.4.14 Thioureas

Oxidations by Co(III)[429a] and Ce(IV)[429b] to the corresponding disulphides require stopped-flow techniques for kinetic study. All oxidations are first-order

TABLE 17

RATE AND ACTIVATION DATA FOR ONE-EQUIVALENT OXIDATIONS OF THIOUREAS

Oxidant	*Thiourea*	*k_2(l.mole^{-1}.sec^{-1})*	*Temperature(°C)*	*E(kcal.mole^{-1})*	*$\Delta S^{\ddagger}$(eu)*
Ce(IV) sulphate	Unsubstituted	163	23.9	7.96±0.2	−32.7±3
	Ethylene-	37.0	26.3	9.61±0.2	−25.4±3
	Diethyl-	403	24.7	8.90±0.5	−28.5±4
	Dimethyl-	490	25.5	9.10±0.5	−29.3±5
	Thiosemicarbazide	227	23.7	9.8 ±0.5	−28.8±3
Co^{3+}	Unsubstituted	7±2	25.0	8.6 ±1.5	−29 ±5
	Ethylene-	20±6	25.0	7.8 ±1.4	−30 ±4
	Diethyl-	36±10	25.0	7.8 ±1.6	−29 ±5
	Dimethyl-	22±4	25.0	7.6 ±1.5	−26 ±4
$CoOH^{2+}$	Unsubstituted	380±20	25.0	9.0 ±0.5	−22 ±3
	Ethylene-	360±30	25.0	10.9 ±0.5	−19 ±3
	Diethyl-	1600±150	25.0	10.6 ±0.4	−12 ±3
	Dimethyl-	3000±150	25.0	9.8 ±0.4	−14 ±3

both in oxidant and substrate. Acidity dependences were determined for the Co(III) oxidations to be of the type

$$k_2 = k_2' + k_2''/[H_3O^+]$$

The two terms are considered to correspond to one-equivalent oxidations by Co^{3+} and $CoOH^{2+}$, respectively, of the monoprotonated thiourea (RSH^+) to give a sulphur radical, *viz.*

$$Co^{3+} + RSH^+ \xrightarrow{k_1} Co^{2+} + \cdot RS^+ + H^+$$

$$CoOH^{2+} + RSH^+ \xrightarrow{k_2} Co^{2+} + \cdot RS^+ + H_2O$$

$$2\,\cdot RS^+ \rightarrow RSSR^{2+}$$

The Ce(IV) oxidation occurs by the first type of path (k_1). Kinetic data are collated in Table 17.

5. Oxidation by Fe(III), Ag(I), Cu(II), Cu(I), Np(V) and Mo(V)

These are thermodynamically relatively weak oxidants (Table 18) and their action is relatively restricted, for example, to inorganic ions of moderate reducing power such as iodide, to polyfunctional organic compounds such as hydroxy-acids, and, in the cases of Ag(I) and Cu(II), to CO and H_2. Fe(III) is particularly affected by hydrolysis and all these oxidants form complexes with suitable ligands. Cyanide ion and 1,10-phenanthroline form strong complexes with Fe(III) which greatly affect its behaviour. Tris-1,10-phenanthrolineiron(III) (ferriin) displays

TABLE 18

REDOX POTENTIALS OF RELATIVELY WEAK ONE-EQUIVALENT OXIDANTS
(phen) = 1,10-orthophenanthroline.

Reaction	*Medium*	π^0 (*Volt*)	*Ref.*
$Fe^{3+} + e^- = Fe^{2+}$	HCl (corrected for $FeCl^{2+}$)	0.772	430
	$HClO_4$	0.772	430
$Fe(CN)_6^{4-} + e^- = Fe(CN)_6^{3-}$	pH > 6 (K^+, counter-ion)	0.355	431
$Fe(phen)_3^{3+} + e^- = Fe(phen)_3^{2+}$	1 *M* H_2SO_4	1.06	19
$Ag^+ + e^- = Ag$	(calculated value)	0.799	19
$Cu^{2+} + 2e^- = Cu$	(calculated value)	0.337	19
$Cu^+ + e^- = Cu$	(calculated value)	0.521	19
$Cu^{2+} + e^- = Cu^+$	(calculated value)	0.153	19
$NpO_2^+ + e^- = Np(IV)$	1.0 *M* HCl	0.737	432
$Mo(CN)_6^{3-} + e^- = Mo(CN)_6^{4-}$	Natural pH, extrapolated to zero ionic strength	0.7260	432a

a quite different reactivity towards compounds such as pinacol, oxalic acid and acetoin from other one-equivalent oxidants[433]. The difference probably stems from the inability of the substitution-inert ferriin to form chelate complexes with these substrates.

Ag(I) and Cu(II) are readily used in neutral solution, although oxidations by Ag(I) are often affected by the deposition of silver metal. Theoretically Cu(II) can behave either as a two- or a one-equivalent reagent, but it usually functions as the latter[6]. The tendency of Cu(I) to disproportionate is normally suppressed by working with solvents such as pyridine which form stable complexes with this oxidation state.

5.1 OXIDATION OF INORGANIC MOLECULES

5.1.1 Halide ions

The kinetics of the oxidation of iodide ion by Fe(III) have been investigated by many groups. The early studies are unreliable as guides to mechanism because of the disregard of the effects of hydrolysis, complexing, ionic strength, *etc.* Fudge and Sykes[434] rectified these errors by using ferric nitrate in moderately strong acid at constant ionic strength. The rate law was determined to be

$$\frac{-\mathrm{d}[\mathrm{Fe(III)}]}{\mathrm{d}t} = \frac{k_1[\mathrm{Fe(III)}][\mathrm{I^-}]^2}{1+k_2[\mathrm{Fe(II)}]/[\mathrm{Fe(III)}]}$$

and a mechanism was proposed compatible with this law, *viz.*

$$Fe(III) + I^- \rightleftharpoons FeI^{2+}$$

$$FeI^{2+} + I^- \rightleftharpoons Fe^{2+} + \cdot I_2^-$$

$$Fe(III) + \cdot I_2^- \rightleftharpoons Fe^{2+} + I_2$$

Specific retardation of this reaction by ions able to form weakly oxidising complexes with Fe(III) has been noted[435].

Identical kinetics have been noted for the oxidation of iodide by ferricyanide ion[436-438]. However, the mechanism cannot be identical because $Fe(CN)_6^{3-}$ is substitution-inert. A one-equivalent oxidation of I^- to an iodine atom would be followed by the rapid reactions

$$I\cdot + I^- \rightleftharpoons \cdot I_2^-$$

$$Fe(CN)_6^{3-} + \cdot I_2^- \rightleftharpoons Fe(CN)_6^{4-} + I_2$$

but this does not lead to the observed kinetics. However, the hypothesis of outer-sphere electron flow is supported by the observed catalysis by added platinum metals[439].

The reaction is specifically catalysed by hydronium ion[440]. This catalysed path shows a rate law

$$-d[Fe(CN)_6^{3-}]/dt = k[Fe(CN)_6^{3-}][I^-][H_3O^+]$$

Indelli and Guaraldi[440] propose the acid-independent path to involve eight-coordinate iron (six CN^- plus two I^- ligands) and the acid-catalysed path to involve seven-coordinate iron, *viz.*

$$\left[H \begin{matrix} \cdots NC \\ \cdots NC \end{matrix} \rangle Fe(CN)_4I \right]^{3-}$$

Specific accelerating effects of certain cations have been noted[438, 440], following a sequence

$$La^{3+} > Cs^+ > NH_4^+ > K^+ > Na^+ > Li^+ > NEt_4^+$$

and a detailed examination of the effect of variation of $[K^+]$ at constant ionic strength reveals a law

$$-d[Fe(CN)_6]/dt = k[Fe(CN)_6^{3-}][I^-][K^+]$$

References pp. 493–509

It is proposed[438] that one pre-equilibrium gives the ion-pair $[KFe(CN)_6]^{2-}$ which may form an ion-triplet in a second fast pre-equilibrium, *viz.*

$$\overset{-}{[I}\overset{+}{K}\overset{3-}{Fe(CN)_6}]^{3-}$$

The ion-triplet oxidises I^- in the slow step to yield $\cdot I_2^-$. The non-participation of any ligand substitution step is confirmed by the absence of any incorporation of activity from added ^{14}C-labelled free cyanide ion into the product ferrocyanide[441]

Specific catalysis by sodium ions is also found in the oxidation of iodide ion by octacyanomolybdate(V) which otherwise shows simple, second-order kinetics[441a] with $E = 5.68 \pm 2.57$ kcal.mole^{-1} and $\Delta S^{\ddagger} = -39 \pm 8.5$ eu and $k_2 = 3.52 \pm 0.13$ l.mole^{-1}.sec^{-1}, at 25.7 °C and $\mu = 0.1$ M. Mo(IV), which is produced stoichiometrically exerts slight retardation upon the reaction. An outer-sphere one-electron transfer is proposed.

Essentially identical results were obtained by a second group[682] with $k_2 = 1.4 \pm 0.3$ l.mole^{-1}.sec^{-1} at 25 °C ($\mu = 0.02$ M). Specific cation effects were also found with rate sequences $K^+ > Na^+ \sim N(C_2H_4OH)_4{}^+ > Li^+$ and positive deviations of the reaction orders above unity were also noted.

The oxidation of iodide ion by Np(V), which exists in solution mainly as NpO_2^+, *viz.*

$$2\,NpO_2^+ + 2\,I^- = 2\,Np(IV) + I_2$$

follows kinetics[442]

$$d[Np(IV)]/dt = k[NpO_2^+][I^-]^{\frac{3}{2}}[H^+]^{\frac{5}{2}}$$

where, at 25 °C ($\mu = [H^+] = 3.29$ M), k equals 1.71×10^{-3} l^4.mole^{-4}.sec^{-1} and E equals 28.3 kcal.mole^{-1}. The fractional orders are not significant and the data are equally well represented by a law

$$d[Np(IV)]/dt = k[NpO_2^+][H^+]^2[I^-] + k_2[NpO_2^+][H^+]^3[I^-]^2$$

This suggests little more than the importance of protonated forms of Np(V).

5.1.2 Pseudohalides; cyanide ion

The oxidation of CN^- ion by ferricyanide follows the course[443]

$$2\,Fe(CN)_6{}^{3-} + CN^- + 2\,OH^- = 2\,Fe(CN)_6{}^{4-} + CNO^- + H_2O$$

The kinetics are complex, but the following features are apparent[437]: (*a*) order in $Fe(CN)_6^{3-}$ between 1 and 2, tending to 1 in presence of $Fe(CN)_6^{4-}$; (*b*) first-order retardation by $Fe(CN)_6^{4-}$; (*c*) order in CN^- between 1 and 2; (*d*) rate independent of pH between pH 9 and 11; (*e*) E zero or negative. The following mechanism was proposed[437]

$$Fe(CN)_6^{3-} + 2\,CN^- \rightleftharpoons Fe(CN)_6^{4-} + \cdot(CN)_2^-$$

$$\cdot(CN)_2^- + xH_2O \rightarrow P + CN^-$$

$$Fe(CN)_6^{3-} + P \rightarrow Fe(CN)_6^{4-} + Q$$

The nature of P and Q remains unestablished.

The stoichiometry of the oxidation of CN^- ion by Cu(II) is[444]

$$8\,CN^- + 2\,Cu^{2+} = 2\,Cu(CN)_3^{2-} + (CN)_2$$

and the rate law in ammoniacal aqueous solution was found to be[444]

$$-\mathrm{d}[Cu(II)]/\mathrm{d}t = k[Cu^{2+}][CN^-]^4$$

A complex was detected in these solutions with the composition $CuCN^+$ but this was not formed in significant quantities under the conditions employed. The role of ammonia molecules could not be assessed, although it was found that the rate showed a strong inverse dependence in ammonia concentration, indicating competition between NH_3 and CN^- for coordination sites.

5.1.3 Pseudohalides; thiocyanate ion

The thermal oxidation of thiocyanate by Fe(III) in a perchlorate medium involves two thiocyanato complexes of Fe(III)[445]. The stoichiometry is thought to be

$$6\,Fe(III) + SCN^- + 4\,H_2O = 6\,Fe(II) + CN^- + SO_4^{2-} + 8\,H^+$$

and the rate law is given as

$$\frac{\mathrm{d}[Fe(II)]}{\mathrm{d}t} =$$

$$\frac{2\,Fe(III)\{k_1 K_1[SCN^-]^2 + k_2 K_1 K_2[SCN^-]^3\}}{\{1 + K_1[SCN^-] + K_1 K_2[SCN^-]^2\}} \cdot \frac{1}{\{1 + k_3[Fe(II)]/k_4[Fe(III)]\}}$$

References pp. 493–509

At 25 °C, $k_1 = 1.8\times10^{-5}$ l.mole^{-1}.sec^{-1}, $k_2 = 1.4\times10^{-4}$ l.mole^{-1}.sec^{-1}, $E_1 = 26.2$ kcal.mole^{-1}, $E_2 = 27.0\pm3.0$ kcal.mole^{-1}, $k_3/k_4 = 3.5\pm1.0$ (independent of temperature), $K_1 = 114$ ($\mu = 1.28$ M) and $K_2 = 20\pm5$ (both in l.mole^{-1} units).

The following mechanism was proposed

$$Fe^{3+} + SCN^- \rightleftharpoons FeSCN^{2+} \qquad (K_1)$$

$$FeSCN^{2+} + SCN^- \rightleftharpoons Fe(SCN)_2^+ \qquad (K_2)$$

$$FeSCN^{2+} + SCN^- \rightarrow Fe^{2+} + \cdot(SCN)_2^- \qquad \text{(slow)} \tag{60}$$

$$Fe(SCN)_2^+ + SCN^- \rightarrow Fe^{2+} + \cdot(SCN)_2^- + SCN^- \qquad \text{(slow)} \tag{61}$$

$$Fe^{2+} + \cdot(SCN)_2^- \rightarrow Fe^{3+} + 2\,SCN^- \tag{62}$$

$$Fe(III) + \cdot(SCN)_2^- \rightarrow Fe^{2+} + (SCN)_2 \tag{63}$$

$$H_2O + (SCN)_2 \rightarrow CN^-, SO_4^{2-}, \textit{etc.} \tag{64}$$

Reaction (61) implies that the transition state contains 3 SCN^- ions. A steady state approximation for $[\cdot(SCN)_2^-]$ leads to the observed rate law. A temperature variation study of K_1 and K_2 is included in this paper. It was also concluded that simple redox breakdown of $FeSCN^{2+}$ is of negligible importance.

5.1.4 Pseudohalides; azide ion

N_3^- is not oxidised by Fe(III)[446], although a statement to the contrary has appeared[447].

5.1.5 Hydrogen peroxide

The rate law of the oxidation by Fe(III) is dependent on the ratio of the concentrations of the reactants. When peroxide is in excess and when the acidity is sufficient to suppress hydrolysis of Fe(III) the rate expression is[448–450]

$$-d[H_2O_2]/dt = k[Fe(III)][H_2O_2]/[H_3O^+]$$

Haber and Weiss[450] proposed a one-equivalent oxidation of HO_2^- to $HO_2\cdot$ later modified by Weiss[451] to

$$Fe(III) + HO_2^- \rightleftharpoons Fe(II) + HO_2\cdot \tag{65}$$

$$HO_2\cdot + H_2O_2 \rightarrow O_2 + H_2O + OH\cdot \tag{66}$$

$$OH\cdot + H_2O_2 \rightarrow H_2O + HO_2\cdot \tag{67}$$

$$OH\cdot + Fe(II) \rightarrow Fe(III) + OH^- \tag{68}$$

However, Andersen[452] noted a new rate law at lower values of $[H_2O_2]/[Fe(III)]$ with a higher order dependence on peroxide.

The reaction scheme as presented is of a non-chain variety provided OH· is removed by Fe(II) rather than by peroxide. Barb *et al.*[448] have included further steps in the scheme, *viz.*

$$Fe(II)+H_2O_2 \rightarrow Fe(III)+OH\cdot+OH^- \tag{69}$$

$$Fe(III)+HO_2\cdot \rightarrow O_2+H^+ +Fe(II) \tag{70}$$

but have eliminated step (66). Application of steady-state kinetics to Fe(II), OH· and $HO_2\cdot$ produces the "normal" rate law at high $[H_2O_2]/[Fe(III)]$ under which conditions reactions (67) and (68) are unimportant. At low $[H_2O_2]/[Fe(III)]$ steady-state treatment yields

$$-d[H_2O_2]/dt = k[Fe(III)]^{\frac{1}{2}}[H_2O_2]^{\frac{3}{2}}$$

Subsequent analysis[448] of Andersen's data confirms a $[H_2O_2]^{\frac{3}{2}}$ dependence but not a $[Fe(III)]^{\frac{1}{2}}$ dependence, and this has been confirmed for these conditions by Barb *et al.*[448] who obtained a first-order dependence on oxidant concentration.

Several groups[453–456] confirmed spectroscopically the existence of an ion-pair complex $Fe^{3+}\cdot HO_2^-$, which may decompose unimolecularly at high $[H_2O_2]/[Fe(III)]$ but become transformed to FeO^{3+} at lower $[H_2O_2]/[Fe(III)]$. FeO^{3+} is attacked by a second molecule of H_2O_2 to provide an alternative route for decomposition, *viz.*

$$Fe^{3+}+HO_2^- \rightleftharpoons [FeHO_2]^{2-}$$

$$[Fe\cdot HO_2]^{2-} \rightarrow FeO^{3+}+OH^-$$

$$FeO^{3+}+H_2O_2 \rightarrow Fe^{3+}+H_2O+O_2$$

A complex of the type $M\cdot HO_2^-$ is also involved in the rather slower haemin-catalysed decomposition of H_2O_2[457].

A detailed physical examination of the purple complex formed in alkaline solution between Fe(III), ethylenediaminetetraacetic acid (EDTA) and peroxide shows it to have a composition $[Fe^{III}(EDTA)O_2]^{3-}$ ($\log K_{1\,3^\circ C} = 4.33$). This complex catalyses decomposition of peroxide, the rate–pH profile going through a maximum at pH 9–10[689].

Ferricyanide catalyses the decomposition of H_2O_2 at pH 6–8. The reaction is first-order in peroxide but the dependence on oxidant concentration is complex[457a]. The Arrhenius plots were curved but averaged values indicated that at pH 7 and 8, respectively, E is 25 ± 2 and 11 ± 1.5 kcal.mole^{-1}. Clearly several processes are contributing to the overall reaction.

References pp. 493–509

Oxidation by the 1 : 1 chelate of Cu(II) with histamine (Hm)

$CH_2CH_2NH_2$

N NH

has been studied[457b] as a model for the catalytic activity of copper-containing proteins. The overall reaction rate at pH 7.0 is given by

$$d[O_2]/dt = k_{obs}[Cu(II)]_T[H_2O_2]_T$$

where the subscripts denote total molar concentrations. Variation of rate with pH indicates that

$$k_{obs} = k_1[CuHm^{2+}][HO_2^-]+k_2[CuHm(HO_2^-)^+][HO_2^-]$$

with $k_1 = 50$ and $k_2 = 45$ l.mole^{-1}.sec^{-1} at 25 °C. The first term probably corresponds to a conventional radical-chain mechanism, *cf.* Fe(III) oxidation, but the second term indicates participation of a molecular mechanism.

5.1.6 Thiosulphate ion

The oxidation by Fe(III)[458], *viz.*

$$2\,Fe(III)+2\,S_2O_3^{2-} = 2\,Fe(II)+S_4O_6^{2-}$$

involves the formation of a spectroscopically detectable[459, 460] but labile complex $FeS_2O_3^+$ which reacts with further thiosulphate giving rise to the observed rate law

$$\text{rate} = \frac{k[FeS_2O_3^+]^{\frac{3}{2}}[S_2O_3^{2-}]^{\frac{1}{2}}}{f[Fe(II)]}$$

The mechanism proposed is

$$Fe(III)+S_2O_3^{2-} \rightleftharpoons FeS_2O_3^+ \qquad \text{(fast)}$$

$$FeS_2O_3^+ + S_2O_3^{2-} \rightleftharpoons FeS_2O_3 + \cdot S_2O_3^- \qquad \text{(slow)}$$

$$FeS_2O_3^+ + \cdot S_2O_3^- \rightarrow Fe(II)+S_4O_6^{2-} \qquad \text{(slow)}$$

The slow oxidation of $\cdot S_2O_3^-$ required to explain the kinetics is most anomalous; radical-anions are normally most powerful reducing agents, reaction being diffusion-controlled. The retardation by Fe(II) is explained in terms of competition with Fe(III) for $S_2O_3^{2-}$.

5.1.7 *Sulphurous acid*

H_2SO_3 is oxidised to two sets of products by both ferric ion and ferricyanide depending upon the reaction conditions, *e.g.*

$$H_2SO_3+2\,Fe(III)+H_2O = 2\,Fe(II)+4\,H^+ +SO_4^{2-}$$

$$2\,H_2SO_3+2\,Fe(III) = 2\,Fe(II)+4\,H^+ +S_2O_6^{2-}$$

When a large excess of ferric ion was employed to promote the sulphate-forming reaction under nitrogen, it was found[461] that a plot of $t_{\frac{1}{2}}$ *versus* $\{[Fe(II)]_0+2\,[H_2SO_3]_0\}$ is a straight line and in general the rate equation

$$[Fe(III)]^2 t_{\frac{1}{2}} = k[H_3O^+]\{[Fe(II)]_0+2\,[H_2SO_3]_0\}+k'[Fe(III)][H_3O^+]$$

expresses the observed data very adequately. Such an equation may be obtained by applying a steady-state approximation to $HSO_3\cdot$ in the following mechanism[462]

$$H_2SO_3 \rightleftharpoons H^+ +HSO_3^-$$

$$Fe(III)+HSO_3^- \rightleftharpoons Fe(HSO_3)^{2+}$$

$$Fe(HSO_3)^{2+} \rightarrow Fe(II)+HSO_3\cdot$$

$$HSO_3\cdot+Fe(II) \rightarrow Fe(III)+HSO_3^-$$

$$HSO_3\cdot+Fe(III) \rightarrow SO_4^{2-}+Fe(II)+H^+$$

Colour changes indicate the role of the bisulphito complex.

Higginson and Marshall[462] have examined the catalytic effect upon this reaction of Cu(II) ions; these oxidise $HSO_3\cdot$ about 10^2 times faster than does Fe(III).

Vepřek-Šiška and Wagnerová[12] have obtained simpler kinetics with ferricyanide, *viz.*

$$-d[SO_3^{2-}]/dt = k_2[Fe(CN)_6^{3-}][SO_3^{2-}]$$

with k_2 strongly dependent upon pH, becoming virtually zero at pH 3.5 but at pH 8.4 being given by $1.4\times10^8\ \exp(-14.0\times10^3/\boldsymbol{RT})$ l.mole^{-1}.sec^{-1}. These authors object to free radical intermediates on the grounds of non-production of dimeric products and propose a scheme

$$Fe(CN)_6^{3-} + SO_3^{2-} \rightleftharpoons Fe(CN)_5SO_3^{4-} + CN^- \quad \text{(slow)}$$

$$Fe(CN)_5SO_3^{4-} + Fe(CN)_6^{3-} \rightleftharpoons Fe(CN)_5SO_3^{3-} + Fe(CN)_6^{4-} \quad \text{(fast)}$$

$$Fe(CN)_5SO_3^{3-} + H_2O \rightleftharpoons Fe(CN)_5^{3-} + H_2SO_4 \quad \text{(fast)}$$

$$Fe(CN)_5^{3-} + CN^- \rightarrow Fe(CN)_6^{4-} \quad \text{(fast)}$$

Swinehart[463] believes that specific ion-pairing effects are important with both Na^+ and K^+ and that the rate law should be written

$$-d[Fe(CN)_6^{3-}]/dt = k_3[Fe(CN)_6^{3-}][SO_3^{2-}][M^+]$$

The ion-pairs $[KFe(CN)_6]^{2-}$ and $[NaSO_3]^-$ are regarded as mechanistically significant and a radical mechanism is preferred, *viz.*

$$Fe(CN)_6^{3-} + SO_3^{2-} \rightarrow Fe(CN)_6^{4-} + \cdot SO_3^-$$

$$\cdot SO_3^- + Fe(CN)_6^{3-} \rightarrow SO_3 + Fe(CN)_6^{4-}$$

$$2 \cdot SO_3^- \rightarrow S_2O_6^{2-}$$

Vepřek-Šiška[463a] ascribes the accelerative effects of "inert" electrolytes to catalysis by trace quantities of Cu^{2+} ions.

Wiberg *et al.*[441] have performed the reaction in the presence of ^{14}C-labelled cyanide ion and find no incorporation of activity into product ferrocyanide. Evidently the reversible ligand displacement proposed by the Czech workers does not take place and the electron-transfer scheme of Swinehart is preferable. Recent spectroscopic studies[463a] indicate that a complex $[Fe(CN)_5(CNSO_3)]^{5-}$ functions as an intermediate in this reaction.

5.1.8 *Hypophosphorous acid*

The oxidations by Cu(II)[464] and Ag(I)[465] are zero-order in oxidant and completely analogous to that by Hg(II) (p. 334). They involve a slow, acid-catalysed tautomerisation of H_2POOH to $HP(OH)_2$ followed by a rapid attack on the latter by the oxidant.

5.1.9 *Persulphate ion*

The reactions of this species with Ag(I) and Cu(II) are, of course, *oxidations* of these ions and are discussed in the appropriate section.

5.1.10 Phosphorothioic acid

H_3PO_3S (denoted PS) is oxidised by ferricyanide to a dimer[466], *viz.*

$$2\,H_3PO_3S+2\,Fe(CN)_6^{3-} \rightleftharpoons H_2PO_3S\text{–}SO_3PH_2+2\,Fe(CN)_6^{4-}+2\,H^+$$

This is an example of a reversible reaction; the standard electrode potential of the 2 PS/PSSP+2e^- couple is zero at pH 7. The oxidation kinetics are simple second-order and the presence of a radical intermediate (presumably PS·) was detected. Reaction occurs in the pH range 5 to 13 with a maximum rate at pH 6.2, and the activation energy above 22 °C is zero. The ionic strength dependence of k_2 afforded a value for $z_A z_B$ of 9 from the Brønsted relation

$$\log k_2 = \log k_0 + 1.02\, z_A z_B \sqrt{\mu}$$

Several of these features remain unexplained but it is clear that here we have an example of a relatively well-behaved reversible electron transfer reaction involving radical intermediates.

5.1.11 Hydrazine

The oxidation by simple salts of Fe(III) has been studied by several groups. Two overall routes are apparent

$$N_2H_4 \xrightarrow{-e^-} 1/2\,N_2 + NH_3 + H^+$$
$$N_2H_4 \xrightarrow{-4e^-} N_2 + 4\,H^+$$

The mechanism put forward is complex[467–469] (Cahn and Powell[467] omit k_{-1}), *viz.*

$$Fe(III)+N_2H_4 \underset{k_{-1}}{\overset{k_1}{\rightleftharpoons}} Fe(II)+N_2H_3\cdot+H^+ \tag{71}$$

$$2\,N_2H_3\cdot \xrightarrow{k_2} N_4H_6 \xrightarrow{\text{fast}} N_2+2\,NH_3 \tag{72}$$

$$2\,N_2H_3\cdot \xrightarrow{k_3} N_2H_2+N_2H_4 \tag{73}$$

$$N_2H_2 \xrightarrow[-(2\,e^-)]{\text{fast}} N_2$$

$$Fe(III)+N_2H_3\cdot \xrightarrow{k_4} N_2H_2+Fe(II) \xrightarrow[-(2\,e^-)]{\text{fast}} N_2 \tag{74}$$

Application of the steady-state approximation to $[N_2H_3\cdot]$ gives

$$\frac{d[Fe(II)]/dt}{d[NH_3]/dt} = 1+\frac{2\,k_3}{k_2}+\frac{4\,k_4[Fe(III)]}{\sqrt{k_2 d[NH_3]/dt}}$$

References pp. 493–509

This expression is closely adhered to in a chloride ion medium[468] with k_3/k_2 (60 °C) = 0.015±0.015. Cahn and Powell[467] find $k_3/k_2 = 0.15$ in a sulphate medium at 50 °C; under similar conditions Higginson and Wright[468] conclude that $k_3/k_2 = 0$ to 0.02 and believe the difference arises from the neglect by Cahn and Powell of the reversibility of the initial step for which $k_1 = 1.33 \times 10^{-4}$ $\text{l.mole}^{-1}.\text{sec}^{-1}$ (sulphate medium, 50 °C).

Addition of cupric ion[467,469] greatly increases the relative amount of 4-equivalent oxidation, both by catalysis of (74) and by reducing the back-reaction in step (71). Analysis of the effect of Cu(II) confirms the importance of this reversibility[469].

The role of di-imine is supported by ^{15}N-labelling of N_2H_4 (ref. 470). The one-equivalent route produces equal quantities of N_2 derived from the same N_2H_4 molecule and from two different N_2H_4 molecules. Each molecule of N_2 produced in the four-equivalent route originates from *one* N_2H_4 molecule.

Hydrazine reduction of ferric ethylenediaminetetraacetate (denoted FeY^-) gives nitrogen in high yield[471]. In the absence of added phenanthroline the reaction, initially rapid, slows down and becomes second order in FeY^-. When ferrous ethylenediaminetetraacetate (FeY^{2-}) is initially present, the reaction is second-order in FeY^- throughout. In the presence of phenanthroline the order in FeY^- is one throughout. Under both sets of conditions the reaction is first-order in hydrazine.

The reaction is discussed in terms of the scheme (71)–(74) for oxidation by simple complexes of Fe(III) except that the one-equivalent oxidation to NH_3 is disregarded. A steady state treatment for $[N_2H_3\cdot]$ leads to

$$\frac{-d[FeY^-]}{dt} = \frac{4\,k_1\,k_4[FeY^-]^2[N_2H_4]}{k_4[FeY^-]+k_{-1}[FeY^{2-}]}$$

In the presence of sufficient FeY^{2-} then $k_{-1}[FeY^{2-}] \gg k_4[FeY^-]$ and the kinetics observed with added FeY^{2-} result. Phenanthroline removes all Fe(II) from solution thereby suppressing the back-reaction in (71) and the change in order is explained. Cu(II) exerts a catalytic effect, as in the oxidation by ferric ion, by oxidising $N_2H_3\cdot$, thereby reducing the importance of the back-reaction.

A few data exist on the oxidation by ferricyanide. This is simple second-order[471] (in oxidant and *neutral* hydrazine), and leads to quantitative production of nitrogen[16,473] in accordance with scheme (71)–(74) with $k_4 \gg k_3$ and k_{-1}. No scrambling occurs during oxidation of ^{15}N-labelled N_2H_4 indicating that all N_2 is formed *via* di-imine[470]. Di-imine so prepared is capable of hydrogenating added unsaturated compounds, for example, phenylpropiolic acid gives *cis*-cinnamic acid[474].

5.1.12 *Hydroxylamine*

The oxidation by Fe(III) chloride is kinetically complex[475]. Only in the presence of excess acid and Fe(II) was a rate law discernible, *viz.*

$$-\mathrm{d}[\mathrm{Fe(III)}]/\mathrm{d}t = k[\mathrm{NH_2OH}]_{\cdot \mathrm{otal}}[\mathrm{Fe(III)}]^2/[\mathrm{H_3O^+}]^2[\mathrm{Fe(II)}]$$

The stoichiometry is represented by

$$4\,\mathrm{Fe(III)}+2\,\mathrm{NH_3OH^+} = 4\,\mathrm{Fe(II)}+6\,\mathrm{H^+}+\mathrm{N_2O}+\mathrm{H_2O}$$

The reaction shows a rather high activation energy of 32.0 kcal.mole^{-1}. Speculation about the mechanism is worthless until more controlled experiments are performed, taking account of hydrolysis, the role of the reaction Fe(II)+NH_2OH, *etc.*
The oxidation by Ag(I) sulphate[476] is complicated by the deposition of metallic silver according to the stoichiometry

$$2\,\mathrm{Ag^+}+2\,\mathrm{NH_2OH} = 2\,\mathrm{Ag}+\mathrm{N_2}+2\,\mathrm{H_3O^+}$$

A dependence of d$[N_2]$/dt upon $[\mathrm{Ag(I)}]^2$ is considered to result from heterogeneous catalysis.

5.1.13 *Carbon monoxide*

The oxidation of CO by Cu(II) aq. to $[\mathrm{Cu{-}CO{-}OH}]^+$ (*plus* H^+) has been proposed[477] as an initial step in the Cu(II)-catalysed oxidation of CO by molecular oxygen in aqueous solution at 120 °C (7–20 atm) which displays kinetics

$$\frac{-\mathrm{d}[\mathrm{O_2}]}{\mathrm{d}t} = \frac{k[\mathrm{O_2}][\mathrm{Cu(II)}][\mathrm{CO}]}{[\mathrm{H^+}]}$$

The rapid insertion of CO, which recalls the Hg(II)–CO reaction (p. 334), is followed by

$$[\mathrm{Cu{-}CO{-}OH}]^+ + \mathrm{O_2} \rightarrow \mathrm{Cu(II)}+\mathrm{CO_2}+\mathrm{HO_2^-} \qquad \text{(slow)}$$

$$2\,\mathrm{H^+} + [\mathrm{Cu{-}CO{-}OH}]^+ + \mathrm{HO_2^-} \rightarrow \mathrm{Cu(II)}+\mathrm{CO_2}+2\,\mathrm{H_2O} \qquad \text{(fast)}$$

The direct reduction of Cu(II) acetate to Cu(I) by CO at high pressures (up to 1360 atm) in aqueous solution at 120 °C shows several kinetic paths, the rate expression being[477a]

$$\frac{-\mathrm{d}[\mathrm{Cu(II)}]}{\mathrm{d}t} = k_1 \frac{[\mathrm{Cu(II)}][\mathrm{Cu(I)}]}{[\mathrm{H_3O^+}]} + k_2 \frac{[\mathrm{Cu(II)}]^2}{[\mathrm{H_3O^+}]} \cdot P_{\mathrm{CO}} + k_3[\mathrm{Cu(II)}]^2 P_{\mathrm{CO}}$$

where, $k_1 = (4.2 \pm 0.4) \times 10^{-6}$ sec^{-1}, $k_2 = (1.4 \pm 0.3) \times 10^{-9}$ sec^{-1} atm^{-1} and $k_3 = (1.15 \pm 0.17) \times 10^{-4}$ l.mole^{-1}.sec^{-1}.atm^{-1} at 119.4 °C ($\mu = 0.050$ *M*, NaOAc). k_2 is dominant and $E_2 = 25.8 \pm 1.0$ kcal.mole^{-1}. The three terms correspond to the reactions

$$(k_1) \qquad \mathrm{Cu(I)\text{–}CO_2H + Cu(II) \rightarrow CO_2 + CuH^+ + Cu(I)}$$

$$(k_2) \qquad \mathrm{Cu(II)\text{–}CO_2H^+ + Cu(II) \rightarrow CO_2 + 2\ Cu(I) + H^+}$$

$$(k_3) \qquad \mathrm{Cu(II)\text{–}CO_2H_2{}^{2+} + Cu(II) \rightarrow CO_2 + 2\ Cu(I) + 2\ H^+}$$

each of which is preceded by equilibria between the metal ion, CO and H_2O, *e.g.*

$$\mathrm{Cu(II) + CO + H_2O \rightleftharpoons (Cu\text{–}CO\text{–}OH)^+ + H^+}$$

5.1.14 *Molecular hydrogen*

Cu(II) and Ag(I), in common with other oxidising ions with d^9 or d^{10} configurations react readily with H_2 in aqueous solution, giving the following rate laws:

(*a*) cupric perchlorate[478]

$$-\mathrm{d}[\mathrm{H_2}]/\mathrm{d}t = k_2[\mathrm{Cu^{2+}}][\mathrm{H_2}] \tag{75}$$

$$k_2 = 1.7 \times 10^{11} \exp(-26.6 \times 10^3/\boldsymbol{RT})\ \text{l.mole}^{-1}.\text{sec}^{-1}$$

(*b*) cupric acetate[479]

$$-\mathrm{d}[\mathrm{Cu(OAc)_2}]/\mathrm{d}t = k_2[\mathrm{Cu(OAc)_2}][\mathrm{H_2}] \tag{76}$$

$$k_2 = 1.6 \times 10^{12} \exp\{-24.2 \pm 0.8) \times 10^3/\boldsymbol{RT}\}\ \text{l.mole}^{-1}.\text{sec}^{-1}.$$

(*c*) silver perchlorate (see also p. 291)[63, 480]

$$-\mathrm{d}[\mathrm{H_2}]/\mathrm{d}t = k_3[\mathrm{Ag^+}]^2[\mathrm{H_2}] + \frac{k_1[\mathrm{Ag^+}]^2[\mathrm{H_2}]}{[\mathrm{Ag^+}] + k_{-1}/k_2[\mathrm{H^+}]} \tag{77}$$

with $k_3 = 6.8 \times 10^7 \exp(-14.7 \times 10^3/\boldsymbol{RT})$ l^2.mole^{-2}.sec^{-1}, $k_1 = 1.2 \times 10^{12}$ exp

$\{(-24\pm2)\times10^3/\boldsymbol{RT}\}$ l.mole^{-1}.sec^{-1} and $k_{-1}/k_2 = 2.3\times10^7 \exp\{(-14\pm4) \times10^3/\boldsymbol{RT})\}$.

Expressions (75) and (77) were deduced from essentially catalytic experiments when the gross oxidant was dichromate and the concentrations of Ag(I) or Cu(II) remained unchanged. Under these conditions the disappearance of Cr(VI) is zero-order. The $[Ag_2H_2]^{2+}$ complex may be attacked rapidly by Cr(VI) but the study with cupric acetate shows that the $[CuH_2]^{2+}$ complex can break down in a redox manner, possibly through attack by further Cu(II). Cr(VI) oxidation of H_2 is catalysed by cupric acetate[481], with an activation energy of 24.6 ± 0.6 kcal.mole^{-1} and with a rate coefficient in reasonable agreement with that for the direct oxidation, indicating a common slow step involving activation of H_2 by $Cu(OAc)_2$. Solvent isotope effects have been determined[66], and are given below.

Oxidant	Cu^{2+}	$Cu(OAc)_2$	Ag^+
k_{H_2O}/k_{D_2O} (±0.1)	1.20 (110 °C)	0.93 (100 °C)	1.23 (50 °C)

The complex term of the Ag^+-catalysed oxidation, which dominates at higher temperatures, is believed to relate to an equilibrium

$$Ag^+ + H_2 \underset{k_{-1}}{\overset{k_1}{\rightleftharpoons}} AgH + H^+ \tag{78}$$

followed by

$$AgH + Ag^+ \xrightarrow{k_2} \text{further intermediates}$$

The occurrence of equilibrium (78) is confirmed by the incorporation of deuterium into molecular hydrogen from deuterated water. A brief study of the direct reaction between Ag^+ and H_2 indicates that both direct and catalysed reactions have the same velocity[482].

The presence of ligands, either in the form of added anions such as acetate or as co-solvents or solvents, such as pyridine, markedly affect the kinetics. In pyridine or dodecylamine solvents[483, 484] the hydrogenation of Ag(I) acetate follows simple second-order kinetics, as does that of Cu(I) acetate. This behaviour is also shown in aqueous solutions by Ag(I) in the presence of acetate ions and by an ethylenediamine complex of Ag(I)[482]. The rate of hydrogenation of Cu(II) acetate, on the other hand, is independent of oxidant concentration. The rate of oxidation of hydrogen by Cu(II) acetate in quinoline is also independent of oxidant concentration[485-487], but does depend on the square of the concentration of cuprous acetate which acts as a catalyst. For further details of these complicating features, reference should be made to the original papers and to Halpern's review[202].

5.1.15 Antimony(III)

Sb(III) is oxidised by alkaline ferricyanide with complex kinetics including variable reaction orders[488]. It is proposed that the reacting form of Sb(III) is dimeric, and, that subsequent to an initial one-equivalent reduction of $Fe(CN)_6^{3-}$, a further form of Sb(III) is involved, *viz.*

$$Sb(III)_2 + Fe(CN)_6^{3-} \rightarrow Fe(CN)_6^{4-} + Sb(IV) + Sb(III)$$

$$Sb(III) \rightarrow Sb(III)'$$

$$Sb(III)' + Fe(CN)_6^{3-} \rightarrow Sb(IV) + Fe(CN)_6^{4-}$$

$$Sb(IV) + Fe(CN)_6^{3-} \rightarrow Sb(V) + Fe(CN)_6^{4-}$$

$Sb(III)_2$, Sb(III) and Sb(III)′, respectively, are thought to be $(HO)_2Sb\text{–}O\text{–}Sb(OH)_2$, $Sb(OH)_4^-$ and $Sb(OH)_6^{3-}$, the last two on the basis of hydroxyl-ion dependences.

5.1.16 Borohydride ion

Oxidation by ferricyanide gives borate[489], *viz.*

$$8\ Fe(CN)_6^{3-} + BH_4^- + 2\ H_2O = 8\ Fe(CN)_6^{4-} + BO_2^- + 8\ H^+$$

Three conflicting accounts of the kinetics exist.

(*a*) In the pH range 9.22 to 12.52 the kinetics are reported by Freund[489] to be

$$-d[Fe(CN)_6^{3-}]/dt = k_2[H_3O^+][BH_4^-]$$

with $k_2 = 5.9 \times 10^{11} \exp(-7.2 \times 10^3/\boldsymbol{RT})$ l.mole^{-1}.sec^{-1}

and he regards ion pairing as the rate-determining step as in hydrolysis, *viz.*

$$H^+ + BH_4^- \rightleftharpoons H^+BH_4^-$$

(*b*) A potentiometric study[490] indicates a different rate law, *viz.*

$$-d[OH^-]/dt = k_2[BH_4^-][Fe(CN)_6^{3-}]$$

with $k_2 = 8.2 \times 10^{-6}$ l.mole^{-1}.sec^{-1} at pH > 12 and room temperature.

(*c*) Mochalov and Khain[491] believe that only the hydrolysis products of BH_4^- can reduce ferricyanide. Solutions of BH_4^- made strongly alkaline to prevent hydrolysis do not react. However, solutions of $Na^+BH_3OH^-$ rapidly reduce ferricyanide although those of $Na^{+}BH_2(OH)_2^{-}$ are inactive. The rate law with borohydride was found to be

$$-\mathrm{d}[Fe(CN)_6^{3-}]/\mathrm{d}t = k_3[H_3O^+][Fe(CN)_6^{3-}][BH_4^-]$$

The authors believe hydrolysis to be rate-determining but do not account for the presence of ferricyanide in the transition state.

5.2 OXIDATION OF ORGANIC MOLECULES

These oxidants are generally too feeble to attack monofunctional compounds except thiols, carbonyl- and nitro-compounds in their enolic forms, phenols and aromatic amines. However, ferric *tris-o*-phenanthroline readily oxidises cyclohexanone.

5.2.1 Thiols

These are oxidised by both Fe(III) and Cu(II) octanoates (denoted Oct) in non-polar solvents at moderate temperatures[492]. 80–90 % yields of the corresponding disulphide are obtained with Fe(III) and this oxidant was selected for kinetic study, the pattern of products with Cu(II) being more complex. The radical nature of the reaction was confirmed by trapping of the thiyl radicals with added olefins. Simple second-order kinetics were observed, for example, with 1-dodecane thiol oxidation by $Fe(Oct)_3$ in xylene at 55 °C (k_2 = 0.24 l.mole^{-1}.sec^{-1}). Reaction proceeds much more rapidly in more polar solvents such as dimethylformamide. The course of the oxidation is almost certainly

$$RSH + Fe(Oct)_3 \rightarrow RS\cdot + HOct + Fe(Oct)_2$$

$$RS\cdot \rightarrow \tfrac{1}{2}\,RSSR$$

3-Mercaptopropionic acid (HRSH) has been oxidised with ferricyanide in aqueous solution to give 3,3′-dithiodipropionic acid in 95 % yield. Whilst individual runs showed second-order disappearance of oxidant, the magnitude of k_2 varied with increasing thiol, oxidant and ferrocyanide concentrations[493], *viz.*

$$k_2 = \frac{k[RSH]}{[Fe(CN)_6^{3-}]+[Fe(CN)_6]^{4-}}$$

Dithiodipropionic acid was without effect on the reaction rate but the reaction went faster at high pH, suggesting the participation of species HRS^- (thiolate) and RS^{2-} (thiolate carboxylate). The retardation by ferrocyanide suggests the following pattern of steps

$$HRS^- + Fe(CN)_6{}^{3-} \rightleftharpoons HRS\cdot + Fe(CN)_6{}^{4-} \quad \text{(fast)}$$

$$HRS\cdot + Fe(CN)_6{}^{3-} \rightarrow HRS^+ + Fe(CN)_6{}^{4-} \quad \text{(slow)}$$

$$HRS^+ + HRSH \rightarrow HRSSRH + H^+ \quad \text{(fast)}$$

A steady-state treatment for the radical (and inclusion of similar reactions incorporating $RS^{2-} \rightarrow \cdot RS^-$) produces the observed rate law.

Aqueous acidic ferricyanide oxidises thioglycollic acid, $HSCH_2COOH$, to the disulphide with identical kinetics, and a similar reversible initial step is proposed[493a].

The aqueous ferricyanide oxidation of 2-mercaptoethanol to the disulphide is also complex kinetically[494]. In the pH range used (1.8–4.1) no complication from ionisation of the thiol is expected. Individual decays of oxidant concentrations are initially second-order but eventually become almost zero-order. For both second- and zero-order paths the rate depends on the first power of the thiol concentration and the former path is retarded by increasing the acidity, an approximately inverse relation existing above pH 3.2. Addition of ferrocyanide transforms the kinetics; the rapid, second-order path is inhibited and the zero-order path is accelerated until, at 10^{-2} *M* ferrocyanide, the whole of the disappearance of oxidant is zero-order. Addition of $Pb(ClO_4)_2$, which removes product ferrocyanide, greatly enhances the oxidation rate and the consumption of oxidant becomes *first*-order. Two routes are considered to co-exist (taking due account of the acidity of ferrocyanic acid), *viz.*

$$Fe(CN)_6{}^{3-} + RSH \rightleftharpoons HFe(CN)_6{}^{3-} + RS\cdot \quad \text{(slow)}$$

$$Fe(CN)_6{}^{3-} + RS\cdot + H^+ \rightarrow HFe(CN)_6{}^{3-} + RS^+ \quad \text{(fast)}$$

$$RS^+ + RSH \rightarrow RSSR + H^+ \quad \text{(fast)}$$

$$Fe(CN)_6{}^{4-} + RSH \rightleftharpoons Fe(CN)_5SR^{4-} + HCN \quad \text{(slow)}$$

$$Fe(CN)_6{}^{3-} + Fe(CN)_5SR^{4-} \rightarrow Fe(CN)_6{}^{4-} + Fe(CN)_5SR^{3-} \quad \text{(fast)}$$

Rather simple kinetics are shown by the ferricyanide oxidation of *n*-octyl mercaptan to disulphide in aqueous acetone[495], *viz.*

$$(\text{pH} > 7) \; -d[Fe(CN)_6{}^{3-}]/dt = k[Fe(CN)_6{}^{3-}][RSH][OH^-]$$

although a strongly specific effect of K^+ ions was discovered. E is 24.0 kcal.mole^{-1}. Potassium ferrocyanide acts only as a source of K^+ ions. Cyanide ions depress the rate markedly but the effect reaches a "ceiling" at $[CN^-] \sim 4 \times 10^{-4}$ *M*.

The mechanism proposed is one of ligand substitution (to explain the effect of

CN^-) followed by a redox process, *viz.*

$$Fe(CN)_6{}^{3-} + RS^- \rightleftharpoons Fe(CN)_5RS^{3-} + CN^- \quad \text{(slow)}$$

$$Fe(CN)_5RS^{3-} \rightarrow Fe(CN)_5{}^{3-} + RS\cdot \quad \text{(fast)}$$

and then further reactions as above. The "residual" reaction at high cyanide concentrations is regarded as a simple non-substitutional redox process.

The displacement of CN^- by RS^- in surprising since ^{14}C-labelled cyanide ion does not exchange with unlabelled ferricyanide[496]. Wiberg *et al.*[441] have shown that no exchange occurs under the conditions of the oxidation and that the effect of added cyanide is to be attributed to nucleophilic addition of cyanide to the thiol, *viz.*

$$RSH + HCN \rightarrow HC(SR){=}NH$$

$$HC(SR){=}NH + RSH \rightarrow HC(SR)_2NH_2$$

The reaction mechanism probably simply involves one-equivalent oxidation of thiolate ion.

5.2.2 *Carbonyl and nitro compounds*

The oxidations of three aldehydes, four ketones and two nitroparaffins by ferricyanide ion were examined by Speakman and Waters[497]. The rate laws are

$$-d[Fe(CN)_6{}^{3-}]/dt = k[\text{substrate}][OH^-][Fe(CN)_6{}^{3-}]^x$$

where $x = 1$ tending to zero at high $[Fe(CN)_6{}^{3-}]$ (*n*-butanal, diethyl ketone); $x = 1$ (propanal, isobutanal, ethylmethyl ketone, diisopropyl ketone, 2-nitropropane); $x = 1$ tending to two at high $[Fe(CN)_6{}^{3-}]$ (acetone, nitroethane). For *n*-butanal the reaction is not strictly first-order in OH^-. Maltz has confirmed the rate law for isobutanal[498].

The alkali-dependence suggests that the enolic forms (or probably their anions) are involved, *viz.*

$$R_2CH{-}CO{-}R' \rightleftharpoons R_2C{=}C(OH){-}R' \overset{OH^-}{\rightleftharpoons} R_2C{=}C(O^-){-}R' + H_2O \quad \text{(fast or slow)}$$

$$R_2C{=}C(O^-){-}R' + Fe(CN)_6^{3-} \longrightarrow R_2C{=}C(O\cdot){-}R' + Fe(CN)_6^{4-} \quad \text{(slow or fast)}$$

$$R_2C{=}C(O\cdot){-}R' \leftrightarrow R_2\dot{C}{-}C({=}O){-}R'$$

$$R_2\dot{C}{-}C({=}O){-}R' + Fe(CN)_6^{3-} \xrightarrow{H_2O} R_2C(OH)C({=}O){-}R' + Fe(CN)_6^{4-} + H^+ \quad \text{(fast)}$$

The enolisation may be rate-determining (to afford the zero-order dependence on oxidant concentration) or the oxidation step may be slower (to give the first-order dependence). The second-order dependence on oxidant concentration for acetone and nitroethane cannot involve slow oxidation of a free radical and no ready alternative explanation is available. Maltz[498] showed that the rate of oxidation of isobutanal equals the rate of enolisation, and that *two* main paths of oxidation are followed subsequent to enolisation leading either to tetramethyldihydropyrazine and a poly-aquocyanoiron(II) species or to isobutyric acid.

An interesting catalytic effect upon the alkaline ferricyanide oxidations of ketones is shown by osmium(VIII) tetroxide, the rate expression being

$$-\mathrm{d}[\mathrm{Fe(CN)_6}^{3-}]/\mathrm{d}t = k[\text{ketone}][\mathrm{OH}^-][\mathrm{Os(VIII)}][\mathrm{Fe(CN)_6}^{3-}]^0$$

k (25 °C) is 44.9 and 11.1 $\mathrm{l^2.mole^{-2}.sec^{-1}}$ for acetone and methylethyl ketone respectively[498a]. A slow two-equivalent attack of Os(VIII) upon enolate ion followed by a fast reoxidation of Os(VI) by $\mathrm{Fe(CN)_6}^{3-}$, *viz.*

$$\mathrm{R_2CH{-}CO{-}R'} + \mathrm{OH^-} \underset{k_{-1}}{\overset{k_1}{\rightleftharpoons}} \mathrm{R_2C{=}C(O^-){-}R'} + \mathrm{H_2O} \qquad (K)$$

$$\mathrm{R_2C{=}C(O^-){-}R'} + \mathrm{Os(VIII)} \xrightarrow{k_2} \text{complex} \xrightarrow[\text{fast}]{k_3} \mathrm{Os(VI)} + \text{products}$$

$$\mathrm{Os(VI)} + 2\,\mathrm{Fe(CN)_6}^{3-} \xrightarrow{\text{fast}} \mathrm{Os(VIII)} + 2\,\mathrm{Fe(CN)_6^{4-}}$$

is proposed. A steady-state treatment gives

$$\frac{-\mathrm{d}[\mathrm{Fe(CN)_6}^{3-}]}{\mathrm{d}t} = \frac{k_2 K[\mathrm{R_2CHCOR'}][\mathrm{OH^-}][\mathrm{Os(VIII)}]}{[\mathrm{H_2O}] + (k_2/k_{-1})[\mathrm{Os(VIII)}]}$$

which reduces to the observed law if $[\mathrm{H_2O}] \gg (k_2/k_{-1})[\mathrm{Os(VIII)}]$.

The oxidation of the non-enolising formaldehyde to formic acid by Cu(II) sulphate in the presence of glycerol follows kinetics[498b]

$$-\mathrm{d}[\mathrm{Cu(II)}]/\mathrm{d}t = k[\mathrm{Cu(II)}][\mathrm{CH_2O}][\mathrm{OH^-}]^2$$

where $k_{35\,^\circ\mathrm{C}} = 3.2 \times 10^{-2}$ $\mathrm{l^3.mole^{-3}.sec^{-1}}$, $E = 25.3$ $\mathrm{kcal.mole^{-1}}$ and $\Delta S^\ddagger =$ 9.2 eu. The *gem*-diol form of the substrate appears to lose both protons before oxidation, possibly in forming a complex with Cu(II).

The ferricyanide oxidation of formaldehyde is also base-catalysed[498c], the rate law being

$$\frac{-\mathrm{d}[\mathrm{Fe(CN)_6^{3-}}]}{\mathrm{d}t} = \frac{2\,k_1 k_2[\mathrm{CH_2O}][\mathrm{OH^-}][\mathrm{Fe(CN)_6^{3-}}]}{k_{-1}[\mathrm{H_2O}] + k_2[\mathrm{Fe(CN)_6^{3-}}]}$$

(excluding an initial rapid reaction). The rate coefficients refer to the steps

$$H_2C(OH)_2 + OH^- \underset{k_{-1}}{\overset{k_1}{\rightleftharpoons}} H_2C(OH){-}O^- + H_2O$$

$$H_2C(OH){-}O^- + Fe(CN)_6^{3-} \overset{k_2}{\rightarrow} H_2C(OH){-}O\cdot + Fe(CN)_6^{4-}$$

$$H_2C(OH){-}O\cdot + Fe(CN)_6^{3-} \rightarrow HCO_2H + Fe(CN)_6^{4-} + H^+ \qquad \text{(fast)}$$

when $[H_2O] \gg k_2[Fe(CN)_6{}^{3-}]/k_{-1}$ then the observed third-order rate coefficient becomes $2\,k_1k_2/k_{-1}[H_2O]$. At 40 °C its value is 1.5×10^{-3} l^2.mole^{-2}.sec^{-1} ($\mu = 2.0\ M$) and $E = 12.0$ kcal.mole^{-1}.

Cupric chloride oxidises isobutanal in 80 % acetone–20 % water according to the stoichiometry[499]

$$(CH_3)_2CHCHO + 2\,CuCl_2 = (CH_3)_2CClCHO + HCl + 2\,CuCl$$

The rate law is

$$-d[Cu(II)]/dt = k_2[Cu(II)][aldehyde] + k_3[H_3O^+][Cu(II)][aldehyde]$$

where, at 30 °C, $k_2 = 2.1 \times 10^{-5}$ l.mole^{-1}.sec^{-1} and $k_3 = 1.4 \times 10^{-3}$ l^2.mole^{-2}.sec^{-1}. k_2 has an activation energy of 23 kcal.mole^{-1}. Addition of LiCl accelerates reaction but the effect levels out at a LiCl:$CuCl_2$ ratio of 2, suggesting $CuCl_4{}^{2-}$ may be an active species.

Rate-determining enolisation is discounted by the authors on the grounds of the appearance of cupric ion in the rate law and of the value of the rate coefficient, which shows oxidation to be faster than enolisation. However, it is known that Cu(II) catalyses enolisation and an intermediate radical $(CH_3)_2\dot{C}CHO$ could abstract chlorine from $CuCl_2$.

The oxidation of *n*-butanal by $CuCl_2$ in dimethylformamide showed simple second-order kinetics in the presence of lithium chloride[500]. At 83 °C, k_2 is 2×10^{-4} l.mole^{-1}.sec^{-1}. α-Monohalogenation occurs in 97 % yield. Cu(II)-catalysed enolisation followed by ligand-transfer is proposed.

α-Halogenation of acetone is accomplished by $CuCl_2$, *viz.*

$$2\,CuCl_2 + CH_3COCH_3 = 2\,CuCl + ClCH_2COCH_3 + HCl$$

The kinetics in aqueous solution are more complex than those for the aldehydes[501]. Initial rates indicate the reaction order in Cu(II) to be one-half and an approximately first-order dependence on chloride ion was noted; cuprous ion retards reaction. The mechanism proposed includes dimerisation of $CuCl_2$ which is not a feature of other reactions of this oxidant.

References pp. 493–509

Tris-1,10-phenanthrolineiron(III) (also called ferriin) is substitution-inert and its ready oxidation of cyclohexanone is of interest as a possible example of an electron-transfer process[433]. The reaction has simple second-order kinetics and the rate is both acid-independent and faster than enolisation. k_H/k_D is 2.2 (15 °C) and the reaction is very sensitive to oxygen but remains second-order. An attempt to differentiate between electron- and hydrogen atom-transfer was made, using α-tritiocyclohexanone. Very little tritium was transferred to the ligand and so an electron-transfer mechanism was preferred, *viz.*

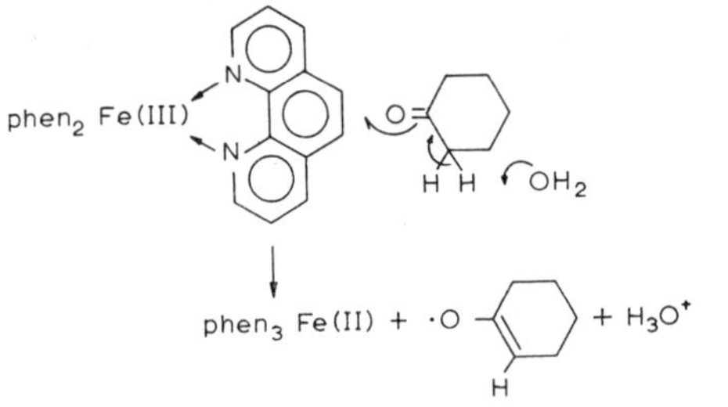

The radical is rapidly oxidised to a carbonium ion which is hydrolysed to 2-hydroxycyclohexanone. Electron transfer is compatible with the observed Arrhenius parameters (E = 12.6 kcal.mole^{-1}, $\Delta S^{\ddagger}$ = −31 eu).

Oxidation of butan-2-one in nitrate media by $[Fe^{III}(H_2O)_2phen_2]$ is reported to proceed much more slowly than enolisation[683]; moreover $Fe(phen)_3^{3+}$ is said to be even less reactive, in total contrast to the results on cyclohexanone.

5.2.3 *Formic acid*

At the natural pH of the solution, sodium formate is oxidised by silver nitrate. The following kinetics have been reported in an early study[177]

$$-d[Ag(I)]/dt = k_3[HCO_2^-][Ag^+]^2$$

A re-investigation of this reaction under controlled conditions of acidity and ionic strength is desirable.

5.2.4 *Unsaturated alcohols*

The homogeneous, anaerobic, oxidation of propargyl alcohol by cupric acetate in buffered pyridine solution[502] is an example of a general reaction

$$2\,RC{\equiv}CH + 2\,Cu(II) = RC{\equiv}C{-}C{\equiv}CR + 2\,Cu(I) + 2\,H^+$$

Individual runs show initial autocatalysis followed by zero-order disappearance of Cu(II) when the substrate is in considerable excess. The reaction is affected by added cuprous acetate, according to

$$-\mathrm{d}[\mathrm{Cu(II)}]/\mathrm{d}t = k_1[\mathrm{Cu(I)}]/(1+k_2[\mathrm{Cu(I)}])$$

and the dependence on the substrate concentration is

$$-[\mathrm{d}(\mathrm{Cu(II)}]/\mathrm{d}t = k_3[\mathrm{RC{\equiv}CH}]/(1+k_4[\mathrm{RC{\equiv}CH}])$$

The reaction is catalysed by added base (piperidine). These individual dependences were combined (at 40 °C) to give

$$-\mathrm{d}[\mathrm{Cu(II)}]/\mathrm{d}t = \frac{8.84\times10^{-4}[\mathrm{RC{\equiv}CH}][\mathrm{base}][\mathrm{Cu(I)}]}{[\mathrm{base}]^{\frac{1}{2}}[\mathrm{CH_3CO_2H}]^{\frac{1}{2}}(1+1.49[\mathrm{RC{\equiv}CH}]+17.3[\mathrm{Cu(I)}])}$$

A mechanism consistent with this rate expression was put forward, *viz.*

$$\mathrm{RC{\equiv}CH + B \rightleftharpoons RC{\equiv}C^- + BH^+} \quad \text{(slow)}$$

$$\mathrm{RC{\equiv}C^- + Cu(CH_3CO_2) \rightleftharpoons RC{\equiv}CCu + CH_3CO_2^-} \quad \text{(slow)}$$

$$\mathrm{RC{\equiv}CCu + Cu(CH_3CO_2) \rightleftharpoons R{-}C{\equiv}C{-}Cu \;(\pi\text{-complexed}\rightarrow Cu(CH_3CO_2))} \quad \text{(fast)}$$

$$\mathrm{RC{\equiv}CCu + Cu(CH_3CO_2)_2 \longrightarrow RC{\equiv}C\cdot + 2\,Cu(CH_3CO_2)} \quad \text{(fast)}$$

$$\mathrm{2\,RC{\equiv}C\cdot \longrightarrow RC{\equiv}C{-}C{\equiv}CR} \quad \text{(fast)}$$

An analogous study has been reported[503] of the oxidation of 2-methyl-but-3-yn-ol by Cu(II) chloride in aqueous ammonia to give 2,7-dimethylocta-3,5-diene-2,7-diol. Simple, second-order kinetics were obtained, but a very sharp increase in rate occurred when the pH was increased from 8 to 10. Addition of ammonium ions retarded reaction but Cu(I) was without effect. The reaction mechanism put forward is similar to that given above.

The methanolic cupric bromide oxidation of propargyl alcohol[500] to *trans*-$\mathrm{BrCH{=}CBrCH_2OH}$ (30 %) and $\mathrm{Br_2C{=}CBrCH_2OH}$ (18 %) and, under other reaction conditions, $\mathrm{Br_2C{=}CBr{-}CH_2OH}$ (93 %) follows simple second-order kinetics with a rate coefficient of 1.5×10^{-2} l.mole^{-1}.sec^{-1} at 64 °C. A mechanism of ligand-transfer in a π-complex is proposed†.

The analogous quantitative oxidation of methanolic allyl alcohol to 2,3-dibromopropan-1-ol shows unusual kinetics[500], *viz.*

$$-\mathrm{d}[\mathrm{Cu(II)}]/\mathrm{d}t = k_2[\mathrm{CuBr_2}]^2[\text{allyl alcohol}]^0$$

† Some of the formulae for the oxidation products (in the original paper) are obviously incorrect.

with k_2 equal to 5.5×10^{-3} l.mole^{-1}.sec^{-1} at 64 °C. This is ascribed to slow reversible dissociation

$$2\ CuBr_2 \rightleftharpoons 2\ CuBr + Br_2$$

followed by fast bromination of the olefinic group. This hypothesis could easily be tested with related compounds.

5.2.5 *Hydroxy-ketones (α-ketols, acyloins)*

The oxidation of acetoin by Fe(III) perchlorate[504] follows the stoichiometry

$$2\ Fe(III) + CH_3COCH(OH)CH_3 = 2\ Fe(II) + CH_3COCOCH_3 + 2\ H^+$$

and the rate law

$$-d[Fe(III)]/dt = k[Fe(III)][acetoin](a + b/[H^+])$$

The two terms correspond to oxidation by Fe^{3+} and $Fe(OH)^{2+}$ which exist in a hydrolytic equilibrium. A radical intermediate $CH_3CO\dot{C}(OH)CH_3$, which suffers rapid oxidation by a second ferric ion, is proposed.

Ferricenium ion oxidation of acetoin in acidic aqueous ethanol (50 % v/v) involves rate-determining enolisation[713], producing a zero-order decay of oxidant. The oxidation rate approximates to that of deuteration under similar conditions.

Several studies have been made of oxidations of ketols by Cu(II). Simple cupric salts deposit cuprous oxide, but this has been overcome by using a citrate complex of Cu(II) or a 1 : 1 (mole) pyridine–water mixture. A recent detailed account is that of Wiberg and Nigh[505], who found the stoichiometry and rate law for the oxidation of α-hydroxyacetophenone by Cu(II) acetate to be, respectively,

$$2\ Cu(II) + C_6H_5COCH_2OH + 2\ OH^- = 2\ Cu(I) + C_6H_5COCHO + 2\ H_2O$$

$$-d[Cu(II)]/dt = k_1[ketol] + k_2[ketol][Cu(II)]$$

The first term was found to correspond to the rate of enolisation (measured by an NMR study of hydrogen–deuterium exchange at the methylene group). The second term predominates at $[Cu(II)] > 10^{-2}\ M$ and is characterised by a primary kinetic isotope effect of 7.4 (25 °C) and a ρ value of 1.24. Addition of 2,2′-bipridyl (bipy) caused an increase in k_2 up to a bipy: Cu(II) ratio of 1 : 1 but at ratios greater than this k_2 fell gradually until the enolisation term dominated. The oxidation of α-methoxyacetophenone is much slower but gives a similar rate expression. These various data suggest the following mechanism

$$R-\overset{O}{\overset{\|}{C}}-CH_2OH + Cu(II) \rightleftharpoons R-\overset{O\cdots Cu^{+}\cdots O}{C \text{———} CH_2} + H^{+} \qquad \text{(fast)}$$

$$R-C(=O\cdots Cu^{+}\cdots O-)C(H)(H)\cdots Py: \longrightarrow R-\overset{O}{\overset{\|}{C}}-\dot{C}HOH + Cu^{+} \qquad \text{(slow)}$$

$$RCO\dot{C}HOH + Cu(II) \longrightarrow RCOCHO + Cu(I) + H^{+} \qquad \text{(fast)}$$

The chelation is suggested by the effect of adding 2,2′-bipyridyl and of methylation of the hydroxyl group; moreover, Cu^{2+} ions catalyse the enolisation of ethyl acetoacetate[506].

In aqueous solution the rate law for oxidation of simple α-ketols, such as acetoin[507] and benzoin[508], and various aldoses and ketoses[507, 509–510], is markedly different from that of Wiberg and Nigh[505] (*vide supra*), *viz.*

$$-\mathrm{d}[Cu(II)]/\mathrm{d}t = k[\alpha\text{-ketol}][OH^-][Cu(II)]^0$$

In several instances an induction period has been observed but this has been shortened by elimination of dissolved oxygen[508]. Deposition of cuprous oxide (Fehling's test) does not appear to influence the zero-order disappearance of Cu(II), but the induction period can be eliminated by adding copper powder (but not Cu_2O)[507].

The rate expression suggests a rate-determining step not involving the oxidant; this is very probably enolisation, *viz.*

$$CH_3CH(OH)COCH_3 + OH^- \rightleftharpoons CH_3\bar{C}(OH)COCH_3 + H_2O \qquad \text{(slow)}$$

$$\left.\begin{array}{c} CH_3\bar{C}(OH)COCH_3 \\ \updownarrow \\ CH_3C(OH){=}\underset{\displaystyle O^-}{\underset{|}{C}}-CH_3 \end{array}\right\} + H_2O \rightleftharpoons CH_3C(OH){=}C(OH)CH_3 + OH^- \qquad \text{(fast)}$$

The rapid oxidation of the enediol then ensues. Marshall and Waters[508] propose that a Cu(I) complex of the enediol is the active reductant in order to explain the induction period (during which O_2 reoxidises Cu(I)), but Wiberg and Nigh[505] believe that it is unlikely that the enediol can displace citrate or tartrate from copper ions. The report of Marshall and Waters that the oxidation proceeds at about one-third of the rate of enolisation is also discounted by Wiberg and Nigh on the grounds of experimental inaccuracy in determining $[OH^-]$ in strongly basic solutions with dioxan–water mixtures.

The faster cupric ion oxidation of ketoses[510] compared with aldoses is due to the more strongly reducing nature of the α-hydroxyketone group. Pre-treatment of an aldose with alkali effects the Lobry de Bruyn transformation to a ketose and an increase in reaction rate is found[510].

Enediol formation is also rate limiting in the oxidation of aldoses and ketoses by alkaline ferricyanide[511–513], the rate expression being

$$-\mathrm{d}[\mathrm{Fe(CN)_6}^{3-}]/\mathrm{d}t = k_2[\text{sugar}][\mathrm{OH^-}][\mathrm{Fe(CN)_6}^{3-}]^0$$

5.2.6 Ascorbic acid

The anaerobic oxidations (to dehydroascorbic acid, A) by several chelate complexes of Fe(III) have a stoichiometry

$$\text{ascorbic acid} + 2\,\mathrm{Fe(III)} = \text{dehydroascorbic acid} + 2\,\mathrm{Fe(II)} + 2\,\mathrm{H^+}$$

and a rate law[513a]

$$\mathrm{d[A]}/\mathrm{d}t = k_2[\mathrm{Fe(III)}][\text{ascorbate ion}]$$

The chelates employed were diethylenetriaminepentaacetic acid (DTPA), 1,2-cyclohexanediaminetetraacetic acid (CDTA), ethylenediaminetetraacetic acid (EDTA) and *N*-hydroxyethylenediaminetriacetic acid (HEDTA). Rate and activation data are given below.

	$10^{-3}\,k_2$(298 °K)($l.mole^{-1}.sec^{-1}$)	E($kcal.mole^{-1}$)	$\Delta S^{\ddagger}$(eu)	π^0(V)
Fe(III)–HEDTA	21.5	5.6±0.2	−21±1	+0.30
Fe(III)–EDTA	4.0	5.1±0.2	−27±1	+0.10
Fe(III)–CDTA	1.30	4.9±0.2	−30±1	+0.08
Fe(III)–DTPA	0.88	4.7±0.2	−32±1	+0.06

The rate sequence is determined by the entropy term and correlates with the oxidation potential of the chelate complex, indicating an outer-sphere electron transfer.

The anaerobic V(IV) oxidation of ascorbic acid (H_2A) in the pH range 1.75–2.85 follows a rate law [513b]

$$\mathrm{d[A]}/\mathrm{d}t = k_3[\mathrm{H_2A}][\mathrm{VO^{2+}}][\mathrm{H_3O^+}]$$

A two-equivalent process is proposed, yielding V(II). In the presence of oxygen the rate law becomes

$$d[A]/dt = k_4[H_2A][VO^{2+}][H_3O^+][O_2]$$

$k_{3(298\ °K)} = (1.8 \pm 0.1) \times 10^2$ l.mole^{-1}.sec^{-1} ($\mu = 0.10$ M KNO_3), $k_{4(298\ °K)} = (4.4 \pm 0.3) \times 10^3$ l.mole^{-1}.sec^{-1}. atm^{-1}; $E_3 = 7.6 \pm 0.5$ kcal.mole^{-1}, $\Delta S_3^\ddagger = -25 \pm 2$ eu, $E_4 = 14.4 \pm 0.7$ kcal.mole^{-1} and $\Delta S_4^\ddagger = 4.1 \pm 0.2$ eu. Complexes VOH_3A^{3+} and $VOH_3A(O_2)^{3+}$ are regarded as breaking down in the rate-determining steps

$$VOH_3A^{3+} \rightarrow V^{2+} + A + H_3O^+$$

$$VOH_3A(O_2)^{3+} \rightarrow VO^{2+} + A + H_2O_2 + H^+$$

Identical kinetics are found for the uranyl ion-catalysed aerobic oxidation of ascorbic acid and a similar mechanism has been put forward[513d]. These results and others afford a sequence of catalytic activity for the *aerobic* oxidation of ascorbic acid[513a-d]

$$Cu(II) > Fe(III) > V(IV) > U(VI)$$

The ferricyanide oxidation of ascorbic acid at pH 1.1 follows kinetics[513e]

$$-d[Fe(CN)_6{}^{3-}]/dt = k[H_2A][Fe(CN)_6{}^{3-}]/[H_3O^+]$$

indicating oxidation of the HA^- species. k (15 °C) $\sim 2.1 \times 10^{-2}$ sec^{-1}, $E = 10.8$ kcal.mole^{-1} and $\Delta S^\ddagger = -39.8$ eu ($\mu = 1$ M KCl).

5.2.7 Phenols

The early work of Porret[514] has been superseded by that of Baxendale *et al.*[515] who found the oxidation of hydroquinone (QH_2) by Fe(III) perchlorate to follow the rate law

$$-d[QH_2]/dt = k_{obs}[Fe(III)][QH_2]$$

k_{obs} is sharply reduced by added ferrous ion and by increased acidity; plots of $k_{obs}{}^{-1}$ *versus* [Fe(II)] are linear with an intercept and the retardation, which is too severe to be ascribed to the Fe(II)+Q back-reaction, is considered to result from re-oxidation by a semi-quinone intermediate, *viz.*

$$Fe(III) + QH_2 \underset{k_{-1}}{\overset{k_1}{\rightleftharpoons}} Fe(II) + QH\cdot + H^+$$

$$Fe(III) + QH\cdot \underset{k_{-2}}{\overset{k_2}{\rightleftharpoons}} Fe(II) + Q + H^+$$

Assuming a steady-state for [QH·] leads to

$$-\mathrm{d}[\mathrm{QH_2}]/\mathrm{d}t = \frac{k_1 k_2[\mathrm{Fe(III)}]^2[\mathrm{QH_2}]}{k_{-1}[\mathrm{Fe(II)}]+k_2[\mathrm{Fe(III)}]}$$

i.e.

$$k_{\mathrm{obs}} = k_1 k_2[\mathrm{Fe(III)}]/k_{-1}[\mathrm{Fe(II)}]+k_2[\mathrm{Fe(III)}]$$

This accords with both [H^+] and [Fe(II)] dependences and gives values for k_1 and k_2/k_{-1}; their variations with temperature yield

$$k_1 = 1.9\times 10^{20} \exp(-25.6\times 10^3/\boldsymbol{RT})\ \mathrm{l.mole^{-1}.sec^{-1}}$$

and

$$k_2/k_{-1} = 1.7\times 10^2 \exp(4.7\times 10^3/\boldsymbol{RT})$$

at unit [H^+]. A study of the reduction of quinone by Fe(II) is included in Section 6.

The ferricyanide oxidation of various phenols has been examined by Waters *et al.*[516] following the pioneering work of Conant *et al.*[517, 518], who observed a correlation between the rates of oxidation by a given oxidant of a number of phenols and amines and the redox potentials of these substrates – a situation which depends critically on the reversible nature of the oxidation

$$\mathrm{M}^{n+} + \mathrm{ROH} \underset{k_{-1}}{\overset{k_1}{\rightleftharpoons}} \mathrm{M}^{(n-1)+} + \mathrm{RO}\cdot + \mathrm{H}^+$$

Waters *et al.*[516] detected retardation by added ferrocyanide and found the reaction to be approximately first-order in hydroxide ion; the order in phenol, while complex, is greater than unity. The isolation of polymeric products is sure indication of the irreversible secondary reactions which follow the initial equilibrium.

An (EMF) study of the ferricyanide oxidation of phloroglucinol revealed kinetics[519]

$$\frac{-\mathrm{d}[\mathrm{Fe(CN)_6}^{3-}]}{\mathrm{d}t} = \frac{k[\mathrm{phloroglucinol}][\mathrm{Fe(CN)_6}^{3-}]}{k'[\mathrm{Fe(CN)_6}^{4-}]+k''}$$

The stoichiometry is initially 1 : 1, but soon becomes 3 or 4 equivalents of oxidant per equivalent of phenol. The initial process is that given above with $k_1 = 1.13$ $\mathrm{l.mole^{-1}.sec^{-1}}$ (25 °C, pH 6.94).

The oxidation of phenol in alcoholic media by a morpholine complex of Cu(II) (as a model for tyrosinase) to give 4,5-dimorpholino-*ortho*-benzoquinone in 64 %

yield, *viz.*

$$\text{PhOH} + 2\,\text{(M)} \xrightarrow[O_2]{Cu(II)} \text{(2,5-bis-M-ortho-benzoquinone)}$$

is kinetically complex[520]. Oxygen is absorbed during the reaction after an induction period and its presence strongly influences the course of reaction as well as the kinetics. The phenol concentration has to exceed a critical value of *ca.* 4×10^{-2} M before the reaction proceeds and the concentration of Cu(II) influences the induction period. Small amounts of H_2O_2 eliminate the induction period. The rate law is complex: the dependence on oxygen pressure is approximately linear: the reaction order in the other reactants (Cu(II), phenol, morpholine) is initially unity but at moderate concentrations decreases to zero or becomes negative. H_2O_2, which is produced during autoxidation of hydroquinones, is considered to play a crucial role in the reaction, and *ortho*-hydroquinone and *ortho*-benzoquinone are the reaction intermediates. The initial step is an attack of Cu(II) on phenol to give phenoxyl radical.

5.2.8 *Amines*

Comparative kinetic studies[680] of the oxidation of triethylamine by a group of oxidants including $Fe(CN)_6^{3-}$, substituted ferriins and $Mo(CN)_8^{3-}$ afford a correlation between $\log k_2$ and π^0. Strong retardation of $Fe(CN)_6^{3-}$ oxidation by $Fe(CN)_6^{4-}$ was noted and the kinetics were further complicated by specific ion-pairing effects. However, an electron transfer mechanism is plausible and the rate coefficients with different oxidants agreed reasonably well (correlation coefficient 0.966) with those calculated from

$$\log k_2 = 11.64\,\pi^0 - 5.78$$

More general forms of this equation incorporating the ionisation potential of the amine also give a good degree of correlation with several aliphatic amines.

The oxidation of *N*, *N*-dimethylaniline by aerated, ethanolic cupric chloride to give a mixture of products including methyl and crystal violets is simple second-order when an excess of amine is used[521]. Presumably Cu(I) is re-oxidised by dissolved oxygen, for otherwise the observed linearity of log [residual amine] *versus* time plots would not be found as Cu(II) disappears. Under nitrogen the kinetics are complex, but a new optical absorption (472 and 1007 nm) appears immediately on mixing the reactants. This absorption decays whilst a new one at 740 nm develops. The latter absorption originates from a 1 : 1 complex formulated

as

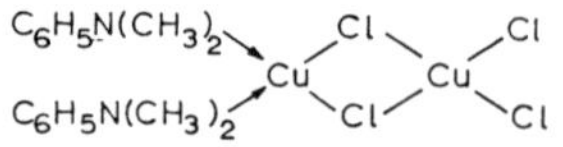

$$\begin{array}{c} C_6H_5N(CH_3)_2 \searrow \quad Cl \quad\quad Cl \\ Cu \quad\quad Cu \\ C_6H_5N(CH_3)_2 \nearrow \quad Cl \quad\quad Cl \end{array}$$

Treatment of the complex with further amine produced the violet dyes. The importance of this complex in the mechanism is suggested by the inability of cupric acetate, nitrate or sulphate to achieve the oxidation.

5.2.9 Phenylhydrazine and its sulphonic acids

The oxidations of these substrates to diazonium salts by acidic ferricyanide[521a]

$$^{-}SO_3C_6H_4NHNH_3^+ + 4\,Fe(CN)_6^{3-} = {}^{-}SO_3C_6H_4N_2^+ + 4\,H^+ + 4\,Fe(CN)_6^{4-}$$

are all first-order in $Fe(CN)_6^{3-}$. With phenylhydrazine and the *ortho*-sulphonic acid the reaction is strictly first-order in the hydrazine, which exists mainly as the protonated form $ArNHNH_3^+$, but with the *meta*- and *para*-sulphonic acids the reaction order is slightly less than one, an effect ascribed to ion-pairing. The acidity dependences are all approximately of the form rate $= a + b/[H_3O^+]$ indicating participation by the, albeit small, quantities of free base. The acid-independent path may involve either or both of the pairs of species, $ArNHNH_3^+ + Fe(CN)_6^{3-}$ and $ArNHNH_2 + HFe(CN)_6^{2-}$, but the former pair only is regarded as significant.

5.2.10 Ortho-aminoazo compounds

The oxidation by Cu(II) in pyridine solution to give benzotriazoles, *viz.*

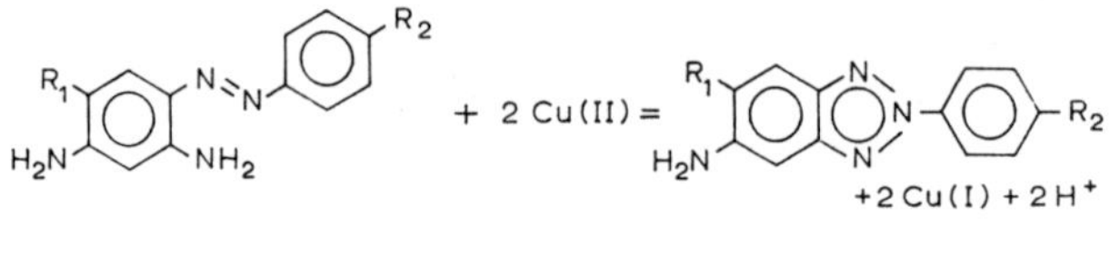

$$\text{(R}_1\text{, H}_2\text{N, NH}_2\text{-substituted aminoazo compound with } R_2\text{-phenyl)} + 2\,Cu(II) = \text{(R}_1\text{, H}_2\text{N-substituted benzotriazole with N-}R_2\text{-phenyl)} + 2\,Cu(I) + 2\,H^+$$

has kinetics[522]

$$d[Cu(I)]/dt = k[\text{aminoazo compound}][Cu(II)]^2$$

The order in oxidant is considered to result from attack of Cu(II) upon a chelate complex of the azo-compound and a second Cu(II) species. Systematic variations of R_1 and R_2 produced Hammett ρ values of -2.50 and -0.946, respectively, but these relate to a combination of equilibrium and reaction constants.

5.2.11 Dichlorophenolindophenol

The reversible reaction with ferricyanide, *viz.*

$$2\,Fe(CN)_6^{4-} + 2\,H^+ + O{=}C_6H_2Cl_2{=}N{-}C_6H_4{-}OH \rightleftharpoons 2\,Fe(CN)_6^{3-} + HO{-}C_6H_2Cl_2{-}NH{-}C_6H_4{-}OH$$

has been examined by means of a temperature-jump technique[523]. The acquisition of two electrons by the indicator takes place consecutively and a semiquinone radical functions as an intermediate. Each stage in the oxidation sequence involves acid-base equilibria; (denoting the indicator as HOx, its semi-reduced form as $\cdot HSq^-$, its fully reduced form as HR^{2-} and adding one H^+ where necessary the mechanism is

$$\begin{array}{ccccc}
Ox^- & & & & \\
\updownarrow & & & & \\
HOx & \underset{k_{-1}}{\overset{k_1}{\rightleftharpoons}} & \cdot HSq^- & & \\
\updownarrow & & \updownarrow & & \\
H_2Ox^+ & \underset{k'_{-1}}{\overset{k'_1}{\rightleftharpoons}} & \cdot H_2Sq & \underset{k_{-2}}{\overset{k_2}{\rightleftharpoons}} & H_2R^- \\
 & & \updownarrow & & \updownarrow \\
 & & \cdot H_3Sq^+ & \underset{k'_{-2}}{\overset{k'_2}{\rightleftharpoons}} & H_3R
\end{array}$$

The following rate coefficients were determined at 30 °C: $k_1 \leqslant 3\times10^5$ l.mole^{-1}.sec^{-1}, $k'_1 = (7.3\pm2)\times10^8$ l.mole^{-1}.sec^{-1}, $k_{-2} = (3.3\pm0.8)\times10^7$ l.mole^{-1}.sec^{-1} and $k'_{-2} = (1.2\pm0.3)\times10^4$ l.mole^{-1}.sec^{-1}.

5.2.12 Ethylenediaminetetraacetic acid (EDTA)

The stoichiometry and the rate law for the oxidation by alkaline ferricyanide are, respectively (EDTA = H_4Y)[524]

$$Y^{4-} + 4\,Fe(CN)_6^{3-} + 4\,OH^- = 4\,Fe(CN)_6^{4-} + H_2O + \text{organic products}$$

$$-d[Fe(CN)_6^{3-}]/dt = k_2[Y^{4-}][Fe(CN)_6^{3-}]$$

where, at 50 °C ($\mu = 0.45$ *M*), $k_2 = 0.57$ l.mole^{-1}.sec^{-1}. k_2 is unaffected by addition of a ten-fold excess of ferrocyanide or 0.1 *M* cyanide ion, but is strongly and specifically affected by added cations, the rate sequence being:

$$Cs^+ > Rb^+ > K^+ > Na^+ > Li^+$$

An irreversible outer sphere electron transfer is thought to be the most likely rate-determining step in view of the lack of retardation by $Fe(CN)_6^{4-}$ and CN^- ions.

Sixteen other nitrogen-containing chelates were examined by Lambert and Jones[524], and a very good correlation is observed between E and $\Delta S^{\ddagger}$ indicating a single effect to predominate in all instances.

5.2.13 *Thiourea and thioacetamide*

Oxidation by aqueous cupric sulphate occurs slowly enough to be studied by conventional means at 25 °C giving a rate law[525]

$$-\mathrm{d}[\mathrm{Cu(II)}]/\mathrm{d}t = k[\mathrm{Cu(II)}][\mathrm{thiourea}]^{\frac{1}{2}}$$

The overall stoichiometry is though to be

$$2\,CuSO_4 + CS(NH_2)_2 + H_2O = Cu_2SO_4 + H_2SO_4 + CO(NH_2)_2 + S$$

The stoichiometries of the oxidations of thiourea and thioacetamide, respectively, by alkaline ferricyanide are[526]

$$8\,Fe(CN)_6^{3-} + CS(NH_2)_2 + 10\,OH^- = CO(NH_2)_2 + SO_4^{2-} + 8\,Fe(CN)_6^{4-} + 5\,H_2O$$

$$8\,Fe(CN)_6^{3-} + CH_3CSNH_2 + 11\,OH^- = CH_3CO_2^- + NH_3 + SO_4^{2-} + 8\,Fe(CN)_6^{4-} + 5\,H_2O$$

The respective rate laws were determined as

$$-\mathrm{d}[Fe(CN)_6^{3-}]/\mathrm{d}t = k_3[\mathrm{thiourea}][Fe(CN)_6^{3-}][OH^-]$$

$$-\mathrm{d}[Fe(CN)_6^{3-}]/\mathrm{d}t = k_2[\mathrm{thioacetamide}][OH^-][Fe(CN)_6^{3-}]^0$$

At 35 °C $k_3 = 72$ l^2.mole^{-2}.sec^{-1} and $k_2 = 0.31$ l.mole^{-1}.sec^{-1}. Also $E_3 = 11.65$ kcal.mole^{-1}, $\Delta S_3^{\ddagger} = -27.3$ eu, $E_2 = 16.7$ kcal.mole^{-1} and $\Delta S_2^{\ddagger} = 21.4$ eu.

The hydroxyl-ion dependences suggest oxidation of substrate anions. Alkali-catalysed enolisation is the slow step of the oxidation of thioacetamide but is a fast pre-equilibrium in the thiourea oxidation.

The reaction between two polydentate complexes of Cu(II), CuY (Y_1H_4 = ethylenediaminetetraacetic acid, Y_2H_4 = hydroxyethylethylenediaminetriacetic acid) and thiourea to give a Cu(I) complex of thiourea (this product was not identified), follows kinetics[526a]

$$-\mathrm{d}[CuY^{2-}]/\mathrm{d}t = k[CuY^{2-}][\mathrm{thiourea}]^2[H_3O^+]/[YH_4]$$

The mechanism postulated involves rapid, acid-catalysed ligand exchanges to give $Cu(thiourea)_2^{2+}$ which breaks down in the slow step. Activation energies are 13.4 and 12.9 kcal.mole^{-1} for CuY_1^{2-} and CuY_2^{2-}, respectively.

6. Reduction

6.1 INTRODUCTION

The term reductant is relative but in this Section it is applied to the metal ion (which becomes oxidised) irrespective of its reduction potential (Table 19). The stages of oxidation may differ by one [Ti(III), Cr(II), Co(II), Cu(II)] by two [Sn(II), Tl(I)] or by one *and* two [V(II), Mn(II)], and in some cases, *e.g.* certain reductions by Fe(II), the issue is in doubt.

The field of reduction is much less well charted than that of oxidation but a substantial literature exists nonetheless and is growing rapidly. Reductions are conveniently classified into (*i*) those involving and initial electron acceptance by the substrate (possibly followed by rapid protonation) and (*ii*) those involving electron acceptance concerted with, or followed very rapidly by, homolysis of the substrate; the latter includes the important Fenton and silver–persulphate reactions, as well as reductions of halogens, hydrazine and possibly NO_3^- and NO_2^-.

TABLE 19

STANDARD ELECTRODE POTENTIALS FOR REDUCING IONS IN AQUEOUS SOLUTION

All values taken from Ref. 19 unless otherwise indicated.

Reaction	π^0 *(Volt)*
$Eu^{2+} = Eu^{3+} + e^-$	+0.43
$Cr^{2+} = Cr^{3+} + e^-$	+0.41
$Ti^{2+} = Ti^{3+} + e^-$	*ca.* +0.37
$V^{2+} = V^{3+} + e^-$	+0.255
$Ti^{3+} + H_2O = TiO^{2+} + 2\,H^+ + e^-$	*ca.* −0.1
$Np^{3+} = Np^{4+} + e^-$	−0.147
$Sn^{2+} = Sn^{4+} + 2\,e^-$	−0.15
$Cu^+ = Cu^{2+} + e^-$	−0.153
$U^{4+} + 2\,H_2O = UO_2{}^{2+} + 4\,H^+ + 2\,e^-$	−0.334
$Fe(CN)_6{}^{4-} = Fe(CN)_6{}^{3-} + e^-$	−0.355[431]
$V^{3+} + H_2O = VO^{2+} + 2\,H^+ + e^-$	−0.361
$Cu = Cu^+ + e^-$	−0.521
$Mo(V) = Mo(VI) + e^-$	−0.53
$Fe^{2+} = Fe^{3+} + e^-$	−0.772
$Pu^{3+} = Pu^{4+} + e^-$	−0.97
$IrCl_6{}^{4-} = IrCl_6{}^{3-} + e^-$	−1.017
$NpO_2{}^+ = NpO_2{}^{2+} + e^-$	−1.15
$Tl^{3+} = Tl^{2+} + e^-$	−1.25
$Mn^{2+} = Mn^{3+} + e^-(H_2SO_4)$	−1.51
$Ce^{3+} = Ce^{4+} + e^-(HNO_3)$	−1.61

Within each group relatively simple molecules will be discussed first. In some cases a given substrate may be reduced by modes (*i*) and (*ii*) depending on the reductant and in other cases the mechanism is unknown, for example the reduction of perchlorate ion may involve either electron-acceptance or oxygen atom transfer.

6.2 SIMPLE ELECTRON ACCEPTANCE BY INORGANIC MOLECULES

6.2.1 *Perchlorate ion*

This ion, thermodynamically a powerful oxidant[19], is reduced by comparatively few reagents. However, both V(II) and V(III) are effective[527], *viz.*

$$8\,V(II) + ClO_4^- + 8\,H^+ = 8\,V(III) + Cl^- + 4\,H_2O$$

$$8\,V(III) + ClO_4^- + 4\,H_2O = 8\,VO^{2+} + Cl^- + 8\,H^+$$

The combined rate expression is

$$-d[V(II)]/dt = k_1[V^{2+}][ClO_4^-] + 2\,k_2[V^{3+}][ClO_4^-]$$

The second term corresponds to the concurrent reduction of ClO_4^- by the product, V(III), to give V(IV), which oxidises further V(II) very rapidly to V(III). k_1 is almost acid-independent but k_2 is doubled by a 15-fold variation in $[H_3O^+]$. At 50 °C and an acidity of 1 *M* ($\mu = 2.5$ *M*), $k_1 = 2.12 \times 10^{-5}$ l.mole^{-1}.sec^{-1} and $k_2 = 4.2 \times 10^{-6}$ l.mole^{-1}.sec^{-1}.

Ti(III) also reduces aqueous acidic ClO_4^- in accordance with the stoichiometry given for V(II)[528, 529] and with kinetics

$$-d[Ti(III)]/dt = [Ti(III)][ClO_4^-](k_1 + k_2[H_3O^+])$$

Chloride ion retards reaction, presumably by forming less reactive chlorocomplexes with Ti(III). At 40 °C ($\mu = 0.5$ *M*), $k_1 = (8.89 \pm 0.9) \times 10^{-2}$ l.mole^{-1}.sec^{-1} and $k_2 = (3.40 \pm 0.3) \times 10^{-2}$ l^2.mole^{-2}.sec^{-2}. $E_1 = 26.0$ kcal.mole^{-1} (± 15 %), $\Delta S_1^\ddagger = 20$ eu (± 15 %), $E_2 = 6 \pm 3$ kcal.mole^{-1} and $\Delta S_2^\ddagger = -46 \pm 23$ eu. Rate and activation data are also listed[529] for the reduction by $TiCl^{2+}$.

Reduction by $Ru(NH_3)_5^{2+}$ is simple second-order and acid-independent[530]. k_2 is 2.5×10^{-2} l.mole^{-1}.sec^{-1} at 25 °C ($\mu = 0.142$ *M*). Reduction by the corresponding hexammine is about 100 times slower.

Reduction by Eu(II) in a perchlorate medium is too fast for conventional study but chloride ion retards the reaction[531].

These various reductions may involve transfer of oxygen from substrate to metal atom, *viz.*

$$Ti^{3+} + ClO_4^- \rightleftharpoons Ti.ClO_4^{2+} \rightarrow TiO^{2+} + ClO_3\cdot \qquad \text{(slow)}$$

followed by rapid reduction of $\cdot ClO_3$. However, alternative schemes are consistent with available data, for example, an electron-transfer mechanism.

6.2.2 *Chlorate and bromate ions*

The stoichiometry of the reduction by V(II) is

$$6\ V(II) + ClO_3^- + 6\ H^+ = 6\ V(III) + Cl^- + 3\ H_2O$$

A stopped-flow examination produced the rate law[532]

$$-d[V(II)]/dt = k_2[V(II)][ClO_3^-]$$

with $k_2 = 4.85 \pm 0.95$ l.mole^{-1}.sec^{-1} at 0.5 °C, $[H^+] = 0.5\ M$, $\mu = 1.6\ M$ ($NaClO_4$). No V(IV) was detected spectroscopically and it is probable, in view of the comparative slowness of the V(II)–V(IV) reaction[533], that V(III) is the immediate product. The acidity dependence of k_2 could not be clearly established. Arrhenius parameters for k_2 were determined to be: $E = 10.5 \pm 0.6$ kcal.mole^{-1}, $\Delta S^\ddagger = -23 \pm 2$ eu. These resemble those for the Cr(II)–ClO_3^- reaction ($E = 11.8$ kcal.mole^{-1}, $\Delta S^\ddagger = -17$ eu) which shows kinetics[534]

$$-d[Cr(II)]/dt = k_2'[Cr(II)][ClO_3^-] = k_3[Cr(II)][ClO_3^-][H_3O^+]$$

where $k_2' = 39.0 \pm 1.4$ l.mole^{-1}.sec^{-1} at 20.0 °C ($[H_3O^+] = 0.55\ M$, $\mu = 2.00\ M$).

The kinetics of the reduction of chlorate ion by Ir(III) have been determined by controlled-potential electrolysis to be[535]

$$-d[IrCl_6^{3-}]/dt = k_2[IrCl_6^{3-}][ClO_3^-][H^+]^0$$

At an acidity of 0.2 M ($\mu = 1.0\ M$), $k_2 = (3.2 \pm 0.1) \times 10^{-4}$ l.mole^{-1}.sec^{-1} at 45 °C and $E = 30$ kcal.mole^{-1}. The reaction rate is acid-independent only at acidities below 0.2 M. k_2 is in good agreement with a value at the same temperature obtained polarographically (3.0×10^{-4} l.mole^{-1}.sec^{-1})[535]. The stoichiometry was found to be

$$6\ IrCl_6^{3-} + ClO_3^- + 6\ H^+ = 6\ IrCl_6^{2-} + Cl^- + 3\ H_2O$$

A simple electron transfer was proposed. Ir(VI) does not retard reaction and evidently the initial species $\cdot ClO_3^{2-}$ must oxidise further Ir(III) very rapidly.

The reduction of bromate by V(IV) shows a stoichiometry

$$2\, BrO_3^- + 10\, VO^{2+} + 4\, H_2O = 10\, VO_2^+ + Br_2 + 8\, H^+$$

and the kinetics in aqueous sulphuric acid (0.05 M) are[535a]

$$-\frac{d[V(IV)]}{dt} = \frac{kK[V(IV)][BrO_3^-]}{1+K[BrO_3^-]}$$

At 20 °C ($\mu = 0.10$ M) $k = 0.125$ sec^{-1} and $K = 88.8$ l.mole^{-1}; $E = 19 \pm 1$ kcal.mole^{-1}. An intermediate $BrO_3^-.VO^{2+}$ is considered to break down *via* oxygen-atom transfer to give $BrO_2\cdot$ and VO_2^+.

The U(IV) reduction of bromate follows two paths in acid perchlorate or nitrate media[535b], the rate law being

$$-d[U(IV)]/dt = k_2[U(IV)][BrO_3^-] + k_4[U(IV)][BrO_3^-][H_3O^+]^2$$

with corresponding activation parameters $E_2 = 27.8 \pm 0.6$ kcal.mole^{-1}, $\Delta S_2^\ddagger = 27 \pm 2$ eu, $E_4 = 23.9 \pm 1.0$ kcal.mole^{-1} and $\Delta S_4^\ddagger = 16 \pm 3$ eu.

6.2.3 *Chlorite ion and chlorine dioxide*

These present an interesting dichotomy in their reductions by *tris*(1,10-phenanthroline)iron(II) (ferroin)[535c]. That of ClO_2 to ClO_2^- is rapid, is first-order in each component ($k_2 = 1.86 \pm 0.13$ l.mole^{-1}.sec^{-1} at 35 °C) and is independent of acidity. Ferriin is the immediate product and an outer sphere electron-transfer is proposed. The reduction of ClO_2^- is much slower, proceeding at the same rate as dissociation of ferroin at high chlorite concentrations and a major product is feriin dimer, possibly $[(phen)_2Fe\text{–}O\text{–}Fe(phen)_2]^{4+}$. Clearly the reaction depends on ligand-displacement followed by an inner-sphere electron transfer.

Reduction of ClO_2^- by citrate, tartrate and EDTA complexes of Fe(II) and by $Cu(NH_3)_4^+$ has been examined polarographically[535d]. All four reactions are first-order in reductant and ClO_2^-. The data for the citrate and tartrate complexes were not reproducible but estimates of k_2 (27 °C) were obtained. The rate data are

Reductant	k_2 (27 °C)(*l.mole*$^{-1}$.*sec*$^{-1}$)	*E*(*kcal.mole*$^{-1}$)	$\Delta S^\ddagger$(*eu*)
Fe(II)citrate	$(6.3 \pm 0.5) \times 10^4$	—	—
Fe(II)tartrate	$(6.2 \pm 0.2) \times 10^4$	—	—
Fe(II)EDTA	$(3.0 \pm 0.3) \times 10^4$	13	$+2.5 \pm 0.2$
$Cu(NH_3)_4^+$	44 ± 3.7	13	-11.2 ± 0.2

6.2.4 *Molecular oxygen*

The reduction of O_2 is a 4-equivalent process, *viz.*

$$4\,M^{n+} + O_2 + 4\,H^+ = 4\,M^{(n+1)+} + 2\,H_2O$$

which can be regarded as a summation of one-equivalent steps, *viz.*

$$O_2 \underset{e^-}{\overset{H^+}{\rightarrow}} \cdot O_2H \underset{e^-}{\overset{H^+}{\rightarrow}} H_2O_2 \overset{e^-}{\rightarrow} OH\cdot + OH^- \underset{e^-}{\overset{H^+}{\rightarrow}} H_2O + OH^-$$

However, the kinetics of the reduction are often simplified in that the first step (production of $HO_2\cdot$) is rate-determining; the ensuing reactions, however, may confer radical-chain behaviour on the system. Hydrolysis of the reductant can also modify the kinetics.

These possibilities are featured in the autoxidation of U(IV) in perchloric acid[536]. The rate law is

$$-\mathrm{d}[U(IV)]/\mathrm{d}t = k[U(IV)][O_2]/[H_3O^+]$$

with $k = 2\times10^{14}\exp(-2.2\pm10^3/\boldsymbol{RT})\ \mathrm{sec}^{-1}$. This is evidence for slow one-equivalent reduction of O_2 preceded by a rapid hydrolysis, *viz.*

$$U^{4+} + H_2O \rightleftharpoons UOH^{3+} + H^+$$

$$UOH^{3+} + O_2 + H_2O \overset{k_1}{\rightarrow} UO_2{}^+ + HO_2\cdot + 2\,H^+ \qquad \text{initiation (slow)}$$

$$\left.\begin{array}{r} UO_2{}^+ + O_2 + H_2O \overset{k_2}{\rightarrow} UO_2{}^{2+} + HO_2\cdot + OH^- \\ HO_2\cdot + UOH^{3+} + H_2O \overset{k_3}{\rightarrow} UO_2{}^+ + H_2O_2 + 2\,H^+ \end{array}\right\} \qquad \text{propagation (fast)}$$

$$UO_2{}^+ + HO_2\cdot + H_2O \overset{k_4}{\rightarrow} UO_2{}^{2+} + H_2O_2 + OH^- \qquad \text{termination (fast)}$$

$$U^{4+} + H_2O_2 \overset{k_5}{\rightarrow} UO_2{}^{2+} + 2\,H^+ \qquad \text{(fast)}$$

Steady-state treatment for the transients ($HO_2\cdot$ and $UO_2{}^+$) leads to the observed rate law. The chain reaction is indicated by (*i*) strong catalysis by Cu^{2+} ions and (*ii*) partial and complete inhibition respectively by added Cl^- and Ag^+ ions. The inhibition by Ag^+ is not indefinite, however, and takes the form of an induction period, during which time metallic silver is deposited.

The effects of additives are accounted for by the following reactions[536]

Cu²⁺ catalysis $Cu^{2+}+UOH^{3+}+H_2O \rightarrow Cu^{+}+UO_2^{+}+3\,H^{+}$

$$Cu^{+}+O_2+H^{+} \rightarrow Cu^{2+}+HO_2\cdot$$

Cl⁻ inhibition $UO_2^{+}+Cl^{-}+2\,H_2O \rightarrow UOH^{3+}+Cl\cdot+3\,OH^{-}$

$$HO_2\cdot+Cl^{-}+H_2O \rightarrow H_2O_2+Cl\cdot+OH^{-}$$

$$UO_2^{+}+Cl\cdot \rightarrow UO_2^{2+}+Cl^{-}$$

$$HO_2\cdot+Cl\cdot \rightarrow O_2+Cl^{-}+H^{+}$$

Ag⁺ inhibition $UO_2^{+}+Ag^{+} \rightarrow UO_2^{2+}+Ag$

$$HO_2\cdot+Ag^{+} \rightarrow O_2+H^{+}+Ag$$

Gordon and Taube[537], however, have found only one oxygen atom of UO_2^{2+} is derived from labelled O_2 and Fallab[538] prefers an initial two-equivalent oxidation to U(VI).

The rate law for the autoxidation of Ti(III) chloride to Ti(IV) in aqueous HCl was found to be[539]

$$-d[O_2]/dt = \frac{k_1[Ti(III)][O_2]}{[H^+]} + k_2[Ti(III)]^{\frac{1}{2}}[O_2]$$

The first term is identical with that observed for autoxidation of U(IV) and a similar mechanism involving hydrolysis is probable for the initial stage, although the authors prefer an initial three-equivalent oxidation of Ti(III) to give Ti(VI), OH· and OH^-.

The autoxidation of V(III) to V(IV) in aqueous $HClO_4$ also displays the kinetics[540]

$$-d[V(III)]/dt = k_1[V(III)][O_2]/[H_3O^+]$$

and VOH^{2+} is regarded as the active reductant. However, it was noted that the results with different stock solutions of V(III) were in poor agreement and a few runs showed an induction period. Investigation of this point revealed strong catalysis by Cu(II) ions, the activation energy falling from 20.1 kcal.mole^{-1} for the uncatalysed reaction to 17.2 for the catalysed process, the rate of which is independent of p_{O_2} and is given by

$$-d[V(III)]/dt = k[V(III)][Cu(II)]/[H_3O^+]$$

This slow process generates Cu(I) which rapidly reduces O_2 to $HO_2\cdot$ (*vide infra*).

The kinetics of autoxidation of ferrous ion depend on the acidity of the medium and on the anion present. At low acidity (pH ~ 6) the autoxidation of ferrous ammonium sulphate follows the kinetics[541]

$$d[Fe(III)]/dt = k[Fe(II)][O_2][OH^-]$$

i.e. Fe(II) resembles U(IV), Ti(III) and V(III) in its mode of reduction and $FeOH^{2+}$ is presumed to be the active species.

The rate expression for the reaction in a perchlorate ion medium at higher acidities is quite different[542], *viz.*

$$-d[Fe(II)]/dt = k[Fe(II)]^2[O_2]/[H_3O^+]^{0.23}$$

(The acidity dependence is regarded as too slight to have any substantial bearing on the mechanism.) Cu(II) is a weak catalyst for the reduction. A mechanism involving attack of ferrous ion upon a Fe(II)–O_2 complex accounts for the kinetics, *viz.*

$$Fe^{2+} + O_2 \rightleftharpoons FeO_2^{2+} \quad \text{(fast)}$$

$$FeO_2^{2+} + Fe^{2+} \rightarrow 2\,Fe^{3+} + O_2^{2-} \quad \text{(slow)}$$

This concerted reduction by *two* ferrous species eliminates $HO_2\cdot$ (or O_2^-) as an intermediate and explains the weak catalysis by Cu(II) (which is strong for V(III) and V(IV) autoxidations). Weiss[542c] has suggested that the species $Fe^{2+}.O_2.Fe^{2+}$ may be a stable intermediate, but Wells[542a] explains the presence of two Fe(II) species in the rate law in terms of a pre-existing dimeric form of Fe(II) containing an H_2O bridge, for which there is evidence[542b]. The reduction is completed *via* the Fenton reaction (*vide infra*). The hydrogen peroxide dianion is probably never free but is protonated whilst complexed to Fe(III).

Both of these rate laws combine in that for the autoxidation of ferrous sulphate at 140–180 C° in dilute sulphuric acid solution[543, 544], *viz.*

$$-d[Fe(II)]/dt = k_1[Fe(II)][O_2] + k_2[Fe(II)]^2[O_2]$$

The activation energies are 13.4±2 and 16.3±2 kcal.mole^{-1}, respectively. At 30 °C only the second term is evident[543, 544], but the variation of rate with sulphate ion concentration reveals the two paths

$$2\,Fe^{2+} + O_2 \rightarrow 2\,Fe(III) + H_2O_2$$

$$FeSO_4 + Fe^{2+} + O_2 \rightarrow 2\,Fe(III) + H_2O_2$$

Pronounced acid-retardation occurs at acidities lower than 10^{-1} *M*[544]. The reaction is catalysed by cupric ions (unlike that in $HClO_4$)[543, 544], *viz.*

$$-d[Fe(II)]/dt = k_2[Fe(II)][Cu(II)]$$

References pp. 493–509

TABLE 20

AUTOXIDATION OF FERROUS IONS IN PRESENCE OF COMPLEXING IONS

Medium	*Acidity*	*Rate expression*	*Other effects*	*Ref.*
Aqueous H_3PO_4–Na_2HPO_4 ($\mu = 1$ *M* in $NaClO_4$)	0.02–0.08 *M*	$k[Fe(II)][O_2][H_2PO_4^-]^2$	Cu^{2+} catalysis	545
Aqueous HCl	4–8 *M*	$k[Fe(II)][O_2]/f[HCl]$ ($E = 14.6$ kcal.mole^{-1})	Charcoal catalysis	546
Aqueous pyrosphosphate	—	$k[Fe(II)][O_2]$	—	547
Aqueous sulphuric acid *plus* fluoride ion	—	$k[Fe(II)][O_2][F^-]$	—	548

Further data on this reaction are summarised in Table 20. The role of complexing anions is clear from the kinetics and also from relative rates. It appears that strongly bound ligands are associated with second-order reduction but that weakly bound ligands such as H_2O result in a third-order reaction. One view[542a] of the third-order term for dilute sulphuric acid (as for aqueous $HClO_4$) is that the active reductant is a bridged species of the type $(Fe^{2+}SO_4^{2-}Fe^{2+})^{2+}$.

It has also been observed[541] that sulphosalicylic acid strongly catalyses autooxidation of Fe(II) at pH 6. A complex of the chelate, Fe(II) and molecular oxygen is believed to be formed and to break down. Ethylenediaminetetraacetic and its analogues behave similarly[548a].

The fast interaction of O_2 with Fe(II)–cysteine complexes to give an oxygen adduct which rapidly undergoes one-electron breakdown to an Fe(III)–cysteine complex and $\cdot O_2^-$ has been examined by stopped-flow spectrophotometry at 570 nm[685]. Subsequent decomposition of the Fe(III) complex to yield Fe(II) and the disulphide, cystine, was much slower. Both mono- and bis-complexes of Fe(II) are involved and the reaction is first-order in both Fe(II) complex and O_2; k (mono) $= (5 \pm 1) \times 10^3$ l.mole^{-1}.sec^{-1} and k (bis) $= (2 \pm 0.5) \times 10^4$ l.mole^{-1}.sec^{-1} at 25 °C, corresponding to factors of 10^5 and 10^7 times faster than the analogous reactions with sulphosalicylic acid complexes[541] of Fe(II), a feature attributed to Fe(II)–S bonding in the cysteine complexes[686].

Autoxidation of $Fe(CN)_6^{4-}$ proceeds only in acidic solution, being first-order in $Fe(CN)_6^{4-}$ but only first-order in acid up to 10^{-5} *M* and rather less than first order at higher acidities (up to 6×10^{-5} *M*); however the acidity dependence can be rationalised in terms of the sum of the concentrations of $HFe(CN)_6^{3-}$ and $H_2Fe(CN)_6^{2-}$. No dependence upon oxygen concentration was obtained[548c].

Autoxidation of ferrous chloride in nonaqueous solvents is much faster than in water. The rate law is [548b]

$$-d[O_2]/dt = k[Fe(II)]^2[O_2]$$

Fe(III) inhibition, normally strong, vanishes in the presence of excess chloride ion and is attributed to competition between Fe(II) and Fe(III) for Cl^-. The relative rate sequence for different solvents is

$$C_2H_5OH\ 20.0 \qquad (CH_3)_2SO\ 2.6 \qquad CH_3OH\ 1.0$$

$$(CH_2OH)_2\ 0.06 \qquad H_2O < 0.02$$

If benzoin is added to an oxidation mixture, it is oxidised to benzil although the rate of oxygen uptake is unaffected. Fe(III) is not produced in the early stages and accordingly a mechanism including Fe(IV) is favoured, *viz.*

$$Cl_2FeOOFeCl_2 \rightarrow 2\ FeOCl_2 \qquad \text{(slow)}$$

$$FeOCl_2 + FeCl_2 \rightarrow 2\ Fe(III) \qquad \text{(fast)}$$

$$FeOCl_2 + C_6H_5CHOHCOC_6H_5 \rightarrow FeCl_2 + C_6H_5COCOC_6H_5 + H_2O \quad \text{(fast)}$$

The kinetics of the autoxidation of Pu(III) in aqueous sulphuric acid resemble[549] those of Fe(II), *viz.*

$$-d[Pu(III)]/dt = k[Pu(III)]^2[O_2][Pu(IV)]^0[H^+]^0$$

with $E = 19.0$ kcal.mole^{-1}. The accumulation of H_2O_2, the initial product of reduction, was prevented by adding ferrous ions. The reaction is strongly dependent on the concentration of sulphate ion but in a complex manner.

The general reluctance of reagents to reduce O_2 in one-equivalent states is further exemplified in the cases of V(II) and Cr(II). Autoxidation of V(II), ultimately to V(III), produces[550] an intermediate dimer, VOV^{4+}, identified spectrally[533] and a known product[533] of the reaction between V(II) and VO^{2+}. Clearly one path involves an initial two-equivalent oxidation of V(II) to V(IV) and Swinehart[550] calculates that 60 % of the oxidation follows this route. Cr(II) perchlorate produces a species containing two Cr(III) species linked by one oxo- or two hydroxo-bridges and Cr(IV) is proposed as the first intermediate[551].

The ease of autoxidation of Cu(I) is a source of the catalytic power of Cu(II) mentioned previously. In a hydrochloric acid medium the rate law determined was[552]

$$-d[Cu(I)]/dt = k[Cu(I)][O_2]f[H_3O^+]$$

At high cuprous ion concentrations the reaction becomes zero-order in reductant[552, 553] and is determined by the rate of passage of O_2 from the gaseous to the liquid phase. A mechanism involving a Cu(I)–O_2 complex is proposed[552], *viz.*

References pp. 493–509

$$O_2(\text{gas}) \underset{k_{-1}}{\overset{k_1}{\rightleftharpoons}} O_2\,(\text{solution})$$

$$Cu(I)+O_2 \underset{k_{-2}}{\overset{k_2}{\rightleftharpoons}} CuO_2{}^+$$

$$CuO_2{}^+ + H^+ \underset{k_{-3}}{\overset{k_3}{\rightleftharpoons}} Cu^{2+} + HO_2\cdot$$

At 25 °C k_2 is 250 l.mole^{-1}.sec^{-1} and in general is given by $10^{11}\exp(-12.0\times10^3)/RT)$ l.mole^{-1}.sec^{-1}. With slurries of cuprous chloride in acetic acid containing excess chloride, however, the following relation is found[553a],

$$-\mathrm{d}[Cu(I)]/\mathrm{d}t = k[Cu(I)]^2 p_{O_2}$$

With perchloric acid–acetonitrile solutions ($\mu = 0.1\ M$) the rate law becomes[553b]

$$\frac{-\mathrm{d}[Cu(CH_3CN)_2{}^+]}{\mathrm{d}t} = \frac{\mathrm{d}[Cu(II)]}{\mathrm{d}t} = \frac{k[Cu(CH_3CN)_2{}^+][O_2][H^+]}{[CH_3CN]^2}$$

where $k = (6.9\pm0.3)\times10^7$ l^2.mole^{-2}.sec^{-1} at 30 °C and $E = 8.4\pm0.4$ kcal. mole^{-1}. The mechanism proposed includes as its rate-limiting step the oxidation step featured in the preceding discussion, *viz.*

$$Cu(CH_3CN)_2{}^+ \rightleftharpoons Cu^+ + 2\,CH_3CN \quad \text{(fast)}$$

$$Cu^+ + O_2 \rightleftharpoons CuO_2{}^+ \quad \text{(fast)}$$

$$CuO_2{}^+ + H^+ \rightarrow Cu^{2+} + HO_2\cdot \quad \text{(slow)}$$

$$Cu^+ + HO_2\cdot \rightarrow Cu^{2+} + HO_2{}^- \quad \text{(fast)}$$

A recent stopped-flow study of the autoxidation of the complexes $Cu(NH_3)_2{}^+$ and Cu(imidazole)$_2{}^+$ in aqueous acetonitrile indicates a common rate law[554]

$$\mathrm{d}[Cu(II)]/\mathrm{d}t = k_3[Cu(I)][O_2][\text{free ligand}][H^+]^0$$

This implies a mechanism (L = either ligand)

$$CuL_2{}^+ + L \underset{k_{-1}}{\overset{k_1}{\rightleftharpoons}} CuL_3{}^+ \quad \text{(fast)}$$

$$CuL_3{}^+ + O_2 \underset{k_{-2}}{\overset{k_2}{\rightleftharpoons}} CuL_3O_2{}^+ \quad \text{(slow)}$$

$$CuL_3O_2{}^+ + L \overset{k_3}{\rightarrow} CuL_4{}^+ + \cdot O_2{}^- \quad \text{(fast)}$$

At 25 °C k_3 is $(5.5\pm0.6)\times10^3$ l^2.mole^{-2}.sec^{-1} (imidazole) and 1.6×10^4 l^2.

$mole^{-2}.sec^{-1}$ (NH_3). E for the autoxidation of the imidazole complex is under 2.0 $kcal.mole^{-1}$.

Examination[554a] by stopped-flow method of the autoxidation of the bipyridyl complex of Cu(I), $Cu(bipy)_2^+$, shows that it is first-order both in O_2 and in the complex, with k_2 (25 °C) = $(6.5 \pm 0.5) \times 10^3$ $l.mole^{-1}.sec^{-1}$. No ^{18}O was incorporated from labelled water into the product H_2O_2, indicating the O–O bond remains intact during reduction. The authors favour a *two*-equivalent reduction on thermodyamic grounds, proposing a rate-determining formation of a Cu(I)–O_2 complex which reacts rapidly with a second Cu(I) species, *viz.*

$$Cu(bipy)_2^+ + O_2 \rightleftharpoons Cu(bipy)_2O_2^+$$

$$Cu(bipy)_2O_2^+ + Cu(bipy)_2^+ + 2\,H^+ \rightarrow 2\,Cu(bipy)_2^{2+} + H_2O_2$$

Co(II) is autoxidised only in the form of certain complexes, for example with the tridentate di-2-amino-ethylene[555]. An aqueous solution of the chelate Co $(trien)^{2+}$ rapidly takes up O_2, *viz.*

$$2\,Co(trien)^{2+} + O_2 + trien \underset{k_{-1}}{\overset{k_1}{\rightleftharpoons}} Co_2(trien)_3.O_2^{3+}$$

A slow redox process follows to give a dimeric peroxo complex of Co(III), ($k_2 = 1.67 \times 10^{-3}$ sec^{-1} at 25 °C), *viz.*

$$[Co_2(II)(trien)_3.O_2]^{4+} \underset{k_{-2}}{\overset{k_2}{\rightleftharpoons}} [Co_2(III)(trien)_3(O_2^{2-})]^{4+}$$

this then breaks down slowly to 2 Co(III) and H_2O_2 with $k = 3.3 \times 10^{-4}$ sec^{-1}.

Bis-(L-histidinato) cobalt(II) (denoted CoL_2^{2+}) reacts rapidly with O_2 in aqueous solution to give $(CoL_2)_2O_2^{2-}$. The results of a stopped-flow examination[556] supports the mechanism

$$CoL_2^{2+} + O_2 \underset{k_{-1}}{\overset{k_1}{\rightleftharpoons}} CoL_2O_2^{2+}$$

$$CoL_2O_2^{2+} + CoL_2^{2+} \underset{k_{-2}}{\overset{k_2}{\rightleftharpoons}} (CoL_2)_2O_2^{4+}$$

where $k_1 = 5.7 \times 10^7 \exp(-5.6 \times 10^3/\boldsymbol{RT})$ $l.mole^{-1}.sec^{-1}$; at 4 °C $k_1 = 1.6 \times 10^3$ $l.mole^{-1}.sec^{-1}$ and $k_1k_2/k_{-1} = 5.5 \times 10^6$ $l^2.mole^{-2}.sec^{-1}$. Addition of various reagents to a solution containing the oxygenated species brings about decomposition at an identical rate, $k_{-2} = 6 \times 10^{17} \exp(-26.0 \times 10^3/\boldsymbol{RT})$ sec^{-1}. The formulation of the complexes is based on physical evidence cited by the authors[556], who have extended their investigations to a number of polyamine complexes[556a]. An analogous study of the formation of a complex between Co(II), glycylglycine and O_2 has been published[557].

Aqueous ammoniacal Co(II) solutions take up oxygen rapidly and reversibly[557a] to give the complex ion $[(NH_3)_5CoOOCo(NH_3)_5]^{4+}$. The pentammine complex is the major oxygen carrier, *viz.*

$$Co(NH_3)_5(H_2O)^{2+} + O_2 \underset{k_{-1}}{\overset{k_1}{\rightleftharpoons}} Co(NH_3)_5O_2^{2+} + H_2O$$

$$Co(NH_3)_5O_2^{2+} + Co(NH_3)_5(H_2O)^{2+} \underset{k_{-2}}{\overset{k_2}{\rightleftharpoons}} (NH_3)_5CoO_2Co(NH_3)_5^{4+} + H_2O$$

At 25 °C ($\mu = 2\,M$ NH_4NO_3) $k_1 = 2.5 \times 10^4$ l.mole^{-1}.sec^{-1}; $E_1 = 4.6$ kcal.mole^{-1} and $\Delta S_1^{\ddagger} = -25$ eu. k_{-2} (25 °C) = 56 sec^{-1}; $E_{-2} = 18.6$ kcal.mole^{-1} and $\Delta S_2^{\ddagger} = 9$ eu. $k_1 k_2 / k_{-1} k_{-2} = 6.3 \times 10^6$ l.^{2}mole^{-2}. ($= K_{\text{overall}}$) and $\Delta H_{\text{overall}}$ = 30 kcal.mole^{-1}. For the hexammine at 30 °C, $k_1 \sim 1.7 \times 10^3$ l.mole^{-1}.sec^{-1}.

A rather trivial case of autoxidation is that of Mo(V)[558] which exists as a diamagnetic dimer[559] at acidities below 2 M. The stoichiometry is

$$2\,Mo(V)_2 + O_2 + 4\,H^+ = 4\,Mo(VI) + 2\,H_2O$$

and the rate law is

$$-d[Mo(V)_2]/dt = k_1[Mo(V)_2][O_2]^0 f[H_3O^+]$$

The reaction rate is identical with that of reduction of iodine[558] and discussion of the mechanism is deferred to that section (p. 468).

Although the kinetic studies summarised here are useful guides to the gross features of mechanism it is evident from apparently closely related autoxidations, *e.g.* those of V(III) and U(IV), that subtle factors operate. Fallab[538] has pointed out that these reductants give similar kinetics and possess similar reduction potentials, yet differ in autoxidation rate by a factor of 3×10^5, and has discussed differences of this type in terms of the stereochemistry of the electron-transfer process in the coordination sphere.

An altogether different interaction is that between chelates of *oxidising*, as well as reducing, metal ions and oxygen at 100 °C in diphenyl ether solution[560–562]. Attack is on the ligand although the precise site remains unestablished; however, deuteration of the 3-position of tris-acetylacetone iron(III), ($Fe(acac)_3$), leaves the rate unchanged. Only chelates with metal ions capable of oxidation or reduction undergo autoxidation but these cover only a relatively small range of reactivity. The products are a complex mixture, thus $Fe(acac)_3$ absorbs 3.86 moles of oxygen to yield CO_2, H_2O, biacetyl and acetic acid as major products and smaller amounts of mesityl oxide and acetylacetone, and a residue is left of empirical formula $FeC_7H_8O_{4.85}$. Other chelates may enhance the rate for Fe(III) compared with acetylacetone, *e.g.* dibenzoylmethane and 3-methylacetylacetone, whilst others suppress reaction, *e.g.* 3-phenylacetylacetone, 3-benzylacetylacetone and dipivaloyl methane. The rate sequence for autoxidation of acetylacetonates is

$$V(III) > Ce(IV) > Ni(II) > Mn(III) > Fe(III) > Co(II) > Th(IV)$$

The kinetics for $Fe(acac)_3$ and tris-3-methylacetylacetoneiron(III) are

$$-d[O_2]/dt = k[\text{chelate}]^{0.5}[O_2]^{0.24}$$

with $E = 22.0$ kcal.mole^{-1}. However, $Mn(acac)_2$ showed a dependence of rate upon the 0.22 power of the chelate concentration. Reaction of $Fe(acac)_3$ is inhibited by benzoyl peroxide, azobisisobutyronitrile and ferrocene, but not by 2,4,6-tri-*tert*-butylphenol[562]. These various observations are accommodated by the mechanism

$$Fe(III)(acac)_3 \underset{k_{-1}}{\overset{k_1}{\rightleftharpoons}} R\cdot + Fe(II)(acac)_2 \quad (K_1) \tag{79}$$

$$R\cdot \underset{k_{-2}}{\overset{k_2}{\rightleftharpoons}} L\cdot \tag{80}$$

$$L\cdot + O_2 \xrightarrow{k_3} LO_2\cdot \tag{81}$$

$$2\,LO_2\cdot \xrightarrow{k_4} P \tag{82}$$

$$LO_2\cdot + Fe(II)(acac)_2 \xrightarrow{k_5} Q \tag{83}$$

$$Fe(II)(acac)_2 \underset{k_{-6}}{\overset{k_6}{\rightleftharpoons}} Fe(II)_{o.c.} \tag{84}$$

$$Fe(II)_{o.c.} + O_2 \xrightarrow{k_7} (acac)_2FeOO\cdot \tag{85}$$

(and subsequent reactions)
R· and L· are radicals in equilibrium (and *not* resonance), *viz.*

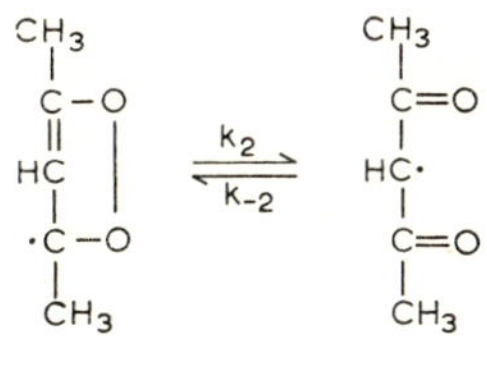

```
CH3              CH3
 |                |
 C-O             C=O
 ||  |    k2      |
 HC  |   ⇌       HC·
 |   |    k-2     |
·C-O             C=O
 |                |
CH3              CH3
(R·)             (L·)
```

Step (84) corresponds to opening of the chelate, *viz.*

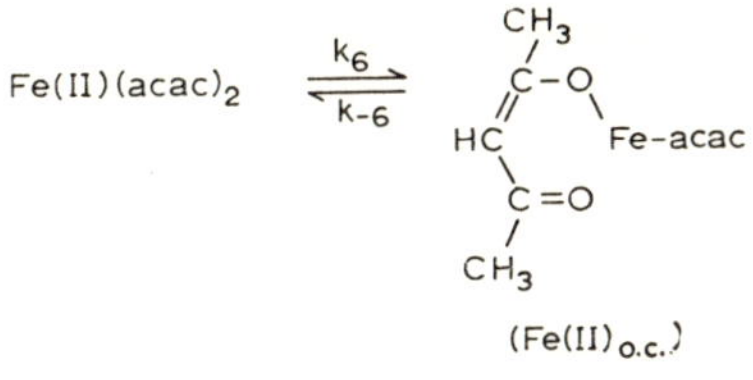

```
                        CH3
                        /
                       C-O
               k6     //   \
Fe(II)(acac)2  ⇌     HC     Fe-acac
               k-6     \
                        C=O
                        /
                      CH3
                   (Fe(II)o.c.)
```

Steady-state treatment for L· leads to a rate law

$$-d[O_2]/dt = k_2(K_1[Fe(acac)_3])^{0.5}$$

One intermediate thought to be the precursor of biacetyl is 2,3,4-pentanetrione which may be formed from 3-peroxyacetylacetone radical.

6.2.5 *Water*

Water is reduced by Ru(II) in the form of $RuCl_4^{2-}$ in deaerated, acidic solution[563]. At 30 °C (pH 1.5, $\mu = 2.5$ *M* KCl) the reductant disappears in first-order fashion with $k_1 = (4.64 \pm 0.15) \times 10^{-5}$ sec^{-1} to give Ru(III). The reaction rate is unaffected by addition of Ru(III), chloride ion, acid or oxygen. E is 26.4 $kcal.mole^{-1}$ and k_{D_2O} (30 °C) $= (3.8 \pm 0.1) \times 10^{-5}$ sec^{-1}. The absence of an appreciable primary isotope effect is regarded as evidence for electron transfer (as opposed to transfer of a hydrogen atom from coordinated water), *viz.*

$$H_2O + Ru(II) \rightarrow e^-_{aq} + Ru(III)$$

$$e^-_{aq} + H_3O^+ \rightarrow \tfrac{1}{2}H_2 + H_2O$$

A further test of the intermediacy of e^-_{aq} would be the reduction of nitrous oxide to nitrogen.

6.2.6 *Sulphur dioxide*

Sulphur dioxide is reduced by pentacyanocobaltate is several stages[564] (*cf.* the reduction of *p*-benzoquinone), *viz.*

$$2\,Co(CN)_5^{3-} + SO_2 \rightarrow \left[(NC)_5Co\text{–}\overset{\overset{O}{|}}{\underset{\underset{O}{|}}{S}}\text{–}Co(CN)_5\right]^{6-} \quad \text{(fast)} \quad \xrightarrow{k_1}$$

$$\text{compound II} + Co(CN)_5OH^{3-} \xrightarrow{k_2} 2\,Co(CN)_5OH^{3-} + \tfrac{1}{2}S_2O_4^{2-}$$

At 25 °C $k_1 = 4 \times 10^{-3}$ sec^{-1} and $k_2 = 5 \times 10^{-5}$ sec^{-1}.

6.2.7 *Xenon trioxide*

Xenon trioxide is reduced in aqueous solution, in a thermal reaction by Pu(III) and photochemically by Np(V). The stoichiometry of the Pu(III) reduction is

$$6\,Pu(III) + XeO_3 + 6\,H^+ = 6\,Pu(IV) + Xe + 3\,H_2O$$

and the kinetics are[565]

$$-\mathrm{d}[\mathrm{Pu(III)}]/\mathrm{d}t = k_2[\mathrm{Pu(III)}][\mathrm{XeO_3}][\mathrm{H_3O^+}]^0$$

with $k_2 = 1.6 \times 10^{-2}$ l.mole^{-1}.sec^{-1} at 30 °C ($\mu = [\mathrm{H^+}] = 2\ M$).

The thermal reduction by Np(V) is a slow reaction of complex kinetics, but it proceeds readily under the influence of light with kinetics[565]

$$-\mathrm{d}[\mathrm{Np(V)}]/\mathrm{d}t = k_1[\mathrm{XeO_3}]$$

Under the conditions employed ($T = 60$ °C, $\mu = 2\ M$) k_1 was 6.3×10^{-6} sec^{-1}. Photoactivation of XeO_3 is considered to be the primary process but no quantum yield was reported.

6.3 SIMPLE ELECTRON ACCEPTANCE BY ORGANIC MOLECULES

A very large number of unsaturated molecules readily accept an electron to give a radical-anion, which may be stable and which can often be identified unequivocally by means of ESR. However, fewer kinetics have been reported for these systems, probably because the most widely used reducing agents are less amenable to kinetic study than their oxidising counterparts. One widely utilised reducing system is sodium dissolved in ammonia but there the active reagent, the electron, is already largely or completely free from the influence of the parent metal atom and the rate of reduction of the substrate is very often diffusion-controlled[567].

One reagent which has received detailed attention, however, is chromous ion, which possibly acts as a model for other divalent reducing metal ions and is the subject of a recent review[568].

6.3.1 Acetylenes

Cr(II) readily reduces a wide range of (but not all) acetylenic compounds to give mainly the corresponding *trans*-olefin. Propargyl alcohol is reduced with kinetics[569]

$$-\mathrm{d}[\mathrm{Cr(II)}]/\mathrm{d}t = k[\mathrm{HC{\equiv}C{\cdot}CH_2OH}][\mathrm{Cr(II)}]^2$$

The activation energy is almost zero, but $\Delta S^{\ddagger} = -60$ eu. The effect of acidity is very slight at constant ionic strength, but $k_{\mathrm{D_2O}}/k_{\mathrm{H_2O}} = 2.86$. The following mechanism was proposed

$$\mathrm{Cr(II)} + \mathrm{RC{\equiv}CR} \rightleftharpoons \left[\begin{matrix}\mathrm{RC{\equiv}CR} \\ \vdots \\ \mathrm{Cr(II)}\end{matrix}\right] \qquad \text{(fast)}$$

$$\left[\begin{matrix}\mathrm{RC{\equiv}CR} \\ \vdots \\ \mathrm{Cr(II)}\end{matrix}\right] + \mathrm{Cr(II)} \rightarrow 2\mathrm{Cr(III)} + \begin{matrix}\mathrm{R} & & \mathrm{H} \\ & \mathrm{C{=}C} & \\ \mathrm{H} & & \mathrm{R}\end{matrix} \qquad \text{(slow)}$$

This accords with the observed sequence of rates, *viz.*

$$HOCH_2C{\equiv}CCH_2OH \sim HC{\equiv}CCH_2OH > CH_3C{\equiv}CCH_2OH >$$
$$> HC{\equiv}CBu^n \gg CH_3C{\equiv}CCH_2CH_3$$

for the ability of the acetylene group to form a π-complex with Ag^+ decreases on alkylation[570]. Chelation of Cr(II) with ethylenediaminetetraacetic acid reduces the reaction velocity and a new rate law is found, *viz.*

$$-d[Cr(II)]/dt = k[Cr(II)]^3[acetylene]^0$$

That some modification of the above mechanism is necessary is apparent from the observation[571] that during the Cr(II) perchlorate reduction of acetylene dicarboxylic acid to fumaric acid at pH 2–3, a light yellow colour is first seen which changes to red-brown within a few minutes. Variation of the relative concentrations reveals this red species to have the composition acetylene dicarboxylic acid: 2 Cr(II). At higher acidity (0.4 M) the red complex is not formed. The red species slowly changes into a second species (II) (λ_{max} 402 and 516 nm) which is also slowly formed in the more acid solution. This second species changes slowly to a third, violet species(III), which finally gives $Cr(H_2O)_6{}^{3+}$. A pH titration showed that the protons are not consumed during the production of the red complex but are consumed thereafter. The reaction giving the red complex has kinetics

$$-d[Cr(II)]/dt = k[acetylene][Cr(II)]^2/[H_3O^+]$$

The reaction of the red complex to give species II is first-order, but the rate is dependent on acidity, *viz.*

$$1/k = a+b/[H_3O^+] \qquad (sec)$$

Polagraphic evidence indicates species II to have a 2 Cr : 1 substrate molecule ratio.

The reaction at pH 2 is four-step, *viz.*

$$2\ Cr(II)+HO_2CC{\equiv}CCO_2H \rightarrow \text{red complex} \tag{86}$$

$$\text{red complex}+n\ H_3O^+ \rightarrow \text{species II} \tag{87}$$

$$\text{species II} \rightarrow \text{species III}+Cr(III) \tag{88}$$

$$\text{species III} \rightarrow Cr(III)+\text{fumaric acid} \tag{89}$$

The inverse acidity dependence of step (86) probably stems from dissociation of the substrate rather than from hydrolysis of Cr^{2+} in view of the slight acidity dependence of the reduction of propargyl alcohol. Stepwise coordination of two Cr(II) species to $HO_2CC{\equiv}CCO_2^-$ would given rise to the observed kinetics, although the authors suggest the second Cr(II) may be sited at the CO_2^- moiety rather than at the triple bond.

This reaction is currently under further examination[571].

6.3.2 *Quinones*

The reduction by Fe(II) is one step of the equilibrium considered in the section on oxidation of phenols by Fe(III)[515], (Q = *p*-benzoquinone), *viz.*

$$\text{Fe(III)} + \text{QH}^- \rightleftharpoons \text{Fe(II)} + \text{QH}\cdot$$

$$\text{QH}\cdot \rightleftharpoons \cdot\text{Q}^- + \text{H}^+$$

$$\text{Fe(III)} + \text{Q}^- \rightleftharpoons \text{Fe(II)} + \text{Q}$$

The kinetics of the reduction step in a perchlorate ion medium are

$$-\mathrm{d}[\mathrm{Q}]/\mathrm{d}t = k_2[\mathrm{Fe(II)}][\mathrm{Q}]$$

with $k_2 = 4.2 \times 10^4 \exp(-7.9 \times 10^3/\boldsymbol{RT})$ l.mole^{-1}.sec^{-1}. This accords well with the other rate coefficient involved and the observed equilibrium constant. Higher orders in Fe(II) have been reported with HCl and H_2SO_4 media, and Fe(II).Q. Fe(II) dimers are postulated[515a]. The effects of adding Cl^- and SO_4^{2-} ions and of introducing alcohols have also been noted[515b, c].

Similar kinetics are reported for the reduction of several quinones by ferrocyanide, but the rate coefficients, k are affected by added ferricyanide[572], *viz.*

$$\frac{1}{k} = \frac{1}{k'} + \frac{C[\mathrm{Fe(CN)_6}^{3-}]}{[\mathrm{Fe(CN)_6}^{4-}]}$$

where C is some function of pH, depending on the quinone employed. $\Delta G^\ddagger$ was obtained using an A factor of 10^{11} and agreed well with calculated values.

Pentacyanocobaltate ion reduces *p*-benzoquinone in several stages[573]. An initial, fast reaction produces the bridged species

$$[(\mathrm{CN})_5\mathrm{CoOC_6H_4OCo(CN)_5}]^{6-}$$

This compound breaks down to $[Co(CN)_5H_2O]^{2-}$ and $[(CN)_5CoOC_6H_4OH]^{3-}$. The latter in turn undergoes an internal redox reaction to give $[Co(CN)_5H_2O]^{2-}$ and hydroquinone. Both first-order steps show general acid catalysis.

Reduction by pentacyanocobaltate(I) ion, $Co(CN)_5H^{3-}$, in alkaline solution ($pH > 9$) proceeds *via* two paths, a slow pH-independent direct reaction ($k_2 = 13.5$ l.mole^{-1}.sec^{-1} at 0 °C) and a fast reaction ($k_2 \sim 10^9$ l.mole^{-1}.sec^{-1}) involving the basic form of the reductant, $Co(CN)_5{}^{4-}$. Both reactions lead to $[(CN)_5CoOC_6H_4OH]^{3-}$ as the sole product[573a].

6.3.3 *Nitro compounds*

The reduction of *m*-nitrochlorobenzene to the corresponding amine by ethanolic stannous chloride follows simple second order kinetics[574]. The reaction is retarded by addition of acid but no effort was made to preserve constant ionic strength.

6.3.4 *Carbonyl compounds*

The reduction of benzaldehyde and cinnamic aldehyde by V(II) has the stoichiometry

$$\mathrm{ArCHO + 2\,V(II) + 2\,H^+ = \begin{array}{c} \mathrm{ArCHOH} \\ | \\ \mathrm{ArCHOH} \end{array} + 2\,V(III)}$$

and in acetone–acetic acid mixture the kinetics of both reductions are of the type[575]

$$-\mathrm{d[V(II)]/d}t = k_2[\mathrm{ArCHO}][\mathrm{V(II)}]f[\mathrm{H_3O^+}]$$

Simple electron transfer followed by rapid protonation would give the ketyl radical which dimerises very rapidly.

Cr(II) reduction of benzaldehyde in aqueous acidic ethanol also yields largely hydrobenzoin although other products were characterised, *e.g.* hydrobenzoin monoethyl ether, and a 1 : 1 stoichiometry is preserved. The rate law in ethanol acidified both with $HClO_4$ and HCl is complex; for example, in HCl–ethanol[700],

$$-\mathrm{d[Cr(II)]/d}t = k_1[\mathrm{Cr(II)}][\mathrm{PhCHO}][\mathrm{H_3O^+}] + k_2[\mathrm{Cr(II)}][\mathrm{PhCHO}]^2[\mathrm{H_3O^+}]$$
$$+ k_3[\mathrm{Cr(II)}][\mathrm{PhCHO}][\mathrm{H_3O^+}][\mathrm{Cl^-}] + k_4[\mathrm{Cr(II)}][\mathrm{PhCHO}]^2[\mathrm{H_3O^+}][\mathrm{Cl^-}]$$

but, in $HClO_4$–ethanol, $k_3 = k_4 = 0$. At 25 °C (HCl–ethanol), $k_1 = 5.9 \times 10^{-5}$ l^2.mole^{-2}.sec^{-1}, $k_2 = 2.4 \times 10^{-4}$ l^3.mole^{-3}.sec^{-1}, $k_3 = 6.3 \times 10^{-5}$ l^3.mole^{-3}.sec^{-3} and $k_4 = 7.6 \times 10^{-4}$ l^4.mole^{-4}.sec^{-1}. The complexity of the individual terms of the rate law is attributed to pre-equilibria giving rise to protonated hydrate, hemiacetal and halohydrin structures, and ethers derived from these,

e.g. (R = H or C_2H_5)

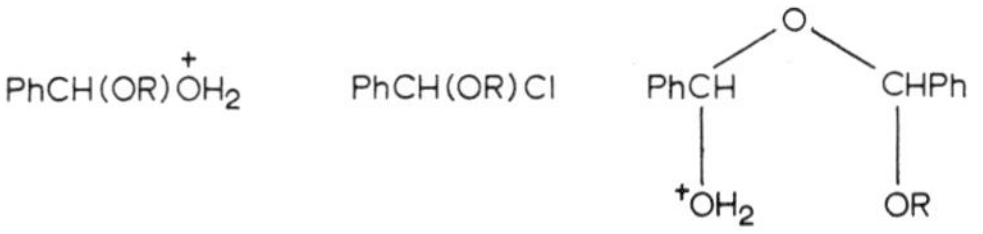

$PhCH(OR)\overset{+}{O}H_2$ $PhCH(OR)Cl$ $PhCH(^+OH_2)-O-CHPh(OR)$

followed by the slow reduction step to give, respectively,

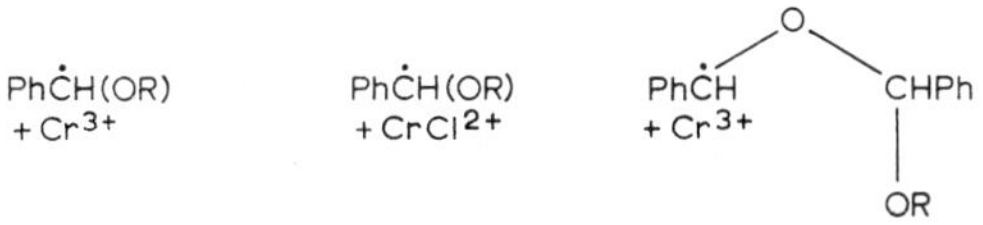

$Ph\dot{C}H(OR) + Cr^{3+}$ $Ph\dot{C}H(OR) + CrCl^{2+}$ $Ph\dot{C}H-O-CHPh(OR) + Cr^{3+}$

The process of ligand transfer to give $CrCl^{2+}$ is well-established in the case of Cr(II) reduction of alkyl halides.

6.3.5 *Unsaturated dicarboxylic acids*

Detailed studies of 1 : 1 complex formation between V^{2+} and maleic and fumaric acids, which precedes reduction to succinic acid, *cis–trans* isomerisation and exchange of the double bond hydrogens, are relevant[692, 693] to the complex kinetics (A = substrate)

$$-d[V^{2+}]/dt = [V^{2+}]^2(k_1[A]+k_2[A][H_3O^+]+k_3[A]^2+k_4[A]^2[H_3O^+])$$

which are also followed when A = citraconic and chloromaleic acids and methyl maleate[701]. The rate coefficients at 23 °C can be summarised (l, mole, sec units):

	Maleic acid	*Fumaric acid*	*Citraconic acid*
k_1	∽0.018	0.003	
k_2	∽0.014	0.005	
k_3	0.420	0.100	0.059
k_4	0.393	0.044	0.178

The mechanism suggested by these kinetics depends on the simultaneous oxidation of *two* V^{2+} ions in substrate–metal ion complexes so that free radicals are not produced. A few data on Cr(II) reduction of these unsaturated acids indicate simple second-order kinetics[702].

6.4 ELECTRON ACCEPTANCE FOLLOWED BY CLEAVAGE

The general equations describing this process are

$$X\text{–}Y + M^{n+} \xrightarrow{k_1} \cdot(X\text{–}Y)^- + M^{(n+1)+}$$

$$\cdot(X\text{–}Y)^- \xrightarrow{k_2} \begin{cases} X\cdot + Y^- \\ X^- + Y\cdot \end{cases}$$

However, no evidence for even a transitory existence of $\cdot(X\text{–}Y)^-$ has been obtained except in the cases of $X = Y =$ halogen or CNS^- (ref. 575a). and it is probable that the breakdown is concerted with reduction. The mole of cleavage appears to be governed by the relative electron affinities of X· and Y·, for example, hypobromous acid[576] and hydroxylamine[577] are cleaved by reducing ions as follows

$$HOBr + Fe^{2+} \rightarrow HO^- + Br\cdot + Fe^{3+}$$

$$NH_2OH + Ti^{3+} \rightarrow HO^- + NH_2\cdot + Ti^{4+}$$

The most celebrated example of this process is the Fenton reaction which is discussed at some length to illustrate the general characteristics of such reductions. An interesting recent example is the reduction of organic halides, *viz.*

$$RX + M^{n+} \rightarrow M^{(n+1)+} + X^- + R\cdot$$

$$R\cdot \rightarrow \tfrac{1}{2}R_2, RH, R\text{–}M^{n+}$$

6.4.1 The Fenton reaction

The overall reaction is

$$2\,Fe(II) + H_2O_2 + 2\,H^+ = 2\,Fe(III) + 2\,H_2O$$

but it has long been known that a reacting mixture is capable of oxidising a wide range of compounds unreactive towards H_2O_2 alone. Several detailed kinetic investigations have been made. Barb *et al.*[448] determined the rate law

$$-d[Fe(II)]/dt = 2\,k[Fe(II)][H_2O_2] \tag{90}$$

where k is the rate coefficient of the step

$$Fe(II) + H_2O_2 \rightarrow Fe(III) + OH\cdot + OH^- \tag{91}$$

(90) holds even in weakly acid solution (pH 3) when excess Fe(II) is present, with k (0 °C) = 12.6±0.3 l.mole^{-1}.sec^{-1} and k (25 °C) = 53.0±0.7 l.mole^{-1}.sec^{-1}. It also holds in very acidic solution even when peroxide is in excess but at lower acidities and in the presence of excess peroxide, deviation occurs as the reaction proceeds. This deviation increases with increasing peroxide concentration, but reaches a limit, and has been shown to be associated with the presence of Fe(III); however, addition of Cu(II) results in deviation from (90) from the outset. Evolution of oxygen takes place with excess peroxide and the volume obtained increases with peroxide concentration, but reaches a limit, which can, however, be increased by adding Fe(III) or Cu(II). However, the deviations from the rate law and the oxygen evolution occasioned by Fe(III) were suppressed by adding fluoride ions. Barb *et al.*[448] modified the original scheme of Haber and Weiss[450] as follows

$$\mathrm{Fe(II)+H_2O_2 \xrightarrow{k_{92}} Fe(III)+OH\cdot+OH^-} \tag{92}$$

$$\mathrm{Fe(II)+OH\cdot \rightarrow Fe(III)+OH^-} \tag{93}$$

$$\mathrm{OH\cdot+H_2O_2 \rightarrow H_2O+HO_2\cdot} \tag{94}$$

$$\mathrm{Fe(II)+HO_2\cdot \rightarrow Fe(III)+HO_2^-} \tag{95}$$

$$\mathrm{Fe(III)+HO_2\cdot \xrightarrow{k_{96}} Fe(II)+O_2+H^+} \tag{96}$$

that is, they have discounted the step

$$\mathrm{HO_2\cdot+H_2O_2 \rightarrow O_2+OH\cdot+H_2O} \tag{97}$$

as a sources of oxygen. k_{92} may be represented as $4.45\times10^8 \exp(-9.4\times10^3/\boldsymbol{R}T)$ l.mole^{-1}.sec^{-1}.

At low peroxide/Fe(II) ratios only (92) and (93) occur. As the ratio increases, OH· increasingly attacks H_2O_2 to give the observed kinetic deviation and evolution of O_2. At high ratios the competition between Fe(II) and Fe(III) for $HO_2\cdot$ dominates and the evolution of O_2 becomes independent of peroxide concentration. The effect of Cu(II) is ascribed to reactions

$$\mathrm{Cu^{2+}+HO_2\cdot \xrightarrow{k_{98}} Cu^++O_2+H^+} \tag{98}$$

$$\mathrm{Cu^++Fe^{3+} \rightarrow Cu^{2+}+Fe^{2+}} \tag{99}$$

where $k_{98} > k_{96}$. The greater prevalence of kinetic deviation and oxygen evolution at higher pH is believed to originate in hydrolysis of Fe^{3+} and ionisation of $HO_2\cdot$. A further possibility noted by the authors, following Bray and Gorrin[578], is the oxidation of Fe(III) to Fe(IV), *viz.*

$$\mathrm{OH\cdot+Fe^{3+} \rightarrow FeOH^{3+} \rightleftharpoons FeO^{2+}+H^+} \tag{100}$$

The main alternative to the modified Haber–Weiss theory is a two-equivalent oxidation of Fe^{2+} by peroxide[578], *viz.*

$$Fe^{2+} + H_2O_2 \rightarrow FeO_2{}^{+} + H_2O$$

$$FeO^{2+} + H_2O_2 \rightarrow Fe^{2+} + H_2O + O_2$$

This has received support from work with ^{18}O-labelled peroxide under conditions when O_2 is evolved[579], but a stopped-flow examination[580] indicated the immediate product to be at least 99 % $Fe^{3+} + FeOH^{2+}$; in contrast, reduction of HOCl and O_3 gave significant quantities of the dimeric $(FeOH)_2{}^{4+}$ formed very probably as follows

$$Fe^{2+} + HOCl \rightarrow Fe(IV) + OH^- + Cl^-$$

$$Fe(IV) + Fe^{2+} \rightarrow [Fe(III)]_2 \qquad \text{(very fast)}$$

The absence of dimer in the Fenton reaction is regarded as evidence for almost complete one-equivalent reduction.

Subsequent studies have been concerned mainly with questions of detail. Wells and Salam[581] have explored the increase of rate with pH at pH > 3 and their data are consistent with $Fe(OH)_2$ as reductant in the region of high pH. The rate with $Fe(OH)_2$ can be summarised as

$$k_2 = 3.7 \times 10^{11} \exp(-12.0 \times 10^3/\boldsymbol{RT}) \text{ l.mole}^{-1}.\text{sec}^{-1}$$

Hardwick[582] has compared his own data with those of several groups[448,583-585] on the rate of the Fenton reaction. The following equations fit most data

$$HClO_4 \quad k_{92} = 5.3 \times 10^8 \exp(-9.45 \times 10^3/\boldsymbol{RT}) \text{ l.mole}^{-1}.\text{sec}^{-1}$$

$$H_2SO_4 \ (0.4\ M) \quad k_{92} = 9.6 \times 10^8 \exp(-9.75 \times 10^3/\boldsymbol{RT}) \text{ l.mole}^{-1}.\text{sec}^{-1}$$

The effects of complexing Fe(II) have been studied. Wells and Salam[586] believe the slightly faster rate in sulphuric acid derives from a faster cleavage of peroxide by $FeSO_4$ compared with Fe^{2+}. Systematic addition of various anions brings about an increase in k_2 although a limiting value is always achieved with a given anion at a fixed temperature[586, 587]. The rate laws suggest the following are the active reductants at 25 °C ($\mu = 1\ M$)

$$FeF_2; \quad k_{92} = 2.5 \times 10^{12} \exp(-14.0 \times 10^3/\boldsymbol{RT}); \quad k\ (25\ °C) = 118 \text{ l.mole}^{-1}.\text{sec}^{-1}$$

$$FeCl^+; \quad k_{92} = 1.2 \times 10^8 \exp(-8.4 \times 10^3/\boldsymbol{RT}); \quad k\ (25\ °C) = 68.5 \text{ l.mole}^{-1}.\text{sec}^{-1}$$

$$\text{FeBr}^+; \quad k_{92} = 8.9 \times 10^8 \ \exp(-9.6 \times 10^3/\boldsymbol{RT}); \quad k\ (25\ °\text{C}) = 78.5\ \text{l.mole}^{-1}.\text{sec}^{-1}$$

$$\text{FePF}_6{}^+; \quad k_{92} = 1.6 \times 10^{10} \ \exp(-11.0 \times 10^3/\boldsymbol{RT}); \quad k\ (25\ °\text{C}) = 124\ \text{l.mole}^{-1}.\text{sec}^{-1}$$

Po and Sutin[688] have disputed both the extent of the catalytic effect of chloride ion reported by Wells and Salam[586] and the formation constant of 5.54 (25 °C, $[Cl^-] = 0.300\ M$, $\mu = 1.00$) for $FeCl^+$ estimated thereby. Wells[687] has replied that the value of k_2 of Po and Sutin at zero chloride concentration is artifically increased because of the presence of stabiliser in their peroxide, consequently masking the catalysis.

Chelates of Fe(II), for example with 2,2′-bipyridyl,9,10–phenanthroline[587] and N-(2-pyridylmethylene)aniline[588].

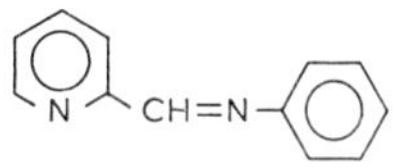

N CH=N

react with hydrogen peroxide at the natural pH of the system with kinetics

$$-\text{d[Fe(II)]}/\text{d}t = k[\text{complex}][\text{H}_2\text{O}_2]^0$$

The slow step is thought to be loss of one ligand from the complex. These chelates and those with a series of ring-substituted Schiff bases undergo acid fission at exactly the same rate as oxidation by peroxide.

The effect of adding 2,2′-bipyridyl is more complicated[589]; it does seem that the mono- and bis-complexes of Fe(II) are capable of reducing peroxide although the rate coefficients can only be given within limits.

U(IV) reduction of acidic H_2O_2 to give U(VI) has a stoichiometry $\Delta[H_2O_2]/\Delta[U(IV)]$ slightly in excess of unity even at high peroxide/reductant ratios[590]. The reaction is second-order with k_2 (25 °C, 2 M $HClO_4$) equal to 0.95 $\text{l.mole}^{-1}.\text{sec}^{-1}$. Slight catalysis by dissolved oxygen was detected and E is of the order of 16 kcal.mole^{-1}. Mild retardation by Cu(II) and Co(II) indicate at least partial chain character.

The Cr(II) reduction of H_2O_2 involves transfer of *one* oxygen atom from the peroxide into the coordination shell of the resulting hexaaquochromium(III)[591]. It would seem that O–O fission is an inner-sphere process, *viz.*

$$\text{Cr(H}_2\text{O)}_6{}^{2+} + \text{H}_2\text{O}_2 \rightleftharpoons [(\text{H}_2\text{O}_2)_5\text{Cr–O–OH}]^+ + \text{H}^+ \qquad \text{(fast)}$$

$$[(\text{H}_2\text{O})_5\text{Cr–O–OH}]^+ \rightarrow [(\text{H}_2\text{O})_5\text{Cr–O}^-]^+ + \text{OH}\cdot \qquad \text{(slow)}$$

$$\downarrow$$

$$[(\text{H}_2\text{O})_5\text{CrOH}]^{2+}$$

In alkaline cyanide solution the kinetic law is found to be[696]

$$-\frac{d[\mathrm{Cr(II)}]}{dt}$$

$$= \frac{2[\mathrm{H_2O_2}]_{\mathrm{total}}[\mathrm{Cr(II)}](k_0K_1[\mathrm{CN^-}]+k_1+k_2K_cK_a/[\mathrm{H_3O^+}][\mathrm{CN^-}])}{(1+K_a/[\mathrm{H_3O^+}])(1+K_1[\mathrm{CN^-}])+K_cK_a[\mathrm{H_2O_2}]_{\mathrm{total}}/[\mathrm{H_3O^+}][\mathrm{CN^-}]}$$

where, at 25 °C ($\mu = 1.0\ M$), $K_1 = (9.55 \pm 0.03)$ l.mole^{-1}, $K_a = 2.15\times10^{-12}$ mole.l^{-1}, $k_0 = (3.29 \pm 0.36)\times10^{2}$ l.mole^{-1}.sec^{-1}, $k_1 = (3.57 \pm 0.16)\times10^{3}$ l.mole^{-1}.sec^{-1}, $K_c = (2.95 \pm 0.19)\times10$ and $k_2 = (2.13 \pm 0.10)\times10$ sec^{-1}, the rate coefficients and equilibrium constants referring to the following set of reactions.

$$[\mathrm{Cr(CN)_5H_2O}]^{3-}+\mathrm{CN^-} \rightleftharpoons \mathrm{Cr(CN)_6^{4-}}+\mathrm{H_2O} \qquad K_1$$

$$\mathrm{H_2O_2} \rightleftharpoons \mathrm{H^+}+\mathrm{HO_2^-} \qquad K_a$$

$$[\mathrm{Cr(CN)_5H_2O}]^{3-}+\mathrm{HO_2^-} \rightleftharpoons [\mathrm{Cr(CN)_4H_2OHO_2}]^{3-}+\mathrm{CN^-} \qquad K_c$$

$$\mathrm{Cr(CN)_6^{4-}}+\mathrm{H_2O_2} \rightarrow \mathrm{Cr(CN)_6^{3-}}+\mathrm{OH\cdot}+\mathrm{OH^-} \qquad k_0$$

$$[\mathrm{Cr(CN)_5H_2O}]^{3-}+\mathrm{H_2O_2} \rightarrow [\mathrm{Cr(CN)_5OH}]^{3-}+\mathrm{OH\cdot}+\mathrm{H_2O} \qquad k_1$$

$$[\mathrm{Cr(CN)_4H_2OHO_2}]^{3-} \rightarrow \text{products} \qquad k_2$$

The final step is

$$\mathrm{Cr(II)}+\mathrm{OH\cdot} \rightarrow \mathrm{Cr(III)}+\mathrm{OH^-} \qquad \text{(fast)}$$

The stoichiometry is the expected $2\Delta[\mathrm{H_2O_2}] = \Delta[\mathrm{Cr(II)}]$. At low $[\mathrm{OH^-}]$ and high $[\mathrm{CN^-}]$ the outer-sphere mechanisms, described by k_0 and k_1, predominate but at high $[\mathrm{OH^-}]$ and low $[\mathrm{CN^-}]$, complex formation becomes significant.

Reduction by pentacyanocobaltate(II) has a stoichiometry[592]

$$2\,\mathrm{Co(CN)_5}^{3-}+\mathrm{H_2O_2} = 2\,\mathrm{Co(CN)_5OH}^{3-}$$

and the rate law, obtained by the stopped-flow technique, is

$$-d[\mathrm{Co(CN)_5}^{3-}]/dt = 2\,k[\mathrm{Co(CN)_5}^{3-}][\mathrm{H_2O_2}]$$

where $k = 7.4\times10^{2}$ l.mole^{-1}.sec^{-1} at 25 °C ($\mu = 0.5\ M$ $\mathrm{NaClO_4}$, pH < 10) and $E = 4.8 \pm 0.5$ kcal.mole^{-1}, $\Delta S^{\ddagger} = -31 \pm 2$ eu. k falls off at pH > 10 in a manner indicating $\mathrm{HO_2}^-$ to be unreactive towards $\mathrm{Co(CN)_5}^{3-}$ (pK_a for $\mathrm{H_2O_2}$ = 11.62)[19]. Addition of iodide ion ($\sim 0.1\ M$) changes the stoichiometry to

$$2\,\mathrm{Co(CN)_5}^{3-}+\mathrm{H_2O_2}+\mathrm{I^-} = \mathrm{Co(CN)_5I}^{3-}+\mathrm{Co(CN)_5OH}^{3-}+\mathrm{OH^-}$$

but not the rate. This suggests reaction steps

$$Co(CN)_5^{3-} + H_2O_2 \rightarrow Co(CN)_5OH^{3-} + OH\cdot \quad \text{(slow)}$$
$$Co(CN)_5^{3-} + OH\cdot \rightarrow Co(CN)_5OH^{3-} \quad \text{(fast)}$$
$$I^- + OH\cdot \rightarrow I\cdot + OH^- \quad \text{(fast)}$$
$$Co(CN)_5^{3-} + I\cdot \rightarrow Co(CN)_5I^{3-} \quad \text{(fast)}$$

The reduction by Cr(III) at high pH is well-known analytically, *viz.*

$$2\,CrO_2^- + 3\,HO_2^- = 2\,CrO_4^{2-} + OH^- + H_2O$$

The rate studies[593] are complicated by an "aging" phenomenon; the velocity falls if the Cr(III) solution has been left to stand. However, the effect is reproducible and "aged" solutions were invariably employed. The best fit with the results is obtained with the equation

$$-\mathrm{d}[Cr(III)]/\mathrm{d}t = k[Cr(III)][HO_2^-]h_-^{\frac{1}{2}}$$

Where h_- is an alkalinity function[594] and k equals $(3.64 \pm 0.07) \times 10^4\ l^{\frac{3}{2}}.mole^{-\frac{3}{2}}.sec^{-1}$ at 25) °C $[OH^-] = 0.50\ M$). Rupture of μ-hydroxo bridges joining chromium atoms in a polymer is viewed as a fast preliminary to the reduction step.

A stopped-flow examination[554a] of the reduction by the bipyridyl complex of Cu(I), $Cu(bipy)_2^+$, reveals the kinetics

$$-\mathrm{d}[Cu(I)]/\mathrm{d}t = k_2[Cu(I)][H_2O_2]$$

with k_2 (25 °C) $= 8.5 \times 10^2$ $l.mole^{-1}.sec^{-1}$. This reaction is involved in the autoxidation of O_2 by $Cu(bipy)_2^+$ (p. 449).

Although Cu^{2+}(aq) is a poor catalyst, it has been established that certain complexes of Cu(II) with a free ligand site can reduce H_2O_2, *i.e.* that the electron transfer is inner-sphere in character[594a–c]. The rate law depends on the other ligands, *e.g.*

(*a*) Cu^{2+}–ethylenediamine(en)[594d]

$$\frac{-\mathrm{d}[H_2O_2]}{\mathrm{d}t} = \frac{k[Cu(en)^{2+}]^2[H_2O_2]}{[H_3O^+]} f([en]_{total})$$

The complex which breaks down in the slow step is portrayed as

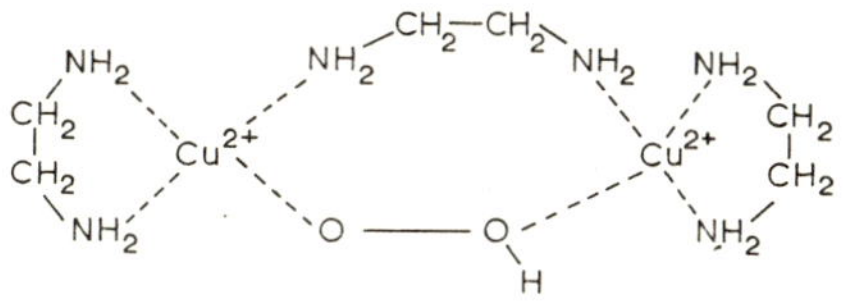

NH2 CH2 CH2 NH2 Cu2+ NH2 CH2–CH2 NH2 Cu2+ NH2 CH2 CH2 NH2 O —— O H

(*b*) Cu^{2+}–diethylenetriamine(dien)[594d] (pH *ca.* 7.5–9.5)

$$-d[H_2O_2]/dt = k[Cu(dien)^{2+}][H_2O_2][H_3O^+]^{-\frac{1}{2}}$$

(*c*) Cu^{2+}-2,2′-bipyridyl (bipy) (pH 5.5–7.5)[594c, e]

$$\frac{-d[H_2O_2]}{dt} = \frac{k[Cu(bipy)^{2+}][H_2O_2]^2}{[H_3O^+]}$$

where $k = 6.9 \times 10^{-4}$ l.mole^{-1}.sec^{-1} at 25 °C. The complex which undergoes internal oxidation–reduction is formulated as

bipy Cu^{2+}(HO–OH)(HO–O) → bipy Cu^{+}–OH + O_2 + H_2O; bipy Cu^{+} → bipy Cu^{2+} + OH^-

Further details of (a)–(c) are given in Sigel's review[594c]. Related mechanisms are proposed in the Cu^{2+} and $Cu(bipy)^{2+}$ catalysed reactions between hydrogen peroxide and hydrazine or hydroxylamine[594f].

6.4.2 *Hydroperoxides*

The stoichiometry of the reduction by Fe(II) of cumene hydroperoxide is 1 : 1[595, 596] (in contrast to reduction of H_2O_2) but the ratio Δ[Fe(II)]/Δ[ROOH] increases greatly in the presence of oxygen. The Arrhenius parameters for reduction of this and related hydroperoxides are quite similar to those of the Fenton reaction (Table 21). The production of acetophenone and ethane in high yield and the simple, second-order kinetics are consistent with the scheme

$$Fe^{2+} + C_6H_5C(CH_3)_2OOH \rightarrow Fe^{3+} + OH^- + C_6H_5(CH_3)_2O\cdot \quad \text{(slow)} \quad (101)$$

$$C_6H_5C(CH_3)_2O\cdot \rightarrow C_6H_5COCH_3 + CH_3\cdot \quad \text{(fast)} \quad (102)$$

$$2\,CH\cdot_3 \rightarrow C_2H_6 \quad \text{(fast)} \quad (103)$$

$$CH\cdot_3 + Fe(III) \rightarrow CH_3^+ + Fe(II) \quad \text{(fast)} \quad (104)$$

$$H_2O + CH_3^+ \rightarrow CH_3OH + OH^- \quad \text{(fast)} \quad (105)$$

Attack of $CH_3\cdot$ upon further hydroperoxide is discounted in view of the constancy of k_2 over a wide range of conditions.

The reaction with a variety of complexes of Fe(II) has been examined (Table 21). The stoichiometry of the ethylenediaminetetraacetic acid (EDTA) complex reduction is 2 Fe(II) : 1 RO_2H in the presence of acrylonitrile but falls to 1 : 1 as the

TABLE 21

ARRHENIUS DATA FOR THE REACTION BETWEEN Fe(II) AND HYDROPEROXIDES

Hydroperoxide	*Fe(II)complex*	*pH*	*A* ($l.mole^{-1}.sec^{-1}$)	*E* ($kcal.mole^{-1}$)	*Ref.*
Cumene (excess O_2)	aquo		3.9×10^{9}	11.1	595
Cumene	aquo		1.07×10^{10}	12.0	596
Cumene	aquo		3.53×10^{8}	9.97	597
Cumene (in D_2O)	aquo		9.25×10^{8}	10.84	597
p-Isopropylcumene	aquo		4.0×10^{9}	10.8	598
p-*tert*-Butylcumene	aquo		1.8×10^{9}	9.90	598
p-Nitrocumene	aquo		8×10^{10}	13.1	599
p-Menthane	aquo		6.3×10^{9}	11.1	599
Phenylcyclohexane	aquo		2.4×10^{9}	10.6	599
Cumene	EDTA	5.36–10	5.0×10^{10}	10.4	600
Cumene	pyrophosphate	4.8	2.0×10^{8}	8.2	600
Cumene	pyrophosphate	6.8	2.7×10^{9}	8.9	600
Cumene	pyrophosphate	8.8	1.6×10^{9}	8.4	600
p-*tert*-Butylcumene	diethylenetriamine	10.0	1.2×10^{12}	16.6	601
p-*tert*-Butylcumene	diethylenetriamine	11.0	7.1×10^{11}	15.0	601
p-*tert*-Butylcumene	diethylenetriamine	11.6	6.7×10^{11}	31.2	601
p-*tert*-Butylcumene	triethylenetriamine	9.8	4.8×10^{13}	16.0	601
p-*tert*-Butylcumene	triethylenetriamine	11.0	3.6×10^{12}	14.3	601
p-*tert*-Butylcumene	triethylenetriamine	11.6	1.4×10^{11}	12.4	601
p-*tert*-Butylcumene	tetraethylenepentamine	9.1	1.0×10^{13}	14.8	601
p-*tert*-Butylcumene	tetraethylenepentamine	10.9	2.7×10^{13}	14.9	601
p-*tert*-Butylcumene	tetraethylenepentamine	11.7	1.3×10^{14}	15.4	601
p-*tert*-Butylcumene	pentaethylenehexamine	10.0	2.8×10^{7}	7.15	601
p-*tert*-Butylcumene	pentaethylenehexamine	11.0	8.0×10^{10}	11.3	601
p-*tert*-Butylcumene	pentaethylenehexamine	11.5	9.0×10^{10}	11.3	601
Cumene	pentaethylenehexamine	10.0	3.2×10^{13}	15.0	601

monomer concentration is decreased, by virtue of a reaction between Fe(II) and growing polymer radical.

The initial step of the anaerobic ferrocyanide reduction of cumene hydroperoxide to acetophenone has kinetics[602]

$$d[Fe(CN)_6^{3-}]/dt = k[Fe(CN)_6^{4-}][ROOH][H_3O^+]$$

The initial step is, in essence, identical with (101).

The cobaltous acetate reduction of *tert*-butyl hydroperoxide in acetic acid yields mainly *tert*-butanol and oxygen; the metal ion stays in the +2 oxidation state because of the reactivity of Co(III) towards hydroperoxides (p. 378)[360]. The rate law is

$$d[O_2]/dt = k[ROOH]^{1.2}[Co(II)]^{1.45}$$

The tendency to an order greater than unity in reductant is explained in terms of complex formation between Co(II) and hydroperoxide. This complex participates in a chain reaction, *viz.*

$$Co(II)\cdot ROOH \rightarrow RO\cdot + OH^- + Co(III)$$

$$Co(III) + ROOH \rightarrow Co(II) + H^+ + RO_2\cdot$$

$$RO_2\cdot + Co(II)\cdot ROOH \rightarrow ROH + RO^- + O_2 + Co(III)$$

$$RO_2\cdot + RO_2\cdot \rightarrow 2\ RO\cdot + O_2$$

$$RO_2\cdot + ROOH \rightarrow RO\cdot + ROH + O_2$$

An analogous study of benzyl hydroperoxide indicates a rate law at 25 °C of

$$-d[ROOH]/dt = 1.15[Co(II)][Co(III)]^{0.5}[ROOH]$$

when steady-state concentrations of Co(II) and Co(III) have been reached[690]. Benzaldehyde is the main product, implying dehydration, rather than reduction, is the course of reaction. Initial rate studies on the Co(II) reaction indicate kinetics

$$d[Co(III)]/dt = k[Co(II)]^2[ROOH]$$

but [Co(III)] reaches a stationary level and the "normal" kinetics then prevail. The mechanism proposed is

$$Co(III) + ROOH \rightleftharpoons Co(III)\cdot ROOH$$

$$Co(III)\cdot ROOH + Co(II) \rightarrow Co(III) + Co(II) + C_6H_5CHO + H_2O$$

$$[Co(III)]_2 \rightleftharpoons 2\ Co(III)$$

which leads to the observed law. The initial rate studies indicate a mechanism

$$Co(II)\cdot ROOH + Co(II) \rightarrow Co(III)OH + [Co(II)(RO\cdot)] \quad \text{(slow)}$$

$$[Co(II)(RO\cdot)] + AcOH \rightarrow Co(III)OH + ROAc \quad \text{(fast)}$$

6.4.3 Halogens, cyanogen iodide, hypohalous acids and hydrogen fluoride

The oxidation of Fe(II) by Cl_2 is fast[580] ($k_2 = 80 \pm 5$ l.mole^{-1}.sec^{-1} at 25 °C, $\mu = 3\ M$) and yields mostly $FeCl^{2+}$ on the several millisecond time-scale of a stopped-flow apparatus. This does not allow differentiation between one- and two-equivalent mechanisms. The analogous oxidation of hypochlorous acid, $k_2 =$

$(3.2 \pm 0.4) \times 10^3$ l.mole^{-1}.sec^{-1}, yields mainly monomeric Fe(III) species but gives some dimeric $(FeOH)_2^{4+}$, which suggests Fe(IV) as an intermediate[580].

A conventional study of the ferrous ion–chlorine reaction in a chloride medium indicates the kinetics[603]

$$d[Fe(III)]/dt = 2\,k_1[Fe(II)][Cl_2] + 2\,k_2[Fe(II)][Cl_3^-]$$

with $k_1 = (9.1 \pm 1.0) \times 10^2$ l.mole^{-1}.sec^{-1} and $k_2 = (1.63 \pm 0.2) \times 10^2$ l.mole^{-1}.sec^{-1} at 30 °C ($\mu = 1\ M$). No inhibition by Fe(III), which is a characteristic of the Fe(II)–bromine reaction, was detected. Catalysis by Cu(II) was found.

The appearance of $FeCl^{2+}$ in the Fe^{2+}(aq) reduction suggests an inner-sphere path. By contrast, the reduction by tris-1 : 10-phenanthrolineiron(II) or ferroin is outer-sphere[603a], for ferriin is formed in high yield. The kinetics are simple second-order with k_2 (25 °C, $\mu = 1\ M$) = 2.2 ± 0.2 (independent of acidity).

The reduction of bromine by Fe(II) at acidities of 0.5 to 0.8 M follows kinetics[604]

$$\frac{-d[Br_2]}{dt} = k_2[Fe(II)][Br_3^-]\left\{\frac{1 + k_{-2}[Br][Fe(III)]}{k_3[Fe(II)]}\right\}^{-1}$$

where the rate coefficients are considered to refer to the reactions

$$Fe(II) + Br_3^- \underset{k_{-2}}{\overset{k_2}{\rightleftharpoons}} Fe(III) + Br^- + \cdot Br_2^-$$

$$Fe(II) + \cdot Br_2^- \xrightarrow{k_3} Fe(III) + 2\,Br^-$$

$k_2 = 3.9 \times 10^7 \exp(-8.4 \pm 0.5 \times 10^3/\boldsymbol{RT})$ l.mole^{-1}.sec^{-1} and $k_{-2}/k_3 = 0.11$ (± 0.01) at 30 °C. Outside this acidity range extra terms are required.

Reduction of Br_2 by U(IV) in perchloric acid has the kinetics[605]

$$-d[U(IV)]/dt = k[U(IV)][Br_2]/[H^+]^2$$

At 25 °C, $k = (5.3 \pm 0.3) \times 10^4$ moles l^{-1}.sec^{-1} and the reaction is strongly catalysed by Fe(III). Br_2 and Br_3^- appear to be equally reactive. The acidity dependence may originate in the pre-equilibrium

$$U^{4+} + 2\,H_2O \rightleftharpoons U(OH)_2^{2+} + 2\,H^+$$

The catalysis presumably involves a reduction of Fe(III) by U(IV).

Reduction of bromine by Tl(I) is important in the bromide-catalysed isotopic exchange between Tl(I) and Tl(III). A potentiometric examination[606] revealed that the reaction was first-order both in Tl(I) and in Br_2 with $k_2 = 7.2 \times 10^3$ l. mole^{-1}.sec^{-1} (25 °C) and acid-independent. This is in passable agreement with a value of k_{-1} of 4.7×10^4 l.mole^{-1}.sec^{-1}, calculated from the equilibrium constant for the reaction

$$\mathrm{Tl(III)+2\,Br^-} \underset{k_{-1}}{\overset{k_1}{\rightleftharpoons}} \mathrm{Tl(I)+Br_2}(K = 2\times10^{-11} \text{ at } 25\,^\circ\mathrm{C}, \mu = 0)$$

and from a value of k_1 of 9.4×10^{-7} l.mole^{-1}.sec^{-1} obtained from the isotopic exchange.

The reduction of iodine by Fe(II) is, of course, the reverse of the ferric ion–iodide ion reaction (p. 408) and it influences the kinetics of the latter. However, the direct reaction has been studied, the rate expression being[607]

$$-\mathrm{d[Fe(II)]/d}t = k\mathrm{[Fe(II)][I_2]}\{1+1/[\mathrm{H^+}][\mathrm{I^-}]\}+k'\mathrm{[Fe(II)][I_3^-]}$$

The reduction of iodine by ferrocyanide[608] is simple second-order with k_2 (25 °C) = $(1.3\pm0.3)\times10^3$ l.mole^{-1}.sec^{-1}. This is the reverse of the oxidation of iodide by ferricyanide (p. 409), but the ratio k(forward)/k(back) does not agree well with the equilibrium constant determined potentiometrically. Addition of I^- strongly retards the reduction and I_3^- was discounted as a reactant, the mechanism suggested being

$$\mathrm{Fe(CN)_6^{4-}+I_2 \rightleftharpoons Fe(CN)^{3-}+\cdot I_2^-}$$

$$\mathrm{Fe(CN)_6^{4-}+\cdot I_2^- \rightleftharpoons Fe(CN)_6^{3-}+2\,I^-}$$

The Mo(V) dimer reduction of iodine has been studied[558] in the pH range 1.65 to 7.20 in phosphate buffer in the presence of excess iodide. The stoichiometry is simple, *viz.*

$$\mathrm{Mo(V)_2+I_3^- = 2\,Mo(VI)+3\,I^-}$$

and the rate law was found to be

$$-\mathrm{d[Mo(V)_2]/d}t = -\mathrm{d[I_3^-]/d}t = k_1\mathrm{[Mo(V)_2][I_3^-]^0}f\mathrm{[H_3O^+]}$$

The reaction is strongly acid-inverse below pH 4 but becomes acid-independent above pH 5.5. The rate and kinetics are identical with those of the autoxidation of $Mo(V)_2$ and it seemd probable that dissociation of the dimer is rate-determining, *viz.*

$$\mathrm{Mo(V)_2} \underset{k_{-1}}{\overset{k_1}{\rightleftharpoons}} \mathrm{2\,Mo(V)} \quad \text{(slow)} \qquad (106)$$

$$\mathrm{Mo(V)+I_3^-} \overset{k_2}{\rightarrow} \mathrm{Mo(VI)+\cdot I_2^-+I^-} \quad \text{(fast)} \qquad (107)$$

$$\mathrm{2\cdot I_2^-} \overset{k_3}{\rightarrow} \mathrm{I_3^-+I^-} \quad \text{(fast)} \qquad (108)$$

However, direct determination of both k_1 and k_{-1} by means of sampling followed by freezing to 77 °K and measurement of ESR absorption intensities (for monomer) gave results inconsistent with step (106). Disproportionation of $Mo(V)_2$ into Mo(IV) and Mo(VI) is also possible but no supporting evidence could be adduced.

The titanous ion reduction of iodine displays the kinetics[690, 610]

$$-d[I_3^-]/dt = \frac{k_1[Ti(III)][I_3^-]}{[H^+][I^-]} + \frac{k_2[Ti(III)][I_3^-]}{[H^+]} + k_3[Ti(III)][H^+]$$

The third term is minor. The first two terms are considered to refer to reactions between $TiOH^{2+}$ and I_2 and I_3^-, respectively. One-equivalent processes involving $\cdot I_2^-$ are favoured.

The V^{2+} ion reductions of iodine, triiodide ion and bromine are all simple second-order[610b], with no acidity dependence. The rate and activation data can be summarised ($\mu = 1.0\ M$) as

Oxidant	$k_{25\,°C}$ (*l.mole*$^{-1}$.*sec*$^{-1}$)	*E* (*kcal.mole*$^{-1}$)	$\Delta S^{\ddagger}$(*eu*)
I_2	$(7.5 \pm 0.5) \times 10^3$	6.6 ± 0.5	-21 ± 2
I_3^-	$(9.7 \pm 0.2) \times 10^2$	9.7 ± 0.2	-14 ± 1
Br_2	$(3.0 \pm 0.4) \times 10^4$	4.1 ± 1.2	-26 ± 4
Cl_2	$> 5 \times 10^2$		

V^{3+} was produced in each oxidation. The low values of E preclude the rate-determining loss of water from the V^{2+} coordination shell found for V^{2+} reductions of several metal complexes[610c]. The kinetics of the Cl_2 reduction were not reproducible.

Cyanogen iodide is reduced by pentacyanocobaltate(II)[592a]

$$2\,Co(CN)_5^{3-} + ICN = Co(CN)_5I^{3-} + Co(CN)_6^{3-}$$

with kinetics

$$-d[Co(CN)_5]^{3-}/dt = 2k[Co(CN)_5^{3-}][ICN]$$

with $k = 9.5 \pm 0.5$ l.mole^{-1}.sec^{-1} at 25 °C ($\mu = 1.0\ M$ $NaClO_4$); $E = 13.3 \pm 0.6$ kcal.mole^{-1} and $\Delta S^{\ddagger} = -12 \pm 2$ eu. The reaction is accordingly very similar to the analogous reductions of H_2O_2 and NH_2OH[592a] and the mechanism given is identical.

The reduction of aqueous HF to hydrogen by Ti(III) is extremely slow at 25 °C ($k_2 \sim 5 \times 10^{-7}$ l.mole^{-1}.sec^{-1}, *i.e.* $t_{\frac{1}{2}} \sim 200$ h at 2 M substrate concentration), although the activation energy is quite small (*ca.* 6 kcal.mole^{-1})[610a]. The reaction is of the first-order both in Ti(III) and undissociated HF when a moderate (up to ten-fold) excess of HF is present. A number of complicating features were found including an induction period of 10–30 h, during which oxidation of Ti(III) is *faster* than during the main course of reaction, and deviation from the rate law for small or large excesses of HF.

Electron-transfer to an HF molecule presumably results in homolysis to H· and F^-.

6.4.4 Hydroxylamine, hydrazine, hydrazoic acid and azide ion

The reduction of NH_2OH is one stage in the reduction of NO_3^- to NH_3· It is formally closely related to the Fenton reaction, the stoichiometry in acidic solution being

$$2\,M^{n+} + HONH_3^+ + 2\,H^+ = 2\,M^{(n+1)+} + NH_4^+ + H_2O$$

Higginson *et al.*[577], have used vinyl monomers to capture the radical intermediate, which turned out to be NH_2· when Ti(III) was employed as reductant. Brown and Drury[611] measured the k_{14}/k_{15} nitrogen isotope effect for the reduction of several molecules containing N–O bonds, obtaining a value of 1.034 ± 0.002 for the reduction of hydroxylamine by Fe(II) at 25 °C. The lack of enrichment of the ^{15}N content of residual NH_2OH during reduction in the presence of added enriched ^{15}N-labelled NH_4^+ rules out a back-reaction involving this ion.

Two studies have been made of the reductions of NH_2OH and its analogues by Cr(II) perchlorate which differ seriously both as regards the results and their interpretation. The results of both studies are combined in Table 22. Wells and Salam[612, 612a] find NH_2OH and N_2H_4 to be much more reactive towards Cr(II) than do Taube *et al.*[612b] (who find no reaction with N_2H_4 in 24 h at 25 °C).

TABLE 22

ARRHENIUS PARAMETERS FOR THE REDUCTION OF SPECIES CONTAINING N–N AND N–O BY CHROMOUS ION ($\mu = 1.0\ M$)

Substrate	k_2 *(25 °C)(l.mole*$^{-1}$*.sec*$^{-1}$*)*	*E (kcal.mole*$^{-1}$*)*	$\Delta S^{\ddagger}$*(eu)*	*Ref.*
NH_2OH	1.00	8.6	−30	612, 612a
NH_3OH^+	0.0141	11.1 ± 0.5	-37 ± 2	612b
$(CH_3)_3NOH^+$	0.66	8.5 ± 0.3	-33 ± 1	612b
$(CH_3)_2C_6H_5NOH^+$	36	6.3 ± 0.2	-33 ± 1	612b
N_2H_4	1.32	10.7	−22	612, 612a
N_2H_4	~0	—	—	612b
HN_3	14	13.6	−7.5	612, 612a
N_3^-	4.2	17.5	3.1	612, 612a

However both groups agree on the stoichiometries, which are all of the type 2 Cr(II) : 1 reductant molecule, and on the rate laws, which are generally

$$\mathrm{d[Cr(III)]/d}t = k_2\mathrm{[Cr(II)][oxidant]}$$

k_2 depends on acidity only for HN_3 above pH 5 when the active oxidant is N_3^-. The efficient transfer of ^{18}O from labelled NH_2OH[612b] to Cr(II) rules out the mechanism proposed by Salam and Wells[612a], *viz.*

$$\mathrm{Cr(II)+NH_3OH^+ = Cr(III)+NH_2\cdot+H^++OH^-\ (or\ \cdot NH_3^++OH^-)}$$

and instead attack on oxygen must occur. In the fast, second step $NH_2\cdot$ or its protonated form must be reduced by Cr(II) without being captured, possibly *via* abstraction of a hydrogen atom from the coordination sphere of Cr(II).

The reduction of NH_2OH by pentacyanocobaltate(II) shows kinetics[592]

$$-\mathrm{d[Co(CN)_5^{3-}]/d}t = 2\,k\,\mathrm{[Co(CN)_5^{3-}][NH_2OH]}$$

with $k = (5.3\pm0.3)\times10^{-3}$ l.mole^{-1}.sec^{-1} at 25 °C ($\mu = 0.2$ *M* KCl); $E = 10.9\pm0.5$ kcal.mole^{-1} and $\Delta S^{\ddagger} = -35\pm2$ eu. The stoichiometry is

$$\mathrm{2\,Co(CN)_5^{3-}+NH_2OH+H_2O = Co(CN)_5OH^{3-}+Co(CN)_5NH_3^{2-}+OH^-}$$

The absence of any effect of added iodide upon the products is in contrast with the reduction of H_2O_2 by pentacyanocobaltate (p. 462) and confirms that $NH_2\cdot$ is the free radical intermediate rather than $OH\cdot$

$$\mathrm{Co(CN)_5^{3-}+NH_2OH \rightarrow Co(CN)_5OH^{3-}+NH_2\cdot}$$

$$\mathrm{Co(CN)_5^{3-}+NH_2\cdot \rightarrow Co(CN)_5NH_2^{3-}}$$

$$\mathrm{Co(CN)_5NH_2^{3-}+H_2O \rightleftharpoons Co(CN)_5NH_3^{2-}+OH^-}$$

6.4.5 Nitrite

Investigations of the reduction to nitric oxide by several reagents have been reported. A recent study[613] is that of reduction by Mo(V) in a chloride medium which displays the stoichiometry

$$\mathrm{2\,NO_2^-+4\,H^++Mo(V)_2 = 2\,Mo(VI)+2\,H_2O+2\,NO}$$

The rate law is

$$-2\,\mathrm{d[NO_2^-]/d}t = -d\mathrm{[Mo(V)_2]/d}t = k\mathrm{[NO_2^-][H_3O^+][Mo(V)_2]^0}$$

and k is unaffected by added Mo(VI) or NO. Activation parameters are: $E = 10.6$ kcal.mole^{-1}, $\Delta S^{\ddagger} = -16.4$ eu. This contrasts with the rate expressions for reaction of Mo(V) with I_2 and O_2. The mechanism proposed involves slow formation of NO^+, *viz.*

$$H_3O^+ + NO_2^- \rightleftharpoons HNO_2 + H_2O \quad \text{(fast)}$$

$$HNO_2 \rightarrow NO^+ + OH^- \quad \text{(slow)}$$

$$NO^+ + Mo(V) \rightarrow NO + Mo(VI) \quad \text{(fast)}$$

Pu(III) in $HClO_4$ and HCl media reduces nitrous acid[614], *viz.*

$$Pu(III) + H^+ + HNO_2 = Pu(IV) + NO + H_2O$$

The rate law and the rate coefficient are the same in the two media; the former being

$$-d[Pu(III)]/dt = k_3[Pu(III)][H^+][HNO_2]$$

At 24 °C k_3 equals 0.30 ± 0.06 l^2.mole^{-2}.sec^{-1}, and E is found to be 6.0 ± 0.3 kcal.mole^{-1}. The mechanism proposed involves a slow oxidation of NO^+ which is produced by

$$HNO_2 + H^+ \rightleftharpoons NO^+ + H_2O \quad \text{(fast)}$$

In nitric acid the rate law includes an additional term, $k_4[Pu(III)][H^+][HNO_2][NO_3^-]$, with k_4 equal to 1.5 ± 0.3 l^3.mole^{-3}.sec^{-1} and E equal to 14.0 ± 0.5 kcal.mole^{-1}. This term dominates at nitrate concentrations in excess of 1 M and is attributed to the further equilibrium

$$HNO_2 + HNO_3 \rightleftharpoons N_2O_4 + H_2O \quad \text{(fast)}$$

followed by

$$N_2O_4 + Pu(III) \rightarrow Pu(IV) + NO_2^- + NO_2\cdot \quad \text{(slow)}$$

Abel *et al.*[615] studied reduction of nitrite to NO by Fe(II) obtaining the rate law

$$d[Fe(III)]/dt = [Fe(II)][HNO_2](k_1 + k_2[H_3O^+] + k_3[HNO_2]/p_{NO})$$

where, at 25 °C, $k_1 = 7.8 \times 10^{-3}$ l.mole^{-1}.sec^{-1}, $k_2 = 0.227$ l^2.mole^{-2}.sec^{-1} and $k_3 = 0.40$ l^2.mole^{-2}.atm^{-1}.sec^{-1}. The three terms are considered to result from the following reaction sequences

(*a*) $Fe(II)+HNO_2 \rightarrow Fe(III)+OH^- + NO\cdot$ (slow)

(*b*) $HNO_2+H^+ \rightleftharpoons NO^+ + H_2O$ (fast)

$Fe(II)+NO^+ \rightarrow Fe(III)+NO\cdot$ (slow)

(*c*) $2\,HNO_2 \rightleftharpoons NO_2\cdot + NO\cdot + H_2O$ (fast)

$Fe(II)+NO_2\cdot \rightarrow Fe(III)+NO_2^-$ (*etc.*) (slow)

The nitrogen kinetic isotope effect for this reaction is 1.034 ± 0.002, which indicates that breaking of the N–O bond occurs in the slow step[611].

The As(III) reduction of nitrite has the kinetics[616]

$$d[As(V)]/dt = k[H_3AsO_3][HNO_2]^2$$

with $k = 9.6 \times 10^{-5}$ $l^2.mole^{-2}.sec^{-1}$ (25 °C). A fast equilibrium is followed by a slow redox process, *viz.*

$$2\,HNO_2 \rightleftharpoons N_2O_3 + H_2O \quad \text{(fast)}$$

$$N_2O_3 + As(III) + 2\,H^+ \rightarrow As(V) + 2\,NO\cdot + H_2O \quad \text{(slow)}$$

6.4.6 Nitrate

The reduction by Ce(III) of nitric acid is a reversible reaction, and in a kinetic investigation[617] it was found necessary to remove the oxides of nitrogen with a stream of nitrogen. The overall reaction is

$$Ce(III) + H_3O^+ + HNO_3 = Ce(IV) + NO_2\cdot + 2\,H_2O$$

At 100 °C and with a substrate concentration range of 12 to 16 *M* the rate law was found to be

$$d[Ce(IV)]/dt = k[Ce(III)]f[HNO_3]$$

which was simplified to

$$d[Ce(IV)]/dt = k[Ce(III)][NO_2^+]$$

with $E = 8.0$ kcal.mole^{-1}. A simple electron-transfer mechanism was proposed.

The reduction by Mo(V) is significant insofar as this element participates in enzymatic reduction[618] of NO_3^- to NO_2^-. In a tartrate buffer and chloride medium the rate law was shown to be[619]

$$-d[Mo(V)_2]/dt = k[Mo(V)_2]^{\frac{1}{2}}[NO_3^-]f[H_3O^+]$$

with $E = 19.6$ kcal.mole^{-1} and $\Delta S^{\ddagger} = -15.4$ eu. The overall stoichiometry is

$$2\,NO_3^- + 3\,Mo(V)_2 = 2\,NO + 6\,Mo(VI)$$

The immediate product of reduction is $NO_2\cdot$ which disproportionates to NO_3^- and NO^+; however, the latter is very rapidly reduced by further Mo(V) to NO.

Only in a tartrate buffer is the rate appreciable and it is significant that only in this buffer does the normally diamagnetic Mo(V) display ESR signals ($g = 1.937$ and $g = 1.945$) indicating the presence of monomeric Mo(V) which is regarded as the active species in the proposed mechanism, *viz.*

$$Mo(V)_2 \rightleftharpoons Mo(V) \qquad \text{(fast)}$$

$$Mo(V) + NO_3^- \rightarrow NO_2\cdot + Mo(VI) \qquad \text{(slow)}$$

$$2\,NO_2\cdot \rightarrow N_2O_4 \rightleftharpoons NO^+ + NO_3^- \qquad \text{(fast)}$$

$$Mo(V) + NO^+ \rightarrow Mo(VI) + NO\cdot \qquad \text{(fast)}$$

$$NO^+ + H_2O \rightarrow HNO_2 + H^+ \qquad \text{(fast)}$$

$$HNO_2 + H^+ \rightarrow NO^+ + H_2O \qquad \text{(fast)}$$

Application of the steady-state approximation leads to the observed kinetics.

It is significant that Mo(VI) is an excellent catalyst for the reduction of NO_3^- by Sn(II) although its precise role has not been elucidated[620].

Re(V) reduction of nitrate in 10 *M* hydrochloric acid to give nitrite and ReO_4^- is rapid and is followed by a slower reaction of Re(V) with NO_2^- of stoichiometry

$$5\,Re(V) + 2NO_2^- = 3ReO_4^- + 2\,[Re^{III}Cl_5NO]^{2-}$$

Both reactions follow simple second-order expressions with k_2(nitrate) $= 2.19 \pm 0.13$ l.mole^{-1}.sec^{-1} and k_2(nitrite) $= 0.644 \pm 0.010$ l.mole^{-1}.sec^{-1}, both at 25 °C[684]. In the NO_3^- reduction, NO_2^+ is considered to enter the inner sphere of the Re(V) complex rapidly to give $[ReCl_4(ONO)]$ which then breaks down to products, following an internal two-electron transfer in the slow step. By analogy, NO^+ is considered to enter the Re(V) complex by displacing H_2O rapidly in the nitrite reduction; slow internal two-electron transfer to give NO^- (or HNO) follows; the latter is then consumed by Re(V).

The Fe(II) reduction of nitrate displays a nitrogen isotope effect k_{14}/k_{15} of 1.075 ± 0.004 at 25 °C[611].

6.4.7 *Peroxodisulphate ion (also called persulphate and peroxydisulphate)*

The reduction by Ag^+ is the most widely investigated example although in recent years several other reductants have been used. The series of oxidations effected by the Ag^+–$S_2O_8^{2-}$ couple are referred to in the section on Ag(II) and Ag(III). In general the rate of disappearance of persulphate is independent of the concentration and, to some extent, the nature of the substrate, *viz.*

$$-\mathrm{d}[S_2O_8^{2-}]/\mathrm{d}t = k_2[S_2O_8^{2-}][Ag^+][\text{reductant}]^0[H_3O^+]^0$$

k_2 is, however, influenced by the charge on the substrate (p. 354).

The various alternative mechanisms for the production of Ag(III) and Ag(II) are given in Section 4.1.1. Reaction *without* added substrate follows the kinetics[621])

$$-\mathrm{d}[S_2O_8^{2-}]/\mathrm{d}t = (k_1 + k_2[Ag^+])[S_2O_8^{2-}]$$

The observation of Fronaeus and Östman[621] that the first-order decay coefficient of persulphate in the presence of Ag^+ is quite unchanged by adding cerous ions (which are oxidised) indicates the decomposition of persulphate and the redox process to have the same rate-determining step. These workers obtain values for k_1 and k_2 of $(8 \pm 3) \times 10^{-7}$ sec^{-1} and $(3.75 \pm 0.17) \times 10^{-3}$ l.mole^{-1}.sec^{-1}, respectively, at 25 °C, and they prefer the mechanism

$$Ag^+ + S_2O_8^{2-} \rightarrow Ag^{2+} + SO_4^{2-} + \cdot SO_4^-$$

It is pertinent that $S_2O_8^{2-}$ accepts an electron generated by pulse radiolysis of water to give optically detectable $\cdot SO_4^-$ within 1.5×10^{-6} sec[622].

The ability of the stable free radical diphenylpicrylhydrazyl (DPPH) to act as an efficient trap for reactive radicals such as $\cdot SO_4^-$ and $OH\cdot$ has been utilised by Bawn and Margerison[623] in their examination of the Ag^+–$S_2O_8^{2-}$ couple. The disappearance of the intensely coloured DPPH gave excellent zero-order kinetics; the rate as a whole was identical with that found by Fronaeus and Östman[621] and k_2 was given by $3.1 \times 10^{11} \exp(-17.9 \times 10^3/RT)$ l.mole^{-1}.sec^{-1}. Sengar and Gupta[624–626] have also determined Arrhenius parameters for this reduction and have compared them with those for some redox processes (Table 23).

The oxidations by persulphate of certain complexes of Ag(I) to *stable* forms of Ag(II) or Ag(III) have recently been examined[626a]. Bipyridyl (bipy) and ethylene-bisbiguanide (enbig) were selected as ligands. The stoichiometry of the oxidation

TABLE 23

KINETIC PARAMETERS FOR REACTIONS OF THE SILVER–PERSULPHATE COUPLE[624]

Reaction	μ	k_2 *(35 °C)(l.mole^{-1}.sec^{-1})*	*E(kcal.mole^{-1})*	*A (l.mole^{-1}.sec^{-1})*	*Ref.*
Oxidation of Mn(II)	0.637	9.00×10^{-3}	15.0	6.45×10^{6}	625
Oxidation of Tl(I)	0.772	9.11×10^{-3}	13.2	4.3×10^{6}	626
Uncatalysed	0.740	9.58×10^{-3}	14.2	2.0×10^{7}	624

of $Ag(bipy)_2^+$ to $Ag(bipy)_2^{2+}$ is $\Delta[Ag(I)]/\Delta[S_2O_8^{2-}] = 0.5$ (as expected) and the reaction is first-order in each of the two reactants, yielding an expression for the rate coefficient in 50 % acetone–water mixture as

$$k_2 = 4.0 \times 10^7 \exp[(19.5 \pm 1.0) \times 10^3/RT] \text{ l.mole}^{-1}.\text{sec}^{-1}$$

The oxidation of $Ag(enbig)^+$ to $Ag(enbig)^{3+}$ is less straightforward, comprising two consecutive stages each of which is first-order in both metal ion and persulphate. The rate data for the steps are

	k_2(20 °C)	*A (l.mole^{-1}sec^{-1})*	*E (kcal.mole^{-1})*
First step	0.0102	5.0×10^{7}	13.1 ± 0.7
Second step	0.119	4.0×10^{9}	14.3 ± 3.8

The two steps are considered to be an oxidation of Ag(I) to Ag(II) followed by an oxidation of the latter to Ag(III). The stoichiometry is unexpected in that one Ag(I) species consumes *two* persulphate ions. The release of $\cdot SO_4^-$ would be expected to result in oxidation of further $Ag(enbig)^+$; the existence of two stages rules out a two-equivalent oxidation directly to $Ag(enbig)^{3+}$.

An isotopic study[626b] of the reaction

$$3\,H^+ + \tfrac{1}{2}\,S_2O_8^{2-} + Co(NH_3)_5OH_2^{3+} = SO_4^{2-} + Co^{2+} + 5\,NH_4^+ + \tfrac{1}{2}\,O_2$$

which proceeds with the customary rate law under the influence of Ag^+ ions and at the same rate as the $Ag(I)–S_2O_8^{2-}$ reduction of Mn^{2+} ions[625], reveals that at least 70 % of the oxygen emanates from the coordinated water molecule. Only a fraction of the persulphate consumed is effective in production of Co^{2+}. The reaction in the absence of Ag^+ (p. 481) produces nitrogen and it is proposed[626b] that neither $OH\cdot$ nor $\cdot SO_4^-$ attack the complex but that the active oxidant is Ag(II) or Ag(III).

Bawn and Margerison[623] have observed cupric ion catalysis of the persulphate–DPPH reaction although the effect is smaller than that with Ag^+. The kinetics are complex and few details are given.

Cu(II) also catalyses the persulphate oxidation of numerous substrates, for example that of oxalate-ion with a rate law[627]

$$-\mathrm{d}[S_2O_8^{\,2-}]/\mathrm{d}t = k[S_2O_8^{\,2-}][\mathrm{Cu(II)}]^{\frac{1}{2}}[\mathrm{oxalate}]^0$$

Cu(III) is believed to be one of the reactive intermediates in a chain mechanism. Both Cu(III) and As(IV) are invoked[628, 628a] in discussion of the oxidation of As(III) by Cu(II)–$S_2O_8^{\,2-}$ which also displays chain character. Two sets of kinetic results have been published. Those of Woods *et al.*[628a] indicate the general reaction orders in Cu(II) and $S_2O_8^{\,2-}$ to be one-half (tending to zero at high [Cu(II)] and one respectively, but the order in As(III) to be zero at low [Cu(II)] ($\leqslant 10^{-3}$ M) but one-half at high [Cu(II)]($> 10^{-3}$ M). The rate coefficient is affected by oxygen, which also produces a shift in the critical concentration at which the order in Cu(II) becomes zero. Bhargava and Gupta[628] used aerated solutions to obtain a general rate law

$$\frac{-\mathrm{d}[\mathrm{As(III)}]}{\mathrm{d}t} = \frac{k[\mathrm{Cu(II)}]^{\frac{1}{2}}[S_2O_8^{\,2-}][\mathrm{As(III)}]}{1+K[\mathrm{As(III)}]}$$

Composite Arrhenius parameters were obtained for k; $E = 24.2$ kcal.mole^{-1} and $\Delta S^{\ddagger} = +10.5$ eu. Mn(II) severely retards reaction.

Both groups invoke $\cdot SO_4^-$ and As(IV) as intermediates in their reaction schemes, *viz.*

(*a*) O_2^- free solutions

$$S_2O_8^{\,2-} \rightarrow 2\cdot SO_4^{\,-}$$

$$\left.\begin{array}{r} \cdot SO_4^{\,-}+\mathrm{As(III)} \rightarrow SO_4^{\,2-}+\mathrm{As(IV)} \\ \mathrm{As(IV)}+\mathrm{Cu(II)} \rightarrow \mathrm{As(V)}+\mathrm{Cu(I)} \\ \mathrm{Cu(I)}+S_2O_8^{\,2-} \rightarrow \mathrm{Cu(II)}+SO_4^{\,2-}+\cdot SO_4^{\,-} \end{array}\right\}\ \text{propagation}$$

$$\left.\begin{array}{r} \mathrm{As(IV)}+\mathrm{Cu(I)} \rightarrow \mathrm{As(III)}+\mathrm{Cu(II)} \\ \mathrm{Cu(I)}+\cdot SO_4^{\,-} \rightarrow \mathrm{Cu(II)}+SO_4^{\,2-} \end{array}\right\}\ \text{termination}$$

(*b*) Aerated solutions (additional steps)

$$\mathrm{As(IV)}+O_2+H^+ \rightarrow \mathrm{As(V)}+HO_2\cdot$$

$$HO_2\cdot+\mathrm{Cu(II)} \rightarrow H^+ +O_2+\mathrm{Cu(I)}$$

$$\mathrm{Cu(I)}+HO_2\cdot+H^+ \rightarrow \mathrm{Cu(II)}+H_2O_2$$

$$\mathrm{Cu(I)}+H_2O_2 \rightarrow \mathrm{Cu(II)}+OH^- +OH\cdot$$

The Indian workers also include a step involving Cu(III)

$$\text{Cu(I).As(III)} + S_2O_8^{2-} \rightarrow \text{Cu(III)} + \text{As(III)} + 2\cdot SO_4^-$$

and an additional chain-breaking step

$$\cdot SO_4^- + \text{As(IV)} \rightarrow SO_4^{2-} + \text{As(V)}$$

The As(III) reduction is also catalysed by Fe(III) with radical-chain kinetics[628a].

The Cu(II)-catalysed oxidation of Sb(III) by $S_2O_8^{2-}$ has also been investigated by Bhargava *et al.*[629b] under essentially anaerobic conditions, the rate law being

$$-\text{d[Sb(III)]}/\text{d}t = k[S_2O_8]^{\frac{1}{2}}[\text{Cu(II)}]^{\frac{1}{2}}[\text{Sb(III)}]^{\frac{1}{2}}[H_3O^+]^0$$

The chain reaction suggested, *viz.*

$$S_2O_8^{2-} \xrightarrow{k_1} 2\cdot SO_4^- \qquad \text{initiation}$$

$$\left.\begin{array}{l} \cdot SO_4^- + \text{Sb(III)} \xrightarrow{k_2} \text{Sb(IV)} + SO_4^{2-} \\ \text{Sb(IV)} + \text{Cu(II)} \xrightarrow{k_3} \text{Sb(V)} + \text{Cu(I)} \\ S_2O_8^{2-} + \text{Cu(I)} \xrightarrow{k_4} SO_4^{2-} + \cdot SO_4^- + \text{Cu(II)} \end{array}\right\} \text{propagation}$$

$$\cdot SO_4^- + \text{Sb(IV)} \xrightarrow{k_5} \text{Sb(V)} + SO_4^{2-} \qquad \text{termination}$$

leads to the expression

$$-\text{d}[S_2O_8^{2-}]/\text{d}t = [k_1 k_2 k_3/k_5]^{\frac{1}{2}}[S_2O_8^{2-}]^{\frac{1}{2}}[\text{Sb(III)}]^{\frac{1}{2}}[\text{Cu(II)}]^{\frac{1}{2}}$$

The temperature dependence of k gives $E = 28.5$ kcal.mole^{-1} and $\Delta S^{\ddagger} = 33.2$ eu. E_1 is 33.5 kcal.mole^{-1} and hence $E_2 + E_3 - E_5 = 11.75$ kcal.mole^{-1}.

Cuprous chloride reduces persulphate with simple second-order kinetics[629]. The first step may involve a short lived complex, *viz.*

$$\text{Cu(I)} + S_2O_8^{2-} \rightarrow CuS_2O_8^- \qquad \text{(slow)}$$

$$CuS_2O_8^- \rightarrow \text{Cu(II)} + SO_4^{2-} + \cdot SO_4^- \qquad \text{(fast)}$$

The reductions by ferrous ion and mono- and bis-bipyridyl complexes of Fe(II) are also simple second-order with (for the Fe^{2+} reaction at zero ionic strength[630]). $k_2 = 1.0 \times 10^{11} \exp(-12.1 \times 10^3/\boldsymbol{RT})$ l.mole^{-1}.sec^{-1}. This reaction generates an intermediate capable of oxidising ethanol[631], but the effect is suppressed by addition of Cl^-, Br^- and acrylonitrile, the latter being polymerised.

An attempt to correlate reduction rate with redox potential of the metal ion

TABLE 24

ARRHENIUS PARAMETERS FOR THE REDUCTION OF PERSULPHATE BY COMPLEXED DIVALENT METAL IONS[632, 633]

Reductant	k_2* (*l.mole⁻¹.sec⁻¹*)	*A* (*l.mole⁻¹.sec⁻¹*)	*E* (*kcal.mole⁻¹*)
Tris-4,4′-dimethyldipyridyliron(II)	6.68 (25 °C)	3.2×10^8	10.6
Tris-2,2′-dipyridyliron(II)	0.59 (25 °C)	5.7×10^8	12.4
Tris-*o*-phenanthrolineiron(II)	0.305 (26.5 °C)	1.5×10^9	13.3
Tris-5-methylphenanthrolineiron(II)	0.112 (17 °C)	1.6×10^8	12.6
Tris-2,2′-dipyridylruthenium(II)	0.0105 (26.5 °C)	5.0×10^8	14.6
Tris-2,2′-dipiridylosmium(II)	52.0 (25.5 °C)	2.8×10^8	9.4

* Extrapolated to zero ionic strength.

was made by Irvine[632, 633] who examined the reductants given in Table 24. Second-order kinetics were followed initially by all these systems. A good correlation was found between k_2 and ΔG°, the standard free-energy change for the reaction

$$2\,M^{2+} + S_2O_8{}^{2-} = 2\,M^{3+} + 2\,SO_4{}^{2-}$$

Raman and Brubaker[633a] have also examined reduction by chelates of Fe(II) and explain the ionic strength dependence in terms of a pre-equilibrium ion-pairing.

The effects of substituents upon the ferroin reduction have also been recorded (Table 25)[634]. A marked correlation between E and log A is found, indicating a single type of cation–anion interaction.

The reduction of persulphate by tris-[α-(2-pyridyl)-benzylideneaniline] iron(II) is, by contrast, independent of persulphate ion concentration[635], and the rates of reaction of several ring-substituted complexes of this type correspond exactly to the rates of acid-catalysed separation of one ligand. Clearly oxidation of the ligand

TABLE 25

ARRHENIUS PARAMETERS FOR THE REDUCTION OF PERSULPHATE ION BY SUBSTITUTED FERROINS

Substituent	*A* (*l.mole⁻¹.sec⁻¹*)	*E* (*kcal.mole⁻¹*)
None	10^9	13.5
5-Nitro	10^{19}	27.5
5-Methyl	10^8	11.9
5-Chloro	10^{14}	20.7
5-Methyl-6-nitro	10^{19}	27
5,6-Dimethyl	10^8	13.3
4,7-Dimethyl	10^5	6.9
3,5,6,8-Tetramethyl	10^6	9.1

follows its release. The reduction by tris-[*N*-(2-pyridylenethylene)aniline] iron(II) has the kinetics

$$-\mathrm{d}[\mathrm{FeL_3}^{2+}]/\mathrm{d}t = k_1[\mathrm{FeL_3}^{2+}]+k_2[\mathrm{FeL_3}^{2+}][\mathrm{S_2O_8}^{2-}]$$

indicating that both kinds of reaction, *i.e.* direct oxidation of complex and oxidation of separated ligand, participate. A dissociative path has also been confirmed for the persulphate oxidation of bis(maleonitriledithiolate)cobaltate-(II)[635].

Ferrocyanide reduces persulphate, the reaction being second-order in a fairly saline medium (0.5 M K_2SO_4)[636] with $k_2 = 3.2\times10^9 \exp(-11.9\times10^3/\boldsymbol{RT})$ l.mole^{-1}.sec^{-1}. The rate is strongly influenced by the presence of potassium ions and this has been shown not to be merely an ionic strength effect[637, 638]. Consideration of all possible modes of ion-pairing led to the conclusion that the two reactants are $[K(Fe(CN)_6]^{3-}$ and $[KS_2O_8]^-$. At zero ionic strength, $E = 9.6$ kcal.mole^{-1} and $\Delta S^\ddagger = -34.7$ eu. Kershaw and Prue[638] have measured the specific effects of many other cations on the rate of this reaction.

The reduction of persulphate by stannous ions is strongly affected by dissolved oxygen[639, 639a]. The anaerobic oxidation is first-order in both oxidant and reductant with $E = 11.9$ kcal.mole^{-1} and $\Delta S^\ddagger = -19.4$ eu[639b].

A total reaction order of two is also shown by the Cr(II) reduction of persulphate[639c]. At 25° $k_2 = (2.5\pm0.3)\times10^4$ l.mole^{-1}.sec^{-1} $[(H^+] = 0.10$ M, $\mu = 1$ M $NaClO_4$). More informative is the product study; one mole of $S_2O_8{}^{2-}$ oxidises two moles of aquochromium(II) to one mole of $CrSO_4{}^+$ and one of Cr^{3+}(aq) over a wide range of conditions. Introduction of bromide ions left both the rate and the yield of $CrSO_4{}^+$ unchanged, but the yield of Cr^{3+}(aq) was depressed, being exactly compensated by the production of $CrBr^{2+}$. These results imply that the reduction of $S_2O_8{}^{2-}$ by Cr^{2+} is of the inner-sphere type to give substitution-inert $CrSO_4{}^+$, *viz.*

$$Cr^{2+}+S_2O_8{}^{2-} \rightarrow CrSO_4{}^+ + \cdot SO_4{}^-$$

The reduction of $\cdot SO_4{}^-$ in the rapid second step is, on the other hand, of the outer-sphere type (for otherwise further $CrSO_4{}^+$ would be found), *viz.*

$$Cr^{2+}+\cdot SO_4{}^- \rightarrow Cr^{3+}+SO_4{}^{2-}$$

Br^- merely reacts with $\cdot SO_4{}^-$ to produce Br·, which is oxidised by Cr^{2+} by an inner-sphere route, possibly in the form of $\cdot Br_2{}^-$.

Both kinetic and product studies have also been made[626b] of the reductions of $S_2O_8{}^{2-}$ by $Co(NH_3)_5OH_2{}^{3+}$ and $Co(NH_3)_6{}^{3+}$. As the concentration of the former reductant is increased the rate coefficient k, defined by

$$k = \frac{d[Co^{2+}]}{dt} / [S_2O_8^{2-}]$$

increases but attains a limiting value at about $[Co(III)] \sim 2 \times 10^{-3}$ M. At lower concentrations of complex a large ($\geqslant 4.5$) isotope effect was apparent when a deuterated complex in D_2O was used, but no similar effect was observed with respect to the "plateau" rate. Even in the plateau region the rate of production of Co^{2+} is linearly dependent on oxidant concentration, and corresponds exactly to the rate of autodecompositition of $S_2O_8^{2-}$ at low acidity[639d]. The stoichiometry is written as

$$3\,H^+ + Co(NH_3)_5OH_2^{3+} + 2\,OH\cdot = Co^{2+} + 4\,NH_4^+ + \tfrac{1}{2}\,N_2 + 3\,H_2O$$

although replacement of $OH\cdot$ by $\cdot SO_4^-$ would not contravene the experimental data. The initial act is probably attack by $\cdot SO_4^-$ on the co-ordinated water molecule, yielding ultimately a Co(IV) species which autodecomposes to Co(II).

Similar stoichiometry and rate saturation phenomena are reported for the reduction by $Co(NH_3)_6^{3+}$ although the rate coefficient is lower at saturation (which requires a higher concentration of Co(III), namely 2×10^{-2} M). The isotope effect for $Co(ND_3)_6^{3+}$ in H_2O is 8–9, suggesting abstraction of a hydrogen atom by $\cdot SO_4^-$ followed by internal electron transfer to give Co(IV). Rapid oxidations ($k_2 \sim 10^7$ l.mole^{-1}.sec^{-1}) by $\cdot SO_4^-$ of molecules such as CH_3OH and HCO_2H have been examined by flash-photolysis[639e]; intermediate formation of $OH\cdot$ (suggested by Tsao and Wilmarth[639f]) is not required and, in view of the *second*-order disappearance of $\cdot SO_4^-$ in absence of added solutes, is quite possibly incorrect.

A few details have been reported on the slow reductions of $S_2O_8^{2-}$ by As(III)[628a] and Tl(I)[639g]. In the anaerobic reductions by As(III) the reaction is first-order in $S_2O_8^{2-}$ and although As(III) certainly catalyses decomposition, the dependence of the rate on [As(III)] is small. Aeration leaves the rate of spontaneous decomposition of $S_2O_8^{2-}$ unaffected, but the As(III)-catalysed route is accelerated by a factor of ten, the kinetic law remaining unchanged. The oxygen effect is interpreted in terms of the chain reaction

$$S_2O_8^{2-} \rightarrow 2\,\cdot SO_4^-$$

$$\cdot SO_4^- + As(III) \rightarrow SO_4^{2-} + As(IV)$$

$$As(IV) + O_2 + H^+ \rightarrow As(V) + HO_2\cdot$$

$$HO_2\cdot + S_2O_8^{2-} \rightarrow H^+ + O_2 + SO_4^{2-} + \cdot SO_4^-$$

$$As(IV) + HO_2\cdot \rightarrow As(III) + H^+ + O_2$$

An alternative termination step involves trace quantities of Cu(II) present in the reaction mixture.

The aerobic reduction by Tl(I) is also first-order in $S_2O_8{}^{2-}$ and zero-order in reductant, and a one-half order in acid concentration is found[639g]. A mechanism identical with that of the As(III) reduction is proposed.

6.4.8 *Peroxomonosulphate ion (Caro's acid)*

The decomposition to sulphate and oxygen is subject to trace metal-ion catalysis[640]. Co(II)[641], Mo(VI)[641] and Mn(II)[642] are particularly effective, but the kinetics could not be resolved unequivocally in all cases. The rate expressions are

Co(II) catalysis (pH 6–6.7)
$$-\mathrm{d}[HSO_5{}^-]/\mathrm{d}t = k_1[HSO_5^-]+k_2[Co(II)]^n[HSO_5{}^-]^2$$

Mn(II) catalysis
$$-\mathrm{d}[HSO_5{}^-]/\mathrm{d}t = k_1[HSO_5^-] + k_2[Mn(III)]^{\frac{1}{2}}[H_2SO_5]^{\frac{1}{2}}$$

Mo(VI) catalysis
$$-\mathrm{d}[HSO_5^-]/\mathrm{d}t = k_1[HSO_5^-]+k_2[Mo(VI)]^{\frac{1}{2}}[HSO_5{}^-]$$

The latter reaction displayed an induction period independent of reactant concentrations and pH. The initial decompositions may be

$$HSO_5{}^- + M^{n+} \rightarrow M^{(n+1)+} + \cdot SO_4{}^- + OH^-$$

Production of the oxidised ion has been observed for Mn(II) catalysis[642].

6.4.9 *Organic halides*

Reductions have been effected with both $Cr(H_2O)_6{}^{2+}$ and $Co(CN)_5{}^{3-}$ to give either or all of the following reactions[643]

$$RX+M^{n+} \rightarrow M^{(n+1)+} + X^- + R\cdot \rightarrow R\text{–}M^{n+}, \tfrac{1}{2}R_2, RH$$

Dihalides are reduced by Cr(II) to the corresponding olefin[643]. Allylic and benzylic halides and polyhalides are reduced more readily than simple alkyl halides, but even the latter are readily reduced by an ethylenediamine complex of Cr(II)[644].

Allyl chloride and α-phenylethyl chloride are reduced by Cr(II) sulphate in aqueous dimethyl formamide (DMF) in a simple second-order process. At 29.7 °C in 1 : 1 solvent mixture and at 7 *M* acidity ($HClO_4$) the second-order rate coefficients are 0.02 and 0.04 $l.mole^{-1}.sec^{-1}$, respectively[643]. The stereochemical course of reaction is illustrated by the reduction of optically inactive α-phenylethyl chloride and bromide to diphenylbutanes of composition 85–90 % *meso* and 10–15 % *d,l*. The mechanism is thought to involve transfer of a halogen atom, *viz*.

$$-\overset{|}{\underset{|}{C}}-X + Cr^{2+}(II) \longrightarrow -\overset{|}{\underset{|}{C}}\cdots X \cdots Cr^{2+} \longrightarrow -\overset{|}{\underset{|}{C}}\cdot + X-Cr^{2+}(III)$$

Halide always appears in the coordination sphere of Cr(III) immediately following reduction. Castro and Kray[643] propose that the organochromium compound is formed by subsequent attack of R· upon further Cr(II).

Similar kinetics are exhibited by the reduction of vicinal dihalides by Cr(II)[645], which proceeds 10–40 times faster than that of the allylic halides. Such activation by a second halogen atom suggests[645] a neighbouring group effect, *viz*.

$$-\overset{X}{\underset{|}{C}}-\overset{|}{\underset{|}{C}}-X + Cr^{2+} \longrightarrow -\overset{X}{\underset{|}{C}}-\underset{|}{C}\cdots X \cdots Cr^{2+} \longrightarrow \overset{X}{C-C} + CrX^{2+} \qquad (109)$$

The dehalogenation of the α-haloalkyl radical is a fast step which can take place by several possible routes[645]. Dibromides are reduced much faster than dichlorides and *trans*-1,2-dibromocylohexane is reduced 100 times faster than the *cis*-isomer. This accords with neighbouring group assistance which bromine seems particularly capable of offering (see subsection 6.4.10).

Attempts have been made[647] to trap the intermediate radical with a monomer, particularly in the reduction of benzyl chloride by Cr(II) to benzylchromium ion (and ultimately to toluene and dibenzyl). The results were ambiguous, however, as benzylchromium ion itself reacts with butadiene and acrylonitrile. This reduction shows second-order kinetics with E = 14.6 $kcal.mole^{-1}$ and $\Delta S^{\ddagger}$ = 14.3 eu. The rate coefficients for benzyl chloride, bromide and iodide follow the expected sequence[647]

$C_6H_5CH_2Cl$	$C_6H_5CH_2Br$	$C_6H_5CH_2I$	
3.2×10^{-3}	4.1×10^{-1}	1.8	k_2 (27 °C) ($l.mole^{-1}.sec^{-1}$)

Reduction of simple alkyl halides to alkanes by ethylenediamine complexes of Cr(II), denoted Cr^{II}(en) occurs readily[691], *e.g.* for isopropyl chloride in aqueous dimethylformamide at 25 °C simple second-order behaviour is found with k_2 dependent on [en]/[Cr(II)] but reaching a limiting value of 1.6×10^{-2} $l.mole^{-1}.sec^{-1}$. Competition studies between a mixture of two alkyl chlorides for Cr(II) was achieved by estimating alkane products by gas–liquid chromatography and

References pp. 493–509

examples of relative rate coefficients at 25 °C obtained in this way are

n-butyl chloride	1	*n*-butyl bromide	140	*n*-butyl iodide	10,000
isopropyl chloride	3.6	isopropyl bromide	1,100	isopropyl iodide	63,000
t-butyl chloride	29	*t*-butyl bromide	5,800	*sec*-butyl iodide	110,000

The mechanism proposed involves halogen atom transfer to give an alkylchromium intermediate which then undergoes hydrolysis, *viz.*

$$RX + Cr^{II}en_2^{2+} \rightarrow R\cdot + Cr^{III}en_2X^{2+}$$

$$R\cdot + Cr^{II}en_2^{2+} \rightarrow RCr^{III}en_2^{2+}$$

$$RCr^{III}en_2^{2+} + H_2O \rightarrow RH + Cr^{III}en_2(OH)^{2+}$$

The hydrolysis step could be followed separately from alkylchromium formation, which was monitored optically.

The reductions in aqueous methanol of a number of alkyl halides[648–649a] by $Co(CN)_5^{3-}$ have simple second-order kinetics. For methyl and benzyl halides (RX) the stoichiometries are[649, 649a]

$$2\,Co(CN)_5^{3-} + RX = Co(CN)_5R^{3-} + Co(CN)_5X^{3-}$$

but for ethyl, isopropyl and *t*-butyl iodides an additional path is apparent[649a] of the type

$$2\,Co(CN)_5^{3-} + C_2H_5I = Co(CN)_5I^{3-} + Co(CN)_5H^{3-} + CH_2{=}CH_2$$

Two α, ω-diiodoalkanes are also reduced[649a], *viz.*

$$2\,Co(CN)_5^{3-} + ICH_2CH_2I = 2\,Co(CN)_5I^{3-} + CH_2{=}CH_2$$
$$2\,Co(CN)_5^{3-} + ICH_2CH_2CH_2I = 2\,Co(CN)_5I^{3-} + \text{cyclopropane}$$

Rate coefficients k_2 defined by

$$-d[Co(CN)_5^{3-}]/dt = 2\,k_2[Co(CN)_5^{3-}][RX]$$

are listed in Table 26. The reduction of methyl and benzyl halides resembles that by Cr^{2+}(aq), *viz.*

$$Co(CN)_5^{3-} + RX \rightarrow Co(CN)_5X^{3-} + R\cdot \qquad (110)$$

$$Co(CN)_5^{3-} + R\cdot \rightarrow Co(CN)_5R^{3-} \qquad (111)$$

TABLE 26

RATE COEFFICIENTS FOR THE REDUCTION OF ORGANIC HALIDES BY PENTACYANOCOBALTATE(II) ION[649a]

Temp., 25.0±0.2 °C; medium, 20 % H_2O–80 % CH_3OH (v/v); pH > 11, μ = 0.02 M ($NaClO_4$).

Halide	k_2 (*l.mole^{-1}.sec^{-1}*)
$C_6H_5CH_2I$	3800
$(CH_3)_3CCl$	9.1
$(CH_3)_2CHI$	1.20
CH_3CH_2I	0.059
$CH_3CH_2CH_2I$	0.043
CH_3I	0.0095
p-$NO_2C_6H_4CH_2Br$	105
p-$BrC_6H_4CH_2Br$	7.5
$C_6H_5CH_2Br$	2.33
$C_6H_5CH_2Cl$	0.00049
ICH_2CH_2I	63
$I(CH_2)_3I$	0.99
$I(CH_2)_3Br$	0.68
$I(CH_2)_4I$	0.30
$I(CH_2)_5I$	0.12

For higher alkyl halides there exist additional paths exemplified by

$$Co(CN)_5^{3-} + CH_3CH_2\cdot \rightarrow Co(CN)_5H^{3-} + CH_2{=}CH_2$$

$$Co(CN)_5H^{3-} + CH_3CH_2I \rightarrow Co(CN)_5I^{3-} + C_2H_6$$

A comprehensive set of rate coefficients for water-soluble halo compounds has been published[649] indicating the same trends apparent in Table 26. Activation parameters for a selection of these compounds fall in a range $E = 6.9 \pm 3.0$ kcal.mole^{-1}, $\Delta S^{\ddagger} = -28 \pm 8$ eu[649].

Bis(glyoximato)cobalt(II) complexes of the types $Co(DH)_2B_2$ and $Co(DH)_2B_2$ (DH = disubstituted glyoxime, B = base, *e.g.* pyridine or triphenylphosphine) reduce benzyl bromide in benzene and acetone solutions[649b]

$$2\,Co(DH)_2B + RX = RCo(DH)_2B + XCo(DH)_2B$$

$$2\,Co(DH)_2B_2 + RX = RCo(DH)_2B + XCo(DH)_2B + 2\,B$$

The reactions are of the first-order in each reactant and, for example, k (25 °C) for the reaction with pyridinatobis(dimethylglyoximato)cobalt(II) in benzene is 0.30 l.mole^{-1}.sec^{-1}. k increases slightly with the basicity of B but is relatively insensitive to changes in DH. A radical mechanism identical with equations (110, 111) is proposed[649b].

Reductions by Co(I) chelates such as vitamin B_{12s} and tributylphosphine–

TABLE 27

RATE COEFFICIENTS FOR THE REDUCTION OF ORGANIC HALIDES BY Co(I) CHELATES[649c]

Temp., 25±2 °C; medium, methanol (0.1 *M* NaOH).

Halide	k_2 (*vitamin* B_{12s}) (*l.mole*$^{-1}$.*sec*$^{-1}$)	k_2 (*tributylphosphine-cobaloxime*$_s$) (*l.mole*$^{-1}$.*sec*$^{-1}$)
CH_3Cl	5.0	8.5×10^{-1}
CH_3Br	1.6×10^{3}	2.2×10^{2}
CH_3I	3.4×10^{4}	2.3×10^{3}
CH_3CH_2Cl	4.7×10^{-2}	9.0×10^{-3}
$(CH_3)_2CHCl$	—	3.2×10^{-4}
CH_3CH_2Br	3.1	1.6
$(CH_3)_2CHBr$	1.1×10^{-1}	1.8
$C_6H_5CH_2Cl$	—	4.4×10^{2}

cobaloximes all proceed with simple second-order kinetics[649c], but the sequence of rates offers a total contrast to the reductions by Cr(II) and $Co(CN)_5{}^{3-}$ (Table 27). In particular the rate sequence

$$CH_3X > CH_3CH_2X > (CH_3)_2CHX$$

is quite the reverse of that found for $Co(CN)_5^{3-}$ and follows instead the dependence expected of a classical S_N2 mechanism, *viz.*

$$Co(I)^- + RX \rightarrow [Co \ldots R \ldots X]^- \rightarrow Co\text{–}R + X^-$$

6.4.10 *β-Substituted alkyl halides*

Closely related to the reductions by Cr(II) of *vic*-dihalides are those of the compounds[650]

$$\underset{\substack{|\quad|\\ X\quad Y}}{\rangle C\text{–}C\langle} + 2\ Cr(II) = \rangle C{=}C\langle + X^- + Y^- + 2\ Cr(III)$$

where X = halogen and Y = OH, CH_3CO_2, NH_2 and Cr(II) refers either to the ethylenediamine complex of Cr(II) or the aquo species. For a series of 3-Y-substituted-7-butyl bromides the rates given in Table 28 were obtained. With the exception of the dibromides the composition of the butenes obtained from a given pair of diastereoisomers is the same, although the *cis*/*trans* ratio depends on Y. Only the dibromide is highly stereospecific, giving up to 97 % *trans* elimination under some conditions, and it appears that only Br is capable of forming the bridged

TABLE 28

RATES OF REDUCTION OF 3-SUBSTITUTED BUTYL BROMIDES BY Cr(II)aq
Temperature, 0 °C; solvent, 85 % DMF+15 % H_2O; $[HClO_4]$ = 0.9 *M*.

Y	$10^4\ k_2$ (*l.mole*$^{-1}$*.sec*$^{-1}$)	
	erythro	*threo*
H	0.14	—
OH	0.56	0.68
CH_3CO_2	2.1	2.0
p-Tosyloxy	7.7	6.0
Cl	51	25
Br	1600	660

species in reaction (109). This is supported by the rate difference both in these reductions and in those of *cis*- and *trans*-dibromocyclohexanes[646] (preceding section).

6.4.11 *Carbon tetrachloride*

The reduction by anhydrous chlorides of Cu(I) and Fe(II) in acetonitrile solution is of the type[650a]

$$\mathrm{Fe(II)} + \mathrm{CCl_4} \rightleftharpoons \underset{\lambda_{max}\ 560\ \mathrm{nm}}{\underset{\text{complex (CT)}}{\text{charge transfer}}} \rightleftharpoons \underset{\lambda_{max}\ 532\ \mathrm{nm}}{\mathrm{Fe(III)Cl}} + \mathrm{CCl_3}\cdot$$

The overall oxidation is first-order in Fe(II), and the main product is tetrachlorethylene, indicating a stoichiometry

$$2\ \mathrm{CCl_4} + 4\ \mathrm{FeCl_2} = \mathrm{C_2Cl_4} + 4\ \mathrm{FeCl_3}$$

At 22 °C the overall second-order coefficient for reductions by Fe(II) and Cu(I) are, respectively, 4×10^{-4} and $\sim 5 \times 10^{-6}$ l.mole^{-1}.sec^{-1}. Products in the presence of olefins demonstrate unequivocally the intermediacy of carbenoid transients, and a complex mechanism is put forward, *viz.*

$$\mathrm{FeCl_4^{2-}} + \mathrm{CCl_4} \rightleftharpoons \mathrm{CT}$$

$$\mathrm{CT} \rightarrow \mathrm{FeCl_4^-} + \mathrm{CCl_3}\cdot + \mathrm{Cl^-} \quad \text{(slow)}$$

$$2\ \mathrm{CCl_3}\cdot \rightarrow \mathrm{C_2Cl_6}$$

$$\mathrm{CCl_3}\cdot + \mathrm{FeCl_4^{2-}} \rightarrow \mathrm{Cl^-} + [\mathrm{FeCl_3(CCl_3)}]^-$$

$$2\ [\mathrm{FeCl_3(CCl_3)}]^- \rightarrow 2\ \mathrm{FeCl_4^-} + \mathrm{C_2Cl_4}$$

$$\text{cyclohexene} + [\mathrm{FeCl_3(CCl_3)}]^- \rightarrow \text{7,7-dichloronorcarane} + \mathrm{FeCl_4^-}$$

6.4.12 Aromatic sulphonyl chlorides

The reduction by Cu(I) in acetonitrile at 110 °C in the presence of styrene was followed dilatometrically[650b]. The mechanism given is

$$\text{Initiation}\quad C_6H_5CH{=}CH_2+2\,CuCl_2 \rightarrow C_6H_5CHClCH_2Cl+2\,CuCl \qquad \text{(fast)}$$

$$\text{Propagation}\quad ArSO_2Cl+CuCl \underset{k_{-2}}{\overset{k_2}{\rightleftarrows}} ArSO_2\cdot+CuCl_2 \qquad \text{(slow)}$$

$$ArSO_2\cdot+C_6H_5CH{=}CH_2 \underset{k_{-3}}{\overset{k_3}{\rightleftarrows}} ArSO_2CH_2\dot{C}HC_6H_5 \qquad \text{(fast)}$$

$$ArSO_2CH_2\dot{C}HC_6H_5+CuCl_2 \overset{k_4}{\rightarrow} ArSO_2CH_2CHClC_6H_5+CuCl \qquad \text{(fast)}$$

$$\text{Termination}\quad ArSO_2\cdot+Cu(I) \underset{k_{-5}}{\overset{k_5}{\rightleftarrows}} ArSO_2^-.\,Cu(II) \rightarrow Cu(II)+\text{products}$$

Application of stationary state treatment for $ArSO_2\cdot$ and $ArSO_2CH_2\dot{C}HC_6H_5$ produces a complex rate law which reduces at low Cu(II) concentrations to

$$-d[ArSO_2Cl]/dt = k_2[CuCl][ArSO_2Cl]$$

This is adhered to experimentally and for $Ar{=}C_6H_5$, $k_2 = 0.0893$ l.mole^{-1}.sec^{-1} under the conditions specified. Results for a set of ring-substituted analogues give a good Hammett plot with $\rho = +0.565$. Certain aspects of this system remain unclear, *e.g.* the non-appearance of polymer at very low Cu(II) concentrations and the exact nature of the Cu(I) species.

7. Redox reactions between radicals and metal ions

The kinetics of reactions between neutral free radicals, either stable or generated thermally or photochemically, and metal ions of variable valence, have been determined. These reactions are generally simple second-order and this will be assumed throughout this section unless stated to the contrary. Although neutral radicals are normally very effective reducing agents, *viz.*

$$R\cdot+M^{n+} \rightarrow R^++M^{(n-1)+}$$

there is evidence that they can function as oxidising agents (*vide infra*), *viz.*

$$R\cdot+M^{n+} \rightarrow R^-+M^{(n+1)+}$$

Reactions of e^-_{aq} and $OH\cdot$ are dealt with later in this series.

7.1 STABLE RADICALS

Diphenylpicrylhydrazyl (DPPH)

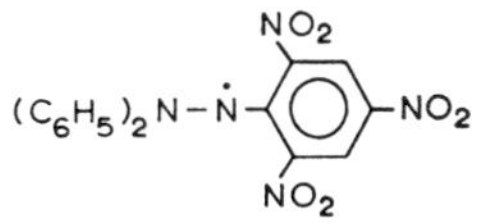

$$(C_6H_5)_2N-\dot{N}-C_6H_2(NO_2)_3$$

is a well-known stable free radical which reduces the acetates of several oxidising metal ions in acetic acid[651], the second-order coefficients at 25 °C revealing remarkably low specificity, particularly towards the two-equivalent oxidant Pb(IV)

Metal acetate	Pb(IV)	Co(III)	Ce(IV)	Fe(III)
k_2 (l.mole^{-1}.sec^{-1})	35	22	55	88

k_2 is always reduced by addition of methanol or benzene, but is increased by addition of water.

The reductions by DPPH of intermediates generated by the Ag(I)– and Cu(II)-persulphate couples are dealt with in section 6.4.7.

DPPH can also function as an oxidant towards Fe(II) in ethanolic solution[652], *viz.*

$$\text{Fe(II)} + C_2H_5OH + \text{DPPH} = [C_2H_5O\text{–Fe(III)}]^{2+} + \text{DPPH}_2$$

($DPPH_2$ represents the corresponding hydrazine). The kinetics are simple second-order with a rate coefficient, k_2, in 1/1(v/v) ethanol–water of 2.73×10^{11} exp $(13.5 \times 10^3/\boldsymbol{RT})$ l.mole^{-1}.sec^{-1}, independent of pH and of ionic strength. Log k_2 is inversely proportional to the dielectric constant of the medium. The small entropy of activation and the rather large activation energy are taken as favouring a hydrogen-atom transfer as opposed to an electron-transfer process.

Ions also reducing DPPH include Sn(II), Cr(II), $Fe(CN)_6{}^{4-}$ and $Mo(CN)_8{}^{4-}$. However, Ag(I), Pb(II), Mn(II), Co(II) and Ce(III) are inert. These data enabled an approximate estimate of 1.1 V for the redox potential of DPPH/$DPPH_2$ to be made[652].

The reduction of 2,2,6,6-tetramethyl-4-piperidol *N*-oxide (RNO·)

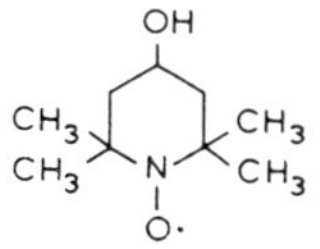

[structure: 4-hydroxy-2,2,6,6-tetramethylpiperidine ring with OH, four CH3 groups, N–O·]

by Fe(II) can be followed by the disappearance of the triplet ESR spectrum[652a]. The reaction is first-order both in Fe(II) and RNO· and Fe(III) does not retard reaction. Anions exert a specific catalytic effect, in the order: $SO_4^{2-} > Cl^- > ClO_4^-$, probably through formation of more reactive complexes of Fe(II) as in the Fenton reaction. The effect of pH is complex, being related to the prevailing anion, but the acidity dependence in perchlorate media is zero. E is *ca.* 7 kcal.mole^{-1} under most conditions.

7.2 GROWING POLYMER RADICALS

Dainton *et al.*[653, 654] have obtained rate and Arrhenius data for both oxidations and reductions in water by the growing polyacrylamide radical by examining the rate of radiation-induced polymerisation as a function of the concentration of added metal ions, which serve to terminate polymerisation. Some examples are given in Table 29.

The following features are discernible: (*i*) the lack of relation between k_2 and π^0 for oxidising ions, (*ii*) the low values of E and (*iii*) the high k_2 for oxidising ions with partly filled d shells and the low k_2 for those with d^{10} configurations. The absence of a solvent isotope effect with Fe^{3+} and Cu^{2+} and the faster rates with

TABLE 29

KINETIC DATA FOR REACTIONS BETWEEN METAL IONS AND GROWING POLYACRYLAMIDE RADICALS

Metal ion	*Acidity*	k_2 (25 °C) (*l.mole*$^{-1}$*.sec*$^{-1}$)	*A*(*l.mole*$^{-1}$*.sec*$^{-1}$)	*E*(*kcal.mole*$^{-1}$)
Fe^{3+}aq	0.0 *M* ($HClO_4$)	2.8×10^3	1.45×10^5	2.35 ± 0.6
Cu^{2+}aq	1.0 *M* ($HClO_4$)	1.17×10^3	1.1×10^7	5.4 ± 1.3
Ce^{4+}aq	1.0 *M* ($HClO_4$)	small	—	—
Ag^+aq	1.0 *M* ($HClO_4$)	zero	—	—
Hg^{2+}aq	0.1 *M* ($HClO_4$)	1.05	4.2×10^4	6.2 ± 1.0
Tl^{3+}aq	0.1 *M* ($HClO_4$)	0.34	21	2.5 ± 0.4
VO_2^+	0.4 *M* (H_2SO_4)	1.1×10^3	—	—
$FeOH^2$	—	2.12×10^4	4×10^4	0.37 ± 0.3
$FeCl^{2+}$	—	8.12×10^4	2.7×10^5	0.7 ± 0.6
$FeCl_2^+$	—	1.70×10^4	—	—
$FeCl_3$	—	1.0×10^6	—	—
$FeBr^{2+}$	—	1.66×10^6	2.3×10^5	-1.2 ± 0.3
FeN_3^{2+}	—	1.56×10^6	3.6×10^6	0.5 ± 0.2
$FeNCS^{2+}$	—	1.36×10^7	2.4×10^5	-2.4 ± 0.9
$Fe(dipy)_3^{3+}$	0.1 *M* (H_2SO_4)	$(0.81 \pm 0.1) \times 10^5$	—	—
$Fe(o\text{-phen})_3^{3+}$	0.1 *M* (H_2SO_4)	$(3.1 \pm 0.1) \times 10^5$	2.2×10^2	-4.3 ± 0.5
$Fe(CN)_6^{3-}$	0.1 *M* (H_2SO_4)	$(8.5 \pm 0.2) \times 10^5$	1.6×10^7	1.2 ± 1.0
Ti^{3+}aq	0.4 *M* (H_2SO_4)	$(5.8 \pm 0.3) \times 10^2$	10^{12}	11.1 ± 2.0
Eu^{2+}aq	0.4 *M* (H_2SO_4)	$(8 \pm 4) \times 10^4$	—	—
V^{2+}aq	0.4 *M* (H_2SO_4)	$(1.1 \pm 0.6) \times 10^5$	—	—
Cr^{2+}aq	0.4 *M* (H_2SO_4)	$(2.8 \pm 1.4) \times 10^5$	—	—
Fe^{2+}aq	0.4 *M* (H_2SO_4)	$0 < k_2 < 1$	—	—
Mo^{3+}aq	0.4 *M* (H_2SO_4)	$(7 \pm 0.7) \times 10^3$	10^8–10^{10}	6.0 ± 1.5

FeX^{2+} suggest an electron-transfer mechanism for the *oxidation* of R·, *viz.*

$$FeX^{2+} + R\cdot \rightleftharpoons [Fe^{3+}X^-R\cdot] \rightleftharpoons [Fe^{2+}X^-R^+] \rightarrow Fe^{2+} + X^- + R^+$$

In the cases of substitution-inert complexes of Fe(III) it is envisaged that R· forms a temporary bond with the ligand through which the electron transport takes place.

Bamford *et al.*[655] have also determined k_2 for the reduction of ferric chloride in dimethylformamide by different R·, *viz.* and regard their results as evidence for an electron-transfer mechanism

R·	acrylonitrile	methacrylonitrile	styrene
k_2 at 60 °C (l.mole^{-1}.sec^{-1})	6.53×10^3	6.15×10^2	5.4×10^4

The Arrhenius parameters reported by Dainton *et al.*[654] for *reduction* of R· are similar to those recorded for reduction of diphenylpicrylhydrazyl by Fe(II) (*vide supra*), and log k_2 is related to the ionisation potentials of the reducing metal ion. It is concluded that hydrogen-atom transfer takes place, *viz.*

$$M^{n+}OH_2 + R\cdot \rightarrow MOH^{n+} + RH$$

7.3 TRANSIENT SIMPLE RADICALS

Measurements of absolute rate constants for the reduction and oxidation of metal ions by e^-_{aq}, H· and OH· has been a prominent achievement of the technique of pulse radiolysis. This subject is too broad to be included in this review and is to be dealt with later in the series. A key reference[567] is given, however, to help cover the interim period.

Kochi *et al.*[656, 657] have made a very detailed investigation of the reactions between salts and complexes of Cu(II) and simple organic free radicals. Measurement of k_2 for R· + Cu(II) is based on a competition between Cu(II) and organic hydrogen donors (denoted DH)[658]. The sequence of reactions is exemplified by the decomposition of valeryl peroxide in aqueous acetic acid by Cu(II) acetate in the presence and in absence of DH

(*a*) $C_4H_9CO\text{–}O\text{–}O\text{–}COC_4H_9 + Cu(I) \rightarrow C_4H_9CO_2\cdot + Cu(II)O_2CC_4H_9$

$$C_4H_9CO_2\cdot \rightarrow C_4H_9\cdot + CO_2$$

$$C_4H_9\cdot + Cu(II)O_2CC_4H_9 \rightarrow CH_3CH_2CH{=}CH_2 + C_4H_9CO_2H + Cu(I)$$

i.e. in toto $(C_4H_9CO_2)_2 = CH_3CH_2CH{=}CH_2 + C_4H_9CO_2H + CO_2$

References pp. 493–509

(*b*) in presence of DH

$$C_4H_9\cdot \begin{cases} \xrightarrow{Cu(II)O_2C_4CH_9} CH_3CH_2CH{=}CH_2 + C_4H_9CO_2H + Cu(I) \\ \xrightarrow{DH} C_4H_{10} + D\cdot \end{cases}$$

The ratio of product yields of but-1-ene and butane gives $k_{R\cdot+Cu(II)}/k_{R\cdot+DH}$ $k_{R\cdot+DH}$ is known for the gas phase reaction[659, 660]

$$CH_3(CH_2)_2CH_2\cdot + CH_3(CH_2)_3CHO \rightarrow C_4H_{10} + CH_3(CH_2)_3CO\cdot$$

and from the known activation energy for this reaction a value of $k_{R\cdot+Cu(II)}$ at 57 °C, the temperature at which the competition was staged, of 1.1×10^8 l.mole^{-1}.sec^{-1} was derived. More generally, for this particular oxidation

$$k_2 = (2.2 \pm 1.8) \times 10^{10} \exp(3.4 \pm 0.5 \times 10^3/\boldsymbol{RT}) \text{ l.mole}^{-1}.\text{sec}^{-1}.$$

k_2 was determined for several analogous reactions, always falling in the range $\log k_2 = 8.2 \pm 0.6$, but correlating roughly with the ionisation potential of the radical. This compares with a value of 1.17×10^3 l.mole^{-1}.sec^{-1} (25 °C) for Cu(II) reacting with growing polyacrylamide radical which is a much larger molecule carrying an electronegative group near the radical site[654].

The reaction between $CH_3\dot{C}HOH$ and Cu(II) has been examined by pulse radiolysis[661]. The reduction of $CuCl_2$ followed the beam pulse shape even down to 2×10^{-4} *M* oxidant, from which it follows that $k_2 > 10^9$ l.mole^{-1}.sec^{-1}. $\cdot CH_2CO_2H$ reduces $Cu(O_2CCH_3)_2$ with a similar rate coefficient but fails to reduce the dimeric $Cu_2(O_2CCH_3)_4$. It was also shown that reaction between ethanolic Cu(I) and $CH_3CH(OH)O_2\cdot$, monitored by pulse radiolysing oxygenated ethanolic Cu_2Cl_2, is diffusion-controlled.

By pulse radiolysis of nitrous oxide-saturated aqueous solutions of ferricyanide (2×10^{-4} *M*) and various alcohols (0.1 *M*), Adams and Willson[694] were able to obtain absolute rate coefficients for the ferricyanide oxidation of the radicals derived from the alcohols by attack of the solvent irradiation product, OH·.

TABLE 30

SECOND-ORDER RATE COEFFICIENTS FOR SOME OXIDATIONS OF FREE RADICALS

Oxidant	*Radical*	*Medium*	*Temp. (°C)*	*k_2 (l.mole^{-1}.sec^{-1})*	*Ref.*
Cu(II)	$CH_3\dot{C}HOH$	98 % ethanol–2 % acetic acid	~20	$>10^9$	661
Cu(II)	*n*-butyl	Aq. acetic acid	57	1.1×10^8	658
Cu(II)	acrylamide (polymer)	1 *M* aq. $HClO_4$	25	1.17×10^3	654
$Fe(CN)_6^{3-}$	acrylamide (monomer)	1 *M* aq. $HClO_4$	~20	$6.8 \pm 0.4 \times 10^6$	662
$Fe(CN)_6^{3-}$	acrylamide (polymer)	0.1 *M* aq. H_2SO_4	25	8.5×10^5	654

The nitrous oxide merely suppresses reduction of ferricyanide by e_{aq}^- by converting the latter to $\cdot O^-$. For a number of alcohols and glycols and also polyethylene oxide of molecular weight $200–2 \times 10^4$, the values of $k(R\cdot + \text{ferricyanide})$ fall in the range $2–5 \times 10^9$ l.mole^{-1}.sec^{-1}.

Table 30 lists some rate coefficients for oxidations of various free radicals.

ACKNOWLEDGEMENTS

Dr. J. S. Littler is thanked for reading the manuscript, for making many usefu comments and for disclosure of unpublished results. The author's wife, Sheila is thanked for her support during the preparation of this chapter. The encouragement of Professor T. C. Waddington, formerly Chairman of the School of Molecular Sciences, University of Warwick, is also acknowledged.

REFERENCES

1 K. B. Wiberg, *Oxidation in Organic Chemistry*, Academic Press, New York, 1965.
2 R. Stewart, *Oxidation Mechanisms; Applications to Organic Chemistry*, Benjamin, New York, 1964.
3 W. A. Waters, *Mechanisms of Oxidation of Organic Compounds*, Methuen, London, 1964.
4 J. P. Candlin, K. A. Taylor and D. T. Thompson, *Reactions of Transition-Metal Complexes*, Elsevier, Amsterdam, 1968, p. 156.
5 T. A. Turney, *Oxidation Mechanisms*, Butterworths, London, 1965.
6 R. E. Kirk and A. W. Browne, *J. Am. Chem. Soc.*, 50 (1928) 337.
7 W. C. E. Higginson, D. Sutton and P. Wright, *J. Chem. Soc.*, (1953) 1380.
8 W. C. E. Higginson and J. W. Marshall, *J. Chem. Soc.*, (1957) 447.
9 P. Saffir and H. Taube, *J. Am. Chem. Soc.*, 82 (1960) 13.
10 R. T. M. Fraser and H. Taube, *J. Am. Chem. Soc.*, 82 (1960) 4152.
11 J. P. Candlin and J. Halpern, *J. Am. Chem. Soc.*, (1963) 2518.
12 J. Vepřek-Šiška and D. M. Wagnerová, *Collection Czech. Chem. Commun.*, 30 (1965) 1390.
13 J. Vepřek-Šiška, D. M. Wagnerová and K. Eckschlager, *Collection Czech. Chem. Commun.*, 31 (1966) 1248.
14 J. Vepřek-Šiška, A. Šolcová and D. M. Wagnerová, *Collection Czech. Chem. Commun.*, 31 (1966) 3287.
15 A. Brown and W. C. E. Higginson, *Chem. Commun.*, (1967) 725.
16 L. Erdey, G. Svehla and L. Koltai, *Anal. Chim. Acta*, 28 (1963) 398.
17 W. C. E. Higginson, *Chem. Soc. London, Spec. Publ.*, 10 (1957) 95.
18 W. C. E. Higginson and D. Sutton, *J. Chem. Soc.*, (1953) 1402.
19 W. M. Latimer, *Oxidation Potentials*, Prentice-Hall, New Jersey, 1952.
20 F. H. Westheimer, *Chem. Rev.*, 45 (1949) 419.
21 Reference 1, p. 74.
22 H. Stamm, *Z. Angew. Chem.*, 47 (1934) 579, 791.
23 J. S. F. Pode and W. A. Waters, *J. Chem. Soc.*, (1956) 717.
24 R. M. Barter and J. S. Littler, *J. Chem. Soc. B*, (1967) 205.
25 J. Y. Tong and R. L. Johnson, *Inorg. Chem.*, 5 (1966) 1902.
26 Y. Sasaki, *Acta Chem. Scand.*, 16 (1962) 719.
27 M. Cohen and F. H. Westheimer, *J. Am. Chem. Soc.*, 74 (1952) 4387.

28 F. Holloway, *J. Am. Chem. Soc.*, 74 (1952) 224.
29 D. G. Lee and R. Stewart, *J. Am. Chem. Soc.*, 86 (1964) 3051.
30 D. G. Lee and D. T. Johnson, *Can. J. Chem.*, 43 (1965) 1952.
31 N. Bailey, A. Carrington, K. A. K. Lott and M. C. R. Symons, *J. Chem. Soc.*, (1960) 290.
32 R. Stewart and M. M. Mocek, *Can. J. Chem.*, 41 (1963) 1160.
33 C. F. Schönbein, *J. Prakt. Chem.*, 75 (1858) 108.
34 C. Benson, *J. Phys. Chem.*, 7 (1903) 1, 356.
35 R. F. Beard and N. W. Taylor, *J. Am. Chem. Soc.*, 51 (1929) 1973.
36 K. E. Howlett and S. Sarsfield, *J. Chem. Soc. A*, (1968) 683.
36a D. C. Gaswick and J. H. Krueger, *J. Am. Chem. Soc.*, 91 (1969) 2240.
37 L. J. Kirschenbaum and J. R. Sutter, *J. Phys. Chem.*, 70 (1966) 3863.
38 M. Bobtelsky and A. Rosenberg, *Z. Anorg. Allgem. Chem.*, 177 (1928) 137.
39 M. Bobtelsky and A. Rosenberg, *Z. Anorg. Allgem. Chem.*, 182 (1929) 74.
40 M. Bobtelsky and A. Glasner, *J. Chem. Soc.*, (1948) 1376.
41 R. Stewart and R. van der Linden, *Can. J. Chem.*, 38 (1960) 2237.
42 T. Freund, *J. Inorg. Nucl. Chem.*, 15 (1960) 371.
43 J. Vepřek-Šiška and V. Ettel, *Chem. Ind. London*, (1967) 548.
44 J. Vepřek-Šiška, V. Ettel and A. Regner, *J. Inorg. Nucl. Chem.*, 26 (1964) 1476.
45 J. Vepřek-Šiška, V. Ettel and A. Regner, *Collection Czech. Chem. Commun.*, 31 (1966) 1237.
45a J. Vepřek-Siška and V. Ettel, *J. Inorg. Nucl. Chem.*, 31 (1969) 789.
46 P. Moore, S. F. A. Kettle and R. G. Wilkins, *Inorg. Chem.*, 5 (1966) 466.
47 M. Orhanović and R. G. Wilkins, *J. Am. Chem. Soc.*, 89 (1967) 278.
48 T. L. Chang, *Science*, 100 (1944) 29.
49 B. Chance, *J. Franklin Inst.*, 229 (1940) 737.
50 G. P. Haight, E. Perchonock, F. Emmenegger and G. Gordon, *J. Am. Chem. Soc.*, 87 (1965) 3835.
51 D. M. Yost and R. Pomeroy, *J. Am. Chem. Soc.*, 49 (1927) 703.
51a K. B. Yatsirmirskii and A. N. Budarina, *Kinetika i Kataliz*, 9 (1968) 422.
52 J. Preer and G. P. Haight, *J. Am. Chem. Soc.*, 87 (1965) 5256.
52a G. P. Haight, M. Rose and J. Preer, *J. Am. Chem. Soc.*, 90 (1968) 4809.
52b G. P. Haight, F. Smentowski, M. Rose and C. Heller, *J. Am. Chem. Soc.*, 90 (1968) 6325.
53 N. R. Dhar, *Ann. Chim. Paris*, 11 (1919) 130.
54 R. F. De Lury, *J. Phys. Chem.*, 7 (1903) 239; 11 (1907) 54.
55 J. O. Edwards, *Chem. Rev.*, 50 (1952) 455.
56 J. G. Mason and A. D. Kowalak, *Inorg. Chem.*, 3 (1964) 1248.
57 I. M. Kolthoff and M. A. Fineman, *J. Phys. Chem.*, 60 (1956) 1383.
58 F. Kessler, *Pogg. Ann.*, 119 (1863) 218.
59 Reference 1, p. 77.
60 G. Just and Y. Kauko, *Z. Physik. Chem.*, 82 (1913) 71.
61 A. C. Harkness and J. Halpern, *J. Am. Chem. Soc.*, 83 (1961) 1258.
62 M. Katz, *Advan. Catalysis*, 5 (1953) 177.
63 A. H. Webster and J. Halpern, *J. Phys. Chem.*, 60 (1956) 280.
64 A. H. Webster and J. Halpern, *Trans. Faraday Soc.*, 53 (1957) 51.
65 G. Just and Y. Kauko, *Z. Physik. Chem.*, 76 (1911) 601.
66 J. F. Harrod and J. Halpern, *J. Phys. Chem.*, 65 (1961) 563
67 Reference 1, p. 69.
68 Reference 1, p. 1.
69 K. B. Wiberg and R. J. Evans, *Tetrahedron*, 8 (1960) 313.
70 F. Mareš and J. Roček, *Collection Czech. Chem. Commun.*, 26 (1961) 2370.
71 J. Roček and F. Mareš, *Collection Czech. Chem. Commun.*, 24 (1959) 2741.
72 J. Roček, *Collection Czech. Chem. Commun.*, 22 (1957) 1519.
73 F. Mareš, J. Roček and J. Sicher, *Collection Czech. Chem. Commun.*, 26 (1961) 2355.
74 W. F. Sager and A. Bradley, *J. Am. Chem. Soc.*, 78 (1956) 1187.
75 K. B. Wiberg and G. Foster, *J. Am. Chem. Soc.*, 83 (1961) 423.
76 A. Streitweiser, *Chem. Rev.*, 56 (1956) 571.

77 H. H. JAFFÉ, *Chem. Rev.*, 53 (1953) 191.
78 H. C. BROWN AND Y. OKAMOTO, *J. Am. Chem. Soc.*, 80 (1958) 4979.
79 I. NECŞOIU, V. PRZEMETCHI, A. GHENCIULESCU, C. N. RENTEA AND C. D. NENITZESCU, *Tetrahedron*, 22 (1966) 3037.
80 K. B. WIBERG, Reference 1, p. 123.
81 R. SLACK AND W. A. WATERS, *J. Chem. Soc.*, (1948) 1666.
82 R. SLACK AND W. A. WATERS, *J. Chem. Soc.*, (1949) 599.
83 Y. OGATA, A. FUKUI AND S. YUGUCHI, *J. Am. Chem. Soc.*, 74 (1952) 2707.
84 H. D. LAW AND F. M. PERKIN, *J. Chem. Soc.*, 93 (1908) 1633.
85 O. H. WHEELER, *Can. J. Chem.*, 38 (1960) 2137.
85a R. C. MAKHIJA AND R. A. STAIRS, *Can. J. Chem.*, 46 (1968) 1255.
86 I. NECŞOIU, A. T. BALABAN, I. PASCARU, E. SLIAM, M. ELIAN AND C. D. NENITZESCU, *Tetrahedron*, 19 (1963) 1133.
87 K. B. WIBERG AND R. EISENTHAL, *Tetrahedron*, 20 (1964) 1151.
88 R. A. STAIRS AND J. W. BURNS, *Can. J. Chem.*, 39 (1961) 960.
89 R. A. STAIRS, *Can. J. Chem.*, 40 (1962) 1656.
90 O. H. WHEELER, *Can. J. Chem.*, 42 (1964) 706.
91 I. P. GRAGEROV AND M. P. PONOMARCHUK, *J. Gen. Chem. USSR*, 32 (1962) 3501.
92 H. C. DUFFIN AND R. B. TUCKER, *Tetrahedron*, 23 (1967) 2803.
93 R. A. STAIRS, *Can. J. Chem.*, 42 (1964) 550.
94 H. C. DUFFIN AND R. B. TUCKER, *Tetrahedron*, 24 (1968) 389.
95 C. N. RENŢEA, I. NECŞOIU, M. RENŢEA, A. GHENCIULESCU AND C. D. NENITZESCU, *Tetrahedron*, 22 (1966) 3501.
96 K. B. WIBERG AND A. S. FOX, *J. Am. Chem. Soc.*, 85 (1963) 3487.
97 C. F. CULLIS AND J. W. LADBURY, *J. Chem. Soc.*, (1955) 555, 1407, 2850.
97a A. RAHMAN, unpublished results quoted in Reference 97b.
97b A. K. ASWATHY AND J. ROČEK, *J. Am. Chem. Soc.*, 91 (1969) 991.
98 K. B. WIBERG AND R. D. GEER, *J. Am. Chem. Soc.*, 88 (1966) 5827.
99 K. B. WIBERG AND K. A. SAEGEBARTH, *J. Am. Chem. Soc.*, 79 (1957) 2822.
100 J. ROČEK AND R. KRUPIČKA, *Chem. Ind. London*, (1957) 1668.
101 J. ROČEK AND R. KRUPIČKA, *Collection Czech. Chem. Commun.*, 23 (1958) 2068.
102 J. ROČEK, F. H. WESTHEIMER, A. ESCHENMOSER, L. MOLDOVÁNYI AND J. SCHREKBER, *Helv. Chim. Acta*, 45 (1962) 2554.
103 F. H. WESTHEIMER AND A. NOVICK, *J. Chem. Phys.*, 11 (1943) 506.
104 F. HOLLOWAY, M. COHEN AND F. H. WESTHEIMER, *J. Am. Chem. Soc.*, 73 (1951) 65.
105 D. G. LEE AND R. STEWART, *J. Org. Chem.*, 32 (1967) 2628.
106 F. H. WESTHEIMER AND N. NICOLAIDES, *J. Am. Chem. Soc.*, 71 (1949) 25.
107 L. KAPLAN, *J. Am. Chem. Soc.*, 77 (1955) 5469.
108 R. BROWNELL, A. LEO, Y. W. CHANG AND F. H. WESTHEIMER, *J. Am. Chem. Soc.*, 82 (1960) 406.
109 H. KWART AND P. S. FRANCIS, *J. Am. Chem. Soc.*, 77 (1955) 4907.
110 J. ROČEK, *Collection Czech. Chem. Commun.*, 25 (1960) 1052.
111 F. H. WESTHEIMER AND Y. W. CHANG, *J. Phys. Chem.*, 63 (1959) 438.
112 W. WATANABE AND F. H. WESTHEIMER, *J. Chem. Phys.*, 17 (1949) 61.
113 K. B. WIBERG AND H. SCHÄFER, *J. Am. Chem. Soc.*, 89 (1967) 455.
113a K. WIBERG AND H. SCHÄFER, *J. Am. Chem. Soc.*, 91 (1969) 927.
113b K. WIBERG AND H. SCHÄFER, *J. Am. Chem. Soc.*, 91 (1969) 933.
114 U. KLÄNING, *Acta Chem. Scand.*, 11 (1957) 1313.
115 U. KLÄNING, *Acta Chem. Scand.*, 12 (1958) 576.
116 A. LEO AND F. H. WESTHEIMER, *J. Am. Chem. Soc.*, 74 (1952) 4383.
117 G. T. E. GRAHAM AND F. H. WESTHEIMER, *J. Am. Chem. Soc.*, 80 (1958) 3030.
118 H. KWART AND P. S. FRANCIS, *J. Am. Chem. Soc.*, 81 (1959) 2116.
119 A. K. ASWATHY, J. ROČEK AND R. M. MORIARTY, *J. Am. Chem. Soc.*, 89 (1967) 5400.
120 J. HAMPTON, A. LEO AND F. H. WESTHEIMER, *J. Am. Chem. Soc.*, 78 (1956) 306.
121 J. J. CAWLEY AND F. H. WESTHEIMER, *J. Am. Chem. Soc.*, 85 (1963) 1771.
122 A. CARRINGTON AND M. C. R. SYMONS, *Chem. Rev.*, 63 (1963) 443.

123 H. C. MISHRA AND M. C. R. SYMONS, *J. Chem. Soc.*, (1962) 4411.
124 J. ROČEK, *Collection Czech. Chem. Commun.*, 22 (1957) 1509.
125 M. C. R. SYMONS, *J. Chem. Soc.*, (1963) 4331.
126 D. G. LEE, W. L. DOWNEY AND R. M. MAASS, *Can. J. Chem.*, 46 (1968) 441.
127 S. WINSTEIN AND M. J. HOLNESS, *J. Am. Chem. Soc.*, 77 (1955) 5562.
128 R. STEWART AND D. G. LEE, *Can. J. Chem.*, 42 (1964) 439.
129 R. P. BELL, *The Proton in Chemistry*, Methuen, London, 1959, p. 183.
130 W. F. SAGER, *J. Am. Chem. Soc.*, 78 (1956) 4970.
131 J. ROČEK, *Collection Czech. Chem. Commun.*, 23 (1957) 833.
131a R. STEWART AND F. BANOO, *Can. J. Chem.*, 47 (1969) 3207.
132 R. STEWART, *J. Am. Chem. Soc.*, 79 (1957) 3057.
133 R. STEWART AND R. van der LINDEN, *Discussions Faraday Soc.*, 29 (1960) 211.
134 R. STEWART AND R. van der LINDEN, *Tetrahedron Letters*, 2 (1960) 28.
134a E. S. LEWIS AND J. K. ROBINSON, *J. Am. Chem. Soc.*, 90 (1968) 4337.
134b C. G. SWAIN, E. C. STIVERS, J. F. REUWER AND L. J. SCHAAD, *J. Am. Chem. Soc.*, 80 (1958) 5885.
134c K. H. HECKNER, R. LANDSBERG AND S. DALCHAU, *Ber. Bunsenges. Physik. Chem.*, 72 (1968) 649.
134d G. STEIN, *J. Chem. Phys.*, 42 (1965) 2986.
135 J. S. LITTLER, *J. Chem. Soc.*, (1962) 2190.
135a F. BANOO AND R. STEWART, *Can. J. Chem.*, 47 (1969) 3199.
136 K. B. WIBERG AND T. MILL, *J. Am. Chem. Soc.*, 80 (1958) 3022.
137 E. LUCCHI, *Chem. Abstr.*, 36 (1942) 6880.
138 E. LUCCHI, *Chem. Abstr.*, 37 (1943) 2252, 4293.
139 J. ROČEK, *Tetrahedron Letters*, 5 (1959) 1.
140 A. C. CHATTERJI AND S. K. MUKHERJEE, *J. Am. Chem. Soc.*, 80 (1958) 3600.
141 T. J. KEMP AND W. A. WATERS, *Proc. Roy. Soc. London Ser. A*, 274 (1963) 480.
142 A. C. CHATTERJI AND V. ANTHONY, *Z. Physik. Chem. Leipzig*, 210 (1959) 103.
143 K. B. WIBERG AND W. H. RICHARDSON, unpublished (quoted in Reference 1, p. 175).
144 K. B. WIBERG AND W. H. RICHARDSON, *J. Am. Chem. Soc.*, 84 (1962) 2800.
145 K. B. WIBERG AND P. A. LEPSE, *J. Am. Chem. Soc.*, 86 (1964) 2612.
146 K. B. WIBERG AND R. STEWART, *J. Am. Chem. Soc.*, 77 (1955) 1786.
147 F. C. TOMPKINS, *Trans. Faraday Soc.*, 39 (1943) 280.
148 J. C. SULLIVAN AND J. E. FRENCH, *J. Am. Chem. Soc.*, 87 (1965) 5380.
149 C. N. HINSHELWOOD AND C. A. WINKLER, *J. Chem. Soc.*, (1936) 368.
150 E. A. ALEXANDER AND F. C. TOMPKINS, *Trans. Faraday Soc.*, 35 (1939) 1156.
151 B. T. ALLEN AND A. BOND, *J. Phys. Chem.*, 68 (1964) 2439.
152 J. ROČEK AND A. RIEHL, *J. Org. Chem.*, 32 (1967) 3569.
153 P. A. BEST, J. S. LITTLER AND W. A. WATERS, *J. Chem. Soc.*, (1962) 822.
154 A. J. GREEN, T. J. KEMP, J. S. LITTLER AND W. A. WATERS, *J. Chem. Soc.*, (1964) 2722.
155 J. ROČEK AND A. RIEHL, *J. Am. Chem. Soc.*, 88 (1966) 4749.
156 K. B. WIBERG AND R. D. GEER, *J. Am. Chem. Soc.*, 87 (1965) 5202.
157 J. S. LITTLER, *J. Chem. Soc.*, (1962) 827.
158 H. C. S. SNETHLAGE, *Rec. Trav. Chim.*, 60 (1941) 877.
159 M. M. MOCEK AND R. STEWART, unpublished results quoted in Reference 1, p. 57.
160 R. P. BELL AND D. P. ONWOOD, *J. Chem. Soc. B*, (1967) 150.
161 D. R. MANN AND F. C. TOMPKINS, *Trans. Faraday Soc.*, 37 (1941) 201.
162 K. B. WIBERG AND R. STEWART, *J. Am. Chem. Soc.*, 78 (1956) 1214.
163 S. M. TAYLOR AND J. HALPERN, *J. Am. Chem. Soc.*, 81 (1959) 2933.
164 H. AEBI, W. BUSER AND C. LÜTHI, *Helv. Chim. Acta*, 39 (1956) 944.
165 F. MAREŠ AND J. ROČEK, *Collection Czech. Chem. Commun.*, 26 (1961) 2389.
166 M. M. WEI AND R. STEWART, *J. Am. Chem. Soc.*, 88 (1966) 1074.
167 D. H. ROSENBLATT, G. T. DAVIS, L. A. HULL AND G. D. FORBERG, *J. Org. Chem.*, 33 (1968) 1649.
167a F. FREEMAN AND A. YERAMYAN, *Tetrahedron Letters*, (1968) 4783.
167b F. FREEMAN, A. YERAMYAN AND F. YOUNG, *J. Org. Chem.*, 34 (1969) 2438.

168 R. Slack and W. A. Waters, *J. Chem. Soc.*, (1949) 594.
169 A. C. Chatterji and S. K. Mukherjee, *Z. Physik. Chem. Leipzig*, 208 (1958) 281.
170 A. C. Chatterji and S. K. Mukherjee, *Z. Physik. Chem. Leipzig*, 207 (1957) 372.
171 A. C. Chatterji and S. K. Mukherjee, *Z. Physik. Chem. Leipzig*, 210 (1959) 166.
172 Y. W. Chang and F. H. Westheimer, *J. Am. Chem. Soc.*, 82 (1960) 1401.
173 J. Roček and F. H. Westheimer, *J. Am. Chem. Soc.*, 84 (1962) 2241.
174 N. D. Heindel, E. S. Hanrahan and R. J. Sinkovitz, *J. Org. Chem.*, 31 (1966) 2019.
175 H. Kwart and D. Bretzger, *Tetrahedron Letters*, (1965) 3985.
176 S. Burstein and H. J. Ringold, *J. Am. Chem. Soc.*, 89 (1967) 4722.
177 N. R. Dhar, *J. Chem. Soc.*, 111 (1917) 707.
178 S. Chandra, S. N. Shukla and A. C. Chatterji, *Z. Physik. Chem. Leipzig*, 237 (1968) 137.
178a G. V. Bakore and C. L. Jain, *J. Inorg. Nucl. Chem.*, 31 (1969) 805.
179 G. G. Rao and R. V. V. Ayyar, *Z. Physik. Chem. Leipzig*, 236 (1967) 17.
179a D. A. Durham, *J. Inorg. Nucl. Chem.*, 31 (1969) 3549.
180 J. W. Ladbury and C. F. Cullis, *Chem. Rev.*, 58 (1958) 403.
181 O. M. Lidwell and R. P. Bell, *J. Chem. Soc.*, (1935) 1303.
182 H. F. Launer and D. M. Yost, *J. Am. Chem. Soc.*, 56 (1934) 2571.
183 H. C. S. Snethlage, *Rec. Trav. Chim.*, 61 (1942) 213.
184 T. J. Kemp and W. A. Waters, *J. Chem. Soc.*, (1964) 3193.
185 G. V. Bakore and S. Narain, *J. Chem. Soc.*, (1963) 3419.
186 T. J. Kemp and W. A. Waters, *J. Chem. Soc.*, (1964) 1192.
186a F. B. Beckwith and W. A. Waters, *J. Chem. Soc. B*, (1969) 929.
186b J. Roček and A. E. Radkowsky, *J. Am. Chem. Soc.*, 90 (1968) 2986.
186c J. H. Espenson, *J. Am. Chem. Soc.*, 86 (1964) 5101.
187 J. M. Pink and R. Stewart, unpublished results quoted in Reference 1, p. 65.
188 T. J. Kemp, D. Phil. Thesis, Oxford, 1963.
189 J. C. Ware and T. G. Traylor, *J. Am. Chem. Soc.*, 85 (1963) 3026.
190 B. Cercek, M. Ebert and A. J. Swallow, *J. Chem. Soc. A*, (1966) 612.
191 S. Ahrland, J. Chatt and N. R. Davies, *Quart. Rev. London*, 12 (1958) 265.
192 R. Criegee, Reference 1, p. 277.
193 A. Peloso and G. Dolcetti, *J. Chem. Soc. A*, (1967) 1944.
194 V. P. Kazakov and M. V. Konovalova, *Zh. Neorgan. Khim.*, 13 (1968) 447.
194a T. Huang and J. T. Spence, *J. Phys. Chem.*, 72 (1968) 4198.
194b T. Huang and J. T. Spence, *J. Phys. Chem.*, 72 (1968) 4573.
195 H. R. Ellison, F. Basolo and R. G. Pearson, *J. Am. Chem. Soc.*, 83 (1961) 3943.
196 A. D. Mitchell, *J. Chem. Soc.*, 119 (1921) 1266.
197 A. C. Harkness and J. Halpern, *J. Am. Chem. Soc.*, 83 (1961) 1258.
198 W. Schoeller, W. Schrauth and W. Essers, *Chem. Ber.*, 46 (1913) 2864.
199 A. B. Fasman, G. G. Kutyukov and D. V. Sokol'skii, *Dokl. Akad. Nauk SSSR*, 158 (1964) 1176.
200 V. A. Golodov, G. G. Kutyukov, A. B. Fasman and D. V. Sokol'skii, *Russ. J. Inorg. Chem. English Transl.*, 9 (1964) 1257.
201 J. Halpern, *J. Phys. Chem.*, 63 (1959) 398.
202 J. Halpern, *Quart. Rev. London*, 10 (1956) 463.
203 G. J. Korinek and J. Halpern, *J. Phys. Chem.*, 60 (1956) 285.
204 G. J. Korinek and J. Halpern, *Can. J. Chem.*, 34 (1956) 1372.
205 J. Halpern, J. F. Harrod and P. E. Potter, *Can. J. Chem.*, 37 (1959) 1446.
206 J. F. Harrod and J. Halpern, *Can. J. Chem.*, 37 (1959) 1933.
207 J. F. Harrod, S. Ciccone and J. Halpern, *Can. J. Chem.*, 39 (1961) 1372.
208 A. Aguiló, *Advan. Organometallic Chem.*, 5 (1967) 321.
209 J. Chatt, *Chem. Rev.*, 48 (1951) 7.
210 R. Criegee, *Ann. Chem.*, 481 (1930) 263.
211 R. Criegee, Reference 1, p. 338.
212 P. M. Henry, *J. Am. Chem. Soc.*, 87 (1965) 990.
213 P. M. Henry, *J. Am. Chem. Soc.*, 87 (1965) 4423.
214 P. M. Henry, *J. Am. Chem. Soc.*, 88 (1966) 1597.

215 L. Balbiano, *Chem. Ber.*, 42 (1909) 1502.
216 B. C. Fielding and H. L. Roberts, *J. Chem. Soc. A*, (1966) 1627.
217 J. C. Strini and J. Metzger, *Bull. Soc. Chim. France*, (1966) 3150.
218 J. Smidt, W. Hafner, R. Jira, J. Sedlmeier, R. Sieber, R. Rüttinger and H. Kojer, *Angew. Chem.*, 71 (1959) 176.
219 P. M. Henry, *J. Am. Chem. Soc.*, 86 (1964) 3246.
220 P. M. Henry, *J. Am. Chem. Soc.*, 88 (1966) 1595.
221 M. N. Vargaftik, I. I. Moiseev and Ya. K. Sirkin, *Dokl. Akad. Nauk SSSR*, 147 (1962) 399.
222 M. N. Vargaftik, I. I. Moiseev and Ya. K. Sirkin, *Isv. Akad. Nauk, SSSR Otd. Khim. Nauk*, (1963) 1148.
223 M. N. Vargaftik, I. I. Moiseev and Ya. K. Sirkin, *Dokl. Akad. Nauk SSSR*, 139 (1961) 1396.
224 I. I. Moiseev, M. N. Vargaftik and Ya. K. Sirkin, *Izv. Akad. Nauk SSSR Otd. Khim. Nauk*, (1963) 1144.
225 J. Smidt, W. Hafner, R. Jira, R. Sieber, J. Sedlmeier and A. Sabel, *Angew. Chem. Intern. Ed. Engl.*, 1 (1962) 80.
226 M. S. Kharasch, R. C. Segler and F. R. Mayo, *J. Am. Chem. Soc.*, 60 (1938) 882.
227 T. Dozono and T. Shiba, *Bull. Japan Petrol. Inst.*, 5 (1963) 8; *Chem. Abstr.*, 59 (1963) 58296.
228 A. P. Belov, G. Yu. Pek and I. I. Moiseev, *Izv. Akad. Nauk SSSR, Ser. Khim.*, (1965) 2204.
229 A. P. Belov, I. I. Moiseev and N. G. Uvarova, *Izv. Akad. Nauk SSSR, Ser. Khim.*, (1965) 224.
230 I. I. Moiseev and M. N. Vargaftik, *Izv. Akad. Nauk SSSR, Ser. Khim.*, (1965) 759.
231 E. W. Stern, *Proc. Chem. Soc.*, (1963) 111.
232 I. I. Moiseev, M. N. Vargaftik and Ya. K. Sirkin, *Dokl. Akad. Nauk SSSR*, 133 (1960) 377.
233 P. M. Henry, *J. Org. Chem.*, 32 (1967) 2575.
234 B. R. James and G. L. Rempel, *Can. J. Chem.*, 46 (1968) 571.
234a A. South and R. J. Ouellette, *J. Am. Chem. Soc.*, 90 (1968) 7064.
234b R. J. Ouelette, R. D. Robins and A. South, *J. Am. Chem. Soc.*, 90 (1968) 1619.
234c R. J. Ouellette, D. Miller, A. South and R. D. Robins, *J. Am. Chem. Soc.*, 91 (1969) 971.
235 M. Lj. Mihailović, Z. Maksiomović, D. Jeremić, Ž. Čeković, A. Milanović and Lj. Lorenc, *Tetrahedron*, 21 (1965) 1395.
236 M. Stefanović, M. Gaślić, Lj. Lorenc and M. Lj. Mihailovič, *Tetrahedron*, 20 (1964) 2289.
237 K. Heusler and J. Kalvoda, *Angew. Chem. Intern. Ed. Engl.*, 3 (1964) 525.
238 W. H. Starnes, *J. Am. Chem. Soc.*, 89 (1967) 3368.
239 W. H. Starnes, *J. Am. Chem. Soc.*, 90 (1968) 1807.
240 S. W. Gilks and G. M. Waind, *Discussions Faraday Soc.*, 29 (1960) 102.
241 W. G. Lloyd, *J. Org. Chem.*, 32 (1967) 2816.
242 J. V. Rund, F. Basolo and R. G. Pearson, *Inorg. Chem.*, 3 (1964) 658.
243 J. Chatt and B. L. Shaw, *Chem. Ind. London*, (1961) 290.
244 R. Criegee, *Chem. Ber.*, 77 (1944) 22.
245 D. Benson and L. H. Sutcliffe, *Trans. Faraday Soc.*, 55 (1959) 2107.
246 H. Hock and H. Kropf, *Chem. Ber.*, 91 (1958) 1681.
247 J. Halpern and S. M. Taylor, *Discussions Faraday Soc.*, 29 (1960) 174.
248 A. R. Topham and A. G. White, *J. Chem. Soc.*, (1952) 105.
249 H. N. Halvorson and J. Halpern, *J. Am. Chem. Soc.*, 78 (1956) 5562.
249a A. Aguiló, *J. Catalysis*, 13 (1969) 283.
250 M. S. Kharasch, H. N. Friedlander and W. H. Urry, *J. Org. Chem.*, 16 (1951) 533.
251 D. Benson, L. H. Sutcliffe and J. Walkley, *J. Am. Chem. Soc.*, 81 (1959) 4488.
251a R. O. C. Norman and M. Poustie, *J. Chem. Soc. B*, (1968) 781.
252 K. Ichikawa and Y. Yamaguchi, *J. Chem. Soc. Japan*, 73 (1952) 415.

253 R. Criegee, Reference 1, p. 305.
254 K. B. Wiberg and W. Koch, *Tetrahedron Letters*, (1966) 1779.
255 H. J. Kabbe, *Ann. Chem.*, 656 (1962) 204.
256 C. A. Bunton, Reference 1, p. 367.
257 R. P. Bell, J. R. Sturrock and R. L. St. D. Whitehead, *J. Chem. Soc.*, (1940) 82.
258 J. P. Cordner and K. H. Pausacker, *J. Chem. Soc.*, (1953) 102.
259 R. P. Bell, V. G. Rivlin and W. A. Waters, *J. Chem. Soc.*, (1958) 1696.
260 R. Criegee and E. Büchner, *Chem. Ber.*, 73 (1940) 563.
261 A. S. Perlin, *J. Am. Chem. Soc.*, 76 (1954) 5505.
262 R. Criegee, L. Kraft and B. Rank, *Am. Chem.*, 507 (1933) 159.
263 F. Baer, J. M. Grosheintz and H. O. L. Fischer, *J. Am. Chem. Soc.*, 61 (1939) 2607.
264 R. Criegee, *Sitzber. Ges. Befoerder, Ges. Naturw. Marburg*, 69 (1934) 25; *Chem. Abstr.*, 29 (1935) 6820.
265 R. Criegee, E. Hoger, G. Huber, P. Kruck, F. Marktscheffel and H. Schellenberger, *Ann. Chem.*, 599 (1956) 81.
266 E. Boyland and G. Wolf, *Biochem. J.*, 47 (1940) 64.
267 R. Criegee, E. Büchner and W. Walther, *Chem. Ber.*, 73 (1940) 571.
268 R. Criegee and H. Zogel, *Chem. Ber.*, 84 (1951) 215.
269 S. J. Angyal and R. J. Young, *J. Am. Chem. Soc.*, 81 (1959) 5251.
270 S. J. Angyal and R. J. Young, *J. Am. Chem. Soc.*, 81 (1959) 5467.
271 H. R. Goldschmid and A. S. Perlin, *Can. J. Chem.*, 38 (1960) 2280.
272 R. C. Hockett and D. F. Mowery, *J. Am. Chem. Soc.*, 65 (1943) 403.
273 N. Dhar, *J. Chem. Soc.*, 111 (1917) 690.
274 G. H. Cartledge, *J. Am. Chem. Soc.*, 63 (1941) 906.
275 V. P. Kazakov, A. I. Mateeva, A. M. Erenburg and B. I. Peshchevitskii, *Russ. J. Inorg. Chem. English Transl.*, 10 (1965) 563.
276 J. A. McMillan, *Chem. Rev.*, 62 (1962) 65.
277 B. M. Gordon and A. C. Wahl, *J. Am. Chem. Soc.*, 80 (1958) 273.
278 D. A. House, *Chem. Rev.*, 62 (1962) 185.
278a N. Venkatasubramanian and A. Sebesan, *Current Sci. India*, 36 (1967) 632.
278b M. C. Agrawal, R. K. Shinghal and S. P. Mushran, *Z. Physik. Chem. Frankfurt*, 62 (1968) 112.
278c G. D. Menghani and G. V. Bakore, *Z. Physik. Chem. Frankfurt*, 61 (1968) 220.
278d L. R. Subbaraman and M. Santappa, *Z. Physik. Chem. Frankfurt*, 45 (1966) 163, 172.
278e T. J. Hardwick and E. Robertson, *Can. J. Chem.*, 29 (1951) 828.
279 W. H. Richardson, Reference 1, p. 246.
280 D. R. Rosseinsky, *J. Chem. Soc.*, (1963) 1181.
281 R. G. Selim and J. J. Lingane, *Anal. Chim. Acta*, 21 (1959) 536.
282 C. F. Wells and G. Davies, *J. Chem. Soc. A*, (1967) 1858.
282a G. Davies, *Coord. Chem. Rev.*, 4 (1969) 199.
283 W. A. Waters and J. S. Littler, Reference 1, pp. 189, 195.
284 J. S. Littler, private communication.
285 F. R. Duke and C. E. Borchers, *J. Am. Chem. Soc.*, 75 (1953) 5186.
286 E. L. King and M. L. Pandow, *J. Am. Chem. Soc.*, 75 (1953) 3063.
287 C. M. Langkammerer, E. L. Jenner ,D. D. Coffman and B. W. Howk, *J. Am. Chem. Soc.*, 82 (1960) 1395.
288 C. F. Wells and D. Mays, *J. Chem. Soc. A*, (1968) 577.
289 M. Bobtelsky and S. Czosnek, *Z. Anorg. Allgem. Chem.*, 205 (1932) 401.
290 K. Julian and W. A. Waters, *J. Chem. Soc.*, (1962) 818.
291 R. G. Yalman, *J. Am. Chem. Soc.*, 75 (1953) 1842.
292 C. G. Swain and K. Hedberg, *J. Am. Chem. Soc.*, 72 (1950) 3373.
293 R. Luther and T. F. Rutter, *Z. Anorg. Chem.*, 54 (1907) 1.
294 J. B. Ramsey, E. L. Colchiman and L. C. Pack, *J. Am. Chem. Soc.*, 68 (1946) 1695.
295 M. H. Boyer and J. B. Ramsey, *J. Am. Chem. Soc.*, 75 (1953) 3802.
295a G. St. Nikolov and D. Mihailova, *J. Inorg. Nucl. Chem.*, 31 (1969) 2499.
296 W. C. Bray and J. B. Ramsey, *J. Am. Chem. Soc.*, 55 (1933) 2279.

296a C. F. Wells and D. Mays, *J. Chem. Soc. A*, (1968) 2740.
296b R. K. Murmann, J. C. Sullivan and R. C. Thompson, *Inorg. Chem.*, 7 (1968) 1876.
296c C. F. Wells and D. Mays, *J. Chem. Soc. A*, (1969) 2175.
297 C. F. Wells and D. Mays, *Inorg. Nucl. Chem. Letters*, 4 (1968) 61.
297a C. F. Wells and D. Mays, *J. Chem. Soc. A*, (1968) 1622.
297b G. Davies, L. J. Kirschenbaum and K. Kustin, *Inorg. Chem.*, 8 (1969) 663.
298 M. A. Suwyn and R. E. Hamm, *Inorg. Chem.*, 6 (1967) 2150.
298a R. C. Thompson, *J. Phys. Chem.*, 72 (1968) 2642.
298b G. Davies and K. Kustin, *J. Phys. Chem.*, 73 (1969) 2248.
298c G. Davies and K. Kustin, *Inorg. Chem.*, 8 (1969) 484.
298d W. A. Waters and I. R. Wilson, *J. Chem. Soc. A*, (1966) 534.
299 J. B. Kirwin, F. D. Peat, P. J. Proll and L. H. Sutcliffe, *J. Phys. Chem.*, 67 (1963) 1617.
300 G. A. Rechnitz and S. B. Zamochnick, *Talanta*, 11 (1964) 713.
301 G. A. Rechnitz and S. B. Zamochnick, *Talanta*, 11 (1964) 1645.
302 G. A. Rechnitz and S. B. Zamochnick, *Talanta*, 12 (1965) 479.
303 A. A. Noyes, C. D. Coryell, F. Stitt and A. Kossiakoff, *J. Am. Chem. Soc.*, 59 (1937) 1316.
304 A. A. Noyes, D. DeVault, C. D. Coryell and T. J. Deahl, *J. Am. Chem. Soc.*, 59 (1937) 1326.
305 H. N. Po, J. H. Swinehart and T. L. Allen, *Inorg. Chem.*, 7 (1968) 244.
306 D. Sen and P. Rây, *J. Indian Chem. Soc.*, 30 (1953) 519.
307 C. E. H. Bawn and A. G. White, *J. Chem. Soc.*, (1951) 331.
308 A. A. Noyes and T. J. Deahl, *J. Am. Chem. Soc.*, 59 (1937) 1337.
309 J. H. Baxendale and C. F. Wells, *Trans. Faraday Soc.*, 53 (1957) 800.
310 K. Jijee and M. Santappa, *Proc. Indian Acad. Sci. Sect. A*, 65 (1967) 155.
311 L. H. Sutcliffe and J. R. Weber, *Trans. Faraday Soc.*, 52 (1956) 1225.
312 L. H. Sutcliffe and J. R. Weber, *J. Inorg. Nucl. Chem.*, 12 (1960) 281.
313 M. Anbar and I. Pecht, *J. Am. Chem. Soc.*, 89 (1967) 2553.
314 H. N. Po, *Dissertation Abstr. Ser. B*, 28 (1967) 1887.
314a C. F. Wells and D. Mays, *Inorg. Nucl. Chem. Letters*, 5 (1969) 9.
315 K. Schug, M. D. Gilmore and L. A. Olson, *Inorg. Chem.*, 6 (1967) 2180.
316 S. Baer and G. Stein, *J. Chem. Soc.*, (1953) 3176.
317 G. Czapski, B. H. J. Bielski and N. Sutin, *J. Phys. Chem.*, 67 (1963) 201.
318 E. Saito and B. H. J. Bielski, *J. Am. Chem. Soc.*, 83 (1961) 4467.
319 B. H. J. Bielski and E. Saito, *J. Phys. Chem.*, 66 (1962) 2266.
320 M. Ardon and G. Stein, *J. Chem. Soc.*, (1956) 104.
321 G. Davies, L. J. Kirschenbaum and K. Kustin, *Inorg. Chem.*, 7 (1968) 146.
322 C. F. Wells and D. Mays, *Inorg. Nucl. Chem. Letters*, 4 (1968) 43.
323 C. F. Wells and D. Mays, *J. Chem. Soc. A*, (1968) 665.
324 G. Veith, E. Guthals and A. Viste, *Inorg. Chem.*, 6 (1967) 667.
325 R. L. Carroll, *Dissertation Abstr.*, 24 (1963) 988.
326 R. L. Carroll and L. B. Thomas, *J. Am. Chem. Soc.*, 88 (1966) 1376.
327 S. K. Mishra and Y. K. Gupta, *J. Inorg. Nucl. Chem.*, 29 (1967) 1643.
328 V. F. Stefanovskiĭ and M. S. Gaukhman, *J. Gen. Chem. USSR*, 11 (1941) 970.
329 E. B. Sandell and I. M. Kolthoff, *J. Am. Chem. Soc.*, 56 (1934) 1426.
330 C. Surasiti and E. B. Sandell, *Anal. Chim. Acta*, 22 (1960) 261.
331 C. Surasiti and E. B. Sandell, *J. Phys. Chem.*, 63 (1959) 890.
332 R. L. Habig, H. L. Pardue and J. B. Worthington, *Anal. Chem.*, 39 (1967) 600.
333 R. D. Sauerbrunn and E. B. Sandell, *Mikrochim. Acta*, (1953) 22.
334 M. Bobtelsky and A. Glasner, *J. Am. Chem. Soc.*, 64 (1942) 1462.
334a S. K. Mishra and Y. Gupta, *J. Inorg. Nucl. Chem.*, 30 (1968) 2991.
335 T. A. Cooper and W. A. Waters, *J. Chem. Soc. B*, (1967) 687.
336 D. R. Rosseinsky and W. C. E. Higginson, *J. Chem. Soc.*, (1960) 31.
337 P. S. R. Murti and S. C. Pati, *Chem. Ind. London*, (1967) 702.
338 P. S. R. Murti and S. C. Pati, *Chem. Ind. London*, (1966) 1722.
338a P. S. R. Murti and S. C. Pati, *J. Indian Chem. Soc.*, 45 (1968) 640.

339 M. S. Venkatachalapathy, R. Ramaswamy and H. V. K. Udupa, *Bull. Acad. Dolon. Sci., Ser. Sci., Chim., Geol. Geograph.*, 6 (1958) 487; *Chem. Abstr.*, 53 (1959) 3853d.
339a K. Sakota, Y. Kamiya and N. Ohta, *Can. J. Chem.*, 47 (1969) 387.
340 C. E. H. Bawn and A. J. Sharp, *J. Chem. Soc.*, (1957) 1854.
341 C. E. H. Bawn and A. J. Sharp, *J. Chem. Soc.*, (1957) 1866.
342 J. H. Baxendale and C. F. Wells, *Discussions Faraday Soc.*, 14 (1953) 239.
343 C. F. Wells, *Trans. Faraday Soc.*, 63 (1967) 156.
343a R. E. van der Ploeg, R. W. de Korte and E. C. Kooyman, *J. Catalysis*, 10 (1968) 52.
343b E. I. Heiba, R. M. Dessau and W. J. Koehl, *J. Am. Chem. Soc.*, 91 (1969) 138.
344 M. Ardon, *J. Chem. Soc.*, (1957) 1811.
345 S. S. Muhammad and K. V. Rao, *Bull. Chem. Soc. Japan*, 36 (1963) 943.
346 J. S. Littler and W. A. Waters, *J. Chem. Soc.*, (1959) 4046.
346a M. Santappa and B. Sethuram, *Proc. Indian Acad. Sci. Ser. A*, 67 (1968) 78.
347 D. G. Hoare and W. A. Waters, *J. Chem. Soc.*, (1962) 965.
348 J. S. Littler, *J. Chem. Soc.*, (1959) 4135.
349 H. L. Hintz and D. C. Johnson, *J. Org. Chem.*, 32 (1967) 556.
350 C. E. H. Bawn and A. G. White, *J. Chem. Soc.*, (1951) 343.
351 C. F. Wells and G. Davies, *Trans. Faraday Soc.*, 63 (1967) 2737.
351a C. F. Wells, C. Barnes and G. Davies, *Trans. Faraday Soc.*, 64 (1968) 3069.
352 W. A. Waters and J. S. Littler, Reference 1, p. 199.
353 J. R. Jones and W. A. Waters, *J. Chem. Soc.*, (1962) 2068.
354 H. Land and W. A. Waters, *J. Chem. Soc.*, (1958) 2129.
355 J. R. Jones and W. A. Waters, *J. Chem. Soc.*, (1962) 2772.
356 D. Greatorex and T. J. Kemp, unpublished observation.
357 D. G. Hoare and W. A. Waters, *J. Chem. Soc.*, (1964) 2552.
358 D. G. Hoare and W. A. Waters, *J. Chem. Soc.*, (1964) 2560.
359 W. A. Waters and J. S. Littler, Reference 1, p. 203.
360 M. H. Dean and G. Skirrow, *Trans Faraday Soc.*, 54 (1958) 849.
361 J. A. Sharp, *J. Chem. Soc.*, (1957) 2026.
362 G. Hargreaves and L. H. Sutcliffe, *Trans. Faraday Soc.*, 51 (1955) 786.
363 P. C. Ford, *Dissertation Abstr. Ser. B.*, 27 (1967) 4303.
364 K. B. Wiberg and W. H. Richardson, unpublished results quoted in Reference 1, p. 262.
364a K. B. Wiberg and P. C. Ford, *J. Am. Chem. Soc.*, 91 (1969) 124.
365 G. Hargreaves and L. H. Sutcliffe, *Trans. Faraday Soc*, 51 (1955) 1105.
366 T. J. Kemp and W. A. Waters, *J. Chem. Soc.*, (1964) 339.
367 T. A. Cooper and W. A. Waters, *J. Chem. Soc.*, (1964) 1538.
368 J. R. Jones and W. A. Waters, *J. Chem. Soc.*, (1963) 352.
369 A. Y. Drummond and W. A. Waters, *J. Chem. Soc.*, (1953) 440.
370 J. Shorter, *J. Chem. Soc.*, (1950) 3425.
371 D. G. Hoare and W. A. Waters, *J. Chem. Soc.*, (1962) 971.
372 J. S. Littler, *J. Chem. Soc.*, (1962) 832.
373 S. Venkatakrishnan and M. Santappa, *Z. Physik. Chem. Frankfurt*, 16 (1958) 73.
374 J. Shorter, *J. Chem. Soc.*, (1962) 1868.
375 J. Shorter and C. N. Hinshelwood, *J. Chem. Soc.*, (1950) 3276.
376 A. Y. Drummond and W. A. Waters, *J. Chem. Soc.*, (1955) 497.
377 J. S. Littler and W. A. Waters, *J. Chem. Soc.*, (1959) 3014.
378 R. Cecil, J. S. Littler and (in part) G. Easton, *J. Chem. Soc. B*, (1970) 626.
379 H. J. Den Hertog and E. C. Kooyman, *J. Catalysis*, 6 (1966) 357.
380 T. A. Cooper and W. A. Waters, *J. Chem. Soc. B*, (1967) 464.
381 A. A. Clifford and W. A. Waters, *J. Chem. Soc.*, (1965) 2796.
381a P. R. Sharan, P. Smith and W. A. Waters, *J. Chem. Soc. B*, (1968) 1322.
381b S. S. Lande and J. K. Kochi, *J. Am. Chem. Soc.*, 90 (1968) 5196.
381c D. Greatorex and T. J. Kemp, *Chem. Commun.* (1969) 383.
382 C. E. H. Bawn and A. G. White, *J. Chem. Soc.*, (1951) 339.
383 T. J. Kemp and W. A. Waters, *J. Chem. Soc.*, (1964) 1610.
384 A. Y. Drummond and W. A. Waters, *J. Chem. Soc.*, (1954) 2456.

385 T. J. KEMP AND W. A. WATERS, *J. Chem. Soc.*, (1964) 1489.
386 H. E. SPENCER AND G. K. ROLLEFSON, *J. Am. Chem. Soc.*, 77 (1955) 1938.
386a D. PAQUETTE AND M. ZADOR, *Can. J. Chem.*, 46 (1968) 3507.
386b L. B. YOUNG, *Dissertation Abstr. Ser. B*, 29 (1968) 128.
386c P. M. NAVE AND W. S. TRAHANOVSKY, *J. Am. Chem. Soc.*, 90 (1968) 4755.
386d G. V. BAKORE AND R. SHANKER, *J. Indian Chem. Soc.*, 6 (1968) 699.
387 F. R. DUKE AND R. F. BREMER, *J. Am. Chem. Soc.*, 73 (1951) 5179.
388 G. G. GUILBAULT AND W. H. MCCURDY, *J. Phys. Chem.*, 67 (1963) 283.
389 F. R. DUKE AND A. A. FORIST, *J. Am. Chem. Soc.*, 71 (1949) 2780.
390 J. S. LITTLER AND W. A. WATERS, *J. Chem. Soc.*, (1960) 2767.
391 G. MINO, S. KAIZERMAN AND E. RASMUSSEN, *J. Am. Chem. Soc.*, 81 (1959) 1494.
392 A. Y. DRUMMOND AND W. A. WATERS, *J. Chem. Soc.*, (1953) 3119.
393 P. LEVESLEY, W. A. WATERS AND A. N. WRIGHT, *J. Chem. Soc.*, (1956) 840.
394 J. R. JONES, W. A. WATERS AND (in part) J. S. LITTLER, *J. Chem. Soc.*, (1961) 630.
395 J. S. LITTLER, I. A. MALLET AND W. A. WATERS, *J. Chem. Soc.*, (1960) 2761.
396 H. LAND AND W. A. WATERS, *J. Chem. Soc.*, (1957) 4312.
396a P. SMITH AND W. A. WATERS, *J. Chem. Soc. B*, (1969) 462.
397 J. R. JONES AND W. A. WATERS, *J. Chem. Soc.*, (1962) 1629.
398 J. HILL AND A. MCAULEY, *J. Chem. Soc. A*, (1968) 1169.
399 A. MCAULEY, *J. Chem. Soc.*, (1965) 4054.
400 B. KRISHNA AND K. C. TEWARI, *J. Chem. Soc.*, (1961) 3097.
401 K. P. BHARGAVA, R. SHANKER AND S. N. JOSHI, *J. Sci. Ind. Res. India, Ser. B*, 21 (1962) 573.
402 S. K. MISRA AND R. C. MEHOTRA, *J. Indian Chem. Soc.*, 44 (1967) 927.
403 P. LEVESLEY AND W. A. WATERS, *J. Chem. Soc.*, (1955) 217.
404 G. V. BAKORE, R. SHANKER AND S. S. DUA, *Z. Physik. Chem. Leipzig*, 236 (1967) 129.
404a R. N. MEHOTRA, *J. Chem. Soc. B*, (1968) 642.
405 J. HILL AND A. MCAULEY, *J. Chem. Soc. A*, (1968) 156.
406 W. F. PICKERING AND A. MCAULEY, *J. Chem. Soc. A*, (1968) 1173.
406a J. HILL AND A. MCAULEY, *J. Chem. Soc. A*, (1968) 2405.
407 W. SCHNEIDER, *Helv. Chim. Acta*, 46 (1963) 1863.
408 A. W. ADAMSON, H. OGATA, J. GROSSMAN AND R. NEWBURY, *J. Inorg. Nucl. Chem.*, 6 (1968) 319.
409 V. H. DODSON AND A. H. BLACK, *J. Am. Chem. Soc.*, 79 (1957) 3657.
410 S. D. ROSS AND C. G. SWAIN, *J. Am. Chem. Soc.*, 69 (1947) 1325.
411 Y. A. EL-TANTWY AND G. A. RECHNITZ, *Anal. Chem.*, 36 (1964) 1774.
412 G. A. RECHNITZ AND Y. A. EL-TANTWY, *Anal. Chem.*, 36 (1964) 2361.
413 H. TAUBE, *J. Am. Chem. Soc.*, 70 (1948) 1216.
414 H. TAUBE, *J. Am. Chem. Soc.*, 69 (1947) 1418.
415 M. A. SUWYN AND R. A. HAMM, *Inorg. Chem.*, 6 (1967) 142.
416 J. R. JONES AND W. A. WATERS, *J. Chem. Soc.*, (1961) 4757.
417 D. N. SATHYANARAYANA AND C. C. PATEL, *Bull. Chem. Soc. Japan*, 37 (1964) 1736.
418 T. J. KEMP AND W. A. WATERS, J. *Chem. Soc.*, (1964) 3101.
419 A. A. CLIFFORD, D. Phil. Thesis, Oxford, 1964.
420 K. K. SENGUPTA AND S. ADITYA, *Z. Physik. Chem. Frankfurt*, 38 (1963) 25.
421 J. I. WATTERS AND I. M. KOLTHOFF, *J. Am. Chem. Soc.*, 70 (1948) 2455.
422 W. A. WATERS AND J. S. LITTLER, Reference 1, p. 233.
422a K. JIJEE, A. MEENAKSHI AND M. SANTAPPA, *Z. Physik, Chem. Frankfurt*, 59 (1968) 206.
423 J. S. LITTLER AND R. CECIL, *J. Chem. Soc. B*, (1968) 1420.
424 G. DOBSON AND L. I. GROSSWEINER, *Trans. Faraday Soc.*, 61 (1965) 708.
425 V. S. MARTEM'YANOV AND E. T. DENISOV, *Zh. Fiz. Khim.*, 41 (1967) 687.
425a G. DAVIES AND K. KUSTIN, *Trans. Faraday Soc.*, 65 (1969) 1630.
425b C. F. WELLS AND L. V. KURITSYN, *J. Chem. Soc. A*, (1969) 2575.
426 T. A. COOPER AND W. A. WATERS, *J. Chem. Soc. B*, (1967) 455.
427 P. J. ANDRULIS, M. J. S. DEWAR, R. DIETZ AND R. L. HUNT, *J. Am. Chem. Soc.*, 88 (1966) 5473.
428 P. J. ANDRULIS AND M. J. S. DEWAR, *J. Am. Chem. Soc.*, 88 (1966) 5483.

429 T. Arantani and M. J. S. Dewar, *J. Am. Chem. Soc.*, 88 (1966) 5479.
429a A. McAuley and U. D. Gomwalk, *J. Chem. Soc. A*, (1969) 977.
429b U. D. Gomwalk and A. McAuley, *J. Chem. Soc. A*, (1968) 2948.
430 W. C. Bray and A. V. Hershey, *J. Am. Chem. Soc.*, 56 (1934) 1889.
431 G. I. H. Hanania, D. H. Irvine and W. A. Eaton, *J. Phys. Chem.*, 71 (1967) 2022.
432 J. J. Katz and G. T. Seaborg, *The Chemistry of the Actinide Elements*, Methuen, London, 1957, p. 225.
432a I. M. Kolthoff and W. J. Tomsicek, *J. Phys. Chem.*, 40 (1936) 247.
433 J. S. Littler and I. G. Sayce, *J. Chem. Soc.*, (1964) 2545.
434 A. J. Fudge and K. W. Sykes, *J. Chem. Soc.*, (1952) 119, and references cited therein.
435 K. W. Sykes, *J. Chem. Soc.*, (1952) 124.
436 C. Wagner, *Z. Physik. Chem.*, 113 (1924) 261.
437 A. W. Adamson, *J. Phys. Chem.*, 56 (1952) 858.
438 Y. A. Majid and K. E. Howlett, *J. Chem. Soc. A*, (1968) 679.
439 M. Spiro, *J. Chem. Soc.*, (1960) 3678, and references cited therein.
440 A. Indeli and G. C. Guaraldi, *J. Chem. Soc.*, (1964) 36.
441 K. B. Wiberg, H. Maltz and M. Okano, *J. Inorg. Chem.*, 7 (1968) 830.
441a M. H. Ford-Smith and J. H. Rawsthorne, *J. Chem. Soc. A*, (1969) 160.
442 N. K. Shastri, J. O. Wear and E. S. Amis, *J. Inorg. Nucl. Chem.*, 24 (1962) 535.
443 A. W. Adamson, J. P. Walker and M. Volpe, *J. Am. Chem. Soc.*, 72 (1950) 4030.
444 F. R. Duke and W. G. Courtney, *J. Phys. Chem.*, 56 (1952) 19.
445 R. H. Betts and F. S. Dainton, *J. Am. Chem. Soc.*, 75 (1953) 5721.
446 D. Bunn, F. S. Dainton and S. Duckworth, *Trans. Faraday Soc.*, 57 (1961) 1131.
447 J. Badoz-Lambling, *Bull. Soc. Chim. France*, (1950) 1195.
448 W. G. Barb, J. H. Baxendale, P. George and K. R. Hargrave, *Trans. Faraday Soc.*, 47 (1951) 462, 591.
449 J. H. van Bertalan, *Z. Physik. Chem.*, 95 (1920) 328.
450 F. Haber and J. Weiss, *Proc. Roy. Soc. London, Ser. A*, 147 (1932) 332.
451 J. Weiss, *Discussions Faraday Soc.*, 2 (1947) 212.
452 V. S. Andersen, *Acta Chem. Scand.*, 2 (1948) 1.
453 M. L. Kremer and G. Stein, *Trans. Faraday Soc.*, 55 (1959) 959.
454 M. L. Huggett, P. Jones and W. F. K. Wynne-Jones, *Discussions Faraday Soc.*, 29 (1960) 153.
455 P. Jones, R. Kitching, M. L. Tobe and W. F. K. Wynne-Jones, *Trans. Faraday Soc.*, 55 (1959) 79.
456 M. G. Evans, P. George and N. Uri, *Trans Faraday Soc.*, 45 (1949) 230.
457 M. L. Kremer, *Trans. Faraday Soc.*, 61 (1965) 1453.
457a K. K. Girdhar and D. V. S. Jain, *J. Inorg. Nucl. Chem.*, 17 (1965) 2633.
457b J. Schubert, V. S. Sharma, E. R. White and L. S. Bergelson, *J. Am. Chem. Soc.*, 90 (1968) 4476.
458 F. M. Page, *Trans. Faraday Soc.*, 56 (1960) 398.
459 F. M. Page, *Trans. Faraday Soc.*, 49 (1953) 635.
460 F. M. Page, *Trans. Faraday Soc.*, 50 (1954) 120.
461 D. G. Karraker, *Phys. Chem.*, 67 (1963) 871.
462 W. C. E. Higginson and J. W. Marshall, *J. Chem. Soc.*, (1957) 447.
463 J. H. Swinehart, *J. Inorg. Nucl. Chem.*, 29 (1967) 2312.
463a J. Vepřek-Šiška and A. Hasnedl, *Chem. Commun.*, (1968) 1167.
463b R. S. Murray, *Chem. Commun.*, (1968) 824.
464 A. D. Mitchell, *J. Chem. Soc.*, 121 (1922) 1624.
465 A. D. Mitchell, *J. Chem. Soc.*, 123 (1924) 629.
466 H. Neumann, I. Z. Steinberg and E. Katchalski, *J. Am. Chem. Soc.*, 87 (1965) 3841.
467 J. W. Cahn and R. E. Powell, *J. Am. Chem. Soc.*, 76 (1954) 2568.
468 W. C. E. Higginson and P. Wright, *J. Chem. Soc.*, (1955) 1551.
469 D. R. Rosseinsky, *J. Chem. Soc.*, (1957) 4685.
470 W. C. E. Higginson and D. Sutton, *J. Chem. Soc.*, (1953) 1402.
471 H. Minato, E. J. Meehan, I. M. Kolthoff and C. Auerbach, *J. Am. Chem. Soc.*, 81 (1959) 6168.

472 E. C. Gilbert, *Z. Physik. Chem. Ser. A*, 142 (1929) 139.
473 P. R. Ray and H. K. Sen, *Z. Anorg. Allgem. Chem.*, 76 (1912) 380.
474 S. Hünig, H. R. Müller and W. Thier, *Angew. Chem. Intern. Ed. Engl.*, 4 (1965) 271.
475 A. D. Mitchell, *J. Chem. Soc.*, (1926) 336.
476 C. P. Lloyd and W. F. Pickering, *J. Inorg. Nucl. Chem.*, 29 (1967) 1907.
477 J. J. Byerley and J. Y. H. Lee, *Can. J. Chem.*, 45 (1967) 3025.
477a J. J. Byerley and E. Peters, *Can. J. Chem.*, 47 (1969) 313.
478 E. Peters and J. Halpern, *J. Phys. Chem.*, 59 (1955) 793.
479 R. G. Dakers and J. Halpern, *Can. J. Chem.*, 32 (1954) 969.
480 A. H. Webster and J. Halpern, *J. Phys. Chem.*, 61 (1957) 1239.
481 E. Peters and J. Halpern, *Can. J. Chem.*, 33 (1955) 356.
482 A. H. Webster and J. Halpern, *J. Phys. Chem.*, 61 (1957) 1245.
483 L. Wright, S. Weller and G. A. Mills, *J. Phys. Chem.*, 59 (1955) 1060.
484 W. K. Wilmarth and A. F. Kapauan, *J. Am. Chem. Soc.*, 78 (1956) 1308.
485 S. Weller and G. A. Mills, *J. Am. Chem. Soc.*, 75 (1953) 769.
486 M. Calvin and W. K. Wilmarth, *J. Am. Chem. Soc.*, 78 (1956) 1301.
487 W. K. Wilmarth and M. K. Barsh, *J. Am. Chem. Soc.*, 78 (1956) 1305.
488 L. Meites and R. H. Schlossel, J. *Phys. Chem.*, 67 (1963) 2397.
489 T. Freund, *J. Inorg. Nucl. Chem.*, 9 (1959) 246.
490 L. S. Hsu, *U.S. At. Energy Comm.*, UCRL-11252 (1964); *Chem. Abstr.*, 61 (1964) 1498b.
491 K. N. Mochalov and V. S. Khain, *Russ. J. Phys. Chem. English Tranl.*, 39 (1965) 1040.
492 T. J. Wallace, *J. Org. Chem.*, 31 (1966) 3071.
493 J. J. Bohning and K. Weiss, *J. Am. Chem. Soc.*, 82 (1960) 4724.
493a R. C. Kapoor, O. P. Kachhwaha and B. P. Sinha, *J. Phys. Chem.*, 73 (1969) 1627.
494 E. J. Meehan, I. M. Kolthoff and H. Kakiuchi, *J. Phys. Chem.*, 66 (1962) 1238.
495 I. M. Kolthoff, E. J. Meehan and Q. W. Choi, *J. Phys. Chem.*, 66 (1962) 1233.
496 H. C. Clark, N. F. Curtis and A. L. Odell, *J. Chem. Soc.*, (1954) 63.
497 P. T. Speakman and W. A. Waters, *J. Chem. Soc.*, (1955) 40.
498 H. Maltz, *Dissertation Abstr., Ser. B*, 28 (1967) 111.
498a V. N. Singh, H. S. Singh and B. B. L. Saxena, *J. Am. Chem. Soc.*, 91 (1969) 2643.
498b U. Shanker and M. P. Singh, *Indian J. Chem.*, 6 (1968) 702.
498c V. N. Singh, M. C. Gangwar, B. B. L. Saxena and M. P. Singh, *Can. J. Chem.*, 47 (1969) 1051.
499 A. Lorenzini and C. Walling, *J. Org. Chem.*, 32 (1967) 4008.
500 C. E. Castro, E. J. Gaughan and D. C. Owsley, *J. Org. Chem.*, 30 (1965) 587.
501 J. K. Kochi, *J. Am. Chem. Soc.*, 77 (1955) 5274.
502 A. A. Clifford and W. A. Waters, *J. Chem. Soc.*, (1963) 3056.
503 A. L. Klebansky, I. V. Grachev and O. M. Kuznetsova, *J. Gen. Chem. USSR, Eng. Transl.*, 27 (1957) 2008.
504 J. K. Thomas, G. Trudel and S. Bywater, *J. Phys. Chem.*, 64 (1960) 51.
505 K. B. Wiberg and W. G. Nigh, *J. Am. Chem. Soc.*, 87 (1965) 3849.
506 K. J. Pedersen, *Acta Chem. Scand.*, 2 (1948) 252.
507 B. A. Marshall and W. A. Waters, *J. Chem. Soc.*, (1960) 2392.
508 B. A. Marshall and W. A. Waters, *J. Chem. Soc.*, (1961) 1579.
509 M. P. Singh, *Z. Physik. Chem. Leipzig*, 216 (1961) 13, 17.
510 M. P. Singh, *Z. Physik. Chem. Leipzig*, 208 (1958) 265, 273.
511 N. Nath and M. P. Singh, *J. Phys. Chem.*, 69 (1967) 2038.
512 N. Nath and M. P. Singh, *Z. Physik. Chem. Leipzig*, 221 (1962) 204.
513 N. Nath and M. P. Singh, *Z. Physik. Chem. Leipzig*, 224 (1963) 419.
513a M. M. Taqui Khan and A. E. Martell, *J. Am. Chem. Soc.*, 90 (1968) 3386.
513b M. M. Taqui Khan and A. E. Martell, *J. Am. Chem. Soc.*, 90 (1968) 6011.
513c M. M. Taqui Khan and A. E. Martell, *J. Am. Chem. Soc.*, 89 (1967) 4176.
513d M. M. Taqui Khan and A. E. Martell, *J. Am. Chem. Soc.*, 91 (1969) 4668.
513e V. S. Mehotra, M. C. Agrawal and S. P. Mushran, *J. Phys. Chem.*, 73 (1969) 1996.
514 D. Porret, *Helv. Chim. Acta*, 17 (1934) 703.
515 J. H. Baxendale, H. R. Hardy and L. H. Sutcliffe, *Trans. Faraday Soc.*, 47 (1951) 963.

515a R. M. DAVYDOV AND L. A. ZAGORUIKA, *Russ. J. Phys. Chem. English Transl.*, 42 (1968) 1400.
515b R. M. DAVYDOV, *Russ. J. Phys. Chem. English Transl.*, 42 (1968) 1563.
515c R. M. DAVYDOV AND G. V. FOMIN, *Russ. J. Phys. Chem. English Transl.*, 42 (1968) 481.
516 C. G. HAYNES, A. H. TURNER AND W. A. WATERS, *J. Chem. Soc.*, (1956) 2823.
517 J. B. CONANT, *Chem. Rev.*, 3 (1926) 1.
518 J. B. CONANT AND M. F. PRATT, *J. Am. Chem. Soc.*, 48 (1926) 3148, 3220.
519 H. N. STEIN AND H. J. C. TENDERLOO, *Rec. Trav. Chim.*, 74 (1955) 905.
520 W. BRACKMAN AND E. HAVINGA, *Rec. Trav. Chim.*, 74 (1955) 937, 1021, 1070, 1100, 1107.
521 J. R. LINDSAY-SMITH, R. O. C. NORMAN AND W. M. WALKER, *J. Chem. Soc. B*, (1968) 269.
521a A. M. A. VERWEY, J. A. JONKER AND S. BALT, *Rec. Trav. Chim.*, 88 (1969) 439.
522 I. ČEPČIANSKY, V. SLAVÍK, L. NĚMEC, H. FINGEROVÁ AND I. NÉMETH, *Collection Czech. Chem. Commun.*, 33 (1968) 100.
523 H. DIEBLER, *Ber. Bunsenges. Physik. Chem.*, 67 (1963) 396.
524 D. G. LAMBERT AND M. M. JONES, *J. Am. Chem. Soc.*, 88 (1966) 4615.
525 S. N. KHODASKAR, R. A. BHOBE AND D. D. KHANOLKAR, *Science and Culture*, 33 (1967) 191.
526 M. C. AGRAWAL AND S. P. MUSHRAN, *J. Phys. Chem.*, 72 (1968) 1497.
526a R. R. DAS, T. R. BHAT AND J. SHANKER, *Indian J. Chem.*, 7 (1969) 240.
527 W. R. KING AND C. S. GARNER, *J. Phys. Chem.*, 58 (1954) 29.
528 G. BREDIG AND J. MICHEL, *Z. Physik. Chem.*, 100 (1922) 124.
529 F. R. DUKE AND P. R. QUINNEY, *J. Am. Chem. Soc.*, 76 (1954) 3800.
530 J. F. ENDICOTT AND H. TAUBE, *Inorg. Chem.*, 4 (1965) 437.
531 D. J. MEIER AND C. S. GARNER, *J. Phys. Chem.*, 56 (1952) 853.
532 G. GORDON AND P. H. TEWARI, *J. Phys. Chem.*, 70 (1966) 200.
533 T. W. NEWTON AND F. B. BAKER, *Inorg. Chem.*, 3 (1964) 569.
534 R. C. THOMPSON AND G. GORDON, *Inorg. Chem.*, 5 (1966) 562.
535 G. A. RECHNITZ AND J. E. MCCLURE, *Anal. Chem.*, 36 (1964) 2265.
535a C. W. FULLER AND J. M. OTTAWAY, *Analyst*, 94 (1969) 32.
535b A. G. RYKOV, V. YA. VASIL'EV, P. G. KRUTIKOV AND G. N. YAKOVLEV, *Khim. Transuranovykh Oskolochnykh Elem., Akad. Nauk SSSR, Otb. Obshch. Tekh. Khim.*, (1967) 177; *Chem. Abstr.*, 69 (1968) 22408r.
535c B. Z. SHAKHASHIRI AND G. GORDON, *J. Am. Chem. Soc.*, 91 (1969) 1103.
535d R. L. BIRKE AND W. F. MARZLUFF, JR., *J. Am. Chem. Soc.*, 71 (1969) 3481.
536 J. HALPERN AND J. G. SMITH, *Can. J. Chem.*, 34 (1956) 1419.
537 G. GORDON AND H. TAUBE, *Inorg. Chem.*, 1 (1962) 69.
538 S. FALLAB, *Angew. Chem. Intern. Ed. Engl.*, 6 (1967) 496.
539 H. A. E. MACKENZIE AND F. C. TOMPKINS, *Trans. Faraday Soc.*, 38 (1942) 465.
540 J. B. RAMSEY, R. SUGIMOTO AND H. DE VORKIN, *J. Am. Chem. Soc.*, 63 (1941) 3480.
541 TH. KADEN, D. WALZ AND S. FALLAB, *Helv. Chim. Acta*, 43 (1960) 1639.
542 P. GEORGE, *J. Chem. Soc.*, (1954) 4349.
542a C. F. WELLS, *J. Inorg. Nucl. Chem.*, 30 (1968) 893.
542b C. F. WELLS AND M. A. SALAM, *Nature*, 205 (1965) 690.
542c J. WEISS, *Experientia*, 9 (1953) 61.
543 R. E. HUFFMAN AND N. DAVIDSON, *J. Am. Chem. Soc.*, 78 (1956) 4836.
544 A. B. LAMB AND L. W. ELDER, *J. Am. Chem. Soc.*, 53 (1931) 137.
545 M. CHER AND N. DAVIDSON, *J. Am. Chem. Soc.*, 77 (1955) 793.
546 A. M. POSNER, *Trans. Faraday Soc.*, 49 (1953) 382, 389.
547 J. KING AND N. DAVIDSON, unpublished data quoted in Reference 545.
548 W. TAYLOR AND J. WEISS, unpublished data quoted in Reference 542c.
548a Y. KURIMURA, R. OCHIAI AND N. MATSUURA, *Bull. Chem. Soc. Japan*, 41 (1968) 2234.
548b G. S. HAMMOND AND C. H. WU, *Advan. Chem. Ser.*, 77 (1968) 186.
548c S. AŠPERGER, I. MURATI AND D. PAVLOVIĆ, *J. Chem. Soc. A*, (1969) 2044.
549 T. W. NEWTON AND F. B. BAKER, *J. Phys. Chem.*, 60 (1956) 1417.
550 J. H. SWINEHART, *Inorg. Chem.*, 4 (1965) 1069.
551 M. ARDON AND R. A. PLANE, *J. Am. Chem. Soc.*, 81 (1959) 3197.
552 H. NORD, *Acta Chem. Scand.*, 9 (1955) 430.

553 G. W. Filson and J. H. Walton, *J. Phys. Chem.*, 36 (1932) 740.
553a P. M. Henry, *Inorg. Chem.*, 6 (1966) 688.
553b R. D. Gray, *J. Am. Chem. Soc.*, 91 (1969) 56.
554 A. Zuberbühler, *Helv. Chim. Acta*, 50 (1967) 466.
554a I. Pecht and M. Anbar, *J. Chem. Soc. A*, (1968) 1902.
555 B. Erdem and S. Fallab, *Chimia*, 19 (1965) 463.
556 J. Simplicio and R. G. Wilkins, *J. Am. Chem. Soc.*, 89 (1967) 6092.
556a F. Miller, J. Simplicio and R. G. Wilkins, *J. Am. Chem. Soc.*, 91 (1969) 1962.
557 C. Tanford, D. C. Kirk and M. K. Chantooni, *J. Am. Chem. Soc.*, 76 (1954) 5325.
557a J. Simplicio and R. G. Wilkins, *J. Am. Chem. Soc.*, 91 (1969) 1325.
558 E. P. Guymon and J. T. Spence, *J. Phys. Chem.*, 71 (1967) 1616.
559 C. R. Hare, I. Bernal and H. B. Gray, *Inorg. Chem.*, 1 (1962) 831.
560 M. A. Mendelsohn, E. M. Arnett and H. Freiser, *J. Phys. Chem.*, 64 (1960) 660.
561 E. M. Arnett, H. Freiser and M. A. Mendelsohn, *J. Am. Chem. Soc.*, 84 (1962) 2482.
562 E. M. Arnett and M. A. Mendelsohn, *J. Am. Chem. Soc.*, 84 (1962) 3821, 3824.
563 G. A. Rechnitz and H. A. Catherino, *Inorg. Chem.*, 4 (1965) 112.
564 A. A. Vlček and F. Basolo, *Inorg. Chem.*, 5 (1966) 156.
565 J. M. Cleveland, *J. Am. Chem. Soc.*, 87 (1965) 1816.
566 J. M. Cleveland and J. Werkema, *Nature*, 215 (1967) 732.
567 M. Anbar and P. Neta, *Intern. J. Appl. Radiation Isotopes*, 16 (1965) 227.
568 J. R. Hanson and E. Premuzic, *Angew. Chem. Intern. Ed. Engl.*, 7 (1968) 247.
569 C. E. Castro and R. D. Stephens, *J. Am. Chem. Soc.*, 86 (1964) 4358.
570 G. K. Helmkamp, F. L. Carter and H. J. Lucas, *J. Am. Chem. Soc.*, 79 (1957) 1306.
571 R. S. Bottei and W. A. Joern, *J. Am. Chem. Soc.*, 90 (1968) 297.
572 S. A. Levison and R. A. Marcus, *J. Phys. Chem.*, 72 (1968) 358.
573 A. A. Vlćek and J. Hanzlík, *Inorg. Chem.*, 6 (1967) 2053.
573a J. Hanzlík and A. A. Vlček, *Inorg. Chem.*, 8 (1969) 669.
574 J. R. Sampey, *J. Am. Chem. Soc.*, 52 (1930) 88.
575 J. B. Conant and H. B. Cutter, *J. Am. Chem. Soc.*, 48 (1926) 1016.
575a J. H. Baxendale, P. L. T. Bevan and D. A. Stott, *Trans. Faraday Soc.*, 64 (1968) 2389.
576 M. G. Evans, J. H. Baxendale and D. J. Cowling, *Discussions Faraday Soc.*, 2 (1947) 206.
577 P. Davis, M. G. Evans and W. C. E. Higginson, *J. Chem. Soc.*, (1951) 2563.
578 W. C. Bray and M. H. Gorin, *J. Am. Chem. Soc.*, 54 (1932) 2124.
579 A. E. Cahill and H. Taube, *J. Am. Chem. Soc.*, 74 (1952) 2312.
580 T. J. Conocchioli, E. J. Hamilton and N. Sutin, *J. Am. Chem. Soc.*, 87 (1965) 926.
581 C. F. Wells and M. A. Salam, *J. Chem. Soc. A*, (1968) 24.
582 T. J. Hardwick, *Can. J. Chem.*, 35 (1957) 428.
583 F. S. Dainton and H. C. Sutton, *Trans. Faraday Soc.*, 49 (1953) 1011.
584 J. H. Baxendale, M. G. Evans and G. S. Park, *Trans. Faraday Soc.*, 42 (1946) 155.
585 T. Rigg, W. Taylor and J. Weiss, *J. Chem. Phys.*, 22 (1954) 575.
586 C. F. Wells and M. A. Salam, *Trans. Faraday Soc.*, 63 (1967) 620.
587 J. Burgess and R. H. Prince, *J. Chem. Soc.*, (1965) 6061.
588 J. Burgess, *J. Chem. Soc. A*, (1967) 955.
589 W. G. Barb, J. H. Baxendale, P. George and K. R. Hargrave, *Trans. Faraday Soc.*, 51 (1955) 935.
590 F. B. Baker and T. W. Newton, *J. Phys. Chem.*, 65 (1961) 1897.
591 L. B. Anderson and R. A. Plane, *Inorg. Chem.*, 5 (1964) 1470.
592 P. B. Chock, R. B. K. Dewar, J. Halpern and L. Y. Wong, *J. Am. Chem. Soc.*, 91 (1969) 82.
593 M. R. Baloga and J. E. Earley, *J. Am. Chem. Soc.*, 83 (1961) 4906.
594 G. Schwarzenbach and R. Sulzberger, *Helv. Chim. Acta*, 27 (1944) 348.
594a H. Sigel and U. Müller, *Helv. Chim. Acta*, 49 (1966) 671.
594b H. Erlenmeyer, U. Müller and H. Sigel, *Helv. Chim. Acta*, 49 (1966) 681.
594c H. Sigel, *Angew. Chem. Intern. Ed. Engl.*, 8 (1969) 167.
594d T. Kaden and H. Sigel, *Helv. Chim. Acta*, 51 (1968) 947.
594e H. Sigel, C. Flierl and R. Griesser, *J. Am. Chem. Soc.*, 91 (1969) 1061.
594f H. Erlenmeyer, C. Flierl and H. Sigel, *J. Am. Chem. Soc.*, 91 (1969) 1065.

595 J. W. L. Fordham and H. L. Williams, *J. Am. Chem. Soc.*, 72 (1950) 4465.
596 J. W. L. Fordham and H. L. Williams, *J. Am. Chem. Soc.*, 73 (1951) 1634.
597 W. L. Reynolds and I. M. Kolthoff, *J. Phys. Chem.*, 60 (1956) 969.
598 R. J. Orr and H. L. Williams, *Can. J. Chem.*, 30 (1952) 985.
599 R. J. Orr and H. L. Williams, *J. Phys. Chem.*, 57 (1953) 925.
600 W. L. Reynolds and I. M. Kolthoff, *J. Phys. Chem.*, 60 (1956) 996.
601 R. J. Orr and H. L. Williams, *J. Am. Chem. Soc.*, 76 (1954) 3321.
602 H. Boardman, *J. Am. Chem. Soc.*, 75 (1953) 4268.
603 J. H. Crabtree and W. P. Schaefer, *Inorg. Chem.*, 5 (1966) 1348.
603a B. Z. Shakhashiri and G. Gordon, *Inorg. Chem.*, 7 (1968) 2454.
604 P. R. Carter and N. Davidson, *J. Phys. Chem.*, 56 (1952) 877.
605 G. Gordon and A. Andrewes, *Inorg. Chem.*, 3 (1964) 1733.
606 L. G. Carpenter, M. H. Ford-Smith, R. P. Bell and R. W. Dodson, *Discussions Faraday Soc.*, 29 (1960) 92.
607 A. V. Hershey and W. C. Bray, *J. Am. Chem. Soc.*, 58 (1936) 1760.
608 W. L. Reynolds, *J. Am. Chem. Soc.*, 80 (1958) 1830.
609 C. E. Johnson and S. Winstein, *J. Am. Chem. Soc.*, 73 (1951) 2601.
610 D. M. Yost and S. Zabaro, *J. Am. Chem. Soc.*, 48 (1926) 1181.
610a J. W. Johnson, K. Shen and W. J. James, *J. Less. Common Metals*, 15 (1968) 177.
610b J. M. Malin and J. H. Swinehart, *Inorg. Chem.*, 8 (1969) 1407.
610c H. J. Price and H. Taube, *Inorg. Chem.*, 7 (1968) 1.
611 L. L. Brown and J. S. Drury, *J. Chem. Phys.*, 46 (1967) 2833.
612 C. F. Wells and M. A. Salam, *Chem. Ind. London*, (1967) 2079.
612a C. F. Wells and M. A. Salam, *J. Chem. Soc. A*, (1968) 1568.
612b W. Schmidt, J. H. Swinehart and H. Taube, *Inorg. Chem.*, 7 (1968) 1984.
613 J. A. Frank and J. T. Spence, *J. Phys. Chem.*, 68 (1964) 2131.
614 E. K. Dukes, *J. Am. Chem. Soc.*, 82 (1960) 9.
615 E. Abel, H. Schmid and F. Pollak, *Monatsh. Chem.*, 69 (1936) 125.
616 E. Abel, H. Schmid and J. Weiss, *Z. Physik. Chem., Ser. A*, 147 (1930) 69.
617 R. W. Johnson and D. S. Martin, *J. Inorg. Nucl. Chem.*, 10 (1959) 94.
618 C. A. Fewson and D. J. D. Nicholas, *Biochim. Biophys. Acta*, 49 (1961) 335.
619 E. P. Guymon and J. T. Spence, *J. Phys. Chem.*, 70 (1966) 1964.
620 G. P. Haight, P. Mohliner and A. Katz, *Acta Chem. Scand.*, 16 (1962) 221.
621 S. Fronaeus and C. O. Östman, *Acta Chem. Scand.*, 10 (1956) 320.
622 D. M. Brown and F. S. Dainton, *Trans. Faraday Soc.*, 62 (1966) 1139.
623 C. E. H. Bawn and D. Margerison, *Trans. Faraday Soc.*, 51 (1955) 925.
624 H. G. S. Sengar and Y. K. Gupta, *J. Indian Chem. Soc.*, 44 (1967) 769.
625 H. G. S. Sengar and Y. K. Gupta, *J. Inorg. Nucl. Chem.*, 9 (1959) 178.
626 H. G. S. Sengar and Y. K. Gupta, *J. Indian Chem. Soc.*, 43 (1966) 223.
626a J. D. Miller, *J. Chem. Soc. A*, (1968) 1778.
626b D. D. Thusius and H. Taube, *J. Phys. Chem.*, 71 (1967) 3845.
627 E. Ben-Zvi and T. L. Allen, *J. Am. Chem. Soc.*, 83 (1961) 4352.
628 A. P. Bhargava and Y. K. Gupta, *Bull. Chem. Soc. Japan*, 41 (1968) 843.
628a R. Woods, I. M. Kolthoff and E. J. Meehan, *Inorg. Chem.*, 4 (1965) 697.
629 O. A. Chaltykyan and N. M. Beileryan, *Izv. Akad. Nauk Armyan SSSR, Khim. Nauki*, 11 (1968) 13; *Chem. Abstr.*, 52 (1958) 18054i.
629b A. P. Bhargava, Y. K. Gupta and K. S. Gupta, *J. Inorg. Nucl. Chem.*, 31 (1969) 777.
630 J. W. L. Fordham and H. L. Williams, *J. Am. Chem. Soc.*, 73 (1951) 4855.
631 I. M. Kolthoff, A. I. Medalia and H. P. Raaen, *J. Am. Chem. Soc.*, 73 (1951) 1733.
632 D. H. Irvine, *J. Chem. Soc.*, (1959) 2977.
633 D. H. Irvine, *J. Chem. Soc.*, (1958) 2166.
633a S. Raman and C. H. Brubaker, *J. Inorg. Nucl. Chem.*, 31 (1969) 1091.
634 J. Burgess and R. H. Prince, *J. Chem. Soc. A*, (1966) 1772.
635 J. Burgess, *J. Chem. Soc. A*, (1968) 497.
636 J. Holluta and W. Hermann, *Z. Physik. Chem. Ser. A*, 166 (1933) 453.
637 R. W. Chlebek and M. W. Lister, *Can. J. Chem.*, 44 (1966) 437.

638 M. R. Kershaw and J. E. Prue, *Trans. Faraday Soc.*, 63 (1967) 1198.
639 J. C. Gupta and S. P. Srivastava, *Z. Physik. Chem. Leipzig*, 216 (1961) 293.
639a J. C. Gupta and S. P. Srivastava, *Z. Physik. Chem. Leipzig*, 277 (1964) 152.
639b J. C. Gupta and S. P. Srivastava, *Z. Physik. Chem. Leipzig*, 277 (1964) 158.
639c D. E. Pennington and A. Haim, *J. Am. Chem. Soc.*, 90 (1968) 3700.
639d I. M. Kolthoff and I. K. Miller, *J. Am. Chem. Soc.*, 73 (1951) 3055.
639e I. Dogliotti and E. Hayon, *J. Phys. Chem.*, 71 (1967) 3802.
639f M. S. Tsao and W. K. Wilmarth, *J. Phys. Chem.*, 63 (1959) 346.
639g R. K. Shinghal, M. C. Agrawal and S. P. Mushran, *Z. Physik. Chem. Frankfurt*, 60 (1968) 34.
640 D. L. Ball and J. O. Edwards, *J. Am. Chem. Soc.*, 78 (1956) 1125.
641 D. L. Ball and J. O. Edwards, *J. Phys. Chem.*, 62 (1968) 343.
642 J. Beltran and R. Ferrus, *An. Real. Soc. Espan. Fiz. Quim. Ser. B*, 63 (1967) 283; *Chem. Abstr.*, 67 (1967) 26113x.
643 C. E. Castro and W. C. Kray, *J. Am. Chem. Soc.*, 86 (1964) 4603.
644 J. K. Kochi and P. E. Mocadlo, *J. Am. Chem. Soc.*, 88 (1966) 4094.
645 W. C. Kray and C. E. Castro, *J. Am. Chem. Soc.*, 86 (1964) 4603.
646 D. M. Singleton and J. K. Kochi, *J. Am. Chem. Soc.*, 89 (1967) 6547.
647 J. K. Kochi and D. D. Davis, *J. Am. Chem. Soc.*, 86 (1964) 5264.
648 J. Halpern and J. P. Maher, *J. Am. Chem. Soc.*, 86 (1964) 2311.
649 J. Halpern and J. P. Maher, *J. Am. Chem. Soc.*, 87 (1965) 5361.
649a P. B. Chock and J. Halpern, *J. Am. Chem. Soc.*, 91 (1969) 582.
649b P. W. Schneider ,P. F. Phelan and J. Halpern, *J. Am. Chem. Soc.*, 91 (1969) 77.
649c G. N. Schrauzer and E. Deutsch, *J. Am. Chem. Soc.*, 91 (1969) 3341.
650 J. K. Kochi and D. M. Singleton, *J. Am. Chem. Soc.*, 90 (1968) 1582.
650a M. Asscher and D. Vofsi, *J. Chem. Soc. B*, (1968) 947.
650b A. Orochov, M. Asscher and D. Vofsi, *J. Chem. Soc. B*, (1969) 255.
651 L. H. Sutcliffe and J. Walkley, *Nature*, 178 (1956) 999.
652 C. E. H. Bawn and D. Verdin, *Trans. Faraday Soc.*, 56 (1960) 519.
652a R. M. Davydov, *Russ. J. Phys. Chem. English Transl.*, 42 (1968) 1397.
653 E. Collinson and F. S. Dainton, *Nature*, 177 (1956) 1224.
654 E. Collinson, F. S. Dainton, B. Mile, S. Tazuke and D. R. Smith, *Nature*, 198 (1963) 26.
655 C. H. Bamford, A. D. Jenkins and R. Johnston, *Proc. Roy. Soc. London, Ser. A*, 239 (1957) 214.
656 J. K. Kochi, *Science*, 155 (1967) 415.
657 H. E. de la Mare, J. K. Kochi and F. F. Rust, *J. Am. Chem. Soc.*, 85 (1963) 1437.
658 J. K. Kochi and R. Subramanian, *J. Am. Chem. Soc.*, 87 (1965) 4855.
659 J. A. Kerr and A. F. Trotman-Dickenson, *J. Chem. Soc.*, (1960) 1602.
660 J. A. Kerr and A. F. Trotman-Dickenson, *Prog. Reaction Kinetics*, 1 (1961) 107.
661 A. Maclachlan, *J. Phys. Chem.*, 71 (1967) 4132.
662 K. W. Chambers, E. Collinson, F. S. Dainton, W. A. Seddon and F. Wilkinson, *Trans. Faraday Soc.*, 63 (1967) 1699.
663 J. N. Cooper, *J. Phys. Chem.*, 74 (1970) 955.
664 J. G. Mason, A. D. Kowalak and R. M. Tuggle, *Inorg. Chem.*, 9 (1970) 847.
665 J. I. Brauman and A. J. Pandell, *J. Am. Chem. Soc.*, 92 (1970) 329.
666 P. M. Nave and W. S. Trahanovsky, *J. Am. Chem. Soc.*, 92 (1970) 1120.
667 M. H. Ford-Smith and J. J. Habeeb, *Chem. Commun.*, (1969) 1445.
668 K. K. Banerji and P. Nath, *Bull. Chem. Soc. Japan*, 42 (1969) 2038.
669 K. S. Gupta and Y. K. Gupta, *J. Chem. Soc. A*, (1970) 256.
670 N. K. Shastri and E. S. Amis, *Inorg. Chem.*, 8 (1969) 2487.
671 N. K. Shastri and E. S. Amis, *Inorg. Chem.*, 8 (1969) 2484.
672 O. L. Kabanova, M. A. Danushenkova and P. N. Paley, *Anal. Chim. Acta*, 22 (1960) 66.
673 M. N. Malik, J. Hill and A. McAuley, *J. Chem. Soc. A*, (1970) 643.
674 C. F. Wells and M. Husain, *J. Chem .Soc. A*, (1969) 2981.
675 C. F. Wells and M. Husain, *J. Chem. Soc. A*, (1970) 1013.
676 S. K. Mishra and Y. K. Gupta, *J. Chem. Soc. A*, (1970) 260.

677 E. I. Heiba, R. M. Dessa and W. J. Koehl, Jr., *J. Am. Chem. Soc.*, 91 (1969) 6830.
678 C. F. Wells and L. V. Kuritsyn, *J. Chem. Soc. A*, (1970) 676.
678a C. F. Wells and L. V. Kuritsyn, *J. Chem. Soc. A*, (1969) 2930.
679 R. N. Mehotra, *Chem. Commun.*, (1969) 1357.
680 L. A. Hull, G. T. Davis and D. H. Rosenblatt, *J. Am. Chem. Soc.*, 91 (1969) 6247.
681 S. V. Singh, O. C. Saxena and M. P. Singh, *J. Am. Chem. Soc.*, 92 (1970) 537.
682 F. Ferranti, *J. Chem. Soc. A*, (1970) 134.
683 V. D. Komissarov and E. T. Denisov, *Russ. J. Phys. Chem.*, 43 (1969) 426.
684 J. A. Casey and R. K. Murmann, *J. Am. Chem. Soc.*, 92 (1970) 78.
685 A. D. Gilmour and A. McAuley, *J. Chem. Soc. A*, (1970) 1006.
686 A. Tomita, H. Hirai and S. Makishima, *Inorg. Chem.*, 7 (1968) 760.
687 C. F. Wells, *J. Chem. Soc. A*, (1969) 2741.
688 H. N. Po and N. Sutin, *Inorg. Chem.*, 7 (1968) 621.
689 C. Walling, M. Kurz and H. J. Schugar, *Inorg. Chem.*, 9 (1970) 931.
690 E. J. Y. Scott, *J. Phys. Chem.*, 74 (1970) 1174.
691 J. K. Kochi and J. W. Powers, *J. Am. Chem. Soc.*, 92 (1970) 137.
692 E. Vrachnou-Astra and D. Katakis, *J. Am. Chem. Soc.*, 89 (1967) 6772.
693 E. Vrachnou-Astra, P. Sakellaridis and D. Katakis, *J. Am. Chem. Soc.*, 92 (1970) 811.
694 G. E. Adams and R. L. Willson, *Trans. Faraday Soc.*, 65 (1969) 2981.
695 I. Baldea and G. Niac, *Inorg. Chem.*, 9 (1970) 110.
696 G. Davies, N. Sutin and K. O. Watkins, *J. Am. Chem. Soc.*, 92 (1970) 1892.
697 R. C. Thompson and J. C. Sullivan, *Inorg. Chem.*, 9 (1970) 1590.
698 J. R. Pladziewicz and T. R. Webb, private communication to Thompson and Sullivan, quoted in Ref. 697.
699 R. Cecil, A. J. Fear and J. S. Littler, *J. Chem. Soc. B*, (1970) 632.
700 D. D. Davis and W. B. Bigelow, *J. Am. Chem. Soc.*, 92 (1970) 5127.
701 E. Vrachnou-Astra, P. Sakellaridis and D. Katakis, *J. Am. Chem. Soc.*, 92 (1970) 3936.
702 A. Malliaris and D. Katakis, *J. Am. Chem. Soc.*, 87 (1965) 3077.
703 T. J. Conocchioli, G. H. Nancollas and N. Sutin, *Inorg. Chem.*, 5 (1965) 1.
704 A. McAuley, M. N. Malik and (in part) J. Hill, *J. Chem. Soc. A*, (1970) 2461.
705 F. Freeman, J. B. Brant, N. B. Hester, A. A. Kamego, M. L. Kasner, T. G. McLaughlin and E. W. Paul, *J. Org. Chem.*, 35 (1970) 982.
706 K. B. Wiberg and F. Freeman, quoted in footnote 51 of ref. 705.
707 F. Freeman and N. J. Yamachika, *J. Am. Chem. Soc.*, 92 (1970) 3730.
708 F. Freeman, P. D. McCart and N. J. Yamachika, *J. Am. Chem. Soc.*, 92 (1970) 4621.
709 F. Freeman and N. J. Yamachika, *Tetrahedron Letters*, (1969) 3615.
710 L. I. Simándi and M. Jáky, *Tetrahedron Letters*, (1970) 3489.
711 I. M. Mathai and R. Vasudevan, *J. Chem. Soc. B*, (1970) 1361.
712 R. A. Sheikh and W. A. Waters, *J. Chem. Soc. B*, (1970) 988.
713 J. Lubach and W. Drenth, *Rec. Trav. Chim.*, 89 (1970) 144.
714 K. K. Banerji, P. Nath and G. V. Bakore, *Bull. Chem. Soc. Japan*, 43 (1970) 2027.
715 M. I. Edmonds, K. E. Howlett and (in part) B. L. Wedzicha, *J. Chem. Soc. A*, (1970) 2866.
716 L. Hin-Fat and W. C. F. Higginson, *J. Chem. Soc. A*, (1970) 2836, 2842.

Chapter 5

Induced Reactions

L. J. CSÁNYI

1. Introduction

1.1 DEFINITIONS

It has been well known for a long time that some reactions taking place at a very slow rate may be markedly accelerated by the simultaneous occurrence of another reaction of measurable velocity. On the suggestion of Kessler[1] this phenomenon is called chemical induction and it is said that the reaction of measurable velocity induces the other slow reaction.

An induced reaction may be represented by the following scheme.

$$A+I = C \tag{1}$$

$$A+Ac = D \tag{2}$$

While in separate solutions only reaction (1) occurs, in a common solution of the three constituents reaction (2) also takes place.

Reaction (1) is called the primary, main, or inducing reaction, which brings about induced reaction (2). Substances A, I and Ac taking part in both reactions (1) and (2) are called actor, inductor, and acceptor, respectively. The extent of the induced change is conventionally expressed by the induction factor F_i, defined as the ratio of the equivalents of the induced reaction to those of the primary reaction.

In the majority of cases both the primary and the induced reactions are oxidation–reduction reactions. In such reactions the actor can have either reducing or oxidizing properties. The chemical characteristics of the inductor and acceptor are always identical and opposite to that of the actor. When the latter is a reducing agent the acceptor and inductor are oxidants and *vice versa*.

Besides induced oxidation–reduction reactions we often speak of induced dissolution, induced precipitation, as well as of induced complex formation; there is even a reference to an induced reaction caused by neutralization. It is only necessary to examine briefly the latter cases.

Induced precipitation is a collective name for processes accompanying the formation of solid phase, such as occlusion, adsorption, compound formation, formation of isomorphous mixtures, mixed crystals, colloidal solutions, etc. In our opinion "induction" occurring at precipitations does not fall within the category of chemical induction since coprecipitation is determined by different crystal-physical and -chemical factors.

As far as we know there is only one reference to induced change caused by neutralization. Raschig[2] observed a strong air oxidation during the conversion of bisulphite–sulphite. We are of the opinion that, here, not the air oxidation of sulphite induced by neutralization but the pH dependence of oxidation is involved.

Induced dissolution is also a well known phenomenon and frequently applied in chemical analysis. To dissolve platinum easily it was suggested by Ropp[3] that the sample should be alloyed with silver or copper, the alloys being easily soluble even in dilute acids. Anhydrous chromic chloride, insoluble in water and dilute acids, becomes easily soluble by adding metallic magnesium or zinc to the dilute acid[4]. In this case the chromium(III) compound is reduced to chromium(II), which will be oxidized by the solvent to water-soluble chromium(III)

$$CrCl_2 + 3\,H_2O + H_3O^+ = [Cr(H_2O)_4Cl_2^+] + \tfrac{1}{2}H_2 \tag{3}$$

Dissolution of tellurium in sodium sulphide can be effected only by simultaneous addition of sulphur, when the tellurium will be oxidized to thiotellurite and dissolved[5].

There are also examples of induced complex formation, an essential step of which is always an oxidation–reduction reaction. Rich and Taube[6] found that the rate of exchange between $PtCl_4^{2-}$ and Cl^- was considerably increased by addition of cerium(IV). In the presence of this oxidizing agent a labile complex of Pt(III) is formed, the chloride of which is easily exchangeable. Exchange of platinum between $PtCl_4^{2-}$ and $PtCl_6^{2-}$ is similarly rapid *via* the intermediate labile $PtCl_5^{2-}$ complex formed by cerium(IV).

Summarizing, it can be said that induced reactions are connected mostly with oxidation–reduction processes and that this is true for the induced complex formation, too. On the other hand, induced precipitation has nothing to do with genuine induced reactions; therefore it is advisable not to use this collective name in order to keep the concept of chemical induction clear.

1.2 TYPES OF INDUCED REACTIONS

In the early years of this century Luther and Schilow[7] attempted to classify induced reactions. This classification was based on the specific nature of substances taking part in the induced reaction. Thus they distinguished two groups: coupled and induced chain reactions (this term was adopted later).

Coupled reactions occur when the primary reaction results in an intermediate which enables the acceptor to react too. The main characteristic of this reaction is that the value of the induction factor is small, not exceeding 2 even under most favourable conditions for the induced change. Plotting F_i against the ratio of the initial concentrations of acceptor and inductor results in a curve having a limiting value.

According to the authors mentioned above, induced chain reactions (Livingston[8]) or induced catalysis (Bray and Ramsey[9]) take place when the very slow reaction between the acceptor and actor is catalyzed by the inductor. However, since the chemical characters of the acceptor and the inductor are the same, the actor reacts with the inductor, too, thus a part of it will be excluded from the catalysis. The principal characteristics of reactions of this type according to Luther and Schilow are that the value of F_i largely exceeds 2 and that the plot of F_i *versus*$([\mathrm{Ac}]/[\mathrm{I}])_0$ rises exponentially.

1.2.1 Coupled reactions

Coupled reactions take place when in the primary process a reactive intermediate is formed which enables the acceptor to react. The coupling intermediate can equally well be formed (*a*) from the actor and (*b*) from the inductor.

(*a*) The actor is converted by inductor into a species of more active oxidizing or reducing properties than the actor itself, and in the further steps the inductor and the acceptor compete for this intermediate.

(*b*) The inductor is converted by the actor into a product which attacks the acceptor. According to Manchot[10] this is the only correct way to interpret induced reactions.

(*a, i*) When the actor is an oxidizing agent and is reduced in the primary reaction by steps $\mathrm{A_{ox}} \rightarrow \mathrm{A_1} \rightarrow \mathrm{A_{red}}$, we can write

$$a\ \mathrm{A_{ox}} + \alpha b\ \mathrm{I_{red}} = a\ \mathrm{A_1} + \alpha b\ \mathrm{I_{ox}} \tag{4}$$

$$\beta a\ \mathrm{A_1} + \beta(1-\alpha) b\ \mathrm{I_{red}} = \beta a\ \mathrm{A_{red}} + \beta(1-\alpha) b\ \mathrm{I_{ox}} \tag{5}$$

$$(1-\beta) a\ \mathrm{A_1} + (1-\beta) c\ \mathrm{Ac_{red}} = (1-\beta) a\ \mathrm{A_{red}} + (1-\beta) c\ \mathrm{Ac_{ox}} \tag{6}$$

where a, b and c are stoichiometric coefficients, α and β are the corresponding conversion factors.

Since $F_i = \Sigma[\mathrm{Ac_{ox}}]/\Sigma[\mathrm{I_{ox}}]$ the value of the induction factor depends on the rate ratio of competing reactions (5) and (6). Thus, on increasing the initial concentration ratio of Ac to I over any limit, the rate ratio w_6/w_5 increases to infinity. In this case the coupling intermediate $\mathrm{A_1}$ is entirely consumed in reaction (6) in the oxidation of $\mathrm{Ac_{red}}$, *i.e.* $\beta = 0$, and F_i reaches its limiting value. The limiting value of F_i is called the coupling index, CI. When $\beta = 0$, the induced reaction can be described by the following overall equation

$$aA_{ox}+\alpha bI_{red}+cAc_{red} = aA_{red}+\alpha bI_{ox}+cAc_{ox} \tag{7}$$

To emphasize the coupling caused by intermediate A_1, the term coupling factor, F_c, was introduced by Medalia[11] and defined as

$$F_c = \frac{1-\beta}{\beta} = \frac{\text{equivalents of Ac reacted in coupling reaction}}{\text{equivalents of I reacted in competing step}}$$

On the basis of the above scheme the value of the coupling index is

$$\text{CI} = \frac{1-\alpha}{\alpha}$$

Since $\beta a+(1-\beta)a = (1-\alpha)b$, we obtain for the induction factor

$$F_i = \frac{(1-\beta)a}{\alpha b+\beta(1-\alpha)b} = \frac{(1-\beta)(1-\alpha)}{\alpha+\beta(1-\alpha)} = \frac{\text{equivalents of Ac reacted}}{\text{sum of equivalents of I reacted}}$$

Knowing F_i and CI, the coupling factor can be expressed as

$$F_c = \frac{F_i(\text{CI}+1)}{\text{CI}-F_i} = \frac{F_i+(F_i/\text{CI})}{1-(F_i/\text{CI})} \tag{8}$$

If $F_i/\text{CI} \rightarrow 1$, then $F_c \rightarrow \infty$.

From the foregoing it can be seen that F_c refers to the rate ratios of competing reactions (6) and (5). A knowledge of F_c makes it possible in certain cases to obtain the relative rate coefficient, k_6/k_5.

Denoting the equivalents of the actor converted in reactions (4), (5) and (6) by x_4, x_5, and x_6, and the initial concentration of the actor (also in equivalents) by c_A, then

$$c_A = x_4+x_5+x_6$$

The rate equations of the competing reactions (6) and (5), assumed second order, are

$$-\frac{dx_6}{dt} = k_6[A_1][Ac]$$

$$-\frac{dx_5}{dt} = k_5[A_1][I]$$

Dividing we obtain

$$F_c = \frac{dx_6}{dx_5} = \frac{k_6[Ac]}{k_5[I]}$$

If the change in concentrations of Ac and I is not great, integrating this equation we obtain

$$\frac{x_6}{x_5} = \frac{k_6[Ac]_{av}}{k_5[I]_{av}}$$

Hence

$$\frac{k_6}{k_5} = \frac{x_6[I]_{av}}{x_5[Ac]_{av}}$$

If the initial concentrations of Ac and I are high so that the amount of substance converted can be ignored, *i.e.* if the concentration of the actor is low, then

$$\frac{k_6}{k_5} \approx F_c \left(\frac{[I]}{[Ac]}\right)_0 \tag{9}$$

The utility of this approximation (9) can be judged by the figures in the last columns of Tables 4, 6, 8, 9 and 10.

In practice, however, it may happen that $w_5 > w_6$, even when concentrations of I and Ac are commensurable. In such cases, even with the greatest ratio of $(Ac/I)_0$ which can be realized experimentally, F_i cannot reach a limiting value, *i.e.* the characteristics of coupled reactions cannot be observed.

(*a, ii*) When the actor is a reducing agent and is oxidized by the inductor in steps $A_{red} \rightarrow A_1 \rightarrow A_{ox}$, the considerations are similar to those in the previous case. Obviously, the problem is more involved when the reduction or oxidation of the acceptor is taking place in two or more steps. Cases are particularly complicated when the inductor also has more than two oxidation states.

From the above it can generally be concluded that the basic condition of the occurrence of coupled reactions – if the coupling intermediate is derived from the actor – is that the actor has at least three (including zero) oxidation states.

Case *a, i* is well illustrated by the arsenite-induced oxidation of manganese(II) by chromic acid, studied by Lang and Zwerina[12]. The overall equation of this induced reaction is

$$2\,Cr(VI)+2\,As(III)+Mn(II) = 2\,Cr(III)+2\,As(V)+Mn(IV) \tag{10}$$

According to eqn. (10), CI = 0.5, *i.e.* for every equivalent of chromate used in the oxidation of arsenic(III), a further half equivalent of chromate is also consumed in the oxidation of manganese(II).

To illustrate case *a, ii* we may refer to the reduction of chlorate by arsenious acid induced by different 1-equivalent oxidizing reagents[13]. The effect of the oxidizing agent is to form arsenic(IV). Chlorate will be reduced by arsenic(IV) which is a stronger reducing agent than arsenic(III).

Manchot's concept[10] that the inductor is converted by the actor into an active intermediate by which the oxidation of the acceptor takes place (case *b*) can be illustrated by the scheme

$$A_{ox} + I_{red} = A_{red} + I_P \qquad (11)$$

$$I_P + Ac_{red} = I_{ox} + Ac_{ox} \qquad (12)$$

Thus, I_P (inductor peroxide, primary oxide) derived from I_{red} is converted by Ac_{red} into the end-product, I_{ox}, which is inactive regarding the induction, *i.e.* coupling cannot turn into catalysis. Moreover, since there are several consecutive steps, and since the direct reaction between actor and acceptor is very slow, it follows that the value of F_i will not be changed by altering the ratio, $([Ac]/[I])_0$. However, in most cases studied the function $F_i = f(Ac/I)_0$ attains a limiting value, or at least varies; therefore it has to be concluded that Manchot's concept cannot be maintained in its original form.

The situation is different when I_{red} can be regenerated according to $I_P \rightleftharpoons I_{red}$. If this process is rapid enough, a considerable amount of the acceptor can be converted even by a few molecules of I_{red}, *i.e.* coupling turns into catalysis. With this modification Manchot's concept can be applied to characterize autoxidation processes belonging to the category of induced chain reactions.

So far possible processes leading to the occurrence of coupled reactions have been indicated. However, with regard to the nature of the reactive intermediates, it has been mentioned only that these, formed either from the actor or from the inductor, are more active than the actor itself. In the case of simpler systems, as was pointed out by Luther and Rutter[15], the knowledge of the coupling index makes it possible to estimate the quality (oxidation number) of the coupling intermediate. If α and ω are the oxidation numbers of the actor before and after the reaction, and x is the oxidation number of the coupling intermediate, and furthermore, if only the acceptor reacts with the intermediate then

$$\text{CI} = \frac{x - \omega}{\alpha - x}$$

Hence

$$x = \frac{\text{CI}\alpha + \omega}{1 + \text{CI}}$$

References pp. 577–580

Applying this to reaction (10) where $\alpha = 6$, $\omega = 3$, and CI = 0.5

$$x = \frac{0.5 \times 6 + 3}{1 + 0.5} = 4$$

If we try to interpret reaction (10) according to Manchot's concept, taking m and n as oxidation numbers of inductor before and after the reaction and y as oxidation number of the "primary oxide", since $(y-n)$ equivalents are consumed in the oxidation of acceptor we get

$$\text{CI} = \frac{y-n}{n-m}$$

and so

$$y = n + \text{CI}(n-m)$$

Since CI = 0.5, $m = 3$ and $n = 5$, $y = 5+0.5\,(5-3) = 6$. According to this, the effect of chromate is to produce a primary oxide containing arsenic(VI), which is in disagreement with experiment.

In the case of more involved systems the study of the mechanism of induced reactions may serve as a diagnostic tool. This study ought to be (as usual) from both the kinetic and thermodynamic point of view.

1.2.2 Induced chain reactions

Induced chain reactions occur when the very slow reaction between acceptor and actor is catalyzed by the inductor. Actually it means that primary reaction between actor and inductor induces a chain reaction between actor and acceptor. That is why this group of reactions is called induced chain reactions.

Assuming that the actor forms its stable end-product through steps $A_{ox} \rightarrow A_1 \rightarrow A_{red}$, and the acceptor is oxidized according to $Ac_{red} \rightarrow Ac_1 \rightarrow Ac_{ox}$, and furthermore, for the sake of simplicity, that the inductor is transformed directly to the end-product according to $I_{red} \rightleftharpoons I_{ox}$ which is a reversible reaction, we can write

$$A_{ox} + I_{red} = A_1 + I_{ox} \tag{13}$$

$$A_1 + I_{red} = A_{red} + I_{ox} \tag{14}$$

$$A_1 + Ac_{red} = A_{red} + Ac_1 \tag{15}$$

$$Ac_1 + I_{ox} = Ac_{ox} + I_{red} \tag{16}$$

The above scheme represents a closed-chain reaction showing that, at a high concentrations of Ac if the chain carrier A_1 reacts only with Ac_{red}, F_i never reaches a limiting value: the plot of F_i *versus* $([Ac]/[I])_0$ is exponential. If the concentration of Ac_{red} is high enough, the competing reaction (14) is practically negligible; thus, in the overall equation of the induced reaction, composed of steps (13), (15) and (16)

$$A_{ox}+Ac_{red} = A_{red}+Ac_{ox} \tag{17}$$

only the acceptor and the actor take part, in contrast to coupled reactions, in the overall process of which the inductor is also involved. This fact is often emphasized as a characteristic difference between the two types of induced reactions. To stress the point is, however, not reasonable, for it is valid only in special cases when rate $w_{15} \gg w_{14}$.

From the above scheme it is evident that the extent of the induced change here as well as in the case of coupled reactions depends on the rate ratios of competing reactions (14), (15) and (16). Thus any factor influencing these rates will also affect the value of the induction factor. According to our experience of investigating induced reactions by titration methods, *e.g.* titrating the common solution of the actor and the acceptor with the solution of the inductor, the delivery rate of the titrant, the dilution of the solution, the stirring rate, the hydrogen ion concentration, the order of addition of the reagents, the temperature, the presence of several, apparently indifferent substances – all considerably influence the value of F_i.

The principal characteristic of induced reactions of this type which have not been stressed so far, is that the extent of the induced change greatly decreases and in most cases reaction even ceases in the presence of chain-breaking substances. The induced reaction can be suppressed by any substances reacting with chain carriers at a higher rate than does the acceptor, and the product of the reaction of the suppressor can easily react with the inductor. Since the concentration of the chain carriers is generally low, the supressors of induced chain reactions exert considerable effect even in small quantity. The effect is particularly pronounced when the suppressor reacts reversibly.

From the above it is obvious that merely the magnitude of the numerical value of F_i and the shape of the plot of F_i *versus* $([Ac]/[I])_0$ do not make it possible to classify induced reactions correctly. It is necessary to learn more about the mechanism of induced reactions. The schemes presented show clearly that a genuine coupled reaction can be regarded as an open-chain, and an induced catalysis as a closed-chain reaction. However, these limiting types of reactions occur only rarely.

The examples mentioned illustrate well the peculiarities of induced reactions, *i.e.* a hardly oxidizible substance can be oxidized when a simultaneous reduction

References pp. 577–580

occurs, and a slow reduction can be accelerated by a simultaneous fast oxidation reaction, because in the primary reaction intermediates more active than the original partners are formed. Thus occurrence of induced reactions can be expected in every case when different numbers of electrons are involved in oxidation–reduction couples taking part in the induced reaction, *e.g.* the Cr(VI)/As(III)/Mn(II) system consists of chromium(VI)/chromium(III), arsenic(V)/arsenic(III), and manganese-(III)/manganese(II) couples involving 3, 2 and 1 electrons. This can easily be understood by considering the principle of equivalent changes due to Schaffer[16]. This principle states that those processes are fast in which the numbers of electrons required by the partners are equal, *e.g.* in general a 2-equivalent oxidant rapidly reacts with a 2-equivalent reducing agent, but the reaction will be slower if the reducing partner is a 1-equivalent reagent or *vice versa.* The kinetic background of this principle is clear. The reaction between 2- and 1-equivalent partners requires a termolecular collision which has a very low probability. If this reaction were to take place by bimolecular steps, a considerable greater energy of activation would be required because of the formation of an unstable intermediate. Therefore, it can be stated that in the case of the reactions of partners of different equivalents, both the relative slowness of the primary step on the one hand, and the formation of unstable intermediates on the other, favour the competing reaction of the acceptor, *i.e.* the occurrence of induced reactions.

It seems to be worthwhile touching on the correlation between induced reaction and catalysis. Defining a catalyst as a relatively small amount of an additional substance which brings about a considerable change in the reaction rate we may speak of catalysis in every case when chain initiating atoms, radicals, or radical ions are introduced into the reacting system. And just this happens in an induced reaction. In the primary reaction an intermediate is formed and initiates longer or shorter chains in which the acceptor is converted. Therefore, a definite difference between catalysis and induced reaction cannot be established. If this were to be accepted, then genuine coupled reactions should be regarded as limiting cases where the "catalyst" (= inductor) undergoes a stoichiometrical change (length of chain = 1). The transition between the limiting cases, *i.e.* between the genuine coupled reactions and induced catalysis is continuous. This fact can be observed not only by comparing different systems, but also with a given system if the experimental conditions are appropriately changed. To show this we refer to the induced oxidation of iodide caused by the iron(II)–chromium(VI) reaction[17]. In this system a coupled reaction occurs if the concentrations of the partners are low and the pH is fairly high. Accordingly, iodine is liberated just until iron(II) ions are present in the solution. By increasing the concentration of the partners and that of hydrogen ions, the rate of reaction between iron(III) and iodide ions increases to such an extent that the reduction of chromate ions by iodide becomes instantaneous even in the presence of very few iron(II) ions. At the end of the reaction iron(II) is recovered, *i.e.* the coupled reaction taking place under the first conditions has turned into a catalytic process. Of course, under such experimental conditions iron(III) is equivalent to iron(II) with respect to the acceleration.

2. Examples of induced reactions

Although in the fifties of the last century it had already been recognized that in several oxidation–reduction reactions the co-existence principle (*i.e.* the assumption that the individual processes take place independently of each other) was not valid and to date many examples of chemical induction have been found, there are only a few cases known where the mechanism of the induced reaction has been satisfactorily elucidated. There are several reasons for this. Some of the induced reactions take place too rapidly to be investigated by conventional kinetical methods; in other cases a thorough investigation was frustrated by the lack of appropriate analytical methods.

Chemical analysis of composite systems is often severely restricted by the invalidity of the co-existence principle (although there are a few cases known in which estimations are made possible just because of the occurrence of chemical induction). Therefore, many efforts have been directed at exploring at least qualitatively the source of errors caused by induced reactions. That is why our present knowledge about such reactions is rather qualitative in nature.

Because of the foregoing we will not attempt in this article to compile every observation recorded so far in the field of induced reactions; only the results where the mechanism is sufficiently clear will be referred to. Also, the wide and technically important field of autoxidation processes will be omitted, because they can be considered more properly later in the series.

Systems selected for discussion will be presented below according to the reactive species involved in the induced reactions.

2.1 CHROMIUM(IV) AND CHROMIUM(V) SPECIES AS COUPLING INTERMEDIATES

The observation of induced reactions involving chromate almost coincided with the discovery of the phenomenon of chemical induction itself. According to the the role of chromate ions in these reactions, two groups can be distinguished:

(*i*) Chromium(VI) plays the role of actor, whose reaction with various inductors listed in Table 1 results in the oxidation of several acceptor ions or molecules.

(*ii*) Chromium(VI) is functioning as inductor in the systems summarized in Table 2.

Unfortunately, only few of the systems listed in Tables 1 and 2 have been studied in detail; there are many where even the value of the induction factor is unknown. Therefore, we shall deal with systems whose mechanisms are clear to some degree.

References pp. 577–580

TABLE 1

INDUCED REACTIONS WITH Cr(VI) AS ACTOR

Inductor	*Acceptor*	F_i	*Ref.*
As(III)	Mn(II)	0.5	12
As(III)	I^-	2	18
As(III)	tartaric acid		1
iso-propanol	Mn(II)	0.5	19
sec-butanol	Mn(II)	0.5	20
n-butanol	Mn(II)		21
sec-hexanol	Mn(II), Ce(III)		21
n-propanol	Mn(II)		22
caprylalcohol	Mn(II)		22
ethyleneglycol	Mn(II)		23
propyleneglycol	Mn(II)		24
2, 3-butyleneglycol	Mn(II)		24
pinacol	Mn(II)		25
benzaldehyde	Mn(II)		26, 27
formaldehyde	Mn(II)		28
formic acid	Mn(II), Ce(III)		29
Fe(II)	I^-		30–33
Fe(II)	Br^-		15
V(IV)	I^-		15
V(III)	I^-		15
V(II)	I^-		15
Ti(III)	I^-		34
Ti(III)	Br^-		34
UO^{2+}	I^-		15, 35, 36
oxalic acid	indigo		17, 37
Sn(II)	tartaric acid		1
Sb(III)	tartaric acid		1
$[Fe(CN)_6]^{4-}$	tartaric acid		1
hydroquinone	oxalic acid; glycerol; lactic acid, citric acid; maleic acid		38

TABLE 2

INDUCED REACTIONS WITH Cr(VI) AS INDUCTOR

Actor	*Acceptor*	*Ref.*
As(III)	O_2	1, 15, 39
As(III)	$HBrO_3$	15
As(III)	Fe(III)	15
As(III)	$HClO_3$	15
As(III)	$H_2S_2O_8$	15
Sn(II)	O_2	15

2.1.1 Reaction between arsenic(III) and chromium(VI)

The reaction between arsenic(III) and chromium(VI) in 0.01–0.05 *M* sulphuric acid was studied by DeLury[40]. On recomputing his data[41] taking into consideration the dichromate–hydrogen chromate equilibrium

$$Cr_2O_7^{2-}+H_2O \rightleftharpoons 2\,HCrO_4^-$$

it was found that the rate of disappearance of chromic acid is proportional to the first power of the concentration of arsenous acid and of hydrogen chromate and approximately to the second power of the hydrogen ion concentration.

Edwards[42] has pointed out that, at low acid chromate concentrations, the plot of $k/[HCrO_4][H^+]^2$ against $1/[H^+]$ is linear but not horizontal. Therefore the rate law

$$\text{rate} = k_a[HCrO_4^-][H_3AsO_3][H^+]+k_b[HCrO_4^-][H_3AsO_3][H^+]^2$$

was recommended. The increase in rate in solutions containing large amounts of acid chromate is a good evidence for general acid catalysis.

Mason and Kowalak[43] have found that, in 0.2 *M* acetate buffer at low concentrations of chromate and high concentrations of arsenous acid, the rate law has the form

$$-\frac{d[Cr(VI)]}{dt} = \frac{A[As(III)][Cr(VI)]}{1+B[As(III)]}$$

which is consistent with the mechanism

$$As(III)+HCrO_4^- \rightleftharpoons As(III).HCrO_4^- \qquad \text{pre-equilibrium, } K$$

$$As(III).HCrO_4^- \rightarrow \text{product} \qquad \text{rate determining, } k$$

Considering also the stoichiometry of the reaction, we find that

$$A = 2kK, \quad B = K, \quad k = 3.76\times10^{-4}\ \text{sec}^{-1}, \quad K = 22.4\ \text{l.mole}^{-1}$$

In the case of high concentrations of chromate and low concentrations of arsenous acid, we must take into account that dichromate ions are also involved in reaction. Thus

$$-\frac{d[Cr(VI)]}{dt} = \frac{A[As(III)][HCrO_4^-]}{1+B[HCrO_4^-]}+k'[Cr_2O_7^{2-}][As(III)]$$

References pp. 577–580

where

$$k' = 2.47 \times 10^{-4} \text{ l.mole}^{-1}.\text{sec}^{-1}$$

The data reported and mechanism proposed permit no conclusions to be drawn concerning the products formed in the rate-determining step. According to the authors the reaction could occur by oxygen transfer, and no production of arsenic-(IV) would be required.

Oxidation of arsenic(III) by chromate in alkaline medium was studied by Kolthoff and Fineman[39]. They found the reaction to be first order with respect to both chromate and arsenic(III). At pH greater then 9.1 the rate coefficient is independent of the hydrogen-ion concentration. The average value of the rate coefficient at 30° in solutions of pH 9.1 and ionic strength 1.75 was found to be $(1.61 \pm 0.08) \times 10^{-3}$ l.mole^{-1}.sec^{-1}.

Reaction between arsenic(III) and chromate may induce the oxidation of several acceptors, such as manganese(II), iodide and bromide ions.

Induced oxidation of manganese(II) salts was studied by Lang and Zwerina[12]. According to their measurements (Table 3) the limiting value of the induction factor is 0.5. It is known that manganese(II) cannot be oxidized by chromium(VI) directly owing to the relative low value of oxidation potentials of chromium(VI)–chromium(III) couple unless complex forming agents such as fluoride or metaphosphate ion etc. are present[44]. In the presence of iodide ions[18,45] the value of CI = 2, *i.e.* the oxidation of every equivalent of arsenic(III) is accompanied by formation of 2 equivalents of iodine (Table 4). The same was found with bromide ions.

TABLE 3

OXIDATION OF MANGANESE(II) INDUCED BY REACTION BETWEEN CHROMIC ACID AND ARSENOUS ACID

Data of Lang and Zwerina[12]. Conditions: 3.14×10^{-3} *M* $K_2Cr_2O_7$, 1.13 *M* H_2SO_4; 0.105 *M* $MnSO_4$; initial volume, 55.4 cm^3.

Initial ratio $\frac{[Mn(II)]}{[As(III)]}$	F_i	F_c
57.3	0.49	73.5
37.1	0.49	73.5
28.8	0.47	23.5
23.0	0.46	13.3
19.2	0.45	13.6
17.7	0.41	6.8
16.4	0.39	5.3
15.4	0.33	2.9
14.4	0.27	1.8
12.9	0.14	0.9

TABLE 4

ARSENITE-INDUCED OXIDATION OF IODIDE BY DICHROMATE ACCORDING TO DE LURY[18]

Conditions: 1.66×10^{-4} M As_2O_3; 1.11×10^{-4} M $K_2Cr_2O_7$; 8.0×10^{-3} M H_2SO_4.

$\left(\frac{[KI]}{[As(III)]}\right)_0$	$(F_i)_{obs}$	$(F_i)_{calc}$	(F_c)†	k_{21}	k_{18}	$\frac{k_{21}}{k_{18}}$	$F_c\left(\frac{[As(III)]}{[KI]}\right)_0$
				$(sec^{-1})\times10^3$			
1.96	1.24	1.15	4.07	4.93	4.05	1.21	2.07
2.94	1.41	1.34	6.11	5.23	3.73	1.40	2.07
3.95	1.41	1.46	8.21	5.28	3.78	1.39	2.08
4.92	1.50	1.54	10.23	5.42	3.67	1.47	2.09
5.90	1.51	1.60	12.27	5.50	3.30	1.66	2.08
7.38	1.77	1.67	15.35	5.67	3.23	1.75	2.07
9.88	1.86	1.74	20.55	5.77	3.13	1.84	2.08
14.77	2.05	1.82	30.72	5.78	2.87	2.01	2.08
24.62	2.12	1.89	51.21	5.98	2.87	2.08	2.08
34.42	2.11	1.92	71.59	5.80	2.78	2.08	2.08
0.0	—	—	—	—	8.93	—	—

† F_c is calculated by eqn. (8) (p. 513) using values of $(F_i)_{obs}$. The value of

$$(F_c)_{calc}\left(\frac{[As(III)]}{[KI]}\right)_{av} \text{ is about 2.1; } \left(\frac{[As(III)]}{[KI]}\right)_{av} = \frac{[As(III)]_0-\frac{1}{2}[As(III)]_{reacted}}{[KI]-\frac{1}{2}[KI]_{reacted}}.$$

Considering the limiting values of the induction factor it may be postulated that in the case of iodide and bromide the induced oxidation is caused by chromium(V), whereas for induced oxidation of manganese(II) chromium(IV) is the coupling intermediate. Therefore, one has to assume that in the course of reaction between arsenic(III) and chromium(VI) both chromium(V) and chromium(IV) intermediates are involved. The mechanism below, proposed by Westheimer[41], seems to be in agreement with experiment.

$$Cr(VI)+As(III) \rightarrow Cr(IV)+As(V) \tag{18}$$

$$Cr(IV)+Cr(VI) \rightarrow 2Cr(V) \tag{19}$$

$$Cr(V)+As(III) \rightarrow Cr(III)+As(V) \tag{20}$$

It was found by DeLury[18] that the overall rate of reduction of chromate is practically unaffected by the concentration of iodide, *i.e.* the sum of the rates of formation of iodine and aresenic(V) is constant and just equal to the rate at which chromate is reduced in a raction mixture containing no iodide (Fig. 1). The rate of oxidation of arsenite at a sufficiently high concentration of iodide decreases to $\frac{1}{3}$ of its original value; this is in accordance with the value of CI = 2 found. Fig. 1 well illustrates the general feature of coupled reactions, that the reaction of the inductor is always inhibited by the acceptor. The induced oxidation of iodide can

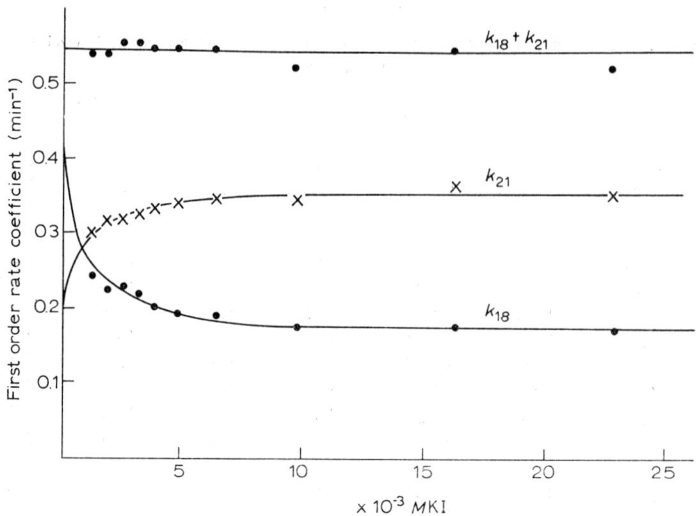

Fig. 1. Arsenic(III)-induced oxidation of iodide by dichromate. Data of DeLury[18].

be described by reactions (18), (19), (21) and (22)

$$Cr(V)+I^- = IO^- + Cr(III) \tag{21}$$

$$IO^- + I^- + 2H^+ = I_2 + H_2O \tag{22}$$

The induced oxidation of manganese(II) can be represented by steps (18) and (23)

$$Cr(IV)+Mn(II) = Cr(III)+Mn(III) \tag{23}$$

followed by stabilization of manganese(III) through disproportionation, *viz.*

$$2Mn(III) = Mn(IV)+Mn(II) \tag{24}$$

From the stoichiometric point of view the above induced reactions can be described equally well if the formation of arsenic(IV) is assumed, *viz.*

$$Cr(VI)+As(III) = Cr(V)+As(IV) \tag{25}$$

At first sight the argument of Westheimer[41], according to which it is unwise to postulate a reaction in which two unstable intermediates are simultaneously produced, seems to be acceptable, especially if we take into consideration that these induced oxidations are not autocatalytic in nature; most of mechanisms involving arsenic(IV), chromium(IV), and chromium(V) lead to the conclusion that the reaction should be autocatalytic. However, omission of step (25) from the mechanism of induced oxidation of manganese(II) leads to the difficulty that reaction (19) involves a positive standard free-enthalpy change[56], and, consequently, will be too slow to be significant.

2.1.2 Reaction between isopropyl alcohol and chromium(VI)

Oxidation of isopropyl alcohol (H_2R) by chromic acid has been studied in det ai by Westheimer and Novick[46], and it was found that acetone (R) is formed nearly quantitatively. The reaction proved to be first order with respect to hydrogen chromate and second order with respect to hydrogen ions. Measurements using 2-deutero-2-propanol under identical conditions as those for the oxidation of ordinary isopropyl alcohol showed the rate of reaction to be $\frac{1}{6}$ of that with the hydrogen compound. This fact is considered to prove that the secondary hydrogen atom is removed in the rate-controlling step and that the assumption of hydride-ion abstraction can be excluded. The data are consistent with the following mechanism

$$HCrO_4^- + H_2R = Cr(IV) + R \tag{26}$$

$$Cr(IV) + H_2R = Cr(II) + R \tag{27}$$

$$Cr(IV) + Cr(VI) = Cr(V) + Cr(III) \tag{28}$$

$$Cr(V) + H_2R = Cr(III) + R \tag{29}$$

in which only chromium intermediates are assumed and neither organic free radicals nor reaction of chromium species with the solvent molecules are taken into account. The absence of organic free radicals seems to be supported by the fact that oxygen has no influence on the oxidation. Reaction between the solvent and the chromium species is unlikely because it would lead to the production of OH radicals and hydrogen peroxide. If hydrogen peroxide were formed, it would cause the oxidizing capacity to decrease, *i.e.* no quantitative formation of acetone could be observed.

A recent study by Lee and Stewart[47] confirms the previous observations that with rather low acid concentrations the oxidation rate of H_2R depends only on the acidity of the medium and not on the nature of the proton-supplying mineral acid. On the contrary, at rather high acid concentrations the rate of oxidation depends not only on the acidity but also on the nature of the acids. This can be explained by the fact that the hydrogen chromate and the acid present interact with each other

$$HCrO_4^- + H^+ + HB = HCrO_3B + H_2O$$

where HB and B represented the acid and acid radical, respectively. The incorpora-

References pp. 577–580

TABLE 5

APPARENT pK_a VALUES OF CHROMIC ACID IN AQUEOUS SOLUTIONS OF VARIOUS MINERAL ACIDS, ACCORDING TO LEE AND STEWART[47]

Acid	*pK_a(H_0 scale)*	*Species formed*
HNO_3	−1.91	$HCrNO_6$
$HClO_4$	−0.83	$HCrClO_7$
H_2SO_4	+0.34	$HCrSO_7$
HCl	+0.52	$HCrClO_3$
H_3PO_4	+1.74	H_3CrPO_7

tion of the acid radical increases the electron-accepting power of chromium. The oxidizing ability of the protonated species, $HCrO_3B$, increases in the order: $H_2CrPO_7 < HCrClO_3 < HCrSO_7 < HCrClO_7 < HCrNO_6$ for a given H_0 value. This is the same order in which the apparent pK_a values of these species vary (Table 5).

The first step in the oxidation of the alcohol is the formation of a chromate ester which probably decomposes unimolecularly to products, *viz.*

$$R_2C(H)\text{–}O\text{–}CrO_2H \xrightarrow{H^+} [R_2C\cdots O\cdots CrO_2H_2^+ \text{ (cyclic, } H\cdots O)] \longrightarrow R_2C{=}O + H_3CrO_3^+$$

At higher acid concentrations, when formation of $HCrO_3B$ is also to be considered, because of the incorporation of the conjugate base the rate of oxidation will depend also on the nature of the acid radical present, *viz.*

$$R_2C(H)\text{–}O\text{–}Cr(O)(O)B \xrightarrow{H^+} [R_2C\cdots O\cdots Cr^+(OH)B \text{ (cyclic, } H\cdots O)] \longrightarrow R_2C{=}O + H_2CrO_2B^+$$

Oxidation of isopropyl alcohol by chromic acid in concentrated acetic acid solution has recently been studied by Wiberg and Schäfer[48,123] spectrophotometrically. At 385 nm a rapid increase in absorbance (with a half life of about 6 sec) due to mono- and diester formation was noted. When the reaction was examined at 510 nm[124], first a rapid increase, then a decrease of the absorbance was found. Since at this wavelength only chromium species can absorb, the intermediate could be chromium(V) or (IV). The ESR spectra of reaction mixtures showed a relatively sharp signal with a $g = 1.9805$ value corresponding to chromium(V). The fact that the relative concentrations of the intermediate determined from the spectral data agree well with the intensity of ESR signals, indicates that the same species is responsible for the both phenomena. It is then clear that the oxidation of isopropyl alcohol proceeds *via* chromium(V).

It is also interesting to note that kinetic data of Wiberg and Schäfer are incompatible with a mechanism involving the direct reaction of acetochromic acid with the alcohol, *i.e.* only the mono- and diesters should be taken into consideration as reactive intermediates.

As to the mechanism of the oxidation, the decomposition of the esters is followed by either the steps

$$H_2R + Cr(IV) \rightarrow H\dot{R} + Cr(III)$$

$$H\dot{R} + Cr(VI) \rightarrow R + Cr(V)$$

$$H_2R + Cr(V) \rightarrow R + Cr(III)$$

or, as was suggested by Watanabe and Westheimer[19],

$$Cr(IV) + Cr(VI) \rightarrow 2Cr(V)$$

$$2\{H_2R + Cr(V) = R + Cr(III)\}$$

At present there is no experimental data to differentiate between the two sets of reactions.

In the presence of manganese(II) ions the rate of oxidation of H_2R by chromic acid decreases[19]. Under favourable experimental conditions (high concentration of alcohol and low concentration of chromate) the diminution of rate is about 50 % which is in accordance with results listed in Table 6, according to which CI = 0.5. The inhibiting effect of manganese(II) on the oxidation of H_2R can be explained by reaction (23) followed by step (24). Therefore the induced oxidation of manganese(II) can be described by reactions (26), (23) and (24).

The mechanism of the induced oxidation of manganese(II) cannot be regarded as sufficiently clear. Thus, the data obtained do not make it possible to decide whether manganese(IV) is formed by interacting with the chromium(V) intermediate

TABLE 6

ISOPROPYL ALCOHOL-INDUCED OXIDATION OF MANGANESE(II) BY CHROMIC ACID

Data of Watanabe and Westheimer[19]. Conditions: 0.166 *M* isopropyl alcohol (H_2R); 0.0162 *M* chromic acid; 0.82 *M* $HClO_4$.

MnSO₄ (*mole.l⁻¹*)	*Milliequiv. chromic acid reduced*	*Milliequiv. Mn²⁺ oxidized*	F_i	F_c	$F_c \left(\frac{[H_2R]}{[Mn^{2+}]} \right)_{av.}$
0.053	0.137	0.036	0.36	3.87	12.7
0.106	0.141	0.040	0.40	6.0	9.9
0.213	0.148	0.047	0.47	23.4	15.5
0.331	0.153	0.050	0.50	∞	—
0.426	0.149	0.048	0.48	36.0	14.0
					Average = 13.3

directly, or indirectly by reacting with chromium(IV) followed by step (24). It was mentioned above that the occurrence of step (19) must be criticized because of its positive standard free-enthalpy change; on the contrary, the disproportionation of chromium(IV) according to

$$2\mathrm{Cr(IV)} \rightarrow \mathrm{Cr(V)} + \mathrm{Cr(III)} \tag{30}$$

is probable. However, it should be considered that reaction (31)

$$2\mathrm{Cr(V)} \rightarrow \mathrm{Cr(VI)} + \mathrm{Cr(IV)} \tag{31}$$

is concurrent with step (30). Therefore, if either chromium(V) or chromium (IV) were formed in the primary reactions, as the results of subsequent steps both chromium intermediates could occur.

There are also other unsolved problems. For instance, we have sufficient knowledge neither about the structures of chromium(V) and chromium(IV) species nor about the mechanism of striping of the covalently-bonded oxygen atoms from the chromate ion.

Furthermore, reaction (27) of Westheimer's scheme seems to be rather arbitrary. If chromium(II) were really formed, the rate of oxidation of alcohol would certainly be influenced by oxygen. However, careful experiments show that there is no oxygen effect at all[19].

2.1.3 Oxidation of other alcohols by chromic acid

The oxidation of several alcohols by chromic acid was studied by Chatterji *et al.*[20-25] and, as a diagnostic tool for identification of intermediates formed, the induced oxidation of manganese(II) was also investigated. However, these data furnish only qualitative information on the chemistry of induced reactions; therefore their results will not be discussed here.

Recently Mosher and Driscoll[125] have noted that the polymerization of acrylonitrile can be observed during the chromic acid oxidation of 2,2-dimethyl-1-phenyl-1-propanol. The polymerization is caused by radicals formed during the oxidation of benzaldehyde (which is one of the cleavage product of phenyl-1-butylcarbinol). The oxidation of benzaldehyde is due to the chromium(IV), most probably, or chromiun(V) intermediates.

In the oxidation of 2-propanol no polymer could be detected. However, when benzaldehyde was added, polymerization occurred to a considerable extent. These data suggest that the intermediate chromium(IV) or chromium(V) species formed in the oxidation of alcohol was responsible for the radical products and that the benzaldehyde was involved in the initiation.

2.1.4 *Oxidation of aldehydes and organic acids by chromium(VI)*

a. Benzaldehyde

The oxidation of benzaldehyde with chromic acid was investigated by Graham and Westheimer[26] using a spectrophotometric method, and the following rate equation was found

$$-\frac{d[\mathrm{Cr(VI)}]}{dt} = [\mathrm{H^+}][\mathrm{HCrO_4^-}][\mathrm{C_6H_5CHO}]\{0.147+0.95[\mathrm{H^+}]\}$$

Lucchi[49], who studied the oxidation of substituted benzaldehyde derivatives found that chlorine atoms in the *meta* and *para* position accelerate the reaction and alkyl groups retard the oxidation. A Hammett plot of Lucchi's data yields a good straight line with the slope $\rho = 1.06$. These data suggest that the reaction proceeds by way of the chromic ester of hydrated benzaldehyde as intermediate, *viz.*

$$\mathrm{C_6H_5CHO+HCrO_4^- +2H^+ \rightleftharpoons C_6H_5\overset{\displaystyle H}{\underset{\displaystyle OH}{\overset{|}{\underset{|}{C}}}}OCrO_3H_2^+} \tag{32}$$

$$\mathrm{C_6H_5\overset{\displaystyle H}{\underset{\displaystyle OH}{\overset{|}{\underset{|}{C}}}}OCrO_3H_2^+ +H_2O \rightarrow C_6H_5CO_2H+H_3O^+ +H_2CrO_3} \quad \text{(rate-determining)} \tag{33}$$

$$\mathrm{Cr(IV)+Cr(VI) \rightarrow 2Cr(V)} \tag{19}$$

$$\mathrm{C_6H_5CHO+Cr(V) = C_6H_5CO_2H+Cr(III)} \tag{34}$$

This mechanism accounts for the positive value of ρ, namely, electron-withdrawing substituents increase the acidity of the hydrogen atom. If the reaction occurred by abstraction of the hydride ion, one would expect a negative value of ρ.

Wiberg and Mill[27] investigated this reaction in acetic acid and found the rate equation

$$\mathrm{R} = k[\mathrm{RCHO}][\mathrm{HCrO_4^-}]h_0$$

where

$$h_0 = a_\mathrm{H}\frac{f_\mathrm{B}}{f_\mathrm{BH^+}}$$

The value of the second-order rate coefficient is considerably decreased by increasing the ionic strength as well as by increasing the water concentration. If the water content of the solvent is more than 25 %, the rate of reaction decreases more than the acidity function.

The rate coefficient was found to be $(5.98 \pm 0.06) \times 10^{-3}$ l.mole^{-1}.sec^{-1} at 30.01 °C and $(4.28 \pm 0.4) \times 10^{-2}$ l.mole^{-1}.sec^{-1} at 59.94 °C, indicating that the reaction has an activation energy of 13.2 ± 0.3 kcal.mole^{-1} and $\Delta S^{\ddagger} = -28$ eu.

To decide whether the reaction involves 1- or 2-electron transfers, *i.e.* chromium-(IV) or chromium(V) is formed first, the induced oxidation of manganese(II) was investigated. When sodium perchlorate was used to maintain a constant ionic strength, the rate of oxidation of benzaldehyde dropped to one-half of the original rate in the presence of manganese(II) ions. On the contrary, when magnesium perchlorate was used as the neutral salt, the rate was reduced to $\frac{1}{3}$ of its original value. This peculiar observation, however, has not been interpreted.

Knowledge of stoichiometry of the induced reaction could help to distinguish whether chromium(V) or chromium(IV) species are involved in the oxidation of benzaldehyde. Thus, the Cr(V) hypothesis predicts that for each molecule of benzaldehyde oxidized two molecules of manganese dioxide should be formed, whereas the Cr(IV) predicts that one molecule of manganese dioxide should be formed for each two molecules of benzaldehyde oxidized. Unfortunately, the attempt to determine the stoichiometry of the induced reaction failed because the oxidized manganese species was not precipitated during the reaction presumably due to formation of acetate complexes in the concentrated acetic acid solution.

b. Formaldehyde

It was found by Chatterji and Mukherjee[28] that the rate law for the oxidation of formaldehyde indicated that the chromic acid was esterified by the aldehyde hydrate formed, although they did not succeed in isolating the ester. The hypothesis of ester formation seems to be supported by the experience that the rate of reaction is increased by addition of pyridine.

Just as for the oxidation of alcohols by chromic acid, the rate of oxidation of

TABLE 7

FORMALDEHYDE-INDUCED OXIDATION OF MANGANESE(II) BY CHROMIC ACID

Data of Chatterji and Mukherjee[28]. Conditions: 9.52×10^{-3} *M* chromic acid; 0.143 *M* $MnSO_4$; 0.166 *M* $HClO_4$; initial volume, 35 cm³.

[*HCHO*] (*M*×10³)	*Equiv. of* Mn^{2+} *oxidized*	*Equiv. of HCHO oxidized*	F_i	F_c	$\frac{k_{HCHO}}{k_{Mn^{2+}}}$
8.20	0.870	5.745	0.15	0.645	26.5
6.56	0.891	4.596	0.19	0.96	18.3
4.92	0.809	3.447	0.23	1.27	21.8
3.28	0.779	2.298	0.34	3.18	13.5
1.64	0.603	1.149	0.51	—	—

aldehyde will be decreased by addition of manganese(II) and cerium(III). The limiting value of F_i is about 0.5, which points to the occurrence of chromium(IV) intermediate (Table 7).

c. Formic acid

The mechanism of the oxidation of formic acid by chromic acid[41] is far from being solved yet. The reaction in many respects reminds one of the oxidation of isopropyl alcohol. The induced oxidation of manganese(II) can also be observed during the reduction of chromium(VI) by formic acid. The stoichiometry of the induced reactions, however, cannot be given, because the oxidized product of manganese cannot be separated from the solution. The rate of oxidation of formic acid is reduced to one-third of its original value by adding manganese(II) in a sufficient quantity; thus it might be assumed that a chromium(IV) species is the active intermediate. The factor by which manganese(II) reduces the rate was explained by Wagner[137], who suggested that manganese(II) ion catalyzes the disproportionation of chromium(IV) or chromium(V) to chromate and chromic ions. The experimental facts can be explained by steps analogous to reactions (18), (19) and (20).

Cerium(III) also proved to be an effective inhibitor of the oxidation of formic acid. As the oxidation of cerium(III) to cerium(IV) is a 1-equivalent process, the inhibition furnishes additional evidence for the chromium(IV) species as intermediate.

d. Other acids

The kinetics of the initial stages of the oxidation of some α-hydroxy-carboxylic such as lactic, malic and mandelic acids by chromic acid have been studied by Bakore and Narain[126]. The initial reaction resembles the oxidation of a secondary alcohol to ketone. The authors concluded that the rate determining step involves C–H bond rupture at the α-carbon atom. The rate of oxidation of these acids is reduced to one-half by the addition of manganous ions, when the concentration of the latter is commensurable with that of the acids.

The oxidation of tartaric[127] and glycollic acid[128] by chromic acid also induces the oxidation of manganous ions. In the presence of higher concentrations of manganese(II) the rate of oxidation of the acids is diminished to about one-third of that in the absence of manganous ions. The decrease of the rate has been attributed to manganese(II) catalysis of the disproportionation of the intermediate valence states of chromium probably chromium(IV).

It was observed by Gopala Rao and Sastri[38] that the reaction between hydroquinone and chromic acid leads to the induced oxidation of oxalic acid, glycerol, lactic acid, glucose, citric acid, and malic acid. If the concentrations of the above acceptors are ten times that of that of the hydroquinone inductor, the values of F_i found are, respectively, 0.51, 0.46, 0.35, 0.27 and 0.17. The numerical values of the induction factor do not permit us to discuss the nature of coupling intermediate.

References pp. 577–580

Gopala Rao and Venkateswara Rao[37] found that the oxidation of indigo to isatin by chromic acid is accelerated by the presence of oxalic acid, and at the same time the extent of the oxidation of oxalic acid by chromic acid is increased. This observation is an example of mutual induction.

2.1.5 *Reaction between iron(II) and chromium(VI)*

Benson[31] has found that, for the reaction between iron(II) and chromate, the following rate law holds

$$-\frac{\mathrm{d[Cr(VI)]}}{\mathrm{d}t} = k\,\frac{[\mathrm{Cr(VI)}]^{1.7}[\mathrm{Fe(II)}]^2[\mathrm{H}^+]^2}{[\mathrm{Fe(III)}]}$$

The order of greater than unity with respect to chromate concentration suggests that here the active oxidizing agent is the dichromate ion. The concentration of this ion must vary as the square of the gross concentration of chromic acid, whenever that concentration is small.

The fact that the rate of reaction is inversely proportional to the concentration of iron(III) is explained by Wagner and Preiss[33] on the basis of an equilibrium between chromium(VI) and iron(II), *viz.*

$$\mathrm{Cr_2O_7^{2-} + Fe(II) \rightleftharpoons Cr_2O_7^{3-} + Fe(III)} \tag{35}$$

$$\mathrm{Cr_2O_7^{3-} + Fe(II) = 2Cr(V) + Fe(III)}\ \text{(rate determining)} \tag{36}$$

$$\mathrm{Cr(V) + Fe(II) = product} \tag{37}$$

In the presence of iodide ions the reaction of iron(II) with chromate induces the formation of iodine. The induced formation of iodine can be represented by reaction (21) and (22).

Relative rate coefficient, k_{37}/k_{21}, has been calculated by Wagner end Preiss from the expression

$$\frac{k_{37}}{k_{21}} = \frac{\frac{2}{3}c-y}{y}\cdot\frac{[\mathrm{I}^-]_{\mathrm{av}}}{[\mathrm{Fe(II)}]_{\mathrm{av}}} = \frac{\frac{2}{3}c-y}{y}\cdot\frac{b-\dfrac{y}{2}}{a-\frac{1}{2}(c-y)}$$

where a, b, and c are the initial concentrations of iron(II), iodide and chromate and x and y represent the equivalents of iodine formed and of iron(II) converted. Values of k_{21}/k_{37} for the initial stages of the reaction, at constant acidity, are satisfactorily constant. In Table 8 both the values of F_c and $F_c([\mathrm{Fe(II)}]/[\mathrm{I}^-])_0$ are listed, and it can be seen that they agree satisfactorily with the k_{21}/k_{37} ratio.

The induced iodine formation can be formally explained by the following reac-

TABLE 8

IRON(II)-INDUCED OXIDATION OF IODIDE BY DICHROMATE

Data of Wagner and Preiss[33]. Conditions: 0.95 M KCl; 1.91×10^{-2} M HCl; 1.91×10^{-2} M KI; 1.91×10^{-3} M $FeCl_3$; 1.58×10^{-4} M $K_2Cr_2O_7$.

$\left(\frac{[I^-]}{[Fe(II)]}\right)_0$	F_i	F_c	$\frac{k_{21}}{k_{37}}$	$F_c\left(\frac{[Fe(II)]}{[I^-]}\right)_0$
1.00	1.299	5.557	5.780	5.557
0.50	0.852	2.225	4.694	4.451
0.33	0.653	1.454	4.587	4.361
0.25	0.534	1.092	4.587	4.377
0.20	0.465	0.909	4.761	4.546
0.166	0.401	0.752	4.716	4.532
0.133	0.356	0.649	5.076	4.883
0.10	0.282	0.492	5.102	4.924

tions

$$\text{Cr(V)}+\text{Fe(III)} = \text{Cr(III)}+\text{Fe(IV)} \tag{38}$$

$$\text{Fe(IV)}+\text{Fe(II)} = 2\,\text{Fe(III)} \tag{39}$$

However, there is no experimental evidence for an iron(IV) intermediate.

2.1.6 *Oxidation of vanadium(II) and vanadium(IV) by chromium(VI)*

Luther and Rutter[15] have observed the induced oxidation of iodide during the reactions between chromic acid and vanadium(IV), vanadium(III), and vanadium(II) ions. In all the three systems CI = 2, therefore it is probable that the coupling intermediates are chromium(V) species, these being, especially the two latter systems, too complicated for a detailed kinetic treatment to be given.

According to spectrophotometric studies of Espenson[50], the oxidation of vanadium(IV) by chromic acid follows the rate law

$$-\frac{\mathrm{d}[\text{Cr(VI)}]}{\mathrm{d}t} = k[\text{VO}^{2+}]^2[\text{HCrO}_4^-]/[\text{VO}_2^+]$$

The mechanism of the reaction is

$$\text{V(IV)}+\text{Cr(VI)} \rightleftharpoons \text{V(V)}+\text{Cr(V)} \tag{40}$$

$$\text{Cr(V)}+\text{V(IV)} \rightleftharpoons \text{V(V)}+\text{Cr(IV)} \quad \text{(rate determining)} \tag{41}$$

$$\text{Cr(IV)}+\text{V(IV)} \rightarrow \text{Cr(III)}+\text{V(V)} \tag{42}$$

TABLE 9

VANADIUM(IV)-INDUCED OXIDATION OF IODIDE BY DICHROMATE

Data of Luther and Rutter[15]. Conditions: 9.16×10^{-4} M $K_2Cr_2O_7$; 9.7×10^{-4} M $VOSO_4$; 7.0×10^{-4} M H_2SO_4.

$\left(\text{Initial ratio } \frac{[I^-]}{[V(IV)]_0}\right)$	F_i †	F_c	$F_c \left(\frac{[(V(IV)]}{[I^-]}\right)_0$
4.11	1.875	45.00	10.948
20.57	1.945	106.09	5.157
41.15	1.962	154.89	3.764
82.30	1.993	854.14	10.378

† Corrected by extrapolation of original plots.

TABLE 10

VANADIUM(II)-INDUCED OXIDATION OF IODIDE BY DICHROMATE

Data of Luther and Rutter[15]. Conditions: 9.16×10^{-4} M $K_2Cr_2O_7$; 4.4×10^{-4} M VSO_4; 7.0×10^{-4} M H_2SO_4.

Initial ratio $\left(\frac{[I^-]}{[V(II)]}\right)_0$	F_i	F_c	$F_c \left(\frac{[V(II)]}{[I^-]}\right)_0$
4.50	1.13†	3.896	0.8657
8.70	1.375†	6.600	0.7586
19.80	1.41†	7.169	0.3620
39.30	1.47†	8.320	0.2117
91.2	1.61	12.384	0.1357
182.4	1.68	15.75	0.0863
459.0	1.87	43.153	0.0940
903.0	1.96	147.0	0.162
1530.0	1.99	597.0	0.3901

† Corrected by extrapolation of original plots.

The value of $k = k_{40}k_{41}/k_{-40}$ is 0.62 ± 0.06 l.mole^{-1}.sec^{-1} at 25 °C for a solution containing 1 M lithium perchlorate and 0.5 M perchloric acid.

Induced oxidation of iodide caused by vanadium(IV) and vanadium(II) presumably involves steps analogous to those in the iron(II)–chromium(VI)–iodide system. Some data obtained by Luther and Rutter[15] are summarized in Tables 9 and 10.

2.1.7 Chromium(VI) as inductor in the induced oxidation of arsenic(III) by molecular oxygen

This reaction was observed first by Kessler[1] and investigated by Kolthoff and Fineman[39]. In the absence of chromate, arsenic(III) is hardly oxidizable (< 1.7 %)

in alkaline medium. The values of F_i (= equivalents of O_2 reduced/equivalents of chromium(VI) reduced) obtained are independent of the concentration of arsenic(III), while inversely proportional to that of chromate. Under favourable experimental conditions the limiting value of F_i is of 1.3 ± 0.2.

By choosing the smallest O_2/Cr(VI) ratio at which the closest agreement with the limiting value of F_i found experimentally is obtained, the following stoichiometric equation can be written

$$7\,As(III) + 2\,Cr(VI) + 2\,O_2 = 7\,As(V) + 2\,Cr(III) + 10(OH^-)$$

The relation between the average F_i observed and the ratio of the average chromate ion concentration to the average pressure of oxygen is

$$1/F_i = A + B\frac{[Cr(VI)]}{p_{O_2}}$$

For the induced reaction Kolthoff and Fineman[39] suggested a mechanism similar to that proposed for isopropyl alcohol, *viz.*

$$Cr(VI) + As(III) = Cr(IV) + As(V)$$

$$Cr(IV) + As(III) = Cr(II) + As(V)$$

$$Cr(II) + Cr(VI) = Cr(V) + Cr(III)$$

$$Cr(V) + As(III) = Cr(III) + As(V)$$

Induced reduction of oxygen can be interpreted by the steps

$$Cr(II) + O_2 = Cr(II) \cdot O_2$$

$$2\,CrO_2 = (CrO_2)_2$$

$$(CrO_2)_2 + 2\,As(III) + 4\,H^+ = 2\,As(V) + 2\,Cr(III) + 2\,OH^- + H_2O_2$$

$$H_2O_2 + As(III) = As(V) + 2\,OH^-$$

According to this scheme the Cr(II) species is responsible for the induced reduction of oxygen, though its occurrence cannot be supported by independent experimental evidence.

Abel[51] has assumed that the reaction between arsenite and molecular oxygen is catalyzed by a chromium intermediate. He suggested that chromium(IV) is converted by oxygen into chromium(VI) which causes the excess oxidation of arsenic(III). However, this mechanism is also devoid of experimental support.

We are of the opinion that the reaction can be explained without assuming the reduction of chromium(IV) to chromium(II) by arsenic(III), although this step is not improbable ($E^\circ_{4/2}$(Cr) $\approx$ 0.7 V; $E^\circ_{5/3}$(As) = 0.56 V). As molecular oxygen is easily reduced by arsenic(IV), the following sequence of reactions is proposed

$$\mathrm{Cr(VI)+As(III) \rightarrow Cr(V)+As(IV)}$$

$$\mathrm{Cr(V)+As(III) \rightarrow Cr(III)+As(V)}$$

$$\mathrm{As(IV)+O_2 \rightarrow As(V)+O_2^-}$$

$$\mathrm{O_2^- +As(III) \rightarrow As(IV)+O_2^{2-}}$$

$$\mathrm{O_2^{2-} +As(III) \rightarrow 2O^{2-}+As(V)}$$

$$\mathrm{Cr(VI)+3\tfrac{1}{2}As(III)+O_2 = Cr(III)+3\tfrac{1}{2}As(V)+2O^{2-}}$$

2.1.8 Properties of the chromium(V) and chromium(IV) intermediates

In the course of reactions of chromium(VI) with different substrates (inductors), depending on whether the inductor is 1- or 2-equivalent reagent, in the primary reaction formation of Cr(V) and Cr(IV), respectively, was assumed. However, no definite statement can be made as to whether the chromium species formed in the primary reaction or another chromium entity produced in a secondary step reacts with the acceptor. The possibility of simultaneous formation of both chromium species in the primary reaction can be excluded.

The reactivity of chromium(V) and chromium(IV) species is uncertain since there are no reliable thermodynamic data, and not much can be said at present about the structure of these species. With respect to the latter some hints can be obtained from the fact that the changeover from chromium(V) to chromium(IV) or *vice versa* in all cases was found to be rate determining, which seems to correlate well with the conclusion of Tong and King[136] that Cr(VI) and Cr(V) have coordination number four, whereas Cr(IV) and Cr(III) have six.

Attempts of Sanko and Stefanovskii[52] to make direct measurements of the oxidation potentials of couples involving these chromium species were not successful. An estimate of the potential of the Cr(VI)/Cr(V) couple was given by Westheimer[41], based on the equilibrium between chromium(VI) and iron(II), *viz.*

$$\mathrm{Cr(VI)+Fe(II) \rightleftharpoons Fe(III)+Cr(V)}$$

Since the reaction is markedly displaced to the left, the standard potential of Fe(III)/Fe(II) couple (0.77 V) must exceed that of the Cr(VI)/Cr(V) couple; therefore $E^\circ_{6/5}$ must be $\leqslant$ 0.6 V. This estimate seems reasonable. Regarding the oxidation potential of chromium(IV) there is an estimate, also by Westheimer, that the Cr(IV)/Cr(III) couple has a value of $E^\circ_{4/3} \geqslant$ 1.5 V. This was based on the fact that manganese(II) can rapidly be converted into MnO_2 by chromium(IV) species. However, it is not known by how much this potential exceeds 1.5 V.

A slightly better estimate can be given based on the arsenite-induced oxidation of cobalt(II) by chromic acid in sulphuric acid medium[53]. According to Johnson and Sharpe[54] the potential of the Co(III)/Co(II) couple in the absence of strong complex forming agents is about 1.95 V. As cobalt(II) can readily be oxidized by chromium(IV), it can be assumed that $E^{\circ}_{4/3} \geqslant 2.0$ V. By using these estimated values the potential diagram[55] of chromium is constructed (Fig. 2). Data obtained for single couples are only of semi-quantitative value because of the roughness of the estimations and the omission of the effect of hydrogen ions. This may cause a few tenths of a volt uncertainty in the potentials.

In spite of the uncertainty in the values of the potentials we are of the opinion that they enable the mechanism of the induced reactions involving chromium to be clarified.

For instance, it was assumed by Westheimer that, in the As(III)–I^-–Cr(VI)

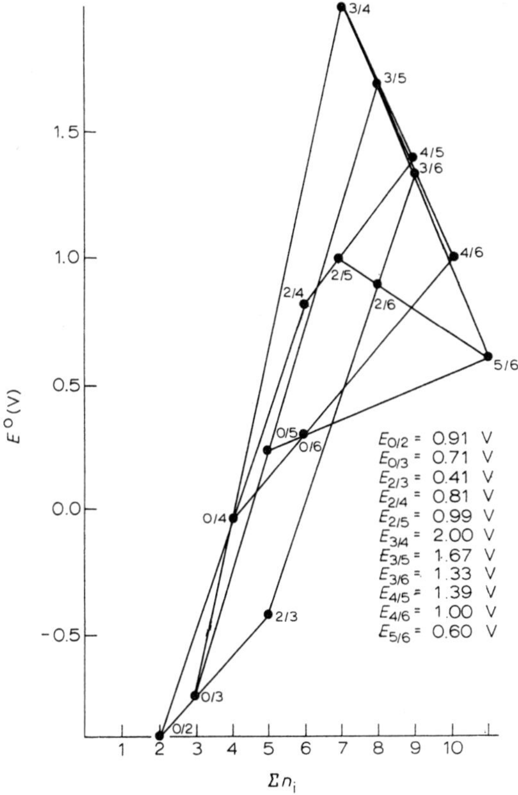

Fig. 2. Potential diagram of the redox couples of chromium. According to Csányi[55].

system, reaction (19) takes place rapidly

$$Cr(IV)+Cr(VI) = 2Cr(V) \tag{19}$$

However, this reaction is not possible thermodynamically[56], because $E^{\circ}_{6/5} \leqslant 0.6$ and $E^{\circ}_{5/4} \approx 1.4$ V; on the contrary, the reverse reaction, *i.e.* the disproportionation of chromium(V), is favoured. This and other objections make it necessary to modify the reaction scheme for this system. It is recommended that steps (18)–(24) be replaced by

Primary reactions

$$Cr(VI)+As(III) \rightleftharpoons Cr(VI).As(III)$$

$$Cr(VI).As(III)+Cr(VI) \rightarrow 2Cr(V)+As(V)$$

$$2Cr(V) \rightarrow Cr(VI)+Cr(IV)$$

$$Cr(V)+As(III) \rightarrow Cr(III)+As(V)$$

$$Cr(IV)+As(III) \rightarrow Cr(III)+As(IV)$$

$$Cr(VI)+As(IV) \rightarrow Cr(V)+As(V)$$

Induced reactions

$$Cr(V)+Ac_{red} \rightarrow Cr(III)+Ac_{ox}$$

$$Cr(IV)+Ac'_{red} \rightarrow Cr(III)+Ac'_{ox}$$

This scheme accounts for both the induced oxidation of iodide (where CI = 2) and that of manganese(II) (where CI = 0.5) without including step (19). Furthermore it can be seen that in the presence of iodide the rate of disappearence of chromate will not be altered, whilst the rate of oxidation of arsenic(III) will be reduced to one-third of its original value, as found experimentally.

It should be mentioned that problems of chromic acid oxidations are discussed in detail in recent excellent reviews by Wiberg[129] and Stewart[130].

2.2 INDUCED REACTIONS CAUSED BY ARSENIC(IV) INTERMEDIATES

2.2.1 *Iron(II)–arsenic(III)–peroxydisulphate system*

In the absence of oxygen

The reaction between arsenic(III) and peroxydisulphate is very slow, but is markedly accelerated in the presence of iron(II) as inductor. According to Kolthoff *et al.*[57] the reduction of peroxydisulphate in the presence of arsenic(III) and iron(II) is first order with respect to peroxydisulphate, the apparent rate coefficient being proportional to the iron(II) concentration. Thus

$$-\frac{d[S_2O_8^{2-}]}{dt} = k[S_2O_8^{2-}] = k'[\text{Fe(II)}][S_2O_8^{2-}]$$

$k' = 30$ l.mole^{-1}.sec^{-1} at 25 °C. The induction factor, F_i (= equivalents of arsenic(III) oxidized/equivalents of iron(II) oxidized), depends considerably on the experimental conditions.

(*a*) F_i decreases on increasing the acid concentration.

(*b*) Addition of iron(III) ions results in the increase of F_i to infinity. Copper(II) ions have a similar „catalytic" effect but, their activity does not depend on the acid concentration of the solution.

(*c*) Increasing the iron(II) concentration decreases F_i.

(*d*) Fluoride was found to reduce the induction factor to values approaching zero. However, the rate of the iron(II)–peroxydisulphate reaction is not influenced by fluoride ions.

(*e*) The value of F_i is independent of the arsenic(III) concentration, when $[\text{As(III)}]/\text{Fe(II)}] \geqslant 2$.

The results can be explained by the following mechanism

$$\text{Fe(II)} + S_2O_8^{2-} \rightarrow \text{Fe(III)} + SO_4^{2-} + SO_4^{-} \quad (43)$$

$$\text{Fe(II)} + SO_4^{-} \rightarrow \text{Fe(III)} + SO_4^{2-} \quad (44)$$

$$\text{As(III)} + SO_4^{-} \rightarrow \text{As(IV)} + SO_4^{2-} \quad (45)$$

$$\text{As(IV)} + \text{Fe(III)} \rightarrow \text{As(V)} + \text{Fe(II)} \quad (46)$$

$$\text{As(IV)} + \text{FeOH}^{2+} \rightarrow \text{As(V)} + \text{Fe(II)} \quad (46')$$

$$\text{As(IV)} + \text{Fe}^{3+} \rightarrow \text{As(V)} + \text{Fe}^{2+} \quad (46'')$$

$$\text{As(IV)} + \text{Fe(II)} \rightarrow \text{As(III)} + \text{Fe(III)} \quad (47)$$

According to this mechanism only iron(II) reacts with peroxydisulphate and gives an SO_4^{-} radical. With respect to reactions of the SO_4^{-} radical, reaction (44) becomes insignificant compared with (45) at [As(III)]/[Fe(II)] ratios greater than two, and then F_i will no longer depend on the arsenic(III) concentration.

The relative rate coefficient, k_{46}/k_{47}, can be determined by assuming a steady-state with respect to arsenic(IV). According to the data given in Table 11 the value of k_{46}/k_{47} depends on the ionic strength and the pH. Dependence on the pH can be explained by the fact that arsenic(IV) reacts with FeOH^{2+} ions, present at low acid concentrations, 90 times faster than with Fe^{3+} ions.

TABLE 11

VARIATION OF INDUCTION FACTOR WITH VARYING ACIDITY AND IONIC STRENGTH FOR THE IRON(II)–ARSENIC(III)–PEROXYDISULPHATE SYSTEM

Data of Woods *et al.*[57]. Conditions: $[Fe^{2+}] = 4.56 \times 10^{-5}$ *M*; $[As(III)] = 10^{-2}$ *M*; $[S_2O_8] = 2.5 \times 10^{-5}$ *M*.

$[HClO_4]$ (M)	μ	F_i	k_{46}/k_{47}
0.002	0.50	1.32	3.8
0.002	0.002	1.54	5.1
0.005	0.50	1.07	2.6
0.010	0.50	0.88	1.8
0.010	0.01	1.23	3.3
0.10	0.50	0.32	0.27
0.50	0.50	0.23	0.14

The effect of copper(II) is essentially similar to that of iron(III). In reaction (48) copper(I) is formed which results either indirectly by reaction (49) or directly by reaction (50) in the reduction of a peroxydisulphate ion

$$As(IV)+Cu(II) \rightarrow As(V)+Cu(I) \tag{48}$$

$$Cu(I)+Fe(III) \rightarrow Cu(II)+Fe(II) \tag{49}$$

$$Cu(I)+S_2O_8^{2-} \rightarrow Cu(II)+SO_4^{2-}+SO_4^{-} \tag{50}$$

In the presence of oxygen

The induced reaction occurring in the system iron(II)–arsenic(III)–peroxydisulphate is greatly influenced by the presence of oxygen[59]. While in the absence of oxygen the sum of the equivalents of iron(II) and arsenic(III) oxidized is equal to the equivalents of peroxydisulphate reduced, in the presence of oxygen $\frac{1}{2}X_{Fe}+X_{As} > 1$; where X_{Fe} = moles of iron(II) oxidized/moles of peroxydisulphate reduced, and X_{As} = moles of arsenic(III) oxidized/moles of peroxydisulphate reduced; under favourable conditions this sum may be as large as 18. This indicates that the iron(II)–peroxydisulphate reaction induces the oxidation of arsenic(III) by molecular oxygen. In the chain reaction in the presence of oxygen, if the chain is long the stoichiometry approximates to

$$2Fe(II)+As(III)+O_2 = 2Fe(III)+As(V) \tag{A}$$

corresponding to a ratio $X_{Fe}/X_{As} = 2$.

Both iron(III) and copper(II) inhibit the induced chain oxidation and, when present in sufficient quantity, $\frac{1}{2}X_{Fe}+X_{As}$ becomes 1.0, the effect of oxygen is completely eliminated and only the induced oxidation of arsenic(III) by peroxydisulphate occurs, *viz.*

$$As(III)+S_2O_8^{2-} = As(V)+2SO_4^{2-} \quad (B)$$

The initiation step in the chain oxidation, reaction (43), is not affected by the presence of oxygen. SO_4^- radicals formed in (43) give arsenic(IV), reaction (45), initiating the following propagation cycle which leads to the reformation of As(IV)

$$As(IV)+O_2+H^+ \rightarrow As(V)+HO_2 \quad (51)$$

$$Fe(II)+HO_2+H^+ \rightarrow Fe(III)+H_2O_2 \quad (52)$$

$$Fe(II)+H_2O_2 \rightarrow Fe(III)+OH^-+OH \quad (53)$$

$$As(III)+OH \rightarrow As(IV)+OH^- \quad (54)$$

The chain termination steps are (55) and (56)

$$Fe(III)+HO_2 \rightarrow Fe(II)+O_2+H^+ \quad (55)$$

$$Cu(II)+HO_2 \rightarrow Cu(I)+O_2+H^+ \quad (56)$$

followed by either

$$Cu(I)+Fe(III) \rightarrow Cu(II)+Fe(II) \quad (49)$$

or

$$Cu(I)+S_2O_8^{2-} \rightarrow Cu(II)+SO_4^{2-}+SO_4^- \quad (50)$$

If the initiation and termination steps are excluded because the chain is long, the sum of the propagation reactions gives the stoichiometric relationship (A).

At sufficiently high concentration of iron(III) and copper(II), the induced oxidation by oxygen is eliminated because all the HO_2 radicals are oxidized by steps (55) or (56). In such cases F_i approaches infinity and X_{As} becomes equal to 1, *i.e.* arsenic(III) is oxidized according to equation B. Consequently, iron(II) is reformed at the same rate as it is oxidized.

The extent of induced oxidation increases with increasing hydrogen ion concentration. This is a consequence of the fact that both iron(II) and iron(III) are present in the solution and compete for HO_2 radicals, reactions (52) and (55). The rate of reaction (52) increases as the acidity is increased, whereas that of (55) is independent of hydrogen ion concentration.

The extent of the oxygen-induced oxidation passes through a maximum at a certain [As(III)]/[Fe(II)] ratio. When the arsenic(III) concentration is low with respect to that of iron, the induction is decreased because of the enhanced role of

References pp. 577–580

the chain-breaking reaction

$$Fe(II)+OH \rightarrow Fe(III)+OH^- \quad (57)$$

Under these conditions the ratio, X_{Fe}/X_{As}, becomes significantly greater than 2. When [As(III)] > [Fe(II)], there is a competition between arsenic(III) and iron(II) for hydrogen peroxide and the reaction

$$As(III)+H_2O_2 = As(V)+H_2O \quad (58)$$

also results in chain termination. In this case X_{Fe}/X_{As} becomes smaller than 2.

Photoreduction of iron(III) in the absence of oxygen may induce the iron(III)–arsenic(III) reaction. The initiating steps (59) and (60)

$$\text{primary step } Fe^{3+}.OH^- \xrightarrow{h\nu} Fe^{2+}OH \quad (59)$$

$$\text{back reaction } Fe^{2+}OH \rightarrow Fe^{3+}.OH^- \quad (59')$$

$$\text{secondary step } Fe^{2+}OH \rightarrow Fe^{2+}+OH \quad (60)$$

$$\text{back reaction } Fe^{2+}+OH \rightarrow Fe^{2+}OH \quad (60')$$

are followed by reactions (54), (46) and (47).

In the presence of oxygen the arsenic(IV) formed is oxidized by the step (51) and reactions (52), (53), (54) and (58) take place as in the reaction initiated by the iron(II)–peroxydisulphate reaction.

2.2.2 Iron(II)–hydrogen peroxide–arsenic(III) system

The reaction between arsenic(III) and hydrogen peroxide is rather slow; at 25 °C $k = 10^{-2}$ l.mole^{-1}.sec^{-1}. However, it is induced by the fast reaction between iron(II) and hydrogen peroxide (k = 53 l.mole^{-1}.sec^{-1})[60].

The stoichiometry of the induced reaction depends, as in the Fe(II)–$S_2O_8^{2-}$ system, on the iron(II)/iron(III) ratio and on the pH. Therefore, it can be expected that under identical experimental conditions (actor, inductor, and hydrogen ion concentration) the induction factors for the two systems should be identical. The data obtained show that this expectation is fulfilled. For the photo-induced oxidation of arsenic(III) the value of k_{54}/k_{57} was found to be 2, while in the present system $k_{54}/k_{57} = 4$. (Comparing these values with the value of $k_{45}/k_{44} = 21$, it can be concluded that the SO_4^- radical, formed by reaction (43), is not removed by the reaction

$$SO_4^- + H_2O \rightarrow OH + HSO_4^-$$

to any appreciable extent.)

In the absence of oxygen the mechanism of the induced oxidation of arsenic(III) involves the steps

$$\mathrm{Fe(II)+H_2O_2 \rightarrow Fe(III)+OH^- + OH} \quad (53)$$

$$\mathrm{As(III)+OH \rightarrow As(IV)+OH^-} \quad (54)$$

$$\mathrm{As(IV)+Fe^{3+} \rightarrow As(V)+Fe^{2+}} \quad (46'')$$

$$\mathrm{As(IV)+FeOH^{2+} \rightarrow As(V)+Fe(II)} \quad (46')$$

$$\mathrm{As(IV)+Fe(II) \rightarrow As(III)+Fe(III)} \quad (47)$$

In the presence of oxygen the chain-oxidation of arsenic(III) consists of

$$\mathrm{Fe(II)+H_2O_2 \rightarrow Fe(III)+OH^- + OH} \quad (53)$$

$$\mathrm{As(III)+OH \rightarrow As(IV)+OH^-} \quad (54)$$

$$\mathrm{As(IV)+O_2+H^+ \rightarrow As(V)+HO_2} \quad (51)$$

$$\mathrm{Fe(II)+HO_2+H^+ \rightarrow Fe(III)+H_2O_2} \quad (52)$$

together with the termination reactions

$$\mathrm{Fe(III)+HO_2 \rightarrow Fe(II)+O_2+H^+} \quad (55)$$

$$\mathrm{Fe(II)+OH \rightarrow Fe(III)+OH^-} \quad (57)$$

$$\mathrm{As(III)+H_2O_2 = As(V)+H_2O} \quad (58)$$

2.2.3 Arsenic(III)–peroxydisulphate reaction catalyzed by iron(III) and copper(II)

The slow reaction between arsenic(III) and peroxydisulphate is first order with respect to peroxydisulphate[61]. In the presence of oxygen the rate of reaction increases. Values of the first order rate coefficient

$$k = -\frac{\mathrm{d} \ln [\mathrm{S_2O_8^{2-}}]}{\mathrm{d}t}$$

are listed in Table 12. It has been observed that the extent of the oxygen effect was reduced by the addition of allyl acetate and acrylonitrile.

The rate of reaction between arsenic(III) and peroxydisulphate is markedly increased by iron(III) ions. In the absence of air

$$k = 7.6 \times 10^{-5} \frac{[\mathrm{Fe(III)}]^{\frac{1}{2}}}{([\mathrm{H^+}]+K_a)^{\frac{1}{2}}} \ \mathrm{sec}^{-1}$$

TABLE 12

RATE OF REDUCTION OF PEROXYDISULPHATE

Data of Woods *et al.*[61]. Conditions: 10^{-3} *M* $S_2O_8^{2-}$; 10^{-2} *M* $HClO_4$; $\mu = 0.1$ ($NaClO_4$).

[*As(III)*] (*M*)	*k*(*sec*$^{-1}$)$\times 10^7$
Absence of oxygen	
0.0	0.9
10^{-3}	3.0
10^{-2}	6.0
2×10^{-2}	6.7
Air-saturated	
0.0	0.9
10^{-3}	35
4×10^{-3}	35
10^{-2}	41, 35, 40
2×10^{-2}	46

In the presence of air

$$k = 4.3\times10^{-5}\frac{[\mathrm{Fe(III)}]^{\frac{1}{2}}}{([\mathrm{H}^+]+K_a)^{\frac{1}{2}}}\ \mathrm{sec}^{-1}$$

where [Fe(III)] is the gross concentration of iron(III) and $K_a = [\mathrm{FeOH}^{2+}][\mathrm{H}^+]/[\mathrm{Fe}^{3+}]$ which has the value[62] of 2.8×10^{-3} mole.l^{-1}, at $\mu = 0.1$.

The effectiveness of copper(II) catalyst can be shown by the following values of the rate coefficient. In the absence of air

$$k = 4.2\times10^{-3}[\mathrm{Cu(II)}]^{\frac{1}{2}}\mathrm{sec}^{-1}$$

In the presence of air

$$k = 2.8\times10^{-2}[\mathrm{Cu(II)}]^{\frac{1}{2}}\mathrm{sec}^{-1}$$

The thermal decomposition of peroxydisulphate proceeds by two parallel paths[63], *viz.*

$$S_2O_8^{2-} \rightarrow 2SO_4^- \tag{61}$$

$$H^+ + S_2O_8^{2-} \rightarrow SO_4 + HSO_4^- \tag{62}$$

$k = k_{61}+k_{62}\ [\mathrm{H}^+]$, and $k_{61} = 1.3\times10^{-8}$ sec^{-1} at 25 °C the activation energy being 33.5 kcal.mole^{-1}. The overall rate coefficient at 25 °C was found to be $k = 9\times10^{-8}$ sec^{-1} at a hydrogen ion concentration of 0.01 *M* and an ionic strength of 0.1, giving 7.7×10^{-6} l.mole^{-1}.sec^{-1} for k_{62} under the same conditions.

In the absence of oxygen the addition of arsenic(III) has only a slight effect on the rate of reduction of peroxydisulphate. In the presence of air the rate of reduction of persulphate increases nearly fortyfold (Table 12). The oxidation of arsenic(III) by SO_4 from reaction (62) is not a chain process, thus it need not be considered in the iron(III)- and copper(II)-catalyzed reaction between peroxydisulphate and arsenic(III).

In the presence of air the chain reaction consisting of steps (61), (45), (51) and (63) occurs.

$$S_2O_8^{2-} \rightarrow 2\, SO_4^- \tag{61}$$

$$As(III) + SO_4^- \rightarrow As(IV) + SO_4^{2-} \tag{45}$$

$$As(IV) + O_2 + H^+ \rightarrow As(V) + HO_2 \tag{51}$$

$$HO_2 + S_2O_8^{2-} \rightarrow H^+ + O_2 + SO_4^{2+} + SO_4^- \tag{63}$$

If termination occurred by either of reactions (64) and (65)

$$SO_4^- + HO_2 \rightarrow SO_4^{2-} + O_2 + H^+ \tag{64}$$

$$2\, SO_4^- \rightarrow S_2O_8^{2-} \tag{65}$$

the rate of reaction would not be independent of the concentration arsenic(III), contrary to experiment. However, if termination occurred by the reaction,

$$As(IV) + HO_2 \rightarrow As(III) + O_2 + H^+ \tag{66}$$

the overall process would be of first order in peroxydisulphate and zero order in arsenic(III) as found experimentally, and half order in oxygen.

Catalysis by iron(III)

In the presence of iron(III) ions and in the absence of oxygen, reaction (61) is followed by steps (45), (46) or (46′) and by reaction (43) with chain termination by reaction (47), *viz.*

$$S_2O_8^{2-} \rightarrow 2\, SO_4^- \tag{61}$$

$$As(III) + SO_4^- \rightarrow As(IV) + SO_4^{2-} \tag{45}$$

$$As(IV) + O_2 + H^+ \rightarrow As(V) + HO_2 \tag{51}$$

$$As(IV) + Fe(III) \rightarrow As(V) + Fe(II) \tag{46}$$

$$As(IV) + FeOH^{2+} \rightarrow As(V) + Fe(II) \tag{46'}$$

$$Fe(II) + S_2O_8^{2-} \rightarrow Fe(III) + SO_4^{2-} + SO_4^- \tag{43}$$

$$As(IV) + Fe(II) \rightarrow As(III) + Fe(III) \tag{47}$$

In the presence of oxygen, reactions (46) and (46′) will be replaced in the chain propagation cycle by reactions (63) and (55); consequently, there will be no change in the overall stoichiometry of the reaction. The chain reactions will be terminated by steps (52), (53), (54) and (58)

$$HO_2 + S_2O_8{}^{2-} \rightarrow H^+ + O_2 + SO_4^{2-} + SO_4^- \quad (63)$$

$$Fe(III) + HO_2 \rightarrow Fe(II) + O_2 + H^+ \quad (55)$$

$$Fe(II) + HO_2 + H^+ \rightarrow Fe(III) + H_2O_2 \quad (52)$$

$$Fe(II) + H_2O_2 \rightarrow Fe(III) + OH^- + OH \quad (53)$$

$$As(III) + OH \rightarrow As(IV) + OH^- \quad (54)$$

$$As(III) + H_2O_2 = As(V) + H_2O \quad (58)$$

Assuming a steady state for the chain carriers, in the absence of oxygen we obtain

$$-\frac{d[S_2O_8^{2-}]}{dt} = \left\{k_{61}\,k_{43}\,\frac{k_{46'}}{k_{47}}\,K_a\right\}^{\frac{1}{2}} \cdot \frac{[Fe(III)]^{\frac{1}{2}}}{([H]^+ + K_a)^{\frac{1}{2}}}\,[S_2O_8^{2-}]$$

while in the presence of air

$$-\frac{d[S_2O_8^{2-}]}{dt} = \frac{k_{61}\,k_{43}\,k_{55}([H^+]+K_a)^{\frac{1}{2}}}{k_{52}} \cdot \frac{[Fe(III)]^{\frac{1}{2}}}{([H^+]+K_a)^{\frac{1}{2}}}\,[S_2O_8^{2-}]$$

Catalysis by copper(II)

In the presence of copper(II) ions and in the absence of oxygen, steps (48), (50) and (67)

$$As(IV) + Cu(II) \rightarrow As(V) + Cu(I) \quad (48)$$

$$Cu(I) + S_2O_8^{2-} \rightarrow Cu(II) + SO_4^{2-} + SO_4^- \quad (50)$$

$$As(IV) + Cu(I) \rightarrow As(III) + Cu(II) \quad (67)$$

will replace reactions (46), (43) and (47), respectively. By a steady-state treatment we obtain

$$-\frac{d[S_2O_8^{2-}]}{dt} = (k_{61}\,k_{50}\,k_{48}/k_{67})^{\frac{1}{2}}[Cu(II)]^{\frac{1}{2}}[S_2O_8^{2-}]$$

In the presence of air, reactions (55), (43), (52) and (53) are replaced by reactions (56), (50), (68) and (69), *viz.*

$$Cu(II)+HO_2 \rightarrow Cu(I)+O_2+H^+ \quad (56)$$

$$Cu(I)+S_2O_8^{2-} \rightarrow Cu(II)+SO_4^{2-}+SO_4^- \quad (50)$$

$$Cu(I)+HO_2+H^+ \rightarrow Cu(II)+H_2O_2 \quad (68)$$

$$Cu(I)+H_2O_2 \rightarrow Cu(II)+OH^-+OH \quad (69)$$

The rate equation is

$$-\frac{d[S_2O_8^{2-}]}{dt} = (k_{61}k_{50}k_{56}/k_{68})^{\frac{1}{2}}[Cu(II)]^{\frac{1}{2}}[S_2O_8^{2-}]$$

At high copper(II) concentrations the rate of reduction of peroxydisulphate becomes independent of copper(II) concentration and is unaffected by the presence of oxygen. This can be explained by the occurrence of the termination reaction

$$Cu(I)+SO_4^- \rightarrow Cu(II)+SO_4^{2-} \quad (70)$$

As the concentration of copper(II) is increased the steady-state concentration of copper(I) is also increased, and the SO_4^- radical reacts with copper(I) rather than with arsenic(IV). In such a case the rate of reduction of peroxydisulphate, taking $k_{50}[Cu(I)]/k_{61} \gg 1$, can be given by

$$-\frac{d[S_2O_8^{2-}]}{dt} = (k_{61}k_{50}k_{45}/k_{70})^{\frac{1}{2}}[As(III)]^{\frac{1}{2}}[S_2O_8^{2-}]$$

To complete the chemistry of arsenic(IV), and of the induced reactions involving this intermediate, we will mention some further results.

2.2.4 Polarographic behaviour of the system containing peroxydisulphate, arsenic(III) and copper(II)

Copper(II) ions in the presence of chloride ions are reduced at the dropping mercury electrode (DME) in two steps, Cu(II) → Cu(I) and Cu(I) → Cu(O) producing a double wave at +0.04 and 0.22 V *versus* SCE half-wave potentials. In the presence of peroxydisulphate[64], when the chloride concentration is large enough, two waves are also observed; the first limiting current corresponds to the reduction of the Cu(II) to Cu(I) *plus* reduction of a fraction of peroxydisulphate and the total diffusion current at a more negative potential is equal to the sum of the diffusion currents of reduction of Cu(II) to Cu(O) and of the peroxydisulphate. There is evidence that peroxydisulphate is not reduced at the potential of the first wave because of the adsorption of the copper(I) chloride complex at the electrode surface. In the presence of thiocyanate a similar effect was observed. An adsorbed layer of copper(I) complex at the DME also prevents the electro-reduction of hydrogen peroxide[65].

The kinetic current, i_k, of peroxydisulphate exalting the Cu(II) to Cu(I) wave is caused by a regenerative reaction consisting of the following steps

$$Cu(II)+e \rightarrow Cu(I) \tag{71}$$

$$Cu(I)+S_2O_8^{2-} \rightarrow Cu(II)+SO_4^{2-}+SO_4^{-} \tag{50}$$

$$Cu(I)+SO_4^{-} \rightarrow Cu(II)+SO_4^{2-} \tag{70}$$

$$SO_4^{-}+e \rightarrow SO_4^{2-} \tag{72}$$

where (50) is the rate-determining step. Using the relationship derived by Koutecky[66] for a bimolecular regenerative process, we can write

$$\frac{i}{i_d} = 0.812(\alpha k_{50}\, t[S_2O_8^{2-}])^{\frac{1}{2}}\gamma H(u_i)$$

where i_d is the first diffusion current of copper(II), $i = i_d+i_k$ is the current found experimentally, α is the stoichiometric factor which is 2 in this case, t is the drop time,

$$\gamma = \left(1+\frac{2D_1[Cu(II)]}{3\alpha D_2[S_2O_8^{2-}]}\right)^{\frac{1}{2}}, \quad u_i = \frac{\left(\dfrac{k_{50}\, TD_1[Cu(II)]^2}{\alpha D_2[S_2O_8^{2-}]}\right)^{\frac{1}{2}}}{\gamma},$$

D_1 and D_2 are the diffusion coefficients of copper(II) and peroxydisulphate. The function $H(u_i)$ has been tabulated by Koutecky[66]. Plotting $i/\{i_d\gamma H(u_i)\}$ against $[S_2O_8^{2-}]^{\frac{1}{2}}$ a satisfactory straight line is obtained, the slope of which gives the value of $k_{50} = 1.4\times 10^3$ l.mole^{-1}.sec^{-1}.

The reduction wave of peroxydisulphate at DME starts at the potential of the anodic dissolution of mercury. The current–potential curve exhibits certain anomalous characteristics under various conditions. At potentials more negative than the electrocapillary maximum, a current minimum can be observed; this is due to the electrostatic repulsion of the peroxydisulphate ion by the negatively charged electrode surface. The current minimum depends on the concentration and nature of the supporting electrolyte, and can be eliminated by the adsorption of capillary active cations of the type NR_4^+.

A minimum different from the above can be observed on the polarographic wave of peroxydisulphate at about +0.1 Volt in the presence of both copper(II) and arsenic(III) which does not occur with either constituent alone[67]. The decrease in the current due to peroxydisulphate, i_k, is kinetic in nature and caused by the copper(II)-catalyzed reaction between arsenic(III) and the peroxydisulphate. The reaction between arsenic(III) and peroxydisulphate is initiated by step (61) and followed by reactions (45), (48) and (50)

$$S_2O_8^{2-} \rightarrow 2\,SO_4^{-} \tag{61}$$

$$As(III)+SO_4^{-} \rightarrow As(IV)+SO_4^{2-} \tag{45}$$

$$As(IV)+Cu(II) \rightarrow As(V)+Cu(I) \tag{48}$$

$$Cu(I)+S_2O_8^{2-} \rightarrow Cu(II)+SO_4^{2-}+SO_4^{-} \tag{50}$$

or, when oxygen is present, by steps (51), (56) and (50)

$$As(IV)+O_2+H^+ \rightarrow As(V)+HO_2 \tag{51}$$

$$Cu(II)+HO_2 \rightarrow Cu(I)+O_2+H^+ \tag{56}$$

The chain length of this reaction proved to be as large as 27,000 with 27 m*M* Cu(II) and 20 m*M* As(III). In the case of electroreduction the chain is initiated by the reaction

$$Cu(II)+e \rightarrow Cu(I) \tag{71}$$

which is followed by reactions (50), (45) and (48).

In the presence of sufficient arsenic(III) reaction (45) is so rapid that the termination reactions (72) and (70) can be left out of consideration. The decrease of the peroxydisulphate current, i_k, should be proportional to the rate of reaction (50), *viz.*

$$-\frac{d[S_2O_8^{2-}]}{dt} = k_{50}[Cu(I)]_e[S_2O_8^{2-}]_e$$

where the subscript *e* refers to concentrations on the electrode surface. At potentials at the beginning of the reduction of Cu(II) to Cu(I) the concentration of the copper(I) ions at the surface of DME is proportional to the copper(II) concentration, and thus

$$i_k \approx [Cu(II)]_e[S_2O_8^{2-}]_e$$

Considering only termination by reaction (72) and assuming that stationary conditions are attained, the rate of reduction of peroxydisulphate in the chain process is given by

$$-\frac{d[S_2O_8^{2-}]}{dt} = (k_{45}\,k_{50}\,k_{71}/2\,k_{72})^{\frac{1}{2}}[As(III)]_e^{\frac{1}{2}}[Cu(II)]_e^{\frac{1}{2}}[S_2O_8^{2-}]_e^{\frac{1}{2}}$$

in accordance with experiment. The mechanism correctly predicts the absence of any influence of oxygen on i_k.

It should be mentioned that there is no decrease in the peroxydisulphate current in the presence of both iron(III) and arsenic(III). Presumably, the iron(II)–peroxydisulphate reaction is too slow to compete with the reduction of peroxydisulphate at the DME at the given concentration. However, iron(III) reduces the kinetic current in the presence of copper(II) and arsenic(III). This can be accounted for by the termination reactions

$$\mathrm{Fe(II)+SO_4^- \rightarrow Fe(III)+SO_4^{2-}} \qquad (44)$$

$$\mathrm{Fe(II)+As(IV) \rightarrow Fe(III)+As(III)} \qquad (47)$$

the iron(II) being formed by the electroreduction of iron(III).

Further convincing evidence was found by Catherino[134,135] for the formation of arsenic(IV) during the electro-oxidation of arsenic(III) in perchloric acid solution.

2.2.5 The induced reduction of chlorate by arsenic(III)

It was observed by Gleu[68] that chlorate ions can be rapidly reduced by arsenic(III) during the reaction between arsenic(III) and cerium(IV). Csányi and Szabó[69] have established that induced reduction can be carried out by other 1-equivalent reagents, *e.g.* cobalt(III), manganese(III), permanganate; while 2-equivalent reagents, *e.g.* bromine, chlorine, periodate, proved to be inactive.

According to Sanko and Stefanovskii[70] the reaction between arsenic(III) and chlorate is autocatalytic in character and at constant hydrogen ion concentration the rate is given by

$$-\frac{\mathrm{d[ClO_3^-]}}{\mathrm{d}t} = k_a\mathrm{[ClO_3^-][As(III)]} + k_b\mathrm{[ClO_3^-][Cl^-]}$$

Rutter[71] and McDougall[72] have pointed out that during the reaction chlorine dioxide is formed. Because of this observation either reaction (73) or (74)

$$\mathrm{ClO_3^- + As(III) \rightarrow ClO_2 + As(IV)} \qquad (73)$$

$$\mathrm{ClO_3^- + As(III) \rightarrow ClO_2^- + As(V)} \qquad (74)$$

can be assumed to be the rate-determining step, followed by the fast disproportionation of chlorite[132]

$$\mathrm{4\,ClO_2^- + 2H^+ = 2\,ClO_2 + Cl^- + ClO_3^- + H_2O} \qquad (75)$$

and by the fairly rapid reaction

$$HClO_2+HClO_3 = 2\,ClO_2+H_2O \tag{76}$$

As the disproportionation of chlorite is faster than reaction (76), at higher chlorate concentrations both reactions, and at lower concentrations rather reaction (75), are to be considered.

Reaction between arsenic(III) and chlorate is fairly slow. Although the reaction can be markedly accelerated by osmium tetroxide as catalyst[73], the quantitative reduction of chlorates takes nearly an hour. In the case of the induced reaction it was assumed that arsenic(III) is oxidized to arsenic(IV) by 1-equivalent oxidizing agents. Chlorate is reduced to chlorine dioxide by the arsenic(IV) intermediate, *viz.*

$$As(IV)+ClO_3^- \rightarrow As(V)+ClO_2 \tag{77}$$

Then, in fast consecutive steps, chlorine dioxide is reduced to chloride. The induced reduction of chlorate is quantitative within a few minutes. At the beginning of the titration not only the special smell but also the faint-yellowish colour of chlorine dioxide can be observed after the complete disappearance of the cerium(IV) reagent.

The induced reduction of chlorate can be inhibited by iodide, bromide and chloride ions. The effectiveness of these ions is about 400 : 10 : 1 in the given order. The order and the magnitude of the effect agree fairly well with the catalytic activity of these ions in the arsenic(III)–cerium(IV) reaction. This inhibition by halides is presumably connected with the opening of a new two-electron route for the arsenic(III)–cerium(IV) reaction.

During the induced reduction of chlorate a considerable oxygen effect was observed. The air oxidation of arsenic(III) is also an induced reaction, the extent of which decreases with increasing acid concentration and is increased by decreasing the rate of primary reaction. The induced oxidation caused by air also can be observed[69] during the osmium tetroxide-catalyzed chlorate–arsenic(III) reaction.

It may be assumed that the induced air oxidation is caused by reactions (51), (78) and (58), *viz.*

$$As(IV)+O_2+H^+ \rightarrow As(V)+HO_2 \tag{51}$$

$$HO_2+HO_2 \rightarrow H_2O_2+O_2 \tag{78}$$

$$As(III)+H_2O_2 \rightarrow As(V)+H_2O \tag{58}$$

since the latter is very fast in the presence of OsO_4.

If this were really the case, then during the reaction between arsenic(III) and

References pp. 570–580

cerium(IV) catalyzed by OsO_4 a considerable oxygen effect should also be observed However, no air oxidation occurs in the system, presumably because reaction (79)

$$HO_2 + Ce(IV) \rightarrow O_2 + H^+ + Ce(III) \tag{79}$$

which competes with reaction (78), is at least 6 times faster[132].

It is necessary therefore to seek the cause of the induced air oxidation in the ClO_3^-–Cl^- system. The fact that, during the stepwise reduction of chlorate, chlorine oxide (ClO), which can easily react with molecular oxygen[74]

$$ClO + O_2 \rightleftharpoons ClO_3 \tag{80}$$

is formed, may offer an explanation. However, the formation of ClO is not supported by direct experimental evidence.

According to experiment the induced air oxidation can be suppressed by silver(I) or copper(II) if added in sufficient quantity[69].

2.2.6 *Properties of arsenic(IV) intermediate*

The reactions discussed above show that arsenic(IV) is of redox amphoteric character and a stronger reducing agent than arsenic(III), but at the same time it is a stronger oxidant than arsenic(V). Partners of the oxidation–reduction reactions of arsenic(IV) known so far can be seen in Table 13. It follows from the redox amphoteric character that the oxidation potentials of couples involving arsenic species are in the order

$$E^\circ_{4/3} > E^\circ_{5/3} > E^\circ_{5/4}$$

This order of the potentials indicates that arsenic(IV) is an intermediate in the oxidation of arsenic(III) and reduction of arsenic(V) which is unstable from thermodynamic point of view and disproportionates easily according to

$$2\,As(IV) \rightarrow As(III) + As(V) \tag{81}$$

if there is no appropriate partner present. This is why arsenic(IV) cannot be prepared in quantity in aqueous solutions.

An estimate of the oxidation potential of couples involving arsenic(IV) can be given, based on the reaction

$$As(IV) + O_2 + H^+ \rightleftharpoons As(V) + HO_2 \tag{51}$$

TABLE 13

OXIDATION–REDUCTION REACTIONS OF ARSENIC(IV)

	E^0 (*V*)
As(IV) as reducing agent	
$O_2 \rightarrow HO_2$	−0.13
$Cu^{2+} \rightarrow Cu^+$	0.15
$Fe^3 \rightarrow Fe^{2+}$	0.77
$Ag^+ \rightarrow Ag$	0.79
$VO_3^- \rightarrow VO^{2+}$	1.00
$ClO_3^- \rightarrow ClO_2$	1.15
As(IV) as oxidant	
$HO_2 \rightarrow O_2$	−0.13
$Cu^+ \rightarrow Cu^{2+}$	0.15
$Fe^{2+} \rightarrow Fe^{3+}$	0.77

If it is assumed that equilibrium (51) is displaced at least 10 % to the right, with the aid of the standard potential of the O_2/HO_2 couple (−0.13 V) the potential of the As(V)/As(IV) couple can be estimated as $E^\circ_{5/4} = -0.07$ V, and $E^\circ_{4/3}$ as about 1.2 V. To characterize the reactivity of arsenic(IV) there are some relative rate coefficients compiled in Table 14.

It is striking that reactions of arsenic(III) with 1-equivalent oxidizing agents take place fairly slowly, while reactions with 2-equivalent partners *e.g.* with iodine, are fast. To interpret this behaviour it was suggested by Waters[58] that arsenious acid has the tautomeric forms

$$\begin{array}{ccc} \text{OH} & & \text{OH} \\ | & & | \\ \text{HO–As} & \rightleftharpoons & \text{H–As=O} \\ | & & | \\ \text{OH} & & \text{OH} \\ \text{a} & & \text{b} \end{array}$$

and form b is favoured. The oxidation of the latter by hydrogen abstraction can occur easily, because the heterolysis, such as

$$\text{I–I} + \text{H–AsO(OH)}_2 \rightarrow \text{I}^- + \text{I–H} + {}^+\text{AsO(OH)}_2 \tag{82}$$

requires very little activation energy.

On the other hand, reactive radicals that are capable of abstracting hydrogen atoms from other molecules could react homolytically, *e.g.*

$$\text{HO} + \text{H–AsO(OH)}_2 \rightarrow \text{HO–H} + \cdot\text{AsO(OH)}_2$$

TABLE 14

RELATIVE RATE COEFFICIENTS OF REACTIONS INVOLVING ARSENIC(IV)

$\frac{k_{As(IV)+O_2}}{k_{As(IV)+Fe(II)}} = 40$	$\frac{k_{As(IV)+FeOH^{2+}}}{k_{As(IV)+Fe(II)}} = 9$	$\frac{k_{As(IV)+Fe(III)}}{k_{As(IV)+Fe(II)}} = 0.1$
$\frac{k_{As(IV)+O_2}}{k_{As(IV)+Cu(II)}} = 4.4$	$\frac{k_{As(IV)+Cu(II)}}{k_{As(IV)+Fe(II)}} = 9$	
$\frac{k_{As(IV)+O_2}}{k_{As(IV)+Fe(III)}} = 400$		
$\frac{k_{As(IV)+Cu(II)}}{k_{As(IV)+Fe(III)}} = 90$		

However, this method of oxidation differs considerably from the effect of strong oxidizing ions such as cerium(IV), manganic(III) etc., because these merely gain electrons from the reducing partner.

This proposed structure of arsenic(IV) could also explain why the $FeOH^{2+}$ ion reacts faster with arsenic(IV) than does Fe^{3+}(aq). The former can react by direct transfer of hydroxyl, *viz.*

$$Fe(OH)^{2+} + AsO(OH)_2 \rightarrow Fe^{2+} + HO\text{–}AsO(OH)_2 \tag{46'}$$

while the reaction of the latter will involve an electron transfer followed by the addition of hydroxyl ion, *viz.*

$$Fe^{3+}(aq) + AsO(OH)_2 \rightarrow Fe^{2+}(aq) + {}^{+}AsO(OH)_2 \xrightarrow{OH^-} HO\text{–}AsO(OH)_2 \tag{46''}$$

Arsenic(IV) is a 25 electron radical, the molecular parameters of which can be found in the monograph by Atkins and Symons[133].

2.3 INDUCED REACTIONS INVOLVING HO_2 AND OH RADICALS

2.3.1 Induced reactions occurring in the $H_2S_2O_8$–H_2O_2 system

It is known that the permanganometric or cerimetric determination of hydrogen peroxide in the presence of peroxydisulphate furnishes too low results[75,76]. It has been shown by recent studies[77] that the error is due to the reaction between hydrogen peroxide and peroxydisulphate, *viz.*

$$H_2O_2 + S_2O_8^{2-} = 2\,HSO_4^- + O_2 \tag{83}$$

induced by the hydrogen peroxide–permanganate (or cerium(IV) sulphate) reaction. In the induced reaction hydrogen peroxide is the actor, peroxydisulphate the acceptor, while the oxidizing agent is the inductor.

Details of the kinetics of this induced reaction are not yet clear, and only a qualitative presentation can be given. The essential features of the induced reaction can be summarized as follows.

(*i*) Induced reduction of peroxydisulphate takes place only when hydrogen peroxide is oxidized by a 1-equivalent reagent.

(*ii*) On keeping the total oxidizing capacity of the system at a constant value and changing the $[H_2O_2]/[H_2S_2O_8]$ mole ratio within wide limits, the value of F_i goes through a maximum. This maximum is reached when $[H_2O_2]/[H_2S_2O_8] = 1$, showing that, in the induced reaction, the partners are consumed according to eqn. (83).

(*iii*) The value of F_i is greater with permanganate than with ceric systems. With the latter, the value of F_i is lower for perchloric acid than for sulphuric acid media.

(*iv*) The plot of F_i *versus* $([S_2O_8^{2-}]/[H_2O_2])$ results in a curve rising to a constant value. The limiting value of F_i does not exceed 1 even under the most favourable conditions.

(*v*) On changing the pH, the value of F_i goes through a maximum in the case of cerium(IV) ions. The maximum value of F_i was found to be in the pH range from 1–2 with both perchloric and sulphuric acid media. For permanganate systems the value of F_i increases continuously up to pH 4, but it was not possible to determine the pH of any maximum because of the slowness of the reaction between hydrogen peroxide and permanganate at pH's greater than 4.

(*vi*) On increasing the rate of titration, the value of F_i decreases.

(*vii*) On increasing the rate of stirring of the titrated solution, the value of F_i increases and reaches a limiting value.

(*viii*) On diluting the solution to be titrated, the other parameters being kept constant, the value of F_i increases.

(*ix*) In the presence of arsenious acid (only for cerimetric measurements) the error in H_2O_2 determined decreases to a minimum and is replaced by an error in the estimated As(III) concentration.

(*x*) During the induced reaction a slight polymerization of added acetanilide and biphthalate can be observed. Acrylonitrile, although it also exerted an inhibiting effect, did not undergo any observable polymerization.

(*xi*) During the induced reaction intermediates of strong oxidizing properties are formed, which partially or wholly destroy the rather resistent ferroin indicator.

(*xii*) When a solution containing cerium(IV) sulphate and peroxydisulphate was titrated with hydrogen peroxide (inverse titration), the error due to the induced reaction fell to a minimum.

(*xiii*) The induced reaction is influenced by a number of substances:

(*a*) The extent of the induced change is considerably increased by copper(II), silver(I), iron(III), and fluoride ions. The copper(II) ions catalyze the induced reaction without modifying its characteristics.

(*b*) The induced reaction is inhibited by iodide, bromide and chloride, the magnitude of the effect being in the order given, as well as by acetate, manganese(II), cerium(III), and uranyl ions. Cations of groups I, II and III also exert a slight inhibiting effect.

To understand this induced reaction it is necessary to examine the reaction between hydrogen peroxide and peroxydisulphuric acid. The rate equation based on the careful experiments of Tsao and Wilmarth[78] is

$$-\frac{d[S_2O_8^{2-}]}{dt}$$

$$= \frac{[S_2O_8^{2-}]}{\left\{\frac{9.5\times 10^6}{[H_2O_2]}+\frac{7.6\times 10^7[S_2O_8^{2-}]}{[H_2O_2]}+7.9\times 10^8+2.9\times 10^{10}[S_2O_8^{2-}]\right\}^{\frac{1}{2}}}$$

By varying the hydrogen peroxide and peroxydisulphuric acid concentration the rate expression approaches the limiting forms

(*a*) 0.001–0.005 M H_2O_2 and 0.001–0.025 M $K_2S_2O_8$

$$-\frac{d[S_2O_8^{2-}]}{dt} = k_a[S_2O_8^{2-}][H_2O_2]^{\frac{1}{2}}$$

(*b*) 0.025 M H_2O_2 and 0.002–0.01 M $K_2S_2O_8$

$$-\frac{d[S_2O_8^{2-}]}{dt} = k_b[S_2O_8^{2-}]$$

(*c*) 1.0 M H_2O_2 and 0.1–0.25 M $K_2S_2O_8$

$$-\frac{d[S_2O_8^{2-}]}{dt} = k_c[S_2O_8^{2-}]^{\frac{1}{2}}$$

Chain initiation is due to reaction (61), *i.e.* the thermal decomposition of peroxydisulphate into sulphate radicals, *viz.*

$$S_2O_8^{2-} \rightarrow 2\,SO_4^- \tag{61}$$

and the chain propagation steps are

$$SO_4^- + H_2O \rightarrow H^+ + SO_4^{2-} + OH \tag{84}$$

$$OH + H_2O_2 \rightarrow HO_2 + H_2O \tag{85}$$

$$HO_2 + S_2O_8^{2-} \rightarrow O_2 + HSO_4^- + SO_4^- \tag{63}$$

$$HO_2 + H_2O_2 \rightarrow O_2 + H_2O + OH \tag{86}$$

Of the six possible bimolecular termination reactions which HO_2, SO_4^-, and OH radicals might undergo only reactions (87), (88), (64) and (65) appear to be of importance under the experimental conditions covered

$$HO_2 + OH \rightarrow O_2 + H_2O \quad (87)$$

$$SO_4^- + OH \rightarrow HSO_5^- \quad (88)$$

$$HO_2 + SO_4^- \rightarrow O_2 + HSO_4^- \quad (64)$$

$$SO_4^- + SO_4^- \rightarrow S_2O_8^{2-} \quad (65)$$

In deriving the rate law, it was assumed that $k_{86}[H_2O_2] \gg k_{-84}[H^+][SO_4^{2-}]$. In principle there is a competition for the HO_2 radical between peroxydisulphate and hydrogen peroxide [reactions (63) and (86)]; however, when the stoichiometry is 1 : 1 reaction (86) can be neglected. Assuming that the chain length is large, with the usual steady-state approximation, we obtain the following rate equation:

$$-\frac{d[S_2O_8^{2-}]}{dt}$$

$$= \frac{[S_2O_8^{2-}]}{\left(\frac{k_{87}}{k_{61}k_{85}k_{63}[H_2O_2]} + \frac{k_{88}[S_2O_8^{2-}]}{k_{61}k_{84}k_{85}[H_2O_2]} + \frac{k_{64}}{k_{61}k_{84}k_{63}} + \frac{k_{65}[S_2O_8^{2-}]}{2k_{61}k_{84}^2}\right)^{\frac{1}{2}}}$$

By using the numerical constants obtained experimentally the following relations between the different rate coefficients are found

$$k_{87}k_{65}/k_{88}k_{64} = 9.2$$

$$k_{84}/k_{63} = 0.0136(k_{65}/k_{64}) = 0.12(k_{88}/k_{87})$$

$$k_{84}/k_{85} = 1.3 \times 10^{-3}, \; (k_{65}/k_{88}) = 1.2 \times 10^{-2}(k_{64}/k_{87}),$$

$$k_{63}/k_{85} = 0.1(k_{64}/k_{88})$$

From these relations it can be concluded that $k_{85} > k_{63} > k_{84}$, which implies that the concentrations of SO_4^- and HO_2 radicals are relatively large compared to that of OH. It is striking, however, that for kinetic reasons this mechanism does not involve termination steps (78) and (89) which are widely favoured in the chemistry of HO_2 and OH radicals

$$HO_2 + HO_2 \rightarrow H_2O_2 + O_2 \quad (78)$$

$$OH + OH \rightarrow H_2O_2 \quad (89)$$

However, the relations

$$\frac{[OH]}{[HO_2]} = \frac{k_{63}[S_2O_8^{2-}]}{k_{85}[H_2O_2]} \text{ and } \frac{[OH]}{[SO_4^-]} = \frac{k_{84}}{k_{85}[H_2O_2]}$$

indicate that OH should be the dominant radical at sufficiently large value of $[S_2O_8^{2-}]/[H_2O_2]$ and $1/[H_2O_2]$, *i.e.* a situation would arise in which step (89) would compete with reactions (87) and (88) and, in the limit, the rate equation would become

$$-\frac{d[S_2O_8^{2-}]}{dt} = \left(\frac{2\,k_{61}\,k_{85}^2}{k_{89}}\right)[S_2O_8^{2-}]^{\frac{1}{2}}[H_2O_2]$$

which is not in agreement with experiment.

A feature of this mechanism is that reaction (90) is omitted

$$SO_4^- + H_2O_2 \rightarrow HO_2 + H^+ + SO_4^{2-} \quad (90)$$

According to the thermochemical data, reaction (90) is considerably more likely energetically than step (84) as a process removing the SO_4^- radical. Previous experiments[60] indicated that reaction (84) is not too rapid, although the neglect of reaction (90) requires it to be very fast.

This mechanism does not account for the pH dependence of the reaction rate. According to our experiments[79] the rate of reaction between hydrogen peroxide and peroxydisulphate strongly depends on the acidity and at about pH = 5 a maximum can be observed.

It is worth mentioning that an attempt was made by Tsao and Willmarth[80] to determine the acid dissociation constant of HO_2. The reaction between hydrogen peroxide and peroxydisulphate was used for the generation of the HO_2 radical. However, these experiments, like others where the HO_2 radical is studied under steady-state conditions, could yield only a value of acidity constant multiplied by a coefficient consisting of a ratio of kinetic parameters. Unfortunately, in this case there are no independent data for the kinetic coefficient, and the value of K_{HO_2} cannot be evaluated. Considering the kinetic analogue of the titration curve it can be stated only that ionization of HO_2 becomes important in the pH range from 4.5–6.5. The value of acidity constant of HO_2 obtained by Czapski and Dorfman[81] is $(3.5 \pm 2.0) \times 10^{-5}$ mole.l.$^{-1}$.

Returning to the explanation of induced reactions, we can say the following. Friend's proposal[75], according to which the error in the H_2O_2 determination is caused by reaction (83) catalyzed by manganese(II) or cerium(III) formed in the primary reaction between hydrogen peroxide and permanganate or cerium(IV) cannot be accepted. The reaction between the ions mentioned and peroxydisulphate at room temperature is very slow, and, furthermore, the increase in acidity – in contrast to its effect on the induced reaction – promotes the oxidation. There is another important objection; induced reaction is suppressed by these ions. Similarly, there are not any observations which could support the assumption of Skrabal and Vacek[76], who regarded the induced reaction between peroxydisulphate and permanganate as the source of error.

A better explanation[77] can be given by considering the fact that the induced reduction of peroxydisulphate is observed only when hydrogen peroxide reacts with 1-equivalent oxidizing agents. This reaction takes place in steps

$$H_2O_2 + Ox \rightarrow HO_2 + H^+ + Red \quad (91)$$

$$HO_2 + Ox \rightarrow O_2 + H^+ + Red \quad (79')$$

The HO_2 radical formed being a strong reducing agent ($E_0 = -0.13$ V[82]) attacks the peroxydisulphate present according to

$$HO_2 + S_2O_8^{2-} \rightarrow HSO_4^- + O_2 + SO_4^- \quad (63)$$

The sulphate radical formed either reacts directly with H_2O_2 or, according to the assumption of Tsao and Willmarth[80], is converted to OH (step 84) which reacts with hydrogen peroxide, the HO_2 radical being reformed, *viz.*

$$OH\ (\text{or } SO_4^-) + H_2O_2 \rightarrow HO_2 + H_2O\ (\text{or } SO_4^{2-}) \quad (85)$$

Therefore, a chain reaction initiated by step (91), occurs in which a considerable amount of hydrogen peroxide and peroxydisulphate will be converted according to a 1 : 1 stoichiometry.

The value of F_i depends on the rate ratio of competing steps (79) and (63). The smaller this ratio, w_{79}/w_{63}, the greater is the value of F_i. Therefore, any factor which decreases the rate of reaction (79) effects an increase in the induced change. This accounts for the variation of F_i on changing the speed of titration. At low delivery rates only a slight amount of oxidizing agent is added to the solution at once, and thus the local concentration of the reagent is lower than during a more rapid titration. Consequently, reaction (79) takes place to a smaller extent and the value of F_i increases. Increasing the dilution of the titrated solution, and enhancing the stirring rate likewise decrease the local concentration of the reagent, resulting in an increasing disappearance of peroxydisulphate and hydrogen peroxide.

The effect on the induced reaction of the acidity cannot be satisfactorily considered with the kinetic data available. It was mentioned that the rate of reaction between hydrogen peroxide and peroxydisulphate is at maximum at about pH 5. In contrast, the value of F_i obtained cerimetrically goes through a flat maximum in the pH range from 1–2. This maximum should be regarded as an apparent one because the hydrolysis of cerium(IV) is considerable at pH's higher

References pp. 577–580

than 2. Measurements using permanganate confirm this assumption; the value of F_i was found to increase on increasing the pH. Unfortunately, measurements cannot be carried out at pH's higher than 4 because the rate of the hydrogen peroxide–permanganate reaction becomes too slow for the titration method. The rate maximum found for the hydrogen peroxide–peroxydisulphate reaction at about pH 5 indicates that the amphoteric character of the HO_2 plays an important role.

According to Bielski and Allen[83], since the HO_2 radical is of amphoteric character, it will be either protonated

$$HO_2 + H^+ \rightleftharpoons H_2O_2^+ \qquad K_{H_2O_2^+}$$

or dissociated

$$HO_2 \rightleftharpoons H^+ + O_2 \qquad K_{HO_2}$$

depending on the pH. Since $pK_{H_2O_2^+}$ and pK_{HO_2} are 1.0 ± 0.4 and 4.5, respectively, at pH < 1 reaction (63′) and at pH > 4 reaction (63″) is more probable than reaction (63)

$$H_2O_2^+ + S_2O_8^{2-} \rightarrow O_2 + HSO_4^- + H^+ + SO_4^- \tag{63'}$$

$$O_2^- + S_2O_8^{2-} \rightarrow O_2 + SO_4^{2-} + SO_4^- \tag{63''}$$

In view of the behaviour of the induced reaction it can be assumed that $k_{63'} < k_{63} > k_{63''}$.

The chain character of the induced raction is supported by the strong inhibiting effect of halide ions. Halides easily react with OH (or SO_4^-) radicals breaking the chain and causing the value of F_i to drop markedly. It was observed that the induced reaction is inhibited not only by halide ions but even by halogen molecules, since these oxidize the HO_2 radicals. Radical scavengers will also remove HO_2. Acetanilide and biphthalate strongly reduce the value of F_i and at the same time polymerization of these substances takes place. The assumption that reaction (63) occurs seems to be supported by the fact that during the induced reaction such a stable molecule as the ferroin indicator is partly or wholly destroyed.

The presence of arsenous acid causes a considerable change in the induced reaction: the error in the H_2O_2 determination decreases to a minimum and an As(III) error appears, while the $S_2O_8^{2-}$ error remains practically unchanged. Though reaction between arsenic(III) and peroxydisulphate is about ten times as rapid as that between hydrogen peroxide and peroxydisulphate, the extent of the induced reduction of peroxydisulphate remains practically unchanged. This indicates that, in the induced chain oxidation, reaction (85), is replaced by the more rapid reaction

$$SO_4^- (\text{or OH}) + \text{As(III)} \rightarrow SO_4^{2-} (\text{or OH}) + \text{As(IV)} \tag{45}$$

Assuming that arsenic(IV) reacts with peroxydisulphate according to

$$As(IV)+S_2O_8^{2-} \rightarrow As(V)+SO_4^{2-}+SO_4^- \quad (92)$$

and that chain carried by arsenic(IV) and sulphate radicals is nearly as long as that by HO_2 and OH (or SO_4^-) radicals, one can understand that the H_2O_2 error decreases, while the $S_2O_8^{2-}$ error remains the same as it was in the absence of arsenic(III), but an As(III) error also appears.

If reaction (92) occurred during the cerimetric (or permanganometric) titration of arsenic(III) in the presence of peroxydisulphate negative errors in concentration determinations should also appear contrary to experience. Therefore step (92) is presumably replaced by a reaction in which the As(IV) radical is oxidized by the oxidant according to

$$As(IV)+Ox \rightarrow As(V)+Red \quad (93)$$

The occurrence of reaction (93) results in a shortening of the chain causing the $S_2O_8^{2-}$ error to be considerably reduced.

Considering, however, that the solution contains oxygen from the beginning of the titration and oxygen is evolved during the reaction as well, the effect of oxygen should not be overlooked. In fact, it was found that the extent of induced reaction differs considerably in the presence or absence of oxygen. When oxygen was removed by bubbling N_2 through the solution, the H_2O_2 error remained the same in the presence as in the absence of arsenic(III). However, in the presence of oxygen, as was mentioned above, the H_2O_2 error was decreased by adding arsenic(III). This observation indicates that reaction (51) takes place

$$As(IV)+O_2+H^+ \rightarrow HO_2+As(V) \quad (51)$$

accounting for the variation in the H_2O_2 error and the constancy of the As(III) error.

In connection with the role of arsenic the reaction

$$As(III)+HO_2+2\,H^+ = As(V)+OH+H_2O \quad (94)$$

must also be ruled out, because in the H_2O_2–As(III)–Ce(IV) system an induced reaction does not occur. Lastly, if hydrogen peroxide were attacked by the arsenic(IV) intermediate according to

$$As(IV)+H_2O_2 \rightarrow As(V)+OH^-+OH \quad (95)$$

then the error in the H_2O_2 determination should be doubled in comparison with the As(III) and $S_2O_8^{2-}$ errors. However, if reaction (51) occurs, the HO_2 radical formed sustains the chain leading to the reduction of peroxydisulphate by arsenic(III).

The results of the inverse titrations can be interpreted as follows. On titrating cerium(IV) sulphate with hydrogen peroxide in the presence of peroxydisulphate, owing to the great excess of cerium(IV) present a much larger proportion of HO_2 radicals are removed by reaction (79) than in the case of direct titration; therefore the value of F_i considerably decreases. F_i is decreased further because there will be no hydrogen peroxide present to remove SO_4^- radicals, which oxidise cerium(III) ions, formed during titration, to cerium(IV). Thus the amount of cerium(IV) increases on adding a charge of peroxydisulphate. However, this increase is significant only in the presence of copper(II) ions, being due to the oxidation reaction of peroxydisulphate markedly accelerated by the latter. Halides, just as in direct titrations, strongly inhibit the induced reaction.

It was found that the value of F_i is markedly increased by ions which are effective catalysts of oxidation reactions of peroxydisulphate. These are silver(I) copper(II), and iron(III). Cobalt(II) and nickel(II) ions, although they are good catalysts for the decomposition of hydrogen peroxide, exert their effect merely as inert electrolytes in the induced reaction. Therefore it can be concluded that, in this process, activation of the rather less reactive $S_2O_8^{2-}$ is more important than that of hydrogen peroxide[84].

The induced change is decreased by ions of groups I, II and III. The effect of these ions in decreasing F_i can be ascribed to an increase in the rate of primary reaction (91)[85].

The inhibiting effect of halide ions and other easily oxidizable and reducible substances can be interpreted by assuming that these substances scavenge the chain carriers, HO_2 and SO_4^- (or OH) radicals. The role of uranyl ions is different, a weak 1 : 1 adduct with hydrogen peroxide is formed, which reacts with permanganate more slowly[86]. Owing to this slow reaction the local concentration of the oxidant will be increased, therefore the removal of HO_2 radicals by reaction (79′) will be favoured. In the presence of great excess of uranyl ions the induced change practically ceases.

Summarizing our findings, the induced reaction occurring in the H_2O_2–$H_2S_2O_8$–$KMnO_4$, or $Ce(SO_4)_2$, system can be illustrated by the following reaction scheme

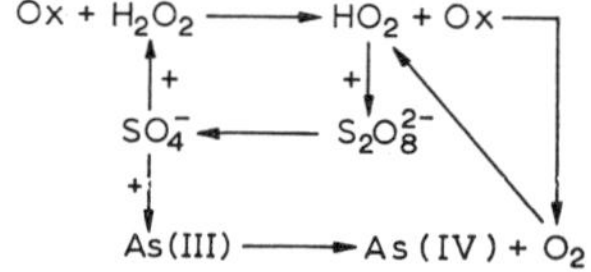

2.3.2 *Induced reactions involving other peroxy compounds*

According to our investigations on the H_2O_2–H_2SO_5–MnO_4^- [87], the H_2O_2–CH_3COOOH–$Ce(SO_4)_2$[88], the H_2O_2–H_3PO_5 (and $H_2P_2O_8$)–$Ce(SO_4)$[88,89], the H_2O_2–$MoOH_3$ (and WO_3, V_2O_5)–MnO_4^- [90] and the H_2O_2–Tl(III)–$Ce(SO_4)_2$[91] systems, an induced reaction occurs, similar in nature to that observed previously. But, since only qualitative data are available, a detailed discussion of these reactions will not be given here.

The induced reaction observed in the H_2O_2–OsO_4–MnO_4^- (or $Ce(SO_4)_2$) system[92] will be dealt with briefly. When hydrogen peroxide is titrated with permanganate in the presence of osmium tetroxide, a considerable negative H_2O_2 error occurs, the magnitude of which depends on the osmium tetroxide concentration (Table 15). In acid medium the decomposition of hydrogen peroxide by osmium tetroxide takes place only slowly, thus it is reasonable to assume that the H_2O_2 error is caused by an induced reaction. The value of F_i considerably decreases with an increase in the acid concentration, while it increases with a decrease in the rate of titration and with the dilution of the solution, as well as with an increase in the stirring rate. The limiting value of F_i under the most favourable experimental conditions is 7.5. The effect of foreign ions is similar to that observed in the H_2O_2–$H_2S_2O_8$ system so it will not be considered.

Induced decomposition of hydrogen peroxide can be interpreted as follows. The HO_2 radical formed in reaction (91) reduces OsO_4, which will be re-oxidised

TABLE 15

EFFECT OF CHANGES IN THE CONCENTRATION OF OSMIC ACID ON THE INDUCED REACTION BETWEEN H_2O_2, OsO_4 AND $KMnO_4$

Data of Csányi *et al.*[92].

$[OsO_4](M \times 10^6)$	$0.1\ N\ H_2O_2(ml)$		$\Delta H_2O_2\ (ml)$
	Taken	*Found*	
—	9.30	9.30	0.00
0.01	9.30	9.15	−0.15
0.10	9.30	8.96	−0.34
1.0	9.30	8.15	−1.15
1.0	9.30	8.15	−1.15
5.0	9.30	6.20	−3.10
5.0	9.30	6.22	−3.08
10.0	9.30	5.06	−4.24
10.0	9.30	5.12	−4.18
50.0	9.30	3.45	−5.85
50.0	9.30	3.47	−5.83
100.0	9.30	2.99	−6.31
100.0	9.30	2.91	−6.39
500.0	9.30	1.83	−7.47
500.0	9.30	1.86	−7.44

to Os(VIII) by the hydrogen peroxide present, *viz.*

$$HO_2+Os(VIII) \rightarrow Os(VII)+O_2+H^+ \quad (96)$$

$$Os(VII)+H_2O_2 \rightarrow Os(VIII)+OH^-+OH \quad (97)$$

The OH radical formed in reaction (97) forms a HO_2 radical again in reaction (85); thus, without any extra oxidizing agent being present, a considerable amount of hydrogen peroxide is decomposed.

Dependence of the value of F_i on the titrating rate, on the dilution and on the concentration of the reactants can be interpreted in a similar way as for the H_2O_2–$H_2S_2O_8$ system.

2.3.3 The Fenton reaction

The oxidation of organic compounds by hydrogen peroxide in the presence of iron(II) salts is called the Fenton reaction[93–95]. A great number of compounds, which ordinarily are attacked very slowly, if at all, by hydrogen peroxide, are readily oxidised by Fenton's reagent (a solution containing both iron(II) and hydrogen peroxide). The Fenton reaction is a typical example of induced chain reactions, and is of importance not only from the preparative point of view but plays a significant role in redox polymerization reactions[96] and many biochemical oxidation processes. In addition it is important in analytical chemistry, as the determination of organic peroxides and hydroperoxides by iron(II) is a widely used method.

The reaction between hydrogen peroxide and iron(II) has been studied in detail by many authors after the pioneering works of Haber and Weiss[97]. Without striving for completeness we should like to refer to the papers of Barb *et al.*[98]. and of Hardwick[99].

The mechanism of the Fenton reaction can be represented by the steps

$$Fe(II)+H_2O_2 \rightarrow Fe(III)+OH^-+OH \quad (53)$$

$$Fe(II)+OH \rightarrow Fe(III)+OH^- \quad (57)$$

$$OH+H_2O_2 \rightarrow HO_2+H_2O \quad (85)$$

$$Fe(II)+HO_2 \rightarrow Fe(III)+HO_2^- \quad (52')$$

$$Fe(III)+HO_2 \rightarrow Fe(II)+O_2+H^+ \quad (55)$$

At a low ratio of $[H_2O_2]/Fe^{2+}]$ only reactions (53) and (57) take place, oxygen is not evolved and a second order rate law is operative. In dilute perchloric acid solution (pH = 1.35) and at 25 °C, k_{53} was found to be 53.0 ± 0.7 $l.mole^{-1}$.

sec^{-1}. However, according to the recent measurements of Wells and Salam[100], the value of k_{53} is somewhat lower. The activation energy of the reaction is 9.4 $kcal.mole^{-1}$.

As the concentration ratio $[H_2O_2]/[Fe^{2+}]$ increases, the process will become governed by reactions (57) and (85) competing for the OH radicals. At even higher ratios of $[H_2O_2]/[Fe^{2+}]$, steps (52′) and (55), competing for the HO_2 radical, become dominant. As hydrogen peroxide does not take part in these latter steps, the rate of the evolution of oxygen becomes independent of the concentration of hydrogen peroxide. The relative rate coefficient $k_{55}/k_{52'}$ at about 25 °C and at pH = 1.35 is 0.14, while at pH = 2.65 it is 1.0, showing that the partially hydrolyzed $Fe(OH)^{2+}(aq)$ ion is much more reactive than the $Fe^{3+}(aq)$. The value of k_{85}/k_{57} is 0.11 ± 0.02 at 24.6 °C. A more recent and detailed study on the pH dependence of the reaction has shown[101] that on increasing the pH the rate of reaction increases to a limiting value. This increase in rate is connected with the formation of a $Fe(OH)_2(aq.)$ species. Likewise, the rate of reaction increases on addition of sulphate, selenate, trimetaphosphate, and halide ions. The accelerating effect of the anions is due to the fact that complexes of 1 : 1 composition are formed with iron(II) ions. The appropriate thermodynamical data are to be found in the papers of Wells and Salam[100,102].

As has already been mentioned, during the iron(II)–hydrogen peroxide reaction a number of organic compounds which do not react, or react only slowly with hydrogen peroxide, are readily oxidizable. In the induced oxidation of organic compounds, hydrogen peroxide plays the role of the actor and iron(II) is the inductor.

Induced oxidation of alcohols by hydrogen peroxide was studied by Kolthoff and Medalia[103]. According to their measurements the value of F_i increases with the increase in the concentration of ethanol, while it decreases with increase in the acid concentration (see Table 16). In acetic acid medium the value of F_i is considerably lower. Chloride ions effectively suppress the induced oxidation of alcohols. The main product of the oxidation of ethanol is acetaldehyde which can be further oxidized to acetic acid. The data on the induced oxidation of alcohol (H_2A) can be interpreted by reactions (53), (98), (99) and (57).

$$Fe(II)+H_2O_2 \rightarrow Fe(III)+OH^-+OH \qquad (53)$$

$$OH+H_2A \rightarrow HA+H_2O \qquad (98)$$

$$HA+Fe(III) \rightarrow A+H^++Fe(II) \qquad (99)$$

$$Fe(II)+OH \rightarrow Fe(III)+OH^- \qquad (57)$$

The sum of steps (53) and (99) gives the direct reaction of the organic radical with hydrogen peroxide, *viz.*

$$HA+H_2O_2 \rightarrow A+OH+H_2O \qquad (100)$$

TABLE 16

IRON(II)-INDUCED OXIDATION OF ETHANOL BY HYDROGEN PEROXIDE

Data of Kolthoff and Medalia[103]

Initial concentrations		*Initial*			
Ethanol (M)	*Iron(II) (M×10³)*	$\frac{[Fe^{++}]}{[H_2O_2]}$	H_2SO_4 *(M)*	$\frac{[Fe^{++}]\ reacted}{[H_2O_2]\ taken}$	F_i
	0.52	3.8	0.75	1.98	0.01
1.0×10^{-4}	0.52	3.8	0.75	1.64	0.22
1.0×10^{-3}	0.52	3.8	0.75	0.54	2.7
1.0×10^{-3}	0.52	7.6	0.75	0.56	2.6
1.0×10^{-3}	0.52	1.9	0.75	0.58	2.5
1.0×10^{-2}	0.52	3.8	0.75	0.15	12.3
1.0×10^{-2}	0.90	0.85	0.75	0.19	9.5
1.0×10^{-2}	2.0	2.0	0.75	0.21	8.5
1.0×10^{-3}	2.0	4.1	0.75	0.70	1.86
1.0×10^{-3}	2.0	3.8	0.027	0.49	3.08

The suppressing effect of the acetic acid can be interpreted by reactions (101) and (102)

$$OH+CH_3COOH \rightarrow CH_3COO+H_2O \tag{101}$$

$$CH_3COO+Fe(II) \rightarrow CH_3COO^-+Fe(III) \tag{102}$$

Baxendale *et al.*[104] have observed that in the presence of organic substances the iron(II) consumption in the reaction with hydrogen peroxide is increased by

TABLE 17

IRON(II)-INDUCED OXIDATION OF VARIOUS ORGANIC SUBSTANCES BY HYDROGEN PEROXIDE IN THE PRESENCE OF AIR

Data of Kolthoff and Medalia[105]. $[H_2SO_4] = 0.75\ M$.

Substance added	*Concentration (M)*	*Initial* $[Fe^{++}]$ $(M\times10^3)$	*Initial* $\frac{[Fe^{++}]}{[H_2O_2]}$	$\frac{[Fe^{++}]\ reacted}{[H_2O_2]\ taken}$
—	—	1.0	3.75	2.04
Methanol	10^{-4}	1.0	3.75	2.22
Methanol	10^{-2}	1.0	3.75	2.10
Methanol	10^{-2}	1.3	9.28	3.21
Acetic acid	10^{-4}	1.0	3.75	2.07
Acetic acid	10^{-2}	1.0	3.75	2.41
Acetone	10^{-4}	1.0	3.75	2.09
Acetone	10^{-2}	1.0	3.75	3.04
Ethanol	10^{-4}	1.0	4.16	2.19
Ethanol	10^{-2}	2.0	4.24	2.06

oxygen. This phenomenon has been studied by Kolthoff and Medalia[105]. The increase in the amount of iron(II) reacted is caused by autoxidation of the organic compound present. Some relevant data are included in Table 17. All the organic substances studied, unlike their effects in the absence of oxygen, show similar behaviour. Chloride ion, however, proved to be an effective suppressor of the induced reaction in the presence as in the absence of oxygen. The induced reaction occurring in the presence of oxygen can be described by the following scheme

$$\mathrm{Fe(II)+H_2O_2 \rightarrow Fe(III)+OH^- + \dot{O}H} \quad (53)$$

$$\mathrm{H_2A+\dot{O}H \rightarrow H\dot{A}+H_2O} \quad (98)$$

$$\mathrm{H\dot{A}+O_2 \rightarrow HA\dot{O}_2} \quad (103)$$

$$\mathrm{HA\dot{O}_2+H^+ +Fe(II) \rightarrow HAO_2H+Fe(III)} \quad (104)$$

$$\mathrm{HAO_2H+Fe(II) \rightarrow HA\dot{O}+Fe(III)+OH^-} \quad (105)$$

$$\mathrm{HA\dot{O}+H_2A \rightarrow HAOH+H\dot{A}} \quad (106)$$

$$\mathrm{HA\dot{O}_2+H_2A \rightarrow HAO_2H+H\dot{A}} \quad (107)$$

$$\mathrm{HA\dot{O}+Fe(II)+H^+ \rightarrow HAOH+Fe(III)} \quad (108)$$

$$\mathrm{Fe(II)+\dot{O}H \rightarrow Fe(III)+OH^-} \quad (57)$$

Induced reactions involving hydrogen peroxide can be observed with hydrogen peroxide derivatives, as well. For instance, the reaction between cumene hydroperoxide and iron(II), in the absence of oxygen, results in a considerable induced decomposition of the peroxy compound, while, in the presence of oxygen, a marked oxidation of iron(II) takes place[106].

2.4 INDUCED REACTIONS INVOLVING SULPHATE RADICALS

Reaction between iron(II) and peroxydisulphate ions, if the iron(II) is added rapidly, occurs according to

$$\mathrm{S_2O_8^{2-}+2\,Fe(II) = 2\,SO_4^{2-}+2\,Fe(III)}$$

On the other hand, Kolthoff *et al.*[107] established that the ratio [Fe(II)] reacted/ [$S_2O_8^{2-}$] reacted in acidic, neutral and alkaline medium is the same but much less than 2, if iron(II) is added at very low speed. The decrease in the oxidizing capacity can be explained by the induced decomposition of peroxydisulphate, though the detection of oxygen as a product was not successful.

In the presence of organic compounds, according to the observations of Merz

References pp. 577–580

and Waters[108], reaction between the organic compound and peroxydisulphate is induced by the iron(II)–peroxydisulphate reaction, resulting in the value of the $[Fe(II)]/[S_2O_8^{2-}]$ ratio being much less than 2. The induced reaction can be explained by equations (43), (44), (109) and (110), *viz.*

$$Fe(II)+S_2O_8^{2-} \rightarrow Fe(III)+SO_4^{2-}+SO_4^{-} \quad (43)$$

$$Fe(II)+SO_4^{-} \rightarrow Fe(III)+SO_4^{2-} \quad (44)$$

$$SO_4^{-}+H_2A \rightleftharpoons HA+HSO_4^{-} \quad (109)$$

$$HA+S_2O_8^{2-} \rightarrow A+HSO_4^{-}+SO_4^{-} \quad (110)$$

According to this mechanism the value of the consumption ratio

$$\frac{-d[H_2A]}{-d[Fe(II)]} = \frac{k_{109}[H_2A]}{2\,k_{44}[Fe(II)]} \quad (111)$$

must be independent of the rate of addition of peroxydisulphate to the reaction mixture. On the other hand, if iron(II) is added to an alcohol–peroxydisulphate mixture, the extent of the oxidation of alcohol is determined by the rate of addition of iron(II). If the instantaneous concentration of iron(II) is very low – which can be attained by very slow addition – the value of $-d[H_2A]/-d[Fe^{2+}]$ should become infinite. This is supported experimentally. With ethanol the value of k_{109}/k_{44} was found to be 0.006 by Merz and Waters[108] and 0.015 by Kolthoff *et al.*[107]. The difference was ascribed by the latter to traces of oxygen, as the value of k_{109}/k_{44} decreases in the presence of O_2.

Dilution of the reaction mixture with water or acid results in a decrease in the consumption ratio. This result cannot be interpreted by the above mechanism because the induced oxidation depends on the relative and not on the absolute concentrations. But assuming that reaction (109) is reversible we obtain for the consumption ratio

$$\frac{-d[H_2A]}{-d[Fe(II)]} = \frac{k_{109}[H_2A]}{2\,k_{44}[Fe(II)]}\left(\frac{k_{110}/k_{-109}[S_2O_8^{2-}]}{1+k_{110}/k_{-109}[S_2O_8^{2-}]}\right) \quad (112)$$

If $k_{110} \gg k_{-109}$, or the concentration of the oxidizing agent is high, eqn. (112) reduced to eqn. (111). On the contrary, with a very low concentration of oxidizing agent the consumption ratio becomes dependent on the concentration of the oxidizing agent.

During the induced reactions involving peroxydisulphate in the presence of oxygen, the induced oxidation of iron(II) by oxygen can be observed just as with the hydrogen peroxide–iron(II) system. Chloride and particularly bromide ions are effective inhibitors in the iron(II)–peroxydisulphate system.

By comparing the values of k_{109}/k_{44} and k_{98}/k_{57}(for the hydrogen peroxide system), it can be seen that the OH radical is much more reactive than the SO_4^- radical (see Table 18).

TABLE 18

RELATIVE RATE OF OXIDATION OF ALCOHOLS BY SO_4^- AND OH RADICAL, RESPECTIVELY

Data of Merz and Waters[108].

		CH_3OH	C_2H_5OH	n-C_3H_7OH	iso-C_3H_7OH
k_{109}/k_{44}	(SO_4^-)	0.0022	0.0062	0.0061	0.0089
k_{98}/k_{57}	(OH)	1.14	2.05	1.30	1.77

2.5 INDUCED REACTIONS INVOLVING INTERMEDIATES PRODUCED BY PARTIAL OXIDATION OF THIOCYANATE

Recently an interesting observation was described by Pungor *et al.*[109]. They found that in moderately acid solution in the presence of hydrogen peroxide not only was peroxysulphuric or peroxyacetic acid reduced by thiocyanate ions, but also the hydrogen peroxide, although H_2O_2 reacts only slowly with thiocyanate under the given conditions. This phenomenon was interpreted in terms of formation of dipole adducts between peroxy acids and hydrogen peroxide, called solvated peroxy acids. However, there are some observations[110] which are not in agreement with this interpretation. For example, it has to be assumed that increase of temperature gives rise to an increase in the amount of "solvated" compound and that the change in the order of addition of peroxy acid and hydrogen peroxide can considerably alter the quantity of the "adduct" formed. This seems unlikely.

These and similar results can be explained if the simultaneous reduction of hydrogen peroxide is due to an induced reaction. To show the characteristic features of this reaction some results are presented in Table 19 and Table 20. The procedure for these measurements was as follows. The solution of peroxy compounds given in columns 1 and 2 was made up to 20 ml and the pH was adjusted to the given value. Then potassium thiocyanate solution was added and, after the reaction time noted, the process was quenched by adding potassium iodide solution (0.3 g KI). After 5 sec the solution was acidified with 1 ml 2 *N* sulphuric acid; then using, molybdate catalyst solution, the iodine liberated was titrated with standard thiosulphate.

The data in Tables 19 and 20 show that both peroxysulphuric and peroxyacetic acid give the induced reduction of hydrogen peroxide by thiocyanate. The induced reduction depends strongly on the pH of the solution and on the concentration of thiocyanate. At pH > 3 both peroxy acids reacts more slowly

References pp. 577–580

TABLE 19

DEPENDENCE OF INDUCED REDUCTION OF HYDROGEN PEROXIDE ON EXPERIMENTAL CONDITIONS IN THE H_2O_2–H_2SO_5–SCN^- SYSTEM

Data of Csányi and Solymosi[110]. Initial amounts taken: 2.83 ml 0.01 *N* H_2SO_5, 7.46 ml 0.01 *N* H_2O_2. Total oxidising capacity taken: 10.29 ml 0.01 *N*.

pH	*0.1 N KSCN (ml)*	*Time of reaction (sec)*	*Total oxidising capacity found x ml 0.01 N*	*Difference 10.29−x*
3.00	0.5	15	8.30	1.99
2.50	0.5	15	7.90	2.39
2.10	0.5	15	7.65	2.64
1.30	0.5	15	6.98	3.31
0.90	0.5	15	6.86	3.43
3.6	0.5	15	8.70	1.59
3.6	0.5	45	8.06	2.23
3.6	0.5	120	7.35	2.94
3.6	0.5	180	6.90	3.39
3.0	0.5	15	8.30	1.99
3.0	1.5	15	7.50	2.79
3.0	2.0	15	7.14	3.15
3.0	3.0	15	6.86	3.43
3.0	4.0	15	6.86	3.43

(the reactivity of peroxyacetic acid is lower than that of peroxysulphuric acid); consequently at the end of the short reaction time there is no quantitative reduction of peroxy acids. However, it should be mentioned that, under such unfavourable conditions for the induced change, not only the peroxy acids but also a part of hydrogen peroxide will be reduced.

On the basis of these results it can be stated that the fast oxidation of thiocyanate by peroxy acids gives rise to the induced reduction of hydrogen peroxide. In order to elucidate the mechanism of this interesting reaction let us have a look at reactions of thiocyanate with peroxy compounds of different types.

The reaction between peroxysulphuric acid and thiocyanate ions was investigated by Smith and Wilson[111–113] by a stopped-flow conductance method. Their results can be summarized as follows. If l and m are the numbers of moles of cyanate and sulphur dicyanide produced per mole of thiocyanate consumed, the overall stoichiometric equation is

$$\begin{aligned} &SCN^- + (3-2m+l)\,HSO_5^- + (1-2m)\,H_2O \\ &= (4-3m+l)\,SO_4^{2-} + (4-4m-2l)\,H^+ + l\,OCN^- + (1-2m+l)\,HCN \\ &\qquad + m\,S(CN)_2 \end{aligned}$$

TABLE 20

DEPENDENCE OF INDUCED REDUCTION OF HYDROGEN PEROXIDE ON THE EXPERIMENTAL CONDITIONS IN THE H_2O_2–CH_3COOOH–SCN^- SYSTEM

Data of Csányi and Solymosi[110].

Taken		*pH*	*0.1 N KSCN (ml)*	*Time of reaction (sec)*	*Total oxidizing capacity*		
CH_3COOOH (ml 0.01 N)	*H_2O_2 (ml 0.01 N)*				*Taken*	*Found (ml 0.01 N)*	*Difference*
4.97	6.73	3.05	0.25	30	11.70	8.38	3.32
4.97	6.73	3.05	0.50	30	11.70	6.90	4.80
4.97	6.73	3.05	0.75	30	11.70	5.93	5.77
4.97	6.73	3.05	1.00	30	11.70	4.45	7.25
4.97	6.73	3.05	1.50	30	11.70	3.78	7.92
4.97	6.73	3.05	2.00	30	11.70	3.10	8.60
4.97	6.73	3.05	2.50	30	11.70	2.93	8.77
4.97	6.73	3.05	3.00	30	11.70	2.93	8.77
4.03	4.03	3.34	0.5	30	7.23	4.80	2.34
4.03	4.03	3.23	0.5	30	7.23	4.70	2.53
4.03	4.03	3.10	0.5	30	7.23	4.23	3.00
4.03	4.03	2.92	0.5	30	7.23	3.40	3.83
4.03	3.03	2.75	0.5	30	7.23	3.00	4.23
4.03	4.03	2.62	0.5	30	7.23	2.82	4.41
4.03	4.03	2.54	0.5	30	7.23	2.70	4.53
4.03	4.03	2.10	0.5	30	7.23	2.30	4.93
4.03	4.03	1.70	0.5	30	7.23	2.18	5.05
4.03	4.03	1.27	0.5	30	7.23	2.15	5.08
4.97	6.73	3.05	0.5	30	11.70	7.17	4.53
4.97	6.73	3.05	0.5	45	11.70	6.15	5.55
4.97	6.73	3.05	0.5	60	11.70	5.17	6.53
4.97	6.73	3.05	0.5	90	11.70	4.40	7.30
4.97	6.73	3.05	0.5	120	11.70	4.25	7.45
4.97	6.73	3.05	0.5	180	11.70	4.24	7.46
4.97	6.73	3.05	0.5	240	11.70	4.24	7.46

Schematically, the reaction involves the pathways

k_b — SCN^- → $(SCN)_2$
k_c — HSCN → $(SCN)_2$ → products
HSO_5^- + SCN^- $\overset{k_a}{\rightleftharpoons}$ X → disproportionation → products
k_d — SO_5^{2-} → products
k_e — CN^- → $S(CN)_2$

The rate of reaction depends in a fairly complex way upon concentrations. In the pH range from 2.3–3.7 (which is the interesting range from the aspect of the induced reaction), the initial rates were shown to obey the relation

$$R = (k_b + k_c[H^+])[SCN^-]_0 + k_d K[H^+]^{-1}[HSO_5]_0 \cdot \frac{[HSO_5^-]_0[SCN^-]_0}{[HSO_5^-]_0 + [SCN^-]_0}$$

References pp. 577–580

where $K = (5.0 \pm 1.6) \times 10^{-10}$ mole.l^{-1} is the formal ionization constant for HSO_5^-.

Since the measurements of conductance change are not directly related to the composition of the solution, as an alternative method numerical integration of the differential rate equations implied by the proposed mechanism was employed. The second order rate coefficients obtained by this method are

	k_b	k_c	k_d	k_e
30.1 °C	2.8	2300	0.0052	0.1
10.1 °C	1.0	1000	0.0040	0.04

To confirm the proposed mechanism calculated conductance–time curves were compared with the experimental ones. The agreement is satisfactory up to at least 50 % reaction over a fairly wide range of experimental conditions.

Not much can be said about the nature of entity of X produced by the rapid initial equilibrium. The formation of X as HOSCN, *viz.*

$$HSO_5^- + SCN^- \rightleftharpoons HOSCN + SO_4^{2-} \tag{113}$$

or as an 1 : 1 addition complex, *viz.*

$$HSO_5^- + SCN^- \rightleftharpoons HSO_5.SCN^{2-} \tag{114}$$

is equally consistent with the initial rate data. The two possibilities (113) and (114) are in principle distinguishable. If (113) is occurring, then increasing the sulphate ion concentration should produce a decrease in the initial rate, while there is no sulphate dependence in the case of step (114). Unfortunately, at present there are no results reliable enough to distinguish between these two possibilities.

The acid-catalyzed reaction between hydrogen peroxide and thiocyanate ions has been investigated by several authors, most thoroughly by Wilson and Harris[114]. The rate law has the form

$$R = k[H^+][SCN^-][H_2O_2]^2/\{[H_2O_2]+\alpha[HCN]\}$$

with $k = 1.05 \times 10^9 \exp(-11{,}000/\boldsymbol{R}T)$ l^2.mole^{-2}.sec^{-1}. Following the reaction merely to about 30 % conversion and using different analytical methods, Csányi and Horváth[116] obtained a simple third order rate law with a value of $k = 1.54 \times 10^9 \exp(-11{,}100/\boldsymbol{R}T)$ l^2.mole^{-2}.sec^{-1}, fairly close to that of Wilson and Harris.

The experimental rate law is consistent with the following mechanism

$$SCN^- + H_2O_2 + H_3O^+ \rightleftharpoons HOSCN + 2\,H_2O \quad (115)$$

$$HOSCN + H_2O_2 \rightarrow HOOSCN + H_2O \quad (116)$$

$$HOSCN + HCN \rightarrow S(CN)_2 + H_2O \quad (117)$$

$$HOOSCN + H_3O^+ \rightarrow H_2SO_3 + HCN + H^+ \quad (118)$$

$$H_2SO_3 + H_2O_2 = H_2SO_4 + H_2O \quad (119)$$

If it is assumed that the rate-determining step is reaction (116), that reactions (118) and (119) are much more rapid than (115) and (116) and that the steady-state hypothesis can be applied, the resulting rate law corresponds exactly to the experimental one, provided that $\alpha = k_{117}/k_{116} = 2$ and $k = k_{115}$. At the beginning of the reaction when [HCN] = 0, the rate law reduces to the simple third order form as found[116].

At higher pH's, reactions (115) and (118) will be replaced by reactions (120) and (121), respectively

$$SCN^- + H_2O_2 \rightarrow HOSCN + OH^- \quad (120)$$

$$HOOSCN + H_2O_2 \rightarrow H_2SO_3 + HOCN \quad (121)$$

steps (120), (116), (121) and (119), together with reaction (122)

$$HOCN + 2\,H_2O = HCO_3^- + NH_4^+ \quad (122)$$

reaction (120) being rate determining, account for the stoichiometry and kinetics of the non-catalyzed reaction.

The reactions of peroxysulphuric acid and of hydrogen peroxide with the thiocyanate are obviously very similar. In both reactions the reactive intermediate is the same or at least has a very similar structure, and the subsequent reactions of this entity are equally fast either with peroxy acid or with hydrogen peroxide. Accordingly it can be said that if both peroxy acid and hydrogen peroxide are present in the solution a competition between these two substances for the intermediate occurs, as a result of which not only the peroxy acid but also the hydrogen peroxide will be reduced.

In solutions of pH > 3, when the rate of reduction of peroxy acid is much lower, the induced disappearance of hydrogen peroxide will be considerably reduced.

2.6 INDUCED REACTIONS EFFECTED BY REDUCTION OF PERMANGANATE IONS

Doroshevskii and Bardt[117] observed that during the reduction of permanganate by iron(II) sulphate water-soluble alcohols are oxidized. A similar oxidation was found by Waters *et al.*[118] in the course of the permanganate–hydrogen peroxide reaction in acid medium. It is known that alcohols in dilute solution are slowly attacked by permanganate; therefore the alcohol oxidation observed during the reduction of permanganate is to be regarded as an induced reaction.

Investigation of the reaction was carried out by Waters *et al.* by titrating an alcoholic solution of iron(II) or hydrogen peroxide with potassium permanganate. If the values obtained are plotted as {equiv. of Fe(II) (or H_2O_2)/moles of $KMnO_4$} against {moles of alcohol/equiv. of Fe(II) (or H_2O_2)} then, as the concentration of the alcohol in the solution is increased, the first ratio progressively declines from 5 to 3 (Fig. 3). This result shows that out of 5 equivalents of oxygen available 2 equivalents are consumed in the oxidation of alcohol in the limit. If the pH of the solution is kept constant, the extent of the conversion of alcohol becomes independent of the dilution of the solution. The extent of the induced reaction does not change even if the titration is carried out inversely, *i.e.*, a solution containing alcohol and permanganate is titrated by iron(II) sulphate or hydrogen peroxide. This observation excludes the possibility that a chain reaction is occurring. On increasing the concentration of sulphuric acid (*a*) the rate of the induced reaction increases and (*b*) there is a slight, but definite, discontinuity in the ratio plots (type of Fig. 3) corresponding to the oxidation ratio (equiv. of Fe^{2+}/moles of $KMnO_4$) = 4.

The addition of manganese(II) salts reduces significantly the extent, while addition of fluoride enhances the initial rate, of the oxidation of the alcohol. In

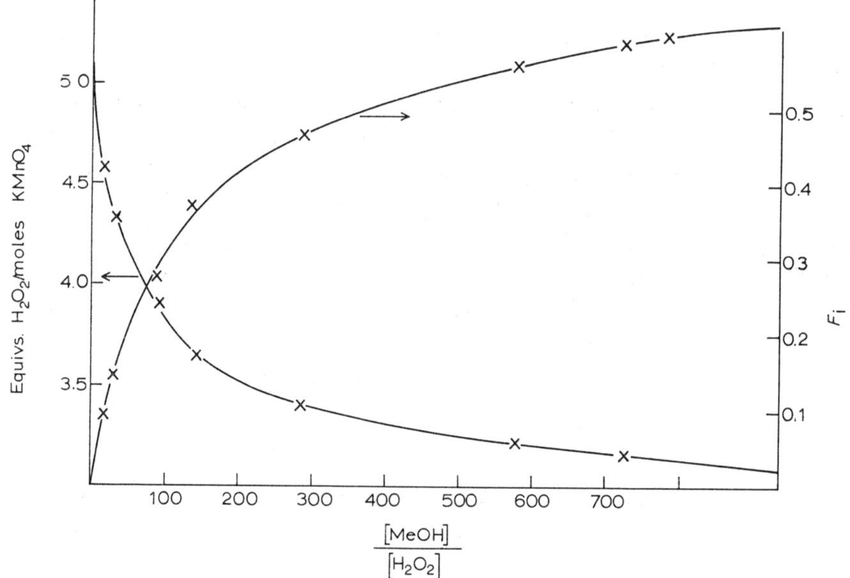

Fig. 3. Induced oxidation of methanol by permanganate in the H_2O_2–$KMnO_4$ system. According to data of Waters *et al.*[118].

the presence of pyrophosphate ions, not only the rate of reduction of permanganate but also the rate of the oxidation of alcohol was decreased.

The data for this induced reaction can be interpreted as follows. In the course of the permanganate–iron(II) or permanganate–hydrogen peroxide reaction a manganese intermediate [presumably maganese(IV)] is formed which reacts both with the inductor (iron(II) or hydrogen peroxide), and with the acceptor (alcohol). The suggestion that manganese(IV) is the active intermediate is supported by the value of CI = 0.6 and by the limiting value of the oxidation ratio, (equiv. of alcohol/mole of $KMnO_4$) = 3, found experimentally. This assumption is also in accordance with the fact that addition of manganese(II), by both displacing the equilibrium

$$\mathrm{Mn(III)+Mn(II)} \rightleftharpoons 2\ \mathrm{Mn(IV)}$$

to the right, and increasing the rate of the primary reaction, greatly decreases the rate of the alcohol oxidation.

The mechanism of the induced oxidation of alcohol is assumed by the authors mentioned to be analogous to that of oxidation of isopropyl alcohol by chromic acid, *viz.*

$$\mathrm{Me_2C(H){-}\bar{O}(H)| + Mn^{4+} \longrightarrow Me_2C(H){-}\overset{+}{\bar{O}}(H)| \longrightarrow Mn^{3+}}$$

$$\downarrow$$

$$\mathrm{Me_2C{=}O + H^+ + Mn^{2+} \longleftarrow Me_2C(H){-}\bar{O}{-}Mn^{3+} + H^+}$$

2.7 INDUCED REACTIONS INVOLVING TIN(III) INTERMEDIATE

The formation of tin(III) intermediate during the oxidation of tin(II) has been postulated earlier in some cases, although direct evidence has not been quoted. Recently a couple of cobalt(III) complexes were supposed by Higginson *et al.*[119] to demonstrate the presence of tin(III).

Generally to detect strongly reducing intermediates it is necessary to have a substrate which is not a strong oxidizing agent, or it is likely to react fairly rapidly with the reduced form of the oxidizing partner under investigation. (*Mutatis mutandis*—the same is valid for detection of oxidizing intermediates.) The suitability of some cobalt(III) complexes, *e.g.* in order of sensitivity: trioxalato- > > aquo-chloro-tetramino- > (hydroxyethylethylenediamine-triacetato)aquo-(Co (YOH)H_2O) > ethylenediamine-tetraacetato cobaltate(III)-(Co(Y)), arises from the fact that they are weak oxidizing agents, and that cobaltous ions produced are inert towards oxidation, or further reduction. The ease of determination of cobalt(III) complexes by spectrophotometry is an additional advantage.

The method proposed is appropriate to show the presence of a strongly reducing intermediate. However, it is usually not possible to identify this entity as tin(III) merely on the evidence of the consumption of cobalt(III) complex present. To this end additional (kinetic) evidence is necessary. Nevertheless, the investigation of the induced reduction of cobalt(III) complexes is useful as a simple means of deciding whether the oxidation of tin(II) involves 1- or 2-equivalent steps.

2.7.1 *Reaction between iron(III) and tin(II)*

The mechanism of this reaction adopted by earlier investigators[120, 121, 122] was

$$\mathrm{Fe(III)+Sn(II) \rightarrow Fe(II)+Sn(III)}$$

$$\mathrm{Fe(III)+Sn(III) \rightarrow Fe(II)+Sn(IV)}$$

In perchlorate media, a marked decrease in the apparent second-order rate coefficients was found as the reaction proceeded if tin(II) was present in excess, whereas only little change occurred when iron(III) was in excess. This behaviour indicates that there is a competition between iron(II) and iron(III) for tin(III). In experiments with added $Co(YOH)H_2O$ a considerable consumption of this complex was found[56].

The fact that the results of kinetic study are in good agreement with the above mechanism, together with the additional evidence obtained by using the cobalt(III) complex, makes it highly probable that tin(III) is, in fact, the only intermediate involved in this reaction.

2.7.2 *Reaction between tin(II) and chromate*

It is interesting to note that the reaction

$$\mathrm{2\ Cr(VI)+3\ Sn(II) = 2\ Cr(III)+3\ Sn(IV)}$$

unlike the reaction between arsenous acid and chromate, takes place in 1-equivalent steps. Thus, a considerable consumption of added trioxalato cobaltate(III) complex has been observed[56]. This finding also supports the author's opinion that it is necessary to reconsider some steps in the widely accepted mechanism of chromic acid oxidations.

2.7.3 *Reaction between tin(II) and permanganate*

During the fast reaction between tin(II) and permanganate in hydrochloric acid medium the induced reduction of cobalt(III) complex, $Co(YOH)H_2O$ takes place. According to the meagre kinetic data it seems likely that the fast stage of the reaction is the formation of Mn(V) by a 2-equivalent change. The step

$$Mn(V)+Sn(II) \rightarrow Mn(IV)+Sn(III)$$

is assumed by Wetton and Higginson[56] to be the source of tin(III).

During oxidation of tin(II) ions by hydrogen peroxide, iodine, bromine, mercury(II) and thallium(III) the induced reduction of cobalt(III) complexes cannot be observed. Therefore, it was concluded that these reactions proceed by 2-equivalent changes in the oxidation states of the reactants.

3. Conclusions

This short and far from complete survey shows that the previously obscure field of chemical induction is becoming more and more understood. The accelerating pace of progress has furnished from the forties onwards a great deal of interesting information about the chemistry of unstable intermediates, *e.g.* chromium(V), chromium(IV), arsenic(IV), tin(III), HO_2, OH, SO_4^- radicals. These results were obtained mostly by conventional methods. Therefore, it may be expected that the more extensive application of methods suitable for detection and estimation of short-living entities (*e.g.* resonance methods, fast reaction techniques) will enable our somewhat qualitative knowledge (as it is today) to be put onto a quantitative basis.

REFERENCES

1 F. Kessler, *Pogg. Ann.*, 119 (1863) 218.
2 F. Raschig, *Z. Angew. Chem.*, 17 (1904) 580, 1407.
3 A. von der Ropp, *Z. Anal. Chem.*, 40 (1901) 482.
4 F. Feigl, *Chemistry of Specific Selective and Sensitive Reactions*, Academic Press, New York, 1949, p. 156.
5 F. Feigl, *Chemistry of Specific Selective and Sensitive Reactions*, Academic Press, New York, 1949, p. 159.
6 R. L. Rich and H. Taube, *J. Am. Chem. Soc.*, 76 (1954) 2608.
7 R. Luther and N. Schilow, *Z. Physik. Chem. Leipzig*, 46 (1903) 777.
8 R. Livingston, in S. L. Fries and A. Weissberger, Eds., *Technique of Organic Chemistry*, Vol. VIII, Interscience, New York, 1953, pp. 219 and 224.
9 W. C. Bray and J. B. Ramsey, *J. Am. Chem. Soc.*, 55 (1933) 2279.
10 W. Manchot, *Ann.*, 213 (1882) 312.
11 A. I. Medalia, *Anal. Chem.*, 27 (1958) 1678.
12 R. Lang and J. Zwerina, *Z. Anorg. Allgem. Chem.*, 170 (1928) 389.
13 L. J. Csányi and M. Szabó, *Talanta*, 1 (1958) 359.
14 W. Nernst, *Theoretische Chemie*, 4th ed., ENKE, Stuttgart, 1903, p. 656.
15 R. Luther and T. F. Rutter, *Z. Anorg. Allgem. Chem.*, 54 (1907) 1.
16 P. A. Schaffer, *J. Am. Chem. Soc.*, 55 (1933) 2169; *J. Phys. Chem.*, 40 (1936) 1021.
17 C. F. Schönbein, *J. Prakt. Chem.*, 75 (1858) 108.

18 R. E. De Lury, *J. Phys. Chem.*, 11 (1907) 54.
19 W. Watanabe and F. H. Westheimer, *J. Chem. Phys.*, 17 (1949) 61.
20 A. C. Chatterji and V. Antony, *Proc. Natl. Acad. Sci. India*, 23A (1954) 20.
21 V. Antony and A. C. Chatterji, *Z. Anorg. Allgem. Chem.*, 280 (1955) 111.
22 A. C. Chatterji and V. Antony, *Z. Physik. Chem. Leipzig*, 210 (1959) 50.
23 A. C. Chatterji and S. K. Mukherjee, *Z. Physik. Chem.* Leipzig, 207 (1957) 372.
24 A. C. Chatterji and S. K. Mukherjee, *Z. Physik. Chem.* Leipzig, 208 (1958) 281.
25 A. C. Chatterji and S. K. Mukherjee, *Z. Physik. Chem.* Leipzig, 210 (1959) 103.
26 T. E. Graham and F. H. Westheimer, *J. Am. Chem. Soc.*, 80 (1958) 3030.
27 K. B. Wiberg and Th. Mill, *J. Am. Chem. Soc.*, 80 (1958) 3022.
28 A. C. Chatterji and S. K. Mukherjee, *J. Am. Chem. Soc.*, 80 (1958) 3600.
29 N. R. Dhar, *J. Chem. Soc.*, 111 (1917) 707; *Ann. Chim. Paris*, 11 (1919) 130.
30 W. Manchot and O. Wilhelms, *Ann.*, 325 (1902) 105.
31 C. Benson, *J. Phys. Chem.*, 7 (1903) 1.
32 W. Manchot and O. Wilhelms, *Ann.*, 325 (1902) 125.
33 C. Wagner and W. Preiss, *Z. Anorg. Allgem. Chem.*, 168 (1927) 265.
34 W. Manchot and P. Richter, *Ber.*, 39 (1906) 488.
35 W. Manchot, *Ber.*, 39 (1906) 1352.
36 W. Manchot, *Ber.*, 39 (1906) 3510.
37 G. Gopala Rao and Venkateswara Rao, *Z. Anal. Chem.*, 151 (1956) 247.
38 G. Gopala Rao and T. P. Sastri, *Z. Anal. Chem.*, 152 (1956) 270.
39 I. M. Kolthoff and M. A. Fineman, *J. Phys. Chem.*, 60 (1956) 1383.
40 R. E. DeLury, *J. Phys. Chem.*, 7 (1907) 47.
41 F. H. Westheimer, *Chem. Rev.*, 45 (1949) 419.
42 J. O. Edwards, *Chem. Rev.*, 50 (1952) 472.
43 J. G. Mason and A. D. Kowalak, *Inorg. Chem.*, 3 (1964) 1248.
44 R. Lang, *Z. Anorg. Allgem. Chem.*, 170 (1928) 387.
45 R. E. DeLury, *J. Phys. Chem.*, 7 (1903) 239.
46 F. H. Westheimer and A. Novick, *J. Chem. Phys.*, 11 (1943) 506.
47 D. G. Lee and R. Stewart, *J. Am. Chem. Soc.*, 86 (1964) 3051.
48 K. B. Wiberg and H. Schäfer, *J. Am. Chem. Soc.*, 89 (1967) 455.
49 E. Lucchi, *Gazz. Chim. Ital.*, 71 (1941) 729, 752.
50 J. H. Espenson, *J. Am. Chem. Soc.*, 86 (1964) 1883.
51 E. Abel, *Monatsh. Chem.*, 88 (1957) 711.
52 A. M. Sanko and V. F. Stefanovskii, *J. Gen. Chem. USSR*, 7 (1937) 100.
53 L. J. Csányi, unpublished results.
54 D. A. Johnson and A. C. Sharpe, *J. Chem. Soc.*, (1964) 3490.
55 L. J. Csányi, *J. Chem. Educ.*, 37 (1960) 147.
56 E. A. M. Wetton and W. C. E. Higginson, *J. Chem. Soc.*, (1965) 5890.
57 R. Woods, I. M. Kolthoff and E. J. Meehan, *J. Am. Chem. Soc.*, 85 (1963) 2385.
58 W. A. Waters, *Discussions Faraday Soc.*, 29 (1960) 170.
59 R. Woods, I. M. Kolthoff and E. J. Meehan, *J. Am. Chem. Soc.*, 85 (1963) 3334.
60 R. Woods, I. M. Kolthoff and E. J. Meehan, *J. Am. Chem. Soc.*, 86 (1964) 1698.
61 R. Woods, I. M. Kolthoff and E. J. Meehan, *Inorg. Chem.*, 4 (1965) 697.
62 R. M. Milburn and W. C. Vosburgh, *J. Am. Chem. Soc.*, 77 (1955) 1352.
63 I. M. Kolthoff and I. K. Miller, *J. Am. Chem. Soc.*, 73 (1951) 3055.
64 I. M. Kolthoff and R. Woods, *J. Am. Chem. Soc.*, 88 (1966) 1371.
65 I. M. Kolthoff and R. Woods, *J. Electroanal. Chem.*, 12 (1966) 385.
66 J. Koutecky, *Collection Czech. Chem. Commun.* 22 (1957) 160.
67 R. Woods and I. M. Kolthoff, *Anal. Chem.*, 38 (1966) 1297.
68 K. Gleu, *Z. Anal. Chem.*, 95 (1933) 385.
69 L. J. Csányi and M. Szabó, MTA Kém. Tud. Oszt. Közl., 11 (1959) 71.
70 A. M. Sanko and V. F. Stefanovskii, *J. Gen. Chem. USSR*, 7 (1937) 2385.
71 R. Rutter, *Dissertation*, Leipzig, 1906.
72 A. O. McDougall, *Dissertation*, Leipzig, 1906.
73 K. A. Hofman, *Ber.*, 45 (1912) 3329.

74 Z. G. Szabó, *J. Chem. Soc.*, (1950) 1356.
75 J. A. N. Friend, *J. Chem. Soc.*, 85 (1904) 527.
76 A. Skrabal and J. P. Vacek, *Österr. Chemiker-Ztg.*, 13 (1901) 27.
77 L. J. Csányi, J. Bátyai and F. Solymosi, *Z. Anal. Chem.*, 195 (1963) 9.
78 M.-S. Tsao and W. K. Wilmarth, *Discussions Faraday Soc.*, 29 (1960) 137.
79 L. J. Csányi, unpublished results.
80 M.-S. Tsao and W. K. Wilmarth, *Advan. Chem.*, 36 (1962) 113.
81 G. Czapski and L. M. Dorfman, *J. Phys. Chem.*, 68 (1964) 1169.
82 W. M. Latimer, *The Oxidation States of the Elements and their Potentials in Aqueous Solutions*, 2nd Ed., Prentice-Hall, New York, 1952.
83 B. H. J. Bielski and A. O. Allen, *Proc. 2nd Tihany Symp. Radiation Chemistry*, Akadémiai Kiadó, Budapest, 1967, p. 81.
84 L. J. Csányi, J. Bátyai and F. Solymosi, *Acta Phys. Chem. (Szeged)*, 9 (1963) 106.
85 L. J. Csányi, L. Mucsi and K. Németh, *Acta Phys. Chem. (Szeged)*, 9 (1963) 97.
86 R. Havemann, K. Schiller and W. Schneider, *Z. Physik. Chem. Leipzig*, 233 (1966) 343.
87 L. J. Csányi and F. Solymosi, *Acta Chim. Acad. Sci. Hung.*, 13 (1957) 19.
88 L. J. Csányi and F. Solymosi, *Acta Chim. Acad. Sci. Hung.*, 17 (1958) 69.
89 L. J. Csányi and F. Solymosi, *Acta Chim. Acad. Sci. Hung.*, 13 (1958) 275.
90 E. Nagy, V., *B.Sc. Thesis*, Szeged, 1963.
91 I. Cseh, *B.Sc. Thesis*, Szeged, 1967.
92 L. J. Csányi, S. Kaszai and I. Molnár, *Talanta*, 10 (1963) 449.
93 H. J. H. Fenton, *J. Chem. Soc.*, 65 (1894) 899; 77 (1900) 1294.
94 H. J. H. Fenton, *Chem. News*, (1876) 190; (1881) 110.
95 H. J. H. Fenton and H. Jackson, *J. Chem. Soc.*, 75 (1899) 1.
96 A. I. Medalia and I. M. Kolthoff, *J. Polymer Sci.*, 4 (1949) 377.
97 F. Haber and J. Weiss, *Proc. Roy. Soc. London, Ser. A*, 147 (1934) 332.
98 W. G. Barb, J. H. Baxendale, P. George and K. H. Hargrave, *Trans. Faraday Soc.*, 47 (1951) 462.
99 T. J. Hardwick, *Can. J. Chem.*, 35 (1957) 428.
100 C. F. Wells and M. A. Salam, *Trans. Faraday Soc.*, 63 (1967) 620.
101 C. F. Wells and M. A. Salam, *J. Chem. Soc. A*, (1968) 24.
102 C. F. Wells and M. A. Salam, *J. Chem. Soc. A*, (1968) 308.
103 I. M. Kolthoff and A. I. Medalia, *J. Am. Chem. Soc.*, 71 (1949) 3777.
104 J. H. Baxendale, M. G. Evans and G. S. Park, *Trans. Faraday Soc.*, 42 (1946) 155.
105 I. M. Kolthoff and A. I. Medalia, *J. Am. Chem. Soc.*, 71 (1949) 3784.
106 I. M. Kolthoff and A. I. Medalia, *J. Am. Chem. Soc.*, 71 (1949) 3789.
107 I. M. Kolthoff, A. I. Medalia and H. Parks Raaen, *J. Am. Chem. Soc.*, 73 (1951) 1733.
108 J. H. Merz and W. A. Waters, *Discussions Faraday Soc.*, 2 (1947) 179.
109 E. Pungor, E. Schulek and J. Trompler, *Acta Chim. Acad. Sci. Hung.*, 4 (1954) 417.
110 L. J. Csányi and F. Solymosi, *Acta Chim. Acad. Sci. Hung.*, 15 (1958) 231.
111 R. H. Smith and I. R. Wilson, *Australian J. Chem.*, 19 (1966) 1357.
112 R. H. Smith and I. R. Wilson, *Australian J. Chem.*, 19 (1966) 1365.
113 R. H. Smith and I. R. Wilson, *Australian J. Chem.*, 20 (1967) 1353.
114 I. R. Wilson and G. M. Harris, *J. Am. Chem. Soc.*, 82 (1960) 4515.
115 I. R. Wilson and G. M. Harris, *J. Am. Chem. Soc.*, 83 (1961) 286.
116 L. J. Csányi and Gy. Horváth, *Acta Chim. Acad. Sci. Hung.*, 34 (1962) 1.
117 A. G. Doroshevskii and H. Bardt, *J. Russ. Phys. Chem. Soc.*, 46 (1914) 754.
118 J. H. Merz, G. Stafford and W. A. Waters, *J. Chem. Soc.*, (1951) 638.
119 W. C. E. Higginson, R. T. Leigh and R. Nightingale, *J. Chem. Soc.*, (1962) 435.
120 R. A. Robinson and N. H. Law, *Trans. Faraday Soc.*, 31 (1935) 899.
121 M. H. Gorin, *J. Am. Chem. Soc.*, 58 (1936) 1787.
122 F. R. Duke and R. C. Pinkerton, *J. Am. Chem. Soc.*, 73 (1951) 3045.
123 K. B. Wiberg and H. Schäfer, *J. Am. Chem. Soc.*, 91 (1969) 927.
124 K. B. Wiberg and H. Schäfer, *J. Am. Chem. Soc.*, 91 (1969) 933.
125 W. A. Mosher and G. L. Driscoll, *J. Am. Chem. Soc.*, 90 (1968) 4189.
126 G. V. Bakore and S. Narain, *J. Chem. Soc.*, (1963) 3419.

127 G. V. BAKORE AND S. NARAIN, *Z. Phys. Chem.*, 227 (1964) 8.
128 G. V. BAKORE AND A. A. DESHPANDE, *Z. Phys. Chem.*, 227 (1964) 14.
129 K. B. WIBERG, *Oxidation in Organic Chemistry, Part A*, Academic Press, New York, N.Y., 1965.
130 R. STEWART, *Oxidation Mechanism: Application to Organic Chemistry*, W. A. Benjamin, New York, N.Y., 1964.
131 A. O. ALLEN, private communication.
132 C. C. HONG AND W. H. RAPSON, *Can. J. Chem.*, 46 (1968) 2053.
133 P. W. ATKINS AND M. C. R. SYMONS, *The Structure of Inorganic Radicals*, Elsevier, Amsterdam–London–New York, 1967, p. 175, 210.
134 H. A. CATHERINO, *J. Phys. Chem.*, 71 (1967) 268.
135 H. A. CATHERINO, *J. Phys. Chem.*, 70 (1966) 1339.
136 J. Y. TONG AND E. L. KING, *J. Am. Chem. Soc.*, 82 (1960) 3805.
137 C. WAGNER, *Z. Anorg. Allgem. Chem.*,168 (1928) 279

Index

A

B

D

E

F

L

M

O

Q

R

T

U

V